Challenges in analysis of complex natural mixtures

John McIntyre Conference Centre, University of Edinburgh, United Kingdom

13–15 May 2019

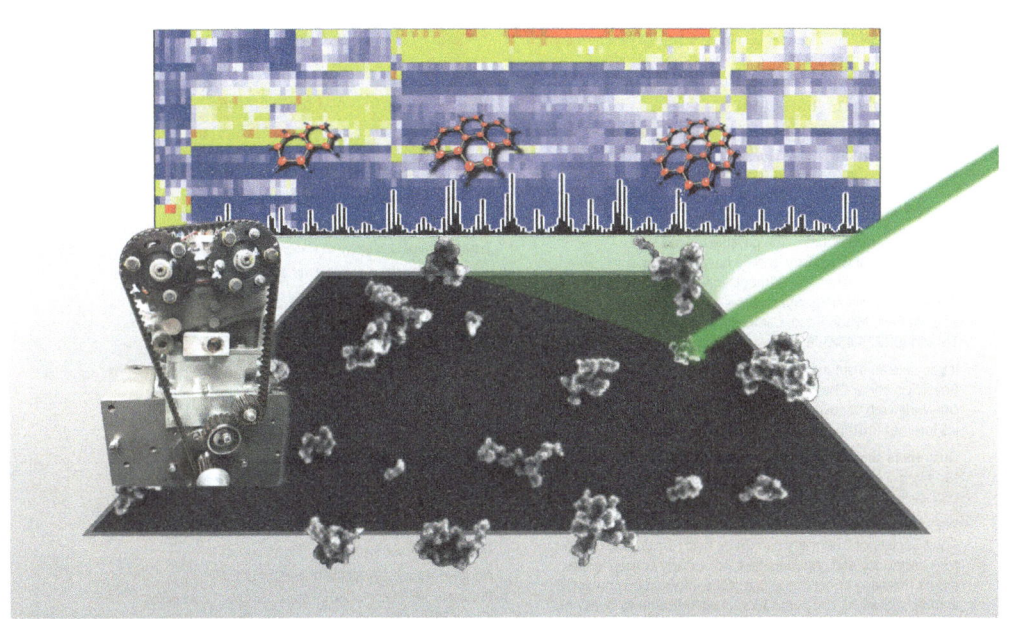

FARADAY DISCUSSIONS
Volume 218, 2019

The Faraday Division of the Royal Society of Chemistry, previously the Faraday Society, was founded in 1903 to promote the study of sciences lying between chemistry, physics and biology.

Editorial Staff

Executive Editor
Richard Kelly

Deputy Editor
Maria Southall

Editorial Production Manager
Claire Darby

Publishing Editors
Suzanne Howson, Ella Wren

Editorial Assistant
Aliya Anwar

Publishing Assistants
David Bishop, Allison Holloway

Publisher
Jamie Humphrey

Faraday Discussions (Print ISSN 1359-6640, Electronic ISSN 1364-5498) is published 8 times a year by the Royal Society of Chemistry, Thomas Graham House, Science Park, Milton Road, Cambridge, UK CB4 0WF.

Volume 218 ISBN 13: 978-1-78801-674-2

2019 annual subscription price: print+electronic £1220 US $2148; electronic only £1162, US $2046. Customers in Canada will be subject to a surcharge to cover GST. Customers in the EU subscribing to the electronic version only will be charged VAT.

All orders, with cheques made payable to the Royal Society of Chemistry, should be sent to the Royal Society of Chemistry Order Department, Royal Society of Chemistry, Thomas Graham House, Science Park, Milton Road, Cambridge, CB4 0WF, UK Tel +44 (0)1223 432398; E-mail **orders@rsc.org**

If you take an institutional subscription to any Royal Society of Chemistry journal you are entitled to free, site-wide web access to that journal. You can arrange access via Internet Protocol (IP) address at **www.rsc.org/ip**

Customers should make payments by cheque in sterling payable on a UK clearing bank or in US dollars payable on a US clearing bank.

Faraday Discussions are unique international discussion meetings that focus on rapidly developing areas of chemistry and its interfaces with other scientific disciplines.

Challenges in analysis of complex natural mixtures

Faraday Discussions

www.rsc.org/faraday_d

A General Discussion on Challenges in analysis of complex natural mixtures was held in Edinburgh, UK on the 13th , 14th and 15th of May 2019.

RSC Publishing is a not-for-profit publisher and a division of the Royal Society of Chemistry. Any surplus made is used to support charitable activities aimed at advancing the chemical sciences. Full details are available from www.rsc.org

CONTENTS

ISSN 1359-6640; ISBN 978-1-78801-674-2

Cover
See Duca *et al.*, *Faraday Discuss.*, 2019, **218**, 115–137.

On the benefits of using multivariate analysis in mass spectrometric studies of combustion-generated aerosols.

Image reproduced by permission of Dumitru Duca from *Faraday Discuss.*, 2019, **218**, 115.

INTRODUCTORY LECTURE

PAPERS AND DISCUSSIONS

CONCLUDING REMARKS

ADDITIONAL INFORMATION

Faraday Discussions

PAPER

Systems chemical analytics: introduction to the challenges of chemical complexity analysis

Philippe Schmitt-Kopplin, [ID] *[ab] Daniel Hemmler, [ID] [ab]
Franco Moritz, [ID] [a] Régis D. Gougeon, [ID] [c] Marianna Lucio, [ID] [a]
Markus Meringer, [ID] [d] Constanze Müller, [ID] [a] Mourad Harir [ID] [a]
and Norbert Hertkorn [ID] [a]

Received 5th June 2019, Accepted 8th July 2019
DOI: 10.1039/c9fd00078j

Understanding complex (bio/geo)systems is a pivotal challenge in modern sciences that fuels a constant development of modern analytical technology, finding innovative solutions to resolve and analyse. In this introductory paper to the Faraday Discussion *"Challenges in the analysis of complex natural systems"*, we aim to present concepts of complexity, and complex chemistry in systems subjected to biotic and abiotic transformations, and introduce the analytical possibilities to disentangle chemical complexity into its elementary parts (*i.e.* compositional and structural resolution) as a global integrated approach termed *systems chemical analytics*.

Introduction/complexity

Over the past few weeks, we have asked friends and colleagues to give their examples of complex systems/phenomena/objects and to describe their definition of "complexity" in one sentence. The answers first looked extremely heterogeneous but after detailed analysis they seemed to converge to a few concepts that are illustrated as an introduction herein.

The "complex" systems, phenomena or objects named were extremely diverse and inspired by our direct environment, such as life, ecosystems, nature, the universe, humans, brains, food, flavours or even wine. Other notions were more

[a]*HelmholtzZentrum Muenchen, German Research Center for Environmental Health, Department of Environmental Sciences, Ingolstädter Landstraße 1, D-85764 Neuherberg, Germany. E-mail: schmitt-kopplin@helmholtz-muenchen.de*

[b]*Technical University Munich, Chair of Analytical Food Chemistry, Maximus-von-Imhof-Forum 2, 85354 Freising Weihenstephan, Germany*

[c]*UMR PAM Université de Bourgogne/AgroSup Dijon, Institut Universitaire de la Vigne et du Vin, Jules Guyot, Dijon, France*

[d]*German Aerospace Center (DLR), Earth Observation Center (EOC), Münchner Straße 20, 82234 Oberpfaffenhofen-Wessling, Germany*

abstract or linked to processes, such as climate, emotions, thinking or health. Even more interesting was that the word "complexity" also stimulated associations with society, languages, religions, mathematics, art, family (teenagers – wonder why?!) and psychology. These answers were mainly related to the background, the field of expertise and the current interests of the respondents. In many cases, the complexity was associated with the description of a "complicated" system in a given quest for knowledge. Complexity is not only related to the scientist's view of natural systems but is also expressed in the dimensions of social sciences, religions, emotions or arts. For example, while a scientific theory of music may describe in mathematical terms the harmony of rhythms and succession of sounds, music evolves more emotionally and thus involves personal levels of enjoyment[1] making this art rather "complex" beyond the combinatorial sound, rhythm, timbre, voices, or choice of instrument. Any art can be projected in a similar way and a song, a sculpture or a painting may be reflected as an emergence resulting from the connections of the elements – art is a complex act of creativity. Thus, it was not surprising to find one of the answers describing "complexity" as: *"Complexity is what we see when we open the eyes, complexity is beauty"* (thank you Régis Gougeon). Bringing a similar dimension to our scientific world, let us remember Feynman's beautiful monologue about art and science concerning the beauty of a flower:

"I have a friend who's an artist and has sometimes taken a view which I don't agree with very well. He'll hold up a flower and say "look how beautiful it is," and I'll agree. Then he says "I as an artist can see how beautiful this is but you as a scientist take this all apart and it becomes a dull thing," and I think that he's kind of nutty. First of all, the beauty that he sees is available to other people and to me too, I believe… I can appreciate the beauty of a flower. At the same time, I see much more about the flower than he sees. I could imagine the cells in there, the complicated actions inside, which also have a beauty. I mean it's not just beauty at this dimension, at one centimeter; there's also beauty at smaller dimensions, the inner structure, also the processes. The fact that the colors in the flower evolved in order to attract insects to pollinate it is interesting; it means that insects can see the color. It adds a question: does this aesthetic sense also exist in the lower forms? Why is it aesthetic? All kinds of interesting questions which the science knowledge only adds to the excitement, the mystery and the awe of a flower. It only adds. I don't understand how it subtracts."

Richard Feynman, Nobel-laureate in physics (from BBC Interview for Horizon 'The Pleasure of Finding Things Out' (https://www.bbc.co.uk/sn/tvradio/programmes/ horizon/broadband/archive/feynman/) animated by Fraser Davidson/https:// vimeo.com/55874553).

These personalized definitions of "complexity" converged in their description of (super)systems and exceeded simple diversity showing various layers of organization and interactions/interconnections between their elements. In these terms, a representative answer and description of complexity, contributed by our friend Nancy Hinman, is *"Complexity is the nexus between parts and processes. If the parts are divided then the processes won't occur. And conversely, if the processes are occurring then the parts cannot be dismantled".*

Further analysis involved text mining by a word cloud generated from almost 120 000 abstracts listed in PubMed, which contained the word "complexity" (Fig. 1A). The concept of complexity is highly relevant in biology and health, which can be concluded from the frequent occurrences of words such as clinics,

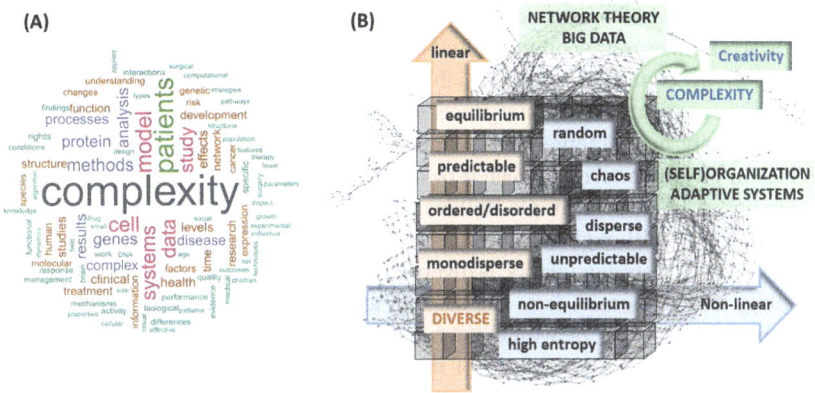

Fig. 1 (A) A word cloud created from abstracts of published articles listed in PubMed and containing the term "complexity". (B) An attempt of a schematic representation of concepts at various levels, as exerted from the definition of complexity within systems theory. While concepts such as diverse, predictable, ordered/disordered, and mono-disperse still show some linearity and, as a consequence, predictability, others such as high entropy, disperse, random, non-equilibrium and chaos may be closer to the concept of "complexity" and thus of the unpredictability of the phenomena. The latter concepts integrate more of the concepts of self-organization, adaptive systems, and network theory as needed and used in modern computational big data processing.

patients, human, risk, cancer, genes, and proteins. Additionally, its great relevance in the field of data science can be deduced from associations with data, systems, study, results, analysis, and network. Complexity is defined in various disciplines and became an important concept in modern social, natural and economic sciences, and "complexity theory" rose out of this need only a few decades ago. The word "complexity" has Latin roots: "*com*" means "*together*" and "*plectere*" means "*to plait*", showing the interconnectivity of the contributing objects/concepts. Some general definitions were given by Darley:[2] "*a typical complex system is one for which at least some of its global behaviors (that result from the interactions between a large number of relatively simple parts) cannot be predicted simply from the rules of underlying interactions*". Complexity thus not only implies interconnectivity but also possible dynamics.

"Complex systems" can be defined as being generally composed of many individual parts interacting with each other, following simple rules, synchronizing without any centralized control. The links between the activities of these parts relate to network approaches and network theory, and their dynamics in the interaction with the environments in cooperation or competition is typical of adaptive systems. This concept is illustrated in Fig. 1B. Out of their linear or non-linear dynamics, patterns can emerge, that can (self)organize and create islands of order within high entropic systems keeping a thermodynamic equilibrium of the whole. In short, as described by Ferreira,[3] "*a complex system is characterized by emergent behavior resulting from the interaction among parts*". The emerging complexity theory is interested in observations of systems and explanations of endogenous behavior, breaking down the complexity of the phenomena,[4] while the related systems theory uses this information to trigger and optimize some functions.[5] The fundamentals of the theory behind complexity have their roots in mathematics and computer sciences, physics, chemistry, biology, ecology, social sciences, engineering, economics and arts, trying to understand or trigger the formation of complex systems such as swarms, social behaviors, business markets, ecosystems or the emergence of life and evolution in general. Even in the arts, creativity may emerge from complexity by reflecting local interactions of objects in global patterns projected by the artists.

Finally, a further example following Feynman's line with a remarkably contemporary – although not recent – representation of such complex systems, where interplay between factors as diverse as ecology, biology, chemistry, physics, sociology and economics is at the basis of an emergent behavior, is the French concept of the terroir of wines. Such emergent behavior, which is constantly challenged,[6] is indeed composed of individual parts, which all bear individual complex patterns, and yet which collectively find coherence, not only from a marketing point of view, but also from a compositional property point of view.[7,8] Richard Feynman wrote "*...But it is true that if we look at a glass of wine closely enough we see the entire universe. There are the things of physics: the twisting liquid which evaporates depending on the wind and weather, the reflections in the glass, and our imagination adds the atoms. The glass is a distillation of the Earth's rocks, and in its composition we see the secrets of the universe's age, and the evolution of stars. What strange arrays of chemicals are in the wine? How did they come to be? There are the ferments, the enzymes, the substrates, and the products. There in wine is found the great generalization: all life is fermentation...*". If a signature of stellar evolution may indeed be found in a glass of wine, individual contributions as subtle as the geographical origin of woods used for barrel aging can emerge in the final composition of a wine.[9]

Complexity in chemistry/complex natural mixtures

The concept of complexity is rather new in chemistry, a field of science that attempts to predict and control rather than simply observe and analyze.[10]

Chemistry defines us and our surroundings; *chemistry is everywhere* from nuclear and organic to polymer chemistry. According to the American Chemical Society (ACS):[11] "*Everything you hear, see, smell, taste, and touch involves chemistry and chemicals (matter). And hearing, seeing, tasting, and touching, all involve intricate series of chemical reactions and interactions in your body. With such an enormous range of topics, it is essential to know about chemistry at some level to understand the world around us*". Chemical complexity thus affects how we sense the world at all levels. Overcoming complexity, adaptation and usage of its intrinsic elements may have led to the emergence of life and further processes may be connected to evolution and progress in time and space. Thus, when defining "chemical complexity" in all environments, one needs to consider various scales in the time/space domain (short to long, tiny to infinitely large) and position these within the chemical evolution. In this context, Fig. 2 is an attempt at illustrating the linear or non-linear chemical evolution in time and space, assuming that each of the elements (presented as *time/space* voxelcubes) is highly entangled with others in

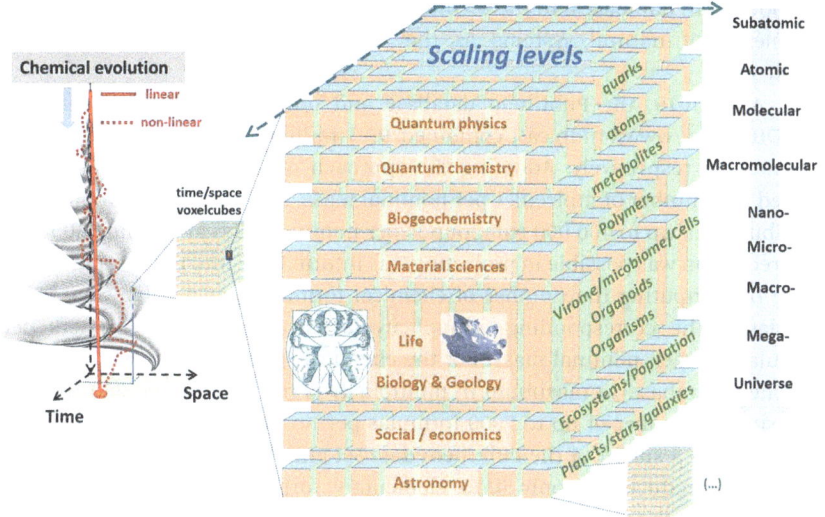

Fig. 2   Schematic representation of the connectivity in time and space of the linear/non-linear chemical evolution occurring at different scales. For example, complex co-evolution in the chemistry of living and non-living systems occurs in highly multifaceted reactivity pathways and the imperfections of its description is due to the extreme limitations in our holistic analytical observations and mathematical projections. "Complexity" is present in every voxelcube at any scale.

this process. Our main problem as humans certainly is how to represent multiple dimensions in our 3D world, and especially in this manuscript on 2D paper. The levels of complexity present in each voxelcube show scaling levels/sizes as well from atoms to molecules across organisms and ecosystems toward planetary dimensions and galaxies with mutual interactions between the respective research disciplines (from quantum physics to astronomy) with strong fractal interdependency. Chemistry itself represents only a limited part within these voxels as it is affected by biotic and abiotic processes in non-living and living systems within the described dimensions.

Chemistry is generally considered to be "complex" and "complicated" and a common strategy in chemistry is to always simplify and linearize.[10] It is generally possible on a reduced scale to explain the processes within small voxelcubes but linear interpolation across all levels and scales shown in Fig. 2 is not possible. In terms of complex chemistry, non-linear processes have not really found the focus yet, because of the limited analytical resolution of our observations leading to a generalization and to averaging if information. The resolution requirements and currently available analytical approaches are briefly discussed in this manuscript and are the focus of this Faraday Discussion on complex natural mixtures.

As illustrated in Fig. 2, scales are extremely important and each voxelcube presents its own chemistry with a high dynamic range in terms of the diversity and abundances of species that are entangled in a network of interactions. We divided the descriptions of diversity into two steps. The first level is considering all mathematically conceivable chemical structures based on the most abundant essential atoms of life (C, H, N, O, P, S) and their comparison with known compounds in databases. In the second part we describe the chemical diversity/ complexity of natural systems as they derive from abiotic and/or biotic synthesis.

In April 2019, the PubChem database and human metabolome database (HMDB, http://www.hmdb.ca/) contained 97.3 million and more than 113 000 compounds, respectively. Adopting the approach of Kauffman,[12] in Fig. 3 we plotted the number of molecules per nominal mass, showing a close to Gaussian distribution of up to 400 000 compounds per nominal mass at around 350 Da. A data reduction was possible in projecting this information into the compositional space by computing the number of chemical formulas per nominal mass and calculating the corresponding multiplicity (average number of isomers per formula) for each nominal mass. For the chemical space (as represented by the PubChem database), a maximum in multiplicity is reached around 350 Da with a maximum number of up to 7000 formulas per nominal mass around 500 Da. The HMDB is a subset of the PubChem data set and showed more than 113 000 compounds reduced to only 9118 elementary formulas having a maximum distribution in the lower mass range around 400 Da. Here, multiplicity is heavily biased by the lipid compounds in the database having an very high number of structural isomers and thus multiplicity. Lipids are among the most diverse metabolites in living systems, and although lipids are composed from a structural combination of only a few defined building blocks, they generate more than several 100 000s of individual lipid structures that vary in chain length, saturation and double bond positions.[13] Lipidomics is a rising complex discipline within the field of metabolite profiling (and metabolomics) and is essentially based on high resolution chromatography hyphenated to tandem mass spectrometry.[14]

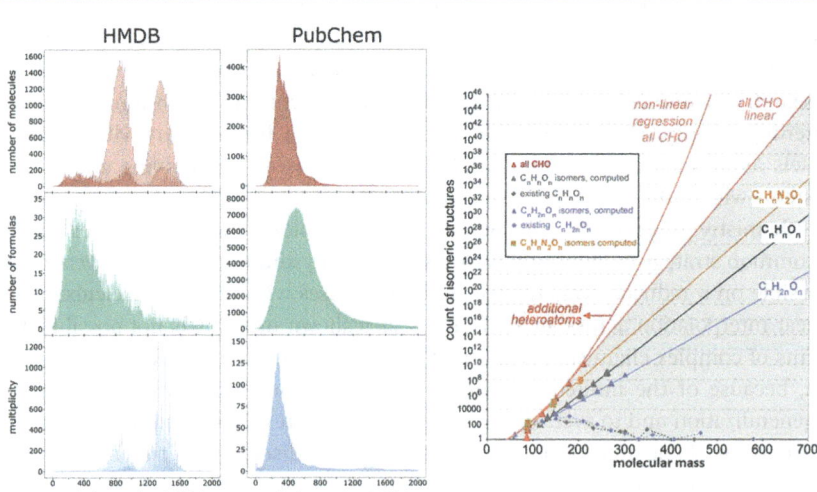

Fig. 3   Left: Abundances of molecules and multiplicity (number of molecules per formula — isomers) of related molecular compositions (formulas) over a wide mass range as found in the metabolite database (HMDB) and general chemical database (PubChem). Right: Diversity and complexity illustrated with the computed number of isomers for given restricted elementary compositions as compared to the corresponding PubChem database. The dotted lines at the bottom depict known molecular structures for the series $C_nH_{2n}O_n$ and $C_nH_nO_n$.

Compositional space, diversity/complexity

When playing with numbers, it is instructive to compare the counts of existing molecular compositions and structures as provided above with mathematically conceivable molecular compositions and isomeric molecular structures that can be computed from given molecular compositions by graph theory.[15,16] Fig. 3 provides the counts of chemically reasonable isomers for compositions $C_nH_{2n}O_n$ (one DBE) and $C_nH_nO_n$ (several DBEs), which can be computed up to $n = 10$, with extrapolation to 700 Da. Molecules of the $C_nH_{2n}O_n$ type feature a single DBE and therefore can carry only a single double bond or a single alicyclic ring; even so, the projected count of isomers at 500 Da and 700 Da reaches $\sim 10^{16}$ and $\sim 10^{22}$, respectively. Molecules of the $C_nH_nO_n$ type feature several DBEs, which give rise to many more conceivable chemical structures, including those with aromatic substructures, which are inaccessible for $C_nH_{2n}O_n$ compositions.

When clusters of nominal masses are recognized, counts of CHO molecular compositions increase moderately with increasing mass,[15] whereas the count of isomeric structures for any of these CHO molecules increases rapidly with the mass. In short, the number of CHO structures per nominal mass (called a cluster) dramatically increases with mass. For example, 18 112 feasible elemental compositions define the compositional space of CHO compounds at m/z 180–700, with H/C $\leq$ 2 and O/C $\leq$ 1.[15] If we assume 260 nominal mass clusters for CHO compounds and 20 mass peaks per cluster, we will observe a total of 5200 mass peaks susceptible to a continually growing number of isomers, ranging from $\sim 10^4$ (at m/z 180) to $\sim 10^{23}$ (at m/z 700) that will be projected on any single mass peak with a cumulative count of $>10^{35}$ conceivable isomers for CHO compounds alone (Fig. 3B). Cumulative counts of CHO compounds define the structural space without additional heteroatoms; the effects of adding heteroatoms, such as N, are exemplified using the $C_nH_nO_nN_2$ formula. The provided isomer counts are vastly underestimated because stereoisomers and several N-bearing functional groups in elevated oxidation states are *not* recognized by the used method of computation. An upward curvature with a considerable progressive size increment is observed after non-linear regression (for the sake of clarity, linear regression is applied here with one exception), which likely reflects continuous evolution of overarching structural motifs, which become accessible only when a certain mass limit is exceeded.[16]

The dotted lines in Fig. 3 depict known molecular structures for the homologous series of type $C_nH_{2n}O_n$ and $C_nH_nO_n$; the discrepancy between the vast theoretical structural space and the space very sparsely occupied by known molecules is immediately obvious. This implies that reliable extrapolations of structures within biogeochemical materials are difficult to confirm from current coverage of databases and contemporary comprehension of biochemistry. Databases overly rely on biological CHO molecules (HMDB) and islands of products from targeted chemosynthesis (PubChem), often produced by combinatorial chemistry.[17]

Natural chemical mixtures

Taking again the definitions of "complex system" from colleagues, one of the answers was "*complex is synergistic interactions of organic molecules and microbial transformations in natural systems*" (thank you Michael Gonsior). This illustrates nicely the setup of complex chemistry involved in the interactions of elements, their adaptation, and their organization to reach homeostasis. The scheme in Fig. 4 integrates these interfaces between *biomes* and *abiomes* in biogeochemical systems and thus sets the challenges for analytical chemistry in describing the dynamic chemistry herein. Living systems (from ubiquitous microbiomes through to higher organisms through to entire ecosystems) are involved in their specific interactions and more general ecosystem adaptations, which are globally transformed on short term geological time scales to complex organic matter and geopolymers. Complexity can be found in interorganismic interactions at the macrolevel of the ecosystems[18] or at the organism level when taking account of supersystems such as holobionts.[19]

Natural chemical mixtures are key challenging samples at the center of focus for the analytical techniques presented during this Faraday Discussion. These cover a wide field of interests in this volume, such as heavy oils, natural/soil/dissolved organic matter, body fluids and plant extracts, to name just a few.

Natural chemical mixtures occur in soil,[20] freshwater,[21] estuarine,[22] marine,[23,24] arid[25] and hydrothermal environments,[26-28] and in the atmosphere,[29] and these

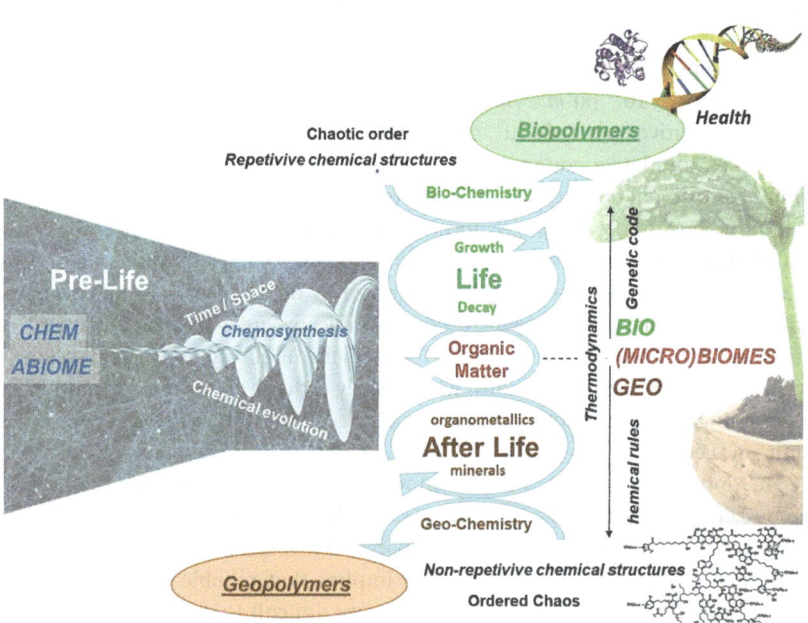

Fig. 4 *"Life in a nutshell"*: natural chemical mixtures are present in all fields of life, pre-life and after-life, with natural organic matter being at the interface of life bioprocesses and abiotic chemical complexity as an entire part of geochemistry. The chemical diversity in biology is mainly a result of and is limited to highly regulated bioprocesses. In comparison, geochemistry involves more diverse interactions including radical reactions and catalysis with metals and mineral phases in the environment and reflecting higher chemical reaction potential leads to a higher molecular diversity. NOM results from biotic and abiotic chemosynthesis and therefore structural assignments are difficult because of non-existing database knowledge. Analytical tools with higher resolution are mandatory for a trustworthy description of abiotic and biogeochemical molecular complexity.

represent exceedingly complex mixtures of organic compounds that collectively exhibit a nearly continuous range of properties (size–reactivity continuum). Their composition and structure in the bio- and geosphere are established and governed according to the rather fundamental constraints of thermodynamics and kinetics. In these intricate materials, the "classical" signatures of the (geogenic or ultimately biogenic) precursor molecules, like lipids, glycans, proteins and natural products, have been attenuated, often beyond recognition, during a succession of biotic and abiotic (e.g. photo and redox chemistry) reactions. Natural organic mixtures incorporate the hugely disparate characteristics of abiotic and biotic complexity.

Natural organic matter (NOM) is a conceptual subset within these complex mixtures and a key component of the global carbon cycle as well as one of the most intricate mixtures of organic compounds on earth.[15] NOM molecular composition and structure follow a dynamic equilibrium that is shaped by ecosystem characteristics with contributions from biochemical and abiotic reactions. Fundamental concepts, such as NOM lability, recalcitrance and persistence, were derived from relative differences in bulk parameters, which are subject to extensive intrinsic averaging and therefore in principle not capable of providing a definite description of NOM molecular and structural features beyond gross oversimplifications. However, massive signal overlaps leading to information projection in complex mixtures does not only apply to bulk measurements including UV, fluorescence and IR spectroscopy but also to more high-resolution techniques such as NMR spectroscopy and FTICR mass spectrometry.

Furthermore, NOM compositional and structural features reflecting the ecosystem characteristics from bulk to spectroscopy level indicate an overall restricted structural diversity determined by biochemistry rather than genuine statistical distributions. This is a result of the continual incorporation of biomolecules contributing towards a dynamic equilibrium of NOM synthesis and degradation. Initial processing of NOM will create chemical bonds between different groups of biomolecules[30] that are rare in biochemistry and hence less susceptible to enzymatic degradation (and difficult to be recognised in standard characterization).

Bio-orthogonal radical reactions (e.g. attack by hydroxyl radicals, lignin and tannin formation, and Fenton and quinone chemistry) initiate large scale production of novel molecules[31] and strongly attenuate biosignatures. Photochemistry recycles a high proportion of otherwise non-(bio)degradable molecules back to the biosphere. While organic matter transformations in an autumn forest soil transform biogenic leaf organic matter in days, subduction of marine sediments initiates organic matter removal and processing on millennial to geological time scales.[32] Adequate orthogonal and integrated combinations of high-resolution analytical methods are required to better understand the chemical structures of these complex organic mixtures and the processes involved in the context of global element cycling.[33] Overall, it is not expected that NOM structural diversity follows pure mathematical probabilities of structure generation. Furthermore, NOM structural diversity will incorporate core biochemical features, however, in a heavily modified manner that cannot be recognized by common analytical techniques because linkage information is either ill-defined or entirely inaccessible.

The abiome

The abiome brings us to all non-enzymatic, non-biological chemistry, such as that observed and described already in the primordial "soup". Our understanding of this ancient Earth chemistry is growing and is illustrated by non-biological systems of reactions that could have formed the network's core for converting carbon dioxide into organic compounds, as one of many seeds in the emergence of life.[34–36] Involved here are complex chemical processes of prebiotic chemistry, such as those found in extreme chemical environments like hydrothermal waters,[26] oceanic hydrothermal vents, deep-reaching tectonic faults, *etc.*[37] The classical surficial *"warm little pond"* concept introduced by Charles Darwin was most closely realized by the experiments conducted by Miller in 1953[38] or later by Wächtershäuser on catalytic surfaces.[39]

Many of these chemical reactions observed by simulations of early Earth conditions, such as hydrolysis, thermochemical sulfate reduction, photochemical, oxidation processes or various chemical conjugations, are found in all fields of life, the environment and food, to cite only a few.

To give a detailed example to illustrate the abiome, we choose the non-enzymatic browning reactions, also termed the Maillard reaction or advanced glycation. These are reactions between reducing sugars and amino compounds (*e.g.* amino acids, peptides, proteins), which lead to a heterogeneous and complex mixture of new compounds.[40,41] In heated foods, such Maillard reaction products (MRPs) are the main contributor to aroma and color formation.[42] Under physiological conditions, analogous reactions can lead to irreversible protein damage and cellular dysfunctions.[43]

The Maillard reaction is not a single type of chemical reaction but rather a superimposition of many simple chemical transformations that eventually can form a huge reaction network. Many of the reactions, independent of the precursor molecules, follow regular reaction patterns,[44] as shown in Fig. 5. Different amino precursors can flow through the shown pathways (Fig. 5A) in the same way. Typical reactions that are observed in the initial and intermediate stage of the reaction cascade are extended dehydration series, redox, and carbonyl cleavage reactions.[44,45] The amino precursor is mainly responsible for the molecular characteristics of the formed reaction products.[44] The almost infinite combining possibilities of precursor molecules, which are available for Maillard or Maillard-type reactions in complex systems (*e.g.* food products), and the same type of underlying chemical reactions ultimately lead to an extraordinary high chemical diversity. In food products, thousands of new compounds can be formed readily from only a few initial precursor molecules (Fig. 5B). As the reaction progresses, compounds with a higher degree of unsaturation and aromaticity are formed. This goes hand-in-hand with a continuous decrease in the available degrees of freedom for bond formation and compilation of atoms in the molecules (see also Fig. 6). Consequently, the compositional properties of reaction products formed from different precursor molecules more and more converge towards "end-products" of similar chemistry.[44] However, additional reactive moieties in the precursor molecules, such as thiols, may strongly influence the reactivity and form a large pool of mostly unpredictable reaction products (Fig. 5C). Nevertheless, an understanding of the entire reaction networks is

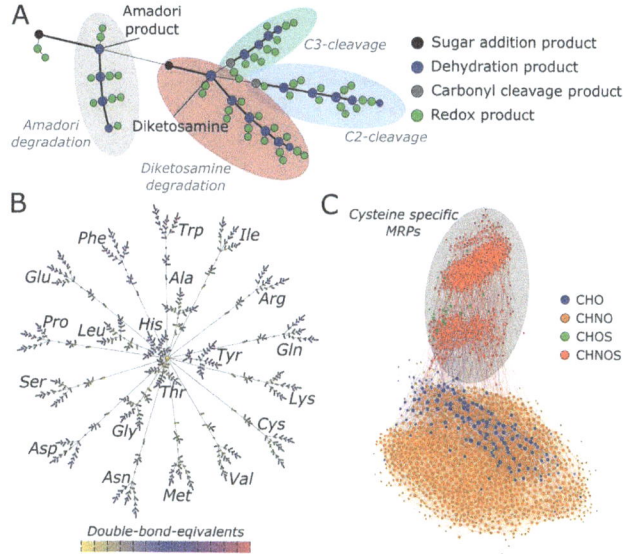

Fig. 5 (A) Schematic Maillard reaction cascade for one amino acid with one reacting sugar, (B) starting with the same sugar and the individual proteogenic amino acids, the series of compounds obtained have a high diversity with no compositional overlap and always the same reaction scaffold, (C) starting with four amino acids in mixtures and one sugar shows cross reactivity and complexity with specific reactivity of sulfur containing Cys.

key to manipulating such complex reaction cascades towards the formation of desired reaction products, while at the same time, the formation of unwanted intermediates and products is avoided. The high chemical diversity observed in the Maillard reaction and also the structural similarity (*e.g.* the formation of isomeric species by carbonyl migration mechanisms) require the highest resolution in several analytical dimensions (*i.e.* separation science, mass spectrometry and exact mass analysis, spectroscopy) to describe the chemistry even in the simplest model systems. Maillard is a great showcase of chemical diversity and complexity in general, and is found in thermal processed foods.[46] To cite here the opposite tendency in clinging to essentials, reducing complexity, "*Note by Note cooking*", as an application of molecular gastronomy, is about reducing complex foods to their elementary compounds: "*Dishes are made entirely from pure compounds or mixtures of pure compounds. No meat, fish, fruits or vegetables are used in the recipes. The aim is not to re-create foods which already exist but to create new foods and potentially new flavours. The shapes, colours, textures, consistency, odours, temperatures and trigeminal stimulation can all be designed by the chef*".[47]

The biome

Life in a nutshell – biology is (bio)chemistry as seen from a molecular perspective. Biotas are islands of high chemical organization at different scales within both theoretical chemical space and their abiotically processed environment. Subcellular organization – compartmentalization – is a living organism's central tool to

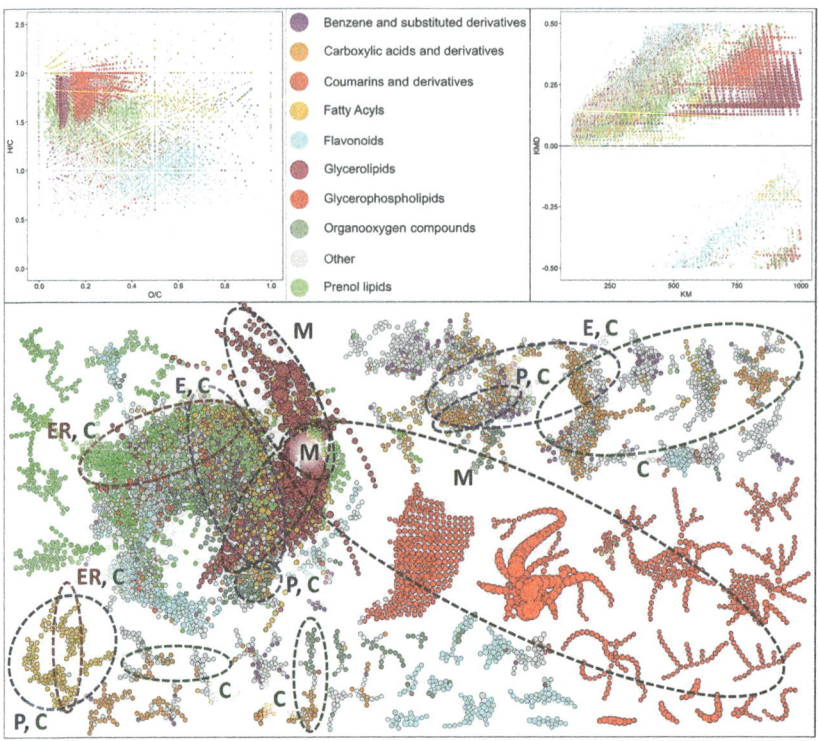

Fig. 6   Upper left: A van Krevelen (VK) diagram visualization of 9118 CHNOPS metabolites from HMDB. All features are colored according to their compound classes (the nine most abundant compound classes are colored in bright colors; all the others are shown in gray). The scatter dot sizes are scaled relative to the number of isomers per formula, as listed in HMDB. Upper right: A Kendrick mass defect plot of the same data. Bottom: A mass difference network of 6654 HMDB molecular formulas (see the main text for more details). The colors are according to the legend in the upper center. The node size is proportional to the count of isomers in HMDB. The data set was mapped against the Recon 2.2 model of the human metabolism. The regions with high annotation densities are cellular compartments C (cytoplasm), E (extracellular), ER (endoplasmic reticulum), P (peroxisome), and M (membrane).

separate its chemical space in ways that support biosynthetic processes. There could be no establishment of electrochemical gradients – the basis for energy production and conduction of neural impulses – without lipidic membrane bilayers, whose building blocks are largely hydrophobic lipids. The basic currency, by which energy is generated and stored, spans highly hydrophilic sugars, carboxylic acids as well as ATP and NADH, whose action needs to be orchestrated both spatially and temporally, and a large number of metabolic intermediates with intermediate log P and varying pK_a values need to be guided towards their appropriate reaction partners to maximize their contribution to an organism's central aim, homeostasis. On the other hand, building blocks of biomembranes may need to be made 'ready for transport' across hydrophilic regions of a cell by means of conjugation to carnitines, which serve as ligand structures for corresponding receptors at mitochondrial membranes. The key language of all these processes is chemistry, and its (bio-)chemical syntax follows rules that are currently too complex for scientists to fit into a single, global model on cells, organs, organisms or their ecology. The most straightforward way that systems biology models this syntax of biochemistry is based on biochemical (metabolic) pathways, signaling pathways and other networks.

The complexity of a living being's organization requires instrumental analytics and scientific methods that can detect, identify, quantify and characterize metabolic building blocks. The best that these intricate analytical methods can do is to either get microscopic, detailed information on the organization of a small number of instances or a macroscopic, generic representation of many instances. Likewise, structural elucidation and quantification of metabolites (generally considered as "metabolomics") must merely cover those metabolites with the highest affinity towards an analytical system (NMR spectroscopy and mass spectrometry are biased towards entirely different subsets of metabolites). A wider range of metabolically relevant features can be detected with ultrahigh resolution mass spectrometry (UHR-MS), but the information obtained is neither quantitative nor complete in terms of structure elucidation. The task to understand and ultimately control complex systems by means of instrumental analysis must therefore always suffer from over-generalization along any of the analytically relevant dimensions (time, sensitivity, costs; see Fig. 8).[48] Fig. 6 shows a possible representation of human metabolism through the eyes of a UHR mass spectrometer with infinite resolution and perfect sensitivity. The upper panels in Fig. 6 show traditional representations of UHR-MS feature spaces, *i.e.* a van Krevelen diagram[49,50] (which requires molecular formulas to be assigned to m/z features) and a Kendrick mass defect plot[51] (KMD plots are based on accurate experimental m/z values).

All representations in Fig. 6 show 9118 unique molecular formulas mined from HMDB 4.0 (released 07/2018). The nine most frequent compound classes are colored, and the others are in gray. The dot and node sizes in the plots are proportional to the number of isomers listed for each represented molecular formula. Fig. 6 shows that metabolites with acetyl-CoA as a fundamental building block donor are the most readily discerned on a van Krevelen diagram and a KMD plot (the colour legend is similar for the whole figure). The corresponding glycerolipids and glycerophospholipids further show the largest numbers of known isomers per molecular formula, which substantiates their prominence: mass spectrometry-based methods for metabolome analysis are biased towards lipids, which are very efficiently ionized by electrospray ionization systems. Their retention time to structure relationships in LC-MS are straightforward and they can even be discerned in shotgun MS on a coarse compound class level. Their rules of formation further facilitate the prediction of structures, leading to an over-representation in metabolic databases.

While flavonoids can be well distinguished from the above mentioned two lipid classes in VK diagrams and KMD plots, most other compound classes appear to be intricately entangled. Better visual and conceptual discrimination of mass spectrometric data can be achieved by Kendrick-analogous mass difference networks (MDiNs),[52,53] as displayed in the lower part of Fig. 6. Molecular formulas with their monoisotopic exact masses are nodes and corresponding differences in element counts and their corresponding exact mass differences are edges in the MDiNs. The presented MDiN was built from the same data as the VK diagram and

KMD plot, and formulas were connected if their corresponding mass differences matched changes in CH_2, H_2 or O. All connections are purely compositional by nature, but they are bound to encompass true chemical reactions. An MDiN on MS data can therefore be seen as a draft metabolic network, *i.e.* a generic framework to mine the syntax of biochemistry from. The MDiN presented here shows the largest 50 sub-networks that result from the above described MDiN reconstruction. It covers 6654 (76%) of HMDB molecular formulas belonging to CHNOPS compositional space. The largest graph component (an isolated sub-network that is not connected to other sub-networks) covers 68% of all corresponding HMDB features. This MDiN articulates the special role of glycerolipids and glycerophospholipids as indicated by their large node sizes, their clear separation and highly ordered net-like patterns. All other compound classes appear entangled intricately once more, yet they show a clearer visual separation on the graph than on the VK diagram and KMD plot. A further mapping of a genome scale metabolic network model of human metabolism (Recon 2.2,[54] extracted from Metexplore,[55] containing 1018 unique molecular formulas) against this HMDB-MDiN led to 769 hits (8.8% of HMDB; 76% of Recon 2.2) and allowed the pinpointing of regions on the graph that were majorly populated by metabolites characteristic of certain cellular compartments. This example shows that the projection of both structural chemical space and even cellular compartments populate distinct regions in compositional space. While compositional space alone does not resolve information on a scale that allowed for mechanistic insights into all processes perturbing a complex biological system, its usefulness as a recommendation system – a coarse guide through complexity – turns out to be apparent.

Analyzing the metabolome certainly is the greatest challenge in modern life science omics. While in genomics, transcriptomics or proteomics the analysis relies on mathematical analysis of repetitive subunits, this is far from being the case in metabolomics. Metabolites are measured from biological samples and we have many application examples related to body fluids, such as urine, saliva, blood plasma, and tissue samples.[56,57] Even simple breathing (exhaled breath condensates) can carry information about the state of health.[58] The methods of choice for targeted or non-targeted analysis are presented in Fig. 7 as a "systems chemical analytics" toolbox.

Towards systems chemical analytics

We have presented natural complex mixtures as originating from biotic and abiotic chemical processes and defined the complex dimensions in the scale of chemical evolution. Chemistry is everywhere and there is no analytical approach to fully characterize the chemistry of a complex natural system as a whole. To quote Goethe at this point: "*One only sees what one looks for. One only looks for what one knows*". Converted to our topic, with any available technology "*one only sees what one looks for with the focus of the analytical system one uses*". Targeted analysis may be appropriate to verify a hypothesis, and non-targeted more holistic approaches may be better suited to generate a new hypothesis in addition within the frame of the analytical observations.

Novel analytical technologies enable a partial description of chemical diversity and a description of biogeosystems at different scales, and some examples are

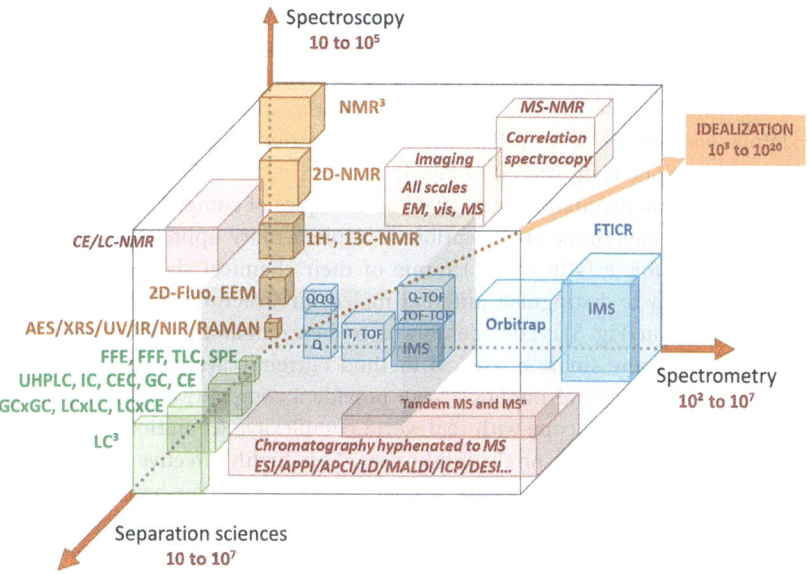

Fig. 7 A schematic representation of the systems chemical analytics space involving dimensions of spectrometry, spectroscopy and separation sciences with increasing resolution. Hyphenations such as LC-NMR, MS-NMR or LC-MS are given on the corresponding side plane projections. Additional resolution or orthogonality is obtained in mass spectrometry with various ionization methods, mass fragmentations and/or ion mobility. The gray cube in the centre covers the analytical methods used in routine applications.

presented in this Faraday Discussion. To name only a few, mass spectrometry is well represented as an orthogonal detection method following chromatography (gas and liquid chromatography (GC and LC)) or in its ultrahigh resolution mode (FTICR-MS) as a direct injection method. Ionization sources, such as electrospray- or photo-ionization (ESI, APPI), ion mobility spectrometry (IMS) and a deuterium exchange approach (specificity on H/D exchange of the analytes), are also presented for the front end of mass spectrometry as possible additional selectivity dimensions. Chromatography is presented as a sample preparation technology (solid phase extraction (SPE), supercritical extraction (SFE)) and as an analytical separation tool (GC, size exclusion chromatography (SEC), reverse phase or hydrophilic interaction chromatography (HILIC)). In these examples, the mode of detection includes spectroscopic (UV/DAD, NMR, FTIR) and mass spectrometric techniques with additional dimensions of information achieved by tandem MS. NMR methods are well represented as well in this volume with novel two or three dimensional setups, combined for an efficient description of structural information out of complex mixtures. Mathematics and statistics offer important modern tools for understanding data and provide context in terms of significant chemical and/or biological information. A few manuscripts also focus specifically on chemometrics as a universal tool to mine correlated information (*e.g.* bioactivity) in multidimensional analytical spectra and data.

For investigations of complex natural mixtures, a systems approach using different bioanalytical approaches[59] with the highest orthogonality appears the most appropriate to reveal the most significant detail. We would call this approach "systems chemical analytics" or SCA. Systems chemical analytics involves selective separation – spectral and/or spectrometric technologies in various resolution setups complemented with extensive mathematical big-data mining (multivariate statistics, neural networks and artificial intelligence, machine learning, *etc.*). An exhaustive representation of possible SCA elements and combinations is illustrated in Fig. 7.

Bringing back metabolomes or NOM as examples of complex natural (super) mixtures, one can think about various complementary approaches that only together assemble a (still coarse) frame of their chemical shape. The mathematical capacity to distinguish different molecular structures of a given molecular composition (Fig. 3) is considerably beyond current awareness and far above the capacity of any single analytical method currently available. Hence, only a combination of several methods[16] can provide a competitive degree of information to successfully cope with real world complexity. Separation sciences, to cite only chromatography or electrophoresis,[60] are highly effective for achieving complexity reduction by the physical separation of molecules based on a wide range of discrimination criteria (size, hydrophobicity, charge, interaction, affinity, *etc.*). High-field NMR spectroscopy provides the capability for quantitative and non-destructive *de novo* determination of chemical environments from any polydisperse and molecularly heterogeneous environmental samples. Quantitative relationships between a number of spins and area (1D NMR spectroscopy) and volume (2D NMR spectroscopy) of NMR resonances operate in the absence of differential NMR spectroscopy. This key feature implies the use of NMR spectroscopy as a quantitative reference for complementary structure-selective analytical methods, like mass spectrometry (which detects gas phase ions and is subject to ionization selectivity) in the case of complex mixtures and fluorescence spectroscopy (which selectively detects fluorescent chemical environments of sp^2-hybridized carbon). To contrast the diverging selectivity of analytical methods: NMR spectroscopy is particularly informative in the description of aliphatic chemical environments which are based on sp^3-hybridized carbon. These are inactive in fluorescence spectroscopy, and the difference in the size of the aliphatic groups will cause rather inconspicuous mass shifts in the FTICR mass spectra: more expansive aliphatic systems will result in higher mass molecules, with somewhat larger H/C and smaller O/C elemental ratios. This characteristic is, however, insufficient to allow reliable conclusions about chemical structures. NMR spectroscopy enables distinction of the size of aliphatic units and also allows for in-depth assessment of the intrinsic chemical environments, like open chain and cyclic arrangements of carbon. The chemical diversity in aliphatic networks as well as typical NMR transverse relaxation rates decrease in the order $C_q > CH > CH_2 > CH_3$ leading to more comprehensible NMR-based constraints for the chemical environments of methyl groups than for carbon deep within branching networks.[61] Mass spectrometry and especially its ultrahigh resolution variants, such as FTICR mass spectrometry,[62] are perfectly suited for the classification of heteroatom-containing molecules directly from complex mixtures, but mathematical elaboration is required to reveal molecular compositions. Traditional spectra (intensity *versus* mass) show useful trends of mass evolution but the intrinsic molecular complexity is caused by isotope-specific mass defects and resides in the internal decimal places, which are not apparent from full width spectra. Various projections of the compositional space separate integer and decimal mass numbers in two dimensions; examples include van Krevelen diagrams, mass-edited H/C ratios, and Kendrick mass defect analysis including KMD/z* diagrams, but also average carbon oxidation state *vs.* mass and other diagrams provide useful relationships between contiguous homologous series and classes of compounds. This kind of data analysis can be advanced to create multidimensional mass difference networks as described above.

As illustrated in Fig. 8, GC-MS and LC-MS are well established in the field of bioanalysis and they benefit from continuously growing databases. In the quest to optimize resolution (for molecular annotation), sensitivity (for trace quantifications) and robustness (for automation), further modern developments involve comprehensive 2D chromatography (LC×LC, GC×GC) with hyphenations to MS in the highest resolution with automated fragmentation strategies. Ideally, one seeks a technology that enables, in a few steps and short time, the integration of the data with the data analyzed in the past, using the growing knowledge over time. Quantitative approaches, such as NMR spectroscopy, enable such data integration with future analysis without complicated transfer functions and the technology can be partially automated. Rapid direct injection ultrahigh resolution mass spectrometry (UHR-MS), such as FTICR-MS and Orbitrap, is a growing field that is combined as a rapid screen (a few minutes) with LC/MS analysis for compound identification. Costs are often rapidly a limiting issue, and resolution limitations should also set the frame of possibilities or provide orientation in the choice of the right orthogonal analytical tools.[63]

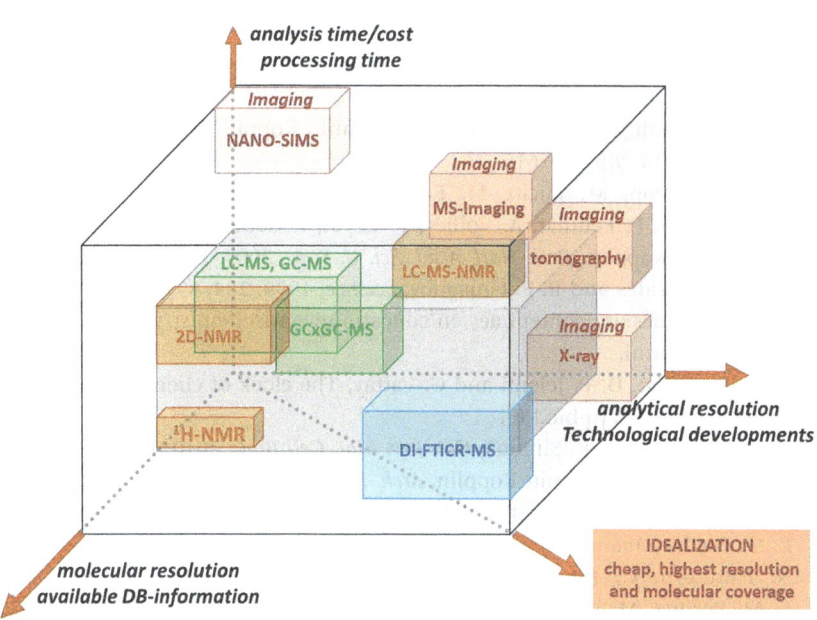

Fig. 8 Schematic representation of the systems chemical analytics as pragmatically and conceptually adapted from project management. Ideally, one aims for limiting costs by short analysis and processing times while maximising resolution in both dimensions (molecular and analytical resolution).

The rapid development of analytical technologies enables even more detailed and sensitive descriptions of complex natural systems, which is perfectly in line with our holistic goals in understanding complex processes and discovering new ones. Systems chemical analytics thus embodies the best combination of approaches for complexity deconstruction and systems simplification. On the other hand, in a pragmatic approach, even in the frame of endless analytical possibilities (in costs, information and resolution), one should never forget the goals within each individual project/experiment to adapt the systems chemical analytics accordingly.

Conflicts of interest

There are no conflicts to declare.

References

1 P. Ball, *Nature*, 2008, **453**, 160–162.
2 D. Bonchev and W. A. Seitz, The concept of complexity in chemistry, in *Concepts in Chemistry: A Contemporary Challenge*, ed. D. H. Rouvray, Chemistry Publisher: Wiley Editors, 1997, ch. 11, pp. 353–381.
3 P. Ferreira, *Tracing complexity theory, ESD 83: Research seminar in Engineering systems*, Massachusetts Institute of Technology MIT, 2001.
4 P. Galanter, Complexism and the role of evolutionary art, in *The art of artificial evolution; a handbook on evolutionary Art and Music*, ed. J. Romero and P. Machado, Springer, Berlin, 2018, pp. 311–332.
5 S. E. Phelan, *Syst. Pract. Action Res.*, 1999, **12**(3), 238.
6 J. McIntyre, *Global Food History*, 2019, **5**(1–2), 1–4.
7 C. Roullier-Gall, L. Boutegrabet, R. D. Gougeon and P. Schmitt-Kopplin, *Food Chem.*, 2014, **152**, 100–107.
8 C. Roullier-Gall, M. Lucio, L. Noret, P. Schmitt-Kopplin and R. D. Gougeon, *PLoS One*, 2014, **9**(5), e97615.
9 R. D. Gougeon, M. Lucio, M. Frommberger, D. Peyron, D. Chassagne, H. Alexandre, F. Feuillat, A. Voilley, P. Cayot, I. Gebefügi, N. Hertkorn and Ph. Schmitt-Kopplin, *Proc. Natl. Acad. Sci. U. S. A.*, 2009, **106**(23), 9174–9179.
10 G. M. Whitesides and R. F. Ismagilov, *Science*, 1999, **284**, 89–92.
11 https://www.acs.org/content/acs/en/education/whatischemistry/everywhere.html.
12 S. A. Kauffman, D. P. Jelenfi and G. Vattay, The clock of chemical evolution, arXiv: 1806.06716 [q-bio.PE].
13 A. Shevchenko and K. Simons, *Nat. Rev. Mol. Cell Biol.*, 2010, **11**, 593–598.
14 M. Witting and P. Schmitt-Kopplin, *Arch. Biochem. Biophys.*, 2016, **589**, 27–37.
15 N. Hertkorn, M. Frommberger, M. Witt, B. P. Koch, P. Schmitt-Kopplin and E. M. Perdue, *Anal. Chem.*, 2008, **80**, 8908–8919.
16 N. Hertkorn, C. Ruecker, M. Meringer, R. Gugisch, M. Frommberger, E. M. Perdue, M. Witt and P. Schmitt-Kopplin, *Anal. Bioanal. Chem.*, 2007, **389**, 1311–1327.
17 J. Liu, S. Wang, T. E. Balius, I. Singh, A. Levit, Y. S. Moroz, M. O. O'Meara, T. Che, E. Algaa, K. Tolmachova, A. A. Tolmachev, B. K. Shoichet, B. L. Roth and J. J. Irwin, *Nature*, 2019, **566**, 224–229.

18 R. Tecon, S. Mitri, D. Ciccarese, D. Or, J. R. van der Meer and D. Johnson, *mSystems*, 2019, **4**(1), e00265-18.

19 T. C. G. Bosch and D. J. Miller, *The Holobionte imperative: Perspectives from Early emerging animals*, Springer, 2016, pp. 1–155.

20 J. Guigue, M. Harir, O. Mathieu, M. Lucio, L. Ranjard, J. Levêque and P. Schmitt-Kopplin, *Biogeochemistry*, 2016, **128**(3), 307–326.

21 M. Gonsior, J. Valle, P. Schmitt-Kopplin, N. Hertkorn, D. Bastviken, J. Luek, M. Harir, W. Bastos and A. Enrich-Prast, *Biogeosciences*, 2016, **13**(14), 4279–4290.

22 N. Hertkorn, M. Harir, K. M. Cawley, P. Schmitt-Kopplin and R. Jaffe, *Biogeosciences*, 2016, **13**, 2257–2277.

23 N. Hertkorn, M. Harir, B. P. Koch, B. Michalke and P. Schmitt-Kopplin, *Biogeosciences*, 2013, **10**, 1583–1624.

24 K. Ksionzek, O. Lechtenfeld, S. L. McCallister, P. Schmitt-Kopplin, J. K. Geuer, W. Geibert and B. P. Koch, *Science*, 2016, **354**(6311), 456–459.

25 D. Schulze-Makuch, D. Wagner, S. P. Kounaves, K. Mangelsdorf, K. G. Devine, J. P. de Vera, Ph. Schmitt-Kopplin, H. P. Grossart, V. Parro, M. Kaupenjohann, A. Galy, B. Schneider, A. Airo, J. Frosler, A. F. Davila, F. L. Arens, L. Caceres, F. Sols Cornejo, D. Carrizo, L. Dartnell, J. DiRuggiero, M. Flury, L. Ganzert, M. Gessner, P. Grathwohl, L. Guan, J. Heinz, M. Hessa, F. Kepplerb, D. Maus, C. P. McKay, R. U. Meckenstock, W. Montgomery, E. A. Oberline, A. J. Probst, J. S. Saenz, T. Sattlerb, J. Schirmack, M. A. Sephton, M. Schloter, J. Uhl, B. Valenzuela, G. Vestergaard, L. Wormerd and P. Zamorano, *Proc. Natl. Acad. Sci. U. S. A.*, 2018, **115**(11), 2670–2675.

26 M. Gonsior, N. Hertkorn, N. Hinman, S. E. M Dvorski, M. Harir, W. J. Cooper and P. Schmitt-Kopplin, *Sci. Rep.*, 2018, **8**(1), 14155.

27 V. La Cono, G. Bortoluzzi, E. Messina, G. La Spada, F. Smedile, L. Giuliano, M. Borghini, Ch. Stumpp, P. Schmitt-Kopplin, M. Harir, W. K. O'Neill, J. E. Hallsworth and M. Yakimov, *Sci. Rep.*, 2019, **9**(1), 1679.

28 R. U. Meckenstock, F. von Netzer, C. Stumpp, T. Lueders, A. M. Himmelberg, N. Hertkorn, P. Schmitt-Kopplin, M. Harir, R. Hosein, S. Haque and D. Schulze-Makuch, *Science*, 2014, **345**(6197), 673–676.

29 P. Schmitt-Kopplin, A. Gelencser, E. Dabek-Zlotozyrnská, G. Kiss, N. Hertkorn, M. Harir, Y. Hong and I. Gebefügi, *Anal. Chem.*, 2010, **82**, 8017–8026.

30 G. A. McKee and P. G. Hatcher, *Geochim. Cosmochim. Acta*, 2010, **74**, 6436–6450.

31 D. C. Waggoner and P. G. Hatcher, *Org. Geochem.*, 2017, **113**, 315–325.

32 P. B. Kelemen and C. E. Manning, *Proc. Natl. Acad. Sci. U. S. A.*, 2015, **112**, E3997–E4006.

33 Y. Li, M. Harir, M. Lucio, M. Gonsior, B. P. Koch, P. Schmitt-Kopplin and N. Hertkorn, *Water Res.*, 2016, **106**, 477–487.

34 A. Ruf, L. L. S. d'Hendecourt and P. Schmitt-Kopplin, *Life*, 2018, **8**, 18.

35 T. Böttcher, *J. Mol. Evol.*, 2018, **86**, 1–10.

36 K. Muchowska, S. J. Varma and J. Moran, *Nature*, 2019, **569**, 104–107.

37 B. Trias, B. Ménez, P. le Campion, Y. Zivanovic, L. Lecourt, A. Lecoeuvrel, P. Schmitt-Kopplin, J. Uhl, S. R. Gíslason, H. A. Alfreðsson, K. G. Mesfin, S. Ó. Snæbjörnsdóttir, E. S. Aradóttir, I. Gunnarsson, J. M. Matter, M. Stute, E. H. Oelkers and E. Gérard, *Nat. Commun.*, 2017, **8**, 1063.

38 S. L. Miller, *Science*, 1953, **117**, 528.

39 G. Wächtershäuser, *Syst. Appl. Microbiol.*, 1988, **10**, 207.

40 J. E. Hodge, *J. Agric. Food Chem.*, 1953, **1**, 928–943.

41 F. Ledl and E. Schleicher, *Angew. Chem., Int. Ed. Engl.*, 1990, **29**, 565–594.

42 M. Hellwig and T. Henle, *Angew. Chem., Int. Ed.*, 2014, **53**, 10316–10329.

43 T. P. Labuza, V. M. Monnier, J. Baynes and J. O'Brien, *Maillard Reactions in Chemistry, Food and Health*, Elsevier, 1st edn, 1998.

44 D. Hemmler, C. Roullier-Gall, J. W. Marshall, M. Rychlik, A. J. Taylor and P. Schmitt-Kopplin, *Sci. Rep.*, 2018, **8**, 16879.

45 D. Hemmler, C. Roullier-Gall, J. W. Marshall, M. Rychlik, A. J. Taylor and P. Schmitt-Kopplin, *Sci. Rep.*, 2017, **7**, 3227.

46 H. This, *Notes Académiques de l'Académie d'agriculture de France (N2AF)*, 2017, vol. 1, pp. 1–12.

47 H. This, *Flavour*, 2013, **2**, 1.

48 S. Forcisi, F. Moritz, B. Kanawati, D. Tziotis, R. Lehmann and P. Schmitt-Kopplin, *J. Chromatogr. A*, 2013, **1292**, 51–65.

49 D. W. Van Krevelen, *Fuel*, 1950, **29**, 269–284.

50 S. Kim, R. W. Kramer and P. G. Hatcher, *Anal. Chem.*, 2003, **75**, 5336–5344.

51 E. Kendrick, *Anal. Chem.*, 1963, **35**, 2146–2154.

52 D. Tziotis, N. Hertkorn and P. Schmitt-Kopplin, *Eur. J. Mass Spectrom.*, 2011, **17**, 415–421.

53 F. Moritz, M. Kaling, J. P. Schnitzler and P. Schmitt-Kopplin, *Plant, Cell Environ.*, 2017, **40**, 1057–1073.

54 N. Swainston, K. Smallbone, H. Hefzi, P. D. Dobson, J. Brewer, M. Hanscho, D. C. Zielinski, K. S. Ang, N. J. Gardiner, J. M. Gutierrez, S. Kyriakopoulos, M. Lakshmanan, S. Li, J. K. Liu, V. S. Martinez, C. A. Orellana, L.-E. Quek, A. Thomas, J. Zanghellini, N. Borth, D. Y. Lee, L. K. Nielsen, D. B. Kell, N. E. Lewis and P. Mendes, *Metabolomics*, 2016, **12**, 109.

55 L. Cottret, C. Frainay, M. Chazalviel, F. Cabanettes, Y. Gloaguen, E. Camenen, B. Merlet, J. C. Portais, S. Heux, N. Poupin, F. Vinson and F. Jourdan, *Nucleic Acids Res.*, 2018, **46**(W1), 495–502.

56 S. Forcisi, F. Moritz, M. Lucio, R. Lehmann, N. Stefan and P. Schmitt-Kopplin, *Anal. Chem.*, 2015, **87**(17), 8917–8924.

57 K. Neth, M. Lucio, A. Walker, B. Kanawati, J. Zorn, P. Schmitt-Kopplin and B. Michalke, *Chem. Res. Toxicol.*, 2015, **28**, 1434–1442.

58 F. Moritz, M. Janicka, A. Zygler, S. Forcisi, A. Kot-Wasik, J. Kot, I. Gebefügi, J. Namiesnik and P. Schmitt-Kopplin, *J. Breath Res.*, 2015, **9**, 027105.

59 *Bioanalytics*, ed. F. Lottspeich and J. Engels, Wiley-VCH, 2018, pp. 1–1110.

60 *Capillary Electrophoresis, Methods in Molecular Biology 1483*, ed. P. Schmitt-Kopplin, Springer, 2nd edn, 2017.

61 N. Hertkorn, M. Harir and P. Schmitt-Kopplin, *Magn. Reson. Chem.*, 2015, **53**, 754–768.

62 *Fundamentals and Applications of Fourier Transform Mass Spectrometry*, ed. B. Kanawati and P. Schmitt-Kopplin, Elsevier, 2019, pp. 1–726.

63 *Metabolomics in Practice, Successful Strategies to Generate and Analyze Metabolic Data*, ed. M. Lämmerhofer and W. Weckwerth, Wiley-VCH, Weinheim, 2013.

PAPER

Combating selective ionization in the high resolution mass spectral characterization of complex mixtures†

Ryan P. Rodgers, [ID] *abc Mmilili M. Mapolelo, [ID] d
Winston K. Robbins, [ID] b Martha L. Chacón-Patiño, [ID] a
Jonathan C. Putman, [ID] c Sydney F. Niles, [ID] c Steven M. Rowland[ab]
and Alan G. Marshall [ID] ac

Received 11th January 2019, Accepted 13th March 2019

DOI: 10.1039/c9fd00005d

Direct "dilute and shoot" mass spectral analysis of complex naturally-occurring mixtures has become the "standard" analysis in environmental and petrochemical science, as well as in many other areas of research. Despite recent advances in ionization methods, that approach still suffers several limitations for the comprehensive characterization of compositionally complex matrices. Foremost, the selective ionization of highly acidic (negative electrospray ionization ((−) ESI)) and/or basic (positive electrospray ionization ((+) ESI)) species limits the detection of weakly acidic/basic species, and similar issues (matrix effects) complicate atmospheric pressure photo-ionization (APPI)/atmospheric pressure chemical ionization (APCI) analyses. Furthermore, given the wide range of chemical functionalities and structural motifs in these compositionally complex mixtures, aggregation can similarly limit the observed species to a small (10–20%) mass fraction of the whole sample. Finally, irrespective of the ionization method, the mass analyzer must be capable of resolving tens-of-thousands of mass spectral peaks and provide the mass accuracy (typically 50–300 ppb mass measurement error) required for elemental composition assignment, and thus is generally limited to high-field Fourier transform ion cyclotron mass spectrometry (FT-ICR MS). Here, we describe three approaches to combat the above issues for (+) ESI, (−) ESI, and (+) APPI FT-ICR MS analysis of petroleum samples. Each approach relies on chromatographic fractionation to help reduce selective ionization discrimination and target either specific chemical functionalities (pyridinic and pyrrolic species (nitrogen) or carboxylic acids (oxygen)) or specific structural motifs (single aromatic core (island) or multi-core aromatics (archipelago)) known to be related to ionization efficiency. Each fractionation method

National High Magnetic Field Laboratory, Florida State University, Tallahassee, FL, USA. E-mail: rodgers@magnet.fsu.edu

bFuture Fuels Institute, Florida State University, 1800 East Paul Dirac Drive, Tallahassee, FL, USA

cDepartment of Chemistry and Biochemistry, Florida State University, Tallahassee, FL, USA

dBotswana Institute for Technology Research and Innovation, Gaborone, Botswana

† Electronic supplementary information (ESI) available. See DOI: 10.1039/c9fd00005d

yields a 2–10-fold increase in the compositional coverage, exposes species that are undetectable using direct "dilute and shoot" analysis, and provides coarse selectivity in chemical functionalities that can both increase the assignment confidence and optimize ionization conditions to maximize compositional coverage.

Introduction

Fossil fuels are immensely complex organic mixtures in terms of the number of chemically distinct components.[1–3] Variations in this complexity are of significant interest, because the composition determines their economic value and behavior. Thus, detailed characterization of molecular structure and chemical functionality in fossil fuels is paramount in all segments of the petroleum industry: exploration, production, and refining.[4] Due to the compositional complexity and inability of GC-based techniques to analyze high-boiling petroleum components, the compositions of most petroleum fractions that boil at temperatures above light gas oil (LGO, ~342 °C) are not yet completely defined.[4] Recent advances in ionization techniques, combined with the performance of high magnetic field FT-ICR mass spectrometry, have led to an increased molecular-level understanding of complex, high-boiling petroleum fractions.[5–7] However, several reports have outlined the problems associated with selective ionization and/or matrix effects that can occur prior to mass spectral detection.[8–12] Those effects limit the accessible compositional information in petroleum-derived samples and other complex systems by direct analysis, and pose a significant hurdle to a comprehensive understanding of petroleum composition and its impact on behavior in exploration, production, and refining. Thus, there is an ever-increasing need for chromatographic fractionation methods designed to both separate suspected chemical functionalities/structural motifs prior to mass spectral analysis and overcome the limitations (selectivity) of current ionization methods. Traditionally, separation prior to analysis is carried out by use of chemical class separation schemes that involve combinations of selective solubility, extractions and column chromatography.[13,14] The schemes for the separations vary, but are most often driven by chemical functionality (i.e. acids, bases and neutrals) followed by chemical group-type (compound class) fractionation. Generally, most separations are designed for ultimate resolution of chemical functionality/group-types without regard to the ionization method used for subsequent mass spectral analysis.[4]

Heteroatom-containing petroleum compounds have long been of interest to the petroleum industry because they are believed to have a significant role in catalyst deactivation, refinery corrosion problems, formation of deposits, and storage instability during upstream and downstream operations.[15,16] Thus, it is of the utmost importance to characterize the heteroatom-containing fractions in petroleum in order to provide insight into the origin of these species with respect to their biological precursors and their compositional dependence/correlation to upstream and downstream problems. Such insight provides an understanding of physiochemical transformations of the biomolecules in maturation into heteroatom-containing species currently found in petroleum and provides invaluable compositional information that can be used for remediation/prevention strategies.[16] It is most commonly termed the "polar fraction", because it is isolated with solvents of higher polarity than heptane and toluene, and it is enriched in nitrogen (N), oxygen (O), and sulfur (S) containing molecular species.[13,17-19] Recently, there has been increased interest in the characterization of heteroatom-containing compounds/fractions in petroleum feedstocks, specifically those containing N, S, and O, as well as asphaltenes (the n-alkane-insoluble, but toluene-soluble fraction of petroleum).[20-29] Petroleum crude oils are typically composed of ~0.1-2% wt of organic nitrogen compounds, ~0.1-6% wt organic sulfur compounds, ~0.1-1.5% wt organic oxygen compounds, and ~0.1 to 19% wt asphaltenes.[30-32] Nitrogen compounds in petroleum-derived streams are mostly alkylated heterocycles with a predominance of neutral nitrogen (pyrrolic) structures over the basic nitrogen (pyridinic) forms.[33,34] Sulfur compounds consist of thiophenic structures (aromatic), linear/cyclic/aromatic sulfides, and mercaptans.[35-37] Oxygen compounds are more functionally diverse, and consist of carboxylic acids, ketones, aldehydes, alcohols, and furanic heterocycles.[38] Asphaltenes are a solubility class of petroleum that is enriched in aromatics and heteroatoms. It is an ultra-complex fraction of crude oil that contains many, if not all, of the chemical functionalities listed above.[39-43]

The higher boiling point of all heteroatom-containing species due to N, S, and O addition, combined with their high compositional complexity, limits characterization by GC-based techniques.[25] Thus, advances in the characterization of these species has been led by high resolution mass spectrometry.[44-51] Despite these advances, many fundamental questions remain unanswered. First, since mass spectral signal-to-noise is not always directly related to the concentration of a specific analyte in the solution (due to ionization efficiency differences and matrix effects), the overall coverage of the total compositional range captured in the mass spectral data is unknown. Recent results suggest that direct "dilute and shoot" analysis of complex petroleum samples can capture as little as 10% of the overall compositional complexity, representing less than 15 wt% of the sample.[52-54] Proper selection of the ionization method can ease the impact of selective ionization, but only subsequent fractionation followed by mass spectral analysis of all fractions provides a measure of the effectiveness. Most researchers in the field do not perform such analyses, due to the additional expense and effort involved. Moreover, the impact of selective ionization within a chemical functionality class (*i.e.* carboxylic acids) has largely been ignored, but can similarly limit the compositional coverage to levels noted above.[52,55] Thus, even within a chemical functionality class, separations are crucial. Finally, as discussed above, aggregation can severely restrict the compositional information attained from direct (whole) sample analysis. Although recently discussed for asphaltene analysis, the topic requires further investigation (see below).

Gas chromatography (GC) separates heteroatom-containing compounds according to boiling point, whereas liquid chromatography (LC) can serve to separate with respect to polarity and/or hydrophobicity (carbon number). Thus, prior separation of heteroatom-containing compounds by LC has the potential not only to simplify mass spectrometric analysis, but also to address selective ionization in the ion source. Schemes for separating nitrogen compounds into well-defined compound classes have been used to eliminate some of the ambiguities, and therefore assist in compound identification by MS.[56] Here, water is added to the oven-dried stationary phase to deactivate highly active sites. The water increases the loading capacity of the stationary phase at the expense of high selectivity to further optimize the separation for ESI mass spectral analysis (the separation of components with widely different chemical functionality or polarity). Clinching evidence of the identification and partial recovery of the lost selectivity of the stationary phase is provided by exploiting the selectivity of ESI. Pyridinic (basic) nitrogen is readily detected as $[M + H]^+$ by protonation, whereas pyrrolic (neutral) nitrogen species are detected as $[M - H]^-$ by deprotonation.[21,24] Thus, polarity-based separation combined with selective ionization facilitates detection and differentiation between the 2 major forms of nitrogen functionalities in petroleum. Hydrophobicity-based separation strategies further assist in selective ionization within a chemical functionality class, because they separate by carbon number (molecular weight).[52,55] Finally, extrography fractionation can separate structural motifs (island (single pericondensed aromatic core) vs. archipelago (multiple alkyl-linked aromatic cores)) in asphaltene samples that exhibit wide variation (~50-fold) difference in ionization efficiency.[54,57,58] All three approaches expand the compositional coverage relative to direct (unfractionated) sample analysis and identify species that were not previously accessible.

Here, we describe three separations that can help overcome selective ionization in the analysis of petroleum samples: (1) a column chromatography method that fractionates deasphalted crude oils into four well-defined fractions that can be classified as nonpolar and polar fractions. The column separation unequivocally separates the two main types of nitrogen compounds found in crude oils. The separation affords chemical class separation of polar species according to aromaticity, polarity, and molecular weight prior to FT-ICR MS analysis. (2) a modified aminopropyl silica (MAPS) fractionation that first retains acidic species and subsequently separates them by hydrophobicity (degree of alkylation/ carbon number) for mass spectral analysis. (3) An extrography fractionation method that employs silica gel as the stationary phase that selectively targets the removal of asphaltene species that have high ionization efficiencies relative to other asphaltene species. Removal of these species facilitates detection of less efficiently ionized species that are not observable in the analysis of the whole asphaltene sample. The ionization efficiency differences are related to molecular structure and tendency to aggregate.[54]

Materials and methods

Samples and preparation

Three crude oils of different geographical origin were used as supplied. The crude oil properties are provided in Table 1. MacKay bitumen is highly degraded with high sulfur (S), moderate nitrogen (N) and high oxygen (O) (naphthenic acid) content.[59] South American crude oils are typically rich in both S and N with complex "bottoms", whereas the Central American "Maya" crude oil is sulfur-rich

Table 1 General properties of the parent crude oils

Sample	Origin	Crude oil (API gravity @ 15 °C, TAN)
Maya crude oil	Central America	21.3°, 0.3 mg KOH per g oil
South American crude oil	South America	25.9°, 0.70 mg KOH per g oil
MacKay bitumen	Canada	8.0°, 2.49 mg KOH per g oil

but has a low nitrogen content.[60,61] The South American crude oil was used for the nitrogen separation, MacKay bitumen for the MAPS fractionation, and Maya crude oil asphaltenes for the extrography separation. HPLC grade n-heptane, toluene, tetrahydrofuran and methanol (J.T. Baker), Whatman No. 1 filter paper (Fisher Scientific), deionized water, silicic acid n-hydrate powder, and anhydrous sodium sulfate (Na_2SO_4) (J.T. Baker), sand (white quartz $-50 + 70$ mesh), chromatographic grade silica gel (100–200 mesh, type 60 Å, Fisher Scientific), glass wool, ammonium hydroxide (NH_4OH) (20–35% ammonia in water), formic acid ($HCOOH$) and tetramethylammonium hydroxide [TMAH (($CH_3)_4N^+OH^-$)] (Sigma-Aldrich) were used as supplied.

Preparation of 10% water-deactivated silicic acid and column packing

100 g of silicic acid n-hydrate powder was oven-dried (activated) at 120–140 °C overnight. The activated silicic acid was allowed to cool in a desiccator. To deactivate the silicic acid, 90 g of the dry silicic acid was combined with 10 mL of deionized water. The deactivated silicic acid was put in a wide-mouthed jar, agitated to eliminate lumps and allowed to equilibrate overnight. Sixty grams of deactivated silicic acid was mixed with 100 mL of n-heptane to make a slurry. The slurry was kept well-mixed and poured into a chromatographic column (25×300 mm) fitted with a 300 mL glass reservoir at its top and a medium glass frit and stopcock at its bottom. The excess n-heptane was allowed to drain through the stopcock until the solvent level reached the top of the stationary phase. Once all of the deactivated silicic acid had settled into the column, the sides of the column were rinsed with n-heptane, leaving several inches of solvent above the deactivated silicic acid stationary phase. Approximately 12 g (1/2 inch) of sand was added on top of the packed deactivated silicic acid stationary phase, followed by an additional wad of glass wool. About 2 mL of n-heptane was left above the top of the glass wool. The column could be dry-packed; however, better resolution and a more consistent behavior can be achieved through the slurry packing technique. The slurry packing eliminates air pockets and uneven flow down the column. The sand and glass wool at the column top allow move even distribution during loading of the sample.

Asphaltene precipitation and isolation

10 g of each of the two crude oils (South American and MacKay bitumen) was deasphalted *via* precipitation of asphaltenes in an excess of saturated hydrocarbon solvent (n-heptane) (40 : 1) prior to chromatographic separation. A Whatman No. 1 filter paper was used to separate and collect the n-heptane insolubles (asphaltenes) from the n-heptane solubles (maltenes). All of the n-heptane solubles washings were combined and the maltene solution reduced to 5 mL. Isolation and extrography fractionation of Maya asphaltenes are discussed below.

Silicic acid sample separation

2–2.5 g of the maltenes were loaded onto the column. The flow rate of the column was adjusted to ~2.5 mL min^{-1}. Components of the crude oils were selectively eluted by use of solvents of increasing polarity (*i.e.*, solvent strength). The most polar components are adsorbed on the active sites of the stationary phase (adsorbent) and the less polar compounds are least retained by the stationary phase. Four fractions were collected in the following elution order; hydrocarbons (non-polar), neutral nitrogen, basic nitrogen, and polars. Hydrocarbons (HC) were eluted with 250 mL of 90 : 10 (vol : vol) n-heptane : toluene. The neutral nitrogen fraction (neutral N) was eluted with 200 mL of 50 : 50 (vol : vol) n-heptane : toluene, whereas the basic nitrogen fraction (basic N) was eluted with 200 mL of 90 : 10 (vol : vol) toluene : methanol. The polars fraction (polars) was eluted with 200 mL of 85 : 15 (vol : vol) methanol : toluene. To overcome overlap between the different fractions, intermediate fractions were collected between fractions when the new eluent (different mobile phase composition) flowed down the column. The methanol used in the elution of the polar fraction did strip some of the water from the column. However, the water was removed by filtration of the fraction through a bed of anhydrous Na_2SO_4. The excess solvent in the collected fractions was reduced by evaporation, and the weights of the dry fractions were recorded (see Table 2). A schematic representation of the separation is shown in Fig. S1.†

MAPS fractionation

Deasphalted MacKay bitumen was isolated as described above and dried under a stream of dry nitrogen gas. The APS and MAPS fractionation were performed as previously described.[52] The first six fractions were analyzed (MA1–MA6).

Asphaltene extrography fractionation

10 g of Maya crude oil was mixed with 400 mL of heptane (dropwise added, with sonication at 60 °C). After 12 h, solids were collected by filtration and Soxhlet-extracted with heptane until the solvent appeared colorless (~84 h). Asphaltenes were recovered *via* dissolution with hot toluene, and dried with N_2. Further asphaltene cleaning was conducted by use of a method reported elsewhere.[62] In short, asphaltene solids were crushed and Soxhlet-extracted with heptane; this cleaning cycle was repeated four times to decrease the concentration of occluded/entrained maltenes. Subsequently, cleaned asphaltenes were adsorbed on silica gel (0.5% mass loading). The resulting solid material was Soxhlet-extracted with acetone, followed by toluene and (1 : 1 vol : vol) toluene/THF. The extraction with each solvent lasted 24 h. Fractions were dried under N_2 and stored in the dark for

Table 2   Weight yield of various fractions from silicic acid chromatographic separation of the parent crude oil^a

Crude oil	South American (g)
Hydrocarbons (HC)	1.4922
Intermediate HC–NN	0.0122
Neutral nitrogen (NN)	0.0874
Intermediate NN–BN	0.0048
Basic nitrogen (BN)	0.0270
Polars (PLR)	0.0287

^{a} All weights are based on loading 2 g of deasphalted parent crude oils onto the column.

MS analyses. The whole asphaltene and extrography fractions are denoted whole, acetone, toluene, and Tol/THF.

Sample preparation for ESI FT-ICR MS

Sample preparation for the analysis of asphaltene, polar acidic, and polar basic species in petroleum and petroleum-derived materials by atmospheric pressure photo-ionization and negative-ion/positive-ion electrospray FT-ICR MS has been previously reported.[53,63] Samples were analyzed at a concentration of 50–500 µg mL^{-1} in a standard ESI spray mix (50 : 50 (vol : vol) toluene : methanol) or toluene (APPI) for mass spectrometric analysis. For (−) ESI nitrogen species analyses, a representative aliquot (1 mL) of each fraction was spiked with 10 µL of 2% TMAH in methanol to facilitate the deprotonation of the acidic species to generate [M − H]$^-$ ions, whereas 10 µL of 2% HCOOH in methanol was used to protonate the basic species to generate [M + H]$^+$ ions. TMAH increases the degree of ionization of slightly acidic nitrogen-containing classes relative to ESI with NH_4OH.[64] Acidic APS and MAPS (MA1–MA4) fractions were analyzed at concentrations of 10–250 µg mL^{-1} in (50 : 50 (vol : vol)) toluene : methanol with 0.1% vol NH_4OH. Due to poor ionization, the concentration of NH_4OH was increased to 0.2% vol for the analysis of fraction MA5, and MA6 performed so poorly even under standard conditions that the solvent system and modifier required further optimization. For MA6, they were changed to 250 µg mL^{-1} in (50 : 50 (vol : vol)) DCM : methanol with 0.125% vol TMAH. Whole Maya asphaltenes and the extrography fractions were diluted in toluene at 200 µg mL^{-1} for positive-ion APPI.

Mass analysis

Each sample was analyzed with a custom-built 9.4 T 22 cm horizontal room temperature bore diameter (Oxford Corp., Oxford Mead, UK) FT-ICR mass spectrometer at the National High Magnetic Field Laboratory.[65] A Predator data system was used to acquire and process ICR data.[65,66] All elemental composition assignments and subsequent data processing were performed with PetroOrg.[67] Mass spectrometer parameters and protocols are similar to those previously described.[52–54,57,68,69]

Results and discussion

Nitrogen separation

LC-MS allows us to address important issues in heavy hydrocarbon characterization, including (i) differentiation of alkyl-aromatics from naphthenoaromatics, (ii) distinction between aromatic hydrocarbons and thiophenes, (iii) analysis of basic and neutral polars, and (iv) correlation of molecular polarity based on elution characteristics in different chromatographic fractions.[56,70] Silicic acid chromatographic separation was developed to fractionate polar compounds with two main types of nitrogen compounds in shale oils, particularly in shale high vacuum gas oils (HVGO), i.e., pyrrole and pyridine benzologs.[14] Mechanistically, a layer of water supported by silicic acid provides an inherently acidic stationary phase surface that separates polar molecules based on differences in their ability to hydrogen-bond. Thus, it retains basic nitrogen functionalities (proton acceptors) more effectively than neutral nitrogen functional groups (proton donors), thereby separating the benzologs of pyrrole from those of pyridine. Highly alkylated compounds, such as N-alkylated pyrroles, exhibit very little H-bonding. They elute as or with the aromatics, because adsorption by nitrogen occurs perpendicularly at the edge of the molecule, and retention is limited by steric accessibility.

Mass spectral data interpretation

Petroleum feedstocks are composed mainly of homologous series, $C_cH_{2c+z}X$, in which c is the carbon number, z is the "hydrogen deficiency" index, and X denotes heteroatoms (N, S, and O) in each molecule, and successive members of the series differ by multiples of CH_2.[71] Although hydrogen deficiency provides a measure of aromaticity, each ring or double bond contributes -2, and $z = +2$ for a fully saturated hydrocarbon. Thus, a more accurate index is double bond equivalents (DBE), defined as the number of rings plus double bonds involving carbon for a molecular elemental composition,[72,73] $C_cH_hN_nO_oS_s$:

$$\text{Double Bond Equivalents (DBE)} = c - \frac{h}{2} + \frac{n}{2} + 1 \qquad (1)$$

The two parameters are related by:

$$z = -2\text{DBE} + n + 2 \text{ or equivalently, DBE} = -z/2 + n/2 + 1 \qquad (2)$$

We highlight the broadband ESI mass spectra of a South American crude oil and its respective fractions to illustrate the molecular weight distribution and compositional complexity. We further analyze the individual heteroatom classes for the separated polar nitrogen compounds. Furthermore, each class is sorted according to DBE and carbon number to reveal patterns of aromaticity and alkylation that help to define their core structures and provide a visual indicator of the compositional coverage achieved for the whole and/or chromatographic fractions. Finally, for all three separation methods, we present monomer ion yields (MIY, a measure of ionization efficiency) to highlight the impact of selective ionization for the analysis of these complex samples.[53] MIY is the inverse of the product of ion accumulation period and analyte concentration required to reach a target number of ions. Thus, a low monomer ion yield value indicates poor ionization efficiency. When possible, all related samples and their respective fractions are analyzed at the same sample concentration.

$$\text{MIY} = 1/[(\text{accum. period})(\text{analyte concentration})] \qquad (3)$$

Basic nitrogen characterization by (+) ESI FT-ICR MS

The crude oil and its respective fractions were analyzed by positive-ion ESI FT-ICR MS to highlight trends in the basic nitrogen (pyridinic) species from the fractionated South American crude oil. Fig. 1 summarizes the DBE and carbon number compositional images for the four major heteroatom classes (N_1, N_1S_1, N_2, and N_1O_1) along with their relative abundances (R.A.) from the (+) ESI FT-ICR MS analysis. Each heteroatom specific image is composed of thousands of basic

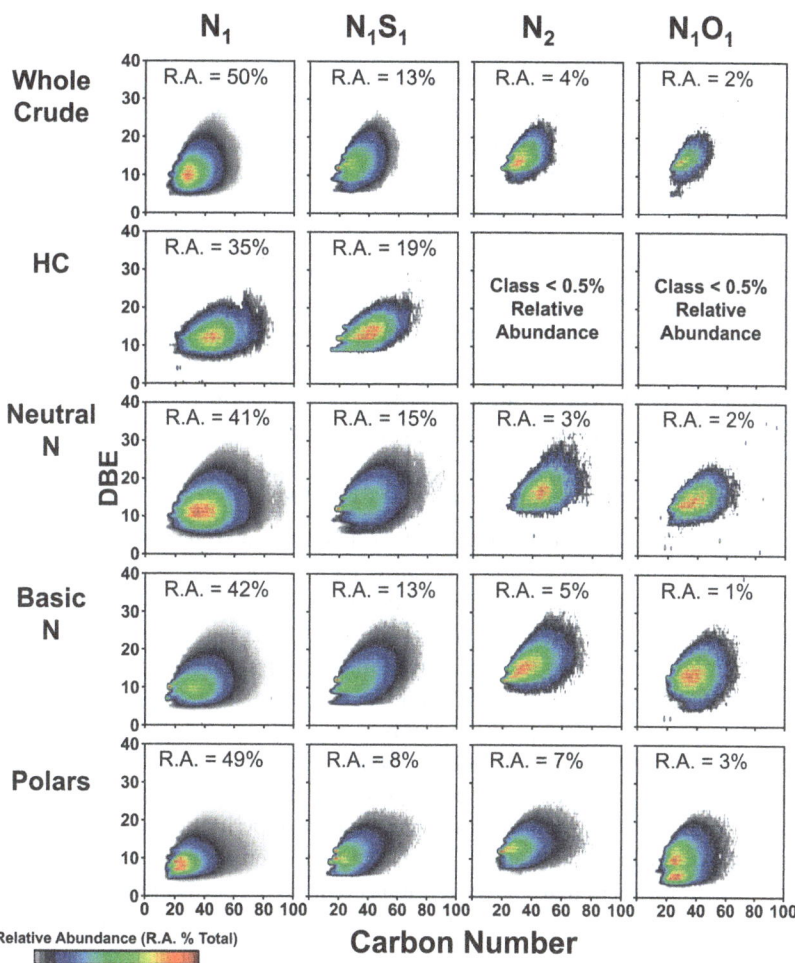

(+) ESI 9.4 T FT-ICR MS

South American Crude Oil

Fig. 1 Isoabundance-contoured isoabundance plots of double bond equivalents (DBE) *versus* carbon number from the (+) ESI FT-ICR MS analysis (pyridinic nitrogen) of the whole deasphalted South American crude oil and its hydrocarbon (HC), neutral nitrogen, basic nitrogen, and polars fractions obtained from silicic acid column chromatography for the four most abundant classes (N_1, N_1S_1, N_2, and N_1O_1) detected in the whole crude oil analysis.

species, predominately nitrogen (N_x species) and nitrogen heteroatom-containing (N_xO_y or N_xS_y) compounds. The fractionation clearly extends the molecular weight distributions of the hydrocarbon and neutral nitrogen fractions (carbon numbers 20–90) beyond the molecular weight distribution of the parent crude (carbon numbers 20–60), and thereby enables the identification of higher molecular weight species unobservable in the parent crude oil. As the fractionation progresses from the HC to the neutral N, basic N, and finally the polars, the DBE and carbon number ranges steadily decline. Thus, the least-retained species have the highest carbon numbers (most sterically hindered) and are the most alkylated N_1 and N_1S_1 species. Unsurprisingly, the molecular weight (carbon number) and DBE ranges for the basic nitrogen and polar fraction more closely resemble those for the parent crude (carbon numbers 20–60), given the monomer ion yield data presented in Fig. 2. The basic N and polars fractions display monomer ion yields between 3- and 10-fold higher than those for the preceding hydrocarbons and neutral N fractions. Thus, without fractionation, the basic N and polars species dominate the ionization process and limit detection of the compositional information found in the hydrocarbons and neutral N fractions (higher carbon number species). Hence, the chromatographic separation gives a broader class composition (carbon number and DBE) and eliminates ambiguities such as isobaric interferences, and somewhat reduces the effects of selective ionization. The preferential ionization of lower molecular weight (lower carbon number) species in complex petroleum samples is a reoccurring theme, as discussed below for the analysis of neutral nitrogen (with this separation method) and acidic (oxygen-containing) species (in MAPS fractionation).

Neutral nitrogen characterization by −ESI FT-ICR MS

The same fractions summarized in Fig. 1 ((+) ESI) were analyzed using (−) ESI FT-ICR MS to highlight trends in neutral nitrogen (pyrrolic) species. Fig. 3 summarizes the compositional information for the whole crude and the HC, neutral N, basic N, and polars fractions for neutral nitrogen (pyrrolic) species and reveals trends similar to those previously observed in the analysis of the pyridinic species using +ESI. Most notably, the HC and neutral nitrogen fraction contain the highest molecular weight (carbon number) species, and, as the fractionation progresses to the basic N and polars fractions, the N_1 and N_1S_1 class species rapidly drop to lower carbon number values but maintain roughly the same DBE range (10 < DBE < 30). As for the previous analysis of pyridinic species (basic N), the compositional information obtained in the whole crude analysis most closely matches the species that are selectively ionized by the respective ESI mode ($\pm$). In the previous discussion, it was the basic nitrogen (pyridinic species) fraction for

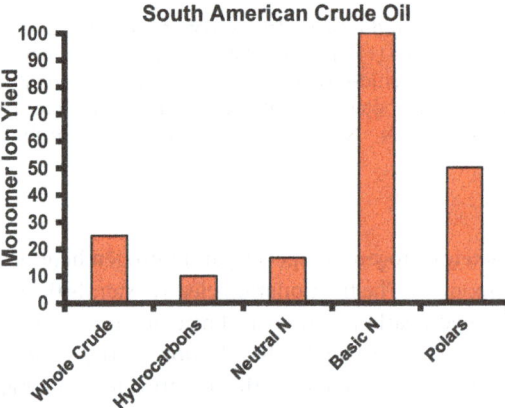

Fig. 2 Monomer ion yield for the whole deasphalted oil and hydrocarbon (HC), neutral nitrogen, basic nitrogen, and polars fractions obtained from silicic acid fractionation of South American crude oil.

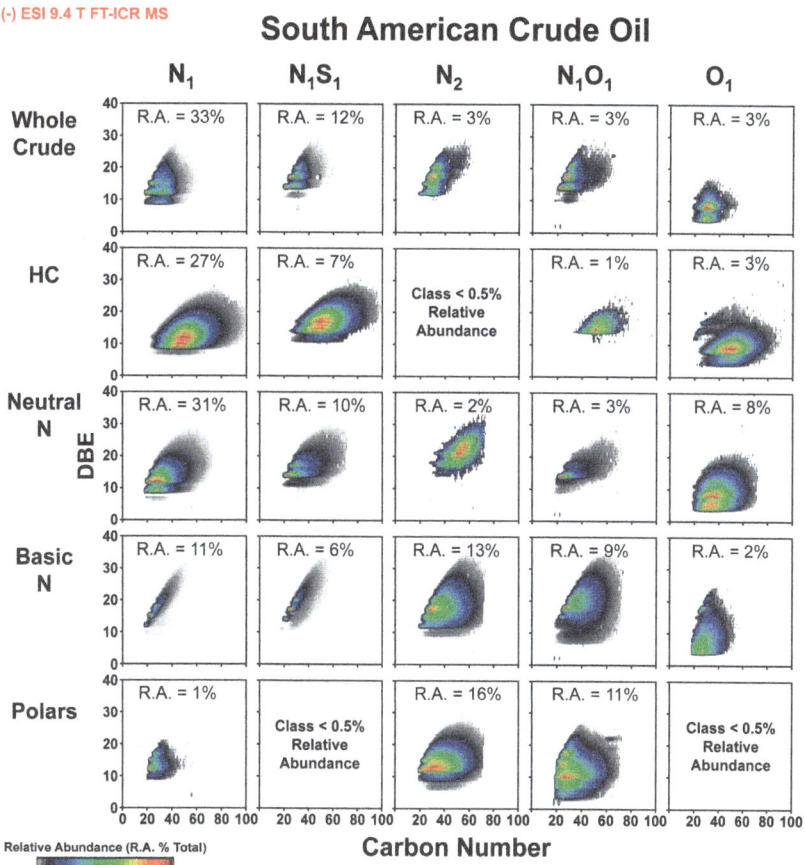

Fig. 3   Isoabundance-contoured plots of DBE *versus* carbon number from (−) ESI FT-ICR MS analysis (pyrrolic nitrogen) of the whole deasphalted South American crude oil and its hydrocarbon (HC), neutral nitrogen, basic nitrogen, and polars fractions obtained from silicic acid column chromatography for the 5 most abundant classes (N_1, N_1S_1, N_2, N_1O_1 and O_1) detected in the whole crude oil analysis.

(+) ESI. Here, it is the neutral nitrogen (pyrrolic species) fraction for (−) ESI. As the fractionation continues to the basic nitrogen fraction, the least sterically hindered (lowest degree of alkylation) pyrrolic species elute (N_1 class) along with their thiophenic sulfur-containing analogues (N_1S_1) class. These 2 classes reside on, or immediately adjacent to the polycyclic aromatic hydrocarbons (PAH) line and are thus bare, or near bare, pyrrolic aromatic cores.[46,74,75] The polar polyfunctional nitrogen species (N_2 and N_1O_1 data) begin to dominate, and account for ∼50% of the total relative ion abundance. Although N_2 and N_1O_1 constitute ∼22% of the total relative ion abundance, other polyfunctional species, such as N_2S_1, $N_1O_1S_1$, N_2O_1, and N_1O_2 contribute to the remainder (data not shown). The trend continues into the polars fraction, for which polar polyfunctional species that contain 2 or more N or O heteroatoms account for nearly 65% of the total relative abundance.

The deactivated silicic acid column chromatography method discussed above enables the coarse separation of the 2 dominant forms of nitrogen (pyridinic and pyrrolic) in petroleum by carbon number, DBE (aromaticity), and polarity, and is therefore better suited for subsequent MS analysis. It achieves the separation through interaction of the nitrogen functionality with the stationary phase. As a result, the most sterically hindered nitrogen-containing species (highest carbon number) elute first and the carbon numbers decline as the separation proceeds. The most strongly retained species elute in the latter 2 fractions (basic N and polars), for which lower-abundance polyfunctional species in the whole crude oil begin to emerge. The greatest difference in this elution window is evident in the mono-functional basic nitrogen species, which continue to elute into the polar fraction. The weaker interaction between the stationary phase and pyrrolic nitrogen species terminates their elution in the basic nitrogen fraction, and polyfunctional species dominate the polars fraction. Analysis of the fractions by both (+) and (−) ESI FT-ICR MS reveals selective ionization, because there are defined monomer ion yield trends and similarity between the whole crude and specific fractions that depend on the ionization mode. Simply, the basic nitrogen fraction from (+) ESI and the neutral nitrogen fraction from (−) ESI are similar to the whole crude results. To further investigate the carbon number dependence of selective ionization, we revisit the modified aminopropyl silica (MAPS) fractionation of MacKay bitumen,[52] which enables separation by molecular weight (carbon number).

Naphthenic acid characterization by MAPS fractionation and (−) ESI FT-ICR MS

The MAPS fractionation procedure first isolates acidic species through their interaction with the amino functionality of the stationary phase. However, once isolated, the solvent system is changed from normal to reversed-phase, and the retained acids then elute by hydrophobicity through their interaction with the propyl functionality of the stationary phase. The retained acids are therefore separated into fractions of increasing carbon number, and thus the separation allows for further investigation of the carbon number dependence of selective ionization (ionization efficiency/monomer ion yield). Fig. 4 summarizes the mass spectral analysis of the whole bitumen APS extract (top), *versus* the MAPS fractions (MA1–MA6, bottom). The MAPS fractionation method clearly extends the detected molecular weight range, as MAPS fractions MA5 and MA6 extend well beyond 1 kDa. The monomer ion yield (ionization efficiency) data for the whole APS bitumen extract and all MAPS fractions are presented in Fig. 5. As the fraction-ation progresses, the molecular weight also increases. However, the monomer ion yield data precipitously decreases and reaches a value 80-fold less for MA5 than for MA1. MA5 and MA6 performed so poorly in the standard NH_4OH modified toluene : methanol solvent systems commonly used for (−) ESI analysis, that MA5 required the use of twice the normal volume of NH_4OH, and MA6 required elimination of toluene and NH_4OH, and replacement with dichloromethane and TMAH for successful analysis. Despite these efforts, the monomer ion yield is still ∼25× less than for the MA1 fraction, and is thus presented for comparison purposes only. The monomer ion yield for the whole bitumen APS extract is roughly the average of that for MA1–MA3, and the mass spectrum spans a similar molecular weight range and chemical composition (dominated by the O_2 class) of these 3 fractions. Thus, as for the pyridinic and pyrrolic nitrogen results, higher molecular weight (higher carbon number) species appear to ionize less efficiently (lower monomer ion yield) than their lower molecular weight forms. The exact

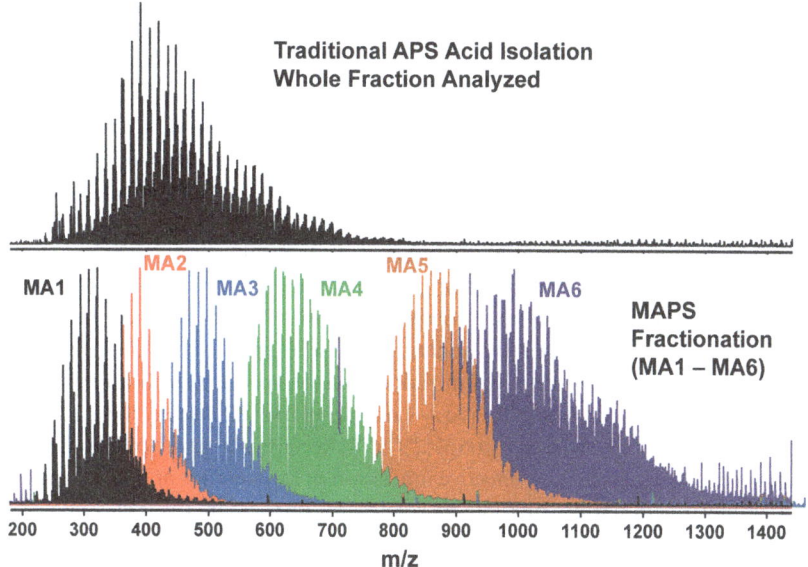

Fig. 4 Broadband (−) ESI FT-ICR mass spectrum of the aminopropyl silica combined acids extract obtained from MacKay bitumen (top). Broadband (−) ESI FT-ICR mass spectra of the modified aminopropyl silica extracts (MA1 (black)−MA6 (purple)) obtained from MacKay bitumen (bottom).

reason is currently unknown, but clues are provided from the analysis of ultra-complex asphaltene samples by APPI FT-ICR MS (see below).

Asphaltene characterization *via* extrography fractionation and +APPI FT-ICR MS

The mass spectral analysis of asphaltenes has been plagued by difficulties for decades.[39,47,53,76–79] Most notably, the H : C ratio of species readily observed based on mass spectrometry (H : C ∼ 0.8) is much lower than that obtained *via* bulk

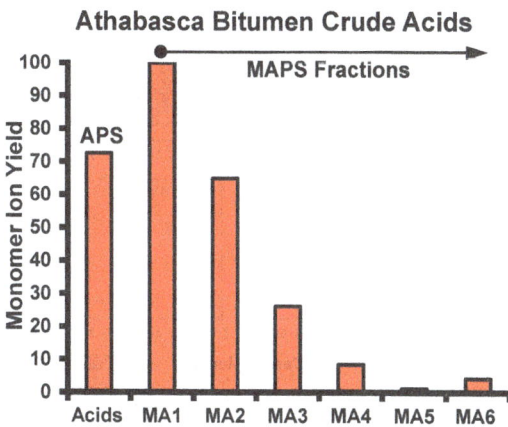

Fig. 5 Monomer ion yield for the seven mass spectra presented in Fig. 4. The whole deasphalted oil APS extract is on the left, and is followed by MAPS fractions MA1–MA6.

analysis (H : C ~1.0–1.1), and the mass spectral decomposition products obtained from tandem MS experiments did not match thermal decomposition products observed in lab-based experiments or readily encountered in normal refinery operations.[47,80–87] Thus, it was obvious that some form of selective ionization influences not only the observed H : C ratio, but also the accessible structures. Briefly, multiple reports of tandem MS results of unfractionated asphaltenes yielded simple dealkylation products (with no change in the precursor ion DBE), strongly suggesting that the dominant asphaltene structural motif is a single, alkyl-substituted aromatic core (known as the island motif).[87–89] However, multiple other reports contradicted those findings and suggested that the dominant motif was composed of multiple aromatic cores linked by alkyl or cycloalkyl groups (known as the archipelago motif).[39,81,90–92] Irrespective of the conflicting reports, it is readily apparent that the mass spectral analysis of asphaltenes suffers from selective ionization, and the species readily ionized have H : C ratios of ~0.8, corresponding to 3–9 ring aromatics with a very low degree of alkyl substitution (0–14 CH_2 groups).[40,47,93,94] Recently, a three-part series of manuscripts investigated the selective ionization of asphaltenes and proposed an extrography fractionation method to reduce its impact on the mass spectral

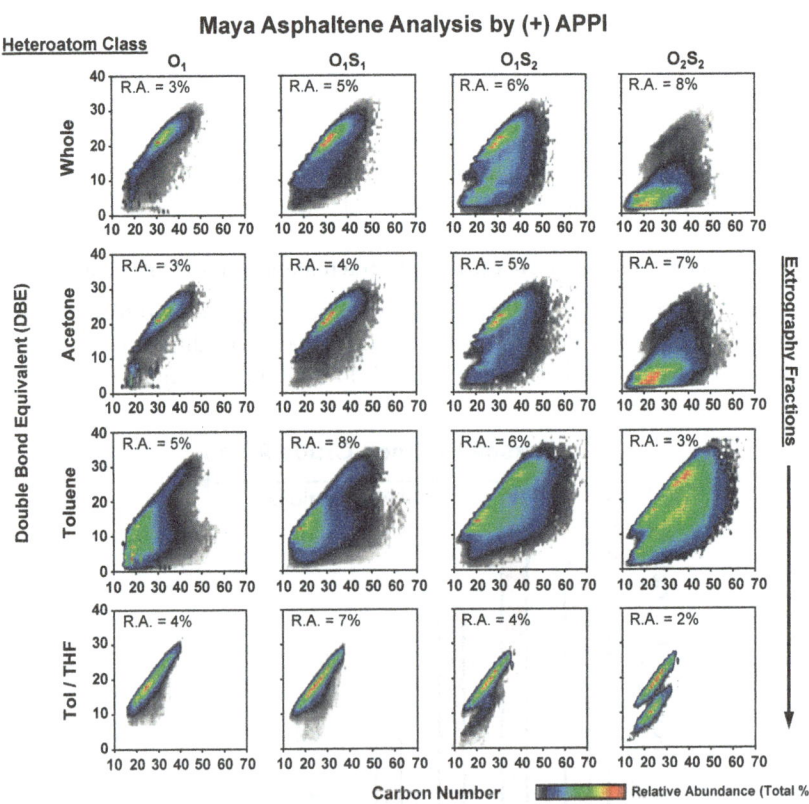

Fig. 6   Isoabundance-contoured plots of DBE *versus* carbon number from the (+) APPI FT-ICR MS analysis of the whole Maya crude oil asphaltene fraction and its acetone, toluene, and toluene/THF fractions obtained *via* extrography fractionation for the 4 heteroatom classes (O_1, O_1S_1, O_1S_2, and O_2S_2).

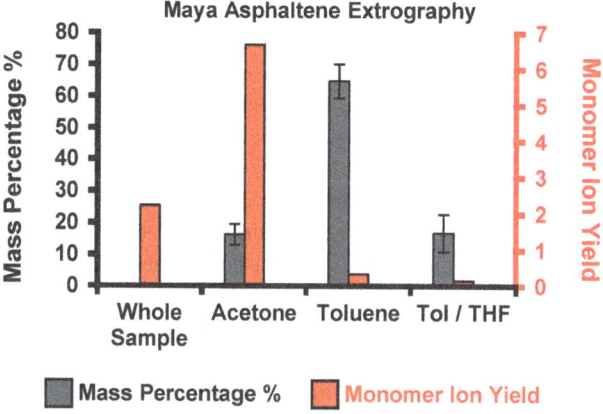

Fig. 7 Mass distribution (left) and monomer ion yield (right) for the extrography asphaltene fractionation for Maya asphaltenes.

analysis.[53,54,57] Fig. 6 provides further evidence of its success for APPI FT-ICR MS analysis of a whole Maya asphaltene and its associated acetone, toluene, and toluene/tetrahydrofuran extrography fractions collected from a simplified version (3 fractions instead of the original 8 fractions) of the previously reported method.[54] Results for the O_1, O_1S_1, O_1S_2, and O_2S_2 classes are presented for the whole asphaltene and 3 extrography fractions. Comparison of the whole asphaltene data to that obtained for the acetone fraction reveals striking similarity and exposes the problem of selective ionization in asphaltene analyses. Simply, as highlighted in Fig. 7, the acetone fraction is only 16% wt of the whole asphaltene, but ionizes 20× more efficiently than the toluene fraction (~65% wt) and 45× more efficiently than the toluene/THF fraction (17% wt). Thus, analysis of the whole asphaltene sample captures the compositional information of only 16% wt of the entire sample, and thereby fails to account for the species highlighted in the toluene and toluene/THF fractions that account for 80+% wt of the asphaltene sample. The proposed reason behind the large differences in monomer ion yield was suggested to be the effects of aggregation, because the precipitated mass *versus* heptane dilution yielded dramatic differences between the two end members in monomer ion yield (acetone and toluene/THF fractions).[54,57] At 40% vol heptane into a toluene solution, only ~5 wt% of the acetone fraction precipitated, whereas more than 90 wt% of the toluene/THF fraction precipitated. The difference in ionization efficiency also appears to be linked to molecular structure, as the dominant structural motif in this and other asphaltene acetone extrography fractions is the island motif. Conversely, the dominant structural motif in this and other asphaltene toluene and toluene/THF extrography fractions is the archipelago motif. Although convincing and correlated with monomer ion yield data for a handful of samples, the impact of aggregation in complex matrices appears to be very important, and warrants future study.

Conclusions

The compositional complexity of crude oils requires a separation method that can generate manageable fractions that can be selectively characterized based on the ionization method, analyte functional group(s), aromaticity, and molecular weight. The silicic acid separation method successfully addresses the separation and targeted ionization of pyridinic ((+) ESI) and pyrrolic ((−) ESI) nitrogen-containing species, but exposes the preferential ionization of lower molecular weight pyridinic and pyrrolic species at the expense of higher molecular weight species. The monomer ion yields for the basic nitrogen fractions, combined with the similarity between the whole sample results and those of the basic nitrogen fraction ((+) ESI) and neutral nitrogen fraction ((−) ESI), strongly suggest that selective ionization is a major problem for the comprehensive characterization of complex samples based on direct "dilute and shoot" analyses. Selective isolation of acidic species by aminopropyl silica and subsequent analysis of the whole acid extract revealed identical selective ionization issues (for acids) compared to the MAPS fractions for the same extract (further separated by hydrophobicity (carbon number)). Ionization favors the lowest molecular weight species over higher molecular weight species, as exemplified by up to 80-fold+ difference in monomer ion yield for the later eluting (higher molecular weight) species. Finally, we revisited selective ionization in an ultra-complex asphaltene fraction isolated from Maya crude oil. Consistent with similar analyses, mass spectral analysis of the whole (unfractionated) asphaltene yielded results nearly identical to those for the acetone extrography fraction, which had a monomer ion yield 20-fold higher than the next eluting fraction (toluene) but comprised only 16% wt of the whole sample. Previous reports have linked this selective ionization to aggregation, and given the present results for nitrogen-containing (pyridinic and pyrrolic), oxygen-containing (carboxylic acids), and oxygen and sulfur-containing (OxSy) species, the link(s) between structure, carbon number, aggregation tendency, and monomer ion yield require(s) further research.

Conflicts of interest

There are no conflicts to declare.

Acknowledgements

This work was supported by the NSF Division of Materials Research through DMR-11-57490 and DMR-1644779, the Florida State University, Future Fuels Institute, and the State of Florida.

References

1 A. G. Marshall and R. P. Rodgers, Petroleomics: The Next Grand Challenge For Chemical Analysis, *Acc. Chem. Res.*, 2004, **37**, 53–59.
2 L. C. Krajewski, R. P. Rodgers and A. G. Marshall, 126264 Assigned Chemical Formulas from an Atmospheric Pressure Photoionization 9.4 T Fourier Transform Positive Ion Cyclotron Resonance Mass Spectrum, *Anal. Chem.*, 2017, **89**, 11318–11324.
3 Y. E. Corilo, B. G. Vaz, R. C. Simas, H. D. Lopes Nascimento, C. F. Klitzke, R. C. L. Pereira, W. L. Bastos, E. V. Santos Neto, R. P. Rodgers and M. N. Eberlin, Petroleomics by ESI(±) FT-ICR MS, *Anal. Chem.*, 2010, **82**(10), 3990–3996.

4 P. L. Grizzle and D. M. Sablotny, Automated Liquid Chromatographic Compound Class Group-Type Separation of Crude Oils and Bitumens Using Chemically Bonded Aminosilane, *Anal. Chem.*, 1986, **58**, 2389–2396.

5 A. M. McKenna, J. M. Purcell, R. P. Rodgers and A. G. Marshall, Heavy Petroleum Composition. 1. Exhaustive Compositional Analysis of Athabasca Bitumen HVGO Distillates by Fourier Transform Ion Cyclotron Resonance Mass Spectrometry: A Definitive Test of the Boduszynski Model, *Energy Fuels*, 2010, **24**(5), 2929–2938.

6 A. M. McKenna, G. T. Blakney, F. Xian, P. B. Glaser, R. P. Rodgers and A. G. Marshall, Heavy Petroleum Composition. 2. Progression of the Boduszynski Model to the Limit of Distillation by Ultrahigh-Resolution FT-ICR Mass Spectrometry, *Energy Fuels*, 2010, **24**(5), 2939–2946.

7 B. M. F. Ávila, B. G. Vaz, R. Pereira, A. O. Gomes, R. C. L. Pereira, Y. E. Corilo, R. C. Simas, H. D. L. Nascimento, M. N. Eberlin and D. A. Azevedo, Comprehensive Chemical Composition of Gas Oil Cuts Using Two-Dimensional Gas Chromatography with Time-of-Flight Mass Spectrometry and Electrospray Ionization Coupled to Fourier Transform Ion Cyclotron Resonance Mass Spectrometry, *Energy Fuels*, 2012, **26**(8), 5069–5079.

8 A. Gaspar, E. Zellermann, S. Lababidi, J. Reece and W. Schrader, Impact of Different Ionization Methods on the Molecular Assignments of Asphaltenes by FT-ICR Mass Spectrometry, *Anal. Chem.*, 2012, **84**, 5257–5267.

9 Y. Cho, J. Na, N. Nho, S. Kim and S. Kim, Application of Saturates, Aromatics, Resins, Asphaltenes Crude Oil Fractionation for Detailed Chemical Characterization of Heavy Crude Oils by Fourier Transform Ion Cyclotron Resonance Mass Spectrometry Equipped with Atmospheric Pressure Photoionization, *Energy Fuels*, 2012, **26**, 2558–2565.

10 G. C. Klein, A. Angstro, R. P. Rodgers and A. G. Marshall, Use of Saturates/Aromatics/Resins/Asphaltenes (SARA) Fractionation To Determine Matrix Effects in Crude Oil Analysis by Electrospray Ionization Fourier Transform Ion Cyclotron Resonance Mass Spectrometry, *Energy Fuels*, 2006, **20**(10), 668–672.

11 D. Giraldo-Dávila, M. L. Chacón-Patiño, J. A. Orrego-Ruiz, C. Blanco-Tirado and M. Y. Combariza, Improving Compositional Space Accessibility in (+) APPI FT-ICR Mass Spectrometric Analysis of Crude Oils by Extrography and Column Chromatography Fractionation, *Fuel*, 2016, **185**, 45–58.

12 B. M. Ruddy, C. L. Hendrickson, R. P. Rodgers and A. G. Marshall, Positive Ion Electrospray Ionization Suppression in Petroleum and Complex Mixtures, *Energy Fuels*, 2018, **32**, 2901–2907.

13 J. D. McLean and P. K. Kilpatrick, Comparison of Precipitation and Extrography in the Fractionation of Crude Oil Residua, *Energy Fuels*, 1997, **11**(3), 570–585.

14 W. K. Robbins and F. C. McElroy, Systematic Evaluation and Development of Adsorption Chromatographic Techniques for Coal Liquid Analyses, *Liq. Fuels Technol.*, 1984, **2**(2), 113–154.

15 R. P. Rodgers, T. Schaub and A. G. Marshall, Petroleomics: MS Returns to Its Roots, *Anal. Chem.*, 2005, **77**(1), 20A–27A.

16 C. S. Hsu, K. Qian and W. K. Robbins, Nitrogen Speciation of Polar Petroleum Compounds by Compound Class Separation and On-Line Liquid

Chromatography - Mass Spectrometry (LC-MS), *J. High Resolut. Chromatogr.*, 1994, **17**, 271–276.

17 M. M. Boduszynski, Composition of Heavy Petroleums. 1. Molecular Weight, Hydrogen Deficiency, and Heteroatom Concentration as a Function of Atmospheric Equivalent Boiling Point up to 1400 °F (760 °C), *Energy Fuels*, 1987, **1**(1), 2–11.

18 M. M. Boduszynski, Composition of Heavy Petroleums. 2. Molecular Characterization, *Energy Fuels*, 1988, **2**(5), 2–11.

19 M. M. Boduszynski and K. H. Altgelt, Composition of Heavy Petroleums. 4. Significance of the Extended Atmospheric Equivalent Boiling-Point (AEBP) Scale, *Energy Fuels*, 1992, **6**(1), 72–76.

20 G. C. Klein, R. P. Rodgers, M. A. G. Teixeira, A. M. R. F. Teixeira and A. G. Marshall, Petroleomics: Electrospray Ionization FT-ICR Mass Analysis of NSO Compounds for Correlation Between Total Acid Number, Corrosivity, and Elemental Composition, *Preprints of Symposia*, American Chemical Society, Division of Fuel Chemistry, 2003, vol. 48, ch. 1, pp. 14–15.

21 C. A. Hughey, R. P. Rodgers, A. G. Marshall, C. C. Walters, K. Qian and P. Mankiewicz, Acidic and Neutral Polar NSO Compounds in Smackover Oils of Different Thermal Maturity Revealed by Electrospray High Field Fourier Transform Ion Cyclotron Resonance Mass Spectrometry, *Org. Geochem.*, 2004, **35**(7), 863–880.

22 M. P. Barrow, J. V. Headley, K. M. Peru and P. J. Derrick, Data Visualization for the Characterization of Naphthenic Acids within Petroleum Samples, *Energy Fuels*, 2009, **23**(5), 2592–2599.

23 G. C. Klein, S. Kim, R. P. Rodgers, A. G. Marshall, A. Yen and S. Asomaning, Mass Spectral Analysis of Asphaltenes. I. Compositional Differences between Pressure-Drop and Solvent-Drop Asphaltenes Determined by Electrospray Ionization Fourier Transform Ion Cyclotron Resonance Mass Spectrometry, *Energy Fuels*, 2006, **20**(9), 1965–1972.

24 Q. Shi, S. Zhao, Z. Xu, K. H. Chung, Y. Zhang and C. Xu, Distribution of Acids and Neutral Nitrogen Compounds in a Chinese Crude Oil and Its Fractions: Characterized by Negative-Ion Electrospray Ionization Fourier Transform Ion Cyclotron Resonance Mass Spectrometry, *Energy Fuels*, 2010, **24**(7), 4005–4011.

25 X. Zhu, Q. Shi, Y. Zhang, N. Pan, C. Xu, K. H. Chung and S. Zhao, Characterization of Nitrogen Compounds in Coker Heavy Gas Oil and Its Subfractions by Liquid Chromatographic Separation Followed by Fourier Transform Ion Cyclotron Resonance Mass Spectrometry, *Energy Fuels*, 2011, **25**(1), 281–287.

26 J. A. Valencia-dávila, M. Witt, C. Blanco-tirado and M. Y. Combariza, Molecular Characterization of Naphthenic Acids from Heavy Crude Oils Using MALDI FT-ICR Mass Spectrometry, *Fuel*, 2018, **231**(February), 126–133.

27 P. V. Hemmingsen, S. Kim, H. E. Pettersen, R. P. Rodgers, J. Sjöblom and A. G. Marshall, Structural Characterization and Interfacial Behavior of Acidic Compounds Extracted from a North Sea Oil, *Energy Fuels*, 2006, **20**(5), 1980–1987.

28 A. C. Clingenpeel, T. R. Fredriksen, K. Qian and M. R. Harper, Comprehensive Characterization of Petroleum Acids by Distillation, Solid Phase Extraction Separation, and Fourier Transform Ion Cyclotron Resonance Mass Spectrometry, *Energy Fuels*, 2018, **32**, 9271–9279.

29 A. Alvarez-majmutov, R. Gieleciak and J. Chen, Modeling the Molecular Composition of Vacuum Residue from Oil Sand Bitumen, *Fuel*, 2019, **241**, 744–752.

30 M. M. Boduszynski, R. J. Hurtubise, T. W. Allen and H. F. Silver, Determination of Hydrocarbon Composition in High-Boiling and Nondistillable Coal Liquids by Liquid-Chromatography Field-Ionization Mass-Spectrometry, *Anal. Chem.*, 1983, **55**(2), 232–241.

31 M. M. Boduszynski, Asphaltenes in Petroleum Asphalts. Composition and Formation, in *Chemistry of Asphaltenes*, American Chemical Society, Washington, D.C., USA, 1982, pp. 119–135.

32 M. M. Boduszynski, R. J. Hurtubise and H. F. Silver, Separation of Solvent-Refined Coal Into Compound-Class Fractions, *Anal. Chem.*, 1982, **54**(3), 375–381.

33 M. Li, S. R. Larter, D. Stoddart and M. Bjoroy, Liquid Chromatographic Separation Schemes for Pyrrole and Pyridine Nitrogen Aromatic Heterocycle Fractions from Crude Oils Suitable for Rapid Characterization of Geochemical Samples, *Anal. Chem.*, 1992, **64**(14), 1337–1344.

34 B. Bennett and G. D. Love, Release of Organic Nitrogen Compounds from Kerogen via Catalytic Hydropyrolysis, *Geochem. Trans.*, 2000, **1**, 61–67.

35 G. F. Bolshakov, Organic Sulfur Compounds of Petroleum, *Sulfur Rep.*, 1986, **5**(2), 103–393.

36 V. V. Lobodin, W. K. Robbins, J. Lu and R. P. Rodgers, Separation and Characterization of Reactive and Non-Reactive Sulfur in Petroleum and Its Fractions, *Energy Fuels*, 2015, **29**(10), 6177–6186.

37 P. Liu, Q. Shi, N. Pan, Y. Zhang, K. H. Chung, S. Zhao and C. Xu, Distribution of Sulfides and Thiophenic Compounds in VGO Subfractions: Characterized by Positive-Ion Electrospray Fourier Transform Ion Cyclotron Resonance Mass Spectrometry, *Energy Fuels*, 2011, **25**(7), 3014–3020.

38 L. R. Snyder, Petroleum Nitrogen Compounds and Oxygen Compounds, *Acc. Chem. Res.*, 1970, **3**(9), 290–299.

39 M. Gray, R. Tykwinski, J. Stryker and X. Tan, Supramolecular Assembly Model for Aggregation of Petroleum Asphaltenes, *Energy Fuels*, 2011, **25**, 3125–3134.

40 J. M. Purcell, I. Merdrignac, R. P. Rodgers, A. G. Marshall, T. Gauthier and I. Guibard, Stepwise Structural Characterization of Asphaltenes during Deep Hydroconversion Processes Determined by Atmospheric Pressure Photoionization (APPI) Fourier Transform Ion Cyclotron Resonance (FT-ICR) Mass Spectrometry, *Energy Fuels*, 2010, **24**(4), 2257–2265.

41 D. Giraldo-Dávila, M. L. Chacón-Patiño, A. M. McKenna, C. Blanco-Tirado and M. Y. Combariza, Correlations Between Molecular Composition and the Adsorption, Aggregation and Emulsifying Behavior of Petrophase 2017 Asphaltenes and Their TLC Fractions, *Energy Fuels*, 2018, **32**(3), 2769–2780.

42 E. Rogel, C. Ovalles and M. Moir, Asphaltene Chemical Characterization as a Function of Solubility: Effects on Stability and Aggregation, *Energy Fuels*, 2012, **26**(5), 2655–2662.

43 E. Rogel and M. Witt, Atmospheric Pressure Photoionization Coupled to Fourier Transform Ion Cyclotron Resonance Mass Spectrometry To Characterize Asphaltene Deposit Solubility Fractions: Comparison to Bulk Properties, *Energy Fuels*, 2016, **30**, 915–923.

44 R. P. Rodgers and A. G. Marshall, Petroleomics: Advanced Characterization of Petroleum-Derived Materials by Fourier Transform Ion Cyclotron Resonance Mass Spectrometry (FT-ICR MS), in *Asphaltenes, Heavy Oils, and Petroleomics*, ed. O. C. Mullins, E. Y. Sheu, A. Hammam and A. G. Marshall, Springer, 2007, pp. 63–93.

45 C. L. Hendrickson, J. P. Quinn, N. K. Kaiser, D. F. Smith, G. T. Blakney, T. Chen, A. G. Marshall, C. R. Weisbrod and S. C. Beu, 21 Tesla Fourier Transform Ion Cyclotron Resonance Mass Spectrometer: A National Resource for Ultrahigh Resolution Mass Analysis, *J. Am. Soc. Mass Spectrom.*, 2015, **26**(9), 1626–1632.

46 Y. Cho, Y. H. Kim and S. Kim, Planar Limit-Assisted Structural Interpretation of Saturates/Aromatics/Resins/Asphaltenes Fractionated Crude Oil Compounds Observed by Fourier Transform Ion Cyclotron Resonance Mass Spectrometry, *Anal. Chem.*, 2011, **83**(15), 6068–6073.

47 A. M. McKenna, A. G. Marshall and R. P. Rodgers, Heavy Petroleum Composition. 4. Asphaltene Compositional Space, *Energy Fuels*, 2013, **27**, 1257–1267.

48 A. M. Wittrig, T. R. Fredriksen, K. Qian, A. C. Clingenpeel and M. R. Harper, Single Dalton Collision-Induced Dissociation for Petroleum Structure Characterization, *Energy Fuels*, 2017, **31**(12), 13338–13344.

49 C. P. Rüger, C. Grimmer, M. Sklorz, A. Neumann, T. Streibel and R. Zimmermann, Combination of Different Thermal Analysis Methods Coupled to Mass Spectrometry for the Analysis of Asphaltenes and Their Parent Crude Oils: Comprehensive Characterization of the Molecular Pyrolysis Pattern, *Energy Fuels*, 2017, **32**(3), 2699–2711.

50 C. P. Rüger, T. Miersch, T. Schwemer, M. Sklorz and R. Zimmermann, Hyphenation of Thermal Analysis to Ultrahigh-Resolution Mass Spectrometry (Fourier Transform Ion Cyclotron Resonance Mass Spectrometry) Using Atmospheric Pressure Chemical Ionization For Studying Composition and Thermal Degradation of Complex Materials, *Anal. Chem.*, 2015, **87**(13), 6493–6499.

51 C. P. Rüger, A. Neumann, M. Sklorz, T. Schwemer and R. Zimmermann, Thermal Analysis Coupled to Ultrahigh Resolution Mass Spectrometry with Collision Induced Dissociation for Complex Petroleum Samples: Heavy Oil Composition and Asphaltene Precipitation Effects, *Energy Fuels*, 2017, **31**, 13144–13158.

52 S. M. Rowland, W. K. Robbins, Y. E. Corilo, A. G. Marshall and R. P. Rodgers, Solid-Phase Extraction Fractionation To Extend the Characterization of Naphthenic Acids in Crude Oil by Electrospray Ionization Fourier Transform Ion Cyclotron Resonance Mass Spectrometry, *Energy Fuels*, 2014, **28**, 5043–5048.

53 M. L. Chacón-Patiño, S. M. Rowland and R. P. Rodgers, Advances in Asphaltene Petroleomics. Part 1: Asphaltenes Are Composed of Abundant Island and Archipelago Structural Motifs, *Energy Fuels*, 2017, **31**(12), 13509–13518.

54 M. L. Chacón-Patiño, S. M. Rowland and R. P. Rodgers, Advances in Asphaltene Petroleomics. Part 2: Selective Separation Method That Reveals Fractions Enriched in Island and Archipelago Structural Motifs by Mass Spectrometry, *Energy Fuels*, 2018, **32**(1), 314–328.

55 G. A. Vasconcelos, R. C. L. Pereira, C. D. F. Santos, V. V. Carvalho, L. V. Tose, W. Romão and B. G. Vaz, Extraction and Fractionation of Basic Nitrogen Compounds in Vacuum Residue by Solid-Phase Extraction and Characterization by Ultra-High Resolution Mass Spectrometry, *Int. J. Mass Spectrom.*, 2017, **418**, 67–72.

56 C. S. Hsu, K. Qian, T. Aczel, S. C. Blum, W. N. Olmstead, L. H. Kaplan, W. K. Robbins, W. W. Schulz and M. A. McLean, On-Line Liquid Chromatography/Mass Spectrometry for Heavy Hydrocarbon Characterization, *Energy Fuels*, 1991, **5**(3), 395–398.

57 M. L. Chacón-Patiño, S. M. Rowland and R. P. Rodgers, Advances in Asphaltene Petroleomics. Part 3. Dominance of Island or Archipelago Structural Motif Is Sample Dependent, *Energy Fuels*, 2018, **32**(9), 9106–9120.

58 L. Nyadong, J. Lai, C. Thompsen, C. J. LaFrancois, X. Cai, C. Song, J. Wang and W. Wang, High-Field Orbitrap Mass Spectrometry and Tandem Mass Spectrometry for Molecular Characterization of Asphaltenes, *Energy Fuels*, 2018, **32**(1), 294–305.

59 O. P. Strausz, A. Morales-Izquierdo, N. Kazmi, D. S. Montgomery, J. D. Payzant, I. Safarik and J. Murgich, Chemical Composition of Athabasca Bitumen: The Saturate Fraction, *Energy Fuels*, 2010, **24**(9), 5053–5072.

60 K. Qian, W. K. Robbins, C. A. Hughey, H. J. Cooper, R. P. Rodgers and A. G. Marshall, Resolution and Identification of Elemental Compositions for More than 3000 Crude Acids in Heavy Petroleum by Negative-Ion Microelectrospray High-Field Fourier Transform Ion Cyclotron Resonance Mass Spectrometry, *Energy Fuels*, 2001, **15**(6), 1505–1511.

61 W. E. Rudzinski, L. Oehlers, Y. Zhang and B. Najera, Tandem Mass Spectrometric Characterization of Commercial Naphthenic Acids and a Maya Crude Oil, *Energy Fuels*, 2002, **16**(5), 1178–1185.

62 M. L. Chacón-Patiño, S. J. Vesga-Martínez, C. Blanco-Tirado, J. A. Orrego-Ruiz, A. Gómez-Escudero and M. Y. Combariza, Exploring Occluded Compounds and Their Interactions with Asphaltene Networks Using High-Resolution Mass Spectrometry, *Energy Fuels*, 2016, **30**(6), 4550–4561.

63 G. C. Klein, S. Kim, R. P. Rodgers, A. G. Marshall and A. Yen, Mass Spectral Analysis of Asphaltenes. II. Detailed Compositional Comparison of Asphaltenes Deposit to Its Crude Oil Counterpart for Two Geographically Different Crude Oils by ESI FT-ICR MS, *Energy Fuels*, 2006, **20**, 1973–1979.

64 V. V. Lobodin, P. Juyal, A. M. Mckenna, R. P. Rodgers and A. G. Marshall, Tetramethylammonium Hydroxide as a Reagent for Complex Mixture Analysis by Negative Ion Electrospray Ionization Mass Spectrometry, *Anal. Chem.*, 2013, **85**, 7803–7808.

65 N. K. Kaiser, J. P. Quinn, G. T. Blakney, C. L. Hendrickson and A. G. Marshall, A Novel 9.4 Tesla FT-ICR Mass Spectrometer with Improved Sensitivity, Mass Resolution, and Mass Range, *J. Am. Soc. Mass Spectrom.*, 2011, **22**(8), 1343–1351.

66 G. T. Blakney, C. L. Hendrickson and A. G. Marshall, Predator Data Station: A Fast Data Acquisition System for Advanced FT-ICR MS Experiments, *Int. J. Mass Spectrom.*, 2011, **306**(2–3), 246–252.

67 Y. E. Corilo, *PetroOrg Software*, Florida State University, All rights reserved, 2013, http://www.petroorg.com.

68 A. C. Clingenpeel, S. M. Rowland, Y. E. Corilo, P. Zito and R. P. Rodgers, Fractionation of Interfacial Material Reveals a Continuum of Acidic Species

That Contribute to Stable Emulsion Formation, *Energy Fuels*, 2017, **31**(6), 5933–5939.

69 A. C. Clingenpeel, W. K. Robbins, Y. E. Corilo and R. P. Rodgers, Effect of the Water Content on Silica Gel for the Isolation of Interfacial Material from Athabasca Bitumen, *Energy Fuels*, 2015, **29**(11), 7150–7155.

70 J. C. Putman, S. M. Rowland, D. C. Podgorski, W. K. Robbins and R. P. Rodgers, Dual-Column Aromatic Ring Class Separation with Improved Universal Detection across Mobile-Phase Gradients via Eluate Dilution, *Energy Fuels*, 2017, **31**, 12064–12071.

71 A. G. Marshall and R. P. Rodgers, Petroleomics: Chemistry of the Underworld, *Proc. Natl. Acad. Sci. U. S. A.*, 2008, **105**(47), 18090–18095.

72 H. Korsten, Characterization of Hydrocarbon Systems by DBE Concept, *AIChE J.*, 1997, **43**(6), 1559–1568.

73 F. W. McLafferty and F. Tureček, *Interpretation of Mass Spectra*, University Science Books, 4th edn, 1993, pp. 27–28.

74 V. V. Lobodin, A. G. Marshall and C. S. Hsu, Compositional Space Boundaries for Organic Compounds, *Anal. Chem.*, 2012, **84**, 3410–3416.

75 C. S. Hsu, V. V. Lobodin, R. P. Rodgers, A. M. McKenna and A. G. Marshall, Compositional Boundaries for Fossil Hydrocarbons, *Energy Fuels*, 2011, **25**, 2174–2178.

76 A. A. Herod, Limitations of Mass Spectrometric Methods for the Characterization, *Rapid Commun. Mass Spectrom.*, 2010, **24**(24), 2507–2519.

77 A. A. Herod, K. D. Bartle, T. J. Morgan and R. Kandiyoti, Analytical Methods for Characterizing High-Mass Complex Polydisperse Hydrocarbon Mixtures: An Overview, *Chem. Rev.*, 2012, **112**(7), 3892–3923.

78 A. M. McKenna, L. J. Donald, J. E. Fitzsimmons, P. Juyal, V. Spicer, K. G. Standing, A. G. Marshall and R. P. Rodgers, Heavy Petroleum Composition. 3. Asphaltene Aggregation, *Energy Fuels*, 2013, **27**(3), 1246–1256.

79 M. L. Chacón-Patiño, C. Blanco-Tirado, J. A. Orrego-Ruiz, A. Gómez-Escudero and M. Y. Combariza, High Resolution Mass Spectrometric View of Asphaltene–SiO$_2$ Interactions, *Energy Fuels*, 2015, **29**(3), 1323–1331.

80 E. Rogel and M. Witt, Atmospheric Pressure Photoionization Coupled to Fourier Transform Ion Cyclotron Resonance Mass Spectrometry to Characterize Asphaltene Deposit Solubility Fractions: Comparison to Bulk Properties, *Energy Fuels*, 2016, **30**(2), 915–923.

81 A. Karimi, K. Qian, W. N. Olmstead, H. Freund, C. Yung and M. R. Gray, Quantitative Evidence for Bridged Structures in Asphaltenes by Thin Film Pyrolysis, *Energy Fuels*, 2011, **25**, 3581–3589.

82 R. I. Rueda-Velásquez, H. Freund, K. Qian, W. N. Olmstead and M. R. Gray, Characterization of Asphaltene Building Blocks by Cracking under Favorable Hydrogenation Conditions, *Energy Fuels*, 2013, **27**, 1817–1829.

83 M. R. Gray, Consistency of Asphaltene Chemical Structures with Pyrolysis and Coking Behavior, *Energy Fuels*, 2003, **17**(6), 1566–1569.

84 M. L. Chacón-Patiño, C. Blanco-Tirado, J. A. Orrego-Ruiz, A. Gómez-Escudero and M. Y. Combariza, Tracing the Compositional Changes of Asphaltenes after Hydroconversion and Thermal Cracking Processes by High-Resolution Mass Spectrometry, *Energy Fuels*, 2015, **29**(10), 6330–6341.

85 P. E. Savage, M. T. Klein and S. G. Kukes, Petroleum Asphaltene Thermal Reaction Pathways, *Preprints of Symposia*, American Chemical Society, Division of Fuel Chemistry, 1985, vol. 30, ch. 3, pp. 408–419.

86 J. Ancheyta, F. Trejo and M. S. Rana, Definition and Structure of Asphaltenes, in *Asphaltene Chemical Transformation during Hydroprocessing of Heavy Oils*, CRC Press, Taylor & Francis Group, Boca Raton, FL, USA, 2010, pp. 1–86.

87 H. Sabbah, A. L. Morrow, A. E. Pomerantz and R. N. Zare, Evidence for Island Structures as the Dominant Architecture of Asphaltenes, *Energy Fuels*, 2011, **25**(4), 1597–1604.

88 M. R. Hurt, D. J. Borton, H. J. Choi and H. I. Kenttämaa, Comparison of the Structures of Molecules in Coal and Petroleum Asphaltenes by Using Mass Spectrometry, *Energy Fuels*, 2013, **27**(7), 3653–3658.

89 T. M. Jarrell, C. Jin, J. S. Riedeman, B. C. Owen, X. Tan, A. Scherer, R. R. Tykwinski, M. R. Gray, P. Slater and H. I. Kenttämaa, Elucidation of Structural Information Achievable for Asphaltenes via Collision-Activated Dissociation of Their Molecular Ions in MSn Experiments: A Model Compound Study, *Fuel*, 2014, **133**, 106–114.

90 Z. Liao, J. Zhao, P. Creux and C. Yang, Discussion on the Structural Features of Asphaltene Molecules, *Energy Fuels*, 2009, **23**(12), 6272–6274.

91 O. P. Strausz, T. W. Mojelsky, E. M. Lown, I. Kowalewski and F. Behar, Structural Features of Boscan and Duri Asphaltenes, *Energy Fuels*, 1999, **13**(2), 228–247.

92 J. D. Payzant, E. M. Lown and O. P. Strausz, Structural Units of Athabasca Asphaltene: The Aromatics with a Linear Carbon Framework, *Energy Fuels*, 1991, **5**(3), 445–453.

93 T. M. C. Pereira, G. Vanini, E. C. S. Oliveira, F. M. R. Cardoso, F. P. Fleming, A. C. Neto, V. Lacerda, E. V. R. Castro, B. G. Vaz and W. Romão, An Evaluation of the Aromaticity of Asphaltenes Using Atmospheric Pressure Photoionization Fourier Transform Ion Cyclotron Resonance Mass Spectrometry – APPI($\pm$)FT-ICR MS, *Fuel*, 2014, **118**, 348–357.

94 J. S. Riedeman, N. R. Kadasala, A. Wei and H. I. Kenttämaa, Characterization of Asphaltene Deposits by Using Mass Spectrometry and Raman Spectroscopy, *Energy Fuels*, 2016, **30**(2), 805–809.

Faraday Discussions

PAPER

Complexity of dissolved organic matter in the molecular size dimension: insights from coupled size exclusion chromatography electrospray ionisation mass spectrometry†

J. A. Hawkes, [ID] *[a] P. J. R. Sjöberg, [ID] [a] J. Bergquist [ID] [a] and L. J. Tranvik [ID] [b]

Received 3rd December 2018, Accepted 25th January 2019
DOI: 10.1039/c8fd00222c

This paper investigates the relationship between apparent size distribution and molecular complexity of dissolved organic matter from the natural environment. We used a high pressure size exclusion chromatography (HPSEC) method coupled to UV-Vis diode array detection (UV-DAD) and electrospray ionisation mass spectrometry (ESI-MS) in order to compare the apparent size of natural organic matter, determined by HPSEC-UV and the molecular mass determined online by ESI-MS. We found that there was a clear discrepancy between the two methods, and found evidence for an important pool of organic matter that has a strong UV absorbance and no ESI-MS signal. Contrary to some previous research, we found no evidence that apparently high molecular weight organic matter is constituted by aggregates of low molecular weight (<1000 Da) material. Furthermore, our results suggest that the majority of apparent size variability within the ESI ionisable pool of organic matter is due to secondary interaction and exclusion effects on the HPSEC column, and not true differences in hydrodynamic size or intermolecular aggregation.

Introduction

Dissolved organic matter (DOM) in aquatic environments is by far the most abundant form of organic matter in natural waters. It is an ultra-complex mixture of phenolic, carboxylic acid rich material that ranges in concentration from >50 mg L^{-1} C^{-1} in terrestrial wetlands down to <0.5 mg L^{-1} C^{-1} in the deep sea. DOM is operationally defined as organic matter that is not retained by filtration,

---

[a]Analytical Chemistry, Department of Chemistry - BMC, Uppsala University, Uppsala, Sweden. E-mail: jeffrey.hawkes@kemi.uu.se
[b]Department of Limnology, Uppsala University, Uppsala, Sweden
† Electronic supplementary information (ESI) available. See DOI: 10.1039/c8fd00222c

typically using pore sizes between 0.2 and 0.7 μm. It may consist of truly dissolved molecules, aggregates and high molecular weight colloids. The size distribution across this wide spectrum is critical to the biogeochemical processing of DOM, including microbial transformation and mineralization, photochemical reactions, and susceptibility to interact with metals, sorb to surfaces or aggregates, and subsequently form aggregates that settle out of the water column.

The biogeochemical utilisation and transformation of DOM along the natural gradient from soils to inland waters to the sea is an important part of the carbon cycle, and slight changes to production and removal processes could lead to significant positive and negative feedbacks in the global climate. There is therefore great interest in the nature and reactivity of DOM, but investigations into its chemistry are always confounded by its extreme complexity.[1,2] The complexity of natural DOM is easily demonstrated by high resolution electrospray ionisation mass spectrometry (HR-ESI-MS), as any aquatic sample contains thousands of molecular peaks with different molecular formulas.[3,4] Each molecular formula is constituted by an unknown number of structural isomers, making compositional analysis extremely challenging.[5,6] Complexity can also be demonstrated using one dimensional chromatography, whereby an unresolvable complex mixture is obtained, no matter which separation mechanism is used.[7-15] It is important to remember that standard compounds do not exist for natural DOM, and natural mixtures are made up of possibly millions of unknown, interacting constituents at trace concentrations.[3,5,16,17]

One of the major debates about the nature of DOM in aquatic environments regards its physical speciation and true molecular size in solution.[18,19] Filtration through different pore size membranes and size exclusion usually suggest that DOM has a large mass range of 0.1–100 kDa,[18,20-24] whereas HR-ESI-MS generally finds molecules restricted to masses of 200–800. One possible explanation for this apparent duality is that naturally occurring molecular aggregates are broken up by sample processing and analysis – *e.g.* during dissolution in organic solvent, ESI and ion trapping steps,[4,25] and natural solutions contain weak aggregates of monomers with masses of 200–800 Da.[18,26-29] Another (less popular) explanation is that negative mode HR-ESI-MS preferentially detects low molecular weight carboxylic acids, and genuine (not aggregated) high molecular weight compounds are masked by the more easily ionisable or volatile lower masses, and are not detected. Either case, or their relative importance, would have major implications for the way that we interpret HR-ESI-MS data with regards to the nature and reactivity of DOM, and the potential effect it can have on topics such as contaminant mobility and drinking water quality.[18,30,31]

High pressure size exclusion chromatography (HPSEC) coupled online with UV-DAD and HR-ESI-MS can be used to investigate this problem. Molecules are separated by hydrodynamic size by HPSEC, and measured according to their UV light attenuation by UV-DAD and according to their mass to charge ratio (m/z) by HR-ESI-MS. Ideally, the estimated HPSEC-UV-DAD size and measured HR-ESI-MS mass would scale together, but molecular shape, charge repulsion and secondary retention effects on the column stationary phase can complicate the chromatography,[19,32] while ionisation efficiency and ionisation effects like adduct formation and disaggregation can still bias ESI-MS, even after simplification of the mixture by separation.[4,25,33] We investigated the HPSEC behaviour of DOM from different sources and compared it with various size standards of differing charge density in order to evaluate the two main theories about the size distribution of DOM presented above. Based on current paradigms about DOM size distribution and aggregation, we expected to find that high molecular weight material measured by HPSEC-UV-DAD would be detected by ESI-MS as monomers with m/z 200–800, however, our results instead provided evidence for the existence of genuinely large, UV active and ESI-MS invisible DOM, and no evidence for the expected molecular aggregates.

Methods

Chemicals and samples

Ultrapure water (18.2 MΩ resistivity) was generated with a MilliQ system (Millipore, Burlington MA, USA). Methanol was high purity hypergrade (LiChroSolv for LCMS, Merck, Kenilworth, NJ, USA). High purity ammonium acetate (NH_4Ac), hydrochloric acid and ammonia were obtained from Fluka (Steinheim, Germany), Merck and Sigma Aldrich (St. Louis, MO, USA), respectively. Model compounds with m/z 369 were purchased from Sigma Aldrich: 2-(4-(2,2-dicarboxy-ethyl)-2,5-dimethoxy-benzyl)-malonic acid (model compound a), 7,8-dihydroxy-6-methoxycoumarin-8-β-D-glucopyranoside (fraxin, model compound b), and isoferulic acid 3-O-β-D-glucuronide (model compound c). Poly(styrene sulfonate) standards (PSS) were purchased from American Polymer Standards Corporation (Mentor, OH, USA) at 6 average weights (218 kDa, 38 kDa, 15 kDa, 4.4 kDa, 1.4 kDa, and 206 Da) and stored at room temperature as powders. Poly(propylene glycol) standards (PPG) were purchased from Aldrich Chemie (Steinheim, Germany) at average masses 1 kDa and 2 kDa, and from Aldrich Chemical Company (Milwaukee, WI, USA) at 425 Da average mass, and stored as liquids at 4 °C.

Suwannee River Fulvic Acid standard (lot 2S101F) was purchased from the International Humic Substances Society (IHSS), and four other samples were prepared as follows:

Soil humic and fulvic acids – 1.42 g of soil from Stadsskogen, Uppsala, Sweden (59.840 N, 17.636 E), was mixed with 10 mL 0.25% ammonia. The mixture was shaken, sonicated and vortexed twice, left at room temperature for 90 minutes, and then centrifuged at 4000 rpm for 10 minutes. One aliquot (2 mL) was taken from the supernatant without further modification and labelled 'Humic acids', and a second aliquot (2 mL) was acidified to pH 2 by adding 40 μL 6 M hydrochloric acid, labelled 'Fulvic acids'. The two aliquots were centrifuged to remove precipitated material and the resulting supernatants were injected onto the SEC column at 100 μL injection volume without further treatment.

Wetland stream reverse osmosis DOM – organic matter was concentrated from a small stream that drains a wetland named Börje Sjö near Uppsala, Sweden (59.918 N, 17.347 E) using a RealSoft PROS/2S portable reverse osmosis system.[34] DOM was concentrated to 700 mg L^{-1} C^{-1}. This sample was stored for >2 years at 4 °C in the dark, and was filtered through glass fibre filter (GF/F 0.7 μm) before injection (20 μL) onto the SEC column.

Leaf cold water extract – 25.96 g of green birch (*Betula* sp.) leaves were taken from a tree beside the Biomedical Centrum, Uppsala, Sweden (59.841 N, 17.634 E), and mixed with 554 g ultrapure water. The mixture was kept at 4 °C in the dark for 18 hours to extract 'rainwater extractable DOM'. A sample (2 mL) was taken to an Eppendorf tube, dried down by vacuum centrifuge (Eppendorf Concentrator

Plus) and redissolved in 100 µL 20 : 80 methanol : water, 25 mM NH_4Ac for injection (80 µL) onto the SEC column.

High pressure liquid chromatography-mass spectrometry

High pressure size exclusion chromatography (HPSEC; Tosoh TSKgel G3000SW 300 × 7.5 mm, 10 µm) was conducted with an Agilent 1100 HPLC system using combinations of water, methanol and NH_4Ac as the mobile phase, with a flow rate of 1 mL min^{-1}. A YMC-Pack-Diol-300 (300 × 8 mm, 5 µm) column was also tested, but found to be much more affected by hydrophobic interactions between analytes and the stationary phase than the Tosoh column. UV light absorbing DOM was detected using a UV-Vis Diode Array Detector (DAD; Agilent 1100) at wavelength 254 nm.

For online ESI-MS analysis, the flow was split (after UV-Vis DAD) approximately 9 : 1 to waste, leaving a flow of ~100 µL min^{-1} for heated ESI-MS (100 °C, sheath gas setting 28, 2.5 kV negative mode). The splitting and heating were to assist with vaporisation of the solvent, which in most cases was largely water. The time delay from UV-Vis DAD detection and ESI-MS detection was determined to be 0.28 minutes using the 206 Da PSS standard, and all ESI-MS times are reported so as to be aligned with UV-Vis DAD detection times. The effect of injection volume was tested by analysing SRFA at four concentrations (0.25–5 mg L^{-1} C^{-1}) with differing injection volumes (1–100 µL) in various combinations (Fig. SI1†), and this had minimal effect on the retention profile determined by UV-DAD and ESI-MS.

The LTQ is a linear ion trap that is used to collect ions before transfer to the Orbitrap analyser, and is also used for collision induced fragmentation (CID) using nitrogen gas. The ion trap was set to trap and transfer ions with masses 200–2000, and the Orbitrap mass analyser was set to fill to 1×10^6 ions (maximum fill time 250 ms) at a resolution setting of '100 000', giving an actual resolution >115 000 at m/z 401. Formulas were assigned in each transient as in our recent papers[35] up to $C_{40}H_{80}O_{30}NS^{13}C$. The formula possibilities were restricted to the following rules: O/C < 1.0, H/C 0.3–2.2, m/z 200–800, double bond equivalence minus oxygen (DBE-O) < 10,[36] N + S + ^{13}C < 2. Formulas were assigned if a peak in the resulting list could be found within 3 ppm ($\Delta m/m \times 1 \times 10^6$) of a measured peak in each individual transient. Isotopologues containing one ^{13}C were assigned but not considered in further data processing (i.e. molecular formulas were not counted twice via isotopologues).

Calibration and method validation was conducted with PSS standards (representing charged polymers), PPG standards (representing uncharged polymers) and three model compounds with 0, 2 and 4 carboxylic groups, but the same m/z (369) and formula ($C_{16}H_{18}O_{10}$). The size standards of PSS and PPG were prepared to 1 mg mL^{-1} in 10 mM NH_4Ac and analysed at a 2 µL injection volume with detection by UV-Vis DAD. After initial testing, the PSS standards were mixed into two mixed standards at equal volumes (A: 218 kDa, 15 kDa, 1.4 kDa and acetone, B: 38 kDa, 4.4 kDa, 206 Da) and injected at a 10 µL volume. The model compounds (a malonic acid derivative, fraxin and an isoferulic acid glucuronide) were prepared to 1 ppm concentration in 50 : 50 water : methanol, and 50 µL of the solution was injected onto the SEC column. They were distinguished by their unique fragments after collision induced dissociation (CID) in the ion trap. This was carried out at a normalised collision energy of 25 eV after selected ion monitoring (SIM) of m/z 368.5–369.5.

In one experiment, chromatographic fractions were collected manually after UV-Vis detection, and fractions (1.5 minutes = 1.5 mL) were dried by vacuum centrifuge (Eppendorf Concentrator Plus) and redissolved in 100 μL 50 : 50 water : methanol. They were then reinjected onto the column at an 80 μL injection volume. This was done to examine potential redistribution of aggregates when size fractions were isolated in solution.

Molecular peaks are later discussed in terms of their polydispersity (PD), which is calculated as:

$$PD = \frac{M_w}{M_n}$$

where the weight averaged molecular weight (M_w) and number averaged molecular weight (M_n) are calculated from the mass spectrum peak intensity (I_k) and HPSEC estimated peak mass (M_k):

$$M_w = \frac{\sum_1^k I_k M_k}{\sum_1^k I_k}$$

$$M_n = \frac{\sum_1^k I_k}{\sum_1^k I_k / M_k}$$

Results and discussion

Development of a HPSEC method for DOM

Coupling HR-ESI-MS to HPSEC imposes important restrictions on the mobile phase used in the chromatography, in that it has to be volatile and high ionic strength should be avoided. Three types of standard were used to evaluate size exclusion behaviour. PSS was used, as in most recent literature, as a charged polymer that is similar in hydrodynamic size to natural humic substances,[17,31,32,37] but the validity of PSS standards or any other model compound in representing the HPSEC retention of DOM must be questioned, as DOM is so chemically diverse.[17,31,32,37] PPG is uncharged and was used to compare the behaviour of analytes that would not be subject to charge exclusion, and three model compounds with identical mass but differing charge in solution were analysed to confirm the differences found in the two polymer series.

We used the PSS polymer series to test combinations of NH$_4$Ac in water and methanol with measured pH > 6, similar to most natural aquatic systems with low DOC. The highest molecular weight standard, 218 400 Da, was always found at ~4.9 minutes, which we suppose is near to the interstitial volume (V_0) of the column. Linearity of response in log(MW) vs. time was only achieved up to the 37 500 Da standard, making this the maximum estimated PSS-equivalent mass. Later, it will be shown that most DOM was retained within the linear range, but some DOM can elute at V_0, making it an unknown size higher than 37.5 kDa in

PSS equivalence (*i.e.* taking charge repulsion into account, as explained below). The total volume of mobile phase in the column was not measured reliably (*e.g.* with D_2O^{32}), and the retention time of acetone (58 Da) varied depending on the extent of hydrophobic interaction imposed by the mobile phase conditions (Fig. 1; see discussion later), meaning that linearity between log(MW) and retention time always broke down at the low mass end.

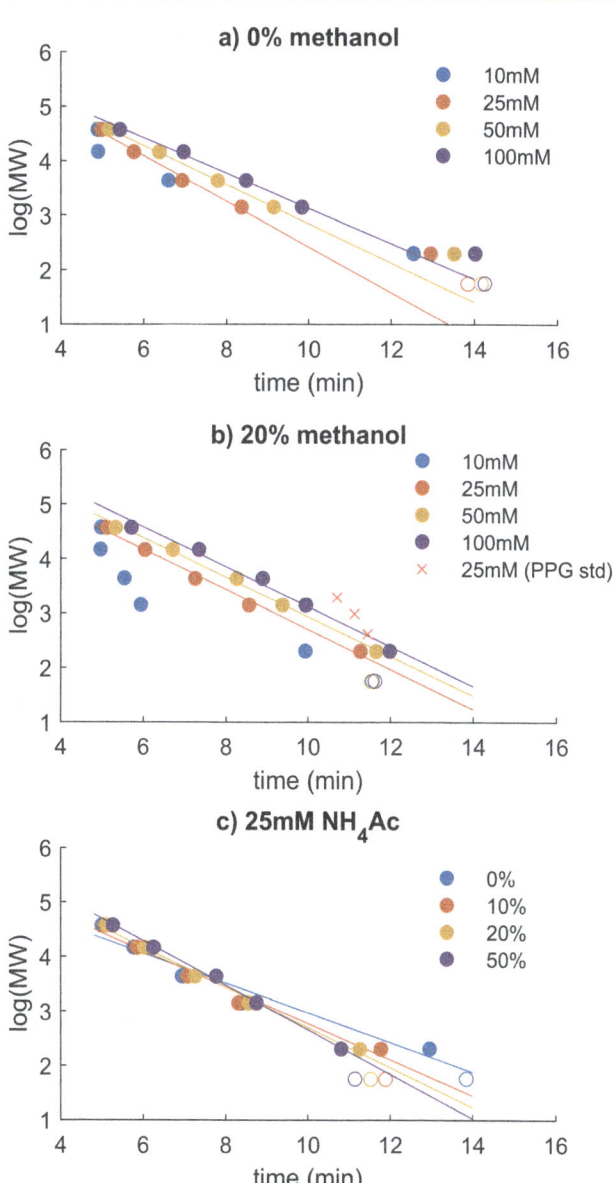

Fig. 1 log(MW) *vs.* retention time relationship of PSS standards (filled circles) in various mobile phases. Note the variation in retention time of acetone, plotted as an unfilled circle. Polypropylene glycol (PPG) standards are plotted as red crosses. (a−c) Different mobile phase conditions, as indicated by the titles and legends.

In aqueous phase, without addition of methanol, calibration curves of log(MW) *vs.* time using PSS were not linear across the entire range, and instead flattened significantly from the 1440 standard to the 206 Da standard and acetone (Fig. 1a). This non-linearity has also been documented using ammonium bicarbonate solutions as the mobile phase.[2,11] An increase in NH_4Ac from 10–100 mM improved retention of the PSS standards due to decreased influence of ionic exclusion, which greatly affects charged molecules in SEC[19,32,38] due to repulsive interactions between charged analytes and the slightly charged stationary phase.[39] The increase in NH_4Ac in pure aqueous solvent did not improve the flattening of the calibration at low masses. The adjusted r^2 of the slopes using all 5 standards up to 37.5 kDa was 0.93–0.98. At the lowest NH_4Ac concentration, no retention was found for the standards >14 800 Da because of ion exclusion, and retention of the 37 500 Da standard began at 25 mM NH_4Ac, similarly to the previous observations.[32,40]

An increase in methanol notably improved linearity by decreasing the retention time of the 206 Da standard and acetone, while not greatly affecting the higher molecular weight standards (Fig. 1b and c). This suggests that the longer time in the column of the low molecular weight standards allows a greater toll to be taken by secondary hydrophobic interaction effects on the stationary phase in aqueous solution. The flattening of the calibration curve away from linearity due to hydrophobic retention is not desirable as it leads to an incorrect assessment of the molecular size of small, hydrophobic constituents, and unpredictable and variable chromatographic resolution at the crucial mass range of 200–2000.[11] An addition of 10% methanol made a large improvement to linearity by decreasing the retention of the smallest standards (adjusted r^2 0.96 compared with 0.93; Fig. 1c), and 20% methanol made the calibration nearly linear (adjusted r^2 0.99) across the range 206–37 500 Da. The methanol also improves the quality of spray observed with ESI, being more volatile and decreasing surface tension, leading to a better signal to noise in the MS detection. We also noticed at the highest NH_4Ac concentration, and particularly at high sample concentrations of DOM, that organic matter began to appear more hydrophobic (longer retention times, data not shown). This may be due to decreased solubility of the DOM in the high ionic strength solution and resulting non-linear chromatography behaviour where the phase equilibria are not linear.[41] In other words, the concentrations of components in the stationary phase at equilibrium are no longer proportional to the concentrations in the mobile phase. This effect was not investigated further, but was not observed for NH_4Ac concentrations $\leq$ 50 mM.

The PPG standards retained much longer than PSS, as expected[32] (Fig. 1b), and the model compounds also varied in retention according to charge density[37] (Fig. 2). These compounds vary in charge density in pH 6.6 solution from 0–4 charges (Fig. SI4†), and represent the range of charge density expected for natural DOM. The neutral compound, fraxin, was retained longer than acetone when methanol was not added to the mobile phase, indicating hydrophobic retention on the column, and was generally comparable with PPG. The least well retained was the malonic acid derivative, due to its high charge density, leading to an overestimated molecular weight based on the calibration using the PSS standards. The isoferulic acid glucuronide (two carboxylic acid groups) was the most comparable in retention to its equivalent mass in the PSS calibration curve. Some

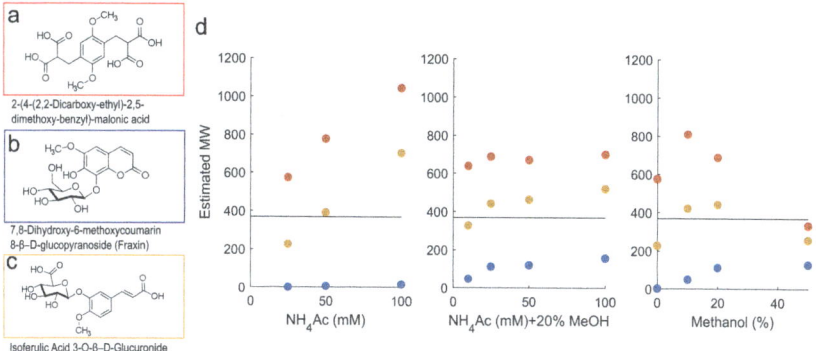

Fig. 2 (a–c) Model compounds with neutral mass 370. (d) Estimated molecular weight of compound (a) (malonic acid derivative; red), (b) (fraxin; blue) and (c) (isoferulic acid glucuronide; yellow). All three compounds have true neutral mass 370, marked with a horizontal line. The compounds were detected as compound-specific fragments after CID fragmentation and the peak position at the apex was converted to mass using the relevant calibration in Fig. 1.

variability in retention time may also be explained by differing hydrodynamic volume, but charge density likely explains the major differences found.

The large range in predicted molecular weights for these three isomers demonstrates the limitations of HPSEC for complex mixtures (Fig. 2). Adding 20% methanol to the mobile phase improved the consistency of the predicted molecular weight across different NH$_4$Ac concentrations, and with 20% meth-anol, the malonic acid derivative never appeared as more than double its true weight. It remains possible that the compound dimerised in solution and was truly present at ~740 Da, but more likely is that charge repulsion was higher for this compound than the PSS standards it is being compared with. It is likely that DOM contains a high occurrence of carboxylic acid groups, according to its reactive chemistry[42,43] and fragmentation patterns when studied by MS,[6,44] but a higher density of carboxylic acids than the malonic acid derivative compound is unlikely. We therefore take comfort that its predicted weight is only two-fold higher than its true value. Also concerning was fraxin, which was retained for far too long compared with the equivalent retention time from the PSS calibration curve, and likely represents all neutral compounds in this sense. Addition of 20% methanol improved the estimated MW for fraxin, and seemed to stabilise the response with respect to NH$_4$Ac concentration for the two carboxylic acids (Fig. 2). At 0% methanol, increasing NH$_4$Ac concentration increased the estimated MW of the carboxylic acids. It might be considered that this high NH$_4$Ac concentration promoted self-aggregation of these standards,[45] but this possibility is difficult to test, as only the monomer mass was detected by ESI-MS.

The final selection of mobile phase has many implications for this study. It is important to have effective retention and resolution by HPSEC, meaning that the NH$_4$Ac concentration must be at least 25 mM and contain at least 10% methanol. The ESI spray efficiency is poor when the methanol content is too low or the salt content is too high, so this favours keeping the salt concentration as low as possible. Sample loading was tested from 2.5–50 μg with no variability in UV retention profile observed (Fig. SI1†), with higher loading giving a better signal and a higher number of assigned peaks by MS. Additionally, the study is improved by making the pH, salt and organic solvent conditions as similar to those of environmental samples as possible, so that the speciation of the DOM by HPSEC is comparable. Buffer capacity and ionic strength change dramatically across landscapes, particularly from terrestrial to marine waters. Marine conditions (0.5 M NaCl) are permissible for HPSEC but incompatible with ESI-MS, so the system was optimised for terrestrial samples of low salinity and buffer strength, usually below 3 meq. Overall, the optimal mobile phase for HPSEC-ESI-MS was chosen as 25 mM NH$_4$Ac in 20% methanol (pH measured at 6.6), as a compromise between effective masking of ionic repulsion effects, limiting hydrophobic interaction and salting out of compounds from solution, optimising spray quality for ESI whilst keeping the salt and alcohol concentration low enough to mimic environmental conditions.

Size distribution of DOM samples according to HPSEC-UV-Vis DAD and HPSEC-HESI-Orbitrap-MS

Suwannee river fulvic acid. Fulvic acids are operationally defined as being soluble in both acid and base. The Suwannee River Fulvic Acid (SRFA) reference sample is a complex mixture of these versatile molecules extracted from river water. The retention profile of SRFA in 25 mM NH$_4$Ac with 20% methanol, measured by HPSEC-UV, is broad, beginning at 6 minutes and ending before 12 minutes, just past the apex of an acetone standard. The bulk of the material elutes between 7–10 minutes with an apex at 8.87 minutes, corresponding to 1331 Da in PSS retention equivalence (Fig. 3a), which is similar to the lower MW range of previous reports.[2,37,40,46,47] Note that the PSS equivalence is only accurate for DOM with similar charge density to PSS, and likely overestimates the MW of DOM with many carboxylic acid groups and underestimates the MW of neutral species, as discussed above (Fig. 2).

The signal from ESI-MS detection was rather different.[48] The total ion count showed two broad, unresolved peaks from 7–9 minutes and 9–12.5 minutes, with apexes at 7.7 and 9.8 minutes, or 3572 and 608 Da PSS. A dot plot (Fig. 3b) of all detected ions m/z 150–2000 shows a broad signal of material with m/z 800–2000 at 7–9 minutes, then starting from 9 minutes, m/z 150–800 were detected with higher intensity. Generally, each m/z in the range 150–800 was detected over a broad retention time range from 9–12 minutes, and all assignments were singly charged, as is typical for DOM.[49]

High resolution mass spectrometry allows accurate formula assignment of the material eluting from HPSEC. The resolution and mass accuracy of the Orbitrap instrument dictates the mass range over which formulas can be assigned (up to about m/z 800). Although this range is limited in the context of this study, multiply charged compounds with higher mass could theoretically be detected and assigned a formula, and ions with m/z up to 2000 can be detected, if not assigned. Fig. 3a shows the chromatogram of the summed intensity of all assigned formulas (total assigned current; TAC), and assigned current was only found in the second hump of material, unlike in some previous results, which did not compare MS assignments to UV data.[29] The high molecular weight material (apex 7.5 minutes, Fig. 3a) with strong UV absorbance was not constituted by aggregated monomers

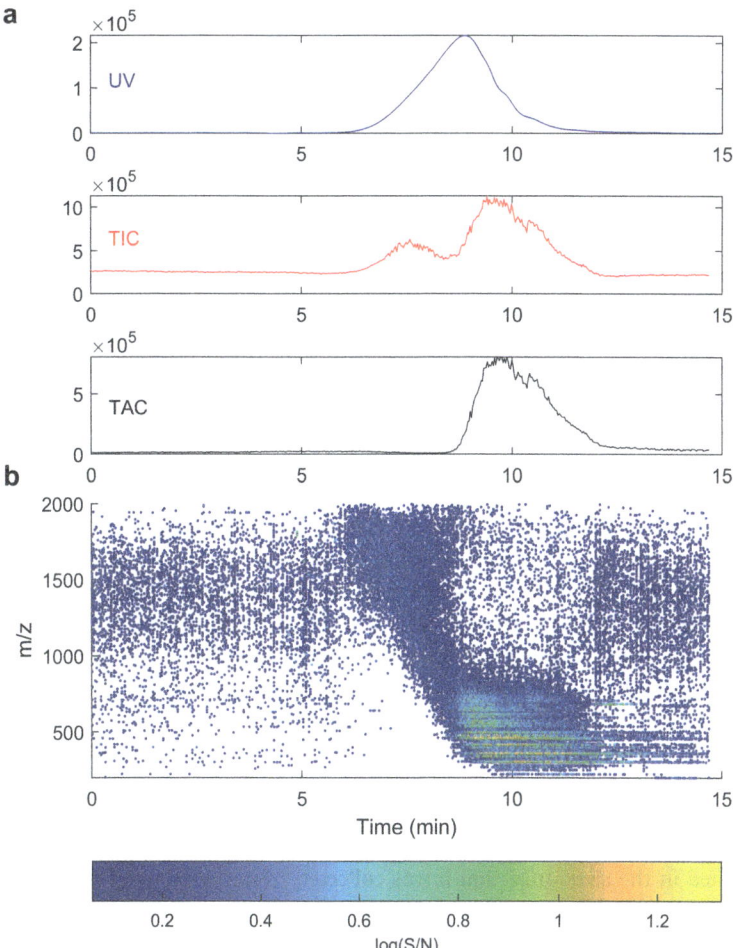

Fig. 3 HPSEC separation of a SRFA sample. (a) Zero-adjusted traces of UV absorbance at 254 nm, total ion current (TIC) and total assigned current (TAC) *vs.* HPSEC retention time. (b) Background-subtracted MS dot plot showing detected masses (see colour scale for log(signal/noise)).

(*i.e.* clusters) that could be dissociated during ESI or with in-source CID (Fig. SI2†), unlike previous results using variable cone voltage (similar to in-source CID) on a triple quadrupole MS.[48] Instead, this data supports the alternative hypothesis that there is genuinely high molecular mass material in the sample that has strong UV absorbance and cannot be dissociated and ionised (either as monomers or multiply charged ions) in the range 200–800 Da. It remains a possibility that the measured compounds aggregate in solution in natural waters,[48] but our data provides no evidence that these aggregates had low retention volumes (high hydrodynamic size) by our HPSEC method.

An extracted ion current of any individual formula shows that molecular peaks can appear at HPSEC retention times higher than predicted for the measured mass. This is likely due to charge density for ESI ionisable carboxylic acids, as discussed above (Fig. 2). The chromatographic peak width can also be greater than the PSS standards, indicating an isomeric diversity in HPSEC response (Fig. 4), similarly to other separation methods.[8,9]

The variability in chromatographic elution time (*i.e.* number averaged molecular weight, M_n) of molecular peaks in SRFA is almost certainly due to variation in charge density, as shown for model compounds in Fig. 2 and for two peaks with similar mass and differing charge density (assumed from the oxygen to carbon ratio (O/C)) in Fig. 4. Inspection of all formulas in a van Krevelen diagram or *m/z vs.* oxygen number diagram shows that charge density (equivalent to oxygen density for unsaturated humic compounds) explains as much variation in obtained peak apex as measured *m/z*, as manifested by the increase in apparent MW with oxygen for any measured *m/z* region (Fig. 5). The polydispersity (apparent range in molecular weight) was generally low (<1.5), particularly for molecular formulas with high charge density and high apparent mass. The most disperse peaks (with broader retention profiles) were mainly low oxygen number peaks with presumably more hydrophobic character and more susceptibility to hydrophobic retention on the column, leading to peak tailing.

Elution fractions were collected, dried down, re-dissolved in the mobile phase and re-injected, in order to investigate whether the high MW material that elutes first in HPSEC originates from monomers that could aggregate, or the dis-aggregation of larger clusters when alone in solution.[50] Fractions were taken at 4–5.5 minutes (blank), 6–7.5 minutes (emerging UV, no MS signal), 7.5–9 minutes (strong UV, weak MS), 9–10.5 minutes (strong UV, strong MS), 10.5–12 minutes (weak UV, strong MS) and 12–13.5 minutes (no UV, weak MS). The fractions are depicted in Fig. 6 as UV signals (a) and MS signals (b). In panels (c) and (d), the signals for the re-injected fractions are shown, along with the summed, re-constituted signal. In general, the molecular distribution in each fraction remained in the size range that it was collected from,[51] especially when considering the poor chromatographic resolution of HPSEC. Certainly; it was not found that fraction 3 (orange) reformed some signal in fraction 1 (sky blue). However, fraction 1 was diminished in magnitude and appeared at lower apparent mass when re-injected, resulting in a slight decrease in the reconstituted total signal at higher apparent masses. This might indicate that some of the higher *m/z* values present in SRFA are supported by self-aggregation due to some form of solubility

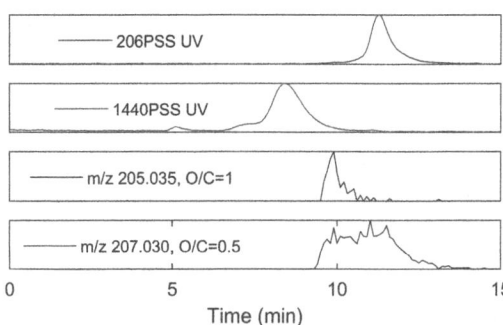

Fig. 4 UV signal for two PSS standards, and extracted ion current (MS) for two molecular masses found in SRFA, with similar actual mass to one PSS standard. O/C means oxygen to carbon ratio of the molecular peak.

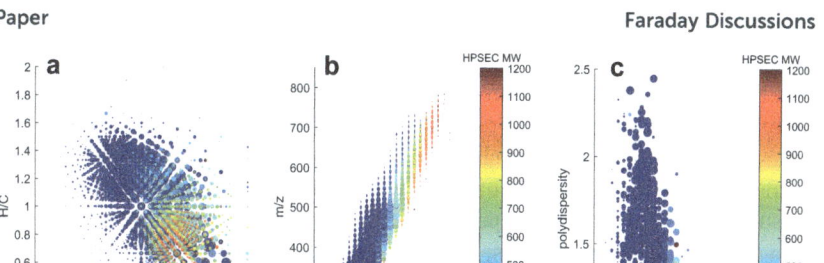

Fig. 5 Distribution of molecular formulas detected in SRFA in apparent MW according to HPSEC (see colour bar). The size of all points is proportional to intensity. (a) van Krevelen (H/C vs. O/C), (b) m/z vs. oxygen (O) number, and (c) polydispersity (M_w/M_n) vs. O number.

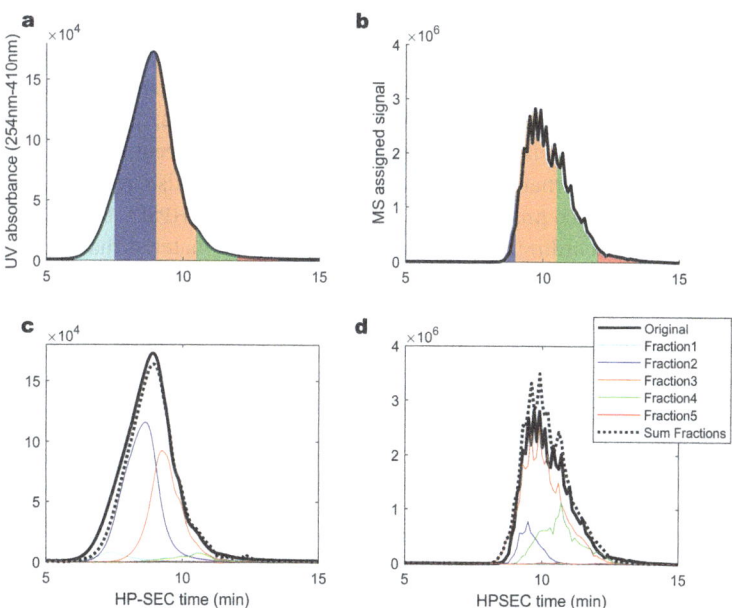

Fig. 6 A HPSEC separation of a SRFA sample. (a and b) Fractions collected for re-injection, and (c and d) signal found in re-injected fractions, including the sum of all fractions as a dotted line and the original spectra as solid lines, adjusted to 80% of the value to account for the incomplete re-injection of material (80 μL of 100 μL). UV signal (a and c), and MS signal (b and d).

limit of the slightly smaller material in fraction 2.[27,45] The MS profile was almost perfectly maintained in the separated fractions, and no signal was drawn from fraction 1 in isolation, contrary to previous results using low resolution MS,[50] and similarly to other previous results that found collected fractions to be stable in retention behaviour over several weeks.[51] This provides further support for the hypothesis that the molecular size distribution of riverine DOM is due to a large range in hydrodynamic sizes, and not dynamic, weak aggregation of small molecules. Overall, our investigations of this complex natural mixture suggest that there is limited aggregation of ionisable molecules (carboxylic acids) in this mobile phase, and variation in retention time for masses detected by ESI-MS can be explained by variation in charge density and hydrophobicity. These non-aggregated ions have recently been shown to have a small cross-sectional area (<250 Å^2) using ion mobility separation following ESI.[15] There is strong evidence for a second pool of compounds, unrelated to the carboxylic acids, which strongly absorb UV light and have a very weak MS signal at m/z from 1000 to an unknown range higher than 2000.

Terrestrial sample comparison. Five samples were compared using the established HPSEC-DAD-ESI-MS method. The samples selected were SRFA, a soil humic acid (HA) extract, an acidified aliquot of the soil HA (fulvic acid (soil FA)), a reverse osmosis (RO) concentrate from a stream draining a wetland, and a cold water extract of birch leaves. The MW profiles in UV-DAD and ESI-MS according to HPSEC are shown in Fig. 7, revealing a wide range in retention behaviour in each and between samples. In these plots, the retention time has been converted to HPSEC MW, inverting the profiles. The wetland stream RO sample was similar to SRFA, but contained a small amount of humic material close to the interstitial volume (V_0) of the column. The soil humic acid had a fairly large UV peak at V_0; this was removed by precipitation after acidification to pH 2, and can be attributed to traditionally defined 'humic acids'. Some low molecular weight hydrophobic constituents that were detected using MS were also removed in this process, possibly by co-precipitation, leaving a very narrow HPSEC MS peak in the soil FA sample. According to our results from SRFA, the material that remains as 'fulvic acids' after precipitation is likely to be dominated by carboxylic acid rich molecules with high charge density. The leaf extract had a wide range in UV absorptivity, and the lowest number averaged molecular weight (M_n) in MS signal of the five samples. The MS signal was centred at an apparent mass similar to acetone, signifying material with much lower charge density, and presumably lower ionisation efficiency.

HPSEC fractions of these samples were isolated *in silico* in the same time windows as described in Fig. 6, and the assigned MS peaks are presented in van Krevelen diagrams in Fig. 8. The river/soil samples were broadly similar in extent across the van Krevelen space, although the SRFA and Stream RO sample had considerably more saturated compounds with H/C > 1.0 and O/C < 0.5 than the soil extracts. This also accounts for the lower abundance of material in the soil extracts in fractions 4–5, where more saturated compounds tend to elute due to their lower charge density (Fig. 5). The leaf extract was very different from the other samples, containing many more labile biomolecules[52] with H/C > 1.5, including likely sugar and glycoside molecules with high oxygen and hydrogen saturation. These eluted across fractions 3–5.

Further discussion

Discrepancy between UV and MS results and the 'true' size distribution of DOM

The diverse chemistry of DOM has an important influence on its detection and characterisation by optical properties and mass spectrometry, and this study has demonstrated that there is a clear distinction in natural terrestrial mixtures between what is optically active and what is efficiently ionised by electrospray ionisation, with

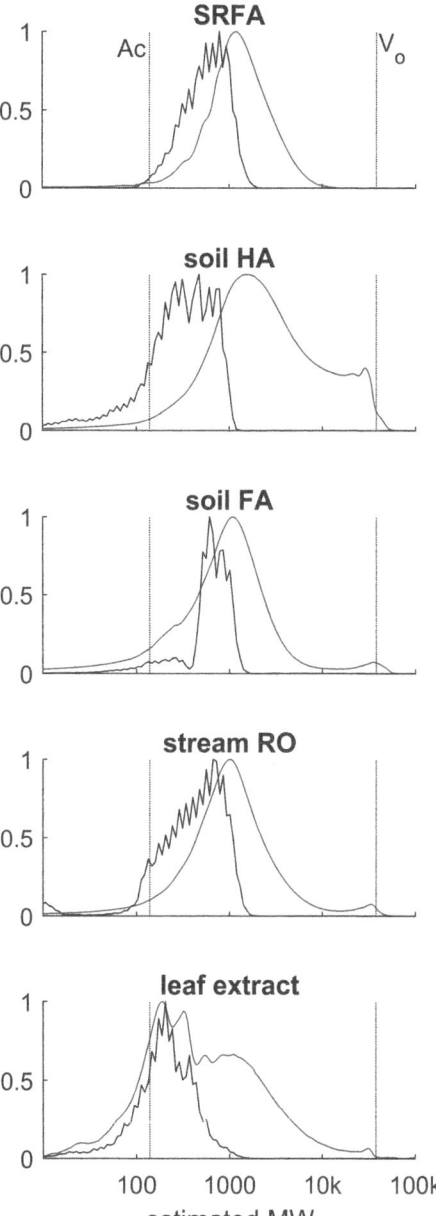

Fig. 7 HPSEC profiles of five samples. Blue: UV absorbance at 254 nm, black: MS assigned intensity, summed in 0.1 minute bins. The interstitial volume (V_O) and acetone (Ac) retention are marked by vertical dotted lines.

only moderate overlap between the two pools. These two pools are not separable with reversed-phase chromatography.[9,35] The compounds that have high light attenuation and low MS signal are not considered in mass spectrometry studies, whereas they probably dominate the results in studies that use optical methods for the characterisation of DOM. The size range difference between optically active and

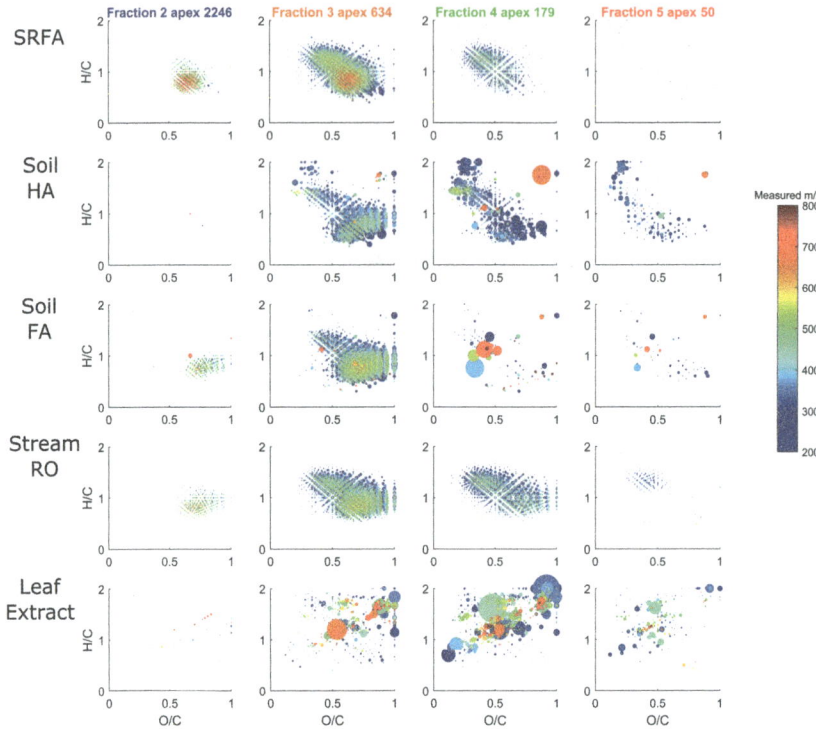

Fig. 8 van Krevelen diagrams depicting H/C and O/C ratios of molecular formulas assigned in each *in silico* fraction for each sample. Essentially no formulas were assigned in the UV-active fraction 1 (not shown). The point size is proportional to the normalised intensity (all points sum to 1×10^6). The assigned mass is shown in colour, and the apex mass according to HPSEC is indicated above the plots. As discussed, the apex mass is mainly related to charge and hydrophobicity, and not actual mass or size.

ionisable DOM probably explains why solid phase extraction, which is poor at extracting high molecular weight components,[53,54] has a minimal effect on ESI-MS results,[54] but a large effect on absorption and fluorescence spectroscopy.[55] Great care should therefore be taken in studies that aim to link the optical and MS character of terrestrial DOM[56–60] or the concentration of DOC and MS signal,[61] as the results being compared are not necessarily representing the same pools of carbon,[47,48,62] and relationships are likely to be coincidental, albeit related.

This larger molecular weight pool, which we refer to as phenolic compounds (Fig. 9), may be the coloured constituents that are most rapidly lost from terrestrial systems along transport to the sea,[63–65] as well as the easiest components to remove during groundwater transport, surface sorption and drinking water treatment.[66,67] This reactivity makes their characterisation highly important, and high resolution ESI-MS appears to be unsuitable for this particular task. Other ionisation techniques, such as atmospheric pressure photoionisation (APPI) and (matrix-assisted) laser desorption ionisation (MA)LDI, and different types of mass spectrometer with higher mass ranges may be useful in future work characterising this 'phenolic' material, but it is unlikely that one technique will be able to cover the full required analytical window due to the specificity and bias of each ionisation source.[3,68,69]

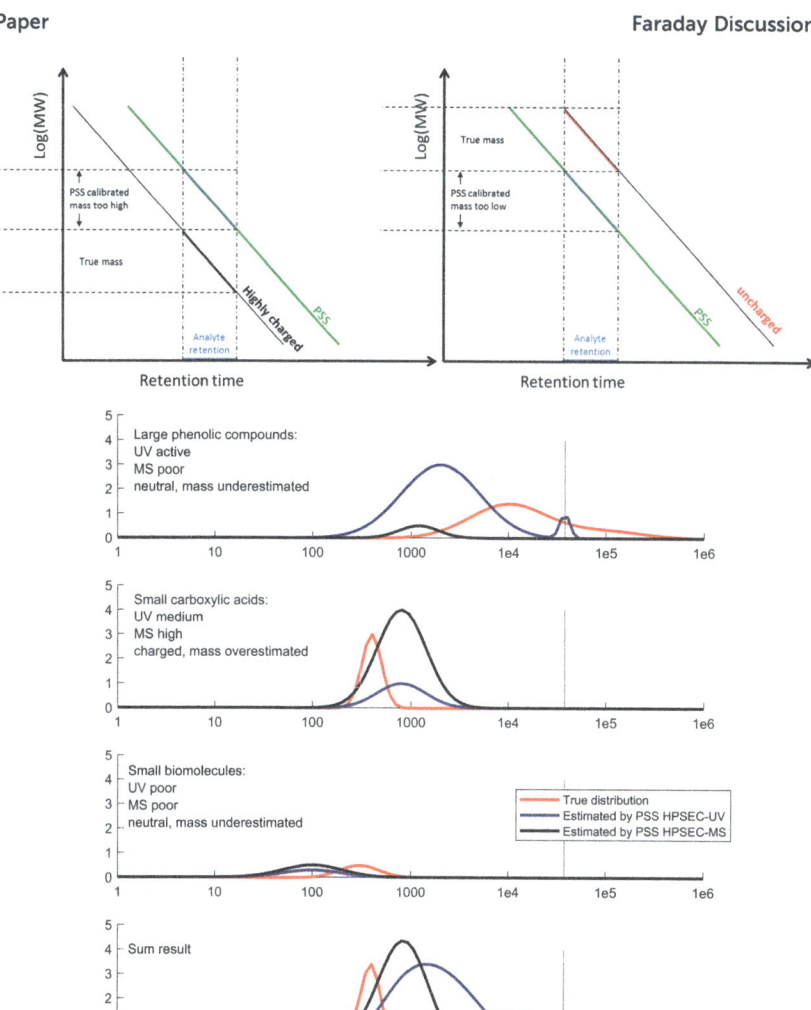

Fig. 9 Conceptual figure summarising our results and conclusions. Top: calibration and apparent mass is highly affected by charge (Fig. 2), meaning a conceptual adjustment needs to be made when considering different analyte types. Bottom: DOM can be grouped into three main classes: large phenolic compounds (top), carboxylic acids (second), biomolecules (third) and the sum of all three (last). The true size distribution of each is plotted in red, and the obtained distribution by UV detection and MS detection is shown in blue and black, respectively. The amplitude of the distributions is affected by the response factor of the detector for that type of molecule, and the size distribution is affected by secondary effects on the column, depending mainly on charge. The direction and extent of bias for each class depends on the polymer used for calibration (in our case, PSS). V_0 is indicated with a vertical dotted line.

The true size distribution of DOM remains elusive using HPSEC due to the important secondary retention and exclusion effects that disrupt the technique. However, knowledge of how model compounds behave, along with information from UV and MS detection, allow us to form a better picture of the likely size distribution of DOM in natural systems based on our results (Fig. 9). We divide the DOM material into three classes, which we loosely name 'phenolic', 'carboxylic' and 'small biomolecules'. These classes clearly have some overlaps, and are meant here only as a guide. We note that previous work using HPSEC coupled to NMR also determined three conceptual classes, in that case named 'carbohydrate + aromatic', 'carboxylic rich alicyclic' and 'linear terpenoids', respectively.[17]

How useful is chemical separation prior to detection in non-targeted complex mixture analysis?

Recent studies have shown that adding a separation to complex mixture analysis brings a new dimension of complexity to the data analysis, often without achieving the goal of separating individual isomers from the mixture.[7,9,11,17,70,71] This may cast doubt on whether the added complexity and time are worthwhile, or simply complicate results that can already be taken from broadband analysis. We feel that the robust chemical information given by separation, especially coupled online to detection, is reason enough for its use. Reversed-phase chromatography can determine reliable solubility partitioning coefficients for the unknown analytes,[72] and chromatography can separate functional classes from complex mixtures with implications for characterisation via NMR[17] fluorescence spectroscopy,[10,73] FT-infrared spectroscopy[2] and mass spectrometry.[8,9,29] Additionally, recent advances have shown that isomeric complexity can be fully resolved in certain cases when methods add enough dimensionality to the separations.[8,16,28]

Unfortunately, there is no clear path towards one method that gives information about every single complex dimension of DOM. In terms of separation, HPSEC seems to be useful for broad analysis of intermolecular speciation and size, while reversed-phase and hydrophilic interaction chromatography are important for functional 'smearing' of material into broad classes, and certainly give clearer information about functionality than HPSEC. Other separation techniques, both old and new,[74] may become useful depending on the particular research question. It may be useful to combine orthogonal chromatographic methods into 2D separations, but this is likely to involve too much complication for the wider community to embrace as a standard technique.

This study makes it clear that the choice of the analysis can have important consequences for the nature and completeness of data obtained. We recently analysed 74 headwater stream samples without prior solid phase extraction using a reversed-phase-high resolution MS technique,[35] and separated material into three chromatographic fractions. Each fraction had a UV and MS response, but the high molecular weight, UV rich material with no MS response was hidden among the three fractions, and was not revealed in that study (Fig. SI3†). The analytical window and inherent biases of any chosen method must be considered in any study of complex mixtures, and the best tools moving forward will include various analytical windows in order to validate assumptions and to thoroughly explore correlations and dependencies within the data.[48,60] Pitfalls to be aware of include limited analytical windows in ionisation and detection methods, interactions between DOM and the stationary phase during chromatography and other separations/extractions, and molecules occurring as aggregates and other intermolecular effects, either naturally or during chromatography and analysis. Considering the diversity of DOM and the different rates at which different fractions of it react in natural waters,[75,76] a full coverage of the molecular diversity of DOM is needed in order to understand its biogeochemistry. Our results support the view that an accurate understanding of DOM and its behaviour in natural environments requires a battery of different techniques, including methods that allow characterisation of molecules larger than those detectable by the currently common ESI-FTICRMS and ESI-Orbitrap MS techniques.

Conflicts of interest

There are no conflicts to declare.

References

1 N. Hertkorn, M. Frommberger, M. Witt, B. P. Koch, P. Schmitt-Kopplin and E. M. Perdue, *Anal. Chem.*, 2008, **80**, 8908–8919.
2 C. Landry and L. Tremblay, *Environ. Sci. Technol.*, 2012, **46**, 1700–1707.
3 N. Hertkorn, M. Frommberger, M. Witt, B. P. Koch, P. Schmitt-Kopplin and E. M. Perdue, *Anal. Chem.*, 2008, **80**, 8908–8919.
4 A. C. Stenson, W. M. Landing, A. G. Marshall and W. T. Cooper, *Anal. Chem.*, 2002, **74**, 4397–4409.
5 J. A. Hawkes, C. Patriarca, P. J. R. Sjöberg, L. J. Tranvik and J. Bergquist, *Limnol. Oceanogr. Lett.*, 2018, **3**, 21–30.
6 M. Zark, J. Christoffers and T. Dittmar, *Mar. Chem.*, 2017, **191**, 9–15.
7 G. C. Woods, M. J. Simpson, P. J. Koerner, A. Napoli and A. J. Simpson, *Environ. Sci. Technol.*, 2011, **45**, 3880–3886.
8 T. A. Brown, B. A. Jackson, B. J. Bythell and A. C. Stenson, *J. Chromatogr. A*, 2016, **1470**, 84–96.
9 C. Patriarca, J. Bergquist, P. J. R. Sjöberg, L. Tranvik and J. A. Hawkes, *Environ. Sci. Technol.*, 2018, **52**, 2091–2099.
10 U. J. Wünsch, K. R. Murphy and C. A. Stedmon, *Environ. Sci. Technol.*, 2017, **51**, 11900–11908.
11 T. Reemtsma and A. These, *Anal. Chem.*, 2003, **75**, 1500–1507.
12 P. Schmitt-Kopplin, A. W. Garrison, E. M. Perdue, D. Freitag and A. Kettrup, *J. Chromatogr. A*, 1998, **807**, 101–109.
13 R. M. B. O. Duarte, A. C. Barros and A. C. Duarte, *J. Chromatogr. A*, 2012, **1249**, 138–146.
14 S. Sandron, P. N. Nesterenko, M. V. McCaul, B. Kelleher and B. Paull, *J. Sep. Sci.*, 2014, **37**, 135–142.
15 K. Lu, W. S. Gardner and Z. Liu, *Environ. Sci. Technol.*, 2018, **52**, 7182–7191.
16 N. Arakawa, L. I. Aluwihare, A. J. Simpson, R. Soong, B. M. Stephens and D. Lane-coplen, *Sci. Adv.*, 2017, **3**, e1602976.
17 G. C. Woods, M. J. Simpson, B. P. Kelleher, M. McCaul, W. L. Kingery and A. J. A. J. Simpson, *Environ. Sci. Technol.*, 2010, **44**, 624–630.
18 A. Piccolo and P. Conte, *Adv. Environ. Res.*, 2000, **3**, 508–521.
19 B. Varga, G. Kiss, I. Galambos, A. Gelencsér, J. Hlavay and Z. Krivácsy, *Environ. Sci. Technol.*, 2000, **34**, 3303–3306.
20 S. A. Huber, A. Balz, M. Abert and W. Pronk, *Water Res.*, 2011, **45**, 879–885.
21 G. Riise, P. Van Hees, U. Lundström and L. Tau Strand, *Geoderma*, 2000, **94**, 237–247.

22 I. Christl, H. Knicker, I. Kögel-Knabner and R. Kretzschmar, *Eur. J. Soil Sci.*, 2000, **51**, 617–625.

23 L. J. Tranvik, *Appl. Environ. Microbiol.*, 1990, **56**, 1672–1677.

24 M. Pivokonsky, J. Safarikova, M. Baresova, L. Pivokonska and I. Kopecka, *Water Res.*, 2014, **51**, 37–46.

25 B. Koch, M. Witt, R. Engbrodt, T. Dittmar and G. Kattner, *Geochim. Cosmochim. Acta*, 2005, **69**, 3299–3308.

26 P. Conte and A. Piccolo, *Chemosphere*, 1999, **38**, 517–528.

27 M. Kerner, H. Hohenberg and S. Ertl, *Nature*, 2003, **422**, 147–150.

28 A. J. Simpson, L. H. Tseng, M. J. Simpson, M. Spraul, U. Braumann, W. L. Kingery, B. P. Kelleher and M. H. B. Hayes, *Analyst*, 2004, **129**, 1216–1222.

29 T. Reemtsma, A. These, A. Springer and M. Linscheid, *Water Res.*, 2008, **42**, 63–72.

30 D. N. Kothawala, R. D. Evans and P. J. Dillon, *Water Resour. Res.*, 2006, **42**, 1–8.

31 M. Hutta, R. Góra, R. Halko and M. Chalányová, *J. Chromatogr. A*, 2011, **1218**, 8946–8957.

32 S. Mori, *Anal. Chem.*, 1989, **61**, 530–534.

33 S. M. Hunt, M. M. Sheil, M. Belov and P. J. Derrick, *Anal. Chem.*, 1998, **70**, 1812–1822.

34 S. M. Serkiz and E. M. Perdue, *Water Res.*, 1990, **24**, 911–916.

35 J. A. Hawkes, N. Radoman, J. Bergquist, M. B. Wallin, L. J. Tranvik and S. Löfgren, *Sci. Rep.*, 2018, **8**, 16060.

36 P. Herzsprung, N. Hertkorn, W. von Tümpling, M. Harir, K. Friese and P. Schmitt-Kopplin, *Anal. Bioanal. Chem.*, 2014, **406**, 7977–7987.

37 J. Peuravuori and K. Pihlaja, *Anal. Chim. Acta*, 1997, **337**, 133–149.

38 M. Gavrilov and M. J. Monteiro, *Eur. Polym. J.*, 2015, **65**, 191–196.

39 P. L. Dubin, *Aqueous Size-Exclusion Chromatography*, Elsevier, 1988.

40 M. Berdén and D. Berggren, *J. Soil Sci.*, 1990, **41**, 61–72.

41 G. Guiochon, A. Felinger, D. G. Shirazi and A. M. Katti, *Fundamentals of Preparative and Nonlinear Chromatography*, Elsevier Academic Press, 2nd edn., 2006.

42 A. Zherebker, Y. Kostyukevich, A. Kononikhin, O. Kharybin, A. I. Konstantinov, K. V. Zaitsev, E. Nikolaev and I. V. Perminova, *Anal. Bioanal. Chem.*, 2017, **409**, 2477–2488.

43 J. A. Leenheer, R. L. Wershaw and M. M. Reddy, *Environ. Sci. Technol.*, 1995, **29**, 393–398.

44 M. Witt, J. Fuchser and B. P. Koch, *Anal. Chem.*, 2009, **81**, 2688–2694.

45 M. Baalousha, M. Motelica-Heino and P. Le Coustumer, *Colloids Surf., A*, 2006, **272**, 48–55.

46 B. C. McAdams, G. R. Aiken, D. M. McKnight, W. A. Arnold and Y. P. Chin, *Environ. Sci. Technol.*, 2018, **52**, 722–730.

47 N. Her, G. Amy, D. Foss and J. Cho, *Environ. Sci. Technol.*, 2002, **36**, 3393–3399.

48 A. These and T. Reemtsma, *Anal. Chem.*, 2003, **75**, 6275–6281.

49 N. R. Novotny, E. N. Capley and A. C. Stenson, *J. Mass Spectrom.*, 2014, **49**, 316–326.

50 A. Piccolo and M. Spiteller, *Anal. Bioanal. Chem.*, 2003, **377**, 1047–1059.

51 M. B. Müller, D. Schmitt and F. H. Frimmel, *Environ. Sci. Technol.*, 2000, **34**, 4867–4872.

52 J. D'Andrilli, W. T. Cooper, C. M. Foreman and A. G. Marshall, *Rapid Commun. Mass Spectrom.*, 2015, **29**, 2385–2401.

53 J. A. Hawkes, C. T. Hansen, T. Goldhammer, W. Bach and T. Dittmar, *Geochim. Cosmochim. Acta*, 2016, **175**, 68–85.

54 J. Raeke, O. J. Lechtenfeld, M. Wagner, P. Herzsprung and T. Reemtsma, *Environ. Sci.: Processes Impacts*, 2016, **18**, 918–927.

55 U. J. Wünsch, J. K. Geuer, O. J. Lechtenfeld, B. P. Koch, K. R. Murphy and C. A. Stedmon, *Mar. Chem.*, 2018, **207**, 33–41.

56 A. M. Kellerman, F. Guillemette, D. C. Podgorski, G. R. Aiken, K. D. Butler and R. G. M. Spencer, *Environ. Sci. Technol.*, 2018, **52**, 2538–2548.

57 E. E. Lavonen, D. N. Kothawala, L. J. Tranvik, M. Gonsior, P. Schmitt-Kopplin and S. J. Köhler, *Water Res.*, 2015, **85**, 286–294.

58 S. Wagner, R. Jaffé, K. Cawley, T. Dittmar and A. Stubbins, *Front. Chem.*, 2015, **3**, 1–14.

59 P. Herzsprung, W. Von Tümpling, N. Hertkorn, M. Harir, O. Büttner, J. Bravidor, K. Friese and P. Schmitt-Kopplin, *Environ. Sci. Technol.*, 2012, **46**, 5511–5518.

60 U. J. Wünsch, E. Acar, B. P. Koch, K. R. Murphy, P. Schmitt-Kopplin and C. A. Stedmon, *Anal. Chem.*, 2018, **90**, 14188–14197.

61 A. Mostovaya, J. A. Hawkes, B. Koehler, T. Dittmar and L. J. Tranvik, *Environ. Sci. Technol.*, 2017, **51**, 11571–11579.

62 A. A. Malik, V. N. Roth, M. Hébert, L. Tremblay, T. Dittmar and G. Gleixner, *Soil Biol. Biochem.*, 2016, **100**, 66–73.

63 G. A. Weyhenmeyer, M. Fröberg, E. Karltun, M. Khalili, D. Kothawala, J. Temnerud and L. J. Tranvik, *Glob. Chang. Biol.*, 2012, **18**, 349–355.

64 T. Dittmar, K. Whitehead, E. C. Minor and B. P. Koch, *Mar. Chem.*, 2007, **107**, 378–387.

65 S. J. Köhler, D. Kothawala, M. N. Futter, O. Liungman and L. Tranvik, *PLoS One*, 2013, **8**, 1–12.

66 C. H. Specht, M. U. Kumke and F. H. Frimmel, *Water Res.*, 2000, **34**, 4063–4069.

67 T. K. Nissinen, I. T. Miettinen, P. J. Martikainen and T. Vartiainen, *Chemosphere*, 2001, **45**, 865–873.

68 D. Cao, H. Huang, M. Hu, L. Cui, F. Geng, Z. Rao, H. Niu, Y. Cai and Y. Kang, *Anal. Chim. Acta*, 2015, **866**, 48–58.

69 J. D. Andrilli, T. Dittmar, B. P. Koch, J. M. Purcell, A. G. Marshall and W. T. Cooper, *Rapid Commun. Mass Spectrom.*, 2010, **24**, 643–650.

70 T. Dittmar, B. Koch, N. Hertkorn and G. Kattner, *Limnol. Oceanogr.: Methods*, 2008, **6**, 230–235.

71 S. Sandron, N. W. Davies, R. Wilson, A. R. Cardona, P. R. Haddad, P. N. Nesterenko and B. Paull, *Chromatographia*, 2018, **81**, 203–213.

72 K. Namjesnik-Dejanovic and S. E. Cabaniss, *Environ. Sci. Technol.*, 2004, **38**, 1108–1114.

73 W. T. Li, S. Y. Chen, Z. X. Xu, Y. Li, C. D. Shuang and A. M. Li, *Environ. Sci. Technol.*, 2014, **48**, 2603–2609.

74 S. Sandron, A. Rojas, R. Wilson, N. W. Davies, P. R. Haddad, R. A. Shellie, P. N. Nesterenko, B. P. Kelleher and B. Paull, *Environ. Sci.: Processes Impacts*, 2015, **17**, 1531–1567.

75 A. M. Kellerman, T. Dittmar, D. N. Kothawala and L. J. Tranvik, *Nat. Commun.*, 2014, **5**, 1–8.

76 B. Koehler, E. Von Wachenfeldt, D. Kothawala and L. J. Tranvik, *J. Geophys. Res.: Biogeosci.*, 2012, **117**, 1–14.

Faraday Discussions

Perspectives on the future of multi-dimensional platforms†

Gino Groeneveld,[a] Bob W. J. Pirok ab and Peter J. Schoenmakers *a

Received 7th December 2018, Accepted 28th January 2019

DOI: 10.1039/c8fd00233a

Two-dimensional liquid chromatography (2D-LC) formats have emerged to help address separation problems that are too complex for conventional one-dimensional LC. There are a number of obstacles to the proliferation of 2D-LC that are gradually being removed. Reliable commercial instrumentation has become available and data analysis software is being improved. Detector-sensitivity and phase-system compatibility issues can largely be solved by using active-modulation strategies. The remaining challenge, developing good and fast 2D-LC methods within a reasonable time, may be solved with smart algorithms. The technology platform that has been developed for 2D-LC also creates a number of other possibilities. Between the two separation stages, all kinds of physical (*e.g.* dissolution) or chemical (*e.g.* enzymatic or light-induced degradation) processes can be made to take place, allowing a wide variety of experiments to be performed within a single, efficient and automated analysis. All these developments are discussed in this paper and a number of critical issues are identified. A practical example, the characterization of polysorbates by high-resolution comprehensive two-dimensional liquid chromatography in combination with high-resolution mass spectrometry, is described as a culmination of recent developments in 2D-LC and as an illustration of the current state of the art.

1 Introduction to two-dimensional liquid chromatography

Analytical instruments are indispensable for modern society (*e.g.* health, food, environment and materials), for industry and trade, and for research and innovation. Almost every chemical or technological innovation in a food product, material or pharmaceutical formulation is accompanied by analytical confirmation of its efficacy and safety, often followed by additional regulatory requirements. To keep pace with the cumulative needs of society to obtain more objective information and to gain more knowledge, analytical methods are continuously

aUniversity of Amsterdam, van 't Hoff Institute for Molecular Sciences, Analytical-Chemistry Group, Science Park 904, 1098 XH Amsterdam, The Netherlands. E-mail: P.J.Schoenmakers@uva.nl; Tel: +31205256642

bTI-COAST, Science Park 904, 1098 XH Amsterdam, The Netherlands

† Electronic supplementary information (ESI) available. See DOI: 10.1039/c8fd00233a

improved. From an analytical perspective, the samples subjected to analysis and the questions asked are increasingly complex. To assess the composition and to characterize various properties of highly complex samples, a multitude of analytical techniques are applied. One of these techniques is liquid chromatography (LC).

In LC, analytes are separated as peaks due the differences in partitioning between two phases that move at different velocities. The number of peaks that can theoretically be baseline separated in a chromatographic experiment is referred to as the peak capacity of the method. In cases where the chromatographic experiment is performed under constant (non-programmed) conditions, the following equation applies.

$$n_p = \frac{\sqrt{N}}{4R_s} \ln\left(\frac{1 + k_{last}}{1 + k_{first}}\right) + 1 \tag{1}$$

Here, k_{first} and k_{last} are the retention factors (or dimensionless retention times), $k = t_R/t_0 - 1$ of the first and last eluting peak, N is the column plate count, and R_s the required resolution. For gradient conditions, the peak capacity is approximately given by

$$n_p \approx \frac{t_G}{4\sigma} \tag{2}$$

where t_G is the duration of the gradient. Literally speaking, n_p is the number of separated peaks that fit in the chromatogram would they be equally distributed. Of course, this is rarely – if ever – the case and in practice the peak capacity must drastically exceed the number of analytes for all of the latter to be separated. According to the statistical overlap theory of Davis and Giddings,[1] the successful separation of 95% of 1000 components would statistically require a peak capacity of 20 000 or more.

The plate numbers and peak capacities offered by LC systems are rather limited when compared with high-resolution techniques, such as gas chromatography (GC) and capillary electrophoresis (CE). Although the performance of LC systems has been improved significantly in recent years thanks to a number of developments, including novel stationary-phase morphologies and ultra-high-pressure technology, the available peak capacities do not suffice to tackle the separation of contemporary complex samples.

While for relatively simple separations data-processing techniques and/or selective detectors may offer refuge, the analysis of truly complex samples with large numbers of analytes becomes extremely challenging. In this case, two-dimensional liquid chromatography (2D-LC) may be fruitful (Fig. 1). In comprehensive 2D-LC (LC $\times$ LC), the entire first dimension effluent is subjected to a second dimension separation, split into a sufficiently high number of fractions so as not too lose (too much of) the separation achieved in the first dimension.[2] Very high peak capacities can be achieved and put to use if two very different ("orthogonal") separation methods are coupled. In principle, the peak capacities of the two individual separation systems may be multiplied to obtain the peak capacity, n_c, of the LC $\times$ LC method.

$$n_c = n_{p,1} \times n_{p,2} \tag{3}$$

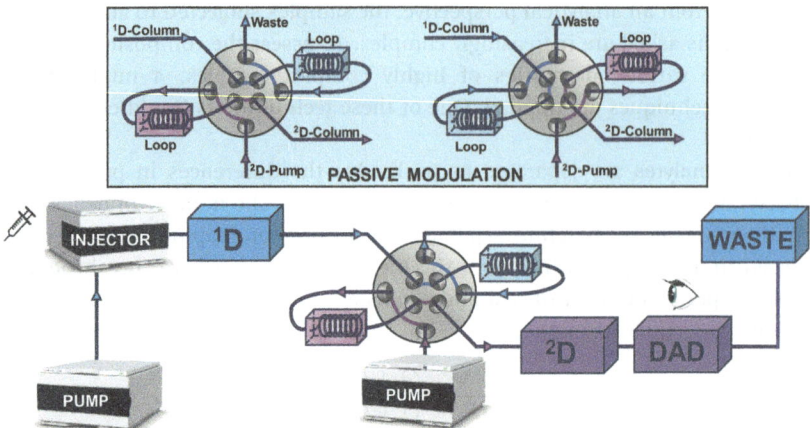

Fig. 1 Schematic overview of a generic comprehensive two-dimensional liquid chromatography (LC × LC) system. The blue colour depicts the first dimension and purple the second dimension. Through the alternating valve-based passive modulation interface, the entire ¹D-effluent can be transferred to the ²D separation dimension.

While the essential concepts of 2D-LC were already introduced in the 1970s by Erni and Frei[3] and worked out in 1990 by Jorgenson and Bushey,[4] it was not until the 21st century that developments really started to pick up. There is a clearly identifiable need for comprehensive two-dimensional liquid chromatography (LC × LC) for the characterization of complex polymers that feature multiple independent distributions. This is illustrated by an example of a separation of industrial surfactants in Fig. 2. This created at least one strong driver for developing and improving the technology.[5,6] Because LC × LC offers peak capacities that are about an order of magnitude higher than those attained in conventional one-dimensional LC, the technique may also be advantageous for the separation of very complex mixtures containing 100 analytes or (many) more.[7] For such samples, the combination of LC with mass spectrometry (LC-MS) and LC × LC are both relevant technologies. For the detailed analysis of extremely complex samples, LC × LC-MS may ultimately be the way to go.[8]

Two-dimensional (2D) LC may be applied in the heart-cut mode (LC-LC), in which one or a few fractions are selected for a detailed separation in the second dimension and comprehensive two-dimensional liquid chromatography (LC × LC), where the entire sample is split into many fractions so as to essentially maintain the first dimension separation. All these fractions are subsequently subjected to a second dimension separation, yielding a two-dimensional colour plot that represents the entire sample. While LC-LC and LC × LC techniques were already introduced in the late 1970s, 2D-LC techniques are still not established in most routine analytical laboratories. Massive developments in terms of dedicated, reliable instrumentation for highly efficient (multiple) heart-cut 2D-LC and for (comprehensive two-dimensional) LC × LC have only emerged in the last decade. 2D-LC now appears to be at a stage of development where LC was in the 1980s. The technique is plagued by the perceptions of a reduced detection-sensitivity, limited applicability, lack of robustness, and impracticality due to its complexity.

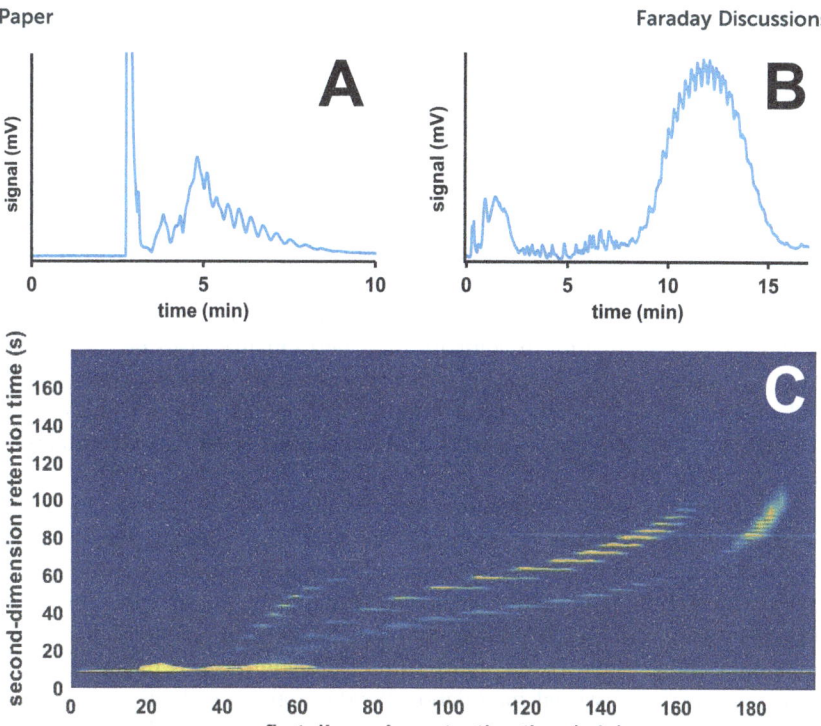

Fig. 2 Separation of a mixture of industrial surfactants using (A) (mixed-mode) ion-exchange chromatography, (B) reversed-phase liquid chromatography, and (C) a comprehensive two-dimensional combination of mixed-mode ion-exchange LC and reversed-phase LC.

We have earlier identified three main challenges as limiting factors for progress in 2D-LC.[9] These are: (i) reduced detection-sensitivity, (ii) limited applicability due to solvent-incompatibility, and (iii) complex and cumbersome method development. Arguably, stationary-phase-assisted modulation (SPAM) has done much to resolve the first issue.[10–12] Significant progress has also been made on the latter two issues.

2 Applicability

In LC $\times$ LC, two vastly different (*i.e.* "orthogonal") retention mechanisms are combined. However, many conceivable combinations of retention mechanisms may potentially suffer from incompatibility issues,[7] because the ^{1}D effluent (mobile phase) negatively affects the ^{2}D separation. Active-modulation techniques help negate these deteriorating effects by altering the mobile-phase composition. Moreover, some active-modulation techniques also facilitate peak sharpening to solve the detection-sensitivity problem. The status of these techniques and future prospects are discussed in this section.

2.1 Looking for a breakthrough in overcoming breakthrough

The last decade has seen a steep increase in the development of active-modulation techniques to address detection-sensitivity issues and solvent incompatibility. Examples include stationary-phase-assisted modulation,[10] vacuum-evaporation modulation,[13] in-column focusing,[14] thermal modulation,[15–17] vacuum-membrane-evaporation and, more recently, although based on earlier developments,[18,19] active-solvent modulation.[20]

While some approaches are still at the pioneering stages, the use of active-modulation techniques is generally seen to be on the rise[2] and this trend is expected to continue in the near future. This will result in improved methods and the exploration of less-common combinations of retention mechanisms.[7]

Because none of the active-modulation techniques has had the time to be fully developed, it is premature to compare their performance or to decide on the strongest options. However, some comments can be made on their current status and on the expected trends and developments heading into the future. The strengths and weaknesses of a number of already implemented active-modulation techniques are listed in Table 1.

The concept of vacuum-evaporation modulation (VEM) and early reports on its functioning are extremely interesting. The technique has specifically been demonstrated for combinations of normal-phase liquid chromatography (NPLC) and reversed-phase liquid chromatography (RPLC).[13] However, it has yet to be demonstrated for LC × LC. Moreover, a thorough investigation into fundamental limitations and possible loss of analytes (discrimination) is very much needed.

Stationary-phase assisted modulation (SPAM) appears to be the most used active-modulation technique[10] and we do not expect this to change rapidly. Incompatible solvents can easily be removed, whilst also improving detection-sensitivity and reducing the analysis time.[7] As a result, we expect more combinations of retention mechanisms to be explored that were previously believed to be incompatible. However, serious attention to the robustness of the traps is

Table 1 Strengths and weaknesses of most used active-modulation techniques

Technique	Strengths	Weaknesses
Stationary-phase-assisted modulation (SPAM)	Eliminates incompatible solvent	Trap robustness
	Improves detection-sensitivity (reduces dilution factors)	Discrimination
	Modulation volume no longer limiting factor	Operation and optimization is sample dependent
		Method development may be challenging
Active-solvent modulation (ASM)	Dilutes incompatible solvent	Modulation volume still a limiting factor
	On-column focusing in the second dimension	Mainly useful with RPLC in second dimension
Vacuum-evaporation modulation (VEM)	Evaporates incompatible solvent	Discrimination: loss of volatile analytes during evaporation not investigated
	Fast operation appears possible (under vacuum conditions)	Some analytes (e.g. polymers) may re-dissolve slowly
		Only demonstrated for heart-cut 2D-LC

required. One fundamental limitation of SPAM is the premise that all analytes must be sufficiently retained by the stationary phase in the trapping cartridge.

Active solvent modulation (ASM) is a recently introduced, powerful alternative.[20,21] The concept is founded on a number of papers that have provided a solid basis for further refinement. Given the ease of implementation, we expect ASM to gain in popularity, possibly, to some extent, at the expense of SPAM. ASM requires very high retention factors for the analytes in the weak eluent (*i.e.* the diluted mobile phase). This is easily realized when the ^{2}D separation is RPLC, where the addition of water drastically increases retention factors of almost all analytes, but it is less suitable for most other ^{2}D mechanisms.

All active-modulation techniques have been developed to reduce solvent-incompatibility effects. One serious effect that may jeopardize chromatographic separations is analyte breakthrough. A schematic representation is shown for reversed-phase LC in Fig. 3 where the blue colour depicts the weak (*i.e.* strongly aqueous) solvent and the purple an organic (*i.e.* hydrophobic) solvent. Upon injection, a too strong solvent inhibits adsorption and partitioning of the analytes on or in the stationary phase. Indeed, hydrophobic analytes will essentially move at the same speed as the hydrophobic solvent (*i.e.* the average mobile-phase velocity). As the plug migrates through the column, it is gradually dispersed, and analytes gradually become more retained. However, as long as a plug with a strong, hydrophobic environment pertains, some of the analyte molecules will remain in this zone, to be ultimately eluted with the solvent peak around the dead time of the column (t_0). In cases of gradient-elution experiments, all highly retained analytes that fall behind the solvent plug essentially wait to be caught up by the gradient. For such analytes, breakthrough may affect quantitation (to an extent depending on the volume of the plug), but the chromatogram usually provides a reasonable impression of the contents of the sample. For less-retained analytes, more severe effects on retention and peak shape may be observed.

Breakthrough effects are most serious for large molecules. If the conditions in the plug of strong solvent (and the pore-size distribution of the column) are such that size-exclusion pertains for large (hydrophobic) molecules, they tend to move ahead of the solvent plug, until they meet the weaker solvent. This means that such molecules become focused before the solvent plug. If a plug of strong solvent pertains throughout the column, such molecules elute as a sharp zone, just before the solvent peak. Only molecules that are at the very end of the solvent plug after injection may be able to "escape" from the zone early in the experiment. Once excluded analytes have moved to the front of the zone, there is no escape possible, as long as a zone of strong solvent pertains. In such cases, most of the high-molecular-weight analytes will be found in the breakthrough peak, with only a small fraction as a regular peak later in the chromatogram.

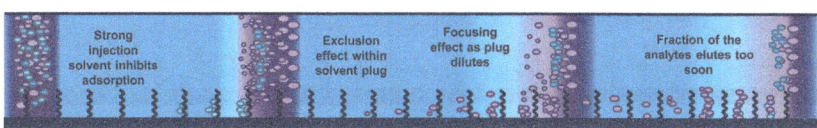

Fig. 3 Schematic representation of the breakthrough phenomena in a reversed-phase column. Here, the blue colour reflects the weak (*i.e.* highly aqueous) eluent, and the purple colour reflects the strong (*i.e.* organic) eluent.

To avoid breakthrough (or any kind of injection-solvent effects) in 1D-LC, the strength of the sample solvent may be lowered, and/or the injection volume may be reduced (Fig. 4). In 2D-LC, the latter can simply be realized by reducing the volume of the modulation loops in any modulator. This may necessitate a lower ^{1}D flow rate, resulting in a longer analysis time and in reduced sensitivity (lower analyte concentrations at the detector). Reducing the strength of the injection solvent is the essence of ASM. By adding a weak ^{2}D eluent to a fraction of modulated ^{1}D strong eluent, the overall solvent-strength is reduced. What is a weak eluent is somewhat dependent on the analytes, but is largely determined by the retention mechanism in the second dimension. To effectively apply these modulation techniques to overcome breakthrough effects, a more fundamental understanding is required.

2.2 Heart-cut 2D-LC to mature more rapidly than comprehensive 2D-LC

We noted in the introduction that the two remaining challenges are solvent-incompatibility and cumbersome and complex method development. The first issue is arguably similar for methods run in heart-cut (2D-LC) and comprehensive (LC × LC) mode. However, method development for heart-cut 2D-LC is much less complex and we believe that heart-cut 2D-LC will more rapidly become established. An investment in dedicated equipment, including a modulation device and a second LC separation, will allow a scientist to couple a familiar 1D-LC separation and select one or a few peaks of interest for further separation.

Heart-cut (2D-LC) has various application domains. It can be used to answer specific questions involving specific analytes, especially in cases where 1D-LC-MS does not suffice. Establishing the purity of an active (pharmaceutical) ingredient is a case in point.[18,22] By separating the product in two orthogonal dimensions, a much better assessment of the purity can be made. Determining enantiomeric ratios in complex mixtures provides another example.[23,24] The complex mixture is first separated, after which the fraction containing the isomers is separated in a second dimension with chiral selectivity. MS cannot usually distinguish between isomers and we believe that chiral LC and chiral supercritical-fluid chromatography (SFC) are, as yet, more powerful than ion-mobility spectrometry (IMS) for such applications.

2.3 Smaller dwell volumes needed

One hardware-related parameter that we believe to be increasingly important is the dwell volume of the chromatographic system. In general, the dwell volume is

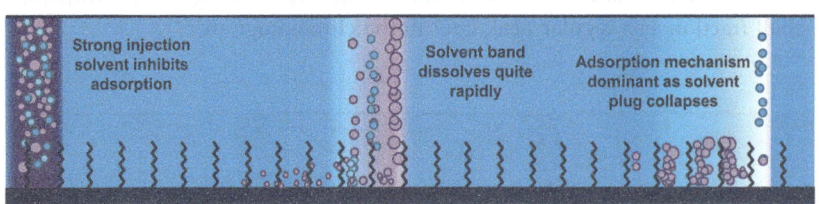

Fig. 4 Schematic illustration of how to prevent breakthrough. The strong sample solvent plug is either sufficiently small or sufficiently weak for the plug to be dispersed to such an extent that the solvent strength falls below a critical level.[56] Analytes will then experience adsorption and will not elute prematurely from the column.

mainly regarded as a nuisance (and a waste of time) for the first dimension separation, due to the low flow rates. Indeed, the flow rates in the ^{1}D separation are generally much lower than those in the second dimension. However, with the introduction of active-modulation techniques, such as SPAM, the restrictions on the ^{1}D flow rate have relaxed. In addition, it should be possible for the dwell volume to be washed away using a high flow rate at the start of the experiment. The volume of liquid may either be flushed through the column prior to injection or diverted away using a valve installed before the column after the injection. The autosampler valve may possibly be used to realize this option.

Ultimately, the ^{2}D dwell volume may be the most important. With the progress of technology (such as SPAM), ^{2}D columns are becoming narrower and ^{2}D flow rates are decreasing. This greatly reduces the solvent volumes needed and the volumes of waste produced. Therefore, we expect this trend to continue. If LC × LC is to be combined with high-resolution MS instruments with interfaces designed to accommodate very low flow rates, such as nano-electrospray ionization (nano-ESI),[10,25] the ^{2}D dwell volume becomes a major bottleneck.

In a well-constructed system, the major contributor to the dwell volume is the mixer, which combines two or more eluent streams into a single homogeneous eluent flow. Efficient mixing is essential, and a certain mixing volume may be fundamentally unavoidable, but mixers should be as small and efficient as possible. Moreover, mixers installed in the eluent delivery system prior to the injector need to withstand very high pressures.

Finally, we would like to emphasize the importance of measuring and knowing the dwell volume for any system. Especially when using computer-aided optimization tools, such as the program for the interpretive optimization of two-dimensional resolution (PIOTR, see Section 3), it is imperative that the dwell volume is accurately known. It can be easily measured by (temporarily) removing the column from the system and measuring the eluent composition in the detector. If necessary, a small amount of a UV-active component, such as acetone, may be added to one of the eluent components for this purpose.

3 (Artificially) intelligent method development

The emergence of robust and reliable instrumentation for 2D-LC and the introduction of active-modulation techniques promise many more successful applications. However, great options, such as variable (e.g. shifting) ^{2}D gradients and SPAM, are arguably making the technique even more complex. Having addressed the issues of detection-sensitivity and applicability, the final issue, method development for advanced 2D-LC applications, looms as an enormous mountain on the paths of chromatographers in academia and industry. In this section we will discuss this Everest of all challenges in 2D-LC and the current state of tackling it.

3.1 Introduction to automated method development

The last few decades have seen an increased understanding of chromatographic theory, which resulted in models relating retention to mobile-phase composition for a large number of retention mechanisms, such as reversed-phase LC, normal-phase LC, ion-exchange and hydrophilic-interaction LC (HILIC). These theoretical models can be fitted to experimental retention times of all analytes of interest from a set of mobile-phase-composition programs. The result will be a retention model for each investigated analyte, from which the retention time can be predicted for any mobile-phase-composition program. By computationally simulating a very large number of methods and assessing the results, the most optimal method can be found.

In a series of papers, we have documented the first steps along what we see as the only way to address the development of complex 2D-LC within a reasonable time (a few days to a week).[26–28] Although significant progress has been made, we can identify a number of further strides that would do much to alleviate the method-development challenge.

3.2 Improving peak-tracking strategies

In computer-aided method development, the peaks do not need to be identified, but we must establish which peaks in the different input chromatograms belong to the same analyte. If this "peak-matching" or "peak-tracking" task is to be performed manually for complex samples, it becomes a painstaking exercise.

We have earlier described our first iteration of a program for automatic peak tracking for use with LC-MS data.[28] During the different stages of testing, our attention became mainly focused on the performance of the program for tracking different isomers present in a sample, which is an indirect indication of the good performance of the system for "easy" analytes (*i.e.* analytes that are substantially different from other ones in the sample). To our surprise, the algorithm also proved to be better than the operators at tracking analytes present at trace-level concentrations. These were frequently missed by the human eye, but unforgivingly spotted by the system. Therefore, we are now more-thoroughly exploring possible routes for testing and using the algorithm as a tool for peak discovery.

Our tracking approach comprises four steps: (i) data preparation, (ii) peak selection (recognition of likely pairs), (iii) comparison, and (iv) evaluation.

The first step encompasses the detection of peaks, followed by removal of noise, background and system peaks. Looking forward, the peak detection and recognition of background signals can still be improved.

In the second step, peaks are classified and pooled as likely or unlikely candidates for matching, depending on the relative retention of peaks in the different input chromatograms. This pre-selection step has mainly been designed and implemented to reduce the time required. To assess all possible peak combinations is quite feasible, but it may be painstakingly slow on regular laptop or desktop computers. All peaks that can be combined in this step do not need to be cross-checked in further steps. However, this step is currently exclusively applicable if the two chromatograms have been recorded using similar chromatographic methods, as is often the case when performing gradient-elution scanning experiments in method-development strategies. One of our aims for the near future is to adapt the algorithm to also allow swift tracking of peaks across chromatograms obtained using very different methods or even across different samples. To accomplish this, the tracking algorithm may be fed with knowledge on the chromatographic mode, the experimental conditions and their presumed (theoretical) effect on the experimental outcome. For example, when comparing an RPLC chromatogram with a HILIC chromatogram (or, more likely when following a systematic optimization strategy, two of each), the system may be aware of the two main parameters influencing retention in the two modes, which are analyte polarity (inducing peak reversal) and analyte size (promoting a constant elution order). Information on molecular size is expressed in the slope (S) of the RPLC retention curve, which is information that may feasibly be used in a smart peak-tracking algorithm.

The third step involves comparing candidate pairs. It is not dependent on the previous step. In essence, pieces of information are gathered for each peak and these "bits of evidence" are compared for two peaks in different chromatograms. The peak information includes the statistical moments (retention time, area, shape and asymmetry) and the mass spectra. All these bits of information are weighed. Since not all applications use a mass spectrometer, we are actively exploring the use of UV-vis spectra as alternative or additional information on peaks. At a later stage, deep-learning pattern-recognition tools may be investigated to produce additional evidence.

We recognize the need to critically assess the value of each of the pieces of evidence, so as to improve their relative weighing. In this context, it must be immediately noted that some information, such as "the 30 most-abundant peaks in the mass spectrum" and the area (zeroth statistical moment) of the peak, can be significantly influenced by the presence of a nearby, partially co-eluting peak.

The final step involves the evaluation. At this point, alternative pieces of information are consulted with the specific aim to check whether a peak has been sensibly coupled to another peak in another chromatogram. For data involving mass spectrometry, we have tailored this step specifically to a search for possible isomers that are eluting nearby. The algorithm may possibly have matched such peaks incorrectly. In short, in this step the most-abundant mass at the apex is compared between the two peaks. Then, the remainder of the extracted ion-current chromatogram (XIC) is scanned for other signals of the same mass. If other signals representing the same mass exist in the XIC, the elution order in the two chromatograms is compared to verify that the retention order is sensible. If this last test or the apex test fails, the pair is rejected.

While we found this approach to be robust in most cases, one weakness was encountered when an isomer was found to co-elute very closely to another peak of a non-isomer that showed a higher intensity. We are currently adapting the algorithm to improve its robustness in such cases.

In the specific case of LC × LC, another concern is the speed of the algorithm. As the number of peaks to be evaluated increases, the speed may decrease drastically. In terms of peak-tracking, however, the matching across different retention mechanisms becomes very much easier, because every peak in the 2D chromatogram is physically connected to two retention mechanisms. When matching two or more 2D chromatograms, the additional dimension provides an additional set of statistical moments to include as evidence. It also provides additional information on the retention pattern. The number of options to consider within the search domain of the gradient is actually reduced in comparison with the 1D-LC situation described above. Moreover, co-elution issues are less common and often less problematic. Consequently, it may be expected that the robustness of the two-dimensional version of the algorithm will be better than that of the one-dimensional version.

Ultimately, we may reach a situation in which the algorithm has been refined and optimized to an extent that the processing speed can no longer be reduced. Currently, the algorithm applies a brute-force evaluation strategy. For practical and fundamental reasons, we want to investigate smart strategies, such as ant-colony and genetic algorithms.

3.3 Polymer separations

Polymers typically feature molecular distributions. There is not a single, unique molecular structure. Almost always, the chain length varies, resulting in a molecular-weight distribution (or molar-mass distribution). In addition, a specific polymer may feature a functionality-type distribution (describing the variation in functional groups or end-groups), a chemical-composition distribution (describing the ratio of monomers in a copolymer molecule), a block-length distribution (in the case of block-copolymers), *etc.*

Therefore, polymer separations involve the separation of envelopes of peaks, which significantly complicates the optimization of polymer separations. Computer-aided optimization, in particular, is complicated, because peak-matching strategies (Section 3.2) have yet to be developed for the envelopes of peaks or "smears" that appear in the two-dimensional chromatograms (or "bananagrams") typically obtained for polymers.[29] One possible direction for such strategies is to use lines to match different smears, including their end points and intermediate points, across different chromatograms (Fig. 5). The PIOTR program may then be adapted to accommodate the optimization of such lines, yielding optimal separations of polymer distributions. Because many, if not most polymeric samples feature more than a single distribution (so called "complex polymers") and because separation dimensions can often be aligned with the limited number of sample dimensions,[30] LC × LC is especially important in the field of polymer separations. Yet, novel developments, such as shifting second dimension gradients and active-modulation techniques, have not yet been embraced. Therefore, there are great opportunities for meaningful improvements, using PIOTR or otherwise.

3.4 Retention mechanism selectors

One frequently raised point of criticism towards the general approach applied in PIOTR is that it requires an *a priori* decision on the selection of retention

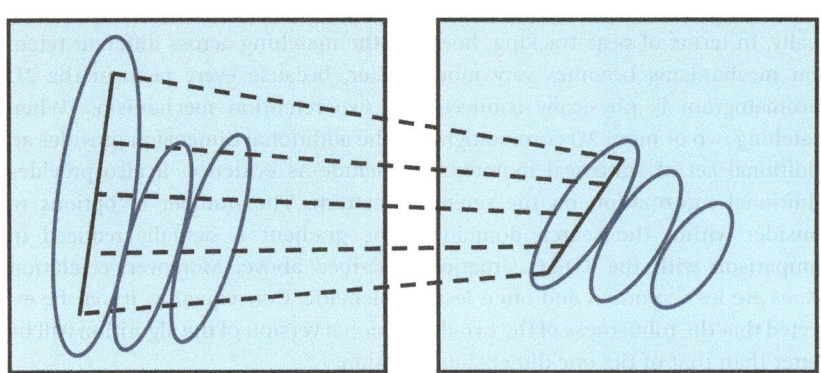

Fig. 5 Polymer distributions usually result in envelopes of peaks in 2D chromatograms. One feasible way to pursue the optimization of polymer separations involves treating them as lines, rather than as individual points (or peaks).

mechanisms. In practice, the selection of suitable and orthogonal retention mechanisms for the sample often involves trial experiments and, therefore, it comprises a significant fraction of the total method-development time for 2D-LC applications.

From a theoretical perspective, the PIOTR approach already contains all the necessary components to combine retention parameters from different 1D-LC retention mechanisms. The user can record 1D-LC data using different retention mechanisms and manually combine these to theoretically assess the efficacy of different combinations through the evaluation of quality descriptions.[7] In its current state, this approach is, however, not effective, because the user must manually match peaks across different mechanisms. This is necessary to use the correct retention parameters for each analyte.

Peak matching is also necessary to perform retention modelling based on experiments carried out with a single retention mechanism (Fig. 6A), but the similarity between the chromatograms thus obtained allows the search window to be narrowed down. For chromatograms obtained using completely different retention mechanisms (Fig. 6B), peak tracking using the approach presented above is still possible, but it is quite slow, because the algorithm cannot focus on a probable retention window, as illustrated in Fig. 6. PIOTR is an efficient program for the optimization of mobile-phase programs. Its usefulness for the selection of retention mechanisms may still be improved.

It is good to realize that peak matching across different dimensions is a non-issue if LC × LC scanning experiments are performed. In that case, peaks in the two dimensions are experimentally matched. Peaks still need to be matched within either mechanism (*i.e.* along the horizontal or vertical axis in the LC × LC chromatogram), but this process may actually be simplified, because the number of candidate peaks may be reduced by information from both axes. The

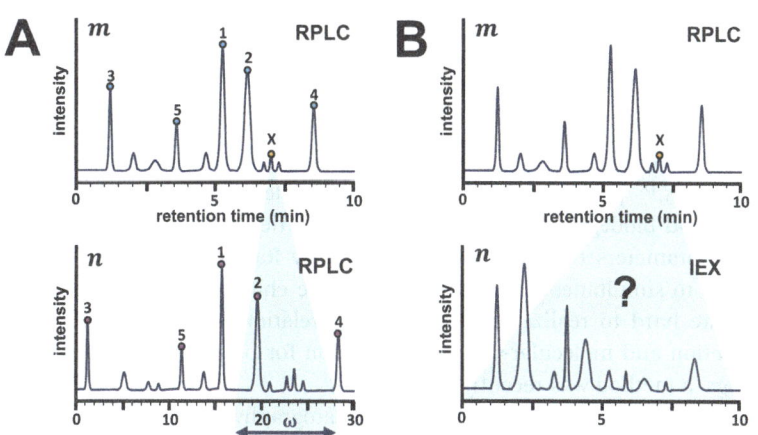

Fig. 6 Tracking peaks across multiple chromatograms is significantly easier when the approximate location of the peak can be predicted. (A) Peak-tracking across chromatograms obtained from similar experiments using identical retention mechanisms, leading to a narrow (initial) search window. (B) Peak tracking of peaks across chromatograms obtained from completely different methods. The search window cannot be narrowed down. The PIOTR approach still works, but more computation time is needed.

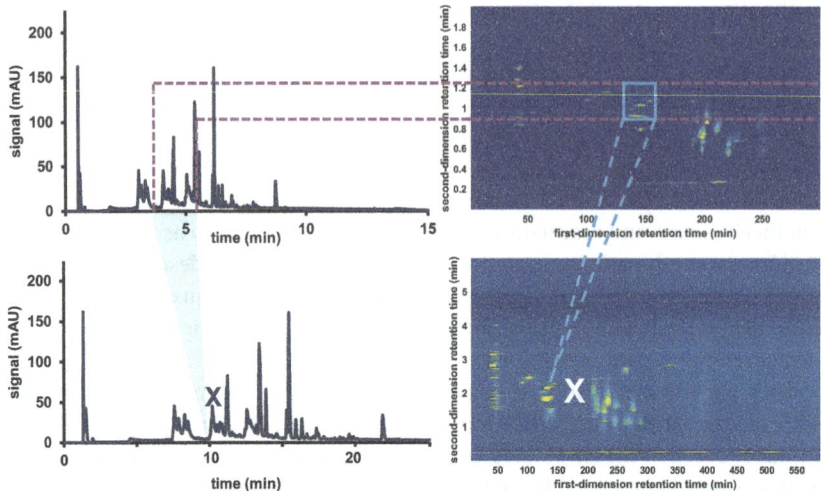

Fig. 7 Even after establishing a narrow search window, plenty of candidate peaks remain during 1D-LC scouting experiments. When using LC × LC for scouting experiments, the search area can be narrowed down in both dimensions, thus significantly reducing the number of candidate peaks.

development of generally applicable LC × LC scouting experiments may be one way to enhance the efficacy of the PIOTR program (Fig. 7).

4   Launching new platforms

The flexible and reliable instrumentation that has emerged for 2D-LC offers a wide range of new possibilities, once it is realized that active-modulation may also include purposeful changes to the sample. Some of the possibilities are discussed in the following sections.

4.1   Physical dissolution

Nanoparticles can feature an array of properties, including their polymer composition,[31] polymer structure,[32] nanoparticle size,[33,34] loading capacity, stability and biodegradability.[31,35,36] These properties can be influenced through several parameters that have been reviewed by Rao and Geckeler.[37] However, methods to simultaneously determine multiple characteristics and their correlation are hard to realize. To establish the relationship between particle-size distribution and molecular-weight distribution for polymeric particles used for coatings, a method was recently developed.[38]

By using very fast size-exclusion-chromatography (SEC) separations[39] and overlapping injections, complete nanoparticle characterizations could be performed in one hour.[38] Fig. 8 displays the configuration used for the first studies on nanoparticles, using a physical-dissolution configuration (PDC), with the inset displaying the situation of the first dimension separation during the initial 12 minutes. Although the setup was successfully used to separate a mixture of polystyrene and polyacrylate particles,[38] there was much room for improvement.

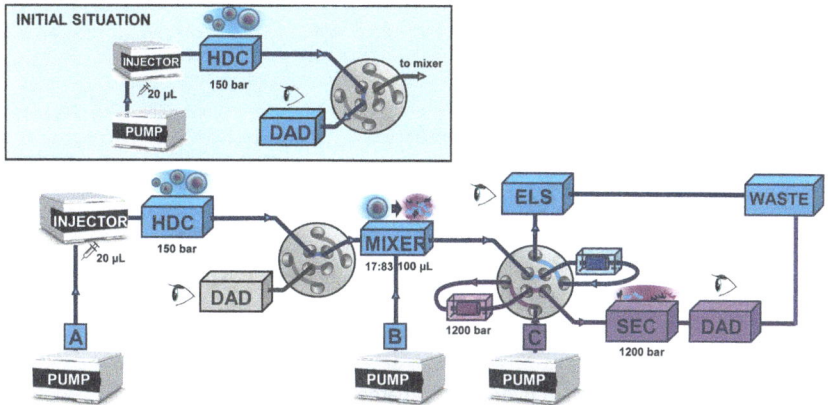

Fig. 8 Example of a physical-dissolution interface, using stationary-phase-assisted modulation traps to reduce dilution effects.

4.1.1 More efficient mixers are needed. One example is the mixing technology. The PDC relies heavily on the efficiency of the employed mixer. At this stage of the study, the band broadening resulting from the very large mixing volumes employed was not critical for proving the concept. However, the development of more efficient mixers is crucial for successful implementation of this technique in industrial practice. This is also true for adapting the PDC to other application fields, such as the analysis of highly complex lignocellulose mixtures or heavy-oil fractions.

Our studies showed a significant influence of the mixing volume.[38] Complete dissolution requires large mixing volumes, which are undesirable from a dispersion perspective. However, only the split-and-recombine approach to mixing was tested. It is recommended to compare other mixing approaches in the search for strategies that require smaller mixing volumes, whilst being more efficient.[40]

4.1.2 Efficient stationary-phase-assisted modulation requires dedicated solutions. A major Achilles heel of the system is its dependence on the stationary-phase-assisted-modulation traps.[38] Sufficient chemical affinity of all analytes for the stationary phase in the trap is important, but this hurdle can usually be overcome. For example, in the setup used for the characterization of poly[(methyl methacrylate)-co-(butyl acrylate)-co-(methacrylic acid)] nanoparticles displayed in Fig. 9, an additional dilution flow was used to improve retention on the trap columns. More critical than the chemistry of the traps is the significant variation in performance that we experienced. This clearly reduced the robustness of the configuration. Typical problems experienced with the traps were clogging and a significant reduction in retention over the course of a few analyses. It is good to realize that SPAM generally utilizes guard columns, which are essentially developed as disposable consumables. At the same time, these guard columns are used at rather extreme conditions from a column-technology perspective, with exposure to (i) elevated temperatures, (ii) large and frequent pressure pulses, (iii) ultra-high-pressure conditions up to 1200 bar, and (iv) potentially aggressive mobile phases in organic second dimension separations. While our published studies mainly featured guard columns from one manufacturer, we have experienced the

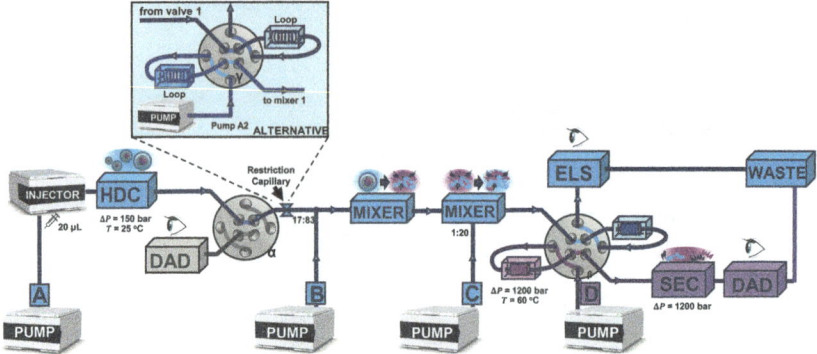

Fig. 9   Setup for the analysis of hydrophilic, charged nanoparticles.

above issues with almost all manufacturers. To reliably use SPAM in routine applications, a dedicated, sustainable solution is needed.

4.2 Enzyme-assisted modulation

4.2.1 Drug-delivery systems. The nanoparticle size, loading capacity, and biodegradability of the polymer are all important for application in controlled-release drug-delivery systems.[31] For example, it is known that small-sized nanoparticles, around 30 nm or below, easily circulate in the human body due to the absence of tissue retention or obstruction,[34] whereas nanoparticles larger than 400 nm may cause hindrances in the vascular system. Capillaries, in particular, may be easily congested by such particles.[34] As explained in a review by Davis *et al.*,[41] particles in the range of 10–100 nm will be optimal for tumour penetration, due to the enhanced permeability and retention (EPR) effect. At the same time, nanoparticles must be sufficiently large to contain a sufficient amount of active ingredient (*e.g.* a 70 nm particle can contain about 2000 small interfering RNA molecules[42]). In essence, it is important that the particle-size distribution (PSD) is well-defined, that the biodegradability is well-characterized, and that the concentration of active ingredient inside the particles is well-known, so that the formulation can be tailored to the application to have the optimal intended effect.

From a chromatographic perspective, correlating particle size with both active-ingredient concentration and polymeric composition comes with another challenge. The particle must be "opened" to release its contents into the eluent. To achieve this inside an organism, nanoparticles created from biodegradable polymers are applied in controlled-release drug-delivery systems.[31,41] Such particles are usually based on poly(lactic-*co*-glycolic acid) (PLGA) and polyethylene glycol (PEG) as PLGA–PEG–PLGA triblocks. PLGA is one of the most successful biodegradable polymers, because its hydrolysis ultimately results in lactic acid and glycolic acid,[31,43] which are both endogenic species. Thus, PLGA is both biodegradable and biocompatible. PLGA also features a number of other attractive properties. It is approved by the Food and Drug Administration (FDA) and the European Medicine Agency (EMA) to be used in drug-delivery systems. Moreover, PLGA formulations and methods for production have been well described. Thus, PLGA is the polymer of choice in many current and future drug-delivery systems.

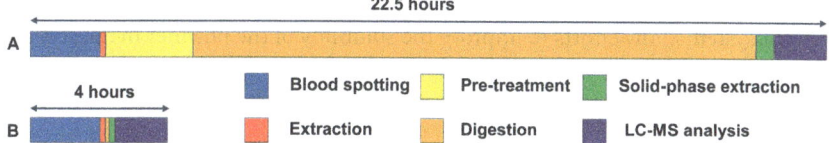

Fig. 10 Comparison of analysis times for dried-blood-spot analysis using (A) in-solution enzymatic degradation, and (B) an immobilized-enzyme reactor (IMER). Using the IMER, the proteins could be digested in 39 s, as compared to 18 h in the case of in-solution degradation. Reproduced with permission from ref. 44.

The biodegradability of the triblock copolymer is largely determined by that of the individual blocks and the overall monomeric composition also plays a role. To characterize the block copolymer, the lengths of the individual blocks must be determined. To this end, physical dissolution does not suffice. However, other forms of reaction modulation may offer a solution.

4.2.2 LC × LC with an immobilized-enzyme reactor (IMER) for nanoparticle analysis. In 2015, Wouters *et al.*[44] investigated the feasibility of enzymes for use in modulators and developed a microfluidic device containing a monolith which was functionalized with immobilized trypsin. Trypsin is an enzyme that is usually used in analytical methods to digest proteins. A trypsin reactor may thus serve as a benchmark for comparison of digestion methods in the field of proteomics. Therefore, initial efforts focused on the degradation of proteins. Wouters *et al.* applied the IMER to the analysis of dried-blood spots and found that it was capable of digesting proteins in 39 s, which compares very favourably with the 18 h required for contemporary in-solution degradation (Fig. 10).[44]

Technically, a similar approach could be applied to polymers. However, the progress in this study was significantly hampered by incompatibility of the enzyme with the typical solvents used to dissolve polymers. One of the first steps in the development of an IMER is the selection of a suitable enzyme. As the interaction of the enzyme with the polymer to be digested can be significantly affected by the morphological state of either, preliminary feasibility studies typically involve testing the enzyme in in-solution-degradation experiments, rather than immobilized in an IMER.

For optimal digestion of a biopolymer, such as PLGA–PEG–PLGA, in solution by an enzyme, (i) the enzyme activity in the solvent must be sufficient, (ii) the polymer must be soluble in the solvent system, and (iii) the enzyme must be able to interact with the polymer (Fig. 11). With some exceptions, most enzymes favour

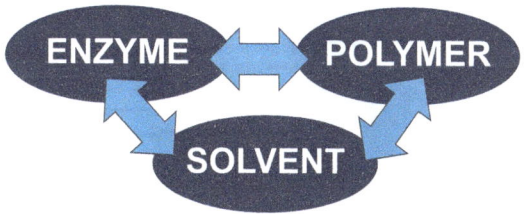

Fig. 11 Compatibility triangle to be taken into consideration during enzyme selection for IMER development.

aqueous systems. However, PLGA–PEG–PLGA polymers are not soluble in such solvent systems. Any efforts to improve the solubility of the polymer in the solvent system, for example by introducing additives, by increasing the organic-modifier content or by adapting the temperature invariably, led to a decrease in the enzyme activity.[45]

In a recent study,[45] nanoparticle dispersions consisting of PLGA–PEG–PLGA polymers were created. Such dispersions are stable in aqueous solvent systems and thus more likely to be compatible with the enzyme. Samples were subjected to various enzymes and were surprisingly well-digested by the enzyme under native conditions in water. The exact mechanism of this process is not yet known. In aqueous solution, PLGA–PEG–PLGA polymers aggregate to form micelles (Fig. 12). The PEG sections of PLGA–PEG–PLGA are relatively hydrophilic and are exposed to the aqueous solvent, whereas the PLGA sections create a hydrophobic core. The investigated enzymes were exclusively able to hydrolyse ester bonds, which exist in PLGA, but not in PEG.[45] This implied that the enzyme somehow managed to penetrate the shell of the particle to break polymer bonds. The energetic interactions between the polar shell and the polar polymers may not be unfavourable. To optimize the IMER-facilitated degradation of PLGA–PEG–PLGA polymers, a better understanding is needed. Ultimately, we want to determine the relationships between the amount of active ingredient, the polymer composition and the particle size in drug-delivery formulations in a single experiment and we are actively investigating possible analytical methods.

Fig. 13A shows one of the setups that was attempted using an IMER in the modulator of an LC × LC system. Using this setup, which is an expansion of the setup shown in Fig. 8, separated particles are subjected to the IMER for degradation. The resulting degradation products may then be trapped in the SPAM cartridges for subsequent analysis by the RPLC second dimension. We are currently testing the feasibility of this system. As a first step, the performance of the IMER, and the viability of trapping and analysing both PLGA–PEG–PLGA-polymer degradation products and the active ingredient using reversed-phase selectivities were investigated. The drug-delivery particles were loaded with

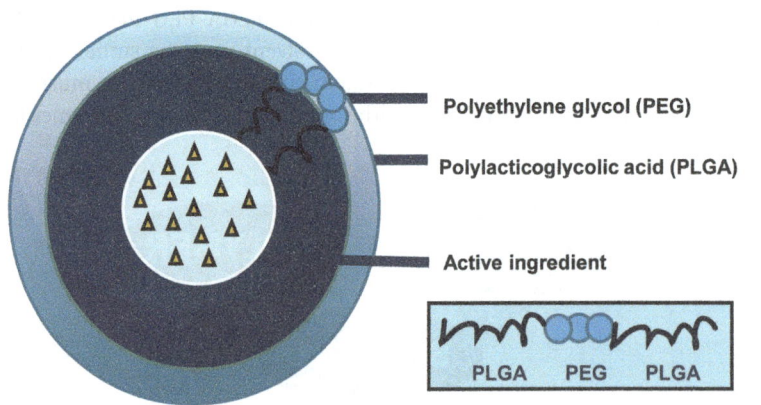

Fig. 12 Illustration of a biodegradable drug-delivery nanoparticle comprised of PLGA–PEG–PLGA polymers and an active ingredient.

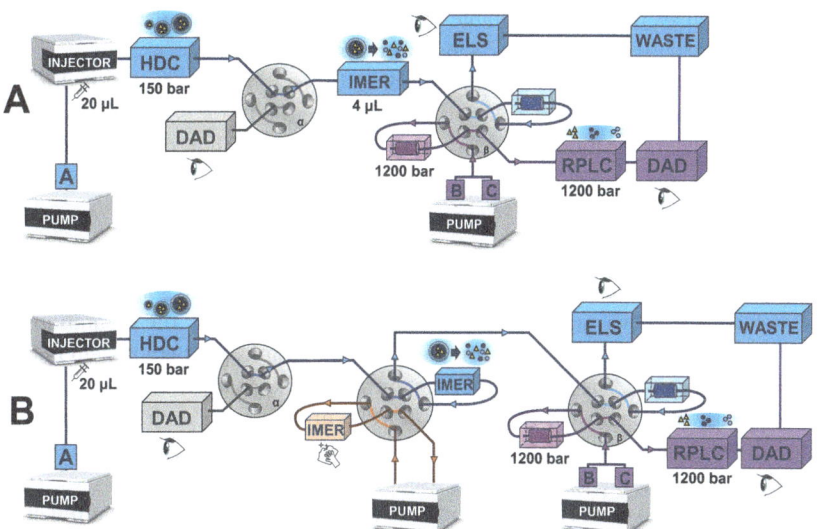

Fig. 13 Two examples of setups explored for on-line reaction modulation of drug-delivery particles using immobilized-enzyme reactors.

a hydrophobic dye. In addition, the HDC column was left out to strictly observe the IMER performance and reduce the number of method variables. An example of a chromatogram resulting from this preliminary study is shown in Fig. 14. As the first dimension HDC column was left out, the first dimension retention time (x-axis) merely represents the peak traveling from the autosampler through the IMER to the valve, whereas this normally would show the particle-size distribution of the undigested particles. The second dimension retention time (y-axis) reflects the reversed-phase LC separation of the degradation products and the active ingredient, which in this case is the dye. Three peaks can be observed, the latter of which was identified as the dye. One problem at this stage is that most of the degradants are PLGA–PEG–PLGA monomers, which are quite hydrophilic in nature. Such monomers are difficult to separate using RPLC. Therefore, current

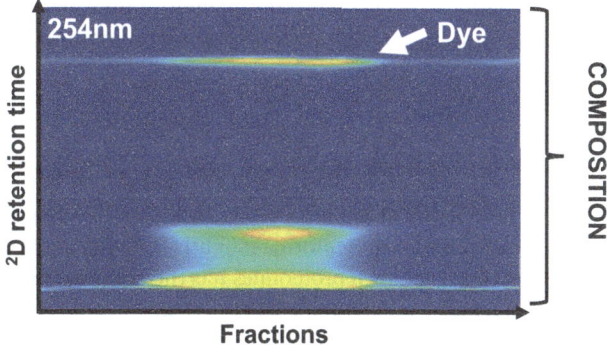

Fig. 14 On-line degradation of dye-loaded PLGA–PEG–PLGA nanoparticles and subsequent RPLC analysis. See text for explanation.

research focuses on finding suitable retention mechanisms, whilst maintaining compatibility with the rest of the separation system.

In addition, there are concerns regarding the stability of the IMER during successive modulations. One solution to regenerate the IMER is the setup shown in Fig. 13B, where the modulation valve can be switched to allow one of the IMERs to be regenerated.

4.3 Dye-degradation mechanisms

Another possibility to manipulate the sample at the modulation stage of an LC × LC experiment is by irradiation. An LC × LC method exists for the separation of dyes.[46] It makes use of a very fast ion-pair chromatography gradient system in the second dimension. The ion-pair agent neutralizes the charged moieties of analytes, greatly increasing their retention in RPLC and allowing their separation in a gradient-elution experiment. In the first dimension, a mixed-mode strong-anion-exchange gradient system was used.

The method has since been expanded to cover virtually all types of natural and synthetic dyes and the selectivity and orthogonality have been optimized using PIOTR.[47] The latter allowed the successful implementation of shifting-gradient assemblies, such that the mobile-phase composition program gradually shifts from (second dimension) run to run. The significant increase in separation power offered by the optimized LC × LC method for separating dyes has led to many more analytes being detected than hitherto possible. A substantial number of these – including analytes present in the extracts obtained from historical cultural-heritage objects – are not identified at this stage.[47] We envisage that the new LC × LC method will make it possible to perform detailed on-line dye-degradation research. A number of off-line strategies exist, but – as explained in Section 4.2 – the implementation of reaction modulation in LC × LC creates exciting new options. For example, a first dimension separation may be used to introduce dyes separately to an exposure cell for light-induced degradation, after which the degradants can be subjected to a second dimension separation for further analysis.

It is anticipated that the matrix, temperature, availability of oxygen, light intensity and duration of the exposure are all important parameters.[48] A well-constructed exposure cell and well-controlled experiments using advanced active-modulation techniques are required for a more thorough, fundamental study into dye-degradation. This should yield qualitative information on degradation pathways and products, and possibly estimates of degradation rate constants.

4.4 Proteoforms

Proteomic studies are typically classified as either top-down or bottom-up proteomics. In the first case, intact proteins are studied, so that information can – in principle – also be obtained on proteoforms (slightly different molecular variations of the same protein). Top-down proteomics has the main advantage that the complete protein sequence can potentially be obtained.

In bottom-up proteomics, the proteins are first digested to peptides. The peptides are then analysed and conclusions are derived about the original protein. This is the most mature and widely used approach for protein identification and characterization. The disadvantage is that only a fraction of the peptide population of a protein is identified and that, consequently, only a limited section of the protein sequence can be obtained. Bottom-up proteomics becomes increasingly more difficult with increasing complexity of the initial protein mixture.

One opportunity is the use of immobilized-enzyme reactors for the analysis of proteins in both their intact form and through their peptide fragments. The intact proteins can first be separated by either RPLC or hydrophobic-interaction chromatography (HIC), after which the eluting, separated proteins are transported to the IMER. In the IMER, the protein is rapidly digested to the constituent peptides. The effluent of the IMER can be injected into a second separation dimension, where the peptides are separated (preferably using RPLC) and finally introduced into a mass spectrometer.

This approach has major advantages. First of all, by separating the proteins prior to digestion, the resulting peptides can be more easily related to the protein that they derived from. Secondly, the individual proteins are digested more efficiently by the IMER, so that potentially all peptides can be detected, rather than just a fraction. This should facilitate a better coverage of the entire sequence and better protein identification.

An alternative opportunity is illustrated by a recent publication of the LC × LC separation of intact histone proteoforms[25] by Gargano et al. Essentially, an intact-protein separation introduces fractions containing resolved proteins into a second dimension separation, where the proteoforms are separated. The authors used HILIC in the second dimension and they found that extremely shallow gradients were required to successfully separate the proteoforms. Such slow gradients are not optimal for second dimension separations. Moreover, the need for such shallow gradients complicates method development. While computer-aided method development for proteoform separations is feasible, the use of spatial devices may eventually also be fruitful.

4.5 Spatial separations

There is actually an egregious waste of time inherent to column-based LC × LC in that all second dimension separations are performed sequentially. If a modest 100 fractions are taken from the ^{1}D effluent and if the cycle time (the analysis time, plus the preparation time in between successive runs, e.g. for column re-equilibration) is 2 min, then the analysis time is already more than 3 h. If we were to include a third dimension to perform comprehensive three-dimensional liquid chromatography (LC × LC × LC) and we cut 100 fractions from every ^{2}D run, we would have a mere one second left to perform the ^{3}D separation. The only group to have performed such a feat is that of Jorgenson,[49] who chose a 5 h size-exclusion chromatography (SEC) separation in the first dimension, a 6 min reversed-phase liquid chromatography separation in the second dimension and a 2 s capillary electrophoresis (CE) separation in the third dimension. For a single experiment to be successfully completed, thousands of CE runs had to be performed in a repeatable and reliable fashion.

An alternative is to resort to spatial LC × LC separations, in which analytes are (at least in the first dimension) not eluted from the column, but separated along the length of a separation medium. A conventional form of a spatial LC × LC separation is 2D thin-layer chromatography. Another example of a spatial 2D separation is 2D-PAGE, where the first dimension separation is based on iso-electric focussing and the second dimension separation on poly(acryl amide) gel electrophoresis. One major advantage of spatial 2D devices is the ability to run all second dimension separations concurrently, almost completely removing the time constraint on the second dimension. Spatial devices have theoretically been shown to be superior in terms of the maximum attainable peak capacity,[50,51] and the implementation of 3D spatial separations is actively pursued by our group.[52]

In spatial 2D-LC, all second dimension separations are normally run with an identical mobile-phase composition program. However, every protein eluting from the intact-protein separation (see Section 4.4) is expected to require a different shallow gradient to obtain a good separation of proteoforms. It is hard to imagine how shifting-gradient assemblies can be implemented in spatial 2D-LC. One alternative may be to "program" the stationary phase. It is conceptually easy to create ^{2}D channels of gradually different lengths, resulting in a version of spatial 2D-LC that is instantly identified as "panflutography" (Fig. 15). Preparing such devices is not more difficult than preparing regular spatial 2D-devices and detection will be only marginally more complicated. Because monolith technology is predominantly used to create stationary phases in spatial separation devices, it may also be possible to "program" the composition of the stationary phase, for example by polymerization at gradually shifting temperatures or by performing grafting reactions with gradually shifting light intensities (in a transparent device).

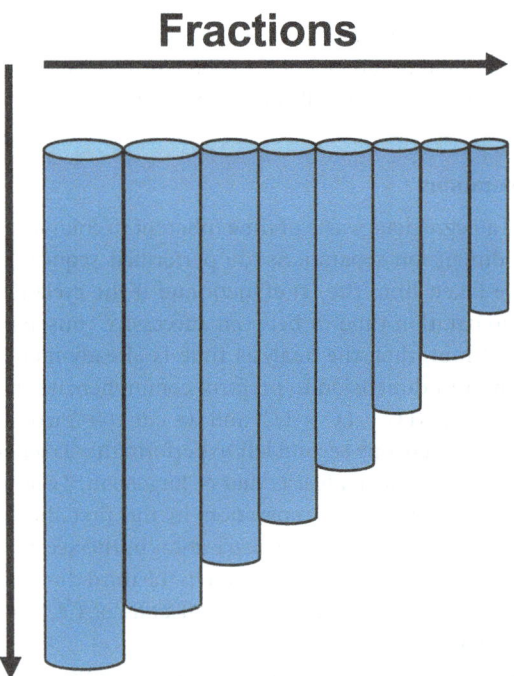

Fractions

Fig. 15   Crude scheme of a spatial 2D-LC device with varying 2D channel lengths. This type of separation is referred to as "panflutography".

5 Polysorbate analysis

Polysorbates are a special kind of non-ionic surfactants used in a variety of application areas, including stabilizing monoclonal-antibody (mAb) formulations, emulsifiers in the food industry, and solubilizers in cosmetics. They are produced by the ethoxylation of sorbitan and isosorbide (derived from the dehydration of sorbitol) and subsequent esterification of the –OH termini by fatty acids.[53] Ultimately, a complex mixture of various structures is formed, including ethoxylated isosorbide mono-/di-esters, ethoxylated sorbitan mono-/di-/tri-/penta-esters, ethoxylated mono-/di-acids and ethoxylated isosorbide and sorbitan. Depending on the fatty acids used during esterification, the complexity of each group is increased. Furthermore, impurities present in the initiator or fatty acid feeds can increase heterogeneities.

Due to hydrolysis in water-rich environments, the polysorbate can lose its surfactant function, which can be a crucial factor in the stability of mAb formulations and emulsions in food.[54] To characterize which compounds are prone to hydrolysis, complete characterization of the polysorbates is necessary. However, this is a challenging task, since polysorbates are simultaneously heterogeneous in molecular weight and chemical composition.[55]

By combining two separation dimensions, multidimensional characterization of the molecular distributions present in polysorbates was achieved. In the ^{1}D separation, HILIC was employed, resolving the chemical structures based on the degree of ethoxylation. The ^{2}D RPLC dimension resolved the chemical features based on hydrophobicity, i.e. ester classes and fatty acid constituents. The HILIC × RPLC separation was hyphenated with high-resolution mass spectrometry to aid the identification of the separated species.

In Fig. 16, the LC × LC-HRMS chromatograms are shown of Tween-20 and Tween-80 (Fig. 16A and B, respectively; structures provided in Fig. 17). For information regarding the chemicals used in the experiments, please see ESI Section S-1.† The identified species using the LC × LC-HRMS data are presented in Table 2. The complexity of the two different samples is clearly presented within the figures, showing group-type clustering of ethoxylated species within the two-dimensional separation space. Overall, ethoxylated isosorbide and sorbitan were detected, as well as ethoxylated mono-/di-acids, isosorbide mono-/di-esters and sorbitan mono-/di-/tri-esters. Within these compound classes, different combinations of fatty acid constituents were detected. These varied between the two Tween samples, which were clearly derived from different fatty acids from different sources. In Tween-20, laurate (12:0), myristate (14:0), palmitate (16:0) and stearate (18:0) species were detected, while myristoleate (14:1), palmitoleate (16:1) and oleate (18:1) species were detected in Tween-80. On average, the isosorbide species have a lower degree of ethoxylation compared to the sorbitan species. This can be explained from the fact that isosorbide has two –OH end-groups available for polymerization, while sorbitan has four –OH end-groups.

Since the two-dimensional separation is highly structured and since the high resolution ensured baseline at many points in the LC × LC chromatogram, very-low-abundant species were observed in both Tween-20 and Tween-80, which could not be detected in 1D-LC chromatograms. In comparison with isosorbide and sorbitan species, these low-abundant species exhibited an even higher degree

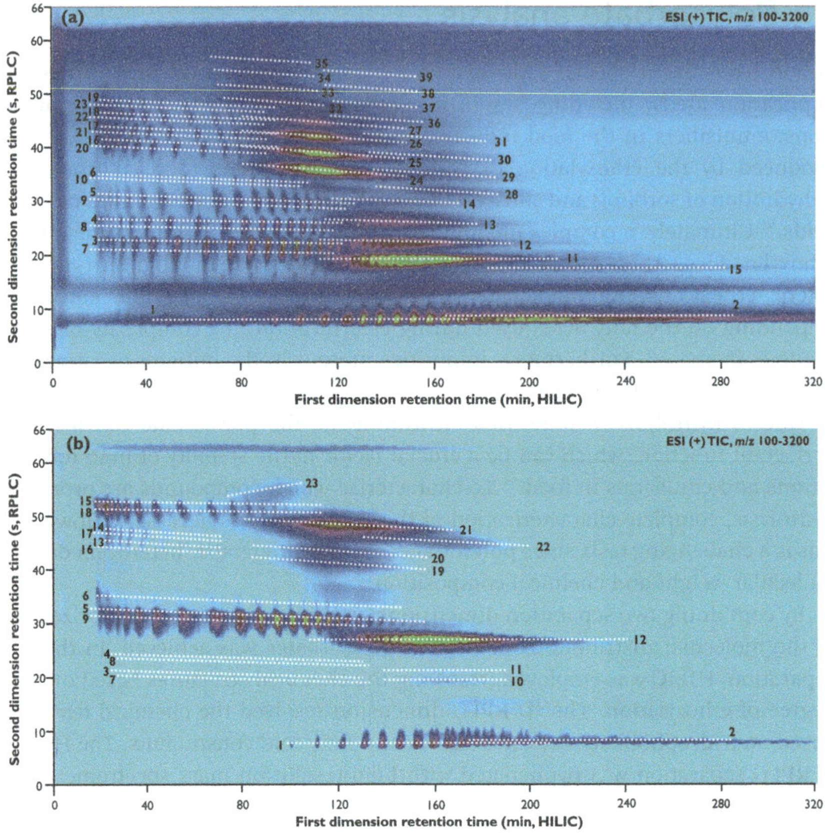

Fig. 16 HILIC × RPLC-HRMS separation of (a) Tween-20 and (b) Tween-80. The degree of ethoxylation is resolved in the ^{1}D HILIC separation, while the ^{2}D resolves the chemical species based on hydrophobicity. Clear group-type separation of isosorbide ethoxylate mono-/di-esters and sorbitan ethoxylate mono-/di-/tri-esters is observed. Within these classes, the species are separated based on the different fatty acid constituents incorporated. Furthermore, sorbitol-initiated species were detected. Identifications are listed in Table 2.

of ethoxylation. Ultimately, the low-abundant species were identified as sorbitol mono-/di-/tri-esters. The sorbitol species may be the result of an impurity left over from the dehydration of the starting compound sorbitol. Such species have not been reported in the literature before, illustrating the power of LC × LC.

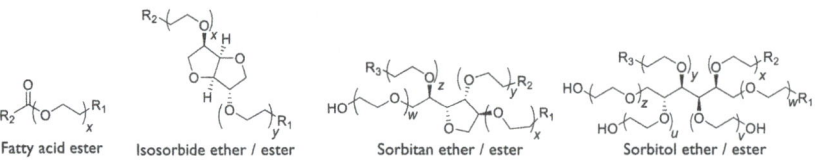

Fig. 17 Generalized structures of the compounds identified in Tween-20 and Tween-80. In Tween-20, $R_x = H$, laurate (C12:0), myrate (C14:0), palmate (C16:0) and/or stearate (C18:0). In Tween-80, $R_x = H$, myristoleate (C14:0), palmitoleate (C16:1) and/or oleate (C18:1).

Table 2 Identified chemical species based on the LC × LC-HRMS data as shown in Fig. 16

Tween-20 (polysorbate)

Class	#	Name
Polyether polyols	1	Isosorbide ethoxylate
	2	Sorbitan ethoxylate
Mono-acid	3	Monolaurate ethoxylate
	4	Monomyristate ethoxylate
	5	Monopalmitate ethoxylate
	6	Monostearate ethoxylate
Mono-ester	7	Isosorbide ethoxylate monolaurate
	8	Isosorbide ethoxylate myristate
	9	Isosorbide ethoxylate monopalmitate
	10	Isosorbide ethoxylate monostearate
	11	Sorbitan ethoxylate monolaurate
	12	Sorbitan ethoxylate myristate
	13	Sorbitan ethoxylate monopalmitate
	14	Sorbitan ethoxylate monostearate
	15	Sorbitol ethoxylate monolaurate
Di-acid	16	Dilaurate ethoxylate
	17	Monolaurate-monomyristate ethoxylate
	18	Monolaurate-monopalmitate ethoxylate
	19	Monolaurate-monostearate ethoxylate
Di-ester	20	Isosorbide ethoxylate dilaurate
	21	Isosorbide ethoxylate monolaurate-monomyristate
	22	Isosorbide ethoxylate monolaurate-monopalmitate
	23	Isosorbide ethoxylate monolaurate-monostearate
	24	Sorbitan ethoxylate dilaurate

Tween-80 (polysorbate)

Class	#	Name
Polyether polyols	1	Isosorbide ethoxylate
	2	Sorbitan ethoxylate
Mono-acid	3	Monomyristoleate ethoxylate
	4	Monopalmitoleate ethoxylate
	5	Monooleate ethoxylate
	6	Monostearate ethoxylate
Mono-ester	7	Isosorbide ethoxylate monomyristoleate
	8	Isosorbide ethoxylate monopalmitoleate
	9	Isosorbide ethoxylate monooleate
	10	Sorbitan ethoxylate monomyristoleate
	11	Sorbitan ethoxylate monopalmitoleate
	12	Sorbitan ethoxylate monooleate
Di-acid	13	Monooleate-monomyristoleate ethoxylate
	14	Monooleate-monopalmitoleate ethoxylate
	15	Dioleate ethoxylate
Di-ester	16	Isosorbide ethoxylate monooleate-monomyristoleate
	17	Isosorbide ethoxylate monooleate-monopalmitoleate
	18	Isosorbide ethoxylate dioleate
	19	Sorbitan ethoxylate monooleate-monomyristoleate
	20	Sorbitan ethoxylate monooleate-monopalmitoleate
	21	Sorbitan ethoxylate dioleate
	22	Sorbitol ethoxylate dioleate
Tri-ester	23	Sorbitan ethoxylate trioleate

Table 2 (Contd.)

Tween-20 (polysorbate)			Tween-80 (polysorbate)		
Class	#	Name	Class	#	Name
	25	Sorbitan ethoxylate monolaurate-monomyristate			
	26	Sorbitan ethoxylate monolaurate-monopalmitate			
	27	Sorbitan ethoxylate monolaurate-monostearate			
	28	Sorbitol ethoxylate dilaurate			
	29	Sorbitol ethoxylate monolaurate-monomyristate			
	30	Sorbitol ethoxylate monolaurate-monopalmitate			
	31	Sorbitol ethoxylate monolaurate-monostearate			
Tri-ester	32	Sorbitan ethoxylate trilaurate			
	33	Sorbitan ethoxylate dilaurate-monomyristate			
	34	Sorbitan ethoxylate dilaurate-monopalmitate			
	35	Sorbitan ethoxylate dilaurate-monostearate			
	36	Sorbitol ethoxylate trilaurate			
	37	Sorbitol ethoxylate dilaurate-monomyristate			
	38	Sorbitol ethoxylate dilaurate-monopalmitate			
	39	Sorbitol ethoxylate dilaurate-monostearate			

6   Concluding remarks

Liquid chromatography is an immensely successful analytical (and preparative) separation technique. Two-dimensional LC (2D-LC) allows better separation of selected peaks in heart-cut mode or much better separations of complex samples in comprehensive (LC × LC) mode. The application field for 2D-LC is much smaller than that for conventional 1D-LC, but 2D-LC (i) provides a much higher peak capacity (by at least an order of magnitude), (ii) provides additional "orthogonal" selectivity, and (iii) may provide structured, readily interpretable chromatograms. These advantages and the rapidly improving hardware and software provide enough indications that 2D-LC is here to stay – and to grow. The latter will be especially the case if we come to realize that there are a myriad of possibilities to manipulate the sample between the two dimensions. This may be "just" aimed at increasing the sensitivity of detection and the efficiency of the separation (peak capacity) through active-modulation, which also opens possibilities to combine highly different, seemingly incompatible separations. We may also manipulate the sample at the intermediate ("modulation") stage between two LC separations. Nanoparticles may be physically dissolved to allow characterization of particles (e.g. the particle-size distribution) and the constituting polymer molecules (e.g. molecular-weight distribution) in single on-line experiments. Macromolecules (e.g. proteins) may be digested to smaller fragments (e.g. peptides) using very fast and efficient immobilized-enzyme reactors and the effects of light on individual sample components (e.g. dyes) may be studied by installing an exposure cell in the modulator. Ultimately, LC × LC technology will greatly benefit from a parallelization of all second dimension separations (spatial 2D-LC) and all modulation reactions.

Highly orthogonal and efficient LC × LC-HRMS methods were applied for the characterization of polysorbates, based on the degree of ethoxylation and on hydrophobicity. Highly structured LC × LC chromatograms were obtained and almost all compounds could be identified. Sorbitol ethoxylate mono-/di-/tri esters were detected in very low concentrations.

Conflicts of interest

The authors declare no conflicts of interest.

Acknowledgements

The MANIAC project is funded by the Netherlands Organization for Scientific Research (NWO) in the framework of the Programmatic Technology Area PTA-COAST3 of the Fund New Chemical Innovations (project 053.21.113). Denice van Herwerden and Rick van der Hurk are acknowledged for their assistance with recording and processing the results shown in Fig. 13 and 14. Melissa Dunkle and Edwin Mes (Dow Benelux B. V., Terneuzen, The Netherlands) are acknowledged for their contributions to the LC × LC-HRMS study of polysorbates. Gino Groeneveld acknowledges the Open Technology Programme (IWT-STW collaboration), project number 14624 (DEBOCS), which is financed by the Netherlands Organization for Scientific Research (NWO).

References

1 J. M. Davis and J. C. Giddings, *Anal. Chem.*, 1983, **55**, 418–424.

2 B. W. J. Pirok, D. R. Stoll and P. J. Schoenmakers, *Anal. Chem.*, 2019, **91**, 240–263.

3 F. Erni and R. W. Frei, *J. Chromatogr. A*, 1978, **149**, 561–569.

4 M. M. Bushey and J. W. Jorgenson, *Anal. Chem.*, 1990, **62**, 161–167.

5 E. Uliyanchenko, P. J. C. H. Cools, S. Van Der Wal and P. J. Schoenmakers, *Anal. Chem.*, 2012, **84**, 7802–7809.

6 X. Jiang, A. van der Horst, V. Lima and P. J. Schoenmakers, *J. Chromatogr. A*, 2005, **1076**, 51–61.

7 B. W. J. Pirok, A. F. G. Gargano and P. J. Schoenmakers, *J. Sep. Sci.*, 2018, **41**, 68–98.

8 G. Groeneveld, M. N. Dunkle, M. Rinken, A. F. G. Gargano, A. de Niet, M. Pursch, E. P. C. Mes and P. J. Schoenmakers, *J. Chromatogr. A*, 2018, **1569**, 128–138.

9 B. W. J. Pirok and P. J. Schoenmakers, *LC-GC Eur.*, 2018, **31**, 242–249.

10 R. J. Vonk, A. F. G. Gargano, E. Davydova, H. L. Dekker, S. Eeltink, L. J. de Koning and P. J. Schoenmakers, *Anal. Chem.*, 2015, **87**, 5387–5394.

11 A. Baglai, M. H. Blokland, H. G. J. Mol, A. F. G. Gargano, S. van der Wal and P. J. Schoenmakers, *Anal. Chim. Acta*, 2018, **1013**, 87–97.

12 A. F. G. Gargano, M. Duffin, P. Navarro and P. J. Schoenmakers, *Anal. Chem.*, 2016, **88**, 1785–1793.

13 H. Tian, J. Xu and Y. Guan, *J. Sep. Sci.*, 2008, **31**, 1677–1685.

14 H. C. Van de Ven, A. F. G. Gargano, S. J. Van der Wal and P. J. Schoenmakers, *J. Chromatogr. A*, 2016, **1427**, 90–95.

15 A. P. Sweeney and R. A. Shalliker, *J. Chromatogr. A*, 2002, **968**, 41–52.

16 M. Verstraeten, M. Pursch, P. Eckerle, J. Luong and G. Desmet, *Anal. Chem.*, 2011, **83**, 7053–7060.

17 M. E. Creese, M. J. Creese, J. P. Foley, H. J. Cortes, E. F. Hilder, R. A. Shellie and M. C. Breadmore, *Anal. Chem.*, 2017, **89**, 1123–1130.

18 D. R. Stoll, E. S. Talus, D. C. Harmes and K. Zhang, *Anal. Bioanal. Chem.*, 2015, **407**, 265–277.

19 Y. Oda, N. Asakawa, T. Kajima, Y. Yoshida and T. Sato, *J. Chromatogr. A*, 1991, **541**, 411–418.

20 D. R. Stoll, K. Shoykhet, P. Petersson and S. Buckenmaier, *Anal. Chem.*, 2017, **89**, 9260–9267.

21 M. Pursch, A. Wegener and S. Buckenmaier, *J. Chromatogr. A*, 2018, **1562**, 78–86.

22 C. J. Venkatramani, S. R. Huang, M. Al-Sayah, I. Patel and L. Wigman, *J. Chromatogr. A*, 2017, **1521**, 63–72.

23 C. J. Venkatramani, M. Al-Sayah, G. Li, M. Goel, J. Girotti, L. Zang, L. Wigman, P. Yehl and N. Chetwyn, *Talanta*, 2016, **148**, 548–555.

24 M. Goel, E. Larson, C. J. Venkatramani and M. A. Al-Sayah, *J. Chromatogr. B*, 2018, **1084**, 89–95.

25 A. F. G. Gargano, J. B. Shaw, M. Zhou, C. S. Wilkins, T. L. Fillmore, R. J. Moore, G. W. Somsen and L. Paša-Tolić, *J. Proteome Res.*, 2018, **17**(11), 3791–3800.

26 B. W. J. Pirok, S. Pous-Torres, C. Ortiz-Bolsico, G. Vivó-Truyols and P. J. Schoenmakers, *J. Chromatogr. A*, 2016, **1450**, 29–37.

27 B. W. J. Pirok, S. R. A. Molenaar, R. E. van Outersterp and P. J. Schoenmakers, *J. Chromatogr. A*, 2017, **1530**, 104–111.

28 B. W. J. Pirok, S. R. A. Molenaar, L. S. Roca and P. J. Schoenmakers, *Anal. Chem.*, 2018, **90**, 14011–14019.

29 F. T. van Beek, R. Edam, B. W. J. Pirok, W. J. L. Genuit and P. J. Schoenmakers, *J. Chromatogr. A*, 2018, **1564**, 110–119.

30 J. C. Giddings, *J. Chromatogr. A*, 1995, **703**, 3–15.

31 L. S. Nair and C. T. Laurencin, *Prog. Polym. Sci.*, 2007, **32**, 762–798.

32 K. Zhang, X. Tang, J. Zhang, W. Lu, X. Lin, Y. Zhang, B. Tian, H. Yang and H. He, *J. Controlled Release*, 2014, **183**, 77–86.

33 J. G. J. L. Lebouille, R. Stepanyan, J. J. M. Slot, M. A. Cohen Stuart and R. Tuinier, *Colloids Surf., A*, 2014, **460**, 225–235.

34 J. G. J. L. Lebouille, L. F. W. Vleugels, A. A. Dias, F. A. M. Leermakers, M. A. Cohen Stuart and R. Tuinier, *Eur. Phys. J. E*, 2013, **36**, 107.

35 K. Herzog, R.-J. Müller and W.-D. Deckwer, *Polym. Degrad. Stab.*, 2006, **91**, 2486–2498.

36 Q. Cai, G. Shi, J. Bei and S. Wang, *Biomaterials*, 2003, **24**, 629–638.

37 J. P. Rao and K. E. Geckeler, *Prog. Polym. Sci.*, 2011, **36**, 887–913.

38 B. W. J. Pirok, N. Abdulhussain, T. Aalbers, B. Wouters, R. A. H. Peters and P. J. Schoenmakers, *Anal. Chem.*, 2017, **89**, 9167–9174.

39 B. W. J. Pirok, P. Breuer, S. J. M. Hoppe, M. Chitty, E. Welch, T. Farkas, S. van der Wal, R. Peters and P. J. Schoenmakers, *J. Chromatogr. A*, 2017, **1486**, 96–102.

40 M. A. Ianovska, P. P. M. F. A. Mulder and E. Verpoorte, *RSC Adv.*, 2017, **7**, 9090–9099.

41 M. E. Davis, Z. Chen and D. M. Shin, *Nat. Rev. Drug Discovery*, 2008, **7**, 771–782.

42 D. W. Bartlett and M. E. Davis, *Bioconjugate Chem.*, 2007, **18**, 456–468.

43 F. Danhier, E. Ansorena, J. M. Silva, R. Coco, A. Le Breton and V. Préat, *J. Controlled Release*, 2012, **161**, 505–522.

44 B. Wouters, I. Dapic, T. S. E. Valkenburg, S. Wouters, L. Niezen, S. Eeltink, G. L. Corthals and P. J. Schoenmakers, *J. Chromatogr. A*, 2017, **1491**, 36–42.

45 B. Wouters, B. W. J. Pirok, D. Soulis, R. C. Garmendia Perticarini, S. Fokker, R. S. van den Hurk, M. Skolimowski, R. A. H. Peters and P. J. Schoenmakers, *Anal. Chim. Acta*, 2018, **1053**, 62–69.

46 B. W. J. Pirok, J. Knip, M. R. van Bommel and P. J. Schoenmakers, *J. Chromatogr. A*, 2016, **1436**, 141–146.

47 B. W. J. Pirok, M. J. den Uijl, G. Moro, S. V. J. Berbers, C. J. M. Croes, M. R. van Bommel and P. J. Schoenmakers, *Anal. Chem.*, 2018, **91**(4), 3062–3069.

48 B. W. J. Pirok, G. Moro, S. V. J. Berbers, C. J. M. Croes, M. R. van Bommel and P. J. Schoenmakers, *J. Cult. Herit.*, 2019, DOI: 10.1016/j.culher.2019.01.003.

49 A. W. Moore and J. W. Jorgenson, *Anal. Chem.*, 1995, **67**, 3456–3463.

50 B. Wouters, E. Davydova, S. Wouters, G. Vivo-Truyols, P. J. Schoenmakers and S. Eeltink, *Lab Chip*, 2015, **15**, 4415–4422.

51 E. Davydova, P. J. Schoenmakers and G. Vivó-Truyols, *J. Chromatogr. A*, 2013, **1271**, 137–143.

52 T. Adamopoulou, S. Deridder, G. Desmet and P. J. Schoenmakers, *J. Chromatogr. A*, 2018, **1577**, 120–123.

53 N. Solak Erdem, N. Alawani and C. Wesdemiotis, *Anal. Chim. Acta*, 2014, **808**, 83–93.

54 Y. Li, D. Hewitt, Y. K. Lentz, J. A. Ji, T. Y. Zhang and K. Zhang, *Anal. Chem.*, 2014, **86**, 5150–5157.

55 G. Vanhoenacker, M. Steenbeke, K. Sandra and P. Sandra, *LC-GC Eur.*, 2018, 360–371.

56 X. Jiang, A. Van Der Horst and P. J. Schoenmakers, *J. Chromatogr. A*, 2002, **982**, 55–68.

Faraday Discussions

PAPER

Collection and identification of an unknown component from *Eugenia uniflora* essential oil exploiting a multidimensional preparative three-GC system employing apolar, mid-polar and ionic liquid stationary phases†

Danilo Sciarrone, [iD] *[a] Antonino Schepis, [iD] [a] Gemma De Grazia, [iD] [b]
Archimede Rotondo, [iD] [c] Filippo Alibrando, [iD] [b]
Roger Raupp Cipriano, [iD] [d] Humberto Bizzo, [iD] [d]
Cicero Deschamps, [iD] [e] Leonard M. Sidisky [iD] [f]
and Luigi Mondello [iD] [abgh]

Received 9th December 2018, Accepted 16th January 2019

DOI: 10.1039/c8fd00234g

The present research deals with the collection and structural elucidation of an unknown component, accounting for about 35% of the essential oil obtained upon distillation of the leaves of *Eugenia uniflora* L., harvested during summer (January, 2017) in Paraná State (Southern Brazil). A multidimensional gas chromatographic preparative system, based on the coupling of three GC systems equipped with apolar, PEG and ionic liquid-based stationary phases, was successfully applied for the isolation of the chromatographic band relative to the unknown molecule. The use of wide-bore columns allowed for an increased sample capacity compared to conventional micro-bore columns, thus the injection of a neat sample was feasible, greatly reducing the total collection time. A higher chromatographic efficiency was afforded by the use of

[a]*Department of Chemical, Biological, Pharmaceutical and Environmental Sciences, University of Messina, Messina, Italy. E-mail: dsciarrone@unime.it; Fax: +39-090-358220; Tel: +39-090-6766463*

[b]*Chromaleont s.r.l., c/o Department of Chemical, Biological, Pharmaceutical and Environmental Sciences, University of Messina, Messina, Italy*

[c]*Department of Biomedical, Dental, Morphological and Functional Imaging Sciences, University of Messina, Messina, Italy*

[d]*Embrapa Agroindústria de Alimentos - Avenida das Américas, 29501 Rio de Janeiro RJ, 23020-470, Brazil*

[e]*Agronomy Departament, Federal University of Paraná, Curitiba, PR 80035-050, Brazil*

[f]*MilliporeSigma, 595 North Harrison Road, Bellefonte, PA 16823, USA*

[g]*Unit of Food Science and Nutrition, Department of Medicine, University Campus Bio-Medico of Rome, via Alvaro del Portillo 21, 00128 Rome, Italy*

[h]*BeSep s.r.l., c/o Department of Chemical, Biological, Pharmaceutical and Environmental Sciences, University of Messina, Messina, Italy*

† Electronic supplementary information (ESI) available. See DOI: 10.1039/c8fd00234g

a multidimensional approach in the heart-cut mode, exploiting the different selectivity of three stationary phases, which ensured the attainment of a highly pure fraction. In only five runs, more than 3 milligrams were collected, with an average purity greater then 95%. Finally, the unknown component was subjected to nuclear magnetic resonance spectroscopy, mass spectrometry and condensed phase Fourier-transform infrared spectroscopy, leading to the identification of 6-ethenyl-6-methyl-3,5-di(prop-1-en-2-yl)cyclohex-2-en-1-one. The presented approach has been demonstrated to be effective for the isolation and structural elucidation of unknown molecules in complex samples, which will allow for further in-depth studies, like biological evaluation or pharmacological tests.

1. Introduction

The investigation of molecules responsible for possible pharmacological effects has always attracted a great deal of attention from researchers and industries. Potential beneficial health effects of natural samples can be related to synergic effects exerted by different molecules, or to specific component(s). As a consequence, it is crucial to acquire full knowledge of a sample composition, and also to be able to study individual compounds, separately from the rest of the sample components. In the latter case, the isolation (physical collection) of specific molecules for deeper study represents a fundamental requirement. In the case of unknown components, identification often requires a separation step before the isolation of the target compound; to this regard, chromatography-based preparative (Prep) systems represent an effective alternative to the common distillation approach.[1,2] Since natural samples are often characterized by a high complexity, with components belonging to different chemical classes and often in a wide range of concentrations, an effective chromatographic separation is required, as prerequisite for a correct structural elucidation of single components. In this regard, the goal of a GC-Prep system is to allow the isolation of specific chromatographic bands, in adequate amounts and with a high purity, to permit further experiments such as NMR, FTIR or MS studies to be carried out prior to biological evaluation. On the other hand, the simultaneous achievement of collecting large amounts of components and obtaining a high degree of purity is challenging. Wide-bore capillaries (0.32–0.53 mm I.D.) with thicker films (1–5 μm) are usually employed to increase the amount of sample manageable by the column, even at the sacrifice of efficiency (almost half if compared to conventional micro-bore columns of 0.10–0.25 mm I.D.).[3] Increasing the sample amount injected allows for the collection of larger quantities per run, but column overloading inevitably compromises the purity of the fractions collected. Heart-cutting multidimensional gas chromatography (MDGC), aiming for a complete separation of target compounds, has proven its capability to increase the efficiency of a separation system, alleviating the issue of the lower efficiency of 0.53 mm I.D. columns when used in a MDGC-Prep system.[1] Heart-cutting MDGC is applied in many fields when dealing with complex samples, particularly when baseline separation of sample components is required.[4] In the last decade, MDGC systems equipped with Deans switch transfer devices have been successfully employed for preparative purposes,[5–13] often followed by [1]H-nuclear magnetic resonance (NMR) experiments.[5-9] For a deeper review of GC-Prep fundamentals, the reader is directed elsewhere.[1,2] The present research is focused on the characterization of an unknown component in a natural sample by means of a three-dimensional GC system. The sample analyzed was a distilled essential oil obtained from the leaves of *Eugenia uniflora* L. from the Myrtaceae family, characterized by shrubs and trees with flowering plants (angiosperms), harvested in January 2017 in Southern Brazil (Paraná State). The leaves, used as a substitute for tea, possess antioxidant activity because of their content in phenolic and flavonoid compounds.[14] *Eugenia uniflora* L. extracts have also been used in popular medicine to treat inflammation, rheumatic pain, fever, diabetes, as a diuretic and in the cosmetics industry.[15] With the aim to investigate the relationship between activity and sample composition, an unknown component, accounting for about 35% of the whole sample, was isolated from the leave extract to undergo further biological evaluation. The MDGC-Prep system exploited enabled the isolation of sufficient quantities of a highly pure compound in a short working time, allowing a reliable structural elucidation by using NMR, condensed phase GC-FTIR and GC-MS analyses.

2.   Materials and methods

2.1. Standard compounds and samples

A distilled essential oil (E.O.) was obtained from the leaves of *Eugenia uniflora* L., during the summer (January, 2017) in Paraná State (Southern Brazil). The distilled E.O. was protected from light and heat, and stored in a refrigerator at 5 °C until it was used. A C_7–C_{30} *n*-alkane mix and *n*-nonane were used for the linear retention index (LRI) measurements and for internal standardization purposes, respectively, kindly provided by Merck Life Science (Merck KGaA, Darmstadt, Germany). The E.O. was diluted 1 : 10 (v/v) in *n*-hexane GC grade from Merck Life Science (Merck KGaA, Darmstadt, Germany) prior to the GC-FID and GC-MS analyses, while it was injected neat in the MDGC-Prep system.

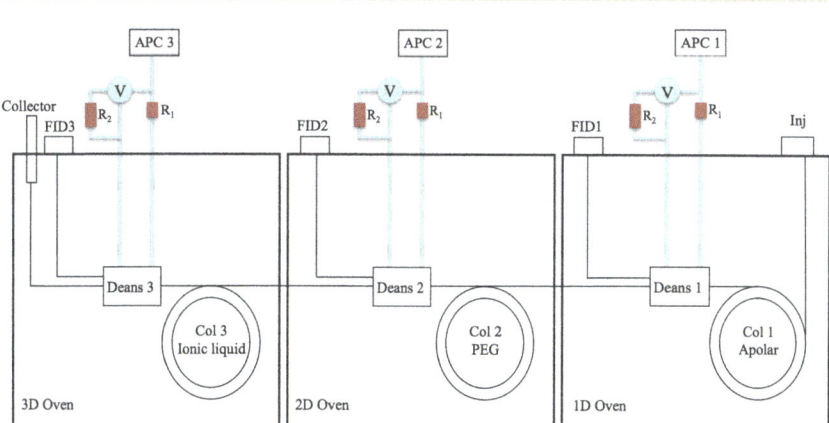

Fig. 1   Scheme of the tridimensional MDGC-Prep system. APC: auxiliary pressure control, R: restrictor, V: valve.

2.2. Multidimensional GC-Prep

The preparative MDGC instrument, illustrated in Fig. 1, consisted of three GC 2010 plus systems (Shimadzu, Kyoto Japan), namely GC1, GC2, and GC3, connected by means of three Deans switch transfer devices (Deans). Each Deans switch element in the three GC systems was connected to an advanced pressure control system (APC1, APC2, and APC3) (Shimadzu, Kyoto Japan) that supplied the carrier gas (He). The system configuration has been described elsewhere; for more details see Sciarrone et al.[5] GC1 was equipped with a split/splitless injector and a flame ionization detector (FID1). GC1 column (^{1}D): equity-5 [poly (5% diphenyl/95% dimethylsiloxane)], 30 m $\times$ 0.53 mm I.D. $\times$ 5 μm d_f from Merck Life Science (Merck KGaA, Darmstadt, Germany), preceded by a 1 m segment of uncoated column of the same I.D. The carrier gas pressure was maintained constant at 140 kPa, while 125 kPa were applied to APC1. Oven temperature program: initial 150 °C, to 280 °C at 3 °C min^{-1}, held for 10 min. FID1 (280 °C) was connected to Deans 1 via 1 m $\times$ 0.22 mm segment of uncoated column. The transfer line between GC1 and GC2 was maintained at 280 °C.

GC2 column (^{2}D): Supelcowax-10 (100% polyethylene glycol, PEG), 30 m $\times$ 0.53 mm I.D. $\times$ 1.0 μm d_f from Merck Life Science (Merck KGaA, Darmstadt, Germany). Oven temperature program: initial 150 °C (held until the end of ^{1}D heart-cut window), to 240 °C at 3 °C min^{-1}, held for 10 min. The APC2 pressure was maintained constant at 105 kPa. FID2 (280 °C) was connected to Deans 2 via a 0.5 m $\times$ 0.25 mm segment of uncoated column. The transfer line between GC2 and GC3 was maintained at 240 °C.

GC3 column (^{3}D) was an SLB-IL60i (custom-made ionic liquid), 30 m $\times$ 0.53 mm I.D. $\times$ 0.42 μm d_f, Merck Life Science (Merck KGaA, Darmstadt, Germany). Oven temperature program: 150 °C (45 min) to 240 °C at 3 °C min^{-1} (10 min). The APC3 pressure was maintained constant at 35 kPa. FID3 (280 °C) was connected to Deans 3 via a 0.6 m $\times$ 0.32 mm I.D. segment of uncoated column.

The detector gases (for FID1, 2, and 3) were H$_2$ at 50.0 mL min^{-1} and air at 400 mL min^{-1}; the sampling rate was 10 Hz. Data were collected by MDGCsolution software (Shimadzu, Kyoto, Japan). A heated (250 °C) aluminum block (11 cm high $\times$ 3 cm wide $\times$ 1.5 cm deep), located inside a modified GC injector port, designed and constructed in the lab, was used as the collector. Two liners in series were used inside the aluminum block: the lower one fixed, with the aim to drive the retention gap to an upper liner. The latter was removable and used to condensate the gas stream. Both the liners were sealed and held in position by means of two nuts: the lower one was to connect the retention gap using a suitable ferrule, while the upper one contained a holed rubber septum, fixing the upper liner used as the collector. The uncoated column was passed through the lower liner and protruded inside the collection liner for about 5 mm, as previously described by Sciarrone et al.[5] After analyte isolation, the collection tube was removed and flushed in a 2 mL vial with 100 μL of deuterated acetone. The resulting solution, containing the collected compound, was then analyzed by GC-MS and GC-FID for qualitative and quantitative purposes, respectively, prior to NMR experiments.

2.3. GC-FID and GC-MS experiments

A Shimadzu GC 2010 gas chromatograph equipped with an AOC-20i series autoinjector, and a GCMS-QP2010 Ultra system mass spectrometer (Shimadzu,

Kyoto, Japan) were used to evaluate the identity, recovery and degree of purity of the collected fraction. Three different 30 m × 0.25 mm I.D. capillary columns were used in order to check the purity of the fractions collected, namely: an SLB-5 ms (0.25 μm d_f) [silphenylene polymer, virtually equivalent to poly(5% diphenyl/95% methylsiloxane)], a Supelcowax-10 (100% polyethylene glycol) (0.25 μm d_f), and an SLB-IL60i (0.20 μm d_f), all from Merck Life Science (Merck KGaA, Darmstadt, Germany). The separation conditions were as follows: oven temperature program, 100 °C to 280 °C at 3 °C min^{-1}; split/splitless injector, 280 °C; injection mode, split (1 : 100 ratio); injection volume, 0.4 μL. The GC-FID conditions were as follows: inlet pressure, 110 kPa; carrier gas, He at a constant linear velocity of 30.0 cm s^{-1}. The FID (310 °C) gases were: H_2 at 40.0 mL min^{-1}; air at 400 mL min^{-1}; make up (N_2) at 40.0 mL min^{-1}; the sampling rate was 10 Hz. Data were acquired by the GCsolution software ver. 2.41 (Shimadzu, Kyoto, Japan). The GC-MS conditions were as follows: inlet pressure, 30.6 kPa; carrier gas, He at a constant linear velocity of 30 cm s^{-1}; source temperature, 200 °C; interface temperature, 250 °C; mass scan range, 40–400 m/z; scan speed, 10 Hz. Data were acquired by GCMSsolution software ver. 2.71 exploiting the FFNSC ver. 3 mass spectral database, for library matching with the additional support of a Linear Retention Indices (LRI) filter (Shimadzu, Kyoto, Japan).

2.4. NMR investigation

^{1}H and ^{13}C{^{1}H} NMR spectra were recorded on an Agilent Propulse 500 MHz spectrometer, equipped with an NMR probe operating at 499.74 (^{1}H) or 125.73 MHz (^{13}C{^{1}H}). After the collection by MDGC-Prep, the sample was dissolved in acetone-d_6 (CD_3COCD_3, 99.9% D, 99.5% purity, from Merck Life Science (Merck KGaA, Darmstadt, Germany)), poured into a 5 mm test tube and analysed after locking on the deuterium lock signal, searching for a good field homogeneity (shimming) and setting the frequency modulation (tuning). The ^{1}H saturation 90° pulse was calculated to be 8 μs at 61 dB of power level, while the protonic spectrum was recorded under 2 s acquisition time, 2 s scan delay and 16 scans. The complete and unambiguous assignment was achieved by processing homonuclear 2D-COSY, TOCSY and ROESY[16] experiments together with the heteronuclear[17] ^{13}C{1H}-HSQC and ^{13}C-HMBC experiments. Calibration was attained using the residual proton signal of the solvent (CD_3COCD_2H quintet at $\delta = 2.05$ ppm) and the ^{13}C solvent septuplets (at $\delta = 49.0$ ppm and $\delta = 29.84$) as internal standards.[18] Data were processed by the vNMRj software and by the ACD/Lab software package (version 2015), which was also used to check for confidence of the structural elucidation.

2.5. Condensed phase GC-FTIR experiments

The GC-FTIR profile was acquired by a DANI Master GC (Dani Instruments, Italy) coupled to a DiscovIR-GC (Spectra Analysis, Inc., Marlborough, UK) system. The GC parameters were as follows: injection volume, 1 μL at 280 °C in split mode (1 : 10); GC column, the same used in the GC-MS experiments; a constant linear velocity of 30 cm s^{-1}; temperature program, 50 °C to 280 °C at 5°C min^{-1} (held for 5 min). The FTIR spectra were acquired from 4000 to 700 cm^{-1}, with a resolution of 4 cm^{-1}. The column eluent was directly deposited on a cryogenically-cooled ZnSe sample disc. The rotation speed and temperature of the disc were 3 mm min^{-1} and −50 °C, respectively; both the transfer line and restrictor temperatures were 280 °C.

3. Results and discussion

In the first step of this research, the essential oil from the leaves of *Eugenia uniflora* L., also known as Brazilian Cherry, was subjected to GC-MS and GC-FID analyses to evaluate its qualitative and quantitative composition. Fig. 2 shows the MS profile acquired, in good accordance with previous data reported in the literature.[19] Thirty components were identified, as reported in Table 1, by means of a twin-filtered mass spectral library, *i.e.* by applying a spectral similarity ≥85% and a linear retention index range of ±5. After GC-MS analysis it was evident that the sample was mainly constituted of sesquiterpenes and their oxygenated components, and different unknown components showed up. A limited number of peaks could be identified, accounting for only 39% of the sample components, while for the rest of the molecules no match was obtained upon searching against a commercial MS database. Remarkably, among these, the most abundant peak (with a linear retention index of 1599), accounting for about 35% v/v (based on GC-FID data) of the oil, was not identified (differently from what was reported elsewhere in the literature[18]). Aiming to achieve a deeper knowledge of the essential oil composition, the sample was subjected to analysis by MDGC-Prep, in order to collect a suitable amount of the unknown molecule for structure elucidation.

3.1. Isolation and collection of the unknown component

When collecting molecules for structural investigation by spectroscopic techniques, such as mass spectrometry, nuclear magnetic resonance and infrared analyses, it is of fundamental importance to isolate the largest sample amount (milligrams), and with a high degree of purity. Obtaining suitable amounts of highly pure compounds may be a challenging task when the collection step is

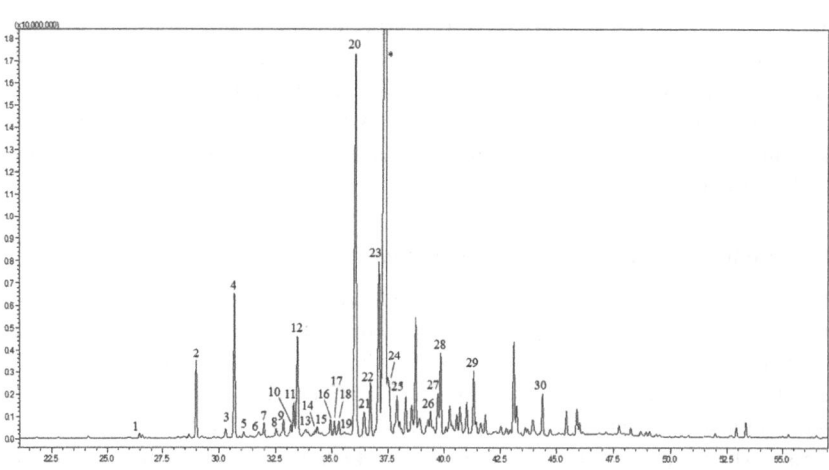

Fig. 2 GC-MS chromatogram of the *Eugenia uniflora* L. essential oil. For peak identification, refer to Table 1.

Table 1 Quali–quantitative profile of *Eugenia uniflora* L. essential oil. *: tentatively identified

ID	Compound name	MS %	LRI exp.	LRI lib.	Area %
1	Elemene(delta-)	87	1335	1335	0.09
2	Elemene(beta-)	95	1385	1390	1.50
3	Caryophyllene((E)-)	95	1424	1424	0.22
4	Elemene(gamma-)	92	1433	1432	2.86
5	Aromadendrene	97	1443	1438	0.16
6	Caryophyllene(9-*epi*-(E)-)	97	1465	1464	0.31
7	Selina-4,11-diene	90	1477	1476	0.23
8	Germacrene D	95	1485	1480	0.39
9	Amorphene(alpha-)	91	1484	1482	0.08
10	Selinene(beta-)	96	1493	1492	0.36
11	Viridiflorene	97	1496	1491	0.76
12	Bicyclogermacrene	95	1500	1497	2.53
13	Germacrene A	86	1512	1511	0.43
14	Cadinene(gamma-)	87	1517	1512	0.06
15	Cadinene(delta-)	94	1523	1518	0.30
16	Selina-4(15),7(11)-diene*	89	1538	1540	0.49
17	Selina-4(15),7(11)-diene*	93	1542	1540	0.38
18	Selina-3,7(11)-diene	92	1547	1546	0.43
19	Elemol(alpha-)	88	1551	1546	0.15
20	Germacrene B	96	1562	1557	10.29
21	Ledol	86	1575	1574	0.81
22	Spathulenol	94	1580	1576	1.37
23	Viridiflorol	90	1592	1594	4.92
	Unknown		1599		35.54
24	Cubeban-11-ol	86	1602	1599	2.20
25	Rosifoliol	95	1613	1609	1.36
26	T-Muurolol	93	1650	1645	0.67
27	Cadin-4-en-10-ol	93	1662	1659	1.18
28	Intermedeol	92	1665	1668	1.83
29	Juniper camphor	95	1700	1696	1.63
30	Cedren-13-ol acetate(8-)	89	1790	1790	1.02

performed right after the GC separation. Small amounts (typically nano–micro-grams) of a sample are usually injected on conventional 0.25 mm I.D. high efficiency columns, not to impair the desirable resolution level. Exploiting this column type is possible to achieve highly purified peaks before the collection step, but, on the other hand, due to the low sample capacity, this approach usually requires tens or hundreds of injections to allow the isolation of an amount close to one milligram. This inevitably leads to very long collection times. Moreover, considering the necessary procedures to remove the condensed fraction from the collector, a large number of collections would cause sample contaminations and losses. Finally, it has to be considered that a certain solvent volume has to be utilized to remove the fraction from the tube, ideally after each collection, causing high dilution of the fraction when a large number of washes are necessary, further requiring a solvent evaporation step that again could expose the fraction to sample loss. A possible alternative could be the use of wide-bore columns, typically 0.32–0.53 mm I.D. Such columns allow the injection of larger sample amounts (micro–milligrams), thanks to the superior sample capacity, thus making this type of column the most suitable for a preparative system. Unfortunately, the cost for increased sample capacity is halving of the efficiency, and thus it is difficult to obtain pure chromatographic bands for the collection of specific fractions, especially in the case of complex samples. Multidimensional chromatography in the heart-cut mode, when the attention is focused on specific sample fractions, has proven to be very useful for improving the separation capability of a system. An effective approach recently developed was based on the use of a three-dimensional GC system, equipped with wide-bore columns with complementary selectivity. Such an approach conjugated the possibility of injecting very large amounts of sample (milligrams) on a wide-bore column, with achieving highly pure fractions through multidimensional GC.[5,7,8,11-13] As such an MDGC-Prep system allows the collection of milligrams of highly-pure chemicals (>90%) in a reduced run time and number of runs, the same approach was selected in the present application.

3.2. MDGC-Prep analysis

The MDGC system employed was based on the coupling of three distinct stationary phases, in detail: a silphenylene polymer, virtually equivalent in polarity to poly(5% diphenyl/95% methylsiloxane) as ^{1}D, a 100% polyethylene glycol as ^{2}D, and a medium-polarity ionic liquid-based column as ^{3}D. The configuration of the Deans switch devices, equipped with a three-restrictor system, is described in Fig. 1. Thanks to the presence of a fixed restrictor (R_1) on the Deans switch inlet side, the back pressure at the column outlet of each dimension is maintained constant during the heart-cut stage, avoiding retention time shifts.[4] Another restrictor (R_2), generating a larger pressure drop with respect to R_1, allows for the heart-cut of selected chromatographic bands by switching of the valve (V). During monodimensional analysis, without any heart-cut window selected (stand-by), no restriction is present on the APC flow path relative to the second dimension, and a slightly higher pressure than FID1 is generated at the second column inlet. R_1 generates a constant pressure drop, driving the diversion of the eluted fraction to the FID1, as in a conventional GC-FID run. When a heart-cut has to be performed, the valve is switched by the software according to the selected heart-cut time, and a slightly lower pressure with respect to FID1 is generated at the head of the second column. In fact, the higher pressure drop generated by R_2, having a higher resistance with respect to R_1, results in the diversion of the fraction to the second column. The same functioning is operated for Deans 2 and Deans 3. In a first stand-by application, 2 μL of neat E.O. were injected in direct mode, theoretically corresponding to about 0.7 mg of the unidentified peak at LRI 1599. During optimization of the method, a series of stand-by (the eluate diverted to the FID) and cut analyses (the eluate directed to the next column/collector) were performed, in order to select the proper heart-cut windows for each chromatographic dimension. Fig. 3 shows the stand-by and cut chromatograms obtained on the three different stationary phases. The resulting first dimension chromatogram, achieved on the apolar stationary phase, is shown in Fig. 3a. From a comparison of the GC-MS profile (Fig. 2) acquired on the high efficiency 0.25 mm I.D. column with that acquired on the 0.53 mm I.D. column used in ^{1}D of the MDGC-Prep system, it is clear that overloading as a result of the large amount of the neat sample (2 μL) injected, greatly exceeds the column

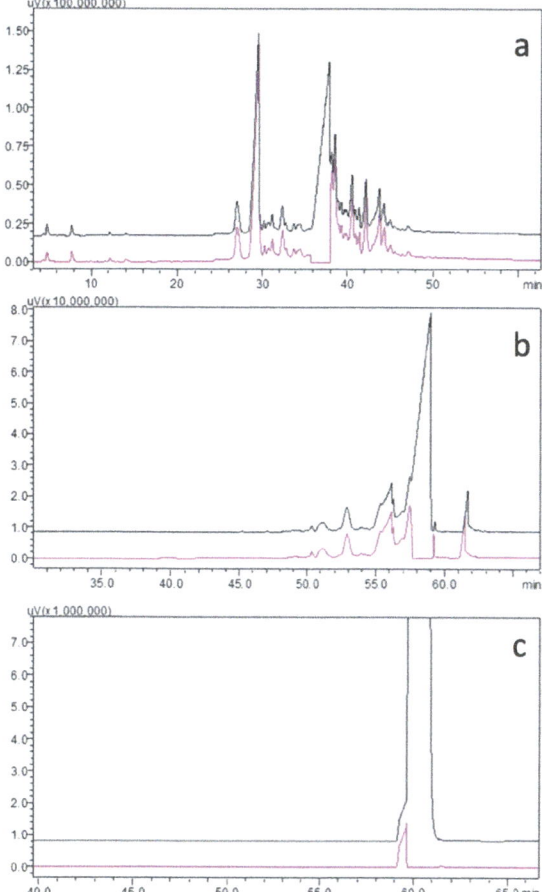

Fig. 3 MDGC-Prep stand-by (black trace) and cut (pink trace) chromatograms relative to the first (a), second (b) and third (c) dimension.

sample capacity. After the first dimension stand-by (black trace), a heart-cut window from 35.5 to 38.0 min was selected in a subsequent run, performed in cut mode (pink trace). As can be appreciated in Fig. 3b, the second dimension stand-by analysis on the PEG stationary phase (black trace) shows a high degree of coelutions around the peak of interest transferred from the first column. Also, in this case, it is evident that the collection of this peak after monodimensional separation (GC-Prep) would lead to an impure component, since it accounted for only 60% of the fraction. With the aim of further purifying the band of interest before the collection, a second heart-cut was selected (pink trace) from 57.4 to 59.0 min; the ^{2}D purification on the PEG column greatly increased the fraction purity. The third purification, operated on the IL-based column, and characterized by a similar polarity to the PEG column but with a different selectivity[5] (Fig. 3c), was effective. A further 5% of impurities were still present (black trace), requiring a further heart-cut step (pink trace) from 59.6 to 61.5 min. After the three chromatographic dimensions, once a highly pure fraction was obtained, the

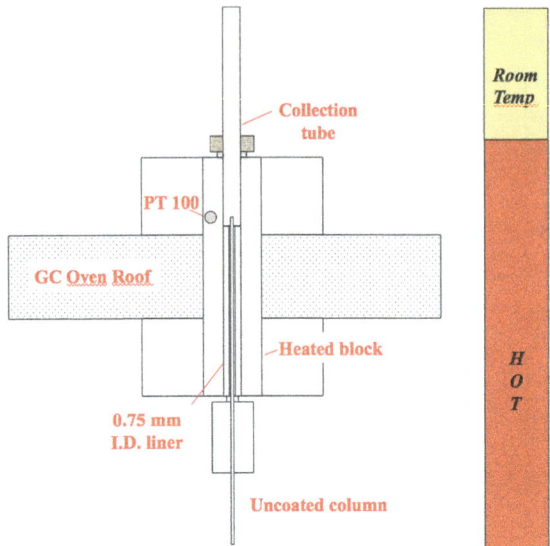

Fig. 4 Lab-made collection system.

third heart-cut diverted the band to the collector, located within the third GC oven roof (Fig. 4). The lab-made collection system included a heated aluminium block, equipped with a PT 100 sensor, directly controlled by the third GC acquisition software. A 7 cm × 0.75 mm I.D. liner was located in the lower part of the block, spanned by the retention gap coming from the third Deans switch device. The last 5 mm of the retention gap protruded inside the collection tube, comprised of an empty straight 3.4 mm I.D. liner, positioned above the 0.75 mm I.D. liner. Due to the shorter length of the latter (75 mm *vs.* 99 mm), 2.5 cm of the collection tube was subjected to the heating of the aluminium block, and the remaining length was exposed to room temperature. As a result, the collection tube received, in its lower part, the gas stream exiting the retention gap in a vapor phase, and subsequent condensation occurred on the glass walls once it reached the room temperature zone. Once the fraction was trapped, the collection tube was immediately removed and flushed in a 2 mL vial with 100 μL of deuterated acetone, compatible with the subsequent NMR analysis, and this was injected in a GC-FID system in order to evaluate the purity of the collected fraction. To ensure complete recovery of the fraction from the collection tube, the latter was flushed again with the same solvent and the obtained solution was injected in a GC-FID system. The resulting chromatogram confirmed the complete removal of the fraction, since no peaks showed up (data not shown). A total of 5 MDGC-Prep collections were performed prior to the NMR analysis for structure elucidation assessment. Afterwards, in order to evaluate the absolute amount collected, 100 μL of the solution obtained after the 5 collections (500 μL) were spiked with an equal volume of a 1 mg mL^{-1} *n*-nonane solution, used as an internal standard, and this was injected in a GC-FID system. The total amount collected was finally calculated referring to a calibration curve built using α-bisabolol, as a representative compound of the oxygenated sesquiterpene family, *versus n*-nonane as the internal standard. A total of 3.3 mg were collected in only 6 hours, with an average collection recovery of about 90% and a purity of 98%. The fraction was then analysed in a GC-MS system: the EI-MS spectrum acquired, relative to the component collected, is reported in the ESI.†

3.3. Structural elucidation

The common strategy for structural characterization and conformational analysis by NMR spectroscopy was successful because of the specific NMR data crossing.[20,21] Briefly, after running [1]D NMR experiments able to detect the proton and ^{13}C resonances of the unknown compound, [2]D HSQC-DEPT experiments were conducted with the aim of connecting the ^{1}H resonances with their ^{13}C parent resonance, where the sign of the peaks indicates the number of attached H atoms per C atom. This analysis evidenced the presence of three vinylic CH$_2$ terminals, one methylene moiety (aliphatic CH$_2$), three methyl groups (CH$_3$) and two olefinic CH groups. In order to clarify the specific connections through the bonds among these chemical groups, a homonuclear 2D-COSY experiment was performed. This gave evidence for the connection between H atoms separated by less than 3 (sometimes 4) bonds, whereas the heteronuclear 2D-HMBC experiment was definitely crucial for the "long-range" connection among ^{13}C resonances and proton resonances coming from nuclei which were separated by 2, 3 or 4 bonds. The structural elucidation reported in Fig. 5, performed on the basis of these data, was further confirmed by a homonuclear 2D-NOESY experiment, which gives evidence for the connections between neighbouring protons, regardless of the specific bonding connections. These through-the-space interactions represent the best way to infer the configurational and conformational arrangement of molecules in solution: in this case, they were crucial in order to define the relative stereochemistry of the two asymmetric (C5 and C6) carbons. Specifically, after supporting the overall assignment, the key "through-the-space" interaction between H4a (namely above the plane) of the methylene endocyclic group and the methyl groups labelled as 12 and 15 (Fig. 5) showed that these chemical groups were clearly on the same side over the mean cyclohexen-one plane. The existence of such a configuration was also strongly supported by the further NOESY vicinity recorded by the vinylic 13-CH nucleus and the endocyclic

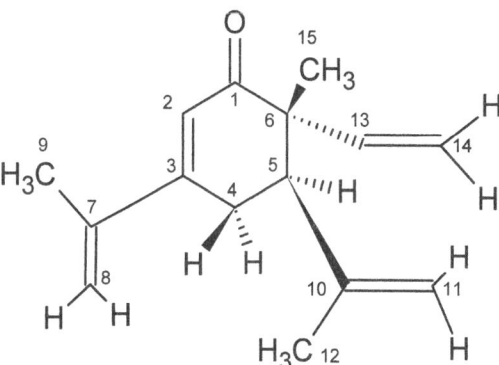

Fig. 5 Structure of the unknown molecule elucidated by NMR spectroscopy after isolation by MDGC-Prep.

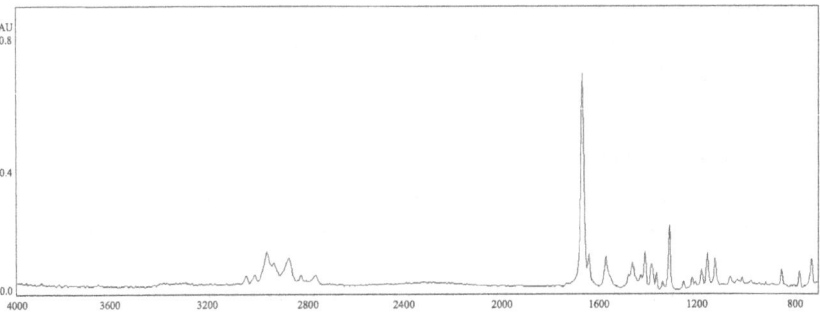

Fig. 6   GC-FTIR spectrum of the unknown molecule.

5b-CH one, both sharing the opposite (below) side of the same mentioned mean plane; again, this was supported by the fact that the terminal 11-CH in the Z position (on the other side with respect to 12-CH$_3$) was close to both protons labelled as 5b and 4b (below the plane). All these evidences were consistent with a 6-ethenyl-6-methyl-3,5-di(prop-1-en-2-yl)cyclohex-2-en-1-one structure (Fig. 5). Because of the presence of two different asymmetric C atoms (namely 5 C and 6 C), four geometric isomers could exist, divided into two enantiomeric couples of diastereomers. Our structural evidences restricted to the same assignment of Chan–Ingold–Prelog conventional configurations so that just two enantiomeric compounds {(5R,6R) and (5S,6S)} could populate the analysed sample. As these compounds presented the same physical properties, it was not possible to understand their mutual relative presence in the analysed sample. For an extended assignment and drawings referring to the structural connections, the reader is directed to the ESI.†

3.4.   Condensed phase GC-FTIR analysis

A condensed-phase FTIR spectrum of the collected molecule was acquired, exploiting a direct micro deposition of the column eluent after GC separation on a cryogenically-cooled ZnSe sample disc. The band, eluted after a GC run on an apolar stationary phase, was first deposited in a concentrated spot onto a rotating IR-transparent disc, by means of a cryogenic temperature control, allowing the deposition on a minimized area. This approach produced a thicker sample layer and therefore a higher intensity in the absorbance spectrum. Afterwards, the disc was rotated, exposing the spot to the infrared beam, and the collected GC-FTIR spectrum is shown in Fig. 6. The most intense signals were detected in the functional groups zone (4000–1350 cm^{-1}): the strongest one was related to the C=C bonds at 1665 cm^{-1}. Stretching signals of C–H, C=C and C–C bonds were detected between 2754 and 3040 cm^{-1}, while bending signals relative to the CH$_2$ and CH$_3$ groups were detected around 1400 cm^{-1}. A complete list of the peaks recorded can be found in the ESI.†

4.   Conclusion

In the present research, the main constituent of *Eugenia uniflora* L. leaves essential oil was isolated from the neat oil in a reasonable collection time and without the need for any sample preparation. The MDGC-Prep configuration allowed for the collection of about 3 mg of pure component, exploiting a lab-constructed collector placed at the outlet of the ^{3}D column. The compound molecular structure was elucidated by means of NMR analysis, and corresponded to 6-ethenyl-6-methyl-3,5-di(prop-1-en-2-yl)cyclohex-2-en-1-one. In addition, the condensed phase FTIR spectra and electron impact mass spectra were acquired for confirmation. The productivity of the system has been improved in terms of larger sample capacity (achieved by using wide-bore columns) and higher separation efficiency (achieved by using multidimensional GC), providing highly pure discrete amounts of the component of interest collected per run. The system can be regarded as a viable alternative to the classical fractional distillation method for the collection of pure components that are not available commercially, and whose content in the matrix is regulated. The characterization of this unknown compound has further expanded the knowledge on the volatile composition of the *Eugenia uniflora* L. leaves essential oil and has made it possible to proceed with performing further biological tests on the isolated molecule.

Conflicts of interest

The authors declare no conflict of interest.

Acknowledgements

The authors gratefully acknowledge Shimadzu Corporation and Merck Life Science.

References

1 D. Sciarrone, S. Pantò, C. Ragonese, P. Dugo and L. Mondello, *TrAC, Trends Anal. Chem.*, 2015, **71**, 65–73.
2 H.-L. Zuo, F.-Q. Yang, W.-H. Huang and Z.-N. Xia, *J. Chromatogr. Sci.*, 2013, **51**, 704–715.
3 E. F. Barry, in *Modern Practice of Gas Chromatography*, ed. R. L. Grob and E. F. Barry, John Wiley & Sons, Inc., Hoboken, New Jersey, 4th edn, 2004, ch. 3, pp. 130–148.
4 P. Q. Tranchida, D. Sciarrone, P. Dugo and L. Mondello, *Anal. Chim. Acta*, 2012, **716**, 66–75.
5 D. Sciarrone, S. Pantò, C. Ragonese, P. Q. Tranchida, P. Dugo and L. Mondello, *Anal. Chem.*, 2012, **84**(16), 7092–7098.
6 C. Ruhle, G. T. Eyres, S. Urban, J.-P. Dufour, P. D. Morrison and P. J. Marriott, *J. Chromatogr. A*, 2009, **1216**, 5740–5747.
7 D. Sciarrone, S. Pantò, A. Rotondo, L. Tedone, P. Q. Tranchida, P. Dugo and L. Mondello, *Anal. Chim. Acta*, 2013, **785**, 119–125.
8 D. Sciarrone, D. Giuffrida, A. Rotondo, G. Micalizzi, M. Zoccali, S. Pantò, P. Donato, R. Goncalves Rodrigues-das-Dores and L. Mondello, *J. Chromatogr. A*, 2017, **1524**, 246–253.
9 G. I. Ball, L. Xu, A. P. Mc Nichol and L. I. Aluwihare, *J. Chromatogr. A*, 2012, **1220**, 122–131.
10 N. Ochiai and K. J. Sasamoto, *J. Chromatogr. A*, 2011, **1218**, 3180–3185.

11 S. Pantò, D. Sciarrone, M. Maimone, C. Ragonese, S. Giofrè, P. Donato, S. Farnetti and L. Mondello, *J. Chromatogr. A*, 2015, **1417**, 96–103.

12 D. Sciarrone, S. Pantò, P. Donato and L. Mondello, *J. Chromatogr. A*, 2016, **1475**, 80–85.

13 D. Sciarrone, S. Pantò, P. Q. Tranchida, P. Dugo and L. Mondello, *Anal. Chem.*, 2014, **86**, 4295–4301.

14 F. N. Victoria, E. J. Lenardão, L. Savegnago, G. Perin, R. G. Jacob, D. Alves, W. Padilha da Silva, A. de Souza da Motta and P. da Silva Nascente, *Food Chem. Toxicol.*, 2012, **50**, 2668–2674.

15 A. C. L. Amorim, C. K. F. Lima, A. M. C. Hovell, A. L. P. Miranda and C. M. Rezende, *Phytomedicine*, 2009, **16**, 923–928.

16 A. E. Derome, *Modern NMR techniques for chemistry research*, Bergamon Press, Exeter, 2013.

17 W. Willker, D. Leibfritz, R. Kerssebaum and W. Bermel, *Magn. Reson. Chem.*, 1993, **31**(3), 287–292.

18 H. E. Gottlieb, V. Kotlyar and A. Nudelman, *J. Org. Chem.*, 1997, **62**(21), 7512–7515.

19 R. M. Melo, V. F. S. Corrêa, A. C. L. Amorim, A. L. P. Miranda and C. M. Rezende, *J. Braz. Chem. Soc.*, 2007, **18**(1), 179–183.

20 A. Rotondo, R. Ettari, M. Zappalà, C. De Micheli and E. Rotondo, *J. Mol. Struct.*, 2014, **1076**, 337–343.

21 A. Rotondo, R. Ettari, S. Grasso and M. Zappalà, *Struct. Chem.*, 2015, **26**(4), 943–950.

Faraday Discussions

PAPER

On the benefits of using multivariate analysis in mass spectrometric studies of combustion-generated aerosols†

D. Duca*, [ID] [a] C. Irimiea, [ID] [b] A. Faccinetto, [ID] [c] J. A. Noble, [ID] ‡[a]
M. Vojkovic, [ID] [a] Y. Carpentier, [ID] [a] I. K. Ortega, [ID] [b] C. Pirim [ID] [a]
and C. Focsa [ID] [a]

Received 10th December 2018, Accepted 5th March 2019
DOI: 10.1039/c8fd00238j

The intricate chemistry of the carbonaceous particle surface layer (which drives their reactivity, environmental and health impacts) results in complex mass spectra. In this respect, detailed molecular-level analysis of combustion emissions may be challenging even with high-resolution mass spectrometry. Building on a recently proposed comprehensive methodology (encompassing all stages from sampling to data reduction), we propose herein a comparative analysis of soot particles produced by three different sources: a miniCAST standard generator, a laboratory diffusion flame and a single cylinder internal combustion engine. The surface composition is probed by either laser or secondary ion mass spectrometry. Two examples of multivariate analysis, Principal component analysis and hierarchical clustering analysis proved their efficiency in both identifying general trends and evidencing subtle differences that otherwise would remain unnoticed in the plethora of data generated during mass spectrometric analyses. Chemical information extracted from these multivariate statistical procedures contributes to a better understanding of fundamental combustion processes and also opens to practical applications such as the tracing of engine emissions.

1 Introduction

Multivariate analysis (MVA) methods are powerful tools for unravelling trends in complex databases. They have been successfully applied to identify drug metabolites in biological fluids,[1] to evaluate profiles of volatile compounds present in mainstream tobacco smoke,[2] and, to assess surface water quality.[3]

[a]Univ. Lille, CNRS, UMR 8523, PhLAM – Laboratoire de Physique des Lasers Atomes et Molécules, F-59000 Lille, France. E-mail: dumitru.duca@univ-lille.fr

[b]ONERA – The French Aerospace Laboratory, F-91123 Palaiseau, France

[c]Univ. Lille, CNRS, UMR 8522, PC2A – Laboratoire de Physico-Chimie des Processus de Combustion de l'Atmosphère, F-59000 Lille, France

† Electronic supplementary information (ESI) available. See DOI: 10.1039/c8fd00238j

‡ Current address: CNRS, Aix Marseille Université, PIIM, UMR 7345, 13397 Marseille cedex, France.

Among the MVA methods commonly used[4] are principal component analysis (PCA) and hierarchical clustering analysis (HCA). The former is used to reveal hidden patterns in databases, by emphasising the variance between samples and thus highlighting their differences and similarities,[5] whereas the latter searches for patterns in a database by grouping the observables into distinct clusters. Their capability to distinguish various complex samples, as exemplified for a while now in the field of biology, has recently led to their consideration for unravelling the chemical composition of multifaceted samples of environmental interest.

Atmospheric aerosols are airborne particles consisting of an intricate mixture of chemical constituents whose nature varies greatly depending upon their emission source and evolution within the atmosphere. Carbonaceous particles account for a significant fraction of atmospheric particulate matter in urban areas (typically 30–50% by mass[6–8]). They are mainly formed of soot, *i.e.* particles generated by the incomplete combustion of hydrocarbon-based fuels or biomass. Accordingly, soot particles possess a multitude of chemical compounds derived from various sources (remnant of fuels, combustion and/or post-oxidation products, *etc.*) that may have been further transformed (aged) by the time they are analysed due to their continuous interaction with environmental elements (solar rays, water molecules, pollutants, *etc.*). Soot particles are therefore considered as complex mixtures that often need a concerted analytical scheme to be fully resolved.

Mass spectrometry (MS) based techniques have significantly contributed to better understanding soot chemistry over the years. They are generally robust techniques that do not require extensive sample preparation, and are hence preferred for the analysis of such complex samples. Furthermore, the amount of particulate matter required to perform MS analysis is relatively small. MS based techniques mostly differ by the way the ions transferred to the mass spectrometer are created (*e.g.* soot particle aerosol mass spectrometry (SP-AMS),[9] two-step laser mass spectrometry (L2MS),[10] and time-of-flight secondary ion mass spectrometry (ToF-SIMS)[11,12]), which often condition their specificity to provide information on either bulk or surface chemical composition. Ultra high resolution mass analyzers such as Orbitrap, Fourier transform ion cyclotron resonance (FT-ICR) and high resolution quadrupole time of flight MS can reach a resolving power higher than 90 000.[13,14] These techniques were developed mainly for proteomics and pharmaceutical analyses, but lately their application has been extended to many other fields, including starting being used and adapted to atmospheric aerosols.[15,16] However, ultra high resolution mass spectrometry is still very rarely applied to the analysis of combustion products, with only a few examples to date.[17] Ultra high resolution mass analyzers are powerful analytical tools, however they still need validation of the sampling protocols. For instance, the sample transfer into the instrument is based on nanospray desorption electrospray using a polar solvent for Orbitrap, followed more recently by laser desorption for FT-ICR and atmospheric pressure chemical ionization (APCI) for APCI-Orbitrap.[13,16,17] Let us also emphasize that in directed energy (laser and ion beam) desorption methods, besides the analyzer performances, the condensed-gas phase transfer itself plays a critical role in the maximum achievable mass resolution and the total number of detected signals, through, *e.g.*, the sample/substrate roughness or conductive properties. We therefore stress the need for a thorough evaluation (and optimization) of the entire analysis chain, from sample collection/deposition on suitable substrates, to sample transfer/ionization into gas phase, ion mass separation and detection, and finally powerful data treatment and interpretation.[18,19]

The mass spectra of soot particles can be very complex, featuring hundreds and even thousands of mass peaks, which quickly renders the interpretation of mass spectra difficult and therefore limits the potentiality of MS to resolve complex mixtures. Accordingly, resolving sample complexity in MS databases is currently tackled using two main approaches. The first is based on the identification of marker species, *i.e.* compounds that are directly linked to a source/process and that can thus be considered as their fingerprints, while the second approach relies on statistical methods. In particular, the use of MVA methods in conjunction with MS is a creative combination to exploit all of the information given by a multitude of peaks within a great variety of sample sets. Both approaches are widely used in the analysis of mass spectra obtained with aerosol mass spectrometers (AMS),[20–22] proton transfer reaction mass spectrometers (PTR-MS),[23,24] and laser-based MS techniques.[19,25,26] Discrimination using marker species was applied to samples of various sources, proving its effectiveness when comparing soot emitted from wood combustion,[20,27] on-road vehicles,[25] aircrafts,[22–24,28,29] ships[30] or other ambient aerosols.[21] However, since some marker species may not remain stable over the aerosols' life span, especially upon atmospheric ageing,[6] this method may misdirect with regards to the origin of samples *a priori* unknown. To circumvent this limitation, MVA approaches are chosen, as they can discriminate samples regardless of their provenance or evolution. Therefore, MVA can uncover trends and features even in samples of unknown/mixed origins,[28,31] which is particularly interesting when analysing natural aerosols.

In constant interaction with their surroundings, aerosol surfaces drive their overall reactivity, and therefore, set their evolution path within the atmosphere (sedimentation, formation of secondary organic aerosols, nucleation, *etc.*). It is hence imperative to uncover their complex surface composition in order to assess their impact on both human health and the environment.[32,33] For example, some polycyclic aromatic hydrocarbons (PAHs), often found adsorbed on the surface of soot particles, are known to be toxic and to have mutagenic effects.[34,35] In addition, the chemical composition of aerosol surfaces determines their hygroscopicity[36] and therefore their ability to act as condensation nuclei, potentially influencing climate forcing, cloud cover and precipitations.

Our group has been addressing this issue of untangling surface chemical compositions of field-collected or laboratory-generated combustion aerosols for over a decade.[10,18,19,26,29,30,37–40] We recently described an original and comprehensive experimental methodology[18] that we later implemented in combining statistical-based approaches with compound classification techniques.[19] This latter systematic study by Irimiea and coworkers[19] was undertaken to characterise over 100 samples collected from different flames. In this work, we developed a comprehensive protocol that allowed significant progress towards the fundamental understanding of soot nucleation and growth. Laboratory flames or standard soot generators are often used to produce soot particles with similar physico-chemical properties to the ones produced by "real world" combustion sources.[41] Laboratory soot particles offer the advantages of a reproducible, easy-access and low-cost production, which is of great importance when testing the robustness of a protocol. Therefore, this necessary step is of paramount importance for further refinements in field-collected combustion-generated particle analyses.

2 Experimental

In this section, the choice of the combustion conditions, the sampling approach and the experimental techniques used to characterised the samples are described. In particular, L2MS and SIMS are used in parallel to obtain information on the chemical composition of combustion generated aerosols.

2.1 Soot samples

Soot samples are generated in different combustion conditions (fuel, burner and sampling method) in order to test the ability of our data treatment protocols to reveal differences and similarities between samples. The sampling procedure, including the substrate choice and its preparation, is optimised according to our previous experience.[18] In particular, the sample-substrate reactivity can lead to the formation of a large number of byproducts that clutter the mass spectrum and make the identification of individual compounds much more difficult. A short description of all analysed samples (summarised in Table 1) is given below. The following soot samples have been used:

• Soot produced by a miniCAST generator (5201c) from Jing Ltd., which is currently proposed as a means of obtaining "standard" soot easily comparable to other studies.[41-43] The main difference between the miniCAST set points is the oxidation flow ($1.50 \rightarrow 1.15 \rightarrow 1.00$ L min^{-1}) resulting in three different combustion conditions ($C_1 \rightarrow C_2 \rightarrow C_3$).[41-43] The hereby generated particles are subsequently deposited on quartz fibre filters.

• Soot produced by laboratory turbulent diffusion flames supplied with two different liquid fuels: diesel (D1–5) and kerosene (K1–5). Soot particles are sampled from the flame at different heights above the burner (HAB) and deposited by impaction on Si wafers. Sampling at various HAB is a means of investigating soot particles of different maturity.[38]

• Soot produced by a gasoline single cylinder internal combustion engine (ICE). Operating conditions of this engine (e.g. injection and ignition crank angle, applied load) could be easily changed, thus allowing exhausts sampling at various working regimes. The following operating points were used:

– Normal engine operation, i.e. engine optimised in terms of high efficiency and low particle emissions, with medium (GOM) and high (GOH) applied loads, which simulate different driving regimes;

– Malfunction simulation with a medium load applied: low air/fuel ratio resulting in a high-sooting regime (GEF) and an addition of oil to the combustion chamber (GEO).

Soot particles are sampled using a cascade impactor (NanoMOUDI) to allow for size selection during sampling, and deposited on Al foils. We analysed the particles collected on the last five stages, having diameters in the range 10–180 nm (Table 1).

Off-line analysis of soot particles requires a careful choice of the deposition substrate, not only to minimise the risk of contaminating the samples, but also to

Table 1 Soot samples used to put in evidence the proposed methodology

Name	Fuel	Source	Substrate	Description	Analysing technique
C_1	Propane	miniCAST	Quartz fibre filters	1.5 L min^{-1} oxidation flow	L2MS+
C_2				1.15 L min^{-1} oxidation flow	
C_3				1.0 L min^{-1} oxidation flow	
D1	Diesel	Diffusion flame	Si wafer	HAB = 6 mm	SIMS±
D2				HAB = 12 mm	
D3				HAB = 14 mm	
D4				HAB = 18 mm	
D5				HAB = 24 mm	
K1	Kerosene	Diffusion flame	Si wafer	HAB = 6 mm	SIMS±
K2				HAB = 12 mm	
K3				HAB = 14 mm	
K4				HAB = 18 mm	
K5				HAB = 24 mm	
GOM1	Gasoline	ICE, optimal conditions, medium load	Al foil	Ø100–180 nm	SIMS±
GOM2				Ø56–100 nm	
GOM3				Ø32–56 nm	
GOM4				Ø18–32 nm	
GOM5				Ø10–18 nm	
GOH1	Gasoline	ICE, optimal conditions, high load	Al foil	Ø100–180 nm	SIMS±
GOH2				Ø56–100 nm	
GOH3				Ø32–56 nm	
GOH4				Ø18–32 nm	
GEF1	Gasoline	ICE, low air/ fuel ratio	Al foil	Ø100–180 nm	SIMS±
GEF2				Ø56–100 nm	
GEF3				Ø32–56 nm	
GEF4				Ø18–32 nm	
GEO1	Gasoline	ICE, addition of oil	Al foil	Ø100–180 nm	SIMS±
GEO2				Ø56–100 nm	
GEO3				Ø32–56 nm	
GEO4				Ø18–32 nm	

ensure that a high mass resolution can be achieved. In particular, among other factors, the mass resolution is directly linked to the surface roughness of the substrate, and can be maximised by depositing the samples on ultra-flat surfaces such as Si or Ti wafers. Furthermore, the sample-substrate reactivity can lead to the formation of reaction byproducts that may heavily interfere with the assignment of sample-specific signals. Therefore, the careful characterization/choice of the deposition substrate is mandatory and the comprehensive identification of its possible reactivity byproducts is necessary for a valid analytical protocol.[18,19] Regardless of its nature, the substrate should undergo a series of preparation steps before it can be used to collect particulate matter.

2.2 Two-step laser mass spectrometry (L2MS)

This laser-based MS technique has been extensively used by our group to characterise the chemical composition of combustion byproducts during the last decade.[10,18,26,29,30,37–39] The main advantages of L2MS are its high sensitivity and selectivity with regards to specific classes of compounds thanks to resonant ionisation processes that can be tuned to reach for instance the sub-fmol limit for the detection of PAHs.[10,37] In addition, the controlled laser desorption process ensures a soft removal of molecules adsorbed on the particle surface (typically sub-monolayer regime), and thus avoids/limits either their fragmentation or the in-depth damaging of the underlying carbon matrix.[37] This qualifies L2MS as a surface-sensitive analysis technique, comparable in limit of detection ($\sim 10^{-6}$ monolayers) with static-mode secondary ion mass spectrometry (SIMS, see below), but with much lower analyte fragmentation. However, our previous L2MS studies were limited by a mass resolution of $m/\Delta m \sim 1000$, significantly lower than the one achievable in SIMS (up to $m/\Delta m \sim 10\,000$, depending on the deposition substrate[18,19]). In the current work, we take the benefit of the recent implementation of a new mass spectrometer (Fasmatech S & T) that combines ion cooling, Radio Frequency (RF) guiding and a Time of Flight (ToF) analyser to reach a mass resolution of about $m/\Delta m \sim 15\,000$. In this new experimental setup, the sample, placed under vacuum (10^{-8} mbar residual pressure), is irradiated at a 30° angle of incidence by a frequency doubled Nd : YAG laser beam (Quantel Brilliant, $\lambda = 532$ nm, 4 ns pulse duration, ~ 50 mJ cm^{-2} fluence, 10 Hz repetition rate) focused to a 0.3 mm^2 spot on the surface. The desorbed compounds form a gas plume expanding in the vacuum normally to the sample surface, and are ionised by an orthogonal UV laser beam (Quantel Brilliant, $\lambda_i = 266$ nm, 4 ns pulse duration, 10 Hz repetition rate, ~ 0.3 J cm^{-2} fluence). At this ionisation wavelength, a high sensitivity is achieved for PAHs through a resonance enhanced multiphoton ionisation process 1 + 1 REMPI.[44–46] Care must be taken on the coupling of the desorption and ionisation steps in this laser-based MS technique.[47–49] Moreover, by changing the ionisation wavelength, one can target different classes of compounds. The generated ions are then RF-guided to a He collision cell for thermalisation and subsequently mass analysed in a time of flight mass spectrometer (ToF-MS).

2.3 Secondary ion mass spectrometry (SIMS)

In addition, the samples are characterised using a commercial IONTOF ToF-SIMS[5] secondary ion mass spectrometer with a maximum resolving power of $m/\Delta m \sim 10\,000$. In short, samples are placed in the analysis chamber with a residual pressure of $\sim 10^{-7}$ mbar. The surface of the sample is bombarded using a 25 keV $(Bi_3)^+$ ion beam with a current of 0.3 pA in static mode. A small fraction of the ejected atoms/molecules are ionised (secondary ions) and can thus be analysed using a time-of-flight tube (V mode). Mass spectra are recorded in both positive and negative polarities, to obtain the maximum amount of information on the sample.[18,19]

3 Data analysis methodology and examples of applications

The data presented below is analysed following an approach structured in three main points that include: mass defect analysis for the identification of unknown compounds (Section 3.1), multivariate analysis for the reduction of the number of dimensions of the dataset (Section 3.2) and eventually mass peak grouping to uncover hidden trends and highlight correlations between different classes of compounds (Section 3.3). This section details the proposed data treatment protocol. The mass spectra of the previously described samples have been used to demonstrate its advantages, including its universal character (the ability to be used with mass spectra of various samples, obtained with different experimental techniques). Mass spectra were recorded with either L2MS or SIMS in multiple regions of the sample surface, to ensure the consistency of the method and to build a database allowing a more advanced statistical analysis. Once all of the peaks coming from the substrate are removed, the data is ready to be processed.

3.1 Mass defect analysis

Mass defect analysis is used to assign a molecular formula to the recorded accurate mass.[50,51] By convention, the mass defect of ^{12}C is defined as zero, therefore the mass defect of every other existing isotope is either positive or negative, depending on its relative nuclear binding energy to ^{12}C. Since each nuclide has a unique mass defect, molecules with different isotopic composition have unique exact masses. For example, while a resolving power of around 5000 is sufficient to completely separate $C_{14}H_{10}^{+}$ and $C_{13}H_{6}O^{+}$, for closely spaced ions, the required resolving power can easily increase up to 10^5 or even higher. As the m/z increases, the number of combinations of different elements resulting in the same nominal mass grows very fast. This experimental limitation is already tackled in Irimiea *et al.*[19] when discussing the role of oxygen containing compounds. Nevertheless, a lower mass resolution mass spectrum can provide some helpful information. In particular, in the investigation of soot particles sampled from laboratory flames, C, H, and O are the major contributors to the total mass of soot, and therefore the mass analysis of peaks with a high signal-to-noise ratio (SNR) can be reasonably limited to $C_mH_nO_p^{+}$ ions. Identification within 5 ppm, often but not necessarily assumed as "certain",[52] in our work is possible up to $m/z \approx 150$–200. *A priori* knowledge of the samples and experimental conditions can extend this range up to $m/z \approx 500$–550 and lead to self-consistent results and coherence with many other works in the literature.

The mass defect analysis can also be used to simplify the visualisation of complex mass spectra (*e.g.* Fig. S1 and S2†). This is generally achieved by plotting the mass defects of all peaks *versus* their nominal mass. The resulting graph (mass defect plot, Fig. 1 and S3†) enables the visualisation of complex databases in one single plot, and highlights trends that are often invaluable to identify unknown species. For instance, aliphatic, aromatic or polycyclic aromatic hydrocarbons are aligned on different positive slopes corresponding to the addition of H atoms. When analysing samples containing hydrocarbons with different degrees of alkylation, the Kendrick mass defect can be used as an alternative way of presenting the mass defect data.[50,51] The Kendrick mass defect is calculated from the re-normalised mass of a repeating molecular fragment to an integer value as shown in eqn (1) for the common case of CH_2 ($m = 14.01565$):

$$m^{\text{Kendrick}} = m^{\text{IUPAC}} \frac{14.0000}{14.01565} \tag{1}$$

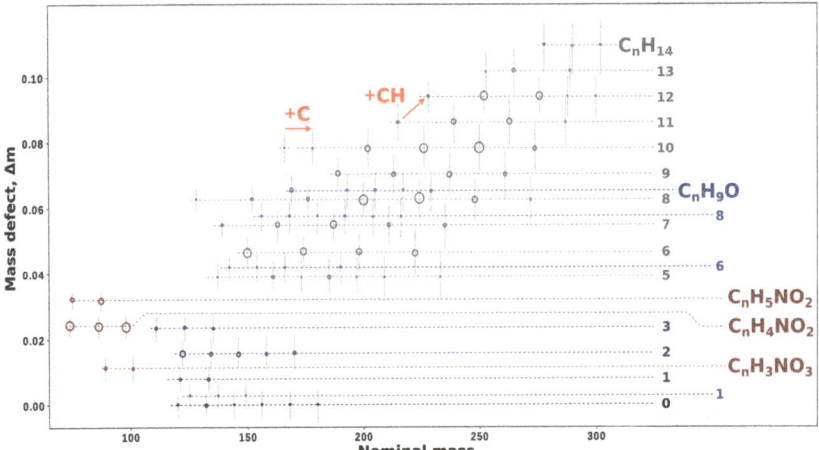

Fig. 1   Mass defect plot obtained from the L2MS mass spectrum of miniCAST soot, C_2 sample. The data points represent the assigned accurate mass. The size of the data points is proportional to the corresponding peak integrated area normalised to the total ion count after background subtraction. Molecular formulas of homologous species are displayed. The error bars show the uncertainty on the accurate mass calculated from the obtained mass resolution.

After this conversion, homologous series that contain the repeating fragment have identical Kendrick mass defect and are found aligned on horizontal lines, making their identification even easier.[50,53] This is useful when dealing with repeating alkyl groups, for instance, since their mass defect increases regularly with their molecular weight and makes their association to a certain series less intuitive when represented on conventional mass defect plots.[50] The most convenient approach (conventional or Kendrick) heavily depends on the nature of the sample. If the sample is dominated by a variety of different species, the use of the conventional mass defect is more advisable. However, when the mass spectrum contains many species that only differ by a repeating unit such as aliphatic chains, for instance (Table S1†), the Kendrick mass defect is more advantageous (Fig. S4†).

In this work, mass defect analysis is applied to the data obtained from L2MS and SIMS to demonstrate its effectiveness when dealing with a variety of mass spectrometric data. Fig. 1 shows the mass defect plot obtained from sample C_2 analysed using L2MS. The suggested representation merges into one graph important information extracted from the raw mass spectra that include the peaks mass defect (y-axis), nominal mass (x-axis) and relative abundance (dot size). Species that line up in the mass defect plots typically contain a repeating unit. Additionally, the detection of a series of homologous species can help the identification of unknown peaks. This is especially helpful for species with high molecular masses, where the attribution of a chemical formula can be rather delicate.

As PAHs exhibit a high thermodynamic stability,[54] they appear in great abundance in all mass spectra and this is amplified by the high sensitivity of the analysis technique to these specific compounds (Fig. S1†). Since the H/C ratio of PAHs is low compared to other hydrocarbons, they have a relatively small mass defect and are thus easily distinguishable from other hydrocarbons. For instance, aromatic hydrocarbons that contain the same number of hydrogen atoms and a progressively increasing number of carbon atoms (*e.g.* $C_{10}H_8 \rightarrow C_{12}H_8 \rightarrow C_{14}H_8 \rightarrow \dots \rightarrow C_{22}H_8$) can be found on the same horizontal line. Besides hydrocarbons, all samples contain oxygen and nitrogen organic derivatives to some extent. As a rule of thumb, in the mass defect plot of combustion generated aerosols, oxygen containing hydrocarbons are often found below the corresponding hydrocarbons due to the large negative mass defect of oxygen. Nitrogen containing hydrocarbons show distinct behaviours. For instance, organic amines are often found mixed with their corresponding hydrocarbons due to the nucleophilicity of nitrogen that results in their tendency to bind one additional hydrogen atom post-ionisation. Organic nitrates, on the other hand, tend to be found at lower mass defect due to the presence of oxygen.

The Kendrick mass defect can be used to emphasise some less obvious patterns as shown in Fig. S4,† in which CH ($m = 13.007\ 825$) is used as the base unit.

3.2 Statistical analysis

In this section we detail the chemometric techniques, based on commonly used statistical tools like multivariate analysis, that were adopted by our group to extract chemical information from mass spectrometric data. A mass spectrometry database can contain an extremely variable number of mass spectra (observations), and each of them typically contain up to thousands of peaks (variables). This database structure should be taken into consideration when choosing the most appropriate statistical methods.

3.2.1 Principal component analysis.
PCA is a powerful statistical tool that can be used to classify samples and reveal trends and patterns in databases,[5] and is often used to increase the readability of very complex data.[55] PCA applied to mass spectrometry is especially useful when many mass spectra are being compared, since it reduces the dimensionality of the database while preserving most of the original information. PCA is a non-parametric analysis, *i.e.* its output is independent of any hypotheses about data distribution.[56] In this work, PCA is performed on a matrix containing the integrated peaks (variables) against the samples (observations). Before applying PCA, data obtained from mass spectrometry should undergo a special preparation procedure[56,57] that includes calibration, baseline removal, construction of a peak list, peak integration and standardisation. PCA applied to data with no normalisation/standardisation is mostly affected by the largest raw variance, which can skew the overall interpretation of the dataset. Therefore, normalisation techniques are applied to mass spectra prior to PCA analysis when there are differences in the sample weights, volumes or other properties that may result in additional sources of variance. The most popular and generally recommended normalisation method is normalisation to the total ion count (TIC), *i.e.* the integrated ion count over a given mass range.[18,58,59]

Care has to be taken when building the peak list as it should only contain species representative of the sample. Minor-abundance isotopes are usually excluded from the peak list, thus allowing a focus on the major-abundance isotopic species.[58] Peaks coming from the substrate and/or originating from the sample-substrate reactivity should also be disregarded. Identifying these peaks, especially the ones corresponding to reaction products, can be a difficult task. One approach to their identification involves comparing the mass spectra of the sample deposited under the same experimental conditions but on different substrates (e.g. Si and Ti wafers).[18] Another possibility relies on the use of PCA: species coming from the sample-substrate reactivity become less prominent as the substrate coverage increases and is less available for the reaction, and are thus likely to be found all clustered in the same principal component.

Each principal component (PC) accounts for a defined percentage of the total variance within the data set; they are represented in a scree plot that is used to select the PCs to take into consideration. The loadings represent the weights of each variable used to calculate the PCs, and are used to understand the contribution of each variable to the selected PC. The distance of an observation from a PC is represented on the score plot. Scores are obtained for each observation in the database and for each principal component, and are often used as a base to display and classify the samples. In the score plot, similar observations group together and are separated from dissimilar observations. The clustering of the scores is strongly related to the values of the loadings, and they are discussed as a whole. The most challenging part of PCA is the interpretation of individual PCs and their contribution to the investigated processes. To this purpose, there is a vast amount of literature providing general guidelines that should be followed.[5,60-62]

To illustrate the potential of this technique, we show below some applications to mass spectrometric data of various combustion generated aerosol samples.

3.2.1.1   MiniCAST soot, L2MS. When L2MS mass spectra of miniCAST soot samples are examined, PC1 and PC2 account for ~96% of the total variance, and are therefore only considered for the data interpretation. The three samples are well separated in the PC2 *vs.* PC1 scores plot (Fig. 2). Sample C_1 is highly influenced by $C_{14}H_8$, $C_{14}H_{10}$ and $C_{16}H_{10}$ (high positive PC1 scores) whereas C_2 and C_3 are dominated by higher mass aromatic compounds (negative PC1 scores). It can be observed that PC2 (~10%) allows for better discrimination between the samples than PC1, especially C_2 and C_3.

3.2.1.2   Flame and ICE soot, SIMS. PCA is applied to the ensemble of SIMS mass spectra obtained in positive polarity from soot samples generated by the gasoline engine and the laboratory flame (diesel and kerosene fuels). PC1 and PC2 together account for 73.3% of the total variance. Two main groups are observed in the score plot of both positive and negative ions (Fig. 3 and S5†). While it was not possible to clearly associate a phenomenon to PC1 (51.7% of total variance), the samples are well separated by the different emission sources (engine, GOM, and flame, D and K) in PC2 (21.6% of total variance). At this level of the analysis, PCA cannot distinguish soot generated by burning the two different liquid fuels (diesel and kerosene) in laboratory flames, which appear mixed together in negative PC2.

PC1 is mainly associated with high H/C fragment ions (negative contribution, red dots in the loadings plot (Fig. 3)), and low H/C fragment ions probably resulting from the dissociation of large aromatic hydrocarbons (positive contribution, green dots in the loadings plot). The main contributions to PC2 come from aromatic species (positive contribution, blue dots on the loadings plot), and to a smaller extent, from high H/C fragment ions. Therefore, the contribution of high H/C fragment ions, possibly related to the dissociation of aliphatic

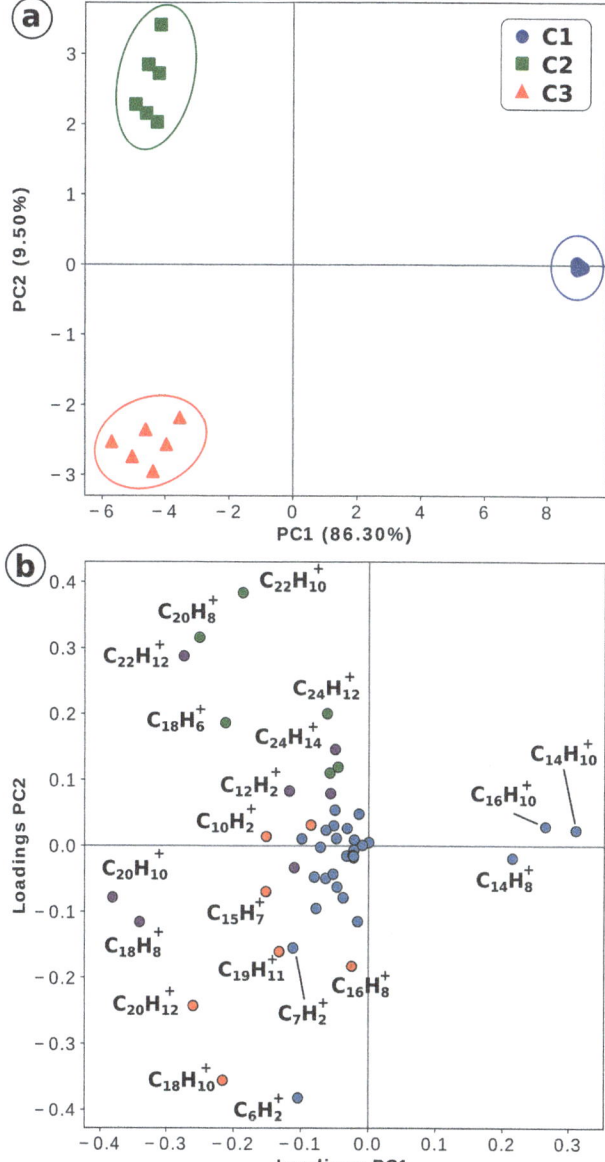

Fig. 2 Score plots of PC2 vs. PC1 for miniCAST soot samples obtained with L2MS (a). The ellipses highlight data points coming from different samples and are added for visual purposes only. (b) The corresponding loadings plot of PC2 vs. PC1. Several homologous series are highlighted: $C_{n+8}H_n$ – red, $C_{n+10}H_n$ – purple, and $C_{n+12}H_n$ – green.

hydrocarbons, depends less on the fuel and more on the combustion conditions (engine *vs.* controlled laboratory flames).

Going a step further, PCA is applied to gasoline soot samples obtained in different engine regimes in order to determine their impact on the chemical composition (Fig. 4). There is an obvious separation between normal engine

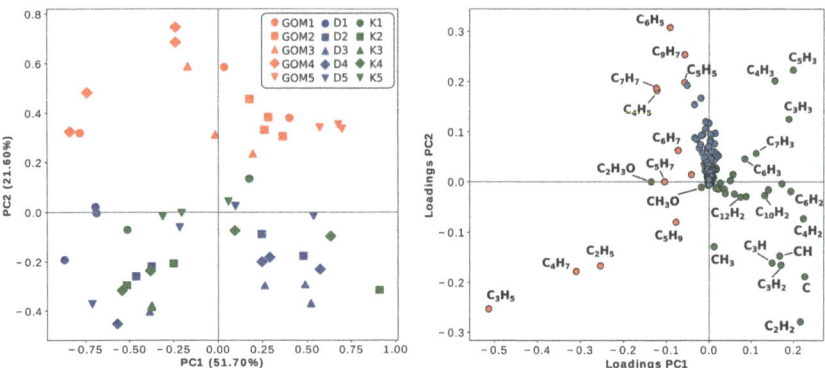

Fig. 3 Score plot of PC2 *vs.* PC1 for positive ions of soot samples obtained from gasoline engine and laboratory flames (left panel). Corresponding loadings plot of the first two principal components (right panel). For sample description, see Table 1.

operation regimes (GOM and GOH) and the ones that simulate a malfunction (GEF and GEO). A good discrimination is achieved with only the first two components that account for ~98% of the total variance. PC1 alone (~91%) allows the separation of regimes, based on the abundance of aliphatic fragment ions (positive contribution to PC1, marked in red in Fig. S6†). Consequently, samples that simulate a malfunction (GEF and GEO) are characterised by a higher relative contribution from aliphatic fragment ions compared to optimised engine regimes (GOM and GOH). PC2 is linked to the contribution of aliphatic fragment ions and aromatic species (positive PC2 value), however some aliphatic fragment ions (C_5H_7, C_5H_9, C_3H_7, C_4H_7) show a contribution to negative PC2. The data points corresponding to optimal engine regimes form a smaller cluster. This implies that soot produced under conditions simulating engine malfunction shows a much larger variability in chemical composition.

At this point of the analysis, it is clear that the two regimes that simulate a malfunction (GEF and GEO) exhibit similarities, while being well separated from the optimised regimes (upper panel of Fig. 4). This implies that the variance of a certain principal component for them is much smaller than the one responsible for the separation between optimised and non-optimised regimes. Consequently, each group should be analysed independently, thus uncovering even smaller contributions to the variance. To demonstrate this concept, the same statistical method was applied a second time to the two non-optimised regimes, and their comparison lead to discrimination between the two main contributors to particulate emissions of the internal combustion engine: fuel and oil, Fig. 4. In this case, PC1 (~71%), accountable for the separation of the two regimes, is linked to the contribution of hydrogen-rich hydrocarbons on one side (negative contribution) and of fragment ions and aromatic species on the other (positive contribution). This reveals that oil-related soot particles feature more hydrogen-rich hydrocarbons, while an excess of gasoline leads to the production of more aromatic species, Fig. S6.† The increase of the contribution of fragment ions in the latter is probably linked to the increase in the aromatic contribution, since the majority of fragment ions can be related to dissociation reactions of PAHs.[63] PC2 (~20%) is associated with the presence of aromatic hydrocarbons

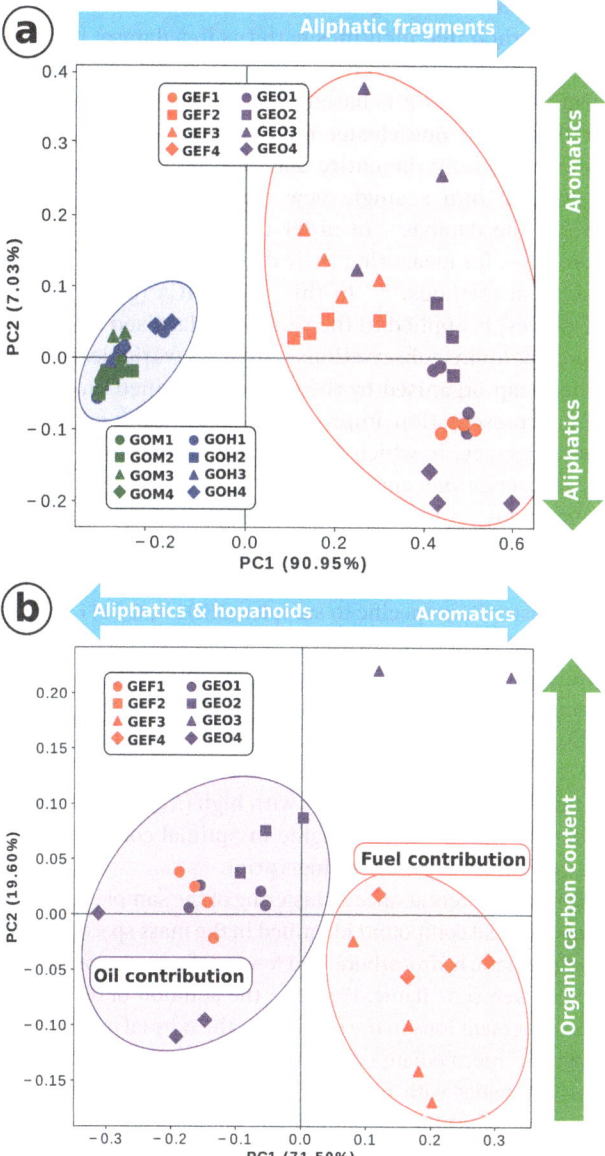

Fig. 4 Score plots of the first two principal components for soot samples produced by a single cylinder engine. Upper panel – discrimination between different engine regimes, lower panel – particle source discrimination. The ellipses highlight clusters of data points and are for visual purposes only. For sample description, see Table 1.

(blue dots in Fig. S6†). One can also notice that samples corresponding to the engine regime with a low air/fuel ratio (GEF1) surprisingly lie in the oil-excess region, while samples GEO3 appear far from the oil-excess region (Fig. 4). It is likely that the specific behaviour observed for these samples relates to their particle size (Table 1) but correlating size to chemical composition is out of the scope of this paper and will be addressed in a future work.

3.2.2 Hierarchical clustering analysis. Hierarchical clustering analysis (HCA) is a MVA method that identifies patterns in a dataset by creating groups of observations called clusters. Unlikely PCA, HCA accounts for the total variance in the database.[60,62] HCA is based on a simple approach for building the clusters that starts with one cluster for each observation and finishes with a single cluster containing the entire database. At each step, the two closest clusters are merged into a single new cluster resulting in a dendrogram representative of the database. In order to decide which clusters to merge, different approaches for measuring their distance can be used and give rise to several hierarchical methods.[61,62] In this work, HCA (group average method, Euclidean distances) is applied to the same standardised matrix used for PCA analysis, on both columns (observations) and rows (variables). The HCA output is built in a heatmap organised by the clusters obtained on observations and variables. This representation improves the visualisation of clusters in the multidimensional space, in which each tile represents the value of the correlation between observations and variables.

The heatmap obtained for the samples analysed in SIMS positive polarity is shown in Fig. 5. HCA groups the samples in three main clusters (C_1, C_2 and C_3) at distance d_1 function of the characteristics of the five clusters of variables (R_1, R_2, R_3, R_4 and R_5). Cluster C_1 is specific to samples GEO1-4, GOM4 and D1 due to the high contribution of compounds with H/C > 1 and identified in the C_{1-1} cluster. C_{1-2} is dissimilar from C_{1-1} due to the presence of aromatic hydrocarbons and other compounds with low H/C ratio. Soot collected from the gasoline engine in optimal conditions and after the addition of oil are dominated by R_5, while there is a shift to R_1 and R_2 for soot collected from the diesel flame. Contrary to C_1, C_2 has a high contribution of fragment ions with high (R_4) and low (R_1) H/C ratio. C_2 shows that soot collected from the engine in optimal conditions with high and medium load have similar chemical fingerprint.

This representation offers at once a clustering of the sample function of the three main classes of chemical compound identified in the mass spectra. For instance, the high content of aromatic hydrocarbons and low H/C fragment ions is specific to soot collected from the kerosene flame. Basically, the addition of oil increases the fraction of high H/C fragment ions in the emissions, the normal operation conditions of the engine have an intermediate content of high H/C fragment ions and a slight contribution of aromatics with four and five aromatic rings, while kerosene soot contains the highest contribution of aromatic compounds and low H/C fragment ions. HCA is also applied to L2MS and SIMS negative polarity data as detailed in the ESI.† In this work, HCA is applied to the raw data corresponding to the selected mass spectra but its usefulness can be extended to more compact data after using another statistical method for sorting the input variables and observations. One of the advantages of this method is that it does not require the raw data set. Moreover, HCA can be used to visualise clusters that form in the principal component space, after applying PCA, or it can group samples according to other properties (mass defect, contribution from different classes of compounds, *etc.*).

3.3 Mass peak grouping into chemical classes

A detailed description of the soot chemical composition is certainly desirable and can lead to important clues on the soot formation, growth, ageing and reactivity.

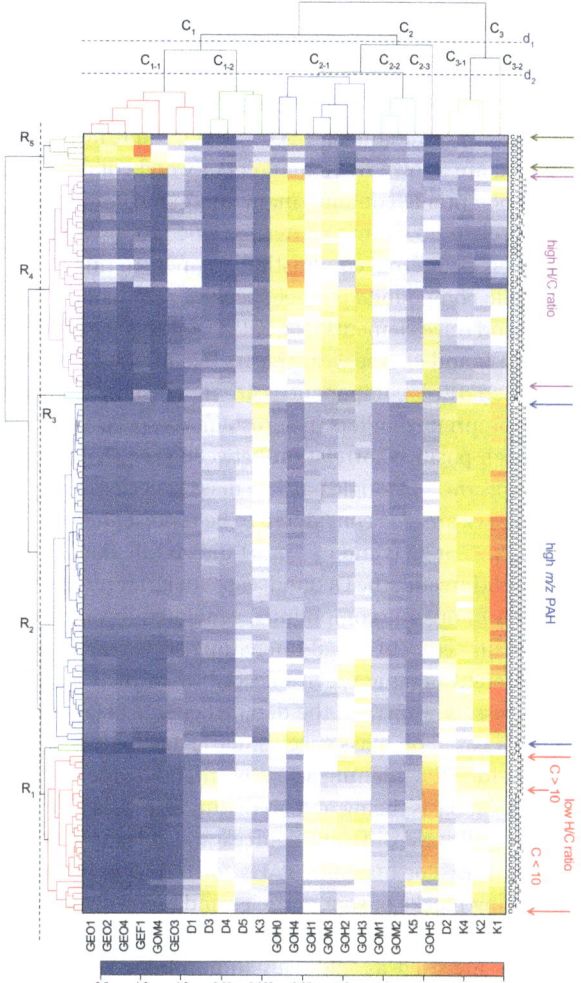

Fig. 5 Two-way hierarchical clustering heat-map for positive ions of gasoline, diesel and kerosene soot obtained with SIMS. Each column corresponds to the averaged mass spectra obtained for a soot sample. The contribution of each mass in individual samples is expressed as relative value and is represented by the cell colour.

However, this can rapidly turn into a very cumbersome task, especially if many different samples are analysed. For the sake of simplicity, most of the time, and especially when long time-series of field-collected data are to be treated, individual compounds are grouped in classes (*e.g.* aliphatics, aromatics, oxygenated, sulphur-containing hydrocarbons, and so on). This grouping of mass peaks into appropriate classes allows easier comparison with other experimental measurements (*e.g.* OC/EC[29]) and facilitates the interaction with modellers that use the data as inputs for various scale simulations. Moreover, this grouping of peaks is also useful when mass spectra of several samples are compared to each other in order to reveal general trends in their chemical composition.

When it comes to the chemical composition of combustion generated aerosols, three non-specific indicators are often considered: amount of ash components (inorganic compounds, IC), amount of carbon associated to the carbonaceous matrix (elemental carbon, EC), and amount of carbon found in organic compounds (organic carbon, OC).[64] IC alone can sometimes help identify the main source of the emissions. For instance, K^+, Na^+, K_2Cl^+ and $K_3SO_4^+$ in the positive polarity mass spectra and Cl^-, SO_3^-, HSO_4^- and KCl_2^- in the negative polarity mass spectra are known to be markers of wood combustion.[65] Generally speaking, since IC potentially contains many inorganic compounds, it can and should be further broken down into source specific groups when characterising complex systems such as internal combustion engines. In this case, the accepted grouping of inorganic compounds is: fuel specific (compounds that are coming from fuel additives and trace elements (Na, K)[11,66]), oil specific (detergent and anti-wear additives (P, Ca)[67] and engine wear tracers (Fe, Al, Cr)[30,67,68]). For addressing the elemental carbon (EC) component, carbon clusters C_n^- ($n = 2$–4) are considered to be appropriate markers in aerosol mass spectrometry.[64] This is also confirmed by the high positive correlation between C_2^-, C_3^- and C_4^- signals in the recorded mass spectra.[26] In single particle mass spectrometry, carbon clusters with even higher masses are also considered to be representative of the elemental carbon (C_5^- at 60 u, C_6^- at 72 u and C_7^- at 84 u).[11] While the handling of IC and EC is relatively straightforward, the OC landscape looks far more complex, with an overwhelming variety of organic compounds, generated in various processes and being themselves main actors of broad-range time-scale reactivity. A subsequent classification of different organic species according to their functional group(s) seems therefore necessary. However, the detailed chemical analysis of a complex mixture of chemicals based on mass spectrometric data only is still an important challenge that requires the identification of the individual ion dissociation patterns. On a practical basis, being able to distinguish these compounds is very important since they all have different sources and roles in the soot formation and ageing mechanisms. For instance, PAHs form during combustion and are well known as building blocks of soot particles and are generally seen as reliable markers of the overall OC content.[29] Organic hydroxyl groups are linked to alcohols that are commonly used as additives in gasoline. The presence of many compounds containing carbonyl groups has been proposed as a marker to distinguish fresh emissions from soot particles aged in the atmosphere.[69]

A combination of previously described mass peak classification methods is shown in Table 2 along with chemical formula assignments.[63,71] Detailed classification of molecular ions by functional groups remains difficult by MS alone, however it can be achieved in combination with complementary techniques (e.g. FTIR).[26] Also, for the sake of simplicity, Table 2 displays only the nominal masses, but the peak assignment is based on the exact mass (see mass defect analysis, Section 3.1). The discussion below is based on this grouping of mass peaks.

Depending on the studied samples, the analysis will focus on specific classes from Table 2. For soot samples obtained with the miniCAST standard generator, one may want to address the impact of the oxidation flow. A possible focus is therefore on the evolution of the oxygenated species vs. PAHs (linked to the OC content). Since miniCAST soot is a well-studied standard, it also allows the comparison of mass spectrometric results with the ones reported in the literature based on other experimental techniques. In the present case, Fig. 6 clearly shows an increase of the oxygenated species abundance with the oxidation flow, however a low oxidation flow (C_2 and C_3) leads to the formation of more PAHs, which

Table 2 Grouping of mass peaks into chemical classes

Category	m/z	Formula	m/z	Formula	m/z	Formula	m/z	Formula
Aliphatics	15	CH_3^+	54	$C_4H_6^+$	71	$C_5H_{11}^+$	99	$C_7H_{15}^+$
(alkynes, alkenes,	27	$C_2H_3^+$	55	$C_4H_7^+$	81	$C_6H_9^+$	109	$C_8H_{13}^+$
alkyl, *etc.*)	29	$C_2H_5^+$	57	$C_4H_9^+$	83	$C_6H_{11}^+$	111	$C_8H_{15}^+$
	41	$C_3H_5^+$	67	$C_5H_7^+$	85	$C_6H_{13}^+$	113	$C_8H_{17}^+$
	43	$C_3H_7^+$	68	$C_5H_8^+$	95	$C_7H_{11}^+$		
	53	$C_4H_5^+$	69	$C_5H_9^+$	97	$C_7H_{13}^+$		
Aromatics	26	$C_2H_2^+$	64	$C_5H_4^+$	152	$C_{12}H_8^+$	216	$C_{17}H_{12}^+$
	38	$C_3H_2^+$	74	$C_6H_2^+$	154	$C_{12}H_{10}^+$	228	$C_{18}H_{12}^+$
	39	$C_3H_3^+$	75	$C_6H_3^+$	166	$C_{13}H_{10}^+$	252	$C_{20}H_{12}^+$
	40	$C_3H_4^+$	76	$C_6H_4^+$	178	$C_{14}H_{10}^+$	276	$C_{22}H_{12}^+$
	50	$C_4H_2^+$	78	$C_6H_6^+$	266	$C_{21}H_{14}^+$	278	$C_{22}H_{14}^+$
	51	$C_4H_3^+$	91	$C_7H_7^+$	190	$C_{15}H_{10}^+$		
	63	$C_5H_3^+$	128	$C_{10}H_8^+$	202	$C_{16}H_{10}^+$		
O-containing (carbonyls,	31	CH_3O^+	69	$C_4H_5O^+$	87	$C_5H_{11}O^+$	129	$C_7H_{13}O_2^+$
acids, ethers,	33	CH_5O^+	71	$C_4H_7O^+$	89	$C_5H_{13}O^+$	137	$C_{10}HO^+$
alcohols, *etc.*)	43	$C_2H_3O^+$	73	$C_3H_5O_2^+$	97	$C_6H_9O^+$	142	$C_{10}H_6O^+$
	45	$C_2H_5O^+$	73	$C_4H_9O^+$	97	$C_5H_5O_2^+$	156	$C_{11}H_8O^+$
	47	$CH_3O_2^+$	75	$C_3H_7O_2^+$	101	$C_6H_{13}O^+$	166	$C_{12}H_6O^+$
	47	$C_2H_7O^+$	75	$C_4H_{11}O^+$	105	$C_7H_5O^+$	169	$C_{11}H_9O^+$
	53	$C_4H_5^+$	81	$C_5H_5O^+$	109	$C_7H_9O^+$	180	$C_{13}H_8O^+$
	55	$C_3H_3O^+$	83	$C_5H_7O^+$	111	$C_6H_7O_2^+$	205	$C_{14}H_9O^+$
	57	$C_5H_5O^+$	85	$C_5H_9O^+$	111	$C_7H_{11}O^+$		
	59	$C_3H_7O^+$	85	$C_4H_5O_2^+$	119	$C_8H_7O^+$		
	61	$C_2H_5O_2^+$	87	$C_5H_{11}O^+$	123	$C_7H_7O_2^+$		
	61	$C_3H_9O^+$	87	$C_4H_7O_2^+$	125	C_9HO^+		
N-containing	26	CN^+	46	CH_4NO^+	60	$C_2H_6NO^+$	89	$C_2H_3NO_3^+$
	29	CH_3N^+	55	$C_3H_5N^+$	74	$C_2H_4NO_2^+$	98	$C_4H_4NO_2^+$
	44	CH_2NO^+			87	$C_2H_5NO_2^+$	121	$C_8H_{11}N^+$
S-containing	32	S^+	44	CS^+	46	CH_2S^+		
Unclassified	28	$C_2H_4^+$	56	$C_4H_8^+$	84	$C_6H_{12}^+$	112	$C_8H_{16}^+$
Hydrocarbons	42	$C_3H_6^+$	70	$C_5H_{10}^+$	98	$C_7H_{14}^+$		

confirms previous observations on the same set-points of the miniCAST generator.[43,70]

Even though examining trends for specific groups can be very informative, when it comes to complex mass spectra containing a multitude of peaks that can be separated in many different ways, not all of the groups feature useful trends. It is therefore advisable to first identify the species of interest, groups or individual compounds that can be linked to variations in the chemical composition of the samples. This information can be retrieved from PCA and HCA as discussed in Sections 3.2.1 and 3.2.2, respectively. Based on the statistical analysis of positive polarity SIMS mass spectra of gasoline, diesel and kerosene soot samples, three groups of interest are chosen for further analysis as shown in Fig. 6: low-mass and low H/C ions (from the dissociation of aromatic species[63]), low-mass and high H/C ions (from the dissociation of aliphatic species), and finally large aromatic ions (mostly PAHs, stable enough to be detected as molecular ions). Gasoline soot shows higher content of large aromatic compounds, with high and almost constant contribution to all considered particle sizes. Gasoline soot also features

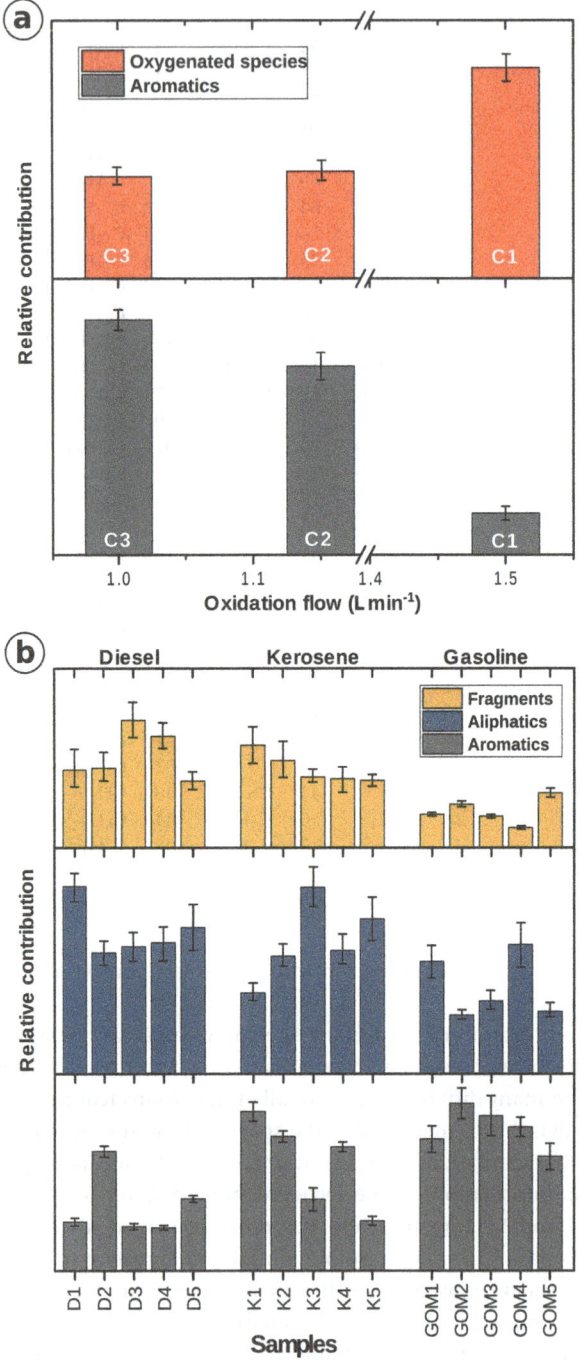

Fig. 6   Several trends retrieved from the mass spectra of: (a) miniCAST soot (L2MS), and (b) gasoline, diesel and kerosene soot (SIMS).

the least fragmentation that is highly consistent with the higher contribution of large aromatics if compared to diesel and kerosene soot. For the other two fuels, different zones of the flame, corresponding to different stages in the soot formation process, were probed, therefore the variation in aromatic content looks more pronounced. It is clear that the aliphatic content alone cannot be used to discriminate between soot coming from combustion of different fuels, just like it was concluded from PCA. However, it still provides valuable information about different soot maturity. For example, for diesel soot the contribution of aliphatics gradually increases with the sampled HAB (HAB $\geq$ 12 cm). On the other hand, HCA on the negative polarity of SIMS is much easier to interpret because the results clearly discriminate the laboratory flame soot from the one produced with the gasoline engine. The samples belonging to the latest category are clearly evidenced by the presence of sulphur and oxygen containing compounds while the soot from the flames contains mainly OC and EC. Generally speaking, the trends that are shown herewith are very useful when interpreting the data. However, they are almost impossible to notice in the raw mass spectra. Being able to follow the contribution of a group of related molecules hidden in a much larger ensemble of signals is a powerful feature used to uncover trends that would have remained hidden to a more basic analysis. The fact that PCA and HCA are able to separate the selected samples into categories dependent on their unique pattern of chemical signatures proves that mass spectrometry and MVA provide useful insights into their properties. The usefulness of this approach allows for an easier identification and traceability of combustion generated particles with unknown sources.

4 Conclusions

Our recently developed comprehensive methodology (based on mass defect analysis, PCA/HCA multivariate methods)[18] dedicated to the chemical analysis of combustion-generated aerosols is applied here to the study of 30 soot samples generated by three different sources using four different fuels. Laser and secondary ion mass spectrometry techniques are used to probe their surface chemistry. A few examples on the performances of this methodology are provided, showcasing its ability to clearly discriminate samples according to various parameters, such as combustion source, soot maturity, or engine operating conditions. The correlations evidenced by the MVA methods were used for peak clustering to highlight the evolution of grand chemical classes with the combustion conditions. These trends, along with detailed molecular-level information, can further help constrain the processes involved in particulate matter emissions and predict the impact of soot particles on the environment and human health. Moreover, aiming for a standardised (generally accepted) methodology in treating complex mass spectrometry data in aerosol science would certainly allow easier intercomparison and the building of extensive shared databases for further specific developments. An appealing perspective is the possible application of neural networks to this type of big data, which would lead to great advances in automated real-time processing of large dataflows.

Conflicts of interest

There are no conflicts to declare.

Acknowledgements

This work was supported by the French National Research Agency (ANR) through the PIA (Programme d'Investissement d'Avenir) under contract ANR-10-LABX-005 (CaPPA – Chemical and Physical Properties of the Atmosphere), the European Commission Horizon 2020 project PEMs4Nano (H2020 Grant Agreement #724145), and the CLIMIBIO project *via* the Contrat de Plan Etat Région of the Hauts-de-France region. We thank N. Nuns from the Regional Surface Analysis Platform for assistance with the SIMS measurements.

References

1 R. S. Plumb, C. L. Stumpf, J. H. Granger, J. Castro-Perez, J. N. Haselden and G. J. Dear, *Rapid Commun. Mass Spectrom.*, 2003, **17**, 2632–2638.

2 M. Brokl, L. Bishop, C. G. Wright, C. Liu, K. McAdam and J. F. Focant, *J. Chromatogr. A*, 2014, **1370**, 216–229.

3 S. Shrestha and F. Kazama, *Environ. Model. Softw.*, 2007, **22**, 464–475.

4 W. K. Härdle and L. Simar, *Applied Multivariate Statistical Analysis Course*, Springer, 2015.

5 H. Abdi and L. J. Williams, *English*, 2010, **2**, 433–470.

6 J. L. Jimenez, M. R. Canagaratna, N. M. Donahue, A. S. Prevot, Q. Zhang, J. H. Kroll, P. F. DeCarlo, J. D. Allan, H. Coe, N. L. Ng, A. C. Aiken, K. S. Docherty, I. M. Ulbrich, A. P. Grieshop, A. L. Robinson, J. Duplissy, J. D. Smith, K. R. Wilson, V. A. Lanz, C. Hueglin, Y. L. Sun, J. Tian, A. Laaksonen, T. Raatikainen, J. Rautiainen, P. Vaattovaara, M. Ehn, M. Kulmala, J. M. Tomlinson, D. R. Collins, M. J. Cubison, E. J. Dunlea, J. A. Huffman, T. B. Onasch, M. R. Alfarra, P. I. Williams, K. Bower, Y. Kondo, J. Schneider, F. Drewnick, S. Borrmann, S. Weimer, K. Demerjian, D. Salcedo, L. Cottrell, R. Griffin, A. Takami, T. Miyoshi, S. Hatakeyama, A. Shimono, J. Y. Sun, Y. M. Zhang, K. Dzepina, J. R. Kimmel, D. Sueper, J. T. Jayne, S. C. Herndon, A. M. Trimborn, L. R. Williams, E. C. Wood, A. M. Middlebrook, C. E. Kolb, U. Baltensperger and D. R. Worsnop, *Science*, 2009, **326**, 1525–1529.

7 C. Fountoukis, A. G. Megaritis, K. Skyllakou, P. E. Charalampidis, C. Pilinis, H. A. C. Denier Van Der Gon, M. Crippa, F. Canonaco, C. Mohr, A. S. H. Prévôt, J. D. Allan, L. Poulain, T. Petäjä, P. Tiitta, S. Carbone, A. Kiendler-Scharr, E. Nemitz, C. O'Dowd, E. Swietlicki and S. N. Pandis, *Atmos. Chem. Phys.*, 2014, **14**, 9061–9076.

8 M. Crippa, F. Canonaco, V. A. Lanz, M. Äijälä, J. D. Allan, S. Carbone, G. Capes, D. Ceburnis, M. Dall'Osto, D. A. Day, P. F. DeCarlo, M. Ehn, A. Eriksson, E. Freney, L. H. Ruiz, R. Hillamo, J. L. Jimenez, H. Junninen, A. Kiendler-Scharr, A. M. Kortelainen, M. Kulmala, A. Laaksonen, A. A. Mensah, C. Mohr, E. Nemitz, C. O'Dowd, J. Ovadnevaite, S. N. Pandis, T. Petäjä, L. Poulain, S. Saarikoski, K. Sellegri, E. Swietlicki, P. Tiitta, D. R. Worsnop, U. Baltensperger and A. S. Prévôt, *Atmos. Chem. Phys.*, 2014, **14**, 6159–6176.

9 T. B. Onasch, A. Trimborn, E. C. Fortner, J. T. Jayne, G. L. Kok, L. R. Williams, P. Davidovits and D. R. Worsnop, *Aerosol Sci. Technol.*, 2012, **46**, 804–817.

10 A. Faccinetto, P. Desgroux, M. Ziskind, E. Therssen and C. Focsa, *Combust. Flame*, 2011, **158**, 227–239.

11 U. Kirchner, R. Vogt, C. Natzeck and J. Goschnick, *J. Aerosol Sci.*, 2003, **34**, 1323–1346.

12 N. Mayama, Y. Miura, K. Misawa, A. Takami, T. Sakamoto and M. Fujii, *Anal. Sci.*, 2013, **29**, 479–482.

13 F. Aubriet and V. Carré, *Anal. Chim. Acta*, 2010, **659**, 34–54.

14 S. Eliuk and A. Makarov, *Annu. Rev. Anal. Chem.*, 2015, **8**, 61–80.

15 K. Wang, Y. Zhang, R. J. Huang, J. Cao and T. Hoffmann, *Atmos. Environ.*, 2018, **189**, 22–29.

16 C. Zuth, A. L. Vogel, S. Ockenfeld, R. Huesmann and T. Hoffmann, *Anal. Chem.*, 2018, **90**, 8816–8823.

17 J. Cain, A. Laskin, M. R. Kholghy, M. J. Thomson and H. Wang, *Phys. Chem. Chem. Phys.*, 2014, **16**, 25862–25875.

18 C. Irimiea, A. Faccinetto, Y. Carpentier, I. K. Ortega, N. Nuns, E. Therssen, P. Desgroux and C. Focsa, *Rapid Commun. Mass Spectrom.*, 2018, **32**, 1015–1025.

19 C. Irimiea, A. Faccinetto, X. Mercier, I.-K. Ortega, N. Nuns, E. Therssen, P. Desgroux and C. Focsa, *Carbon*, 2019, **144**, 815–830.

20 M. F. Heringa, P. F. DeCarlo, R. Chirico, T. Tritscher, J. Dommen, E. Weingartner, R. Richter, G. Wehrle, A. S. Prévôt and U. Baltensperger, *Atmos. Chem. Phys.*, 2011, **11**, 5945–5957.

21 J. L. Jimenez, *J. Geophys. Res.*, 2003, **108**, 8425.

22 M. T. Timko, T. B. Onasch, M. J. Northway, J. T. Jayne, M. R. Canagaratna, S. C. Herndon, E. C. Wood, R. C. Miake-Lye and W. B. Knighton, *J. Eng. Gas Turbines Power*, 2010, **132**, 061505.

23 W. B. Knighton, T. M. Rogers, B. E. Anderson, S. C. Herndon, P. E. Yelvington and R. C. Miake-Lye, *J. Propul. Power*, 2007, **23**, 949–958.

24 M. T. Timko, S. C. Herndon, E. De La Rosa Blanco, E. C. Wood, Z. Yu, R. C. Miake-Lye, W. B. Knighton, L. Shafer, M. J. Dewitt and E. Corporan, *Combust. Sci. Technol.*, 2011, **183**, 1039–1068.

25 M. Bente, M. Sklorz, T. Streibel and R. Zimmermann, *Anal. Chem.*, 2008, **80**, 8991–9004.

26 S. Gilardoni, L. M. Russell, A. Sorooshian, R. C. Flagan, J. H. Seinfeld, T. S. Bates, P. K. Quinn, J. D. Allan, B. Williams, A. H. Goldstein, T. B. Onasch, D. R. Worsnop, J. Pagels, D. D. Dutcher, M. R. Stolzenburg, P. H. Mcmurry, M. E. Gälli, D. S. Gross, J. C. Chow, J. G. Watson, D. Crow, D. H. Lowenthal, T. Merrifield, T. Ferge, E. Karg, A. Schröppel, K. R. Coffee, H. J. Tobias, M. Frank, E. E. Gard, R. Zimmermann, J. W. Sammon, U. Kirchner, R. Vogt, C. Natzeck, J. Goschnick, O. B. Popovicheva, C. Irimiea, Y. Carpentier, I. K. Ortega, E. D. Kireeva, N. K. Shonija, J. Schwarz, M. Vojtíšek-Lom, C. Focsa, H.-P. Ewinger, J. Goschnick, H. J. Ache, D. Kilic, B. T. Brem, F. Klein, I. El-Haddad, L. Durdina, T. Rindlisbacher, A. Setyan, R. Huang, J. Wang, J. G. Slowik, U. Baltensperger and A. S. Prevot, *J. Geophys. Res.: Atmos.*, 2017, **34**, 401–409.

27 A. Kortelainen, Ph.D. thesis, University of Eastern Finland, 2016.

28 M. Abegglen, B. T. Brem, M. Ellenrieder, L. Durdina, T. Rindlisbacher, J. Wang, U. Lohmann and B. Sierau, *Atmos. Environ.*, 2016, **134**, 181–197.

29 D. Delhaye, F. X. Ouf, D. Ferry, I. K. Ortega, O. Penanhoat, S. Peillon, F. Salm, X. Vancassel, C. Focsa, C. Irimiea, N. Harivel, B. Perez, E. Quinton, J. Yon and D. Gaffie, *J. Aerosol Sci.*, 2017, **105**, 48–63.

30 J. Moldanová, E. Fridell, O. Popovicheva, B. Demirdjian, V. Tishkova, A. Faccinetto and C. Focsa, *Atmos. Environ.*, 2009, **43**, 38–44.

31 C. Giorio, A. Tapparo, M. Dall'Osto, R. M. Harrison, D. C. Beddows, C. Di Marco and E. Nemitz, *Atmos. Environ.*, 2012, **61**, 316–326.

32 S. S. Lim, T. Vos, A. D. Flaxman, G. Danaei, K. Shibuya, H. Adair-Rohani, *et al.*, *Lancet*, 2012, **380**, 2224–2260.

33 V. Samburova, B. Zielinska and A. Khlystov, *Toxics*, 2017, **5**, 17.

34 R. Niranjan and A. K. Thakur, *Front. Immunol.*, 2017, **8**, 1–20.

35 T. Petry, P. Schmid and C. Schlatter, *Chemosphere*, 1996, **32**, 639–648.

36 D. A. Knopf, P. A. Alpert and B. Wang, *ACS Earth Space Chem.*, 2018, **2**, 168–202.

37 A. Faccinetto, C. Focsa, P. Desgroux and M. Ziskind, *Environ. Sci. Technol.*, 2015, **49**, 10510–10520.

38 R. Lemaire, A. Faccinetto, E. Therssen, M. Ziskind, C. Focsa and P. Desgroux, *Proc. Combust. Inst.*, 2009, **32**, 737–744.

39 Y. Bouvier, C. Mihesan, M. Ziskind, E. Therssen, C. Focsa, J. F. Pauwels and P. Desgroux, *Proc. Combust. Inst.*, 2007, **31 I**, 841–849.

40 P. Parent, C. Laffon, I. Marhaba, D. Ferry, T. Z. Regier, I. K. Ortega, B. Chazallon, Y. Carpentier and C. Focsa, *Carbon*, 2016, **101**, 86–100.

41 F. X. Ouf, P. Parent, C. Laffon, I. Marhaba, D. Ferry, B. Marcillaud, E. Antonsson, S. Benkoula, X. J. Liu, C. Nicolas, E. Robert, M. Patanen, F. A. Barreda, O. Sublemontier, A. Coppalle, J. Yon, F. Miserque, T. Mostefaoui, T. Z. Regier, J. B. Mitchell and C. Miron, *Sci. Rep.*, 2016, **6**, 1–12.

42 J. Yon, F.-X. Ouf, D. Hébert, J. Mitchell and N. Teuscher, *Combust. Flame*, 2018, **190**, 441–453.

43 R. H. Moore, L. D. Ziemba, D. Dutcher, A. J. Beyersdorf, K. Chan, S. Crumeyrolle, T. M. Raymond, K. L. Thornhill, E. L. Winstead and B. E. Anderson, *Aerosol Sci. Technol.*, 2014, **48**, 467–479.

44 R. Zimmermann, M. Blumenstock, H. J. Heger, K. W. Schramm and A. Kettrup, *Environ. Sci. Technol.*, 2001, **35**, 1019–1030.

45 O. P. Haefliger and R. Zenobi, *Anal. Chem.*, 1998, **70**, 2660–2665.

46 K. Thomson, M. Ziskind, C. Mihesan, E. Therssen, P. Desgroux and C. Focsa, *Appl. Surf. Sci.*, 2007, **253**, 6435–6441.

47 A. Faccinetto, K. Thomson, M. Ziskind and C. Focsa, *Appl. Phys. A: Mater. Sci. Process.*, 2008, **92**, 969–974.

48 C. Mihesan, M. Ziskind, E. Therssen, P. Desgroux and C. Focsa, *Chem. Phys. Lett.*, 2006, **423**, 407–412.

49 C. Mihesan, M. Ziskind, E. Therssen, P. Desgroux and C. Focsa, *J. Phys.: Condens. Matter*, 2008, **20**, 25221.

50 L. Sleno, *J. Mass Spectrom.*, 2012, **47**, 226–236.

51 C. A. Hughey, C. L. Hendrickson, R. P. Rodgers and A. G. Marshall, *Anal. Chem.*, 2001, **73**, 4676–4681.

52 A. G. Brenton and A. R. Godfrey, *J. Am. Soc. Mass Spectrom.*, 2010, **21**, 1821–1835.

53 R. Hilbig and R. Wallenstein, *Appl. Opt.*, 1982, **21**, 913–917.

54 S. E. Stein and A. Fahr, *J. Phys. Chem.*, 1985, **89**, 3714–3725.

55 T. Adam, R. R. Baker and R. Zimmermann, *J. Agric. Food Chem.*, 2007, **55**, 2055–2061.

56 Y. Tanaka, *Comm. Statist. Theory Methods*, 1988, 37–41.

57 R. E. Peterson and B. J. Tyler, *Atmos. Environ.*, 2002, **36**, 6041–6049.

58 P. Cejnar, S. Kuckova, A. Prochazka, L. Karamonova and B. Svobodova, *Rapid Commun. Mass Spectrom.*, 2018, **32**, 871–881.

59 T. Alexandrov, *BMC Bioinf.*, 2012, **13**, S11.

60 L. Pei, G. Jiang, B. J. Tyler, L. L. Baxter and M. R. Linford, *Energy Fuels*, 2008, **22**, 1059–1072.

61 P. Reitz, S. R. Zorn, S. H. Trimborn and A. M. Trimborn, *J. Aerosol Sci.*, 2016, **98**, 1–14.

62 R. C. Alvin, *Methods of multivariate analysis* Wiley - Interscience, 2nd edn, 2001, pp. 1–727.

63 F. W. McLafferty and F. Tureek, *Interpretation of Mass Spectra*, University Science Books, Mill Valley, CA, 1993.

64 J. Pagels, D. D. Dutcher, M. R. Stolzenburg, P. H. Mcmurry, M. E. Gälli and D. S. Gross, *J. Geophys. Res.: Atmos.*, 2013, **118**, 859–870.

65 J. Pagels, M. Strand, J. Rissler, A. Szpila, A. Gudmundsson, M. Bohgard, L. Lillieblad, M. Sanati and E. Swietlicki, *J. Aerosol Sci.*, 2003, **34**, 1043–1059.

66 T. R. Dallmann, T. B. Onasch, T. W. Kirchstetter, D. R. Worton, E. C. Fortner, S. C. Herndon, E. C. Wood, J. P. Franklin, D. R. Worsnop, A. H. Goldstein and R. A. Harley, *Atmos. Chem. Phys.*, 2014, **14**, 7585–7599.

67 E. S. Cross, A. Sappok, E. C. Fortner, J. F. Hunter, J. T. Jayne, W. A. Brooks, T. B. Onasch, V. W. Wong, A. Trimborn, D. R. Worsnop and J. H. Kroll, *J. Eng. Gas Turbines Power*, 2012, **134**, 72801.

68 K. Aras, *Atmos. Environ.*, 1994, **28**, 1385–1391.

69 S. Gilardoni, L. M. Russell, A. Sorooshian, R. C. Flagan, J. H. Seinfeld, T. S. Bates, P. K. Quinn, J. D. Allan, B. Williams, A. H. Goldstein, T. B. Onasch and D. R. Worsnop, *J. Geophys. Res.: Atmos.*, 2007, **112**, 1–11.

70 J. Yon, A. Bescond and F.-X. Ouf, *J. Aerosol Sci.*, 2015, **87**, 28–37.

71 D. Kilic, B. T. Brem, F. Klein, *et al.*, *Environ. Sci. Technol.*, 2017, **51**(7), 3621–3629.

DISCUSSIONS

Dealing with complexity: general discussion

Carlos Afonso, Mark P. Barrow, (iD) Nicholle G. A. Bell, Antony N. Davies,
Dumitru Duca, (iD) Cristian Focsa, (iD) Caroline Gauchotte-Lindsay,
Pierre Giusti, Ruth Godfrey, Royston Goodacre, (iD)
Jeffrey A. Hawkes, (iD) Norbert Hertkorn, Jeroen J. Jansen,
Donald Jones, William Kew, (iD) Adrien Le Guennec, Anneke Lubben,
Ljiljana Paša-Tolić, Ryan P. Rodgers, Christopher P. Rüger,
Philippe Schmitt-Kopplin, Peter J. Schoenmakers, Danilo Sciarrone, (iD)
Stephen Summerfield, Dušan Uhrín and Fleur H. M. van Zelst

DOI: 10.1039/C9FD90055A

Ryan P. Rodgers opened discussion of the introductory lecture by Philippe Schmitt-Kopplin: Emphasizing the importance of data integrity and metadata, with the idea that the data will be/can be mined by future scientists, seems logical, but with science policy/efforts evermore focused on advancing as fast as possible, where do you see the funding coming from for such efforts? How does this idea gain momentum so that it matures?

Philippe Schmitt-Kopplin answered: Many funding agencies in Europe require the main data to be saved on accessible depositories and these guidelines are followed very strongly in some scientific communities. The main limitation, however, is the harmonization of the data set format and the availability of the right metadata (sample information, instrumental setup, *etc.*) to enable use by the community in the future.

Peter J. Schoenmakers said: You described some challenging analyses, but in one sense they are still relatively easy. Most of your examples involved low molecular weight analytes. High molecular weight analytes pose additional challenges, especially when solubility becomes an issue. In one of your slides you referred to lignin structures, which are a prime example of highly cross-linked, essentially insoluble materials. I do not expect an instant solution for such materials, but I would appreciate your comments.

Philippe Schmitt-Kopplin responded: In the presentation, I tried to show that complexity (not only restricted to personal impressions such as "complicated" or "easy" that are rather subjective and related to the expertise of the individuals) covers a wide range of biotic and abiotic chemical continuums in time and space, in incessant dynamic interaction. The modern challenges of analytical chemistry are in the analysis of these complex super-mixtures (organic, organometallic and metallic/mineral) in a size continuum scale from small soluble molecules to macromolecules, nanoparticles and even larger sized objects. Thus, analytical chemistry in a larger sense involves challenges in surface chemistry or the analysis of particle size distributions, which actually reflects all projections of our environment on our five senses. Taking small molecules as an example was on purpose, because in the analysis of these "small and simple" targets one faces the problem of non-repetitive high chemical diversity making their analysis extremely difficult (non-linearity, extreme number of structural and stereoisomers, *etc.*). For biology-derived higher molecular weight chemical families (proteins, DNA, *etc.*), solving the structural problem means looking at the regular composition of successional elementary building blocks. And indeed, as you mentioned, even if lignin is based on elementary phenolic building blocks, its complexity derives from the non-linear and highly cross-linked structures, leading to partially soluble to mainly insoluble high molecular fractions (a reason for their high stability to biodegradation). The chemical analysis and structural description of super-mixtures, such as lignin or natural organic matter, need adapted chemical descriptors involving different scales of bulk chemical descriptions.

Adrien Le Guennec opened discussion of the paper by Ryan P. Rodgers: From the tens of thousands of compounds identified, how many are identified, and how many are monitored for a study?

Ryan P. Rodgers replied: As mentioned in the talk, the elemental compositions are determined with a high confidence. It is the preferential ionization of the source used, or sample fractionation, that provides insight into the chemical functionality of the elemental composition assigned.

Royston Goodacre asked: If diversity and/or complexity are related to your ability to separate chemicals, with respect to their functionality, polarity, *etc.*, prior to mass spectrometry, why spend considerable money and resources on a 21 T magnet, rather than concentrating on better fractionation and expanding the types of ionization techniques?

Ryan P. Rodgers answered: Because, if successful, the inherent complexity of the samples will still require a 21 T scan speed and resolution for the analysis, see Fig. 1.

Ljiljana Paša-Tolić asked: Is there any value in doing direct infusion ESI-MS analyses on complex mixtures? If yes, in which context should it be used?

Ryan P. Rodgers responded: There is value, but it depends largely on the chemical diversity of the sample. We have documented significant differences (50×) in the ionization efficiency within chemical "families" of molecules within complex mixtures. Thus, even within a single chemical functionality, you can have significant differences in the ionization efficiency.

Norbert Hertkorn enquired: Is the amplitude/signal per ion different for small and large molecules? What is, in your eyes, the primary reason for the different

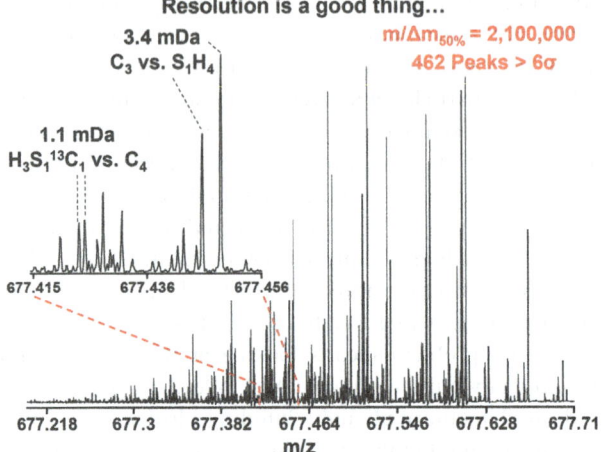

Fig. 1 Mass scale expanded segment of the (+) APPI FT-ICR mass spectrum (200 time-domain averaged transients) of an asphalt volcano sample at *m/z* 677 after ion trap isolation of a 20 Da segment. The inset illustrates the need for ultra-high mass resolving power in petroleum samples, and specifically those that contain abundant sulfur-containing species. The common mass splits for correct sulfur speciation (1.1 and 3.4 mDa) are highlighted. The figure is a modified version of one published in *Anal. Chem.*, 2018, **90**, 2041–2047.

ionisation efficiencies of molecules in complex mixtures (*e.g.* volatility, surface tension, aggregation, ion–molecule reactions, *etc.*)?

Ryan P. Rodgers replied: Our current understanding is that it is due to aggregation.

Stephen Summerfield remarked: I am interested in defining petrochemical substances as a consultant for industry, for the regulation of chemicals such as the EU REACH regulations. What is causing the ecotoxicology and toxic effects? Are you telling me that I am not actually seeing a large amount of what is actually present? This is adding orders of complexity.

Ryan P. Rodgers answered: Yes, the complexity currently exceeds our analytical capabilities.

Dušan Uhrín said: People here are analysing different matrices, but it seems that they are shooting in the dark. Do you have any advice so that we do not spend lots of money and time to discover that we have been looking in the wrong place?

Ryan P. Rodgers responded: It all depends on the matrix and the goal of the experiment. For complex small molecule mixtures (less than 2 kDa), we have had success with efforts to characterize them by differences in the structure and chemical functionality. This helps combat selective ionization, as all (or most) of the molecules have the same chemical functionality, but, as highlighted in the paper (DOI: 10.1039/c9fd00005d), even these efforts led to a humbling conclusion. Even within a chemical functionality, ionization differences between molecules of different molecular weight/structure can be more than $50\times$. Thus, depending on the mass distribution between these "types" of molecules, you can "see" as little as 10% of the species by direct injection analyses.

Nicholle G. A. Bell communicated: In the mass analysis section (DOI: 10.1039/c9fd00005d) you state "Mass spectrometer parameters and protocols are similar to those previously described". Can you please provide these so that one can repeat the experiment or utilize them?

Ryan P. Rodgers communicated in reply: Please see ref. 1 for the parameters and protocols.

1 M. L. Chacón-Patiño, S. M. Rowland and R. P. Rodgers, *Energy Fuels*, 2017, **31**(12), 13509–13518.

Jeffrey A. Hawkes said: In my study, I found it difficult to improve the signal to noise ratio of higher molecular weight material, and I had a very poor signal for this early eluting material in size exclusion chromatography using the tune settings that were optimised for the low molecular weight material. Did you change the tune settings in your study in order to optimise the analysis of the MA1–MA6 samples?

Ryan P. Rodgers replied: Other than the concentration and solvent system/modifier, there is no optimization required. A simple pre-analysis by a time-of-flight or ion trap system can give valuable information for the high resolution analysis.

Cristian Focsa opened discussion of the paper by Jeffrey A. Hawkes: You propose an interesting comparison between UV optical absorption and electro-spray ionisation efficiency. However, I wonder how these can be comparable, as previous experience shows that even comparing optical absorption with resonance-enhanced multi-photon ionisation (REMPI) can be a difficult task, partly because of the presence of isomers (see ref. 1, for example). Is there a possible physical link between high UV absorbance and effective electrospray ionisation? Moreover, it would be interesting to combine both electrospray and multi-photon ionisation methods on the same instrument.

1 A. Faccinetto, P. Desgroux, M. Ziskind, E. Therssen and C. Focsa, *Combust. Flame*, 2011, **158**, 227–239.

Jeffrey A. Hawkes answered: Our approach was not strictly to compare UV and ESI responses, but to use the UV profile to determine what the retention profile measured by ESI-MS should be, under the assumption that all dissolved organic matter (DOM) has some sort of chromophore. As you point out, this is not strictly a valid assumption. However, it is true that where chromophores were detected, we are certain that material exists, so an ESI-MS response may be expected. This is a very diverse mixture, so it is not ideal to categorise material into small groups, but we may take the liberty and say that we found three "groups" that were light-absorbing but not ESI-MS active, active in both detection methods, and ESI-active but transparent. These three groups can not be separated by other methods, such as reversed-phase HPLC. To answer the question more directly: no, we have no reason to believe that there is a link between sensitivity in UV absorption and electrospray ionisation, and our study actually demonstrates that these detection methods are quite orthogonal in natural DOM.

Peter J. Schoenmakers queried: Why did you choose size exclusion chromatography (SEC) in this case? In the case of perfect SEC (no interaction effects), the information obtained is essentially the same as that obtained from your mass spectrum. A high-resolution separation (for example, a reversed-phase LC gradient) would be more sensible.

Jeffrey A. Hawkes responded: You are right that an orthogonal method would be better suited for the most effective separation of the mixture. There are a few publications on this topic, including 2D methods. However, in this study, our objective was to investigate the proposed tertiary structure of DOM as macro-aggregates that can be dissociated into monomers in an ESI source. For this investigation, SEC is a useful tool, under the assumption that aggregates are maintained in the column, eluting early for investigation by ESI-MS in isolation. We did not find evidence for the dispersal of this material into monomers, though, and concluded that ESI does not break up aggregates. Instead, aggregates are either stable, or the large material is genuinely high molecular weight and is not detected by our mass spectrometer for reasons like its ionisation potential or its mass to charge ratio being out of the tuned range.

Pierre Giusti commented: For all the spray-based ionisation techniques, and more specifically ESI, aggregation tendencies can be enhanced during the desolvation process. What do you think about using fraction collection and measuring the elemental composition in each fraction in order to validate that you are not missing some black matter with mass spectrometry?

Jeffrey A. Hawkes replied: This is a good point. Currently, we rely on light absorption for the detection of organic matter. A more quantitative technique would be better, and we are looking into online quantitation methods for future work. We also collected some fractions recently and analysed the dissolved organic carbon (DOC) concentration – see Fig. 2, which shows a profile somewhat between that of the UV absorption and ESI-MS, suggesting that one or both of these techniques respond to the bulk of the DOM. The only time bin when the DOC response was not between the responses of UV and ESI-MS was in the last point, when transparent and ESI-MS poor neutrals may be present.

Ljiljana Paša-Tolić noted: Generally, it is challenging to detect larger analytes in the presence of smaller analytes with ESI-MS due to differences in the ionization efficiencies, dilution of the MS signal among many channels in the case of larger analyses, *etc.* Hence, SEC can be rather useful to get a more uniform response from a mixture containing analytes ranging in size (mass), as has been demonstrated for proteins for instance.

William Kew said: The UV trace showed a normal distribution, as does the typical mass spectra for these DOM samples, however the total ion current (TIC)

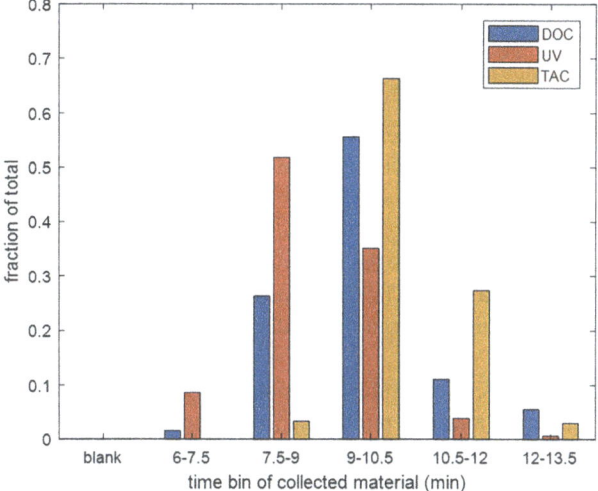

Fig. 2 Abundance analysis by three detection methods (TAC = total assigned current by ESI-MS, UV = UV absorbance at 254 nm, DOC = dissolved organic carbon abundance), presented as a fraction of the total amount detected. The DOC analysis was conducted on collected fractions, analysed by Shimadzu TOC analyser. The UV and ESI-MS signals were summed in the equivalent time bins.

spectrum showed almost a bimodal distribution. Can you comment on the nature and cause of this bimodal distribution?

Jeffrey A. Hawkes answered: The chromatographic trace for UV absorption is indeed quite normal, and this is actually a logarithmic normal distribution as "size" increases exponentially with the inverse of retention time (see Fig. 7 in DOI: 10.1039/c8fd00222c). The TIC is quite difficult to interpret because of mass spectrometer tuning issues and ESI suppression. We have noticed a somewhat wave-like pattern in the signal intensity with our Orbitrap over the broad mass range, and it may be the case that intermediate mass material around 1000 Da has a lower response due to the ion optics, but we have not investigated this deeply. Ion suppression is another issue that is difficult to investigate – possibly, as the lower mass material comes in (represented by the total assigned current (TAC) graph), the higher mass material (1000–2000+ Da) is suppressed, leading to a slightly lower average response. Again, I am not confident about this at this stage.

Anneke Lubben commented: You mentioned that you are unsure whether or not the electrospray ionisation is resulting in the disaggregation of your clusters, eluting off the SEC. One way to determine this would be to collect fractions from the high molecular weight fraction and the low molecular weight fraction, and then analyse them under varying source voltage conditions to see if you see any changes with voltage. For the high molecular weight fraction of aggregates, if you find that with low voltages you see the higher molecular weight, but at higher voltages these transition to only low MS ions, then it suggests that the ESI source is indeed causing disaggregation. It is worth a try, and should be fairly straight forward to do.

Jeffrey A. Hawkes responded: We did not modify the ESI voltage much, but we did look into adding source collision induced dissociation voltage (see Fig. SI2, DOI: 10.1039/c8fd00222c). Adding this voltage should have dispersed weak aggregates, leading to an increased signal of the lower masses, but we saw no evidence of this, supporting the overall conclusion of the study – that monomer carboxylic acids are probably not present in solution as weak aggregates in the ESI spray.

Ruth Godfrey asked: What is the repeatability of the calibration data provided for describing the separation? The chromatographic resolution is not fantastic, so what is the error on the predicted molecular weight?

Jeffrey A. Hawkes replied: Indeed, the resolution of the standard peaks (the calibration standards and our purchased compounds) is low. Because the calibration turned out to be very poor at predicting the molecular weight for known compounds, due to the issues discussed with charge density, we did not put a lot of effort into quality control of the calibration, and instead ended up using elution time and "apparent molecular weight" as a guide, and not a definitive value. The variation of the apex elution time for the standards was actually rather small, but we did not publish an analysis of this and we feel that this is not a main message of the paper. However, in Fig. SI1 (DOI: 10.1039/c8fd00222c) you can see the very low variability in the retention profile of the DOM in Suwannee River Fulvic Acid (SRFA) at different loading volumes. We do not recommend this high pressure size exclusion chromatography (HPSEC) method for the separation of DOM in order to estimate its size/mass, but only as a means of broadly separating small and large material.

Nicholle G. A. Bell communicated: In your methods section (DOI: 10.1039/c8fd00222c), you mention that you have extracted FA and HA from soil samples using ammonia. Can you explain why you decided to use ammonia, as the standard method for HA extraction uses NaOH? The DOM you mention was stored in the dark for >2 years. How do you know that the sample has not degraded/changed with time? We see microbial changes to DOM samples in a matter of days.

Jeffrey A. Hawkes communicated in reply: We used ammonia to ensure full compatibility with ESI-MS during detection (as it is volatile). The "Stream RO" sample is a reverse osmosis concentrate from a wetland stream that we use for method development in our lab. As you suggest, this sample is mainly used to represent non-labile, terrestrial DOM, and it is likely that it has degraded further over time. For a fresh, labile sample, we used the leaf extract. The soil HA and FA fractions could also be considered fresh, as they were prepared directly before analysis.

Mark P. Barrow opened a general discussion of Ryan's and Jeffrey's papers: We are seeing examples of both off-line fractionation and online separation. Do you regard one method as being preferential for the future? What will be the bottlenecks? For example, will it be time spent in the laboratory or will it be the data analysis afterwards?

Ryan P. Rodgers answered: Currently, it is data analysis and consumption. However, I do not think this will last long. The issue when moving to separations prior to mass analysis (for complex systems) will be finding a suitable separation method to deal with complex, polyfunctional mixtures.

Jeffrey A. Hawkes responded: In my opinion, online methods are superior because of the ability to set up a run overnight and have relatively little human influence on the data collection. It also takes less time and effort overall. In my paper, we found that collecting fractions for re-analysis gave very similar results to the online method, which was highly encouraging. However, we were using an isocratic method, so you would not expect a change in response by re-dissolution in the same solvent. We could have improved the signal by re-dissolution in a better ESI solvent, and in gradient runs this could be pretty useful if the objective is to see as many peaks as possible. The bottleneck is always in the data analysis, not data collection.

Ruth Godfrey enquired: Based on the impact of ionisation efficiency and suppression, are you putting forward an argument that we need better ionisation sources that cover a range of techniques and, if so, which techniques do you think would cover the range of chemistries required?

Jeffrey A. Hawkes replied: A more universal ionisation source would make a dramatic improvement to the mass spectrometry analysis of DOM. It is possible that a new ionisation technique will be invented, but another solution may be to have a combined source that incorporates more than one technique. Techniques that have been found to improve coverage of DOM include photo-ionisation (APPI), chemical ionisation (APCI, to a lesser extent than APPI), and deposition-based techniques, like paper spray and laser desorption (LDI).

Ryan P. Rodgers answered: Better, brighter ion sources are clearly desirable. However, it appears that self-association, and methods to deal with it prior to analysis, are equally important.

Danilo Sciarrone asked: Do you have some protocol involving the use of multidimensional chromatographic techniques? Since the samples are very complex, the use of a higher separation power before the MS step could help to achieve better results in an easier way

Ryan P. Rodgers responded: Yes, we separate by structure (number of aromatic rings) and by chemistry (hydrocarbons *vs.* sulfidic species) through the use of two separate columns.

Jeffrey A. Hawkes replied: We do not have any multidimensional techniques operating in our group. I agree that this would be useful, but the data acquisition speed may become a problem with the relatively low resolving power of our mass spectrometer (assuming that the second chromatographic dimension is fairly fast).

Adrien Le Guennec queried: In both studies, do you think having NMR data on top of the other data would or could give you additional information that would be difficult or impossible to access with mass spectrometry?

Jeffrey A. Hawkes answered: It depends on the objectives of the study. For understanding a small number of complex mixtures (as in this study), the more information the better, and NMR data would most definitely be useful. However, if the study aims to compare hundreds of samples across the landscape or over time, you can end up with diminishing returns by adding more and more information – often a message becomes quite clear with a single analytical technique – and having an ionisation bias (ESI, for example) is not a problem as long as you are clear about what you are comparing among the samples.

Ryan P. Rodgers responded: Yes, specifically in samples that self-associate.

Donald Jones opened discussion of the paper by Peter J. Schoenmakers: What are the advantages of GC-GC over LC-IMS-MS?

Peter J. Schoenmakers replied: Whether we consider heart-cutting GC-GC (sending a single fraction to the second dimension) or comprehensive two dimensional GC × GC, the pertaining answer is that these techniques may be used whenever it is allowed by the analytes (and possibly disturbing matrix components). If volatile samples are considered and GC can be used, it has many advantages. Standard GC analyses are more efficient, highly reliable, and flame-ionization detection (FID) can be used for approximate quantitation without the need for calibration. There are fantastic free libraries to aid in the identification of components characterized by GC-MS, and GC × GC is a robust and reliable technique. There are trends towards the increased use of tandem MS (MS/MS) systems and high-resolution (HRMS) systems in combination with GC. The combination of GC with ion-mobility spectrometry and MS is quite feasible, but not common. If GC-MS is inadequate for a given complex (volatile) sample, in my opinion GC × GC-MS is fundamentally and practically the most attractive next step.

Liquid chromatography (LC) or, in some cases, supercritical-fluid chromatography (SFC) can be used for samples that are not compatible with GC. A direct comparison of GC × GC and LC-IMS-MS belongs in the category of apples and oranges.

William Kew said: The manuscript discusses the difficulties with coupling the eluent from the first LC to the inlet of the second, with conflicting solvent types. One solution discussed is vacuum evaporation – similarly, could you use supercritical fluid chromatography as the first stage, as the supercritical CO_2 readily evaporates allowing for easy dissolution for the second stage? What might be the benefits and pitfalls of this approach?

Peter J. Schoenmakers answered: SFC can absolutely be used in the first dimension and/or in the second dimension. The latter is attractive because SFC separations can be very fast. A disadvantage is that SFC is not as broadly applicable as LC. Analytes with multiple ionisable sites, such as peptides and proteins, or ones with high molecular weights (polymers) are not readily amenable with SFC.

Ruth Godfrey asked: You discuss an algorithm to optimise the conditions in your paper – have you tested it for compound quantification? Also, can it be used to determine and optimise the number of data points across a chromatographic peak? Is there a repeatable peak area for the calibration line? Over a linear range, do you get sufficient data points?

Peter J. Schoenmakers responded: Our optimization concerns the chromatographic separation, including the number of cuts per first dimension peak. We also consider the detection sensitivity. This is a stage before calibration and validation of the quantitative analysis. Generally, if resolution and sensitivity are optimized, better quantitative results are to be expected.

Antony N. Davies queried: Your results look like they can be used to help design microfluidic systems, are you also working in that direction?

Peter J. Schoenmakers replied: Microfluidic devices for chromatography have a long history, but their acceptance is still modest. Progress is still being made, though. For example, developments towards low-thermal-mass GC and pillar-array columns for LC are quite exciting. In our group we are pursuing microfluidic devices for LC × LC modulators and modulation reactors. The latter offers a huge range of new possibilities. The paradigm change from column-based LC × LC to spatial LC × LC and, eventually, comprehensive three dimensional spatial LC (LC × LC × LC) hinges on the successful development of microfluidic devices and we have a large effort dedicated to designing such devices and to their fabrication using state-of-the-art additive-manufacturing (3D-printing) techniques.

Ruth Godfrey enquired: You mentioned forensic applications and facial recognition, could you see machine learning increase throughput for this algorithm as a peak detection process?

Peter J. Schoenmakers answered: There is great promise in the application of artificial intelligence (AI) in analytical separations and in analytical chemistry in general. However, it all hinges on the types and amounts of data available. We have just started exploring AI techniques for finding information (in our case, peaks) in very large amounts of data (mainly noise) in a large consortium, which also features applications as diverse as gravitational waves, bird migration, radio signals from space and gene regulation networks. However, analytical chemists often find themselves in a position where we have to extract the most and best possible information from limited data (*e.g.* one or a few samples). Conventional (deterministic) algorithms are already helpful to establish the presence of peaks in such situations.[1]

1 B. W. J. Pirok, S. R. A. Molenaar, L. S. Roca and P. J. Schoenmakers, *Anal. Chem.*, 2018, **90**, 14011–14019.

Dušan Uhrín remarked: I am not that familiar with different LC techniques. Is 2D LC hardware readily available? How about software to optimise the method, is that available too?

Peter J. Schoenmakers responded: Good hardware for LC × LC is now available from several manufacturers. These systems are adequate for routine LC × LC operations and they offer advanced options, such as variable (shifting) second dimension gradients. The limiting factor has now become the software for data analysis and, especially, method development. Without the aid of smart computer programs, it is virtually impossible to establish truly optimal separations. Several research groups are developing such software. We are working towards making a version of our own software[1] available to the public.

1 B. W. J. Pirok, S. Pous-Torres, C. Ortiz-Bolsico, G. Vivó-Truyols and P. J. Schoenmakers, *J. Chromatogr. A*, 2016, **1450**, 29–37.

Antony N. Davies asked: With regard to the need for a fast second dimension, have you run with UHPLC or UHPLC-UHPLC?

Peter J. Schoenmakers replied: We first demonstrated the use of UHPLC conditions (pressures exceeding 100 MPa, particle size smaller than 2 μm) some years ago.[1] We now routinely use such conditions in two dimensional LC separations. Generally speaking, the best results (most efficient separations, highest overall peak capacities) can be obtained when the highest pressures are utilized in both dimensions. We have also demonstrated the use of core–shell particles to achieve fast second dimension separations.[2]

1 E. Uliyanchenko, P. J. C. H. Cools, S. van der Wal and P. J. Schoenmakers, *Anal. Chem.*, 2012, **84**, 7802–7809.
2 B. W. J. Pirok, P. Breuer, S. J. M. Hoppe, M. Chitty, E. Welch, T. Farkas, S. van der Wal, R. Peters and P. J. Schoenmakers, *J. Chromatogr. A*, 2017, **1486**, 96–102.

Donald Jones queried: Have you got any differential pressure between the columns and does that cause any problems?

Peter J. Schoenmakers answered: One of the attractive features of comprehensive two dimensional liquid chromatography (LC × LC) is that we have independent control of the mobile phase compositions and the flow rates in the two dimensions. Thus, it is entirely possible to operate the second dimension column at a higher inlet pressure than the first dimension column. This situation is opposite to the one we encounter in comprehensive two dimensional gas chromatography (GC × GC), where one uninterrupted stream of carrier gas flows through both columns.

Jeffrey A. Hawkes commented: In 2D chromatography, particularly in some of the methods highlighted in your paper, the second dimension is often completed on the scale of several seconds or 1–2 min. What kind of chromatographic resolution do you need for mass analysis (*i.e.* what frequency of analysis), and can you assign an accurate formula with your mass spectrometer, which in my experience requires at least a mass resolving power of 30 000 at m/z 400? Do you think that analysis speed of mass spectrometers will prove to be an obstacle for the widespread use of 2D methods?

Peter J. Schoenmakers responded: This is certainly an issue. In our LC × LC experiments, we have taken the second dimension separation time down to 20 s.[1] If we have a peak capacity of 20, that implies an average peak width of 1 s, which should allow a high resolution mass spectrum to be obtained. We just acquired an Orbitrap QE plus system with a maximum scanning speed of 20 Hz, so that a compromise between speed and resolution can be struck. The situation is more challenging in GC × GC, where the entire second dimension separation may indeed be completed in a few seconds. In GC × GC-MS, time-of-flight mass spectrometers are used most often, but not exclusively. Other types of mass spectrometers have become faster in recent years.

1 E. Uliyanchenko, P. J. C. H. Cools, S. van der Wal and P. J. Schoenmakers, *Anal. Chem.*, 2012, **84**, 7802–7809.

Norbert Hertkorn opened discussion of the paper by Danilo Sciarrone: What are the most promising opportunities to increase your analytical window from your existing analytical set-up?

Danilo Sciarrone replied: A number of additional features have been implemented and published as well, extending the analytical possibilities of the set-up described:

(1) The use of front-end LC pre-separation may increase the absolute sample amount introduced in the system, thanks to the injection of 10× (at least) volumes, exploiting the much higher sample capacity of a packed LC column.

(2) The use of an additional valve between the FID and the Deans switch device in the first and second chromatographic dimensions allows for greater flexibility, *i.e.* for the collection of purified components eluted from any of the three separation steps.

(3) The use of a chiral stationary phase in the final chromatographic dimension would extend the range of collectable compounds, to include pure enantiomers based on the column selectivity.

Caroline Gauchotte-Lindsay said: I am interested in the transfer between the two GCs and at what temperature this happens. Would you be able to detail this?

Danilo Sciarrone answered: The heart-cut process, based on the use of a Deans switch device, is generated upon pressure balancing. In detail, the Deans switch device is equipped with a three-restrictor system. Thanks to the presence of a fixed restrictor (R1) on the Deans switch inlet side, the back-pressure at the column outlet is maintained constant during the heart-cut stage, avoiding retention time shifts. Another restrictor (R2), generating a higher pressure drop with respect to R1, allows for the heart-cut of selected chromatographic bands according to a valve (V) switching. During the stand-by analysis, since no restriction is present on the auxiliary pressure control flow path relative to the second dimension, a slightly higher pressure is generated at the second column head with respect to FID1 where, on the contrary, R1 generates a constant pressure drop, driving the diversion of the eluted fraction to FID1, as in a conventional GC-FID run. When a heart-cut has to be performed, the valve is switched by the software according to the selected heart-cut time, generating a slightly lower pressure at the head of the second column with respect to FID1. In fact, a higher pressure drop is generated by R2, characterized by a higher resistance with respect to R1, producing the diversion of the fraction to the second column. On the other hand, temperature control is not a prerequisite for the transfer to occur. The use of a lower temperature in the second oven can be used to achieve a focusing of the band transferred, by compensating the band broadening effects occurred, eventually.

Caroline Gauchotte-Lindsay remarked: We have been looking at designing a GC prep system too, however, we work with extract samples from the environment and the compounds we are looking to collect are at trace concentrations. I was wondering if the collection "inlet" could be used to trap a smaller amount, maybe in a solvent?

Danilo Sciarrone responded: The collection system can be successfully used to trap trace components, however the data shown in my presentation refer to neat samples where trace components are still present in known absolute amounts. On the other hand, in extracted samples, as in your case, dilution by an unknown factor occurs in the extraction medium, and thus the original absolute amounts of the sample components are greatly reduced. The use of front-end LC pre-separation could help in such cases, by increasing the absolute amounts introduced in the system. This is possible thanks to the injection of $10\times$ (at least) volume of the extracted sample, exploiting the much higher sample capacity of the packed LC column.

Antony N. Davies commented: It is interesting to see you use GC-FTIR, I have not seen that in years. What do you do about the differences in sensitivities for the detection systems, aren't they orders of magnitude different?

Danilo Sciarrone replied: FTIR spectroscopy is clearly characterized by a lower sensitivity with respect to MS or other detectors. On the other hand, in the present application, we isolated a large sample amount allowing the acquisition of the GC-FTIR data. The highly informative spectrum has been included in the information acquired for the characterization of the unknown molecule together with MS and NMR spectra.

Dušan Uhrín said: You isolated 3 mg of a compound that constituted 30% of the mixture. How much of the starting material have you used? How many repeat runs were required? What is your general view on using GC as a preparative chromatography method for less abundant compounds?

Danilo Sciarrone answered: Each injection, consisting of 2 μL of the neat sample, provided, once collected, about 0.6 mg of the component (based on its 30% amount in the sample). In order to reach a total of 3 mg, 5 consecutive runs were done. Regarding the collection of less abundant components, front-end LC pre-separation may be exploited to improve the productivity of the system, since the feasibility for higher column loading will greatly reduce the analysis time needed to collect milligram amounts.

Ljiljana Paša-Tolić remarked: You collected and characterized the most abundant species. How applicable is this approach if you are looking for the less abundant species? What is the dynamic range one can achieve?

Danilo Sciarrone responded: We demonstrated the capability of the system to collect components accounting down to about 5% of the neat samples in an easy way, exploiting the three dimensional configuration. Pre-separation operated with a front-end HPLC system proved to be effective to pre-concentrate the fraction of the sample containing the compound of interest before its further injection into the MDGC system. This approach could provide a great reduction in the collection time of 1–5% components.

Nicholle G. A. Bell communicated: You determined the structure of the unknown compound using NMR spectroscopy. Please could you provide the obtained spectra and/or the obtained chemical shifts/couplings? These should be reported for unknown compound structure determination.

Danilo Sciarrone communicated in reply: All the data can be found in the electronic supplementary information (DOI: 10.1039/c8fd00234g).

Pierre Giusti enquired: Regarding the quantification and, more precisely, the absolute quantification of the unknown, are you considering unspecific isotope dilution of carbon after combustion (through CO_2) as a relevant tool (in combination with molecular weight determination by HR-MS)? Because the response factor in FID can vary quite a lot (0.8–1.2) especially when containing a lot of oxygen.

Danilo Sciarrone replied: For sure, unspecific isotope dilution of carbon after combustion (to CO_2) may be successful for absolute quantification. However, less sophisticated and cheaper methods are also capable of affording a unitary response factor, such as the use of methanizer-like systems, converting all the organic matter to CH_4 before FID.

Antony N. Davies commented: This is high level research that we can deploy in industry – the ultimate resolution is not what we require for industry (not the main factor), we need stability, reproducibility and robustness.

Danilo Sciarrone responded: As reported in the literature, the system has proved to be stable with very good reproducibility and robustness.[1,2]

1 D. Sciarrone, S. Pantò, C. Ragonese, P. Dugo and L. Mondello, *Trends Anal. Chem.*, 2015, **71**, 65–73.
2 D. Sciarrone, S. Pantò, C. Ragonese, P. Q. Tranchida, P. Dugo and L. Mondello, *Anal. Chem.*, 2012, **84**, 7092–7098.

Jeroen J. Jansen opened discussion of the paper by Dumitru Duca: You use a three-step approach to resolve compounds – step 1 is calculating the mass defects, then you assign peaks to several compounds, and then the next step is doing principal component analysis (PCA). Why don't you use the information from the first step to help with the second step? This would give more power and better interpretability.

Dumitru Duca answered: Having an approach with independent steps is advantageous, as it allows us to test the assignment obtained from the mass defect analysis with statistical methods (based on covariance and correlation).

This independent approach has better chances to isolate a possible erroneous assignment from the first step (which can happen, for instance with a low signal to noise ratio or mass calibration issues) in the second step.

Jeroen J. Jansen said: PCA is not unsupervised – you aim for linear correlation between fragments of the same ion. Every peak will have its own measurement error, and you will get rid of this error when you cluster the peaks together that belong to the same compound.

Dumitru Duca responded: PCA does not require a training set of data for clustering, which means that we do not impose any hypothesis on the chemical composition of the sample, leaving only the mathematical requirement of the linear correlation between peaks. For laser-based techniques, the fragmentation is reduced by adjusting the laser fluence (desorption and ionization), however this does not mean that the mass spectra do not contain "redundant" information, *i.e.* the mass peaks that are highly correlated (*e.g.* isotopic peaks, species with the same source, *etc.*). The error, in this case, can indeed be reduced by clustering them. The advantage of our methodology is the use of the mass defect analysis in conjunction with several statistical methods that rely on both correlation and covariance, thus allowing us to precisely identify groups of common source peaks and then perform a second multivariate analysis on the simplified data (*e.g.* another PCA or hierarchical clustering on principal components (HCPC)).

Christopher P. Rüger commented: Deploying state-of-the-art offline approaches, such as LC/GC-MS, will not only back-up the assignments presented in the paper but also allow for further structural information. Ultimately, structural information is needed for understanding the combustion process in detail.

Dumitru Duca answered: This is indeed true, and has been done in some of our previous work, *e.g.* ref. 1.

1 A. Faccinetto, P. Desgroux, M. Ziskind, E. Therssen and C. Focsa, *Combust. Flame*, 2011, **158**, 227–239.

Cristian Focsa replied: We note that the mentioned techniques, providing structural information, usually require a significantly larger amount of sample material (in the range of mg per sample) to provide reliable results. In some cases (*e.g.* transient regimes of internal combustion engines or emissions of specific, expensive engines, such as airplanes[1]), collecting this amount of sample material can be a difficult (if not impossible) task. Our technique provides an excellent limit of detection (sub-fmol) for most of the analytes related to combustion processes,[2] which allows us to work with samples of very low mass.

1 D. Delhaye, F.-X. Ouf, D. Ferry, I. K. Ortega, O. Penanhoat, S. Peillon, F. Salm, X. Vancassel, C. Focsa, C. Irimiea, N. Harivel, B. Perez, E. Quinton, J. Yon and D. Gaffie, *J. Aerosol Sci.*, 2017, **105**, 48–63.
2 A. Faccinetto, C. Focsa, P. Desgroux and M. Ziskind, *Environ. Sci. Technol.*, 2015, **49**, 10510–10520.

Christopher P. Rüger said: The contribution nicely points out the benefits and the analytical power of deploying multivariate analysis in the mass spectrometric analysis of combustion-generated aerosols. When performing photoionization mass spectrometry, the mass spectrometric response when varying the wavelength (perhaps *via* an OPO) is highly interesting and might give you the chance to further assess the complexity of the aerosol or improve the statistical data analysis. The results of the photoionization, in particular the assignment of chemical formula, should be backed-up with offline state-of-the-art techniques in aerosol chemistry, such as GC- and LC-MS protocols. This would definitely contribute to the validation of this approach and study.

Cristian Focsa responded: Both techniques, chromatography and desorption/ionization mass spectrometry, have their own advantages; however, the study of soot particles usually involves very small amounts of sample material and in this context mass spectrometry (the surface sensitive desorption/ionization approach) is preferred and can represent a more suitable solution. Laser-based mass spectrometric techniques have a very small detection limit, tens of attomoles, which makes them most suitable for the characterization of samples that are only available in limited amounts (ng or even less) for reasons such as cost (*e.g.* soot produced by aeronautic engines) and collection difficulty (*e.g.* soot produced in transient engine regimes). Although the isomer discrimination is not (easily) obtained, the assignment of chemical formulas to photoionization products is nevertheless robust on a mass range defined by the mass resolution and accuracy of the analyzer. A combination of both techniques is, of course, desirable, when the collected sample mass allows it.

Dumitru Duca replied: Using multiple ionization wavelengths is indeed interesting as it can provide complementary information and, in some cases, help to differentiate isomers. We routinely use several UV wavelengths (266, 157 and 118 nm) in order to target species belonging to different chemical classes (*e.g.* ref. 1), but this is not showcased in this current paper since its main focus is the multivariate analysis of mass spectrometric data. Concerning the use of continuously tunable radiation (*e.g.* from an OPO), we do have the technical capability to do it, as we have fs/ps radiation covering the 0.2–20 μm spectral range, but a systematic study on a multitude of chemical species would be very time consuming and has not yet been performed. We note, however, that some work has been published in the literature (*e.g.* ref. 2), but this is on limited spectral ranges.

1 O. B. Popovicheva, C. Irimiea, Y. Carpentier, I. K. Ortega, E. D. Kireeva, N. K. Shonifa, J. Schwarz, M. Vojtíšek-Lom and C. Focsa, *Aerosol Air Qual. Res.*, 2017, **17**, 1717–1734.
2 O. P. Haefliger and R. Zenobi, *Anal. Chem.*, 1998, **70**, 2660–2665.

Royston Goodacre asked: Is there any reason why for secondary ion mass spectrometry (SIMS) you used a Bi_3^+ ion gun for the analysis of soot rather than say buckyballs (C_{60}^+) or water, which are softer in their ionization and cause less fragmentation? This may make the downstream data analysis easier.

Dumitru Duca answered: The ToF-SIMS instrument used in our analysis does not have the ability to use buckyballs (C_{60}^+) or water for ionization. Another reason why highly energetic Bi_3^+ ions were used is to ensure the ionization of the majority of species present in the sample, even with the highest ionization potential.

Antony N. Davies queried: You use small amounts – what techniques have you got in place to show these are actually representative of the population as a whole? Do you use an average from various positions?

Dumitru Duca responded: Mass spectra from multiple zones on the same sample (filter) are recorded and then compared to ensure that the samples are homogeneous. Spectra obtained from the different zones are not averaged, instead they are used as input for multivariate statistical analysis techniques. This approach can help identify some possible issues (*e.g.* if one spectrum is not clustered with other spectra from the same sample), thus ensuring that the extracted information is representative of the sample.

Nicholle G. A. Bell communicated: Table 1 (DOI: 10.1039/c8fd00238j) shows the different soot samples analysed. However, what is not clear is why the substrates used were not the same for all samples and why you used two-step laser mass spectrometry (L2MS) for the miniCAST samples but SIMS for all the others?

Dumitru Duca communicated in reply: We describe a comprehensive methodology for analyzing combustion-generated aerosols, providing three examples of applications to prove its universal character. Since in practice these aerosols can be sampled on a variety of substrates and analysed with a variety of techniques, the proposed methodology should be able to encompass all these experimental arrangements. We therefore show that regardless of the substrate and analysis method (L2MS or SIMS), it is possible to extract the maximum amount of chemical information from each mass spectrum.

William Kew said: The statistical techniques employed here – principal component analysis (PCA) and hierarchical clustering analysis (HCA) – are well known and well used techniques. However, there are modern data dimensionality and clustering techniques that might provide insights that PCA cannot provide, such as uniform manifold approximation and projection (UMAP) or t-distributed stochastic neighbour embedding (t-SNE). Have you tried to use these techniques, and, if so, do they yield interesting results?

Secondly, the distance metric used in HCA here was Euclidean, however many other metrics exist. Was the choice of Euclidean deliberate, and do other metrics offer any potential advantages or disadvantages?

Dumitru Duca answered: Mass spectra of combustion generated aerosols generally contain mass peaks that are linearly correlated, and in this case PCA and HCA are very effective in unraveling their complexity. Other more modern techniques have also been tested and provided similar results. However, they require some more optimization in order to be reliable and give reproducible results when applied to mass spectra obtained with a variety of analyzing techniques and from different substrates. This is something that our group is currently working on.

For HCA, several metrics were tested (*e.g.* Euclidian and Manhattan distance) as well as different linkage criteria, and it was found that, in our case, they do not influence the result in a significant manner. Even though the shape of the dendrogram changes (due to different distances), the final clusters remain the same.

Ruth Godfrey opened a general discussion of Peter's, Danilo's and Dumitru's papers: Do we have an idea in terms of specificity? Do we understand what the false positive rate is in terms of classification and in determining the optimum conditions? Do you have a control matrix to test your algorithm?

Peter J. Schoenmakers responded: The algorithm that we have published concerns the development of chromatographic methods, striking a balance between retention time, resolution and, ultimately, robustness. This is a stage before the application and validation of the method for quantitative purposes. In general, we may expect that the quantitative performance is improved if resolution is enhanced.

Dumitru Duca replied: Our formula assignment starts with the mass defect analysis and is then tested with several unsupervised statistical methods (PCA, HCA). Using these two steps independently provides a control for isolating possible false assignments from the first step in the second step.

Norbert Hertkorn remarked: There are fundamental relationships between resolution, bandwidth and the time you need to get an optimal combination of these – we cannot overcome this. Better resolutions in the x- and y-directions are both important for sample discrimination and inevitably need a longer acquisition time.

Fleur H. M. van Zelst responded: A higher resolution indeed calls for longer acquisition times. In 1D NMR spectroscopy this is hardly an issue, however, the time to acquire an FID is only a small part of the entire experiment time. In 2D NMR spectroscopy, however, longer acquisition times indeed call for recording more experiments and thus substantially longer measurement times. This can be alleviated, however, by non-uniform sampling (NUS). Furthermore, given sufficient sensitivity, it is possible to use the approach developed by Frydman and coworkers to record an entire 2D spectrum in a single scan (ultrafast 2D).[1]

1 L. Frydman, T. Scherf and A. Lupulescu, *PNAS*, 2002, **99**, 15858–15862.

Ryan P. Rodgers commented: In complex mixture analysis one often focuses on the resolution required for broadband analysis. However, if mass isolation is performed, the resolution requirements often increase considerably due to the increase in the dynamic range of the mass isolated segment. We have seen the complexity per nominal mass exceed 500 peaks in petroleum analyses. Thus, when MS/MS is performed with mass isolated windows, how confident are you that you know the elemental compositions in the mass isolated segment prior to fragmentation?

Carlos Afonso answered: Calibration of the mass spectra of highly complex mixtures is indeed an important step. Before recording the mass spectra from an isolated *m/z* range, the recording of a broadband mass spectrum is most likely required to be able to make a confident signal assignment. Internal calibration can be done on confidently assigned signals (from known sample types) but this may require the addition of internal calibrants in certain cases.

Philippe Schmitt-Kopplin remarked: Any analytical tool will give only a limited chemical projection of the sample depending on the sample's own properties. While biological samples can be considered as a quantized mixture of highly diverse single molecules over a range of sizes and properties that can be separated with chromatographic tools and assigned *m/z* with mass spectrometry, heavy oils and natural organic matter are rather high dynamic – molecular assembling systems that can behave as pseudocolloids as a function of their solution properties (dilution, pH, type of solvent) – and thus they bring all tools to their limits. These different types of samples, in terms of molecular mixtures, absolutely need to be kept in mind when applying our analytical tools and interpreting the results.

Norbert Hertkorn commented: MS/MS spectra of mixtures largely produce information about (the loss of) functional groups rather than how they are distributed across "complex" skeletons. Hence, you get better information about something you already know a bit about. This is the primary outcome of MS/MS at present – you get an idea about functional groups but not structure. I wonder what the prospects are for improving the coverage of molecular skeletons in molecules of considerable functionalization by MS/MS experiments.

Conflicts of interest

There are no conflicts to declare.

PAPER

Online supercritical fluid extraction mass spectrometry (SFE-LC-FTMS) for sensitive characterization of soil organic matter†

Yufeng Shen,[ac] Rui Zhao,[b] Nikola Tolić, [ID][b] Malak M. Tfaily, [ID][bd] Errol W. Robinson, [ID][a] Rene Boiteau, [ID][be] Ljiljana Paša-Tolić [ID][*b] and Nancy J. Hess [ID][*b]

Received 16th January 2019, Accepted 19th March 2019

DOI: 10.1039/c9fd00011a

We report a novel technical approach for subcritical fluid extraction (SFE) for organic matter characterization in complex matrices such as soil. The custom platform combines on-line SFE with micro-solid phase extraction, nano liquid chromatography (LC), electrospray ionization and Fourier transform mass spectrometry (SFE-LC-FTMS). We demonstrated the utility of SFE-LC-FTMS, including results from both Orbitrap and FTICR MS, for analysis of complex mixtures of organic compounds in a solid matrix by characterizing soil organic matter in peat, a high-carbon soil. For example, in a single experiment, >6000 molecular formulas can be assigned based upon FTICR MS data from 1–50 μL of soil samples (roughly 1–50 mg of soil, dependent on soil density), nearly twice that typically obtained from direct infusion liquid solvent extraction (LSE) from an order of magnitude larger volume of the same soil. The detected species consisted predominately of lipid-like, lignin-like and protein-like compounds, based on their O/C and H/C ratios, with predominantly CHO and CHONP molecular compositions. These results clearly demonstrate that SFE has the potential to effectively extract a variety of molecular species and could become an important member of a suite of extraction methods for studying SOM and other natural organic matter. This is especially true when comprehensive coverage, minimal sample volumes, and high sensitivity are required, or when the presence of organic solvent residue in residual soil is problematic. The SFE based extraction protocol could potentially enable spatially resolved characterization of organic matter in soil with a resolution of ~1 mm^3 to facilitate studies probing the spatial heterogeneity of soil.

[a]Biological Sciences Division, Pacific Northwest National Laboratory, Richland, WA 99354, USA

[b]Environmental Molecular Sciences Laboratory, Pacific Northwest National Laboratory, Richland, WA 99354, USA. E-mail: Ljiljana.PasaTolic@pnnl.gov; Nancy.Hess@pnnl.gov

[c]CoAnn Technologies, Richland, WA 99354, USA

[d]Department of Soil, Water and Environmental Science, University of Arizona, Tucson, AZ 85721-0038, USA

[e]College of Earth, Ocean and Atmospheric Sciences, Oregon State University, Corvallis, OR 97331, USA

† Electronic supplementary information (ESI) available. See DOI: 10.1039/c9fd00011a

1. Introduction

High resolution mass spectrometry (MS) is becoming a preferred method for characterizing organic molecules in complex matrices such as dissolved or soil organic matter (D/SOM) or petroleum.[1–6] Recent applications of this approach have shed new light on the molecular composition of natural organic matter (NOM) and the closely coupled microbial community composition and function, which both change in response to a wide variety of environmental factors including surface adsorption, photochemistry, redox conditions, *etc.*[7–14] Such studies provide mechanistic insight into the source, reactivity, and degradation pathways of the carbon stocks on our planet and will ultimately inform and advance models of land–atmosphere interactions on larger scales.

One of the grand challenges in this field is improving the comparisons of organic matter samples (*e.g.* across ecosystems). This requires the development of robust analytical procedures that capture a comprehensive array of organic analytes while also minimizing the effects of the sample matrix on analytical results. Characterizing SOM using mass spectrometry (MS) generally necessitates extraction of the organic molecules from the rock or soil matrix prior to analysis. This is typically accomplished using liquid solvents, such as water, inorganic acids/bases, organic solvents, and their combinations.[15] Recent work from our group and others demonstrated the effectiveness of liquid solvent extraction (LSE) for isolating SOM from a variety of ecosystems. However, the polarity of the solvent has a strong bias towards extracting organic molecules with similar chemistry.[16,17] As a result, overall organic matter extraction efficiencies are highly dependent on matrix composition and are typically low, which hampers comparisons between samples exhibiting different chemical compositions. Additionally, different ionization techniques – electrospray ionization (ESI), atmospheric pressure photoionization (APPI), atmospheric pressure chemical ionization (APCI), or laser desorption ionization (LDI) – will preferentially ionize specific classes of molecule and further bias analytical results. It has been generally recognized and acknowledged that multiple approaches are required for the robust assessment of natural organic matter composition. However, limited resources, including the amount of material available, usually result in partial characterization, most often using LSE followed by ESI coupled to ultra-high performance Fourier transform (FT) MS, such as Orbitrap or Fourier transform ion cyclotron resonance (FTICR) MS.

Expanding the arsenal of tools available for mass spectrometric natural organic matter characterization, herein we investigate the use of a subcritical fluid extraction (SFE) system to further explore the chemical selectivity for specific classes of SOM molecule and improve the yield from small volumes of soil to allow for the more sensitive measurement needed for *e.g.* characterization of soil spatial heterogeneity. SFE has been widely explored as an environmentally friendly and selective technique for obtaining valuable analytes from many different matrices primarily in large-scale industrial applications.[18,19] More recently, SFE has been applied as an unconventional sample preparation technique for processing polymers, pharmaceuticals, and specialty chemicals with reduced consumption of organic solvents and increased yields.[20,21] Supercritical fluids (SCFs) have ten to a hundred times faster molecular diffusion rates than liquid solvents, and lower surface tension.[22] While many different solvents have been employed as SCFs, CO_2 is the most commonly used because it is chemically stable, safe, cheap and easily available, as well as removable simply by relieving pressure. CO_2 also does not leave chemical residues in either the final extraction products or in the sample matrix. This results in a solvent-free, pure, concentrated, and intact extract allowing for improved analysis recovery and sensitivity compared to conventional LSE. One of the major disadvantages, however, is the limited ability to extract polar analytes. Previous work has demonstrated that altering SCF operation pressure and/or temperature, or modifying SCF composition with the addition of polar solvents (*e.g.* ethanol, methanol, ethyl lactate, *etc.*), can modulate SFE selectivity to enable sequential SFE extraction of compounds with selected molecular properties.[18,22]

Current SFE analytical practices are limited to extracting and analyzing simple, targeted (mixtures of) compounds. Due to their unique physical properties, SCFs exhibit enhanced penetration into porous soil matrices to extract contaminants and thus potentially achieve high extraction efficiencies (relative to LSE). SFE has been widely investigated for remediating environmental matrixes of organic contaminants such as polycyclic aromatic hydrocarbons.[23] However, the utility of SFE for comprehensive molecular characterization of intact organic compounds in soils has yet to be demonstrated, especially when coupled with advanced MS methods such as FTMS.

Since SFE provides clean extracts, it is often directly compatible with downstream analytics. For instance, coupling SFE with gas chromatography (GC) or supercritical fluid chromatography (SFC) can be rather simple due to the compatibility of the chromatographic mobile phases employed. On the contrary, mismatches between the SFE, which produces a large volume of gas, and gas-free liquid-phase ESI-FTMS represent a significant technical hurdle. This difficulty can be resolved using a liquid or solid phase trap for deposition of the extract following SC-CO_2 decompression. To couple SFE to ESI-FTMS for molecular level SOM characterization, we have implemented a solid phase extraction (SPE) trap column.[24] In this way, SFE extracts are trapped on the column and the SCFs are displaced with aqueous solvents for introduction into the mass spectrometer by ESI.[25]

We have further combined these techniques with online high-performance liquid chromatography (HPLC). An online coupling of SFE to LC-FTMS minimizes sample losses and maximizes coverage and sensitivity. This advancement enables the analysis of smaller sample volumes with fully automated operations compared with offline implementation. The automated online SFE-microSPE-nanoLC-FTMS platform enabled characterization of the many thousands of analytes present in soil. We observed an unprecedented diversity of analytes – more than 24 000 LCMS features (*i.e.*, analytes) – from only a few milligrams of soil. This diversity of analytes indicates the wide compositional range of molecules that are observable using SFE-LCMS for the deep characterization of SOM. Furthermore, the data showed that SFE has its own chemical selectivity for extracting specific SOM components that is complementary to a suite of LSE using a range of polar to non-polar solvents.[16,26] These results clearly demonstrate the potential of SFE to expand the chemical characterization of SOM and also assess the composition of much smaller soil samples that are attainable by conventional LSE approaches.

2. Materials and methods

Chemicals and soil sample manipulations

All chemicals and H_2O used in the experiments were HPLC-grade materials from Fisher Scientific (Hanover Park, IL). The soil samples analyzed in this study consisted of peat containing 51% carbon by weight, collected from Northern Minnesota at a depth of 75 cm. We also used low-C sandy-silt soil collected from a wetland at the Savannah River Site (SRS) in South Carolina. It contained 4% carbon by weight. In all cases, the soils were dried and ground to a fine powder prior to analysis.

Design and operation of the automated online SFE-LCMS system

The system constructed for this study is illustrated in Fig. 1. The SFE-grade CO_2 (Praxair, Danbury, CT) was placed into a 100 mL high-pressure syringe pump with a pressure limit of 10k psi (ISCO 100DX, Lincoln, NE). It was maintained at 4 °C using a recirculating cooler (Pharmacia Biotech, Manasquan, NJ). After filling, the system was closed with an on/off valve (VICI, Huston, TX) and detached from the CO_2 cylinder for convenient SFE coupling to different mass spectrometers. Each SFE consumed only about 0.3 mL liquid CO_2. Hence, a single filing of the syringe pump is enough for hundreds of SFE experiments. The CO_2 was purified using a 100 × 4.6 mm ID LC column packed with 3 μm porous graphitic carbon (Thermo Scientific, Waltham, MA) and a 200 × 4.6 mm ID LC column containing 5 μm C18-bonded silica particles (Phenomenex, Terrence, CA) prior to extraction.

SC-CO_2 flow was controlled with an automatic four-port switch valve (V1, VICI). We tested several different volumes of extraction vessels ranging from 1 to 50 μL. Unless otherwise indicated, the results presented here were derived using the 50 μL stainless steel extraction vessel specially manufactured by VICI. The dried and ground soils were transferred into the extraction vessel and placed in an oven

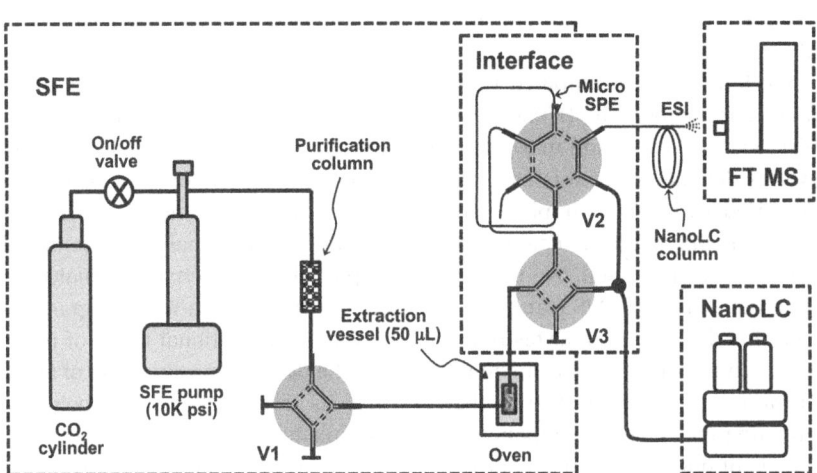

Fig. 1 Schematic of the online SFE-microSPE-nanoLC-ESI-FTMS system implemented for MS characterization of a complex mixture of organic compounds from solid matrices. The soil samples were placed in the extraction vessel.

(Quincy Lab, Chicago, IL) at 31 °C for the duration of the extraction procedure (1 hour). The extraction products were then transferred to an in-house manufactured microSPE fused silica capillary column (20 cm × 150 μm ID) containing 5 μm C18 particles (Phenomenex) through a 50 cm × 50 μm ID fused silica capillary tube with a volume of ~1 μL. The microSPE was connected to an in-house made nanoLC column (50 cm × 75 μm ID fused capillary column packed with 3 μm Phenomenex C18) with the "backflush" mode through a six-port valve (V2, VICI).[27] Unless otherwise indicated, SFE was performed at 3000 psi and 31 °C for 1 hour, with a liquid CO_2 flow of 4–6 μL min^{-1} as indicated by the ISCO pump controller.

After extraction, the microSPE was washed with LC mobile phase A at 5 μL min^{-1} for 10 min to remove residual SCF bubbles and then switched to the nanoLC column for gradient separation of the extracts (controlled by a 4-port valve V3, VICI). Acidic mobile phases (H_2O for A and ACN for B with 0.1% formic acid, v/v) and neutral mobile phases (H_2O for A and ACN for B with 0.1% ammonia acetate, v/v) were applied for negative mode ESI-FTMS. The LC platform previously reported[28] was used for nanoLC separation with gradient from A to 60% B in 120 min with a flow rate of 300 nL min^{-1}. The nanoLC column outlet was fitted with an emitter (3 cm × 20 μm ID fused silica tube with etched tip) for ESI to either of two different mass spectrometers. All packed capillary columns used for both microSPE and nanoLC were manufactured in-house using methods previously reported.[28]

Mass spectrometry experiments

We used two MS platforms for this study: a Thermo Exactive Orbitrap (Thermo Fisher Scientific, Waltham, MA) and a 12 Tesla Bruker solariX Fourier transform ion cyclotron resonance (FTICR) MS (Billerica, MA). The Exactive Orbitrap mass spectrometer was used for evaluating SFE performance under the following operating conditions: −2.3 kV for ESI, 100 < m/z < 2000 range at a resolution of 100 000 (at m/z = 400), an AGC of 3 × 10^6, and a single microscan for data acquisition. 12T FTICR operated with an ESI voltage of −4.4 kV and 150 low m/z cutoff for Q1. The calibration was performed as recommended by the vendor. Mass measurement accuracy < 1 ppm was obtained for singly charged ions within a 200 < m/z < 1200 range.

Data analysis

For FTICR datasets, we used an in-house developed Visual Basic for Applications (VBA) script utilizing the Bruker Data Analysis automation engine to average 5 minute non-overlapping LCMS segments and compile a list of peaks using the "FTMS" peak picker with the signal-to-noise parameter set to 7. Peaks from 20 averaged spectra were internally calibrated using a linear regression function. The list of calibration peaks included fatty acids and humic acid homologous series over the 200 < m/z < 800 range. We submitted peaks from spectra to the Compound Identification Algorithm (CIA) developed by Kujawinski and Behn.[29] A mass measurement accuracy of 1 ppm and a formula propagation with CH_2, H_2 and O building blocks were used for assigning the formula. In the case of ambiguous assignments, the formula with a lower count of heteroatoms (N + S +

P) was reported, or in the case when heteroatom count was the same, the formula with a lower mass error.

For Exactive datasets, in-house developed software packages (Decon-Tools and VIPER[30]) were used to pick peaks and to average m/z measurements over chromatographic peaks. We internally calibrated the resulting list of m/z values using the same calibration peaks with initial mass tolerance of 5 ppm. We submitted the calibrated peaks to the CIA for formula assignment with peak mass tolerances of 2 ppm and a formula propagation with CH_2, H_2 and O building blocks.

3.   Results

Performance of the online SFE-LC-FTMS for characterization of SOM from a low-C soil

The performance of the SFE-LC-FTMS system was optimized on an Exactive Orbitrap mass spectrometer. The microSPE interface used to couple SFE and nanoLCMS produced stable ESI, as evidenced by the smooth chromatogram baselines shown in Fig. 2. Additionally, we visually confirmed a stable ESI spray using the microscope on the spectrometer. The background intensity of the system blank was on a 10^5 level (Xcalibur Software). In comparison, the intensities of chromatographic peaks observed for 1 mg of the SRS soil was two orders of magnitude higher. Increasing the sample size from 1 to 20 mg resulted in more chromatographic peaks, but further increases in sample size up to 35 mg did not lead to significant changes in the chromatogram. Both the number of LCMS features detected and the molecular formulas assigned followed the same sample trends as chromatograms (Fig. S1, ESI†).

Molecular formulas were assigned to the analytes extracted by SFE using accurate masses (within 2 ppm mass error achievable by Exactive Orbitrap) as summarized in Fig. 3. Similar composition distribution was obtained for the SRS soil samples ranging in size from 1 to 35 mg with the maximum number of formula assignments (1618) attained using 20 mg of soil. Approximately 50% of the LCMS features remained unassigned, regardless of the size of the sample. CHO compounds comprised ∼20% of the total LCMS features detected; CHON and CHOS comprised ∼10% each; CHOP, CHNOP, CHONS and CHOSP compounds constituted the majority of the remaining 15% of the LCMS features detected.

508 MS peaks were detected in the system blank and 266 of these peaks, or approximately 52%, could be assigned a molecular formula. The compounds assigned were limited to several alkyl elongations, according to their O/C and H/C ratios.[31] In contrast, the smallest soil sample assessed (1 mg) yielded 820 formula assignments and only 157 were also found in the system blank, indicating that the system background is relatively low.

The extraction recovery of the SCFE system was investigated for the 20 mg sample described above *via* repeated extractions of residual soil (each lasting one hour). The intensities of chromatographic peaks were reduced by 75% after the second extraction and by greater than 99% after the third extraction as shown in ESI Fig. S2.† It is important to note that such recovery cannot be attributed to the total C compounds present due to the chemical selectivity of SFE-CO_2 (see results below). Increasing the extraction time may improve the extraction recovery for

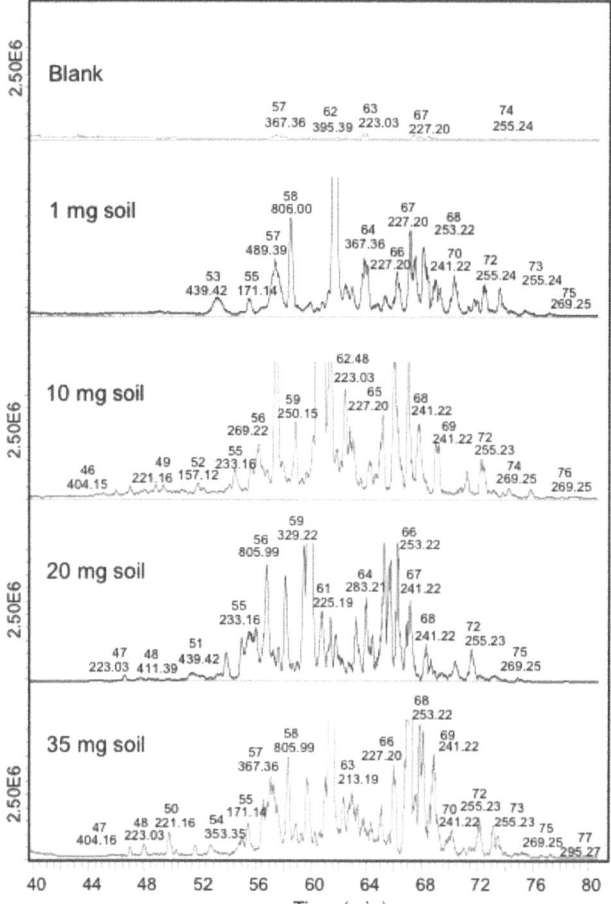

Fig. 2 Optimization of the SFE-microSPE-nanoLC-ESI-FTMS system for characterization of SOM from low C soil employed a Thermo Exactive Orbitrap. NanoLC-ESI-FTMS base peak chromatograms of SOM extracted by SFE from 1–35 mg of low C soil from SRS.

SFE-CO_2 extractable species, but potentially results in some extracts washing off of the microSPE column during the longer collection time.

Performance of the online SFE-LC-FTMS for characterization of SOM from high-C soil

Findings from the Exactive Orbitrap study (described above) guided FTICR experiments. A 12T FTICR mass spectrometer was used for detection of SOM extracted by SFE from peat soil. We have previously reported extensive characterization of the same peat soil using LSE with multiple solvents by direct infusion ESI on the same 12T FTICR mass spectrometer.[16] This study serves as a reference for evaluating the novel analytical methodology introduced herein.

In a single SCFE-LC-FTICR MS analysis using 8 mg (50 µL) of peat soil, we have detected 24 476 LCMS features (including ^{13}C and other minor isotopologues), of which 6414 distinct formulas were assigned as shown in Fig. 4 and Table S1, ESI.†

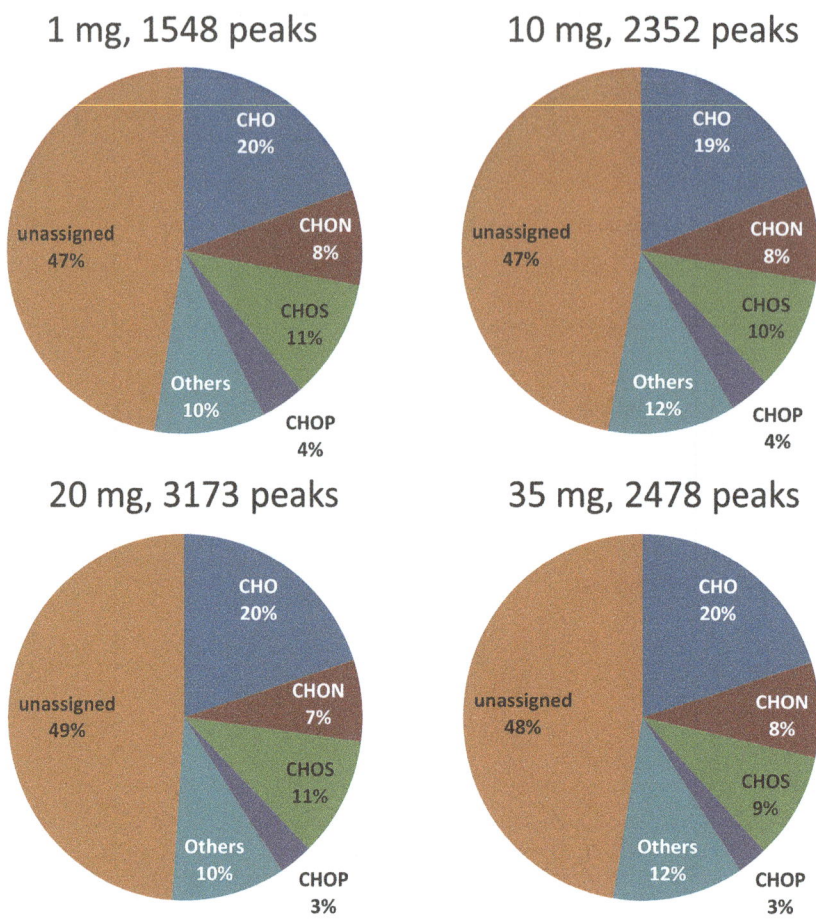

Fig. 3 Compositional distribution of the assigned formulas from SFE-microSPE-nanoLC-ESI-FTMS of SOM from SRS soil of different sample sizes.

This compares favorably to the 5800 total formulas that were assigned for the same sample on the same mass spectrometer using LSE separately with four solvents (MeOH, ACN, H_2O and hexane).[16]

Identifying more features using an LCMS-based method compared to using a direct infusion ESI method is expected because the separation reduces ion suppression effects. However, note that we obtained such a high coverage of SOM using nearly an order of magnitude smaller soil sample (e.g., 50 μL or 8 mg for SFE versus 100 mg for LSE). This demonstrates the high sensitivity of the new SFE based extraction protocol, potentially facilitating probing the spatial heterogeneity of soil with spatial resolution of ~1 mm³. Additionally, though more molecular formulas were assigned, a significant fraction (>60%) of LCMS features detected from the SFE extracts remained unassigned, even with the high mass accuracy (<1 ppm) provided by the 12T FTICR MS. This indicates that perhaps compounds extracted by SFE are not included in existing chemical databases used for formula assignment or include elements not considered in common formula assignment algorithms.

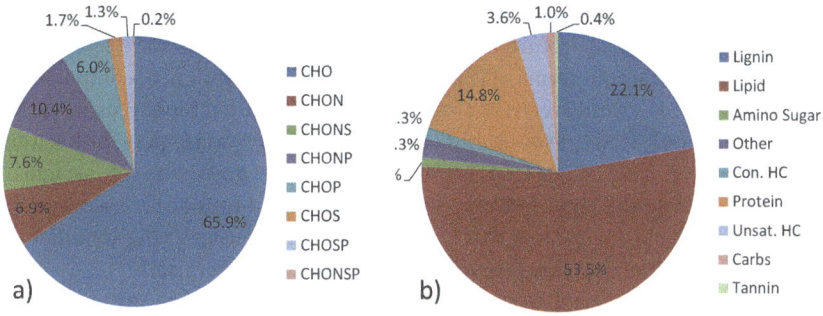

8 mg peat, 6414 peaks

Fig. 4 Performance of the SFE-microSPE-nanoLC-ESI-FTMS system for characterization of SOM from high C peat soil. 8 mg of peat soil was used for SFE-CO$_2$ with ammonia acetate buffers as nanoLC mobile phases. (a) Composition distribution; and (b) compound classes of formulas assigned using FTICR mass spectra.

We found that CHO compounds were the most represented category assigned for peat soil using SFE (66% of all assignments), followed by CHONP compounds (10% of all assignments), with 6414 formulas assigned in total (Fig. 4, Table S1†). The van Krevelen diagram[32,33] shows that most formulas assigned have O/C < 0.5 and H/C ratios between 0.7 and 2.2 (Fig. 5). Note that 3431 formulas (54% of all assignments) are lipid-like compounds according to their O/C and H/C ratios as defined by Sleighter and Hatcher[34] (see the overlay in Fig. 5). This percentage of lipid-like compounds is similar to what was previously obtained with LSE using hexane.[16] Coverage for protein-like compounds (962 formulas, or 15% of all assignments) was higher than obtained using LSE with any solvent except hexane, which suggests enhanced SFE selectivity for this category of compounds. The representation of lignins (~15%) was significantly higher than for LSE–hexane,

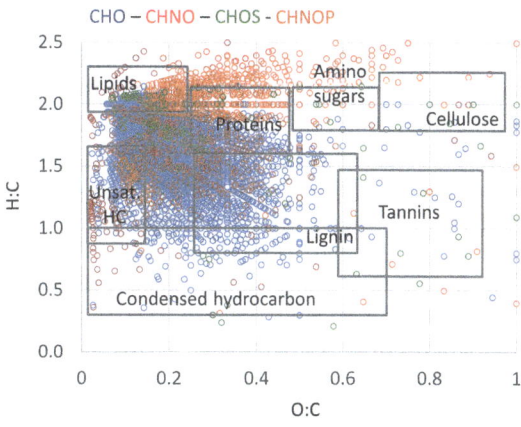

Fig. 5 A van Krevelen diagram showing the distribution of CHO, CHON, CHOS and CHONP compounds detected using the SFE-microSPE-nanoLC-ESI-FTMS system in high-C peat soil.

but lower then for LSE–water, and similar to that for LSE–ACN or LSE–MeOH. However, condensed hydrocarbons, tannins and amino sugars were less represented in SFE extracts in terms of the number of formulas assigned, compared to LSE using ACN, H_2O and MeOH.[16] The peat datasets reported for LSE using ACN, H_2O and MeOH (total 5800 formulas assignments)[16] were pooled for the complementary role of various extraction methods. In terms of assigned formulas, SFE had the largest overlap with LSE with hexane and the least overlap with LSE with water, suggesting that water might be the most effective modifier to SFE (SC-CO_2) to increase the diversity of SOM extractable from soil.

The normalized mass distribution of the molecules identified using the two techniques shows that while the distribution of MS peaks is quite similar for the two techniques, more molecules are detected using SFE in the $700 < m/z < 900$ range, whereas LSE detects more species with $m/z < 300$ (see ESI, Fig. S3†). These observations support the chemical selectivity suggested by Fig. 4, namely that SFE appears to be more selective for lipid-like, lignin-like and protein-like compounds than LSE (with results for all solvents combined) or that the chromatography improved the detection of less polar ions that were suppressed with direct infusion. Although not tested in this study, the higher proportion of lipid-like, lignin-like and protein-like compounds in SFE may also suggest that these compounds reside in pore spaces only accessible using SFE.

4. Discussion

The SFE-based MS characterization of natural organic matter demonstrates a significantly richer composition than attained previously using an LSE-based approach, highlighting the utility of this method for obtaining improved coverage of soil organic matter compound classes from smaller sample sizes. The greater depth of coverage compared to conventional LSE can be attributed to (1) the greater extraction efficiency of organic matter using SFE and (2) reduction of ion suppression due to incorporation of the liquid chromatography separation step.

The high performance of the online SFE-microSPE-nanoLCMS was only achieved by considering several design and technical factors discussed here. First, the high purity of the SCF applied as the extraction solvent was essential for achieving optimal performance. The contaminants in SFE-grade CO_2 can be enriched by the online microSPE, which can result in a high MS background. Purifying the SFE-grade CO_2 using a graphitic carbon LC column prior to soil extraction considerably reduced the background (Fig. 2). We also tested porous silica and silica-bonded C18 packed columns for purification of the SFE-grade CO_2, but a significant background was still observed (Fig. S2†). Second, the dimension of the microSPE affected the sample capacity for SFE and subsequent LC performance. We selected a 20 cm × 150 μm ID (i.e., 3.5 μL) column for the microSPE component of the system as the most effective for handling small (≤20 mg) soil samples (in a 50 μL SFE vessel) (Fig. 2 and S1†). Increasing the microSPE column inner diameter was found to diminish the coupled nanoLC efficiency, whereas lengthening the microSPE column could increase the wash time needed to remove gas bubbles left by SFE, which is critical for effective ESI-MS, thus decreasing the overall throughput.

We incorporated the nanoLC in the system to minimize ion suppression during ESI and maximize sensitivity and coverage. Chromatography reduces ion suppression by separating species in time, such that less heterogeneous mixtures are delivered to the analyzer allowing for lower abundance compounds or compounds that are difficult to ionize to be more easily detected, thanks to the simplified matrix and lower competition for charge. Furthermore, chromatography can separate species with very similar masses that may not be fully resolved even using the 12T FTICR MS. These benefits may explain why a greater proportion of analytes were detected using SFE based LCMS compared to LSE using several solvents with subsequent direct infusion MS. Increased resolving power mass measurement accuracy as well as faster data acquisition, such as those achievable at 21T,[1,35] would certainly further improve the detection of species within complex soil organic matter regardless of the extraction method used.

Our comparison of compound classes of SOM extracted using SFE and LSE methods on high C peat soil based on O/C and H/C ratios indicates that SFE-LCMS has the most similarity to the nonpolar hexane solvent showing selectivity for lipid-like and protein-like compounds (Table S1†). The addition of H_2O or MeOH as a co-solvent to the online SFE method could potentially allow tuning of the selectivity for other classes of SOM. The representation of lignin-like compounds appears to be significantly higher for the SFE than the LSE regardless of the solvent used (Table S1†). We also note that SFE is selective for CHNOP chemical compounds that are similarly represented in the LSE–H_2O analyses but does not appear to extract CHOP or CHNOSP compounds. Since formula assignment algorithms should be independent of the extraction techniques, we believe these results suggest a difference in extraction chemistry. It is likely that the observed increased coverage results in part from the known benefit of LC to decrease charge suppression compared to direct infusion ESI. However, one can speculate that the ability of SC-CO_2 to penetrate into the small pore spaces of the peat soil is likely to play a role in improved extraction efficiency. The resulting high sensitivity and comprehensive chemical characterization on small sample volumes could suggest new applications in the spatially resolved characterization of organic molecules contained in complex matrices, such as soil with ~1 mm^3 resolution. SOM exhibits tremendous heterogeneity not only in terms of its composition, but also in terms of its distribution within the soil.[36] Spatially resolved molecular characterization of organic matter in soil is critical for identifying biologically active sites, so-called hotspots. By enabling the analysis of a small volume of soil, SFE-LC-FTMS provides a means to obtain insights into the localization and distribution of organic matter within the soil mineral matrix (e.g., within soil aggregates or on mineral surfaces). A recent report on the direct analysis of small amounts (500 μg) of unprocessed soil samples using laser desorption ionization coupled with ultrahigh MS provides an attractive alternative approach for assessment of soil heterogeneity.[37]

Lastly, the results presented herein were obtained employing low temperature (31 °C) and neat SC-CO_2 for extraction of intact SOM. However, the SFE-LCMS system described herein is designed for operation at higher temperatures (up to 150 °C) and with modifiers such as water, methanol or isopropyl alcohol. These modifiers could extend SFE extraction coverage and modify chemical selectivity.

Online coupling with HPLC enables additional experimental control through modification of LC mobile phases, such as pH or solvent composition.

5. Conclusions

Online SFE-microSPE-nanoLC-FTMS is a sensitive approach that enables remarkable sensitivity in the molecular characterization of SOM and yields the assignment of >6000 molecular formulas from as little as a few milligrams of soil. The sensitivity provided by the online SFE system enables analyses of limited samples and investigations into the spatial heterogeneity of soil on the cubic-millimeter scale. The selectivity of SFE with neat CO_2 for extraction of SOM is similar to that of LSE with hexane and provides rich information for lipid-like, lignin-like and protein-like compounds based on their O/C and H/C ratios. SFE of SOM tannins, cellulose and amino sugars potentially requires addition of polar solvents, such as water, as modifiers. These results suggest that SFE could have an important role as an efficient extraction method for characterization of SOM, potentially as a complementary method or re-extraction of previously LSE-extracted samples to enrich datasets of SOM currently acquired using LSE methods. To our knowledge, this is the first report of the use of SFE, a widely recognized green sample preparation technique, for organic matter character-ization in complex matrices such as soil.

Conflicts of interest

There are no conflicts of interest to declare.

Acknowledgements

We acknowledge support from the U.S. Department of Energy (DOE) Office of Science and its Office of Biological and Environmental Research (BER). We thank Dr Daniel Kaplan for providing soil samples from SRS, South Carolina. MMT thanks the BER Terrestrial Ecosystem Science program for support. This research was performed using EMSL, a DOE Office of Science User Facility sponsored by BER and located at Pacific Northwest National Laboratory (PNNL). PNNL is operated by Battelle for DOE.

References

1 J. B. Shaw, T. Y. Lin, F. E. Leach III, A. V. Tolmachev, N. Tolic, E. W. Robinson, D. W. Koppenaal and L. Pasa-Tolic, 21 Tesla Fourier Transform Ion Cyclotron Resonance Mass Spectrometer Greatly Expands Mass Spectrometry Toolbox, *J. Am. Soc. Mass Spectrom.*, 2016, **27**(12), 1929.

2 D. F. Smith, D. C. Podgorski, R. P. Rodgers, G. T. Blakney and C. L. Hendrickson, 21 Tesla FT-ICR Mass Spectrometer for Ultrahigh-Resolution Analysis of Complex Organic Mixtures, *Anal. Chem.*, 2018, **90**(3), 2041.

3 J. A. Hawkes, T. Dittmar, C. Patriarca, L. Tranvik and J. Bergquist, Evaluation of the Orbitrap Mass Spectrometer for the Molecular Fingerprinting Analysis of Natural Dissolved Organic Matter, *Anal. Chem.*, 2016, **88**(15), 7698.

4 E. B. Kujawinski, Electrospray Ionization Fourier Transform Ion Cyclotron Resonance Mass Spectrometry (ESI FT-ICR MS): Characterization of Complex Environmental Mixtures, *Environ. Forensics*, 2002, **3**(3–4), 207–216.

5 A. C. Stenson, A. G. Marshall and W. T. Cooper, Exact masses and chemical formulas of individual Suwannee River fulvic acids from ultrahigh resolution electrospray ionization Fourier transform ion cyclotron resonance mass spectra, *Anal. Chem.*, 2003, **75**(6), 1275.

6 N. Hertkorn, M. Frommberger, M. Witt, B. P. Koch, P. Schmitt-Kopplin and E. M. Perdue, Natural organic matter and the event horizon of mass spectrometry, *Anal. Chem.*, 2008, **80**(23), 8908.

7 A. Bhattacharyya, A. N. Campbell, M. M. Tfaily, Y. Lin, R. K. Kukkadapu, W. L. Silver, P. S. Nico and J. Pett-Ridge, Redox Fluctuations Control the Coupled Cycling of Iron and Carbon in Tropical Forest Soils, *Environ. Sci. Technol.*, 2018, **52**(24), 14129.

8 E. M. Bottos, D. W. Kennedy, E. B. Romero, S. J. Fansler, J. M. Brown, L. M. Bramer, R. K. Chu, M. M. Tfaily, J. K. Jansson and J. C. Stegen, Dispersal limitation and thermodynamic constraints govern spatial structure of permafrost microbial communities, *FEMS Microbiol. Ecol.*, 2018, **94**(8), fiy110.

9 J. C. Stegen, T. Johnson, J. K. Fredrickson, M. J. Wilkins, A. E. Konopka, W. C. Nelson, E. V. Arntzen, W. B. Chrisler, R. K. Chu, S. J. Fansler, *et al.*, Influences of organic carbon speciation on hyporheic corridor biogeochemistry and microbial ecology, *Nat. Commun.*, 2018, **9**(1), 585.

10 Q. Yao, Z. Li, Y. Song, S. J. Wright, X. Guo, S. G. Tringe, M. M. Tfaily, L. Pasa-Tolic, T. C. Hazen, B. L. Turner, *et al.*, Community proteogenomics reveals the systemic impact of phosphorus availability on microbial functions in tropical soil, *Nat. Ecol. Evol.*, 2018, **2**(3), 499.

11 N. Gueneli, A. M. McKenna, N. Ohkouchi, C. J. Boreham, J. Beghin, E. J. Javaux and J. J. Brocks, 1.1-billion-year-old porphyrins establish a marine ecosystem dominated by bacterial primary producers, *Proc. Natl. Acad. Sci. U. S. A.*, 2018, **115**(30), E6978.

12 C. M. Kallenbach, S. D. Frey and A. S. Grandy, Direct evidence for microbial-derived soil organic matter formation and its ecophysiological controls, *Nat. Commun.*, 2016, **7**, 13630.

13 R. M. Cory, C. P. Ward, B. C. Crump and G. W. Kling, Carbon cycle. Sunlight controls water column processing of carbon in arctic fresh waters, *Science*, 2014, **345**(6199), 925.

14 C. P. Ward, S. G. Nalven, B. C. Crump, G. W. Kling and R. M. Cory, Photochemical alteration of organic carbon draining permafrost soils shifts microbial metabolic pathways and stimulates respiration, *Nat. Commun.*, 2017, **8**(1), 772.

15 K. Mopper, A. Stubbins, J. D. Ritchie, H. M. Bialk and P. G. Hatcher, Advanced instrumental approaches for characterization of marine dissolved organic matter: extraction techniques, mass spectrometry, and nuclear magnetic resonance spectroscopy, *Chem. Rev.*, 2007, **107**(2), 419.

16 M. M. Tfaily, R. K. Chu, N. Tolic, K. M. Roscioli, C. R. Anderton, L. Pasa-Tolic, E. W. Robinson and N. J. Hess, Advanced solvent based methods for molecular characterization of soil organic matter by high-resolution mass spectrometry, *Anal. Chem.*, 2015, **87**(10), 5206.

17 C. Steelink, Investigating humic acids in soils, *Anal. Chem.*, 2002, **74**(11), 326A.

18 M. Herrero, J. A. Mendiola, A. Cifuentes and E. Ibanez, Supercritical fluid extraction: Recent advances and applications, *J. Chromatogr. A*, 2010, **1217**(16), 2495.

19 J. W. King, Modern supercritical fluid technology for food applications, *Annu. Rev. Food Sci. Technol.*, 2014, **5**, 215.

20 H. W. Yen, S. C. Yang, C. H. Chen, Jesisca and J. S. Chang, Supercritical fluid extraction of valuable compounds from microalgal biomass, *Bioresour. Technol.*, 2015, **184**, 291.

21 A. D. P. Sanchez-Camargo, F. Parada-Alonso, E. Ibanez and A. Cifuentes, Recent applications of on-line supercritical fluid extraction coupled to advanced analytical techniques for compounds extraction and identification, *J. Sep. Sci.*, 2019, **42**, 243–257.

22 M. McHugh and V. J. Krukonis, Supercritical fluid extraction: principle and practice, *Supercritical fluid extraction: principle and practice*, 1994.

23 A. P. Wicker, D. D. Carlton Jr, K. Tanaka, M. Nishimura, V. Chen, T. Ogura, W. Hedgepeth and K. A. Schug, On-line supercritical fluid extraction-supercritical fluid chromatography-mass spectrometry of polycyclic aromatic hydrocarbons in soil, *J. Chromatogr. B: Anal. Technol. Biomed. Life Sci.*, 2018, **1086**, 82.

24 A. D. Sanchez-Camargo, F. Parada-Alfonso, E. Ibanez and A. Cifuentes, On-line coupling of supercritical fluid extraction and chromatographic techniques, *J. Sep. Sci.*, 2017, **40**(1), 213.

25 R. Batlle, H. Carlsson, E. Holmgren, A. Colmsjo and C. Crescenzi, On-line coupling of supercritical fluid extraction with high-performance liquid chromatography for the determination of explosives in vapour phases, *J. Chromatogr. A*, 2002, **963**(1–2), 73.

26 M. M. Tfaily, R. K. Chu, J. Toyoda, N. Tolic, E. W. Robinson, L. Pasa-Tolic and N. J. Hess, Sequential extraction protocol for organic matter from soils and sediments using high resolution mass spectrometry, *Anal. Chim. Acta*, 2017, **972**, 54.

27 Y. Shen, R. J. Moore, R. Zhao, J. Blonder, D. L. Auberry, C. Masselon, L. Pasa-Tolic, K. K. Hixson, K. J. Auberry and R. D. Smith, High-efficiency on-line solid-phase extraction coupling to 15–150 μm-i.d. column liquid chromatography for proteomic analysis, *Anal. Chem.*, 2003, **75**(14), 3596.

28 Y. Shen, R. Zhao, S. J. Berger, G. A. Anderson, N. Rodriguez and R. D. Smith, High-efficiency nanoscale liquid chromatography coupled on-line with mass spectrometry using nanoelectrospray ionization for proteomics, *Anal. Chem.*, 2002, **74**(16), 4235.

29 E. B. Kujawinski and M. D. Behn, Automated analysis of electrospray ionization Fourier transform ion cyclotron resonance mass spectra of natural organic matter, *Anal. Chem.*, 2006, **78**(13), 4363.

30 M. E. Monroe, N. Tolic, N. Jaitly, J. L. Shaw, J. N. Adkins and R. D. Smith, VIPER: an advanced software package to support high-throughput LC-MS peptide identification, *Bioinformatics*, 2007, **23**(15), 2021.

31 A. G. Marshall and R. P. Rodgers, Petroleomics: chemistry of the underworld, *Proc. Natl. Acad. Sci. U. S. A.*, 2008, **105**(47), 18090.

32 S. Kim, R. W. Kramer and P. G. Hatcher, Graphical method for analysis of ultrahigh-resolution broadband mass spectra of natural organic matter, the van Krevelen diagram, *Anal. Chem.*, 2003, 75(20), 5336.

33 D. van Krevelen, *Fuel*, 1950, **29**, 269.

34 R. L. Sleighter and P. G. Hatcher, The application of electrospray ionization coupled to ultrahigh resolution mass spectrometry for the molecular characterization of natural organic matter, *J. Mass Spectrom.*, 2007, **42**(5), 559.

35 C. L. Hendrickson, J. P. Quinn, N. K. Kaiser, D. F. Smith, G. T. Blakney, T. Chen, A. G. Marshall, C. R. Weisbrod and S. C. Beu, 21 Tesla Fourier Transform Ion Cyclotron Resonance Mass Spectrometer: A National Resource for Ultrahigh Resolution Mass Analysis, *J. Am. Soc. Mass Spectrom.*, 2015, **26**(9), 1626.

36 A. P. Smith, B. Bond-Lamberty, B. W. Benscoter, M. M. Tfaily, C. R. Hinkle, C. Liu and V. L. Bailey, Shifts in pore connectivity from precipitation *versus* groundwater rewetting increases soil carbon loss after drought, *Nat. Commun.*, 2017, **8**(1), 1335.

37 N. N. Solihat, T. Acter, D. Kim, A. F. Plante and S. Kim, Analyzing Solid-Phase Natural Organic Matter using Laser Desorption Ionization Ultrahigh Resolution Mass Spectrometry, *Anal. Chem.*, 2019, **91**(1), 951–957.

Faraday Discussions

Structural investigation of coal humic substances by selective isotopic exchange and high-resolution mass spectrometry†

Alexander Zherebker, [iD] *[ab] Irina V. Perminova, [iD] [b] Yury Kostyukevich,[ac]
Alexey S. Kononikhin,[cd] Oleg Kharybin[a] and Eugene Nikolaev[a]

Received 6th January 2019, Accepted 25th January 2019
DOI: 10.1039/c9fd00002j

Here, we report the application of a selective liquid-phase hydrogen/deuterium exchange (HDX) coupled to ultra-high resolution FTICR MS for structural investigations of individual constituents of humic substances (HS) isolated from three coal samples of different geographical origin. Selectivity was achieved by conducting reactions in DCl or NaOD solutions for catalyzing HDX in aromatic ring and side-chain positions with enhanced C–H acidity, respectively. FTICR MS analysis showed a significant overlap of molecular compositions in the HS samples under study, with 2000 common formulae. Using HDX, we demonstrated that the determined common formulae are presented by different structural isomers. We found that aromatic compounds varied both in the substitution pattern and the number of aromatic protons. Depending on the sample, lignin components with the same molecular formulae were composed of coumaryl, coniferyl or sinapyl moieties. Enumeration of HDX series for the 800 most abundant compounds showed that the results of HDX agreed well with the model structures suggested for humic components occupying a van Krevelen plot. In addition, we explored chemical transformations, which could connect individual constituents of coal HS. These transformations included hydrolysis of a guaiacyl moiety and reduction of a catechol unit, which corresponds to the conversion of a coniferyl fragment into a coumaryl unit. The obtained results were supportive of the hypothesis of the reducing humification pathway suggested for lignin transformation in the environment. The conclusion was made that the molecular ensemble of coal HS is composed of individual constituents produced at different humification stages.

[a]Skolkovo Institute of Science and Technology, Skolkovo, Moscow Region, 143025, Russia. E-mail: a.
zherebker@skoltech.ru
[b]Department of Chemistry, Lomonosov Moscow State University, Leninskie Gory 1-3, Moscow, Russia
[c]Moscow Institute of Physics and Technology, Dolgoprudnyi, 141700 Moscow Region, Russia
[d]V.L. Talrose Institute for Energy Problems of Chemical Physics, Russian Academy of Science, Moscow, Russia
† Electronic supplementary information (ESI) available: Additional information on the molecular compositions of coal CHM and a list of HDX results for the 800 most abundant compounds in CHM-Pow. See DOI: 10.1039/c9fd00002j

1. Introduction

Humic substances (HS) are the products of microbial and abiotic oxidation of remnants of living organisms. They are ubiquitous in the environment and constitute the major fraction of organic matter of caustobioliths, in particular of low grade coal (lignite and leonardite), peat, and oil shale.[1] Unlike petroleum, which consists mostly of hydrocarbons and aromatic heterocycles, coalified humic matter is enriched with oxygen-containing compounds carrying carboxylic and phenolic groups.[2] These compounds represent a complex mixture arranged into a supramolecular ensemble.[3] An analysis of different humification models enables the designation of various aromatic molecules with aliphatic side-chains and functional groups, and polyphenols with C–C and C–O–C ether bonds, as essential components of coal HS.[4] Aromatic constituents possess a pronounced biological activity.[5–7] It was shown that coal HS demonstrated the highest activity among HS from different sources.[8] In addition, the acidic character of coal HS is responsible for their metal binding properties, which are of particular importance for agricultural applications:[9] they can be used for plant nutrition in the form of water-soluble complexes with microelements.[10] At the same time, the application of coal humic materials is hindered by the extreme molecular heterogeneity of HS. Only high-resolution analytical techniques are capable of characterising their major structural moieties.[11]

NMR spectroscopy is the most powerful method for structural studies. In the case of HS, only partial structures can be identified due to substantial peak overlap.[12,13] The development of multidimensional NMR techniques enabled much deeper investigation of the aliphatic moieties of HS. Carboxylic rich alicyclic moieties (CRAM) were discovered.[14] Isotopic labeling extends NMR spectroscopy opportunities. The introduction of $^{13}CH_3$ groups in phenolic and carboxyl groups allowed Bell *et al.* to reveal the substitution pattern of aromatic rings and to suggest several possible structures of individual compounds in HS.[15] However, a higher resolution is required to investigate all constituents in complex organic mixtures. Thus, ultrahigh resolution FTICR mass spectrometry has become an indispensable tool for molecular exploration of HS.[16]

Due to its unprecedented high resolution power and mass accuracy, FTICR MS enables identification of thousands of molecules in complex mixtures.[17] The conventional approach is to plot FTICR MS data for HS in a van Krevelen diagram, which is a projection of molecular formulae on O/C and H/C atomic ratio coordinates.[18] Furthermore, van Krevelen plots can be binned into regions corresponding to typical molecular compositions of major HS precursors: lignins, flavonoids, tannins, carbohydrates, *etc.*[19,20] It performs classification tasks[21] but conveys only general information on the structure. In fact, the molecular composition may correspond to a number of different structural isomers, which cannot be resolved by FTICR MS.[22]

To avoid this limitation of FTICR MS, selective labeling of individual components of HS was introduced.[23] Each component undergoes a certain amount of chemical modifications according to its structure. This process is determined as a mass-shift from the parent ion in the mass-spectrum. H/D exchange (HDX) is the simplest technique of chemical labeling. Dissolution of an organic compound in D_2O results in a fast exchange of all labile protons with close to normal distribution of peak intensities in the mass-spectrum.[24] HDX was applied for comparison of HS and their fractions.[25,26] Analysis of HDX series revealed a number of labile protons in the molecules observed in the FTICR mass-spectrum. The power of HDX mass spectrometry is discussed in detail in the corresponding review article.[27] Nevertheless, simple HDX lacks selectivity. Moreover, previously we demonstrated that due to the acceleration of chemical reactions in the charged microdroplets of electrospray, HDX with D_2O can result in additional exchanges of skeletal protons in the moieties with enhanced C–H acidity.[28] Recently, we developed a technique for selective deuteromethylation of carboxylic groups coupled to FTICR MS, which drastically reduces the ambiguity of HDX.[29] Its application to the fractions of HS from different sources (including coal) has shown the similarity of distributions of carboxyl-carrying constituents over van Krevelen diagrams in all fractions studied. The identified constituents were in accordance with the model structures of lignins, condensed tannins and carboxylic rich unsaturated oxygenated compounds. Still, this technique enables the exploration of functional groups, but not carbon skeleton structures.

In our previous work, we applied NaOD and DCl catalyzed exchange reactions[30] to synthetic humic-like substances to facilitate selective HDX of backbone protons in their individual constituents. This enabled elucidation of the exact structures of six products of oxidative condensation.[31] The objective of this study was to apply selective liquid-phase HDX coupled to FTICR MS to three HS samples from different varieties of lignite in order to identify the structural motifs of their common molecular constituents and to determine the exact structural differences of the determined isomers.

2. Materials and methods

2.1. Materials and reagents

Coal hymatomelanic acid (CHM) was isolated *via* exhaustive ethanol extraction in Soxhlet apparatus of freshly precipitated humic acids obtained by acidification of potassium humates produced from two leonardites (Powhumus, Germany and "Gumat-80", Irkutsk, Russia) and from a lignite deposit in Buryatia (Russia). The samples were designated CHM-Pow, CHM-Irk, and CHM-Gl, respectively. The solvents and other reagents used in this study were of analytical grade. Ethanol and methanol for HPLC (Lab-Scan) were used for elution and dissolution. High-purity distilled water was prepared using a Millipore Simplicity 185 water purifying system. Bond Elut PPL (Agilent Technologies) cartridges (100 mg, 3 mL) were used for isolation of labeled CHM samples.

2.2. Deuterium labeling of HS samples

HDX of CHM was performed according to the previously reported procedure.[31] A mixture of 5 mg of CHM with 300 μl of 4 M NaOD (or 16% DCl) in D_2O was heated at 120 °C for 40 hours in a sealed tube. After this step, the labeled material was desalted using solid-phase extraction on a Bond Elut PPL cartridge according to the protocol described by Dittmar *et al.*[32] Labeled CHM solutions were adjusted to pH 2 (in the case of DCl, the solvent was evaporated under vacuum followed by addition of 0.1 NaOH for CHM dissolution) and passed through the activated cartridge. To assure a back-exchange of labile protons, a washing step using

0.01 M HCl was repeated 3 times. The final solution was obtained *via* methanol elution. Native samples were redissolved and extracted following the same procedure. Each solution was analyzed using FTICR MS.

2.3. Fourier transform ion cyclotron resonance mass spectrometry

All experiments were performed on an FTICR MS Bruker Apex Ultra spectrometer with a harmonized cell[33] equipped with a 7 T superconducting magnet and an electrospray ion source (ESI). Prior to analysis, all HS samples were diluted with methanol to 100 mg L^{-1} and then injected into the ESI source using a microliter pump at a flow rate of 90 μL h^{-1} with a nebulizer gas pressure of 138 kPa and a drying gas pressure of 103 kPa. A source heater was kept at 200 °C to ensure rapid desolvation in the ESI droplets. Mass-spectra were first externally calibrated using a carboxylated polystyrene standard made in house.[34] Internal calibration was systematically performed using the residual peaks of fatty acids,[35] reaching accuracy values < 0.2 ppm. The spectra were acquired within a time domain of 4 megawords in ESI(−) and 300 scans were accumulated for each spectrum. The resolving power was 530 000 at $m/z = 400$. The formulae assignment was processed using the lab-made "Transhumus" software designed by A. Grigoriev, which is based on the total mass difference statistics algorithm.[36] All molecular formulas were detected as singly-charged negative ions. The generated formulas were validated by setting sensible chemical constraints (O/C ratio ≤ 1, H/C ratio ≤ 2, element counts (C ≤ 120, H ≤ 200, O ≤ 60, N ≤ 2) and mass accuracy window < 1 ppm). Sulfur and phosphorus containing formulae were excluded from consideration due to their low content.[37] The assigned CHNO formulas were further plotted into van Krevelen diagrams.[18] Based on the O/C and H/C atomic ratios and the aromaticity index (AI),[38] formulae were grouped into 8 compound classes: "saturated" (H/C ≥ 1.5, O/C ≤ 0.3), "aliphatics" (H/C ≥ 1.5, O/C > 0.3), low-oxidized unsaturated (H/C < 1.5, AI ≤ 0.5, O/C ≤ 0.5), highly-oxidized unsaturated (H/C < 1.5, AI ≤ 0.5, O/C > 0.5), low-oxidized aromatic (0.5 < AI ≤ 0.67, O/C ≤ 0.5), highly-oxidized aromatic (0.6 < AI ≤ 067, O/C > 0.5), low-oxidized condensed aromatic (AI > 0.67, O/C ≤ 0.5) and highly-oxidized condensed aromatic (AI > 0.67, O/C > 0.5). AI was calculated according to the following equation:[38]

$$AI = \frac{1 + C - O - S - 0.5N - 0.5H}{C - O - S - N - P} \tag{1}$$

2.4. HDX data treatment

The data from the labeling experiments were processed following an algorithm that was described in our previous work.[31,39] It implies extraction of peaks related to HDX series of individual CHM constituents from the full mass spectrum. Those series are produced by peaks with m/z difference of 1.006277, which corresponds to the substitution of a proton with a deuteron. These HDX series were manually determined for the 800 most abundant peaks of CHM-Pow and some common formulae of all samples under study to identify the number of exchangeable skeletal protons in each parent ion, as shown in Fig. 2. Due to the extreme complexity of mass-spectra and limitations in resolution power, automatic determination of HDX series was impossible. Accordingly, the number of analyzed peaks was limited to 800 for labor- and time-saving reasons: they were represented by peaks with S/N ratio > 10 and a relative magnitude > 1%.[40] If the parent ion could not be detected in the mass spectrum of the labeled CHM, its position was marked by a red mark and HDX series were determined relative to the monodeuterated ion. This prevented misinterpretation. The error constraint was set to 0.0005 m/z.

3. Results

3.1. Comparison of molecular composition of HS from the different coal varieties

The obtained results of FTICR MS and elemental analysis data are shown in Table 1. For all samples, we determined more than 3 thousand molecular compositions. The CHM-Gl lignite sample was characterized by the maximum content of nitrogen exceeding 3%. Both leonardite samples (CHM-Irk and CHM-Pow) contained 1.5% and 1% of N, respectively. In accordance with elemental analysis, the intensity contributions of CHON species decreased from 50% in the case of CHM-Gl to 20% in the case of CHM-Pow. The average values of the FTICR MS parameters listed in Table 1 also varied among the HS samples used in this study and agreed well with the data of elemental analysis. The value of the number-averaged aromaticity index (AI)[38] was lower for the leonardite samples (0.5) as compared to that for lignite (0.6). CHM-Pow and CHM-Irk were also characterized with relatively reduced structures with a number-averaged $O/C_n \approx 0.4$, while CHM-Gl contained ~50% of oxygen by mass and $O/C_n = 0.5$.

To compare the HS samples from different coal varieties on a molecular level, we plotted all of their determined molecular formulae into van Krevelen diagrams (Fig. 1B–D). The latter were further binned into specific areas with respect to the value of aromaticity indexes and of atomic ratios, as described in the Materials and methods section. The intensity-weighted histogram of compound classes is presented in Fig. 1A. The HS samples used in this study possessed quite similar molecular distributions over van Krevelen diagrams. The most pronounced difference between the samples was the relative depletion with saturated compounds along with the higher content of oxidized condensed compounds with AI > 0.67 in CHM-Gl as compared to both CHM-Irk and CHM-Pow samples. At the same time, for all samples, the major components were attributed to compounds with AI > 0.5 and O/C < 0.5. Another common feature of the coal HS samples was the high contribution of low-oxidized unsaturated compounds with atomic ratios H/C < 1.5 and O/C < 0.5. These compounds are usually referred to as polyphenols and lignin, in particular,[19,20] which is considered to be the major precursor for coal humic substances.[41] The obtained results were also in agreement with our previous findings on the comparison of two HS samples isolated from different lignites, which were characterized by the higher contribution of oxidized species as compared to leonardite HS.[42]

The HS sample similarity is clearly visualized in a Venn diagram (Fig. S1†) presenting molecular composition overlap. We found that the three samples studied possessed almost 2000 common formulas (present in all three samples) and 2176 shared formulas (present in two samples). From all common formulae, 669, 765 and 508 could be attributed to lignin, aromatic and condensed

Table 1 The values of the FTICR MS average parameters and elemental compositions of the HS samples used in this study

| Sample | FTICR MS data | | | | | | | | | | Elemental analysis | | | |
	Total formulae	CHO	CHON	CHO, % intensity	CHNO, % intensity	M_n^a	AI_n^a	O/C_n^a	H/C_n^a		C, %	H, %	O, %	N, %
CHM-Gl	3658	1616	2042	49.51	50.49	511.22	0.61	0.50	0.67		42.76	2.92	50.98	3.34
CHM-Irk	5443	3210	2233	69.87	30.13	493.19	0.55	0.43	0.80		55.73	3.58	39.19	1.5
CHM-Pow	3668	2386	1282	79.83	20.17	462.73	0.53	0.42	0.84		50.07	2.17	46.75	1.01

^{a} n stands for the number-average values calculated according to the following equation: $X_n = \dfrac{\sum X_i I_i}{\sum I_i}$, where $X = M$, AI, O/C or H/C, and I = peak intensity.

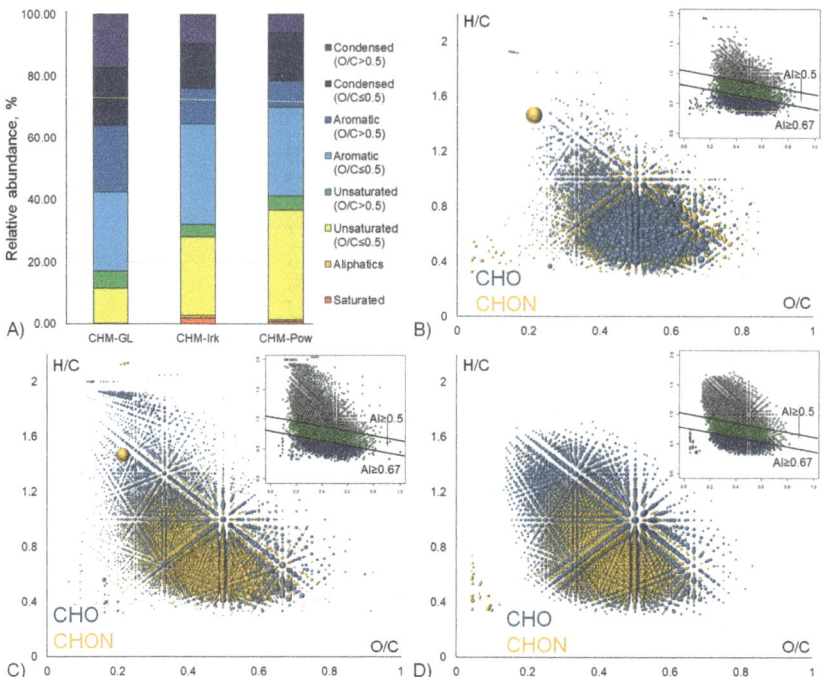

Fig. 1 Characteristics of molecular composition of the three coal HS under study: (A) the relative contribution of different molecular classes based on aromaticity index and O/C ratio; (B–D) van Krevelen diagrams for CHM-Gl, CHM-Irk, and CHM-Pow, respectively. The insets show the AI threshold for all samples: AI ≥ 0.67 (blue), 0.5 ≤ AI < 0.67 (green) and AI < 0.5 (gray).

compounds, respectively. This highlights the importance of polyphenols in the formation of coal organic matter.

The discovered substantial similarity of the molecular composition of the HS samples under study may lead to a wrong conclusion about the similarity of their origin and structural match of major components. In fact, a possible source of organic carbon for coal formation is terrestrial plant biomass, which underwent intensive transformations.[43] It has been reported that the coal originating from East Siberia and represented by CHM-Irk and CHM-Gl was formed during the Cretaceous period,[44] while coal from the German mine is significantly younger, originating from the Middle Miocene.[45] These geological periods were characterized by different vegetation and, therefore, different biomolecular precursors for coal formation. This should have resulted in the structural variance of the major components of the samples under study behind the common molecular compositions.

3.2. H/D exchange of skeletal protons with basic or acidic catalysis

For comparison of the coal HS used in this study on the structural level, we performed acid/base catalyzed H/D exchange of aromatic protons. A reaction scheme is presented in Fig. 2. In DCl, HDX proceeds exclusively *via* electrophilic substitution in the aromatic ring.[30] Adverse HDX of side-chain protons occurs

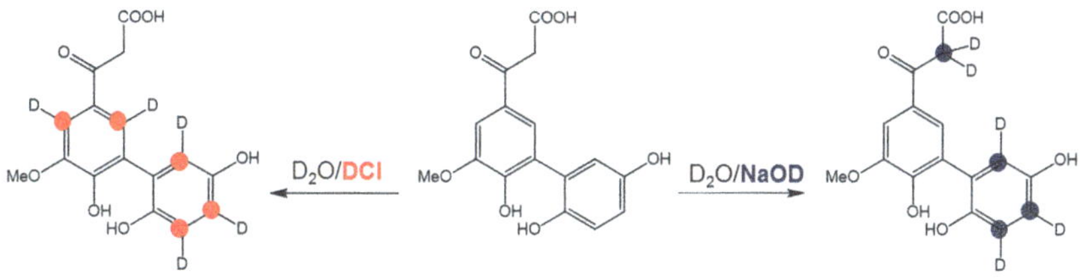

Fig. 2 Scheme of HDX, illustrating reaction regioselectivity under basic and acidic catalysis.

only under supercritical conditions, in which the D_2O/DCl mixture behaves as a super-acid.[46] In the case of NaOD, both aromatic and side-chain protons undergo HDX. In the aromatic ring, HDX is facilitated by keto–enol tautomerism of the phenolic group. So, only protons in *ortho-* and *para-*positions to the phenolic group undergo HDX. Side-chain protons undergo HDX in molecular sites with extended C–H acidity, such as benzylic and α-CH_n moieties.[30] Another advantage of the chosen HDX techniques is the low impact of reagents on the sample integrity. Unlike hydroiodic acid, DCl isn't able to cleave ether bonds at temperatures below 200 °C.[46] Also, a lack of oxidation and condensation processes in NaOD without continuous air flow was shown in our previous paper on the HDX of model humic compounds.[28]

Due to the high analytical capabilities of FTICR MS, parent and deuterated ions were resolved in mass-spectra. Using a custom R-script, we first extracted peaks, which may compose HDX series of the parent ions with nominal mass shift $\Delta m = 1.00628$ (for singly charged ions) as is schematically shown in Fig. 3 on the example of a neutral molecule $C_{19}H_{17}O_8$ identified in the CHM-Pow sample. Furthermore, we manually selected peaks with binomial distribution, which is a well-fitted approximation for HDX.[47] It is noteworthy that incomplete HDX would significantly confuse the data interpretation. Previously, we applied ^2H NMR spectroscopy and FTICR MS to evaluate regioselectivity and completeness of HDX reaction on the model humic and lignin monomers.[28] The chosen HDX technique facilitated a high deuteration degree, which resulted in the absence of the parent ion in the mass-spectrum of the labeled material and enabled determination of all exchangeable protons in molecules. Therefore catalytic HDX provides a high degree of deuteration, which combined with regioselectivity allowed us to reliably attribute exchangeable protons to the particular structural fragments of individual compounds.[31]

The determined lengths of HDX series obtained under acidic or basic conditions for 6 common formulae are presented in Table 2. Evaluation of the data enabled specification of the differences between the samples under study. We found that compounds with common molecular compositions possessed different amounts of specific exchangeable protons. It was necessary to perform both acidic and basic catalysis to explore structural differences in more detail. For example, for molecular composition $C_{17}H_{10}O_7$, we observed four HDX in NaOD for all samples. However, for the same composition in DCl, CHM-Pow, CHM-Irk and CHM-Gl possessed 5, 6, and 4 exchangeable protons, respectively. This indicates the presence of different isomers comprising CHM samples.

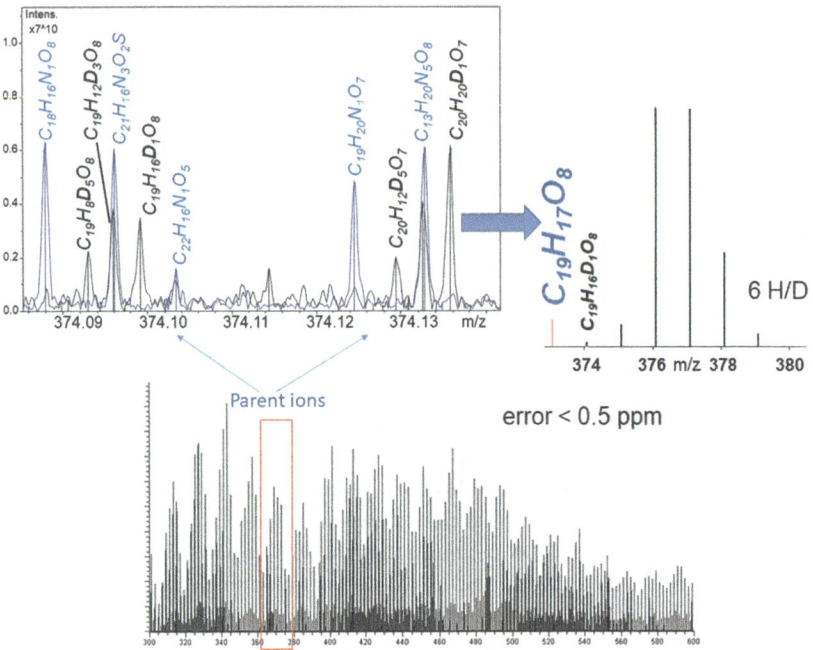

Fig. 3 FTICR mass-spectrum of D-labeled CHM-Pow with a magnified fragment at the nominal $m/z = 374$. Residual parent and deuterated ions are designated in blue and black colors, respectively. The inset on the right shows the mathematically extracted mass-spectrum fragment for determining the HDX series for $C_{19}H_{18}O_8$ molecular composition.

The regioselectivity of HDX and knowledge of the major humification pathways allowed us to suggest which structural motifs represent the differences in the number of exchanges. Given that under acidic conditions, only aromatic protons undergo HDX, the length of HDX is equal to the degree of aromatic rings substitution. This is particularly important for the lignin components of the coal HS under study, which are essentially composed of three model phenylpropanoic units: *p*-coumaryl, coniferyl and sinapyl moieties. They differ by the number of methoxyl substituents in aromatic ring as is shown in Fig. 4A.

The molecular formulae $C_{17}H_{20}O_7$ and $C_{16}H_{16}O_7$ are characterized by lignin-like compositions and the double bond equivalent (DBE) 9 and 8, respectively. We detected these molecules in negative ESI mode, so they likely contain carboxylic

Table 2 The number of exchangeable protons under acidic or basic (in parentheses) conditions determined for six common components of the coal HS samples under study

Formula	DBE	CHM-Pow	CHM-Irk	CHM-Gl
$C_{16}H_{16}O_7$	9	4(5)	3(3)	2(3)
$C_{17}H_{20}O_7$	8	4(5)	3(3)	1(3)
$C_{17}H_{10}O_7$	13	5(4)	6(4)	4 (4)
$C_{20}H_{14}O_6$	14	5(5)	4(1)	2(3)
$C_{14}H_{10}O_9$	10	0(0)	4(4)	3(3)
$C_{16}H_{10}O_{10}$	12	1(1)	4(4)	4(6)

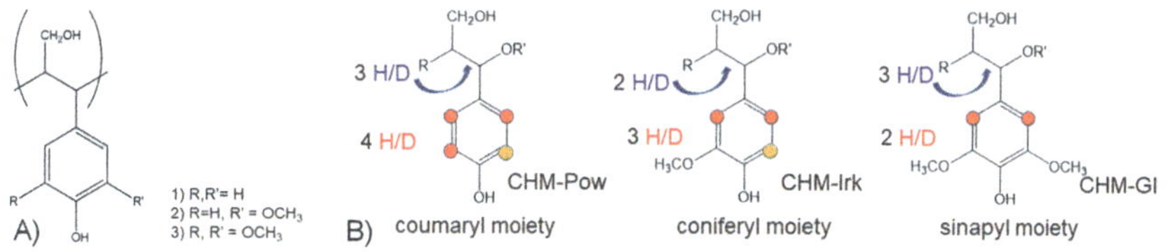

Fig. 4 A) Aromatic moieties of p-coumaryl (1), coniferyl (2) and sinapyl (3) alcohols; (B) phenylpropanoic fragments determined in isomers of $C_{16}H_{16}O_7$ by HDX of three samples of coal HS used in this study. The red, blue and orange circles correspond to protons subjected to HDX in DCl, NaOD and in both cases, respectively.

groups,[48] which facilitate ionization in the negative mode. This might be indicative of the presence of one aromatic ring in both $C_{17}H_{20}O_7$ and $C_{16}H_{16}O_7$. In this case, the length of HDX series under acidic conditions specifies the structure and origin of lignin compounds in the coal HS samples. For CHM-Pow, CHM-Irk, and CHM-Gl, we obtained four, three, and two HDX for $C_{16}H_{16}O_7$, which are attributed to p-coumaryl, coniferyl and sinapyl moieties, respectively, represented in Fig. 4B. Similar results were obtained for $C_{17}H_{20}O_7$, only in the case of the CHM-Gl sample, we observed one exchange instead of two. This could be explained by the formation of a five-membered dihydrofuran cycle between the vacant *ortho*-position in the aromatic ring and side-chain, which is typical for lignin.[49]

Application of HDX under basic conditions enables the comparison of side-chain structures for the common molecular compositions identified in the coal HS samples under study. Based on acidic HDX, we proposed the presence of p-coumaryl moieties for $C_{17}H_{20}O_7$ and $C_{16}H_{16}O_7$ molecules in the case of CHM-Pow. Under basic conditions, we observed 5 exchanges for these molecules. Two of the five exchanges could be attributed to C–H in *ortho*-positions to the phenolic group, which undergo HDX in alkaline solutions. Therefore, three of the five exchanges belong to the side-chain (Fig. 4B). In the case of CHM-Gl, all three exchanges (Table 2) belong to the side-chain, since the presence of methoxyl-groups prevents HDX in the aromatic ring. In the case of CHM-Irk, we observed 3 exchanges under basic conditions for $C_{16}H_{16}O_7$ and $C_{17}H_{20}O_7$. Based on acidic HDX, coniferyl moieties can be proposed for these molecules: they possess C–H protons at an *ortho*-position to the phenolic group, which are exchangeable in NaOD. Therefore, we suggest that in the case of CHM-Irk, both $C_{16}H_{16}O_7$ and $C_{17}H_{20}O_7$ possess 2 exchangeable protons in the side-chain (Fig. 4B).

Consideration of HDX results for low-oxidized aromatic compounds also reveals structural differences in the common components of the three coal HS samples under study. It can be suggested that these aromatic compounds can be attributed to tannins – the major HS precursors, which undergo oxidative condensation in nature.[50] Their aromatic structures have no side-chains and, hence, all HDX are taking place out in aromatic rings. For $C_{17}H_{10}O_7$, we observed 5, 6 and 4 exchanges in DCl for CHM-Pow, CHM-Irk, and CHM-Gl, respectively. The number of exchanges indicates the different degree of aromatic ring substitution. For $C_{17}H_{10}O_7$ in CHM-Pow, we may suggest a flavonoid-like structure with 5 non-substituted aromatic protons subjected to acidic HDX (Fig. 5A). Moreover, in NaOD we observed 4 exchanges for $C_{17}H_{10}O_7$. Proton 2 (Fig. 5A) is in a *meta*-position to the phenolic group, which explains the absence of HDX in

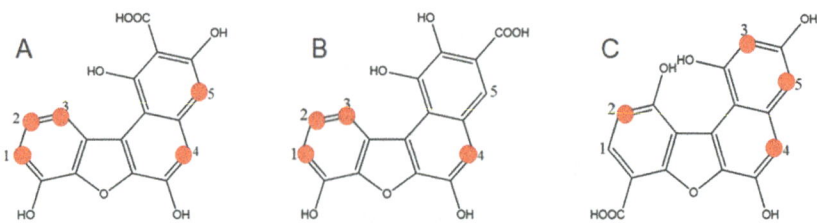

Fig. 5 Flavonoid-like structures proposed for low-oxidized aromatic compounds in coal hymatomelanic acids as revealed by HDX FTICR MS: (A) CHM-Pow, (B and C) CHM-Gl. The numbers indicate C−H bonds subjected to HDX. The red circles designate protons subjected to HDX in DCl.

NaOD. Collectively, HDX under acidic and basic conditions supports the structure depicted in Fig. 5A.

Variation in the order and position of the substituents in the aromatic ring affects the reactivity of the aromatic ring. For $C_{17}H_{10}O_7$ in CHM-Gl, we observed 4 exchanges in DCl. In this case, we suggested two possible structures depicted in Fig. 5B and C. The carboxylic substituent prevents electrophilic substitution and acidic HDX in positions 5 and 1 in the middle and right structures, respectively. In NaOD, we observed 4 HDX, which is in accordance with both suggested structures. The obtained results indicate that aromatic flavonoid-like components of coal HS may vary in substituent positions.

3.3. Structural motifs of different constituents of coal HS

The determination of HDX series for a number of ions determined in the coal HS samples under study enables comparison of different constituents in terms of their reactivity. Also, the obtained information may be used for attribution of molecular compositions to the particular classes of organic compound based on their projection in the van Krevelen diagram.[51] For this purpose, HDX series were determined for the 800 most abundant ions in the FTICR mass-spectrum of CHM-Pow (Table S1†). The results were plotted in a van Krevelen diagram with color designation of HDX series length for both DCl and NaOD (Fig. 6). It is seen that

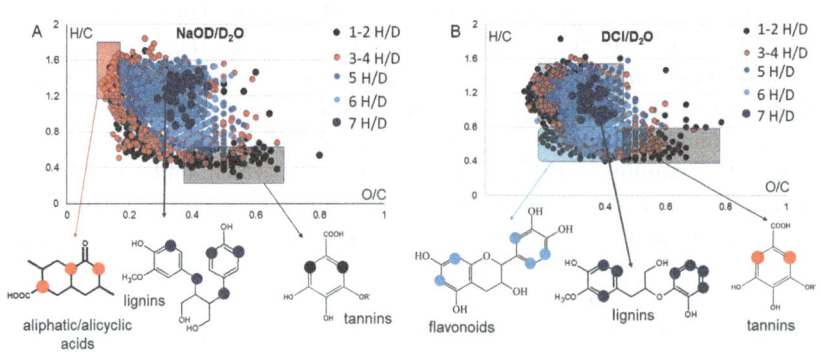

Fig. 6 Color-coded van Krevelen diagram for the 800 most abundant compounds in CHM-Pow with HDX series lengths obtained in (A) NaOD and (B) DCl, as well as model structures.

different components were characterized by different HDX lengths. In particular, the low-oxidized unsaturated components gave rise to the highest amount of exchangeable protons in both DCl and NaOD. Aliphatic compounds demonstrated HDX only in NaOD, while aromatic compounds underwent HDX in both DCl and NaOD. The obtained results are in agreement with the model structures of lignin, flavonoids, tannins and aliphatic/alicyclic carboxylic acids constituting HS.[52] So, oxidized aromatic compounds with H/C < 0.6 and O/C > 0.5 were characterized by a minimum amount of exchangeable protons (1-2 HDX). This agrees well with a gallic acid core proposed for tannins.[53] Low-oxidized aromatic compounds with H/C < 0.8 and O/C < 0.5 possessed up to 6 exchangeable protons in DCl and up to 4 in NaOD. These results are in good agreement with the model flavonoid-like structures shown in Fig. 6B. Low-oxidized unsaturated compounds with H/C > 1.2 and O/C < 0.2 possessed 3–4 exchangeable protons only in NaOD (Fig. 6A). Since HDX occurs in C–H bonds with increased acidity,[54,55] we assign these compounds to the alicyclic carboxylic acids, which undergo HDX in α-positions. The presence of alicyclic carboxyls was suggested previously using gradient fractionation of coal HS[56] and using NMR spectroscopy of pyrogenic soil HS.[57]

Polyphenolic compounds with O/C < 0.5 are of particular interest for structural determination due to their key role in HS biological activity.[8,58] In Fig. 6A, we see that the number of exchanges in NaOD increased with O/C from 5 to 7. An increase in O/C ratio could be accounted for by the presence of acceptor groups in the side-chain of molecules[29] and for hydroxylation of aromatic rings, which consequently creates new positions subjected to HDX. In typical lignin β-O-4 structures, there is a lack of non-substituted side-chain protons for HDX.[49] At the same time, phenolic fragments of phenylpropanoic units could undergo only two exchanges in NaOD as discussed above. Therefore, we suggest dibenzyl-butyrolactol-like (or dibenzyl-butyrolactone-like) structures,[59] as shown in Fig. 7, with a number of benzylic protons that are easily subjected to HDX due to enhanced C–H acidity.

In DCl, we observed a similar trend with the extension of HDX series lengths from 3 to 7 followed by DBE and O/C ratio increases (Fig. 6B). According to our previous results, these compounds are composed mostly of unconjugated carboxylic acids.[56] The absence of carboxylic groups in the aromatic ring

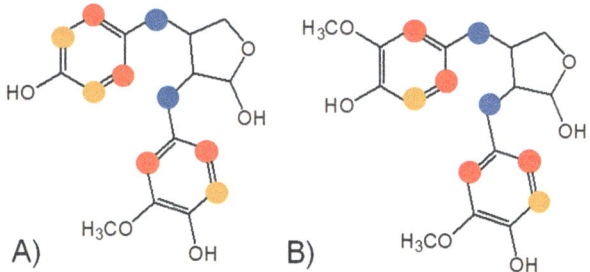

Fig. 7 Model dibenzyl-butyrolactol-like compounds suggested for (A) $C_{19}H_{22}O_5$ and (B) $C_{20}H_{24}O_6$ molecular compositions. The red, blue and orange circles correspond to protons subjected to HDX in DCl, NaOD, and in both cases, respectively.

results in high reactivity in electrophilic substitution, which leads to the efficient HDX.[30] Given that the amount of acidic HDX indicates the degree of aromatic ring substitution, we may compare structures of the different constituents of the coal HS samples under study. For example, $C_{19}H_{22}O_5$ and $C_{20}H_{24}O_6$ possessed 7 and 6 exchangeable protons under acidic conditions, respectively. The first molecule is likely to be composed of p-coumaryl and coniferyl moieties (Fig. 7A), while the second one includes 2 coniferyl fragments (Fig. 7B).

3.4.   Unveiling lignin humification processes using HDX FTICR MS

Considering the structures of lignin-like polyphenols, we could explain an extension of the HDX series in NaOD along with an increase in the O/C ratio by incorporation of acceptor groups and oxidation processes. At the same time, the number of exchangeable protons correlated with the value of DBE, in particular, for the DCl catalyzed reaction. The observed trends could be related to chemical transformations, which occur during lignin humification. It was shown that in addition to oxidation and condensation processes, lignin humification is accompanied by hydrolysis of ether bonds.[41,60] In the case of coniferyl and sinapyl moieties, hydrolysis leads to the generation of a new phenolic group and consequently to an increase in O/C ratio, and to activation of aromatic protons for HDX under basic conditions. It is shown in Fig. 8A on the example of transformation of a guaiacyl unit into a coniferyl fragment. Hydrolysis of methyl ether results in the formation of a catechol moiety with an increase in the O/C ratio and DBE value due to elimination of the methyl group. Catechol undergoes additional HDX in NaOD in the ortho- and para-position to the new phenolic group. The further loss of a phenolic group (Fig. 8B) adds new aromatic protons for HDX in DCl. For example, according to HDX results, a compound with molecular formula $C_{20}H_{28}O_8$ in CHM-Pow possesses six aromatic protons, which corresponds to two coniferyl moieties. Its OCH_2 homologue – $C_{19}H_{26}O_7$ – possesses seven exchangeable aromatic protons, which could correspond to formation of a coumaryl moiety due to methyl ether hydrolysis. Therefore, application of selective HDX enables identification of different phenylpropanoic units within the same coal HS sample and determination of their chemical connectivity reflected in the general humification pathways.

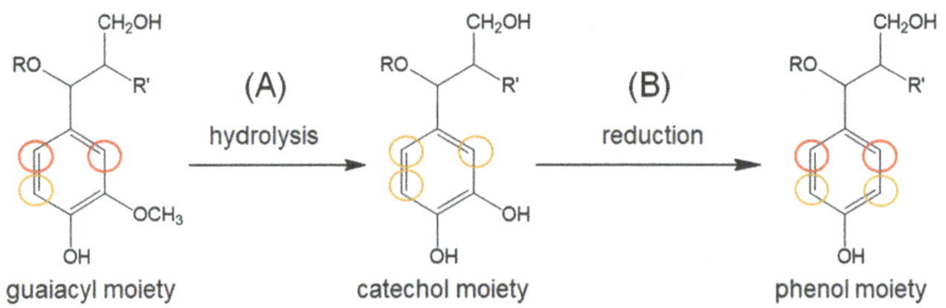

Fig. 8   General scheme of reducing lignin humification pathway: (A) hydrolysis of methyl ether, and (B) reduction of phenolic group.[60] The red and orange circles designate exchangeable protons in DCl, and in both DCl and NaOD, respectively.

4.   Conclusions

The application of HDX enabled enumeration of specific exchangeable protons in the individual components of the three coal HS samples used in this study. Due to the regioselectivity of HDX achieved through liquid-phase catalysis, only sites with enhanced C–H acidity and aromatic protons underwent labeling in NaOD and DCl, respectively. According to the FTICR MS molecular profiles, the three samples under study contained more than 2000 common formulae. The clear differences in their structures were revealed using HDX. Under the same conditions, the so-called common components possessed different amounts of exchangeable protons. We suggested that aromatic constituents varied by the positioning of and by the number of aromatic protons. The degree of aromatic ring substitution, which was observed in DCl, indicated coniferyl, coumaryl and sinapyl moieties as composing lignin components of coal HS. The combination of HDX in DCl and NaOD enabled suggestion of the phenolic group positions in low-oxidized aromatic compounds. The generalization of the obtained data as a distribution of exchangeable protons in the van Krevelen diagram plotted for the 800 most abundant ions determined in FTICR mass-spectrum of the leonardite HS showed that HDX results agree well with the model structures of lignin, tannins, flavonoids and alicyclic carboxylic acids. In addition, HDX enabled exploration of the chemical transformations, which could connect individual lignin constituents of coal HS. Transformations included hydrolysis of the guaiacyl moiety followed by reduction of the catechol unit, resulting in the loss of a phenolic group in the aromatic ring. This transformation corresponded to the reducing humification pathway suggested for lignin in the environment. Hence, the molecular ensemble of coal HS is composed of individual constituents produced at different humification stages, which can be used as an evolution parameter of coal humic matter. The obtained information can be used for fine fractionation and isolation of organic components with the known structures from coal. This can be of great use for the non-fuel chemical industry by enabling production of new chemicals and materials from coal.

Conflicts of interest

There are no conflicts to declare.

Acknowledgements

This work was supported by Russian Science Foundation grant no. 18-79-10127 (FTICR MS and HDX studies). The isolation of coal hymatomelanic acids and their primary characteristics were conducted under support of Russian Foundation for Basic Research grant no. 16-04-01753. AZ and EN acknowledge support from the European's Horizon 2020 Research and Innovation Program under grant agreement No. 731077.

References

1 R. S. Swift. Organic Matter Characterization, in *Methods of soil analysis. Part 3. Chemical methods*, ed. D. L. Sparks, A. L. Page, P. A. Helmke, R. H. Loeppert,

P. N. Soltanpour, M. A. Tabatabai and C. T. Johnson, SSSA Book Series 5, SSSA, Madison, WI, 1996, p. 1036.

2 A. Nebbioso and A. Piccolo, Basis of a Humeomics Science: Chemical Fractionation and Molecular Characterization of Humic Biosuprastructures, *Biomacromolecules*, 2011, **12**(4), 1187–1199.

3 A. Piccolo, The Supramolecular Structure of Humic Substances, *Soil Sci.*, 2001, **166**(11), 810–832.

4 S. E. Cabaniss, G. Madey, L. Leff, P. A. Maurice and R. Wetzel, A Stochastic Model for the Synthesis and Degradation of Natural Organic Matter Part II: Molecular Property Distributions, *Biogeochemistry*, 2007, **86**(3), 269–286.

5 N. A. Kulikova, E. V. Stepanova and O. V. Koroleva, Mitigating Activity of Humic Substances: Direct Influence on Biota, in *Use of humic substances to remediate polluted environments: from theory to practice; Springer*, 2005, pp. 285–309.

6 A. Steinbüchel and R. H. Marchessault. *Biopolymers for Medical and Pharmaceutical Applications*, Wiley-VCH, John Wiley, 2005.

7 R. Klöcking and B. Helbig, Medical Aspects and Applications of Humic Substances, *Biopolymers For Medical & Pharmaceutical Application*, 2005, 3–16.

8 Y. V. Zhernov, S. Kremb, M. Helfer, M. Schindler, M. Harir, C. Mueller, N. Hertkorn, N. P. Avvakumova, A. I. Konstantinov, R. Brack-Werner, *et al.*, Supramolecular Combinations of Humic Polyanions as Potent Microbicides with Polymodal Anti-HIV-Activities, *New J. Chem.*, 2016, **41**(1), 212–224.

9 M. Villén, J. J. Lucena, M. C. Cartagena, R. Bravo, J. García-Mina and M. I. M. De La Hinojosa, Comparison of Two Analytical Methods for the Evaluation of the Complexed Metal in Fertilizers and the Complexing Capacity of Complexing Agents, *J. Agric. Food Chem.*, 2007, **55**(14), 5746–5753.

10 K. Kovács, V. Czech, F. Fodor, A. Solti, J. J. Lucena, S. Santos-Rosell and L. Hernández-Apaolaza, Characterization of Fe-Leonardite Complexes as Novel Natural Iron Fertilizers, *J. Agric. Food Chem.*, 2013, **61**(50), 12200–12210.

11 R. Spaccini and A. Piccolo, Molecular Characterization of Compost at Increasing Stages of Maturity. 1. Chemical Fractionation and Infrared Spectroscopy, *J. Agric. Food Chem.*, 2007, **55**(6), 2293–2302.

12 N. Hertkorn and A. Kettrup, Molecular Level Structural Analysis of Natural Organic Matter and of Humic Substances by Multinuclear and Higher Dimensional NMR Spectroscopy, in *Use of Humic Substances to Remediate Polluted Environments: From Theory to Practice*, Springer-Verlag, Berlin, 2005, pp. 391–435.

13 I. V. Perminova, E. A. Shirshin, A. I. Konstantinov, A. Zherebker, V. A. Lebedev, I. V. Dubinenkov, N. A. Kulikova, E. N. Nikolaev, E. Bulygina and R. M. Holmes, The Structural Arrangement and Relative Abundance of Aliphatic Units May Effect Long-Wave Absorbance of Natural Organic Matter as Revealed by 1H NMR Spectroscopy, *Environ. Sci. Technol.*, 2018, **52**(21), 12526–12537.

14 N. Hertkorn, M. Harir, B. P. Koch, B. Michalke and P. Schmitt-Kopplin, High-Field NMR Spectroscopy and FTICR Mass Spectrometry: Powerful Discovery Tools for the Molecular Level Characterization of Marine Dissolved Organic Matter, *Biogeosciences*, 2013, **10**(3), 1583–1624.

15 N. G. A. Bell, A. A. L. Michalchuk, J. W. T. Blackburn, M. C. Graham and D. Uhrín, Isotope-Filtered 4D NMR Spectroscopy for Structure Determination of Humic Substances, *Angew. Chem., Int. Ed. Engl.*, 2015, **54**(29), 8382–8385.

16 N. Hertkorn, C. Ruecker, M. Meringer, R. Gugisch, M. Frommberger, E. M. Perdue, M. Witt and P. Schmitt-Kopplin, High-Precision Frequency Measurements: Indispensable Tools at the Core of the Molecular-Level Analysis of Complex Systems, *Anal. Bioanal. Chem.*, 2007, **389**(5), 1311–1327.

17 A. G. Marshall, C. L. Hendrickson and G. S. Jackson, Fourier Transform Ion Cyclotron Resonance Mass Spectrometry: A Primer, *Mass Spectrom. Rev.*, 1998, **17**(1), 1–35.

18 S. Kim, R. W. Kramer and P. G. Hatcher, Graphical Method for Analysis of Ultrahigh-Resolution Broadband Mass Spectra of Natural Organic Matter, the van Krevelen Diagram, *Anal. Chem.*, 2003, **75**(20), 5336–5344.

19 E. B. Kujawinski and M. D. Behn, Automated Analysis of Electrospray Ionization Fourier Transform Ion Cyclotron Resonance Mass Spectra of Natural Organic Matter, *Anal. Chem.*, 2006, **78**(13), 4363–4373.

20 W. C. Hockaday, J. M. Purcell, A. G. Marshall, J. A. Baldock and P. G. Hatcher, Electrospray and Photoionization Mass Spectrometry for the Characterization of Organic Matter in Natural Waters: A Qualitative Assessment, *Limnol. Oceanogr.*, 2009, **7**, 81–95.

21 R. L. Sleighter and P. G. Hatcher, *Fourier Transform Mass Spectrometry for the Molecular Level Characterization of Natural Organic Matter: Instrument Capabilities, Applications, and Limitations*, INTECH Open Access Publisher, 2011.

22 N. Hertkorn, M. Frommberger, M. Witt, B. P. Koch, P. Schmitt-Kopplin and E. M. Perdue, Natural Organic Matter and the Event Horizon of Mass Spectrometry, *Anal. Chem.*, 2008, **80**(23), 8908–8919.

23 T. Solouki, M. A. Freitas and A. Alomary, Gas-Phase Hydrogen/Deuterium Exchange Reactions of Fulvic Acids: An Electrospray Ionization Fourier Transform Ion Cyclotron Resonance Mass Spectral Study, *Anal. Chem.*, 1999, **71**(20), 4719–4726.

24 Y. Kostyukevich, A. Kononikhin, I. Popov and E. Nikolaev, Simple Atmospheric Hydrogen/Deuterium Exchange Method for Enumeration of Labile Hydrogens by Electrospray Ionization Mass Spectrometry, *Anal. Chem.*, 2013, **85**(11), 5330–5334.

25 A. C. Stenson, B. M. Ruddy and B. J. Bythell, Ion Molecule Reaction H/D Exchange as a Probe for Isomeric Fractionation in Chromatographically Separated Natural Organic Matter, *Int. J. Mass Spectrom.*, 2014, **360**, 45–53.

26 Y. Kostyukevich, A. Kononikhin, I. Popov, O. Kharybin, I. Perminova, A. Konstantinov and E. Nikolaev, Enumeration of Labile Hydrogens in Natural Organic Matter by Use of Hydrogen/Deuterium Exchange Fourier Transform Ion Cyclotron Resonance Mass Spectrometry, *Anal. Chem.*, 2013, **85**(22), 11007–11013.

27 Y. Kostyukevich, T. Acter, A. Zherebker, A. Ahmed, S. Kim and E. Nikolaev, Hydrogen/Deuterium Exchange in Mass Spectrometry, *Mass Spectrom. Rev.*, 2018, **37**, 811–853.

28 A. Zherebker, Y. Kostyukevich, A. Kononikhin, V. A. Roznyatovsky, I. Popov, Y. K. Grishin, I. V. Perminova and E. Nikolaev, High Desolvation Temperature Facilitates the ESI-Source H/D Exchange at Non-Labile Sites of Hydroxybenzoic Acids and Aromatic Amino Acids, *Analyst*, 2016, **141**(8), 2426–2434.

29 A. Zherebker, Y. Kostyukevich, A. Kononikhin, O. Kharybin, A. I. Konstantinov, K. V. Zaitsev, E. Nikolaev and I. V. Perminova, Enumeration of Carboxyl Groups Carried on Individual Components of Humic Systems Using Deuteromethylation and Fourier Transform Mass Spectrometry, *Anal. Bioanal. Chem.*, 2017, 1–12.

30 J. Atzrodt, V. Derdau, T. Fey and J. Zimmermann, The Renaissance of H/D Exchange, *Angew. Chem., Int. Ed. Engl.*, 2007, **46**(41), 7744–7765.

31 A. Y. Zherebker, D. Airapetyan, A. I. Konstantinov, Y. I. Kostyukevich, A. S. Kononikhin, I. A. Popov, K. V. Zaitsev, E. N. Nikolaev and I. V. Perminova, Synthesis of Model Humic Substances: A Mechanistic Study Using Controllable H/D Exchange and Fourier Transform Ion Cyclotron Resonance Mass Spectrometry, *Analyst*, 2015, **140**(13), 4708–4719.

32 T. Dittmar, B. Koch, N. Hertkorn and G. Kattner, A Simple and Efficient Method for the Solid-Phase Extraction of Dissolved Organic Matter (SPE-DOM) from Seawater, *Limnol. Oceanogr.: Methods*, 2008, **6**, 230–235.

33 Y. I. Kostyukevich, G. N. Vladimirov and E. N. Nikolaev, Dynamically Harmonized FT-ICR Cell with Specially Shaped Electrodes for Compensation of Inhomogeneity of the Magnetic Field. Computer Simulations of the Electric Field and Ion Motion Dynamics, *J. Am. Soc. Mass Spectrom.*, 2012, **23**(12), 2198–2207.

34 A. Zherebker, A. V. Turkova, Y. Kostyukevich, A. Kononikhin, K. V. Zaitsev, I. A. Popov, E. Nikolaev and I. V. Perminova, Synthesis of Carboxylated Styrene Polymer for Internal Calibration of Fourier Transform Ion Cyclotron Resonance Mass-Spectrometry of Humic Substances, *Eur. J. Mass Spectrom.*, 2017, **23**(4), 156–161.

35 R. L. Sleighter, G. A. McKee, Z. Liu and P. G. Hatcher, Naturally Present Fatty Acids as Internal Calibrants for Fourier Transform Mass Spectra of Dissolved Organic Matter, *Limnol. Oceanogr.: Methods*, 2008, **6**(6), 246–253.

36 E. V. Kunenkov, A. S. Kononikhin, I. V. Perminova, N. Hertkorn, A. Gaspar, P. Schmitt-Kopplin, I. A. Popov, A. V. Garmash and E. N. Nikolaev, Total Mass Difference Statistics Algorithm: A New Approach to Identification of High-Mass Building Blocks in Electrospray Ionization Fourier Transform Ion Cyclotron Mass Spectrometry Data of Natural Organic Matter, *Anal. Chem.*, 2009, **81**(24), 10106–10115.

37 B. P. Koch, T. Dittmar, M. Witt and G. Kattner, Fundamentals of Molecular Formula Assignment to Ultrahigh Resolution Mass Data of Natural Organic Matter, *Anal. Chem.*, 2007, **79**(4), 1758–1763.

38 B. P. Koch and T. Dittmar, From Mass to Structure: An Aromaticity Index for High-Resolution Mass Data of Natural Organic Matter, *Rapid Commun. Mass Spectrom.*, 2006, **20**(5), 926–932.

39 Y. Kostyukevich, A. Kononikhin, A. Zherebker, I. Popov, I. Perminova and E. Nikolaev, Enumeration of Non-Labile Oxygen Atoms in Dissolved Organic Matter by Use of $^{16}O/^{18}O$ Exchange and Fourier Transform Ion-Cyclotron Resonance Mass Spectrometry, *Anal. Bioanal. Chem.*, 2014, **406**(26), 6655–6664.

40 R. L. Sleighter, Z. Liu, J. Xue and P. G. Hatcher, Multivariate Statistical Approaches for the Characterization of Dissolved Organic Matter Analyzed by Ultrahigh Resolution Mass Spectrometry, *Environ. Sci. Technol.*, 2010, **44**(19), 7576–7582.

41 P. G. Hatcher and J. L. Faulon, Coalification of Lignin to Form Vitrinite: A New Structural Template Based on an Helical Structure, *J. Am. Chem. Soc.*, 1994, **39**, 7–12.

42 A. Y. Zherebker, Y. I. Kostyukevich, A. S. Kononikhin, E. N. Nikolaev and I. V. Perminova, Molecular Compositions of Humic Acids Extracted from Leonardite and Lignite as Determined by Fourier Transform Ion Cyclotron Resonance Mass Spectrometry, *Mendeleev Commun.*, 2016, **26**(5), 446–448.

43 P. G. Hatcher and D. J. Clifford, The Organic Geochemistry of Coal: From Plant Materials to Coal, *Org. Geochem.*, 1997, **27**, 251–257.

44 V. A. Krassilov, Coal-Bearing Deposits of the Soviet Far East, *Geol. Soc. Am., Spec. Pap.*, 1992, **267**, 263–267.

45 J. Dehmer, Petrographical and Organic Geochemical Investigation of the Oberpfalz Brown Coal Deposit, *Int. J. Coal Geol.*, 1989, **11**(3-4), 273–290.

46 N. H. Werstiuk and G. Timmins, Protium-Deuterium Exchange of Alkylated Benzenes in Dilute Acid at Elevated Temperatures, *Can. J. Chem.*, 1989, **67**(11), 1744–1747.

47 M. Guttman, D. D. Weis, J. R. Engen and K. K. Lee, Analysis of Overlapped and Noisy Hydrogen/Deuterium Exchange Mass Spectra, *J. Am. Soc. Mass Spectrom.*, 2013, **24**(12), 1906–1912.

48 S. M. Shevchenko and G. W. Bailey, The Mystery of the Lignin-Carbohydrate Complex: A Computational Approach, *J. Mol. Struct.*, 1996, **364**(2), 197–208.

49 G. Brunow, I. Kilpelainen, J. Sipila, K. Syrjanen, P. Karhunen, H. Setala and P. Rummakko, Oxidative Coupling of Phenols and the Biosynthesis of Lignin, *J. Am. Chem. Soc.*, 1998, 131–147.

50 W. Flaig and J. C. Salfeld, Nachweis Der Bildung von Hydroxy-*p*-Benzochinon Als Zwischenprodukt Bei Der Autoxydation von Hydrochinon in Schwach Alkalischer L{ö}sung, *Naturwissenschaften*, 1960, **47**(22), 516.

51 R. L. Sleighter and P. G. Hatcher, The Application of Electrospray Ionization Coupled to Ultrahigh Resolution Mass Spectrometry for the Molecular Characterization of Natural Organic Matter, *J. Mass Spectrom.*, 2007, **42**(5), 559–574.

52 A. J. Simpson, D. J. McNally and M. J. Simpson, NMR Spectroscopy in Environmental Research: From Molecular Interactions to Global Processes, *Prog. Nucl. Magn. Reson. Spectrosc.*, 2011, **3**(58), 97–175.

53 P. Delahaye and M. Verzele, Analysis of Gallic, Digallic and Trigallic Acids in Tannic Acids by High-Performance Liquid Chromatography, *J. Chromatogr. A*, 1983, **265**, 363–367.

54 J. V. Castell, L. A. Martinez, M. A. Miranda and P. A. Tárrega, General Procedure for Isotopic (Deuterium) Labelling of Non-Steroidal Antiinflammatory 2-Arylpropionic Acids, *J. Labelled Compd. Radiopharm.*, 1994, **34**(1), 93–100.

55 R. K. Hill, C. Abächerli and S. Hagishita, Synthesis of (2*S*, 4*S*)-and (2*S*, 4*R*)-[5,5,5-^{2}H$_3$] Leucine from (*R*)-Pulegone, *Can. J. Chem.*, 1994, **72**(1), 110–113.

56 A. Zherebker, E. Shirshin, O. Kharybin, Y. Kostyukevich, A. Kononikhin, A. I. Konstantinov, D. Volkov, V. A. Roznyatovsky, Y. K. Grishin, I. V. Perminova, *et al.* Separation of Benzoic and Unconjugated Acidic Components of Leonardite Humic Material Using Sequential Solid-Phase Extraction at Different PH Values as Revealed by Fourier Transform Ion

Cyclotron Resonance Mass Spectrometry and Correlation Nuclear Magneti, *J. Agric. Food Chem.*, 2018, **66**(46), 12179–12187.

57 N. DiDonato and P. G. Hatcher, Alicyclic Carboxylic Acids in Soil Humic Acid as Detected with Ultrahigh Resolution Mass Spectrometry and Multi-Dimensional NMR, *Org. Geochem.*, 2017, **112**, 33–46.

58 E. I. Fedoros, A. A. Orlov, A. Zherebker, E. A. Gubareva, M. A. Maydin, A. I. Konstantinov, K. A. Krasnov, R. N. Karapetian, E. I. Izotova, S. E. Pigarev, *et al.* Novel Water-Soluble Lignin Derivative BP-C*x*−1: Identification of Components and Screening of Potential Targets *in Silico* and *in Vitro*, *OncoTargets Ther.*, 2018, **9**(26), 18578.

59 M. Marcotullio, A. Pelosi and M. Curini, Hinokinin, an Emerging Bioactive Lignan, *Molecules*, 2014, **19**(9), 14862–14878.

60 P. G. Hatcher, H. E. Lerch and T. V. Verheyen, Organic Geochemical Studies of the Transformation of Gymnospermous Xylem during Peatification and Coalification to Subbituminous Coal, *Int. J. Coal Geol.*, 1990, **16**(1), 193–196.

Faraday Discussions

PAPER

Reduced dimensionality hyphenated NMR experiments for the structure determination of compounds in mixtures†

Justinas Sakas [iD] and Nicholle G. A. Bell [iD] *

Received 14th January 2019, Accepted 23rd January 2019

DOI: 10.1039/c9fd00008a

For the structure determination of molecules in mixtures using NMR spectroscopy, the dispersion of ^{13}C chemical shifts provides much needed separation of resonances in the indirectly detected dimension of 2D heterocorrelated NMR experiments. This separation is crucial for establishing networks of coupled spins by hyphenated techniques that combine hetero- and homonuclear polarisation transfers. However, as the sample complexity increases, ^{13}C chemical shifts stop being unique, hindering spectral interpretation. The resulting ambiguities can be removed by adding another dimension to these experiments. However, the spectra obtained from complex samples are riddled with overlapped signals, meaning that another dimension will only reduce the spectral resolution and prevent structure determination. A promising solution is to stay in two dimensions and use the combined ^{13}C and ^{1}H chemical shifts to separate signals. We have developed a suite of (3,2)D reduced dimensionality hyphenated NMR experiments that preserve the information content of 3D spectra but offer all of the advantages of 2D spectra – high resolution and ease of manipulation with only a mild sensitivity penalty. The proposed experiments complement the existing (3,2)D HSQC-TOCSY and include a (3,2)D HSQC-NOESY/ROESY, (3,2)D HSQC-CLIP-COSY and (3,2)D HSQC-HSQMBC. The new experiments represent a set of NMR techniques typically employed in the structure determination of complex compounds and have been adopted here for use on mixtures. The resolving power of these experiments is illustrated on the analysis of hot water extracts of green tea.

Introduction

NMR structure determination of compounds without their purification from mixtures of varying complexity is a challenging task. First of all, the compounds have to reach a concentration threshold to yield sufficient signal intensities to become observable, and secondly, the signals of individual molecules must be tractable in order to identify the investigated molecule or its fragments. This

EaStCHEM School of Chemistry, University of Edinburgh, King's Buildings, Joseph Black Building, David Brewster Road, Edinburgh EH9 3FJ, UK. E-mail: Nicholle.Bell@ed.ac.uk

† Electronic supplementary information (ESI) available. See DOI: 10.1039/c9fd00008a

paper contributes to addressing the second issue by tapping into a large arsenal of NMR techniques developed for the analysis of single molecules. Our aim, similar to others in this field, is to enhance the resolving power of NMR experiments. Different approaches can be employed to achieve this, such as spreading the signals into multiple dimensions,[1] exciting only a fraction of the nuclei,[2] utilising nuclei with larger chemical shift dispersion (usually ^{13}C), reducing the footprint of individual signals by collapsing their multiplets,[3,4] or "separating" the molecules by their size in an NMR tube through molecular diffusion.[5-7] For very complex mixtures, chemical modification in the form of molecular tagging with fully enriched NMR active isotopes seems to be the only option to unambiguously determine structures, although here the main limitation is the potential reach from a given tag and therefore only parts of the molecules can be studied at one time.[8-10]

In this contribution we exploit the possibility of combining ^{1}H and ^{13}C chemical shifts to achieve much needed separation of resonances in the indirectly detected dimension of 2D heterocorrelated NMR experiments. This separation is crucial for establishing networks of coupled spins *via* hyphenated techniques. These techniques combine heteronuclear polarisation transfers, usually in the form of ^{1}H–^{13}C HSQC, with through bond (COSY and TOCSY) or through-space (NOESY or ROESY) homonuclear polarisation transfer steps. We also use this approach in a long-range ^{1}H, ^{13}C correlated experiment.

The basic premise of these techniques is, that as the sample complexity increases, ^{13}C chemical shifts stop being unique at some point, hindering the interpretation of hyphenated heterocorrelated spectra. This ambiguity can, in principle, be tackled by adding another dimension to these experiments, *e.g.* in a form of a 3D HSQC-TOCSY.[11] However, the spectra of complex mixtures are riddled with so many overlapping signals that another dimension, even with non-uniform sampling, will not aid the structure determination process and instead will lower the spectral resolution. A better method is to use the fact that the combined ^{13}C and ^{1}H chemical shifts of a given CH pair are more unique, and use them to distinguish overlapped ^{13}C chemical shifts. Experiments that utilise this approach are commonly referred to as reduced dimensionality NMR experiments.[12] The (3,2)D reduced dimensionality hyphenated NMR experiments preserve the information content of the corresponding 3D spectra but offer all of the advantages of 2D spectroscopy – high resolution and ease of manipulation with only a mild sensitivity penalty.

The described approach has so far been applied to produce (3,2)D HSQC-TOCSY experiments.[13-15] In the work presented here, we have extended it to a suite of complementary experiments, including (3,2)D HSQC-CLIP-COSY, (3,2)D HSQC-NOESY/ROESY and (3,2)D HSQC-HSQMBC. The new experiments, derived from a set of NMR techniques typically employed in the structure determination of pure, but complicated molecules, were adopted here for use on mixtures. The results are illustrated on the analysis of carbohydrates obtained through the hot water extraction of green tea.

Results

Reduced-dimensionality experiments provide equivalent information to that of the corresponding higher dimensionality techniques, *i.e.* they correlate *n* chemical shifts, but they achieve this in a $n - x$ chemical shift space by simultaneous sampling of several chemical shifts during each or some of the incrementable periods of $(n, n - x)$D experiments. In the case of (3,2)D ^{1}H, ^{13}C correlated, hyphenated experiments, the ^{13}C and ^{1}H chemical shifts of directly bonded ^{13}CH$_n$ pairs are sampled simultaneously to record $\Omega_{13C} \pm \kappa\Omega_{1H}$ offset frequencies in a (3,2)D ^{1}H, ^{13}C HSQC part of the experiment. Here, Ω_{13C} (Ω_{1H}) represents the difference between an individual ^{13}C (^{1}H) chemical shift and the respective r.f. carrier frequency; κ is a scaling factor between two frequencies. Such an experiment samples proton chemical shifts twice; the first time during the indirectly-detected period, and the second time during the directly detected period. In between these periods, a polarisation transfer occurs, which can either be homonuclear or, in the case of ^{1}H, ^{13}C long-range correlation, heteronuclear. The displacement of cross peaks by $\pm\kappa\Omega_{1H}$ relative to their position in standard ^{1}H, ^{13}C correlated, hyphenated spectra separates signals with identical ^{13}C chemical shifts, but also allows the determination of the ^{1}H chemical shift of the initial proton in the F_1 dimension, when the direct correlation cross peak is absent.

As a consequence of modulating ^{13}C chemical shifts by ^{1}H frequencies, the number of cross peaks doubles, increasing signal overlap. This issue can however be dealt with easily by recording two data sets while changing the phase of the first 90° ^{1}H pulse in the pulse sequence by 90°. This treatment produces $\cos(2\pi\Omega_{1H}\kappa t_1)$ or $\sin(2\pi\Omega_{1H}\kappa t_1)$ modulated signals in F_1. Consequently, cross peaks in the (3,2)D spectra appear as in phase or antiphase doublets, centred around ^{13}C chemical shifts and separated by $2\kappa\Omega_{1H}$ in F_1, i.e. displaced by $\pm\kappa\Omega_{1H}$ relative to the original HSQC cross peak. The cosine and sine modulated datasets are acquired in an interleaved manner and processed to produce two simplified spectra by the addition or subtraction of the original spectra. The simplified spectra thus contain only one part of the F_1 doublets, each as positive signals with increased intensity. They are referred to here as the $\Omega_{13C} + \kappa\Omega_{1H}$ and the $\Omega_{13C} - \kappa\Omega_{1H}$ spectrum.

Inherent to 2D ^{1}H, ^{13}C HSQC spectra and the spectra of all hyphenated extensions, is the singlet character of cross peaks in F_1. It is crucial that this feature is preserved in the (3,2)D ^{1}H, ^{13}C experiments, as failure to do so would lead to undesirable splitting of signals, increased overlap and reduced sensitivity. In all experiments presented here, the evolution of proton–proton couplings during ^{1}H chemical shift labelling is therefore suppressed by a BIRD^{r,X} pulse[16,17] placed in the middle of this period. The BIRD pulse, in addition to ^{13}C spins, also inverts protons attached to ^{12}C, thus allowing Ω_{1H} labelling of ^{13}C-attached protons, while refocusing the coupling evolution with ^{12}C-attached protons and the directly bonded ^{13}C atom.

The Bruker pulse programs, hsqcedetgpsisp2.3,[18] hsqcdietgpsisp.2,[19] hsqcetgpnosp and hsqcetgprosp[20] were used as the basis for the (3,2)D-based TOCSY, NOESY and ROESY experiments. The DIPSI-2 pulse sequence[21] was applied for the TOCSY transfer, while a low power CW pulse was used for the ROESY spin lock. The (3,2)D BIRD^{r,X}-HSQC-HSQMBC experiment is a modification of the corresponding 3D pulse sequence.[22] The resulting pulse sequences of (3,2)D BIRD^{r,X}-HSQC-(HH/CH-transfer) experiments are shown in Fig. 1.

The proposed experiments are illustrated on the analysis of hot water extracts of green tea. We have recently embarked on a project that uses green tea to assess

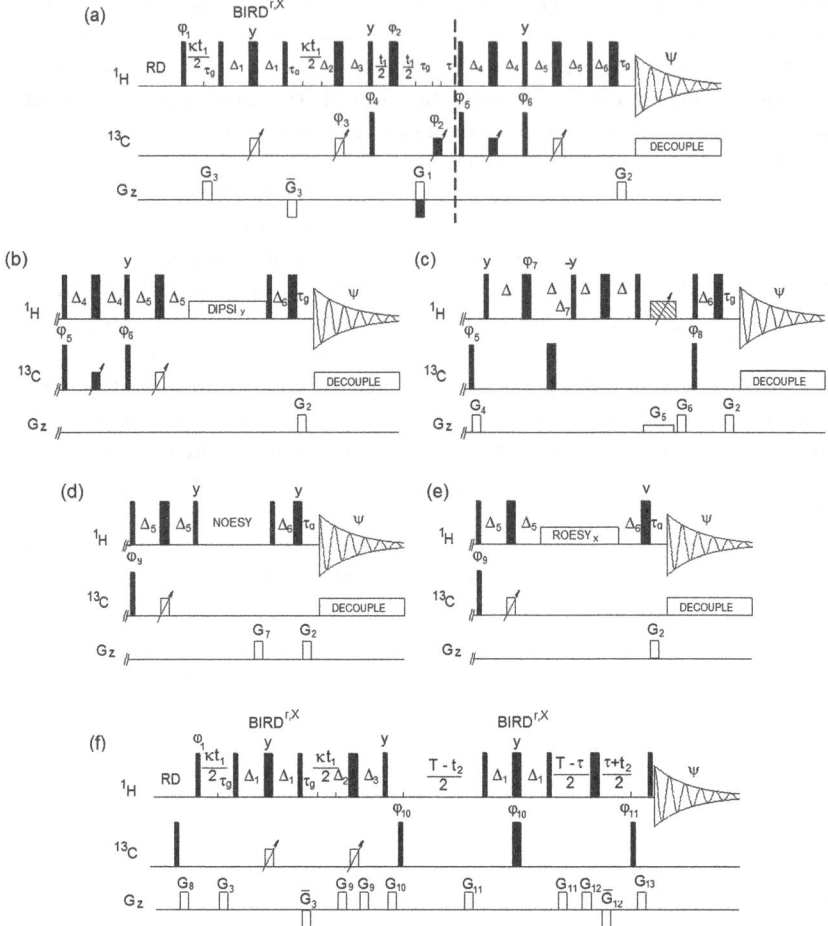

Fig. 1 Pulse sequences of the (3,2)D experiments. (a) (3,2)D BIRD^{r,X}-HSQC. The sequences (b)–(e) start with the first part of the pulse sequence of (a), up to the dashed vertical line; (b) (3,2)D BIRD^{r,X}-HSQC-TOCSY; (c) (3,2)D BIRD^{r,X}-HSQC-CLIP-COSY; (d) (3,2)D BIRD^{r,X}-HSQC-NOESY; (e) (3,2)D BIRD^{r,X}-HSQC-ROESY; (f) (3,2)D BIRD^{r,X}-HSQC-HSQMBC. For explanation of symbols and definition of delays and phases see Experimental.

degradation of organic matter in different soil types. This simple organic mixture acts as a proxy to the degradation of soil organic matter and is used in a citizen science experiment,[23] where members of the public and researchers record the weight loss of buried tea bags after 3 months. Our aim is to provide a molecular signature of the tea degradation process in different soil types as monitored by MS and NMR. Green tea is a well-studied mixture and its hot water extracts contain a number of NMR detectable compounds such as catechins, flavonoids, caffeine, amino acids and carbohydrates.[24] Carbohydrates are the most abundant compound type producing crowded signals in the ^{1}H NMR spectra of green tea (Fig. 2) and as such will be used here to illustrate the resolving power of the proposed experiments.

To illustrate a redistribution of cross peaks, as a result of combined ^{1}H and ^{13}C chemical shift sampling, a 2D ^{1}H, ^{13}C HSQC spectrum of a green tea sample is shown in Fig. 1S,† together with two (3,2)D BIRD^{r,X}-HSQC spectra. Partial (3,2)D BIRD^{r,X}-HSQC-TOCSY spectra of the carbohydrate region are shown in Fig. 3, including the equivalent standard sensitivity-enhanced 2D ^{1}H, ^{13}C HSQC spectrum (Fig. 3a). Fig. 3c and d show separated $\Omega_{13C} + \kappa\Omega_{1H}$ and $\Omega_{13C} - \kappa\Omega_{1H}$ (3,2)D BIRD^{r,X}-HSQC-TOCSY spectra, while Fig. 3b shows an overlay of all three spectra.

The process of identification and assignment of signals from the same molecule is illustrated here on α-D-glucopyranose, a minor component of this mixture. While the chemical shifts of C2α and C5α of α-D-glucopyranose differ by only 8 Hz (and in a lower resolution spectrum would be undistinguishable), the difference in the chemical shifts of H2α and H5α protons leads to a clear separation of their TOCSY traces (Fig. 3b). As discussed previously,[15] the separation of signals in the $\Omega_{13C} + \kappa\Omega_{1H}$ and the $\Omega_{13C} - \kappa\Omega_{1H}$ spectra is different. While an accidental overlap can occur in one of the spectra, separation is usually achieved in the other. This can be seen on the C3α trace of the $\Omega_{13C} - \kappa\Omega_{1H}$ spectrum, where strong sucrose signals interfere, but this is not the case for the $\Omega_{13C} + \kappa\Omega_{1H}$ spectrum, where the signals are well separated. 1D traces through the C1–C5 carbons of α-D-glucopyranose taken from the $\Omega_{13C} - \kappa\Omega_{1H}$ (3,2)D BIRD^{r,X}-HSQC-TOCSY spectra are shown in Fig. 2S.†

The next experiment to be presented is a (3,2)D BIRD^{r,X}-HSQC-CLIP-COSY, which is a modification of a recently published 2D HSQC-CLIP-COSY[25,26] method. Here, the perfect-echo based mixing sequence for in-phase coherence transfer between directly coupled protons is employed in place of an isotropic mixing of a TOCSY. An overlay of the $\Omega_{13C} \pm \kappa\Omega_{1H}$ (3,2)D BIRD^{r,X}-HSQC-CLIP-COSY spectra and a regular HSQC spectrum is presented in Fig. 3e. A comparison with an equivalent presentation of the TOCSY-based experiment (Fig. 3b) indicates that both experiments provide similar information, although the sensitivity of the CLIP-COSY based experiment is lower, as discussed later.

The reduced dimensionality approach was also applied to two through-space proton–proton correlation experiments, HSQC-NOESY and HSQC-ROESY (pulse sequences shown in Fig. 1d and e, respectively). The obtained spectra contain only a few NOESY/ROESY cross peaks. This is to be expected, as the efficiency of NOE-based transfer for small molecules in particular is low on high field instruments, even reaching zero. For carbohydrate structure determination, the most important though-space correlations are those across glycosidic linkages.

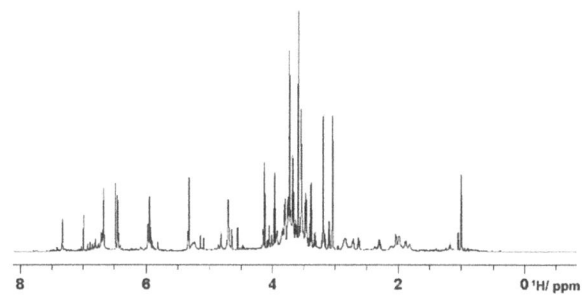

Fig. 2   800 MHz ^{1}H NMR spectrum of hot water extracted green tea.

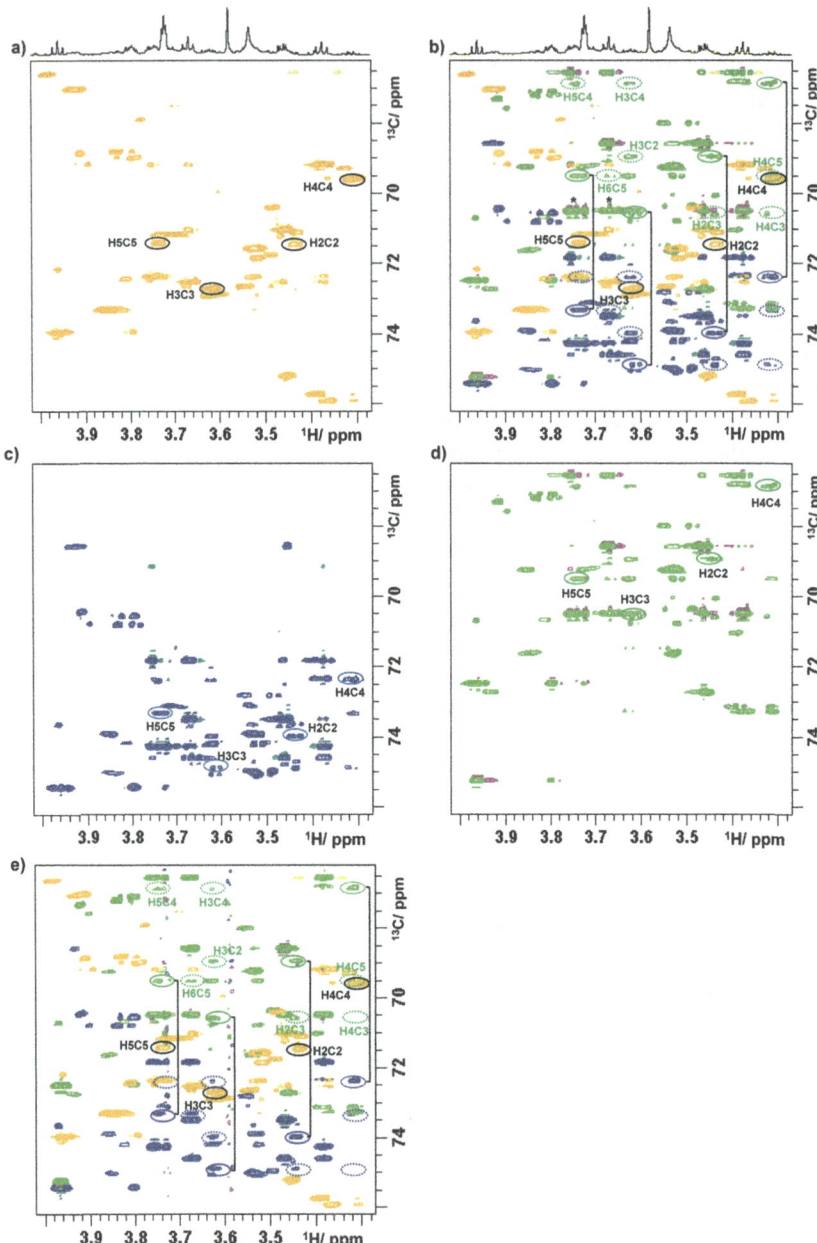

Fig. 3 Partial (3,2)D spectra of green tea. (a) Regular sensitivity-enhanced 2D HSQC spectrum; (b) overlay of spectra (a), (c) and (d); (c) $\Omega_{13C} + \kappa\Omega_{1H}$ (3,2)D BIRD^{r,X}-HSQC-TOCSY spectrum; (d) $\Omega_{13C} - \kappa\Omega_{1H}$ (3,2)D BIRD^{r,X}-HSQC-TOCSY spectrum; (e) (3,2)D BIRD^{r,X}-HSQC-CLIP-COSY spectrum. As a demonstration, the resonances of C2–C5 carbons of α-D-glucopyranose are identified. 1D traces are shown in Fig. 2S.† Intense signals of sucrose that accidentally appear on the same F_1 frequency as those of C3α in the $\Omega_{13C} - \kappa\Omega_{1H}$ spectrum are labelled with an asterisk in (b). Full line ellipsoids represent the displaced cross peaks, while dashed ellipsoids represent the relay cross peaks.

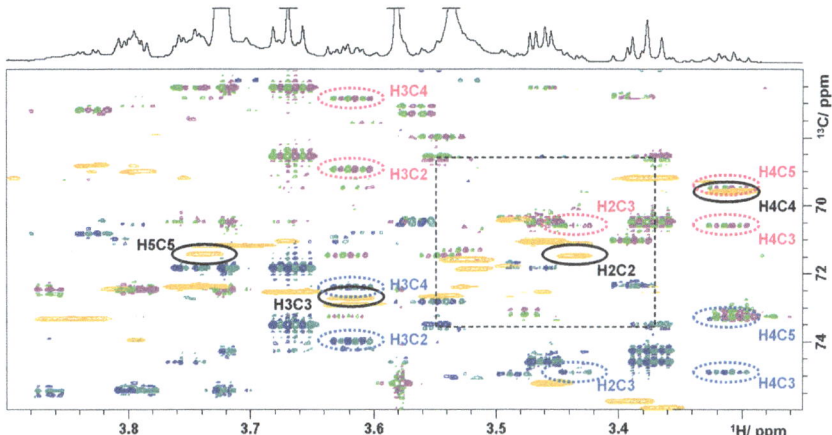

Fig. 4 Overlay of HSQC (gold) and the $\Omega_{13C} \pm \kappa\Omega_{1H}$ (3,2)D BIRD^{r,X}-HSQC-HSQMBC spectra (pink/green and blue/green) of green tea. Resonances of C2–C5 carbons of α-D-glucopyranose are identified. 1D traces are shown in Fig. 4S.†

These have been detected by the (3,2)D BIRD^{r,X}-HSQC-NOESY/ROESY experiments and are presented in Fig. 3S.†

The last experiment discussed is a (3,2)D BIRD^{r,X}-HSQC-HSQMBC technique, which can be viewed as a modification of a 3D HSQC-HSQMBC technique.[22] An expansion identical to that of Fig. 3 shows an overlay of the $\Omega_{13C} + \kappa\Omega_{1H}$ and the $\Omega_{13C} - \kappa\Omega_{1H}$ (3,2)D BIRD^{r,X}-HSQC-HSQMBC spectra together with a regular 2D HSQC spectrum (Fig. 4).

A number of long-range cross peaks are visible in this region, as recognised by their antiphase appearance of a proton–carbon long-range coupling. One-bond correlations are strongly suppressed in the (3,2)D BIRD^{r,X}-HSQC-HSQMBC spectra, nevertheless can appear as four weak cross peaks positioned in the corners of a rectangle separated by $2\kappa\Omega_{1H}$ and $^1J_{CH}$ in F_1 and F_2, respectively. The long-range cross peaks of carbons C2–C5 of α-D-glucopyranose were identified in this spectrum. The F_2 traces through the corresponding cross peaks taken from the $\Omega_{13C} + \kappa\Omega_{1H}$ spectrum are shown in Fig. 4S.† These include correlations of C3 and C5 with H1, which are outside of the spectral region shown in Fig. 4. All expected correlations for C2–C5 carbons were detected by this experiment with good sensitivity.

Discussion

The sensitivity of NMR experiments is always a concern, hence the following discussion focuses on this aspect of the proposed experiments. The overall theoretical sensitivity of a (3,2)D BIRD^{r,X}-HSQC experiment is half of that of a regular 2D HSQC experiment acquired in the same overall time. In practice, additional signal-to-noise reduction occurs, which can be attributed to the $^1J_{CH}$ mismatch and the evolution of proton–proton couplings during the BIRD^{r,X} pulse; signal-to-noise of 30–45% was observed in individual $\Omega_{13C} \pm \kappa\Omega_{1H}$ (3,2)D BIRD^{r,X}-HSQC spectra relative that of a 2D HSQC spectrum, *i.e.* roughly comparable to 3D HSQC-based hyphenated extensions. The spectra are clean and without t_1 noise, with the exception of the CLIP-COSY extension, where strong signals left minor traces, likely due to the use of the gradient z-filter.[27] In all experiments the coherence selection gradient was applied at 20% of the maximum strength (unlike the standard 80%) to minimise diffusion related losses during the long mixing times used in the NOESY, ROESY and also CLIP-COSY experiments. This is not a concern in the (3,2)D BIRD^{r,X}-HSQC-HSQMBC, where gradient coherence selection is not used. Nevertheless, high quality spectra are obtained here as well, mainly due to two G-BIRD blocks[28] that dephase the magnetization of remote protons.

As can be seen from the comparison of corresponding TOCSY and CLIP-COSY $\Omega_{13C} + \kappa\Omega_{1H}$ spectra (Fig. 2d and e), their appearance is very similar. Nevertheless, the sensitivity of the TOCSY transfer is greater, as seen from the comparison of the 1D traces through the $\Omega_{13C} + \kappa\Omega_{1H}$ (3,2)D BIRD^{r,X}-HSQC-TOCSY and CLIP-COSY spectra in Fig. 2S.† We have tested different settings using Δ between 8.33 and 16.7 ms, however the obtained signal-to-noise was between 30 and 50% of that obtained for a 20 ms mixing time TOCSY. This drop of sensitivity has been discussed previously[26] and is caused by the presence of passive proton–proton couplings.

As stated above, the efficiency of the through-space correlation was low, with the ROESY transfer being more efficient than the NOESY transfer. The reduced dimensionality through-space correlation experiments will therefore be most useful for heterogeneous samples with large NOEs such as organic polymers or polysaccharides.

On the other hand, the (3,2)D BIRD^{r,X}-HSQC-HSQMBC designed for establishing long-range correlations of protonated carbons showed a good sensitivity. In this experiment, the active and passive proton–carbon couplings compete for the available magnetisation during the long-range evolution interval T. Even very small couplings, e.g. between C1 and H3 and H5 of α-D-glucopyranose show correlations (see Fig. 4S†). The experiment was repeated twice by optimising the refocusing interval τ for ^{13}CH or ^{13}CH$_2$ pairs. For the CH optimised experiment, low intensity one-bond correlation cross peaks appear in the spectra due to a mismatch of $^1J_{CH}$ couplings with those used to set the τ interval ($0.5/^1J_{CH}$). These cross peaks are characteristically positioned in the four corners of a rectangle separated by $^1J_{CH}$ in F_2 and by $2\kappa\Omega_{1H}$ in F_1, as seen in Fig. 4. For CH$_2$ optimised experiments ($\tau = 0.25/^1J_{CH}$), the intensity of CH one-bond cross peaks increases. It is interesting to note that for CH optimised experiments, very intense one-bond CH$_2$ cross peaks appear in the spectra (the long-range correlation cross peaks are at the same time weak), while for CH$_2$ optimised experiments the intensity of the one-bond CH$_2$ cross peaks drops close to zero (and the long-range correlation cross peaks are more intense). This behaviour is explained in the ESI† using product spin operators. The analysis presented there indicates that refocusing of one-bond couplings before the long-range evolution interval, T, should be considered as a way of removing one-bond correlation cross peaks.

Experimental

The sample was prepared via hot water extraction. 150 mg of Lipton green tea (EAN 87 22700 05552 5) in 40 mL of Milli-Q® water was heated to 80 °C for 30 minutes. The extractant was separated using centrifugation at 8000 rpm and freeze dried. The freeze dried powder (20 mg) was dissolved in D_2O (600 µL, 100% deuterated, Sigma Aldrich®) spun down in a centrifuge. All spectra were acquired on a 4-channel NEO 800 MHz Bruker spectrometer equipped with a 5 mm TCI CryoProbe™ with automated tuning and matching at 300 K. The following parameters were used: 2048 and 2048 complex points in t_2 and t_1, respectively, spectral widths of 9.8 and 160 ppm in F_2 and F_1, yielding t_2 and t_1 acquisition times of 131 and 31.8 ms, respectively. Eight scans were acquired for each t_1 increment using a relaxation time of 1.5 s. The polarisation transfer was optimised for $^1J_{CH}$ = 150 Hz. Forward linear prediction to 4096 points was applied in F_1. A zero filling to 4096 was applied in F_2. A cosine square window function was used for apodization prior to Fourier transformation in both dimensions.

The following parameters are associated with the pulse sequences shown in Fig. 1. Narrow and wide filled rectangles represent 90° and 180° pulses, respectively. Open rectangles with inclined arrows represent 180° ^{13}C CHIRP pulses (p14, 500 µs), while filled rectangles with inclined arrows are 180° composite ^{13}C CHIRP pulses (p24, 2 ms). The following delays were used: $\Delta_1 = 0.5/^1J_{CH} - p14/2$; $\tau_g = 1.2$ ms (1 ms PFG + 200 µs gradient recovery delay); $\Delta_2 = 0.25/^1J_{CH} - \tau_g - p14/2$; $\Delta_3 = 0.25/^1J_{CH} + \tau_g - p14/2 + \kappa t_1(0)$; $\tau = \tau_g + p_{180} \times 2t_1(0)$; $\Delta_4 = 0.125/^1J_{CH} + p14/4$; $\Delta_5 = 0.25/^1J_{CH} - p14/2$; $\Delta_6 = \tau_g - 0.78p_{90} + DE$, where p_{90} and p_{180} are 90° and 180° ^1H pulses, κ is the scaling factor for Ω_{1H} frequencies, $t_1(0)$ is the initial t_1 increment (3 µs) and DE is a pre scan delay (10 µs).

Unless specified otherwise, the pulses were applied from the x axis; $\varphi_1 = x$ or y for cosine and sine modulated signals, respectively; $\varphi_2 = 2x, 2(-x)$; $\varphi_3 = x$; $\varphi_4 = x$, $-x$; $\varphi_5 = 2x, 2(-x)$; $\varphi_6 = 2y, 2(-y)$; $\Psi = x, 2(-x), x$. Phases φ_3, φ_4 and Ψ were increased by 180° simultaneously with t_1 incrementation; the real and imaginary points were acquired by changing the polarity of the G_1 gradient and increasing by 180° the phase φ_5. The gradient G_3 was 600 µs long, while all other gradients were applied for 1 ms. The following relative gradient strengths was used: $G_1 = 20\%$, $G_2 = 5.025\%$, and $G_3 = 70\%$, where 100% represents 53 Gauss per cm.

The (3,2)D BIRD^{r,X}-HSQC-TOCSY spectrum was acquired using identical parameters, a 20 ms mixing time for the TOCSY transfer used a DIPSI-2 pulse sequence[21] ($\gamma B_1/2\pi$ = 5 kHz). The (3,2)D BIRD^{r,X}-HSQC-ROESY spectrum was acquired using a 200 ms CW ROESY spin lock at $\gamma B_1/2\pi$ = 3570 Hz. The (3,2)D BIRD^{r,X}-HSQC-NOESY spectrum was acquired using a 250 ms NOESY mixing time, $G_7 = 15\%$. The phase $\varphi_9 = 4x, 4(-x)$ in the NOESY and ROESY. Phases φ_3, φ_4 and Ψ were increased by 180° simultaneously with t_1 incrementation; the real and imaginary points were acquired by changing the polarity of the G_1 gradient.

For the (3,2)D BIRD^{r,X}-HSQC-CLIP-COSY, $\Delta = 11.4$ ms and $\Delta_7 = 0.25/^1J_{CH}$. A 20 ms CHIRP pulse was applied simultaneously with a G_5 (−5%) PFG followed by a $G_6 = -19.4\%$; $G_4 = 11\%$. Phases were as follows: $\varphi_7 = 2y, 2(-y)$; $\varphi_8 = x, -x$; $\Psi = 2(x, -x), 2(-x, x)$. Phases φ_3, φ_4 and Ψ were increased by 180° simultaneously with t_1 incrementation; the real and imaginary points were acquired by changing the polarity of the G_1 gradient.

For the (3,2)D BIRD^{r,X}-HSQC-HSQMBC spectrum, T represents the long-range evolution interval that was set to 62.5 ms and $\tau = n \times 0.5/^1J_{CH}$, n = 1 for CH and n = 0.5 for all multiplicities. $\varphi_{10} = x, -x$; $\varphi_{10} = 2x, 2(-x)$; $\Psi = x, 2(-x), x$. Phase φ_{10} was incremented in a TPPI manner by 90°, phase Ψ was increased by 180° simultaneously with t_1 incrementation. $G_8 = 10\%$, $G_9 = 11.6\%$, $G_{10} = 26\%$, $G_{11} = 16\%$, $G_{12} = 50\%$, $G_{13} = 22\%$.

The number of scans (NS) was 8, except for the ROESY and the HSQC-HSQMBC, where NS = 12. The overall acquisition time was 7 hours 40 minutes (standard sensitivity-enhance HSQC), 15 hours 38 minutes ((3,2)D HSQC), 15 hours 48 minutes (TOCSY), 17 hours 52 minutes (NOESY), 25 hours 41 minutes (ROESY), 16 hours 5 minutes (CLIP-COSY) and 25 hours 16 minutes (HSQC-HSQMBC).

Conclusions

In conclusion, we have presented a series of reduced dimensionality 2D hyphenated HSQC-based experiments designed to deal with the overlap of carbon resonances encountered in mixtures. In combination with high digital resolution, achievable by non-linear sampling, these experiments are suitable for tracing out individual spin systems of small to medium size molecules embedded in mixtures of considerable complexity.

Conflicts of interest

The authors declare no conflicts of interests.

Acknowledgements

The authors would like to thank Juraj Bella and Dr Lorna Murray for maintenance of the NMR spectrometers. J. Sakas would like to acknowledge the financial support received from an EPSRC Undergraduate Vacation Scholarship. N. G. A. Bell would like to acknowledge the financial support received from a NERC Soil Security Programme Fellowship.

Notes and references

1 C. Griesinger, O. W. Sorensen and R. R. Ernst, *J. Magn. Reson.*, 1987, **73**, 574–579.
2 H. Kessler, H. Oschkinat and C. Griesinger, *J. Magn. Reson.*, 1986, **70**, 106–133.
3 K. Zangger, *Prog. Nucl. Magn. Reson. Spectrosc.*, 2015, **86–87**, 1–20.
4 J. A. Aguilar, S. Faulkner, M. Nilsson and G. A. Morris, *Angew. Chem., Int. Ed.*, 2010, **49**, 3901–3903.
5 W. S. Price, *Concepts Magn. Reson.*, 1997, **9**, 299–336.
6 W. S. Price, *Concepts Magn. Reson.*, 1998, **10**, 197–237.
7 H. Barjat, G. A. Morris, S. Smart, A. G. Swanson and S. C. R. Williams, *J. Magn. Reson., Ser. B*, 1995, **108**, 170–172.
8 N. G. A. Bell, M. C. Graham and D. Uhrin, *Analyst*, 2016, **141**, 4614–4624.
9 N. G. A. Bell, L. Murray, M. C. Graham and D. Uhrin, *Chem. Commun.*, 2014, **50**, 1694–1697.
10 G. A. Bell, A. A. L. Michalchuk, J. W. T. Blackburn, M. C. Graham and D. Uhrin, *Angew. Chem., Int. Ed.*, 2015, **54**, 8382–8385.
11 V. V. Krishnamurthy, *J. Magn. Reson., Ser. B*, 1995, **106**, 170–177.
12 S. Kim and T. Szyperski, *J. Am. Chem. Soc.*, 2003, **125**, 1385–1393.
13 S. M. Pudakalakatti, A. Dubey, G. Jaipuria, U. Shubhashree, S. K. Adiga, D. Moskau and H. S. Atreya, *J. Biomol. NMR*, 2014, **58**, 165–173.

14 A. Singh, A. Dubey, S. K. Adiga and H. S. Atreya, *J. Magn. Reson.*, 2018, **286**, 10–16.

15 N. Brodaczewska, Z. Kostalova and D. Uhrin, *J. Biomol. NMR*, 2018, **70**, 115–122.

16 J. R. Garbow, D. P. Weitekamp and A. Pines, *Chem. Phys. Lett.*, 1982, **93**, 504–509.

17 D. Uhrin, T. Liptaj and K. E. Kover, *J. Magn. Reson., Ser. A*, 1993, **101**, 41–46.

18 L. E. Kay, P. Keifer and T. Saarinen, *J. Am. Chem. Soc.*, 1992, **114**, 10663–10665.

19 J. Schleucher, M. Schwendinger, M. Sattler, P. Schmidt, O. Schedletzky, S. J. Glaser, O. W. Sorensen and C. Griesinger, *J. Biomol. NMR*, 1994, **4**, 301–306.

20 A. Bax and D. G. Davis, *J. Magn. Reson.*, 1985, **63**, 207–213.

21 S. P. Rucker and A. J. Shaka, *Mol. Phys.*, 1989, **68**, 509–517.

22 D. Uhrin, *J. Magn. Reson.*, 2002, **159**, 145–150.

23 J. A. Keuskamp, B. J. J. Dingemans, T. Lehtinen, J. M. Sarneel and M. M. Hefting, *Methods Ecol. Evol.*, 2013, **4**, 1070–1075.

24 Y. F. Yuan, Y. L. Song, W. H. Jing, Y. T. Wang, X. Y. Yang and D. Y. Liu, *Anal. Methods*, 2014, **6**, 907–914.

25 T. Gyongyosi, I. Timari, J. Haller, M. R. M. Koos, B. Luy and K. E. Kover, *ChemPlusChem*, 2018, **83**, 53–60.

26 M. R. M. Koos, G. Kummerlowe, L. Kaltschnee, C. M. Thiele and B. Luy, *Angew. Chem., Int. Ed.*, 2016, **55**, 7655–7659.

27 M. J. Thrippleton and J. Keeler, *Angew. Chem., Int. Ed.*, 2003, **42**, 3938–3941.

28 C. Emetarom, T. L. Hwang, G. Mackin and A. J. Shaka, *J. Magn. Reson., Ser. A*, 1995, **115**, 137–140.

ROYAL SOCIETY
OF CHEMISTRY

PAPER

Unraveling the complexity of complex mixtures by combining high-resolution pharmacological, analytical and spectroscopic techniques: antidiabetic constituents in Chinese medicinal plants†

Yong Zhao, [iD][a] Kenneth Thermann Kongstad, [iD][a] Yueqiu Liu,[a] Chenghua He[ab] and Dan Staerk [iD] *[a]

Received 3rd December 2018, Accepted 17th December 2018

DOI: 10.1039/c8fd00223a

Medicinal plants have been widely used as (poly)pharmacological remedies and constitute a rich source for antidiabetic drug discovery. In the present study, forty medicinal plant samples collected in China were tested for inhibitory activity against α-glucosidase, α-amylase, and protein-tyrosine phosphatase 1B (PTP1B). Crude ethyl acetate extracts of *Dioscorea bulbifera* L., *Boehmeria nivea* Gaudich, *Tinospora sagittata* Gagnep. and *Persicaria bistorta* (L.) Samp. showed dual inhibitory activity towards α-glucosidase and PTP1B, and were chosen for further investigation. Subsequent dual high-resolution α-glucosidase/PTP1B profiling or triple high-resolution α-glucosidase/α-amylase/PTP1B profiling combined with HPLC-HRMS and NMR spectroscopy led to the identification of 28 metabolites with one or more bioactivities. Among these, three new phenanthrenes were identified from *D. bulbifera*, including one new biphenanthrene (**10**) exhibiting promising dual inhibitory activity towards α-glucosidase and PTP1B with IC_{50} values of 2.08 ± 0.19 and 3.36 ± 0.25 µM, respectively. Two triterpenoids and one fatty acid from *B. nivea* and *T. sagittata* as well as some commercially available fatty acids showed strong PTP1B inhibitory activity with IC_{50} values in the range of 4.89 ± 0.38 to 53.77 ± 4.20 µM.

[a]Department of Drug Design and Pharmacology, Faculty of Health and Medical Sciences, University of Copenhagen, Universitetsparken 2, DK-2100 Copenhagen, Denmark. E-mail: ds@sund.ku.dk

[b]College of Veterinary Medicine, Nanjing Agricultural University, Nanjing 210095, China

† Electronic supplementary information (ESI) available: Table S1 with plants tested for antidiabetic activities; Table S2 with HRMS and NMR data of the isolated compounds; Fig. S1 with α-glucosidase and PTP1B IC_{50} curves; Fig. S2–S17 with UV, HRMS, and NMR spectra of compounds **8–10**; Fig. S18 with triple high-resolution α-glucosidase/α-amylase/PTP1B inhibition profile of *P. bistorta*; Fig. S19 with α-glucosidase IC_{50} curves for isolated compounds and acarbose (reference compound); Fig. S20 with PTP1B IC_{50} curves for the isolated compounds and RK682 (reference compound). See DOI: 10.1039/c8fd00223a

Introduction

Diabetes mellitus is a chronic metabolic disease characterized by hyperglycemia and associated long-term renal, retinal, neurovascular and cardiovascular complications.[1] The incidence of diabetes, of which type 2 diabetes (T2D) constitutes approximately 90%, reached an alarming 425 million worldwide in 2017 and is expected to reach 629 million in 2045.[2] Besides being an invalidating disease for the individual patient, the epidemic increase in T2D constitute a large socioeconomic burden worldwide with increasing costs to primary and secondary health care systems.[3] T2D is a multifactorial disease which calls for a multifactorial treatment, and α-glucosidase, α-amylase and PTP1B are three important targets for the management of T2D. α-Glucosidase and α-amylase are both carbohydrate-degrading enzymes, but whereas α-amylase secreted in saliva produces glucose, maltose and oligosaccharides from the random cleavage of α-1,4 glycosidic linkages of amylose and amylopectin, α-glucosidase is located in the brush-border of the small intestine where it hydrolyses terminal $(1 \rightarrow 4)$-linked α-glucose units that are absorbed as monosaccharides into the blood stream.[4,5] Acarbose, miglitol and voglibose are widely used in the clinic as competitive inhibitors of α-glucosidase and/or α-amylase. However, major side effects, such as flatulence, abdominal discomfort and diarrhoea, are associated with their use.[6] Protein tyrosine phosphatase-1B (PTP1B) is a negative regulator of insulin and leptin signalling, which can decrease and/or shorten the action of insulin by dephosphorylating the activated insulin receptor and insulin receptor substrate.[7] There are no clinically approved drugs for PTP1B, and with more than 100 structurally related PTPs identified in humans, the discovery of selective PTP1B inhibitors is a challenge.[8] As a result, new α-glucosidase, α-amylase and PTP1B inhibitors with fewer side effects and higher potency and selectivity are still needed for improved management of T2D.

Nature is a rich source of bioactive constituents, and more than half of today's small-molecule therapeutics in clinical use trace their origin back to natural sources.[9,10] For example, the first-choice drug for treatment of T2D, metformin, a drug that increases insulin sensitivity, is a derivative of galegine, isolated from the medicinal plant *Galega officinalis*.[11] Medicinal plants are used as antidiabetic principles in many traditional treatment systems.[12] This includes traditional Chinese medicine that is a rich source of antihyperglycemic compounds. Many of the bioactive constituents identified in antidiabetic herbs from traditional Chinese medicine inhibit multiple T2D targets and are thus potential poly-pharmacological remedies.[13,14] However, the majority of traditional Chinese medicine plants have not been investigated using a polypharmacological approach, and the antihyperglycemic constituents and their mechanism(s) of action are still largely unknown for the majority of traditional Chinese herbs.

Systematic (poly)pharmacological screening and subsequent structural identification of individual constituents in complex plant extracts, *i.e.* unraveling the complexity of the chemical space made up by one or multiple complex plant extracts, require high-resolution chromatographic, high-resolution pharmacological and high-resolution spectroscopic techniques (Fig. 1).

Aiming at accelerating the separation process but at the same time having enough material available for pharmacological as well as MS and NMR

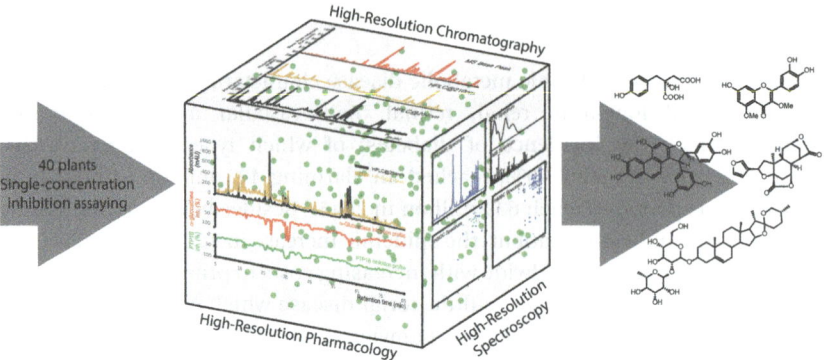

Fig. 1   Single-concentration inhibition percentages used as a filter for the selection of plants and thus for narrowing the chemical space to investigate. The search for drug candidates is accelerated by the combined use of high-resolution chromatographic, high-resolution pharmacological, and high-resolution spectroscopic techniques.

spectroscopic characterization, analytical-scale HPLC is preferred over UHPLC. Thus, analytical-scale HPLC can be used for pharmacological profiling towards one or more pharmacological targets by microplate-based high-resolution inhibition profiling. This is based on repeated time-based microfractionation in one or more 96-well microplates and subsequent bioassaying of the material in each well. The results from each well, expressed as percent inhibition, are subsequently plotted at their respective retention time to provide single,[15] dual,[16,17] triple[18] or even quadruple[19] inhibition profiles. This allows subsequent chemical analysis to be targeted toward constituents with one or more bioactivities, and thereby accelerating the polypharmacological drug discovery process. The same analytical-scale HPLC setup can subsequently be used as part of hyphenated techniques like LC-HRMS or HPLC-PDA-HRMS-SPE-NMR,[20,21] or for the collection of material for NMR spectroscopy.

In this paper, fast single-concentration inhibition percentages of α-glucosidase, α-amylase, and PTP1B were used in combination with LC-HRMS and off-line NMR spectroscopy for the identification of natural products with one or more bioactivities towards T2D targets.

Experimental

Chemicals

α-Glucosidase type I (EC 3.2.20, from *Saccharomyces cerevisiae*, lyophilized powder), *p*-nitrophenol α-D-glucopyranoside (*p*-NPG), α-amylase type VI-B (E.C. 3.2.1.1, from porcine pancreas, lyophilized powder), 2-chloro-4-nitrophenyl-α-D-maltotrioside (CNP-G3), *p*-nitrophenyl phosphate (*p*-NPP), dimethyl sulfoxide (DMSO), NaH_2PO_4, Na_2HPO_4, NaN_3, NaCl, tris(hydroxymethyl)-aminomethane (Tris), bis(2-hydroxyethyl)-imino-tris(hydroxymethylmethane) (bis-Tris), dithiothreitol (DTT), *N,N,N′,N′*-ethylenediaminetetraacetate (EDTA), and HPLC-grade MeCN were purchased from Sigma-Aldrich (St. Louis, MO). Recombinant human protein tyrosine phosphatase 1B (PTP1B) (BML-SE332-0050, EC 3.1.3.48) was purchased from Enzo Life Sciences, Inc., (NY, U.S.A.). Calcium acetate and formic acid were purchased from Merck (Darmstadt, Germany). Water was purified by deionization and 0.22 μM membrane filtration (Millipore, Billerica, MA).

Plant material and extraction

As presented in Table S1,† a total of 40 different species (or parts) of plants were collected in China. Voucher specimens of plants 1, 3, 5–6, 10–15, 17, 20, 22–26, 31, 32, 34–38, and 40 are deposited at Southwest Treasure Herbs (Chendu, China). Voucher specimens for plants 2, 4, 8, 9, 16, 18, 19, 21, 27–30, 33 and 39 are deposited at the lab of Chinese Veterinary Medicine, Nanjing Agricultural University (Nanjing, China). Approximately 1 g of each powdered plant material was dissolved in 10 mL of ethyl acetate, and sonicated for 2 hours to afford 40 crude extracts. Based on previous experience, 50 μg mL^{-1} for each crude extract was used for the initial bioactivity screening. For the four selected plants, powdered material of *Dioscorea bulbifera* L. (12 g), *Boehmeria nivea* (L.) Gaudich (11 g), *Tinospora sagittata* Gagnep. (16 g), and *Persicaria bistorta* (L.) Samp. (21 g) was dissolved in 120 mL, 110 mL, 160 mL and 210 mL of ethyl acetate, respectively, and sonicated for 2 hours at room temperature. The solutions were filtrated and then evaporated to dryness by rotary evaporation (100 mbar, 35 °C), which yielded 83 mg extract of *D. bulbifera*, 74 mg extract of *B. nivea*, 330 mg extract of *T. sagittata*, and 228 mg extract of *P. bistorta*.

Antidiabetic bioassays

The α-glucosidase, α-amylase and PTP1B inhibitory assays were performed in 96-well microplates according to the previously reported procedure.[19] The α-glucosidase assay was performed at 28 °C with a final volume of 200 μL. For each well, samples were dissolved in 10 μL of DMSO and 90 μL of phosphate buffer (0.1 M, pH 7.5, 0.02% NaN$_3$), and subsequently shaken for 10 min. 80 μL of α-glucosidase solution in the same phosphate buffer (2.0 U mL^{-1}) was then added to each well and incubated for 10 min. Then, 20 μL of *p*-NPG solution (10 mM in phosphate buffer) was added to start the enzyme reaction. The absorbance of the 4-nitro-phenol cleavage product was measured at 405 nm every 30 s for 35 min with a Thermo Scientific Multiskan FC microplate photometer (Thermo Scientific, Waltham, MA) coupled to SkanIt ver. 2.5.1 software, to yield the cleavage rate (kinetic measurements) as ΔAU s^{-1}. The inhibition of the enzyme was calculated as follows:

$$\% \text{ Inhibition} = (\text{slope}_{\text{blank}} - \text{slope}_{\text{sample}})/\text{slope}_{\text{blank}} \times 100\%$$

The procedure of the α-amylase assay was similar to the α-glucosidase assay. In brief, samples in each well were dissolved in 10 μL of DMSO and 90 μL of phosphate buffer (0.1 M, pH 6, 0.02% NaN$_3$). 80 μL of α-amylase solution (2.0 U mL^{-1}) was subsequently added to each well and incubated for 10 min at 37 °C. The reaction was started by adding 20 μL of substrate solution (CNP-G3, 10 mM in phosphate buffer). The absorbance was measured at 405 nm every 3 min for 30 min and the percentage inhibition of the enzyme was calculated using the same formula as for the α-glucosidase inhibition assay.

The PTP1B assay was performed at 25 °C in a final volume of 180 μL, using a buffer containing 50 mM Tris, 50 mM bis-Tris and 0.1 M NaCl (pH was adjusted to 7.0 with citric acid). Samples were dissolved in 18 μL of DMSO followed by the addition of 52 μL of EDTA solution (3.46 mM in the above buffer) and 60 μL of substrate solution (1.5 mM p-NPP and 6 mM DTT), and incubated for 10 min at 25 °C. 50 μL of 0.001 μg μL^{-1} PTP1B stock solution was added into each well to start the reaction. The absorbance was measured at 405 nm every 30 s for 10 min and the percentage inhibition of the enzyme was calculated with the same formula as described for the α-glucosidase inhibition assay.

Analytical-scale HPLC separations and high-resolution α-glucosidase, α-amylase and PTP1B inhibition profiling

Analytical-scale HPLC separations for the four selected plant extracts were performed on an Agilent 1200 system (Santa Clara, CA, USA) consisting of a G1367C high-performance autosampler, a G1311A quaternary pump, a G1322A degasser, a G1316A thermostated column compartment, a G1315C photodiode-array detector, and a G1364C fraction collector, all controlled by Agilent ChemStation version B.03.02 software (Agilent, Santa Clara, CA, USA), at 40 °C on a 150 × 4.6 mm i.d. Phenomenex Luna C$_{18}$(2) reversed-phase column (3 μm particle size, 100 Å pore size; Phenomenex, Torrance, CA, USA) with a flow rate of 0.5 mL min^{-1}. The elution solvent system consisted of aqueous eluent A (H$_2$O–MeCN 95 : 5 with 0.1% formic acid) and organic eluent B (H$_2$O–MeCN 5 : 95 with 0.1% formic acid).

For the high-resolution α-glucosidase, PTP1B and/or α-amylase inhibition profiling, four different HPLC methods were developed. For *P. bistorta*, a 10 μL injection of crude extract (40 mg mL^{-1} in MeOH) was separated using the following elution gradient: 0 min, 0% B; 30 min, 30% B; 50 min, 50% B; 51 min, 100% B; 60 min, 100% B. The elute from 8 to 50 min was fractionated into 88 wells (one 96-well microplate) yielding a resolution of 2.10 data points per min. For *D. bulbifera*, a 10 μL injection of crude extract (40 mg mL^{-1} in MeOH) was separated using the following elution gradient: 0 min, 5% B; 18 min, 27% B; 50 min, 35% B; 80 min, 100% B; 90 min, 100% B. The elute from 5 to 85 min was fractionated into 176 wells (two 96-well microplates) yielding a resolution of 2.20 data points per min. For *B. nivea*, a 10 μL injection of crude extract (40 mg mL^{-1} in MeOH) was separated using the following elution gradient: 0 min, 0% B; 20 min, 30% B; 25 min, 70% B; 50 min, 100% B; 60 min, 100% B. The elute from 10 to 50 min was fractionated into 88 wells (one 96-well microplate) yielding a resolution of 2.20 data points per min. For *T. sagittata*, a 10 μL injection of crude extract (40 mg mL^{-1} in MeOH) was separated using the following elution gradient: 0 min, 0% B; 30 min, 100% B; 40 min, 100% B. The elute from 10 to 40 min was fractionated into 88 wells (one 96-well microplate) yielding a resolution of 2.93 data points per min.

The microplates were subsequently evaporated to dryness at 35 °C using a SPD121P Savant SpeedVac concentrator equipped with an OFP400 oil free pump and a RVT400 Refrigerated Vapor Trap (Thermo Scientific, Waltham, MA, USA). This procedure was repeated two or three times, allowing subsequent assays for α-glucosidase, PTP1B and/or α-amylase inhibition as described above. The percentage of inhibition of each well was plotted at its respective chromatographic retention time to yield high-resolution α-glucosidase, PTP1B and/or α-amylase inhibition profiles.

Isolation and purification

Isolation of the individual compounds was performed on the same HPLC system with the same chromatography methods as for high-resolution inhibition profiling. Compounds 1–7, 12–20 were purified from *D. bulbifera* directly after 107 consecutive injections of crude extracts (10 μL injection per separation, 60 mg mL^{-1}). Compound 8 was collected together with compound 7, and the limited amount of mixture did not allow further separation. Compounds 9–10 were obtained by further separation with isocratic elution of 35% B. Similarly, compounds 1, 21–23, 27, and 28 were obtained from *B. nivea* after 98 consecutive injections of crude extracts (10 μL injection per separation, 60 mg mL^{-1}). Compound 1 was obtained from both *D. bulbifera* and *B. nivea*. Compounds 24–26 were further purified with isocratic elution of 70% B. Compounds 27 and 28 were isolated from *T. sagittata* after 50 consecutive injections of crude extracts (10 μL injection per separation, 60 mg mL^{-1}).

HPLC-HRMS and NMR experiments

HPLC-HRMS analyses were performed on an Agilent 1260 HPLC system consisting of a degasser, a quaternary pump, an autosampler, and a photodiode array detector (Agilent, Santa Clara, CA, USA), coupled with a Bruker micrOTOF-Q II mass spectrometer equipped with an electrospray interface (Bruker Daltonik, Bremen, Germany). Separations were performed with the same column, temperature, and solvent gradient as described above. Mass spectra were acquired in both positive and negative mode and a solution of sodium formate cluster was automatically injected to enable internal mass calibration. NMR experiments were performed on a 600 MHz Bruker Avance III instrument (operating frequency of 600.13 MHz) equipped with a cryogenically cooled 1.7 mm TCI probe head and a Bruker SampleJet sample changer (Bruker Biospin, Karlsruhe, Germany). All experiments were acquired in automation (temperature equilibration to 300 K, optimization of lock parameters, gradient shimming, and setting of receiver gain). ^{1}H NMR spectra were acquired with 30°-pulses and 64k data points. 2D homo- and heteronuclear experiments were acquired with 2048 data points in the direct dimension and 128 (HMBC) or 512 (DQF-COSY) or 256 (multiplicity edited HSQC) data points in the indirect dimension. IconNMR ver. 4.2 (Bruker Biospin, Karlsruhe, Germany) was used for controlling the automated sample change and acquisition of NMR data. Topspin ver. 3.5 (Bruker Biospin, Karlsruhe, Germany) was used for acquisition and processing of NMR data.

Results and discussion

Screening samples with antidiabetic potential

In the present study, 40 different samples of medicinal plants traditionally used as antidiabetic remedies in China (ESI Table S1†)[22] were tested for their α-glucosidase, α-amylase and PTP1B inhibitory activity at a concentration of 50 μg mL^{-1}.

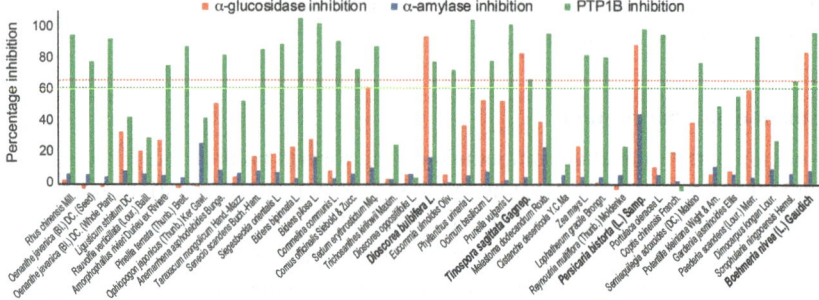

Fig. 2 Percentage inhibition of α-glucosidase, α-amylase and PTP1B measured at a single concentration of 50 μg mL⁻¹ for 40 samples used in traditional Chinese medicine practice. The red and green dashed lines refer to the inclusion criteria of 65% inhibition of α-glucosidase and 60% inhibition of PTP1B – both criteria are based on past experience [96-well microplate assays, UV readings at 405 nm, % inhibition = (slope$_{blank}$ − slope$_{sample}$)/slope$_{blank}$ × 100%].

As shown in Fig. 2 and ESI Table S1,† nine of the extracts showed more than 50% inhibition against α-glucosidase, 30 of them showed more than 50% inhibition against PTP1B, while only one extract showed 45% inhibition against α-amylase. Based on past experience and with an aim of working with samples with inhibitory activity of more than one target, inclusion criteria of more than 65% inhibition of α-glucosidase as well as 60% inhibition of PTP1B were chosen as the threshold levels. This resulted in the inclusion of *Persicaria bistorta* (L.) Samp, *Dioscorea bulbifera* L., *Boehmeria nivea* (L.) Gaudich, and *Tinospora sagittata* Gagnep., showing dual inhibitory activity towards α-glucosidase and PTP1B. The lack of α-amylase inhibitory activity of all the samples except for *P. bistorta* made us choose not to include α-amylase inhibitory activity as an inclusion criterion. The selected crude extracts were assessed by concentration-dependent assays, resulting in IC$_{50}$ values from 17.72 ± 0.88 to 25.12 ± 3.41 μg mL⁻¹ (for α-glucosidase) and 17.43 ± 0.91 to 38.50 ± 4.94 μg mL⁻¹ (for PTP1B), as shown in Table 1 and in ESI Fig. S1.†

High-resolution α-glucosidase/α-amylase/PTP1B profiling of crude extracts

With the aim of identifying individual constituents responsible for the observed α-glucosidase and PTP1B inhibition of these four plants, four different analytical-

Table 1 IC$_{50}$ values of four crude plant extracts against α-glucosidase and PTP1B

Plant	IC$_{50}$ value^a (μg mL⁻¹)	
	α-Glucosidase	PTP1B
Dioscorea bulbifera L.	17.72 ± 0.88	32.21 ± 2.17
Boehmeria nivea (L.) Gaudich	24.60 ± 0.67	20.19 ± 0.94
Tinospora sagittata Gagnep.	25.12 ± 3.41	38.50 ± 4.94
Persicaria bistorta (L.) Samp.	18.14 ± 1.04	17.43 ± 0.91

^{a} Values expressed as mean ± standard error ($n = 3$).

scale HPLC methods with different gradient elution profiles were developed in order to separate as many analytes as possible directly from the crude extracts. Thus, 10 μL injections of crude extract (40 mg mL^{-1} in methanol) were separated by reversed-phase analytical-scale HPLC, and time-based fractionated into one or two 96-well microplates using the described HPLC methods. For *D. bulbifera*, *B. nivea* and *T. sagittata*, this procedure was repeated twice, allowing subsequent inhibitory assays for α-glucosidase and PTP1B. Considering *P. bistorta* also showed weak inhibition of α-amylase, an additional collection for *P. bistorta* was performed and assayed for α-amylase inhibitory activity as well. The inhibitory activities of the material in all wells were plotted according to their respective retention times, providing dual high-resolution α-glucosidase/PTP1B inhibition profiles for *D. bulbifera*, *B. nivea* and *T. sagittata* and a triple high-resolution α-glucosidase/α-amylase/PTP1B inhibition profile for *P. bistorta*.

Identification of metabolites by LC-HRMS and NMR spectroscopy from crude extracts

The *D. bulbifera* extract showed well-separated peaks and several of them were affiliated with α-glucosidase and/or PTP1B inhibition as shown in Fig. 3. The crude extract was analyzed by HPLC-HRMS and twenty well-separated peaks (1–20) were isolated by repeated HPLC separations, allowing subsequent NMR experiments. This resulted in identification of 17 compounds, including 3,4-dihydroxybenzoic acid (**1**), catechin (**2**),[23] eucomic acid (**3**),[24] epicatechin (**4**),[25] 9,10-dihydro-2,4,6,7-phenanthrenetetrol (**5**),[26] dihydropiceatannol (**6**),[27] 3,5-di-*O*-methylquercetin (**7**),[28] cassigarol D (**11**),[29] diosbulbin B (**12**),[30] flavanthrinin (**13**),[31] flavanthridin (**14**),[32] 2,4-dimethoxy-3,7-phenanthrenediol (**15**),[33] nudol (**16**),[34] 5-(2-hydroxyphenethyl)-3,4-dimethoxyphenol   (**17**),[35]   2-(3,4,5-trimethoxyphenethyl)

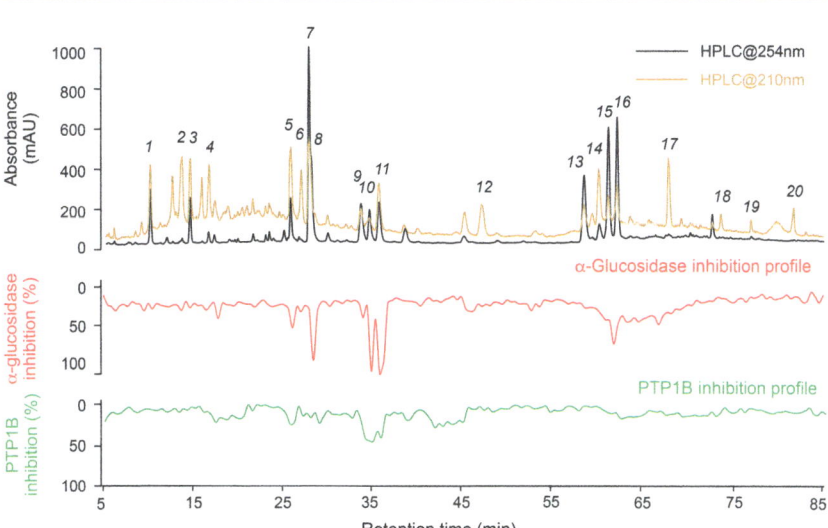

Fig. 3 HPLC chromatogram of the crude *D. bulbifera* ethyl acetate extract monitored at 254 nm (black) and 210 nm (yellow) as well as the high-resolution α-glucosidase inhibition profile (red) and high-resolution PTP1B inhibition profile (green) [40 mg mL^{-1}, 10 μL injection].

phenol (18),[36] trichosanatine (19),[37] and diosgenyl α-L-rhamnopyranosyl-(1 → 2)-β-D-glucopyranoside (20)[38] by comparison of their HRMS, 1D and/or 2D NMR data with data reported in the literature. The chemical structures, HRMS data and ^{1}H NMR data of these compounds are given in ESI Table S2.†

Peak 8 showed a $[M + H]^+$ ion with m/z 243.0649 (calcd for $C_{14}H_{11}O_4{}^+$: 243.0652, ΔM 1.0 ppm), suggesting a molecular formula of $C_{14}H_{10}O_4$. In an attempt to purify peak 8, a mixture of compounds 7 and 8 was obtained in a ratio of 3 : 5, as the limited amount (0.2 mg) did not allow further separation. The structure of compound 8 was elucidated by subtracting the 1D and 2D NMR signals of the known compound 7 from those observed in the spectrum of the mixture (ESI Fig. S2–S7†). The UV absorptions of compound 8 showed absorption maxima at 310, 280 and 265 nm, which is characteristic for a phenanthrene skeleton.[34] The ^{1}H NMR spectrum showed signals for six aromatic protons, including two *ortho*-coupled protons of a tetrasubstituted benzene ring (ring B) at δ_H 7.34 (1H, d, $J =$ 8.8 Hz) and δ_H 7.27 (1H, d, $J =$ 8.8 Hz) and two *meta*-coupled protons of another tetrasubstituted benzene ring (ring A) at δ_H 6.68 (1H, d, $J =$ 2.4 Hz) and δ_H 6.58 (1H, d, $J =$ 2.4 Hz). The remaining two aromatic singlets at δ_H 9.11 (1H, s, H-5) and 7.11 (1H, s, H-8) are typical for the *para* protons on a 1,2,4,5-tetrasubstituted benzene ring (ring C).[29] On the basis of the above analysis, compound 8 was tentatively assigned as a phenanthrene with two *meta* hydroxyl groups on ring A and two *ortho*-positioned hydroxyl groups at C-6 and C-7 of ring C. Analysis of 2D HSQC, HMBC, and COSY spectra confirmed this substitution pattern, and selected HMBC correlations are shown in Fig. 4. This confirmed compound 8 to be 2,4,6,7-phenanthrenetetrol, which is a new compound.

Peak 9 had a $[M + H]^+$ ion with m/z 257.0449 (calcd for $C_{14}H_9O_5{}^+$: 257.0444, ΔM −1.9 ppm), suggesting a molecular formula of $C_{14}H_8O_5$. The ^{1}H NMR spectrum showed – similar to compound 8 – signals for two *ortho*-coupled protons of a tetrasubstituted benzene ring (ring B) at δ_H 7.68 (1H, d, $J =$ 8.2 Hz) and δ_H 7.66 (1H, d, $J =$ 8.2 Hz), and two *para*-coupled protons for a 1,2,4,5-tetrasubstituted benzene ring (ring C) at δ_H 9.26 (1H, s) and δ_H 7.08 (1H, s). The additional unsaturation degree of compound 9 compared to 8 based on the molecular formula, indicated compound 9 should be a phenanthrenedione. The ^{1}H and ^{13}C NMR spectra of compound 9 were closely related to those of previously reported phenanthrenedione (6,7-dihydroxy-2-methoxy-1,4-phenanthrenedione),[39] except for the absence of the methoxy signal. The correlation from H-10 (δ 7.66) to C-1 (δ 187.9) in the HMBC spectrum confirmed one of the carbonyl groups of the 1,4-quinone structure to be at C-1, and the correlations from H-3 (δ 5.43) to C-1 (δ 187.9) and C-4a (δ 128.1) showed the single proton on ring A to be located at C-3.

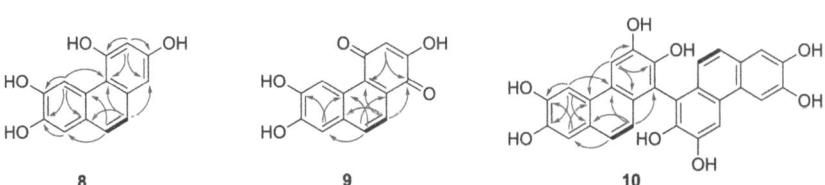

Fig. 4   Key COSY (bold) and HMBC (arrows pointing from H to C) correlations observed for compounds 8–10.

The 1,4-quinone structure was established based on the down-field shift of H-5 (δ 9.25) caused by the carbonyl group in 4-position. However, the signals for C-2 and C-4 could not be detected, indicating that the molecule under the conditions used is in tautomeric equilibrium between its 1,4- and 1,2-quinone forms. HMBC correlations are shown in Fig. 4. This type of compound has previously been reported to be photoactive and form quinone annelations,[40] which is in agreement with our observations that the sample rapidly turned dark red from shallow pink in the NMR tube. Thus, the structure of compound **9** was identified as 2,6,7-trihydroxy-1,4-phenanthrenedione, which is a new compound.

Peak 10 had a $[M + H]^+$ ion with m/z 483.1064 (calcd for $C_{28}H_{19}O_8^+$: 483.1074, ΔM 2.2 ppm), suggesting a molecular formula of $C_{28}H_{18}O_8$. The ^{1}H NMR spectrum only showed signals of 5 aromatic protons, including two *ortho*-coupled protons from a tetrasubstituted benzene ring (ring B) at δ_H 7.22 (1H, d, J = 9.1 Hz) and δ_H 6.83 (1H, d, J = 9.1 Hz), and two *para*-coupled protons from a 1,2,4,5-tetrasubstitution benzene ring (ring C) at δ_H 9.28 (1H, s) and δ_H 7.05 (1H, s). The correlations of the last proton signal H-4 (δ 6.83, 1H, s) to C-4b (δ 127.5) and C-10a (δ 135.8) in the HMBC spectrum suggest a 1,2,3,5,6-substitution pattern of ring A. The molecular formula obtained from HRMS together with the NMR spectra showed that this compound should be a structurally symmetrical biphenanthrene. The correlation from H-10 to C-1 in the HMBC spectrum and the chemical shifts of δ 110.0 for the quaternary C-1, indicated a C–C linkage to the other half of the dimer. Thus, the structure of compound **10** was established to be [1,1′-biphenanthren]-2,2′,3,3′,6,6′,7,7′-octaol, which is a new compound.

Fully assigned NMR spectroscopic data for compounds **8–10** are given in Table 2, selected HMBC correlations are shown in Fig. 4, and UV, HRMS, NMR spectra are given in ESI Fig. S2–S17.†

The dual high-resolution α-glucosidase and PTP1B inhibition profile of *B. nivea* (Fig. 5) showed nine well-separated HPLC peaks correlated with α-glucosidase and/or PTP1B inhibition. These compounds were therefore isolated from the crude extract and identified as dihydroxybenzoic acid (**1**), 3,4-dihydroxybenzaldehyde (**21**), *p*-hydroxybenzaldehyde (**22**),[41] (*E*)-cinnamic acid (**23**), hederagenin (**24**),[42] pomolic acid (**25**),[43] 3-*epi*-pomolic acid (**26**),[44] α-linolenic acid (**27**),[45] and linolenic acid (**28**)[46] by analysis of their HRMS, 1D and/or 2D NMR data – and by comparison with data reported in the literature.

The dual high-resolution α-glucosidase and PTP1B inhibition profile of *T. sagittata* (Fig. 6) showed two well-separated HPLC peaks correlated with α-glucosidase and/or PTP1B inhibition. From the LC-HRMS data, these were identified as α-linolenic acid (**27**)[45] and linolenic acid (**28**),[46] as also identified in *B. nivea*.

P. bistorta did, in addition to α-glucosidase and PTP1B inhibition, show α-amylase inhibitory activity (Fig. 1 and ESI Table S1†). The crude ethyl acetate extract was therefore microfractionated three times to provide the triple high-resolution α-glucosidase/α-amylase/PTP1B inhibition profile (ESI Fig. S18†). The high-resolution inhibition profiles clearly indicate that the observed inhibitory activity towards α-glucosidase, α-amylase and PTP1B can be attributed to the tannins, *i.e.* the hump of broad and closely eluting peaks in the retention time range from 18–50 min. This is in agreement with a recent study showing that tannins are one of the major chemical constituents of *P. bistorta*.[47] Tannins are known to bind nonspecifically to proteins and enzymes – thereby forming

Table 2 ^{1}H NMR (600 MHz) and ^{13}C NMR (150 MHz) data of compounds **8–10** (δ in ppm)

No.	8^a δ_C, type^c	δ_H (J in Hz)	9^a δ_C, type^c	δ_H (J in Hz)	10^b δ_C, type^c	δ_H (J in Hz)
1	104.4, CH	6.68 d (2.4)	187.9, C		110.0, C	
2	155.8, C		n.d.^{d}		158.0, C	
3	103.0, CH	6.58 d (2.4)	108.1, CH	5.43 s	154.0, C	
4	155.7, C		n.d.		102.8, CH	6.83 s
4a	114.6, C		128.1, C		115.4, C	
4b	126.7, C		125.9, C		127.5, C	
5	113.6, CH	9.11 s	110.9, CH	9.26 s	114.0, CH	9.28 s
6	146.2, C		149.0, C		146.0, C	
7	144.8, C		n.d.		145.5, C	
8	112.4, CH	7.11 s	109.5, CH	7.08 s	112.2, CH	7.05 s
8a	127.7, C		134.5, C		127.1, C	
9	127.6, CH	7.34 d (8.8)	128.5, CH	7.68 d (8.2)	127.5, CH	7.22 d (9.1)
10	124.8, CH	7.27 d (8.8)	118.9, CH	7.66 d (8.2)	122.9, CH	6.83 d (9.1)
10a	136.3, C		128.5, C		135.8, C	
1'					110.0, C	
2'					158.0, C	
3'					154.0, C	
4'					102.8, CH	6.83 s
4a'					115.4, C	
4b'					127.5, C	
5'					114.0, CH	9.28 s
6'					146.0, C	
7'					145.5, C	
8'					112.2, CH	7.05 s
8a'					127.1, C	
9'					127.5, CH	7.22 d (9.1)
10'					122.9, CH	6.83 d (9.1)
10a'					135.8, C	

^{a} Spectral data acquired in methanol-d_4. ^{b} Spectral data acquired in DMSO-d_6. ^{c} ^{13}C NMR shifts and assignments were based on 2D HSQC and HMBC. ^{d} n.d. = not determined.

insoluble protein and enzyme complexes – and thus showing false positive inhibitory activity.[48] The crude extract of *P. bistorta* was therefore not investigated further.

Assessment of α-glucosidase inhibitory activity

In the high-resolution α-glucosidase inhibition profile of *D. bulbifera*, peaks 7, 8, 10, and 11 showed more than 90% inhibition and peaks 5 and 15 showed around 50%. The IC$_{50}$ values of these compounds as well as the other isolated bisbenzyls and phenanthrenes (**6**, **14**, and **16–19**) were assessed in an attempt to reveal potential structure–activity relationships (SARs). The IC$_{50}$ values are given in Table 3 and IC$_{50}$ curves are given in ESI Fig. S19.† The IC$_{50}$ values of the new compounds [1,1'-biphenanthren]-2,2',3,3',6,6',7,7'-octaol (**10**) and cassigarol D (**11**) were 2.08 ± 0.19 and 2.39 ± 0.26 μM, respectively, which is in accordance with the strong inhibition observed in Fig. 3. Considering the fact that the dominating compound 7 (3,5-di-O-methylquercetin) only showed moderate inhibition (52% inhibition at a concentration of 100 μM, Table 3), the strong

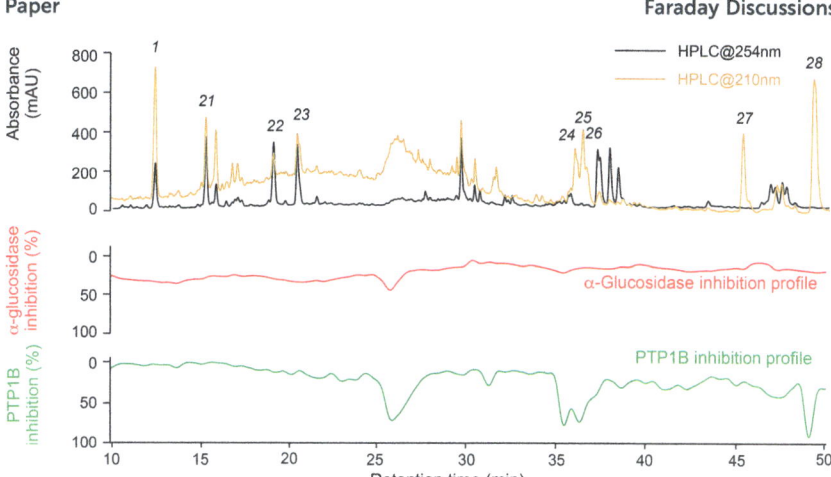

Fig. 5   HPLC chromatogram of the crude *B. nivea* ethyl acetate extract monitored at 254 nm (black) and 210 nm (yellow) as well as the high-resolution α-glucosidase inhibition profile (red) and high-resolution PTP1B inhibition profile (green) [40 mg mL^{-1}, 10 µL injection].

inhibition of peak 7/8 might be attributed to the new compound **8** or a high concentration of **7** in the crude extract. Compounds **5** (9,10-dihydro-2,4,6,7-phenanthrenetetrol) and **6** (dihydropiceatannol) showed moderate inhibition with IC$_{50}$ values of 27.73 ± 2.98 and 46.29 ± 1.81 µM, respectively, whereas other methylated analogues showed much weaker inhibition, indicating that methylation of the hydroxyl group on these compounds may cause a decrease in the α-glucosidase inhibitory activity. Apart from that, a weak inhibition by **28** (linolenic

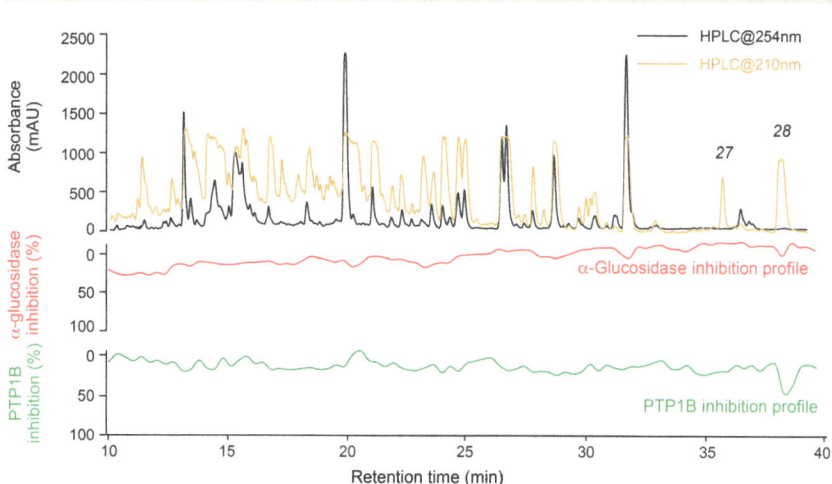

Fig. 6   HPLC chromatogram of the crude *T. sagittata* ethyl acetate extract monitored at 254 nm (black) and 210 nm (yellow) as well as the high-resolution α-glucosidase inhibition profile (red) and high-resolution PTP1B inhibition profile (green) [40 mg mL^{-1}, 10 µL injection].

Table 3 α-Glucosidase inhibitory activity (IC_{50} values) of compounds 5–7, 10, 11, and 13–18

Compound	Structure	IC_{50} value[a] (µM)
5	9,10-Dihydro-2,4,6,7-phenanthrenetetrol	27.73 ± 2.98
6	Dihydropiceatannol	46.29 ± 1.81
7	3,5-Di-O-methylquercetin	>50 (52% inhibition at 100 µM)
10	[1,1'-Biphenanthren]-2,2',3,3',6,6',7,7'-octaol	2.08 ± 0.19
11	Cassigarol D	2.39 ± 0.26
13	Flavanthrinin	>100
14	Flavanthridin	>100
15	2,4-Dimethoxy-3,7-phenanthrenediol	>50
16	Nudol	>100
17	5-(2-Hydroxyphenethyl)-2,3-dimethoxyphenol	>100
18	2-(3,4,5-Trimethoxyphenethyl)phenol	>100
Reference	Acarbose	328.21 ± 25.40

[a] Values expressed as mean ± standard error ($n = 3$).

acid) in *T. sagittata* was also observed. Fatty acids have previously been reported to have moderate inhibitory activity of α-glucosidase,[45] and 28 was therefore not investigated further in the present study.

Characterization of PTP1B inhibitory activity

Peaks 9–11 in the *D. bulbifera* extract showed around 50% inhibition of PTP1B, while peak 5 showed weaker inhibitory activity. However, due to the instability of compound 9, only compounds 5, 10, and 11 were tested for PTP1B inhibitory activity (Table 4 and ESI Fig. S20†). The IC_{50} value of compound 10 was 3.36 ± 0.25 µM, which makes compound 10 the most potent PTP1B inhibitor discovered from these plants. Compounds 5 and 11 showed lower PTP1B inhibitory activity with IC_{50} values of 23.79 ± 1.23 and 13.16 ± 0.63 µM, respectively. For *B. nivea*, four

Table 4 PTP1B inhibitory activity (IC_{50} values) of compounds 5, 10, 11, 25–28 and some commercially available fatty acids

Compound	Structure	IC_{50} value[c] (µM)
5	9,10-Dihydro-2,4,6,7-phenanthrenetetrol	23.79 ± 1.23
10	[1,1'-Biphenanthren]-2,2',3,3',6,6',7,7'-octaol	3.36 ± 0.25
11	Cassigarol D	13.16 ± 0.63
24	Hederagenin	9.53 ± 0.39
25	Pomolic acid	4.89 ± 0.38
26	3-*epi*-Pomolic acid	>100
27	α-Linolenic acid (18 : 3, Δ^9, Δ^{12}, Δ^{15})[a]	>100
28	Linolenic acid (18 : 2, Δ^9, Δ^{12})[a]	13.01 ± 1.70
	Oleic acid (18 : 1, Δ^9)[a]	9.93 ± 1.49
	Linolelaidic acid (18 : 2, Δ^9, Δ^{12})[b]	16.08 ± 1.92
	Ethyl linoleate	>100
	Eicosapentaenoic acid (20 : 5, Δ^5, Δ^8, Δ^{11}, Δ^{14}, Δ^{17})[a]	53.77 ± 4.20
	RK 682	13.15 ± 0.29

[a] All *cis*. [b] All *trans*. [c] Values expressed as mean ± standard error ($n = 3$).

peaks – which were subsequently identified to be three triterpenoids (**24–26**) and one fatty acid (**28**) – showed more than 80% inhibition in the PTP1B inhibition profile (Fig. 5). Similar inhibition was observed for peak 28 in *T. sagittata*, which was attributed to the same fatty acid (Fig. 6). The IC_{50} values of the two triterpenoids **24** and **25** with a 3β-hydroxy group were determined to be 9.53 ± 0.39 and 4.89 ± 0.38 µM, respectively, whereas triterpenoid **26** with a 3α-hydroxy group showed no inhibition of PTP1B. These results were in agreement with a previous study.[46]

Fatty acids have previously been reported to exhibit inhibitory activity against protein tyrosine phosphatase (Shp2).[49] Another study also showed that plant extracts containing different fatty acids possessed potent inhibition of PTP1B.[50] In our study, it is interesting to find that two fatty acids (**27** and **28**) which are structurally closely related exhibit different inhibitory activities of PTP1B, as shown in Table 4. In an attempt to reveal their potential SARs, compounds **27** and **28**, together with four other fatty acids purchased from Sigma-Aldrich (Denmark), were tested for PTP1B inhibitory activity (Table 4 and ESI Fig. S20†). The results show that fatty acids with two (linolenic acid) or one (oleic acid) *cis* carbon–carbon double bonds exhibit strong PTP1B inhibition with IC_{50} values of 13.01 ± 1.70 and 9.93 ± 1.49 µM, respectively. Linolelaidic acid, *i.e.* the linoleic stereoisomer with all *trans* carbon–carbon double bonds, showed a similar level of inhibition (IC_{50} = 16.08 ± 1.92 µM), suggesting that the configuration of the carbon–carbon double bonds are not pivotal for the inhibition of PTP1B. However, with an additional *cis* double bond between C-15 and C-16 (*i.e.* α-linolenic acid), the PTP1B inhibitory activity almost disappears (>100 µM). Similarly, an 18 : 2 fatty acid with esterification of the carboxyl group (ethyl linoleate) shows no PTP1B inhibitory activity, suggesting that the carboxyl group may contribute to the observed inhibition. A C-20 fatty acid (eicosapentaenoic acid) exhibited moderate PTP1B inhibitory activity as well with an IC_{50} value of 53.77 ± 4.20 µM.

Conclusion

In conclusion, assessment of α-glucosidase, α-amylase, and PTP1B inhibitory activity of crude ethyl acetate extracts of 40 Chinese medicinal plant samples combined with dual high-resolution α-glucosidase/PTP1B or triple high-resolution α-glucosidase/α-amylase/PTP1B profiling, allowed fast pinpointing of the HPLC peaks correlated with one or more bioactivities. Subsequent HPLC-HRMS and NMR spectroscopy resulted in identification of 28 metabolites from four selected species, including three new phenanthrenes from *D. bulbifera*. Compound **10** exhibited promising dual inhibition of α-glucosidase and PTP1B with IC_{50} values of 2.08 ± 0.19 and 3.36 ± 0.25 µM, respectively. Two triterpenoids and one fatty acid from *B. nivea* and/or *T. sagittata*, together with several other commercially available fatty acids, exhibited strong PTP1B inhibitory activity, and their SARs were discussed. These results confirmed the potential of traditional Chinese medicines as a source of antidiabetic drug candidates with one or more bioactivities – and demonstrated the advantage of combining high-resolution pharmacologic, high-resolution chromatographic, and high-resolution spectroscopic techniques for accelerating the drug discovery process.

Conflicts of interest

The authors declare no conflicts of interest.

Acknowledgements

Arife Önder is acknowledged for technical assistance. The HPLC equipment used for the high-resolution bioassay profiles was obtained *via* a grant from The Carlsberg Foundation. The HPLC-HRMS and NMR system used in this work was acquired through a grant from "Apotekerfonden af 1991", The Carlsberg Foundation, and the Danish Agency for Science, Technology and Innovation *via* the National Research Infrastructure funds. Y. Z. acknowledges the Chinese Scholarship Council for a PhD scholarship (CSC Grant 201508530222) and the Novo Nordisk Foundation for a postdoc position.

References

1 American Diabetes Association, *Diabetes Care*, 2010, **33**, S62–S69.
2 International Diabetes Federation, *IDF Diabetes Atlas*, 8th edn, ISBN: 978-2-930229-87-4, http://www.diabetesatlas.org/, 2017.
3 S. Chatterjee, K. Khunti and M. J. Davies, *Lancet*, 2017, **389**, 2239–2251.
4 K. Alagesan, P. K. Raghupathi and S. Sankarnarayanan, *Int. J. Pharm. Life Sci.*, 2012, **3**, 1407–1412.
5 L. He, *Diabetes Rev.*, 1998, **6**, 132–145.
6 F. A. Van de Laar, P. L. Lucassen, R. P. Akkermans, E. H. Van de Lisdonk, G. E. Rutten and C. Van Weel, *Cochrane Database Syst. Rev.*, 2005, **18**, CD003639.
7 M. Feldhammer, N. Uetani, D. Miranda-Saavedra and M. L. Tremblay, *Crit. Rev. Biochem. Mol. Biol.*, 2013, **48**, 430–445.
8 T. O. Johnson, J. Ermolieff and M. R. Jirousek, *Nat. Rev. Drug Discovery*, 2002, **1**, 696–709.
9 D. A. Dias, S. Urban and U. Roessner, *Metabolites*, 2012, **2**, 303–336.
10 C. R. Pye, M. J. Bertin, R. S. Lokey, W. H. Gerwick and R. G. Linington, *Proc. Natl. Acad. Sci. U. S. A.*, 2017, **114**, 5601–5606.
11 G. G. Graham, J. Punt, M. Arora, R. O. Day, M. P. Doogue, J. K. Duong, T. J. Furlong, J. R. Greenfield, L. C. Greenup, C. M. Kirkpatrick, J. E. Ray, P. Timmins and K. M. Williams, *Clin. Pharmacokinet.*, 2011, **50**, 81–98.
12 H. Choudhury, M. Pandey, C. K. Hua, C. S. Mun, J. K. Jing, L. Kong, L. Y. Ern, N. A. Ashraf, S. W. Kit, T. S. Yee, M. R. Pichika, B. Gorain and P. Kesharwani, *J. Tradit. Complement. Med.*, 2017, **8**, 361–376.
13 T. T. Zhang and J. G. Jiang, *Expert Opin. Invest. Drugs*, 2012, **21**, 1625–1642.
14 C. S. Jiang, L. F. Liang and Y. W. Guo, *Acta Pharmacol. Sin.*, 2012, **33**, 1217–1245.
15 K. T. Kongstad, C. Ozdemir, A. Barzak, S. G. Wubshet and D. Staerk, *J. Agric. Food Chem.*, 2015, **63**, 2257–2263.
16 Y. Zhao, M. X. Chen, K. T. Kongstad, A. K. Jäger and D. Staerk, *J. Agric. Food Chem.*, 2017, **65**, 4421–4427.
17 S. G. Wubshet, Y. Tahtah, A. M. Heskes, K. T. Kongstad, I. Pateraki, B. Hamberger, B. L. Moeller and D. Staerk, *J. Nat. Prod.*, 2016, **79**, 1063–1072.

18 Y. Tahtah, K. T. Kongstad, S. G. Wubshet, N. T. Nyberg, L. H. Jønsson, A. K. Jäger, Q. Sun and D. Staerk, *J. Chromatogr. A*, 2015, **1408**, 125–132.

19 Y. Zhao, K. T. Kongstad, A. K. Jäger, J. Nielsen and D. Staerk, *J. Chromatogr. A*, 2018, **1556**, 55–63.

20 B. Liu, K. T. Kongstad, Q. Sun, N. T. Nyberg, A. K. Jager and D. Staerk, *J. Nat. Prod.*, 2015, **78**, 294–300.

21 R. Lima, S. Gramsbergen, J. Van Staden, A. K. Jäger, K. Kongstad and D. Staerk, *J. Nat. Prod.*, 2017, **80**, 1020–1027.

22 A. Soumyanath, *Traditional Medicines for Modern Times: Antidiabetic Plants*. CRC Press, 2006.

23 Y. P. Lin, T. Y. Chen, H. W. Tseng, M. H. Lee and S. T. Chen, *Phytochemistry*, 2009, **70**, 1173–1181.

24 M. Okada, S. Park, T. Koshizawa and M. Ueda, *Tetrahedron*, 2009, **65**, 2136–2141.

25 H. M. Abdallah, H. El-Bassossy, G. A. Mohamed, A. M. El-Halawany, K. Z. Alshali and Z. M. Banjar, *Molecules*, 2016, **21**, 251.

26 P. Chaniad, C. Wattanapiromsakul, S. Pianwanit and S. Tewtrakul, *Pharm. Biol.*, 2016, **54**, 1077–1085.

27 T. Tsuruga, Y. T. Chun, Y. Ebizuka and U. Sankawa, *Chem. Pharm. Bull.*, 1991, **39**, 3276–3278.

28 Z. H. Shi, N. G. Li, Y. P. Tang, W. Li, L. Yin, J. P. Yang, H. Tang and J. A. Duan, *Eur. J. Med. Chem.*, 2012, **54**, 210–222.

29 K. Baba, T. Kido, K. Maeda, M. Taniguchi and M. Kozawa, *Phytochemistry*, 1992, **31**, 3215–3218.

30 A. Singh, V. S. Parmar, W. Errington, O. W. Howarth and S. Lawrie, *Acta Crystallogr., Sect. C: Cryst. Struct. Commun.*, 1999, **55**, 559–561.

31 Y. W. Leong, C. C. Kang, L. J. Harrison and A. D. Powell, *Phytochemistry*, 1996, **44**, 157–165.

32 Y. P. Wu, W. J. Liu, W. J. Zhong, Y. J. Chen, D. N. Chen, F. He and L. Jiang, *Nat. Prod. Res.*, 2017, **31**, 1518–1522.

33 P. Tuchinda, J. Udchachon, K. Khumtaveeporn, W. C. Taylor, L. M. Engelhardt and A. H. White, *Phytochemistry*, 1988, **27**, 3267–3271.

34 C. W. Lin, T. L. Hwang, F. A. Chen, C. H. Huang, H. Y. Hung and T. S. Wu, *J. Nat. Prod.*, 2016, **79**, 1911–1921.

35 T. Jarevang, M. C. Nilsson, A. Wallstedt, G. Odham and O. Sterner, *Phytochemistry*, 1998, **48**, 893–896.

36 G. Nesi, N. A. Colabufo, M. Contino, M. G. Perrone, M. Digiacomo, R. Perrone, A. Lapucci, M. Macchia and S. Rapposelli, *Eur. J. Med. Chem.*, 2014, **76**, 558–566.

37 Z. M. Chao and J. M. Liu, *Acta Pharmacol. Sin.*, 1995, **30**, 517–520.

38 J. C. Hernandez, F. Leon, I. Brouard, F. Torres, S. Rubio, J. Quintana, F. Estevez and J. Bermejo, *Bioorg. Med. Chem.*, 2008, **16**, 2063–2076.

39 C. Ma, W. Wang, Y. Y. Chen, R. N. Liu, R. F. Wang and L. J. Du, *J. Nat. Prod.*, 2005, **68**, 1259–1261.

40 H. Suginome, H. Kamekawa, H. Sakurai, A. Konishi, H. Senboku and K. Kobayashi, *J. Chem. Soc., Perkin Trans. 1*, 1994, 471–475.

41 A. A. Dissanayake, B. A. H. Ameen and M. G. Nair, *J. Nat. Prod.*, 2017, **80**, 2472–2477.

42 B. S. Joshi, K. L. Singh and R. Roy, *Magn. Reson. Chem.*, 1999, **37**, 295–298.

43 J. J. Cheng, L. J. Zhang, H. L. Cheng, C. T. Chiou, I. J. Lee and Y. H. Kuo, *J. Nat. Prod.*, 2010, **73**, 1655–1658.

44 X. Chiriboga, G. Gilardoni, I. Magnaghi, P. V. Finzi, G. Zanoni and G. Vidari, *J. Nat. Prod.*, 2003, **66**, 905–909.

45 B. Liu, K. T. Kongstad, S. Wiese, A. K. Jager and D. Staerk, *Food Chem.*, 2016, **203**, 16–22.

46 P. T. Thuong, C. H. Lee, T. T. Dao, P. H. Nguyen, W. G. Kim, S. J. Lee and W. K. Oh, *J. Nat. Prod.*, 2008, **71**, 1775–1778.

47 S. T. Wang, W. Gao, Y. X. Fan, X. G. Liu, K. Liu, Y. Du, L. L. Wang, H. J. Li, P. Li and H. Yang, *RSC Adv.*, 2016, **6**, 27320–27328.

48 Y. Liu, D. Staerk, M. N. Nielsen, N. Nyberg and A. K. Jäger, *Phytochemistry*, 2015, **119**, 62–69.

49 D. Liu, G. Kong, Q. C. Chen, G. Wang, J. Li, Y. Xu, T. Lin, Y. Tian, X. Zhang, X. Yao, G. Feng, Z. Lu and H. Chen, *Bioorg. Med. Chem. Lett.*, 2011, **21**, 6833–6837.

50 G. Kasim, Q. L. Ma, G. Kahar and H. A. Aisa, *Nat. Prod. Res. Dev.*, 2014, **26**, 1835–1838.

Faraday Discussions

PAPER

Characterising polar compounds using supercritical fluid chromatography–nuclear magnetic resonance spectroscopy (SFC–NMR)†

F. H. M. van Zelst,[ab] S. G. J. van Meerten[b] and A. P. M. Kentgens (ID) *[b]

Received 10th December 2018, Accepted 2nd January 2019

DOI: 10.1039/c8fd00237a

To detect and characterise compounds in complex matrices, it is often necessary to separate the compound of interest from the matrix before analysis. In our previous work, we have developed the coupling of supercritical fluid chromatography (SFC) with nuclear magnetic resonance (NMR) spectroscopy for the analysis of nonpolar samples [Van Zelst et al., Anal. Chem., 2018, 90, 10457]. In this work, the SFC–NMR setup was successfully adapted to analyse polar samples in complex matrices. In-line SFC–NMR analysis of two N-acetylhexosamine stereoisomers was demonstrated, namely N-acetyl-mannosamine (ManNAc) and N-acetyl-glucosamine (GlcNAc). ManNAc is a metabolite that is present at elevated concentrations in patients suffering from NANS-mediated disease. With our SFC–NMR setup it was possible to distinguish between the polar stereoisomers. Until now, this was not possible with the standard mass-based analysis techniques. The concentrations that are needed in the SFC–NMR setup are currently too high to be able to detect ManNAc in patient samples (1.7 mM vs. 0.7 mM). However, several adaptations to the current setup will make this possible in the future.

1 Introduction

Polar compounds make up a great part of all molecules and are of special interest in biochemistry. They are present in biological systems, for example in biofluids such as urine. Metabolomics is an emerging field in analytical science, which focusses on characterising small molecules in biological systems. These metabolites form the (intermediate) products of biochemical reactions in our metabolism. Metabolomics can be used to identify biomarkers for many diseases, such as inborn errors of metabolism. The standard workflow in untargeted metabolomics

[a]TI-COAST, Science Park 904, 1098 XH Amsterdam, The Netherlands

[b]Institute for Molecules and Materials, Radboud University, Heyendaalseweg 135, 6525 AJ Nijmegen, The Netherlands. E-mail: A.Kentgens@nmr.ru.nl

† Electronic supplementary information (ESI) available: SFC UV chromatograms of ManNAc and GlcNAC. See DOI: 10.1039/c8fd00237a

is analysis by liquid chromatography–mass spectrometry (LC-MS). In this LC-MS workflow, metabolites are first separated and their mass-to-charge-ratio (m/z) is determined. These m/z-values are then compared to databases, which often result in multiple hits. In this case, the chromatographic retention times and MS/MS-fragmentation patterns will be compared to a reference compound, often leading to the metabolite of interest.[1] However, if the structure cannot be identified in the database, more extensive analyses have to be performed.

Metabolites are very diverse in structure and functional groups, ranging from very hydrophilic to lipophilic. The metabolite composition also varies for different types of biofluids. In urine, very hydrophilic compounds are present, although trace amounts of lipids and fatty acids have also been detected. Serum, on the contrary, contains more lipids. Compared to cerebrospinal fluid and saliva, urine contains more, and more diverse metabolites. Blood contains all the metabolites present in urine, and more, but the concentrations differ substantially. The kidneys filter out some of the metabolites or toxins from the blood, which is why metabolite concentrations in urine are often higher than those in blood.[2] The chemical diversity of the compounds, along with the broad range of concentrations, call for a variety of analytical techniques. As mentioned previously, LC-MS is generally used as the standard technique for metabolomics. Nuclear magnetic resonance (NMR) spectroscopy, which is unbiased and can be used for a broad range of metabolites, can be an attractive alternative. An important advantage of using NMR spectroscopy over MS is that the technique is quantitative. However, NMR spectroscopy is less sensitive than MS and can therefore not be used to identify metabolites at low concentrations (typically the lower limit is around 1 µM).

Due to the complex nature of urine samples, NMR spectra are usually very crowded and therefore especially metabolites present in lower concentrations are difficult to identify from the spectrum. As is the case in the LC-MS workflow, it would be beneficial to separate the sample into less complex fractions by chromatography before NMR analysis. It is however good to note that the sample gets diluted into the mobile phase during chromatography, which can make NMR detection difficult or even impossible. In our group, we have recently developed a new setup, which combines supercritical fluid chromatography (SFC) and NMR spectroscopy.[3] In this work we demonstrated the separation by SFC of a mixture of tocopherol (vitamin E) isomers, followed by in-line concentration and NMR detection. Although tocopherol is a nonpolar molecule and therefore very different from the mostly polar molecules present in urine, we are adapting the setup to analyse polar metabolites in urine. SFC is a type of chromatography that uses supercritical fluids, typically supercritical CO_2, as the mobile phase. Although CO_2 is highly nonpolar, by adding a co-solvent to the supercritical fluid, such as methanol, the mobile phase becomes more polar. In this way, SFC is able to separate nonpolar as well as polar compounds, making it suitable for separating polar metabolites from urine. To be able to detect lower concentrations by NMR spectroscopy, a method for concentrating the sample in-line after SFC separation has been implemented as well. This was achieved by in-line expansion of the supercritical CO_2 ($scCO_2$), thus separating the mobile phase from the sample during transport from the SFC outlet to the NMR probehead.

Another approach to detect smaller amounts in NMR spectroscopy is by miniaturising the detection coil. In this way, smaller sample volumes can be detected at higher sensitivity. The challenge in miniaturisation is to maintain a high resolution. Different designs have been researched, such as tightly wound solenoid coils and planar helical coils.[4-7] Our group has developed a design in which the conventional detection coil is replaced by a high resolution flat stripline radio-frequency inductor. The flat geometry of the stripline allows in-flow detection, by passing a capillary over the stripline.[8]

To see if our SFC–NMR setup can be adapted to analyse polar samples, we have chosen to study N-acetylhexosamines in urine as a concept application. These monosaccharides appear in different diastereomeric forms (shown in Fig. 1), which have equal m/z-values, and are therefore not distinguishable in the standard HPLC-MS workflow described previously. N-Acetylmannosamine (ManNAc) is a biomarker for NANS-mediated disease, an inborn error in the sialic acid metabolism.[9] It is, however, difficult to distinguish N-acetylmannosamine from its stereoisomers, N-acetylglucosamine (GlcNAc) and N-acetylgalactosamine (GalNAc), using MS since they are equal in mass. For identification of ManNAc in body fluids, NMR spectroscopy has been employed.[9] This, however, requires relatively large amounts of sample at a higher concentration compared to MS analyses. Martens et al. developed a new method to study ManNAc in biofluids using infrared ion spectroscopy after MS.[10] This is a sensitive method, requiring only small amounts of sample. However, this technique either requires a reference for comparison, or computer simulations of the IR spectra, which do not always match the experimental spectra. In our approach we will separate the compound of interest (ManNAc) with SFC and shuttle it in-line to the NMR probehead, which is able to detect much smaller volumes (~150 nL in this case) than the standard NMR approach (~500 μL) and does not require a reference for structural characterisation. However, sensitivity remains an issue in NMR analysis.

Due to the variations in urine samples, between different patients but also within patients, metabolite concentrations in urine are usually expressed in μmol per mmol creatinine. Van Karnebeek et al. found that the concentration of ManNAc was between 41 and 98 μmol mmol^{-1} creatinine in adult patients suffering from NANS-mediated disease (<10 μmol mmol^{-1} in the reference

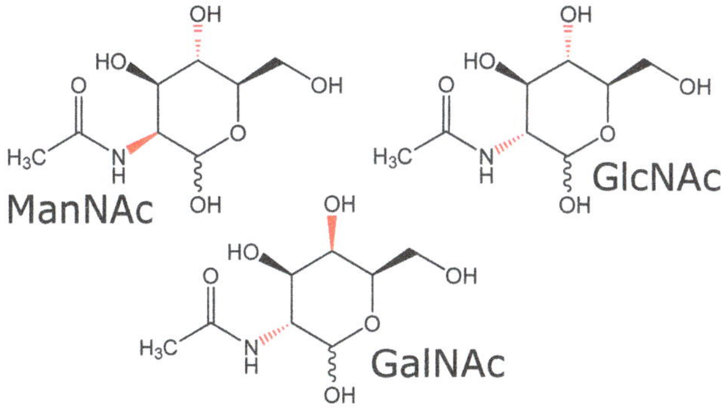

Fig. 1 Molecular structures of the three N-acetylhexosamines. The difference in stereochemistry between the diastereomers is indicated in red. Each diastereomer has an α- and β-stereoisomer, indicated by the wavy bond of the alcohol group at the anomeric carbon atom.

sample).[9] The amount of creatinine within and between patients varies enormously. However according to the guidelines of the World Health Organization (WHO), a valid urine sample ranges between 30 and 300 mg creatinine per dL urine.[11] Taking a value of 150 mg dL^{-1} urine, the concentration of ManNAc in the urine of patients suffering from NANS-mediated disease is estimated to be in the range of 0.3–0.7 mM. This concentration is sufficient in a standard NMR setup, however in our previous SFC–NMR setup we needed samples in the mM range to be able to detect them within 50 minutes.[3] A higher signal can be gained by acquiring over a longer period of time, therefore lower concentrations may also be detectable.

2 Experimental

2.1 Setup

The SFC–NMR setup that was used has been introduced in our previous work[3] and is shown in Fig. 2. An overview of the setup is given here; for further details the reader is referred to our previous paper.

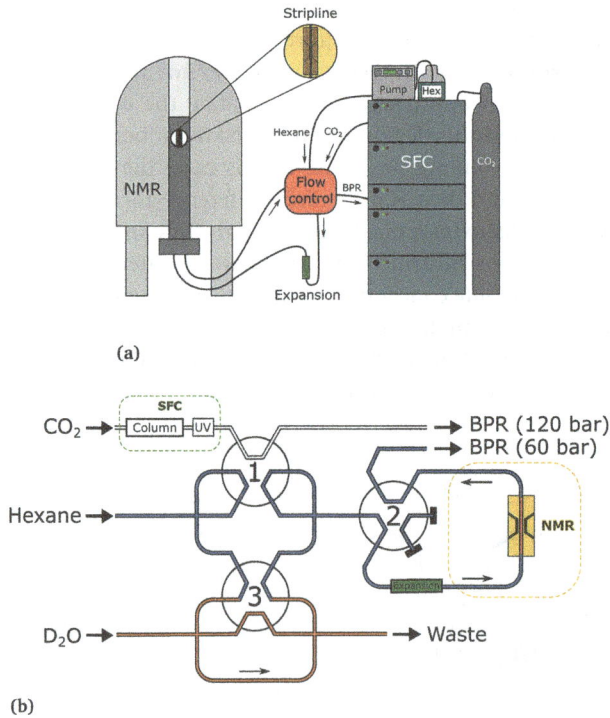

Fig. 2 The SFC–NMR setup (a) and detailed flow control system (b). Adapted from ref. 3. Once a sample of interest is observed in the UV chromatogram, the first valve is switched, letting the sample into the sample loop. When the loop is filled, the first valve switches back. In this way, a sample plug is created in the hexane flow, which is kept at a lower pressure. Simultaneously, the third valve switches to inject a small amount of water in the sample plug. When the sample reaches the middle of the NMR probehead, the second valve switches to stop the flow. In this way, multiple scans can be acquired on the same sample, in order to increase the signal-to-noise ratio.

The SFC is connected to the NMR probe with a flow control system. Once a peak of interest passes the UV detector, the first valve is switched, letting the sample including the mobile phase into a 100 µL sample loop. When the first valve switches back, the sample is injected into a *n*-hexane (AnalaR NORMAPUR, VWR Chemicals) flow-line, which is kept at a lower pressure than the SFC flow (60 bar compared to 120 bar). This pressure is regulated by an additional back-pressure with an operating range between 20 and 103 bar (Vici Jour). Simultaneously, the third valve switches to inject 10 µL of D_2O from the additional sample loop (orange in Fig. 2b), which ends up in the middle of the sample plug. The sample plug is transported in the hexane flow at a rate of 0.1 mL min^{-1} from the SFC to the NMR probe, through PEEK tubing. When the sample reaches the middle of the NMR stripline detector, the flow can be stopped by switching valve 2. In this way, multiple scans can be acquired, to obtain a better signal-to-noise ratio.

The three switching valves are 6-port binary-position valves (Vici Valco). A broad 'expansion tube' made of stainless steel, with an inner diameter of 2.159 mm and length of ~20 cm, is placed in the hexane flow line to allow for the different phases in the sample plug to pass each other during flow. The CO_2 from the trailing edge of the plug can then go past the methanol to the beginning of the plug. All PEEK tubing used in the system has an inner diameter of 0.508 mm, except for the tubing from the 'expansion tube' to the NMR probe, which has an inner diameter of 0.254 mm. In the stripline probe, a fused silica capillary with an inner diameter of 250 µm and outer diameter of 360 µm (Polymicro Technologies) transports the sample over the active detection volume of the chip. After the experiment, the sample flows back from the NMR spectrometer, over the additional backpressure regulator, to the waste.

2.2 Instruments

In the SFC–NMR setup, a Waters Acquity UPC^2 instrument was coupled to a Varian VNMRS spectrometer at 600 MHz Larmor frequency (14.1 T). A home-built stripline probe was used,[8] with a 600 µm wide and 5 mm long chip with split contacts, with an active detection volume of ~150 nL. In-flow experiments were performed with a relaxation delay of 0.2 s and an acquisition time of 0.5 s (in total 0.7 s per spectrum), while stop-flow experiments and the three reference spectra were acquired with a relaxation delay of 10 s and an acquisition time of 1 s. All spectra were recorded with a receiver bandwidth of 10 kHz. The spectra are referenced to the water peak, which was set to 4.79 ppm. All spectra were processed using the ssNake software package.[12]

2.3 SFC chromatography

For the SFC optimisation, in Fig. 3, a sample of 0.78 M N-acetyl-D-mannosamine (>98%, Sigma Aldrich) in ultra pure water was prepared. For the separation in Fig. 4, a sample of 0.035 M N-acetyl-D-mannosamine was dissolved in synthetic urine (Surine Negative Urine Control, Cerilliant, Sigma Aldrich). In both cases, 0.5 µL of the sample was injected onto a 100 mm × 1.7 µm packed 2-PIC SFC column (Waters) at a backpressure of 120 bar, a flow rate of 1.5 mL min^{-1} and a temperature of 60 °C. For the optimal separation, an isocratic mobile phase of

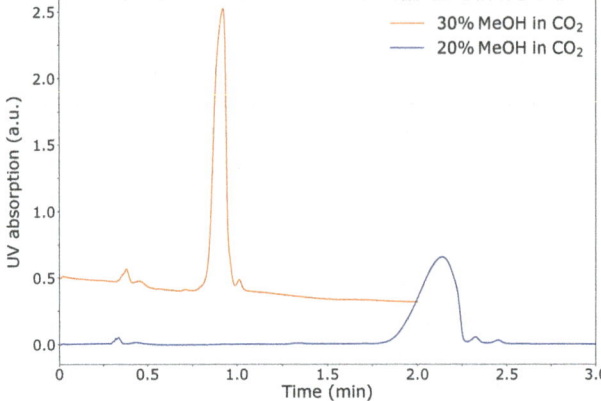

Fig. 3 UV chromatogram of the SFC separation of ManNac in water (0.78 M, 0.5 µL injection volume) on a 2-PIC column (Waters) at 60 °C, 120 bar and a flow rate of 1.5 ml min^{-1}. The top graph was separated with a mobile phase of 30% MeOH in CO_2, and the bottom graph with 20% MeOH in CO_2. The sharpest peak is obtained with 30% MeOH in CO_2.

30% methanol (BioSolve, SFC grade) in CO_2 (Linde Gas Benelux, food grade) was used. All UV chromatograms were recorded at a wavelength of 210 nm.

2.4 (SFC–)NMR

The ^{1}H NMR reference spectra were obtained by dissolving N-acetylglucosamine, N-acetyl-D-mannosamine (>98%) and N-acetyl-D-galactosamine (~98%) (all from Sigma Aldrich) in deuterated water (99.9%; Sigma Aldrich) at concentrations of 0.38 M, 0.38 M and 0.75 M, respectively. The samples were then inserted

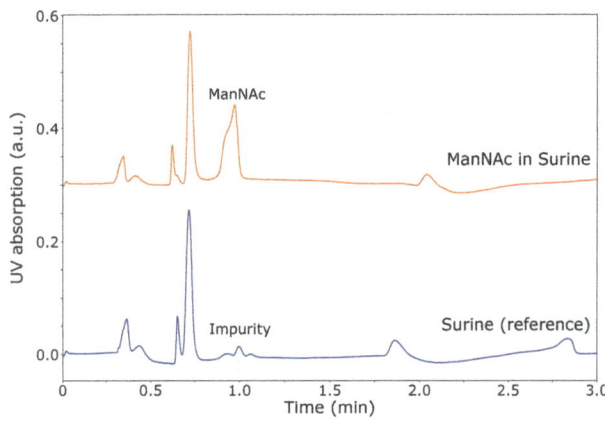

Fig. 4 UV chromatogram of the SFC separation of ManNac in Surine (top) (0.035 M, 0.5 µL injected volume) and pure Surine (bottom) on a 2-PIC column at 60 °C, 120 bar and a flow rate of 1.5 ml min^{-1} for a mobile phase of 30% MeOH in CO_2. A small impurity of a previous separation is present in the chromatogram of pure Surine. Under these conditions, ManNAc can be separated from synthetic urine.

separately into a ∼15 cm long, 250 μm I.D./350 μm O.D. fused silica capillary by capillary suction, which was then sealed at both ends with optical glue. The three reference spectra were acquired using a stripline probe employing a relaxation delay of 10 s and an acquisition time of 1 s, and summing over 6400 scans (19 h 33 min), 13 600 scans (41 h 33 min) and 1500 scans (4 h 35 min), respectively.

For the SFC–NMR spectra, a ManNAc sample of 0.78 M and a GlcNAc sample of 0.73 M in ultra pure water were used, of which 4 μL was injected onto the SFC column (further details in the previous section). The ^{1}H NMR spectra were acquired under the same conditions as the reference spectra, but this time averaging over 4860 scans (14 h 51 min) and 3520 scans (10 h 45 min) for ManNAc and GlcNAc, respectively.

3   Results and discussion

In this research, ManNAc in urine was chosen as a concept application to develop the SFC–NMR setup for polar samples. First, the SFC separation will be optimised for ManNAc in urine. Then, a method for in-line concentration of the sample will be discussed, which is needed to be able to detect the analyte by NMR spectroscopy. Finally, the SFC–NMR analysis of ManNAc (and its stereoisomers) is demonstrated.

3.1   SFC of *N*-acetylhexosamines in synthetic urine

During the optimisation of the SFC separation of ManNAc from urine, we have worked with synthetic urine to eliminate sample variation between batches. Later on, this research could be extended to real urine samples. Several stationary phases were investigated for the SFC separation of ManNAc (BEH, BEH 2-EP and 2-PIC columns from Waters), of which the 2-PIC column showed the best results. The next parameter under investigation was the mobile phase composition. Methanol can be mixed with CO_2 order to make the mobile phase more polar. In Fig. 3, the elutions of ManNAc in water from a 2-PIC SFC column with mobile phases containing 20% and 30% methanol in CO_2 are shown. It is clear that the retention of ManNAc decreases with the increase in the polarity of the mobile phase, which was expected. For 30% methanol, the peak is much sharper. Using a higher concentration of methanol decreases the retention time further and an even sharper peak is obtained. However, 30% methanol in CO_2 was chosen as the optimum, since a higher methanol volume means that the sample will be more diluted in the NMR analysis.

With these optimised parameters, the separation of ManNAc from synthetic urine was performed. In Fig. 4, the UV chromatogram of the SFC separation of ManNAc in synthetic urine, on a 2-PIC column with a mobile phase of 30% methanol in CO_2, is shown. The optimised parameters give a clear separation of ManNAc from synthetic urine. Due to this, the peak from ManNAc can be selected separately from the sample matrix for the SFC–NMR analysis.

3.2   In-line concentration by expansion of scCO$_2$

Since the sample gets diluted in the mobile phase during SFC, it is difficult to detect this low concentration by NMR spectroscopy. This is why concentrating the sample, preferably in-line, after chromatography is required.

In our previous work we have demonstrated that by expanding the sample plug in-line in a controlled fashion, the sample and co-solvent are separated from the mobile phase, CO_2.[3] In this way, the sample is concentrated in a smaller volume, which leads to a higher concentration, which is needed for NMR detection. This method will be employed here as well. However, in our previous work, our transportation medium was water, which did not mix with the nonpolar samples that were investigated. If water is used as the transportation medium in the case of polar samples, the compound will dissolve into the water, thus diluting the sample even more. This is why an alternative, nonpolar transportation solvent was chosen in this work, namely hexane. The sample plug, including CO_2 and co-solvent methanol, is injected into the hexane flow through a valve switching system, and is transported to the NMR probehead. By keeping the hexane flow at a lower pressure, the plug is expected to expand and the sample and co-solvent will separate from the CO_2.

Under the same conditions as employed in our previous work (hexane flow-line at 50 bar, injection of the sample plug at 120 bar), however, no phase separation occurred. No defined sample plug was observed. This was due to the mixing of hexane with CO_2, which both have a similar polarity. Due to this mixing, the expansion did not take place at a pressure of 50 bar. Several pressures were investigated, after which it was found that the sample and methanol co-solvent separated best from the CO_2 at a pressure of 60 bar for the hexane flow-line. Additionally, it was found that methanol has a strong affinity for the tubing walls, which are made out of PEEK (polyether ether ketone). Most of the methanol in the sample plug, and with it the sample, will stick to the tubing and will therefore not reach the NMR detector. In order to transport the methanol in the sample plug to the NMR probe, it is necessary to inject a small volume of water into the sample plug, which collects the methanol and sample and allows them to reach the NMR probe.

To follow the phase separation in the plug in-flow, NMR spectra were recorded each 0.7 s of an expanding mixture of 30% MeOH in CO_2. 100 µL of this mixture at 120 bar, to which 10 µL D_2O is added later on, is injected in the hexane flow-line at 60 bar. The sample plug, as observed in the NMR probe, is shown in Fig. 5. In this figure, the methyl peak of hexane (the transportation medium) and the methyl peak of methanol (part of the plug) were integrated separately in each of the NMR spectra and plotted over time.

The most important result is that methanol and water are mainly collected at the trailing edge of the plug. When a polar sample is present, this will dissolve in methanol and water, not in hexane or CO_2. When the pressure is kept at 120 bar instead of 60 bar, the methanol spreads out over the entire plug (1.55 min until 2.3 min), since it stays dissolved in the CO_2. Without the expansion to 60 bar, the volume in which the sample is dissolved is much larger than when the CO_2 is separated from the methanol, namely approximately 110 µL (CO_2, D_2O and MeOH) of 40 µL (D_2O and MeOH). From Fig. 5, it can be seen that most of the CO_2 is separated from the methanol and therefore the concentration method by in-line expansion is successful. It can be seen, however, that a small amount of methanol is not collected at the trailing edge of the plug, but remains dissolved in the CO_2 at the start of the plug (1.6–2.0 min). This means that a small part of the sample will not be collected at the end, but will remain spread out over this part of the plug. This loss is, however, quite small, since most of the sample will be concentrated at the end.

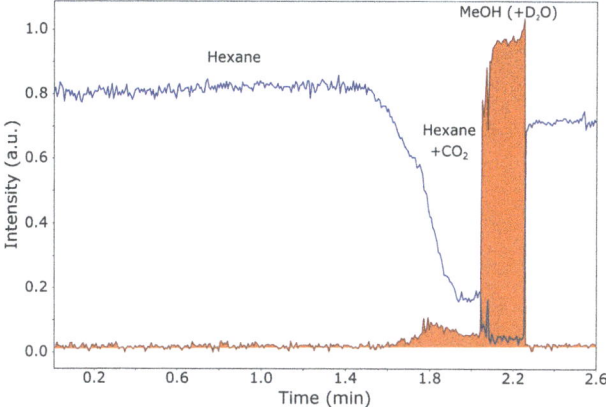

Fig. 5   Sample plug of 30% MeOH in CO_2 injected in a hexane flow, measured in-flow after in-line expansion to 60 bar. ^{1}H NMR spectra were acquired every 0.7 s during the flow of the sample from the SFC towards the NMR probehead. The methyl peaks of hexane (blue) and methanol (orange) were integrated separately in each spectrum over time and nor-malised to the maximum methanol integral to give the graphs. A polar sample will dissolve in methanol/water and will therefore mainly be concentrated in the trailing edge of the plug.

Another interesting phenomenon is that hexane from the flow starts mixing in with CO_2 at the beginning of the plug. This is due to the fact that hexane and CO_2 have a similar polarity. At the trailing edge of the plug, this mixing does not occur since (almost) no CO_2 is present in the highly concentrated methanol. However, a very small amount of hexane is present also in the methanol, probably due to the presence of a small remainder of CO_2 in which the hexane can dissolve. This will be visible in the SFC–NMR spectra.

The in-line concentration method, which is performed by expanding the sample plug in-line to 60 bar and adding a small amount of water to collect the methanol, can be used successfully for polar samples. However, a small sample loss might occur due to the fact that a small amount of methanol remains dissolved in the CO_2 and some methanol sticks to the tubing walls during transportation of the sample to the NMR probe. A small amount of hexane may also be observed in the SFC–NMR spectra that are recorded at the trailing edge of the plug, due to mixing of hexane with a small remainder of CO_2 in the concentrated methanol.

3.3   Stripline NMR analysis of *N*-acetylhexosamines

To show that NMR spectroscopy is able to distinguish between the different isomers ManNAc, GlcNAc and GalNAc, reference ^{1}H NMR spectra were acquired for these three compounds in deuterated water. The three spectra were acquired in the same stripline probe as was used for the in-flow SFC–NMR experiments. The samples were measured in sealed off capillaries. The three spectra are plotted in Fig. 6.

It can be seen that the different diastereomers have different ^{1}H NMR spectra. This is due to the stereochemistry of the molecules. The chemical shift of a proton depends on its chemical environment. When an alcohol group or the amine

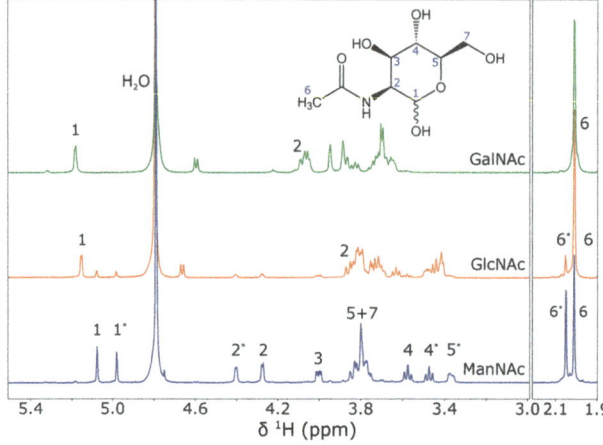

Fig. 6 ^{1}H NMR spectra of GalNac (0.75 M), GlcNAc (0.38 M) and ManNAc (0.38 M) in D_2O (from top to bottom), acquired using the stripline NMR probe. The spectra were scaled to the integrals of the peak(s) around 2 ppm. The structure of ManNAc is shown in the figure for assignment of the peaks. An asterisk (*) next to the peak number indicates the difference between the α- and β-isomer of the same molecule. A full assignment of the spectra can also be found in the Human Metabolome Database (HMDB).[13]

group flips from the top to the bottom of the ring, the chemical environment of the proton attached to that group, but also of the surrounding protons, will change and with it their chemical shifts change. Between the three isomers, the stereochemistry of proton 2 and proton 4 is different, which is observed as a shift in the spectra. As explained before, however, the shifts of the other protons also change. An interesting observation, especially for the spectra of ManNAc and GlcNAc, is that many of the peaks come in pairs. This is due to the fact that within the molecule the alcohol group next to the heteroatom in the ring can change in axial/equatorial position, between α-ManNAc and β-ManNAc. Although the reference compounds that were purchased were 98% pure, in water the position of this alcohol group can change. Also, it can be seen that a small ManNAc contamination is present in the GlcNAc reference spectrum, since all of the peaks match, but are present at much lower intensities.

These reference spectra were acquired at a high concentration, averaging over a substantial number of scans to obtain a high signal-to-noise ratio. For the SFC–NMR analysis, the concentrations will be lower. The most intense peak will be the peak that can be detected in the least amount of time at lower concentrations. In this case, this will be the peak at 2.0 ppm, corresponding to the CH_3 group. However, this peak has the same chemical shift for all three diastereomers and cannot be used to distinguish them. The peak around 5.1 ppm (proton 1) will therefore be the best to make a distinction between the three isomers.

3.4 SFC–NMR analysis of ManNAc and GlcNAc

In the previous sections, the SFC separation was optimised for separating Man-NAc from water or synthetic urine and it was shown that a distinction between the isomers can be made by looking at the ^{1}H NMR spectra. In this section, these two techniques will be combined in-line to obtain an SFC–NMR analysis of these polar stereoisomers. For this, 4 μL of ManNAc (0.78 M) and GlcNAc (0.73 M) in water were injected for SFC separation. The chromatograms can be found in the ESI (S1).† 100 μL of the peak of interest was selected through a valve switching system in a sample loop. 10 μL of D_2O was injected in the middle of this sample loop. The whole plug was then placed into a hexane flow-line at a pressure of 60 bar and transported to the middle of the NMR stripline probe. There, the flow was stopped and multiple scans (taking 10 h 45 min for GlcNAc and 14 h 51 min for ManNAc) were acquired to obtain the SFC–NMR spectra shown in Fig. 7. The reference ^1H NMR spectrum, already shown in Fig. 6, of each corresponding isomer is shown under each SFC–NMR spectrum for comparison.

From the SFC–NMR analysis it can be seen that the different stereoisomers can be distinguished, especially by the peak around 5.1 ppm, corresponding to proton 1 on the anomeric carbon atom next to the heteroatom in the ring. ManNAc can also be identified by the peak around 4.3 ppm, which is not present at this shift for the other two stereoisomers and corresponds to the proton at the carbon atom to which the amine group is attached. Some chemical shifts are slightly different in the SFC–NMR spectra compared to the reference spectra. This is due to the solvent, since the reference spectra were measured in D_2O while the SFC–NMR spectra have a 1 : 3 ratio of D_2O and methanol as the solvent. The latter comes from the SFC separation, since 30% methanol is used as the co-solvent. The water peak at 4.79 ppm and the methanol peak, here at 3.3 ppm, are therefore quite substantial in the spectra. D_2O exchanges protons with the non-deuterated methanol, therefore the residual water peak is more intense than usual. A solution to this would be to use deuterated methanol, but this would be quite expensive since the whole SFC separation would have to be performed with MeOD. However, in the SFC–NMR spectra shown, the solvent peaks do not overlap too much with the peaks of interest. Two extra peaks are present at 0.8 and 1.2 ppm (not shown), which originate from the small amount of hexane that

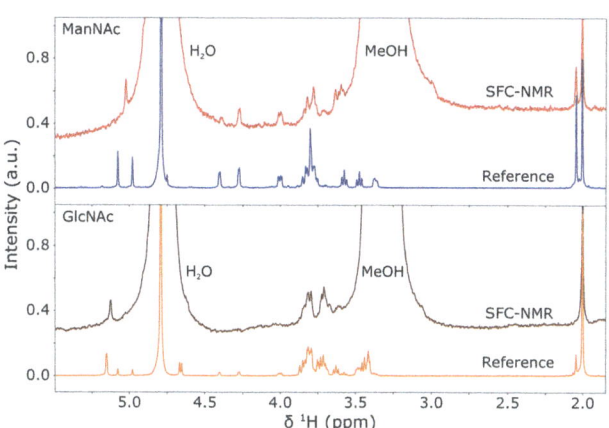

Fig. 7 SFC-^1H NMR spectra of GlcNAc (bottom) and ManNAc (top) in D_2O and MeOH, compared to their respective ^1H NMR reference spectra (shown before in Fig. 6). The spectra were scaled to the integrals of the peak(s) around 2 ppm. A Lorentzian line broadening of 1.5 Hz was applied to the SFC–NMR spectra. The distinction between the two isomers is clearly visible in the SFC–NMR spectra.

is still present in the sample plug (discussed in the previous section). These are, however, low in intensity, namely about the same intensity as the peak at 2.0 ppm, and do not overlap with the peaks of interest. If this causes a problem in the analysis of other compounds, an alternative might be to use fluorinated hexane as the transportation medium, which will not be visible in the ^{1}H NMR spectra.

An interesting feature is the presence of the α- and β-stereoisomers. As can be seen in the figure, the ratios between these two are different in the SFC–NMR spectra compared to the reference spectra. This may have different causes. First, the SFC–NMR spectra contain more methanol than water, which might influence the transition from one stereoisomer into the other. Second, due to the SFC separation, only one part of the peak is selected, which possibly contains only one isomer. However, it would still be expected that the selected isomer transfers into the other stereoisomer, but the timescale of the transition might be longer than the time in which the SFC–NMR analysis took place. This is, however, less likely, due to the long timescale of the experiment.

An important point of interest is to see if this method will be a viable analysis technique for metabolites in urine. For ManNAc, the SFC–NMR analysis started with a 4 μL injection of a sample with a high concentration (0.78 M). The chromatogram of this separation is shown in the ESI (S1).† Although the concentration is high, the amount injected on the column is then only 3.1 μmol. Of this 3.1 μmol, only a part is selected for transportation towards the NMR spectrometer. Since the UV detector was overloaded due to the high concentration that was injected, it is not possible to determine which fraction of the peak was injected. It is, however, possible to determine this from the NMR spectra, since these are quantitative. By comparing the integral of the reference spectrum with a known concentration (0.35 M) to the integral of the SFC–NMR spectrum and correcting for the number of scans, it can be concluded that the ManNAc concentration in the SFC–NMR spectrum is 15 mM. The signal-to-noise (S/N) ratio of the peak at 5.1 ppm is higher than necessary, therefore a lower concentration can be handled. If the minimum S/N-ratio in the spectrum is set to 3, the concentration could be lowered from 15 mM to 1.7 mM. As determined in the introduction, the concentration of ManNAc in the urine of patients suffering from NANS-mediated disease is estimated to be in the range of 0.3–0.7 mM. This is a factor 2.5–5.5 lower than we can currently detect. One of the solutions to this problem is using a larger volume stripline detector. The volume in which our sample is dissolved is 40 μL, however our current NMR probe has an active detection volume of 150 nL. A significant fraction of the sample is therefore not detected. By scaling up the stripline probe by a factor of 10 to 1.5 μL detection volume, a gain of a factor 5 in concentration can be achieved,[14] or a decrease of a factor 25 in experimental time at equal concentration. This will be enough to be able to detect ManNAc in the urine of patients suffering from NANS-mediated disease. Such a large-volume stripline probe is under development.

The calculations above were made assuming that there is no sample loss in the system. The concentration in the SFC–NMR spectrum of ManNAc was determined to be 15 mM. This corresponds to 0.61 μmol in 40 μL of solvent (10 μL D$_2$O and 30 μL MeOH). The entire peak in the SFC chromatogram consisted of 3.1 μmol, of which only a fraction was sent to the NMR spectrometer. If we assume that half of the peak is selected, meaning 1.5 μmol is sent to the NMR detector, this indicates that more than half of the sample gets lost during transport. Probably this sticks to the tubing walls, since methanol binds strongly to the PEEK tubing which couples the SFC machine to the spectrometer. Changing the tubing material to a less hydrophilic material may help in getting a higher concentration to the NMR probe and reducing sample losses during transportation.

4   Conclusions

Our SFC–NMR setup was successfully adapted to analyse polar samples in complex matrices. By controlled expansion of the $scCO_2$, the sample coming from the chromatograph can be concentrated in-line. Concentrating the sample in-line after chromatography is needed to detect the sample with NMR spectroscopy.

SFC separation of N-acetylhexosamine stereoisomers from synthetic urine was optimised, using a 2-PIC column (Waters) as the stationary phase and 30% $MeOH/CO_2$ as the mobile phase. After this, in-line SFC–NMR analysis of two N-acetylhexosamine isomers, namely N-acetyl-mannosamine (ManNAc) and N-acetyl-glucosamine (GlcNAc), was demonstrated. ManNAc is a metabolite that is present at elevated concentrations in patients suffering from NANS-mediated disease. With the SFC–NMR setup it is possible to distinguish between these polar stereoisomers. This is not possible with the current standard analysis techniques, which are based on mass detection.

The concentrations that are needed in the SFC–NMR setup are currently too high to be able to detect ManNAc in patient samples (1.7 mM vs. 0.7 mM) in a reasonable time. However, several adaptations to the current setup will make this possible. First, the detection volume of the stripline NMR probe needs to be increased by a factor 10 to better match the sample volumes coming from the chromatograph. It will then be possible to measure concentrations around 0.3 mM. Second, the tubing connecting the SFC to the NMR probe needs to be replaced with a less hydrophilic material, to decrease sample losses during transportation of the sample towards the NMR spectrometer. With these adaptations it will be possible to detect ManNAc in urine samples from patients suffering from NANS-mediated disease with in-line SFC–NMR. Due to the broad separation range of SFC and its coupling to NMR spectroscopy, it is expected that other polar and non-polar metabolites present at these, or higher, concentrations in biofluids can also be analysed with the same setup. By tuning the polarity of the mobile phase, SFC–NMR can be used to analyse a broad range of hydrophilic and lipophilic metabolites.

Conflicts of interest

There are no conflicts to declare.

Acknowledgements

This research received funding from The Netherlands Organization for Scientific Research (NWO) in the framework of Technology Area COAST (053.21.115). The authors thank Jan van Bentum, Julija Romanuka (Shell), Noud van den Borg (Waters), Peter Schoenmakers (UvA) and Ulrich Braumann (Bruker) for discussions and support, and Hans Janssen (RU) for technical support. The authors would also like to acknowledge Jonathan Martens (RU) for providing the N-acetylhexosamine samples and Arno Hoefnagels for optimising the SFC separations.

Notes and references

1 G. J. Patti, O. Yanes and G. Siuzdak, *Nat. Rev. Mol. Cell Biol.*, 2012, **13**, 263.

2 S. Bouatra, F. Aziat, R. Mandal, A. C. Guo, M. R. Wilson, C. Knox, T. C. Bjorndahl, R. Krishnamurthy, F. Saleem, P. Liu, *et al.*, *PLoS One*, 2013, **8**, e73076.

3 F. H. M. van Zelst, S. G. J. van Meerten, P. J. M. van Bentum and A. P. M. Kentgens, *Anal. Chem.*, 2018, **90**, 10457–10464.

4 D. L. Olson, T. L. Peck, A. G. Webb, R. L. Magin and J. V. Sweedler, *Science*, 1995, **270**, 1967–1970.

5 J. Dechow, A. Forchel, T. Lanz and A. Haase, *Microelectron. Eng.*, 2000, **53**, 517–519.

6 C. Massin, F. Vincent, A. Homsy, K. Ehrmann, G. Boero, P.-A. Besse, A. Daridon, E. Verpoorte, N. F. De Rooij and R. S. Popovic, *J. Magn. Reson.*, 2003, **164**, 242–255.

7 A. P. M. Kentgens, J. Bart, P. J. M. Van Bentum, A. Brinkmann, E. R. H. Van Eck, J. G. E. Gardeniers, J. W. G. Janssen, P. Knijn, S. Vasa and M. H. W. Verkuijlen, *J. Chem. Phys.*, 2008, **128**, 052202.

8 P. J. M. Van Bentum, J. W. G. Janssen, A. P. M. Kentgens, J. Bart and J. G. E. Gardeniers, *J. Magn. Reson.*, 2007, **189**, 104–113.

9 C. D. Van Karnebeek, L. Bonafé, X.-Y. Wen, M. Tarailo-Graovac, S. Balzano, B. Royer-Bertrand, A. Ashikov, L. Garavelli, I. Mammi, L. Turolla, *et al.*, *Nat. Genet.*, 2016, **48**, 777.

10 J. Martens, G. Berden, R. E. Outersterp, L. A. Kluijtmans, U. F. Engelke, C. D. Karnebeek, R. A. Wevers and J. Oomens, *Sci. Rep.*, 2017, **7**, 3363.

11 World Health Organization, *Biological monitoring of chemical exposure in the workplace: guidelines*, World Health Organization, Geneva, 1996.

12 S. G. J. van Meerten, W. M. J. Franssen and A. P. M. Kentgens, *J. Magn. Reson.*, 2019, **301**, 56–66, https://www.ru.nl/science/magneticresonance/software/ssnake/.

13 D. S. Wishart, Y. D. Feunang, A. Marcu, A. C. Guo, K. Liang, R. Vázquez-Fresno, T. Sajed, D. Johnson, C. Li, N. Karu, *et al.*, *Nucleic Acids Res.*, 2018, **46**, D608–D617.

14 J. Bart, J. W. G. Janssen, P. J. M. Van Bentum, A. P. M. Kentgens and J. G. E. Gardeniers, *J. Magn. Reson.*, 2009, **201**, 175–185.

PAPER

Polar mixture analysis by NMR under spin diffusion conditions in viscous sucrose solution and agarose gel†

Pedro Lameiras, [iD] ‡* Simon Mougeolle, François Pedinielli and Jean-Marc Nuzillard [iD] ‡*

Received 3rd December 2018, Accepted 17th December 2018

DOI: 10.1039/c8fd00226f

The use of two new viscous solvents, sucrose solution and agarose gel, is reported for the first time for giving access to the individual NMR spectra of polar and potentially bioactive compounds in a mixture. Under viscous conditions, the tumbling rate of small and mid-sized molecules reduces in solution, so that the longitudinal cross-relaxation encourages the observation of spin diffusion. As a result, all of the resonances of the ^{1}H nuclei within the same molecule tend to correlate together in a 2D NOESY spectrum, thus paving the way to mixture analysis. This work describes the individualization of four structurally close mixed dipeptides: Leu–Val, Leu–Tyr, Gly–Tyr and Ala–Tyr dissolved in each of sucrose solution and agarose gel, by means of spin diffusion in homonuclear selective 1D NOESY, selective 2D NOESY experiments and heteronuclear 2D HSQC-NOESY. Sucrose solution should be preferred to agarose gel for the investigation of mixtures made of small and flexible polar compounds, due to its capability to give rise to more suitable viscous conditions mandatory for efficient spin diffusion, even though agarose gel reveals the benefit of not offering intense residual solvent proton signals due to active transverse relaxation.

Introduction

The identification of organic molecules in mixture is a crucial issue in the vast majority of human activities related to chemistry (healthcare, energy, materials, *etc.*). Transformation processes that lead from natural resources, either from the fossil carbon industry or biomass exploitation, to low-value products in large amounts or to high-value speciality products including pharmaceutical and cosmetic active ingredients, rarely produce chemically pure compounds. In the current state of knowledge, mixture analysis remains a necessity, and to date,

Université de Reims Champagne-Ardenne, Institut de Chimie Moléculaire de Reims, CNRS UMR 7312, SFR CAP-Santé, BP 1039, 51687 Reims Cedex 02, France. E-mail: pedro.lameiras@univ-reims.fr; jm.nuzillard@univ-reims.fr; Tel: +33 3 26 91 82 28; +33 3 26 91 82 10

† Electronic supplementary information (ESI) available. See DOI: 10.1039/c8fd00226f

‡ Contributed equally to the work.

addressing this issue using NMR spectroscopy has been explored in a small number of ways. (i) The translational diffusion coefficient (D) characterizes molecules in solution and reflects their mobility. In theory, sub-spectra of the NMR spectrum of a mixture can be extracted using D as a discriminating factor, so as to attain the spectra of the individual mixture components.[1-3] Nonetheless, D values are poorly discriminant, although experimental tricks enable an increase of the resolving power in special cases: the introduction of various chromatographic solid phases in the sample, the interaction of analytes with soluble polymers or lanthanide shift reagents, and analyte inclusion in micelles.[4-14] (ii) Multi-quantum spectroscopy combined (or not) with broadband homonuclear decoupling, sparse sampling and pure shift data acquisition may be pertinent for individualizing molecules in a mixture. For example, a simple 2D TOCSY spectrum or a series of 1D selective TOCSY spectra points out ^{1}H nuclei coupled together, which may turn out to be useful for molecules composed only of one single spin system.[15-21] (iii) Viscous solvents (or solvent blends) under specific operating conditions lower the tumbling rate of small and mid-sized molecules in solution since the value of the molecular overall correlation time τ_c depends upon the medium viscosity according to the microviscosity theory of Gierer and Wirtz.[22] The longitudinal cross-relaxation regime thus favours the detection of spin diffusion. Consequently, the molecules present a negative nuclear Overhauser effect (NOE) regime and their resonances can be grouped according to their ability to share magnetization through intramolecular diffusion of spin magnetization. All of the resonances of the ^{1}H nuclei within the same molecule tend to correlate together in a 2D NOESY spectrum, thus giving access to the individual ^{1}H NMR spectra of the mixture components. The original idea was implemented in 1981 [23] by means of a perfluorinated polymer solvent and was considered again in 2008;[24] the use of supercooled water to tailor the spin dynamics of small metabolites was published in 2012 by the same author.[25] Our team reported for the first time, in 2011, 2016 and 2017, the use of glycerol and glycerol carbonate, of DMSO/glycerol and of DMSO/water binary solvents, respectively, as viscous media for ^{1}H and ^{19}F spin diffusion promotion.[26-28]

The present work focuses on the assessment of sucrose solution and agarose gel in the individual NMR characterization of four polar structurally close dipeptides: Leu–Val, Leu–Tyr, Gly–Tyr and Ala–Tyr within a single mixture by means of spin diffusion in homonuclear selective 1D NOESY, selective 2D NOESY experiments and heteronuclear 2D HSQC-NOESY.

Dissolving "table sugar" in water yields a viscous solution that enables us to profit from spin diffusion from room temperature to a lower temperature due to the overall decrease in the tumbling rate of molecules in solution. For instance, the viscosity of a sucrose/water (1 : 1, w/w) blend is around 27 cP at 283 K.[29,30] In comparison, that of water at the same temperature is 1.307 cP.[31] At room temperature, the viscosity of the same sucrose/water blend is sufficiently low, such that NMR samples may be prepared and transferred into an NMR tube without any difficulty, in contrast to if using highly viscous solvents such as glycerol (η = 934 cP at 298 K)[32] and glycerol carbonate (η = 85.4 cP at 298 K).[32] Adding sucrose to water (deuterated or not) also allows the possibility of working from room to sub-zero temperature,[30] which is especially suitable for thermally unstable molecules such as bioactive molecules. Spin diffusion may thus happen at a wide range of temperatures, from 243 K [30] to room temperature and even higher. In addition, the sample preparation cost is very low since "table sugar" is a widespread staple. However, the major experimental pitfall is the strong residual proton signals of non-deuterated sucrose that necessitate mandatory elimination. Mid-sized molecules will only require a low amount of sucrose in water, whereas smaller molecules will require more sucrose for driving spin diffusion under temperature control from room temperature to sub-zero temperature.

Agarose gel is commonly used for the separation of biological molecules by electrophoresis, although it has already been proposed as a confinement medium for investigating the molecular confinement effect on the conformational dynamics of apomyoglobin.[33] The main problem in the sample preparation is the addition of molecules of interest to the agarose powder previously dissolved in water at high temperature (363 K for melting regular agarose and around 338 K for low-melting agarose). Peptide or protein solutions may suffer from the high temperature exposition, preventing NMR structural and conformational studies, and premature gelation might occur. In practice, the heat time only takes a few minutes, and thus the sample preparation is compatible with quite small molecules that are reasonably heat resistant, such as nucleic acids, peptides or saccharides. In addition, the NMR signal line sharpness can be dramatically altered in the case of inhomogeneous samples due to partially unmelted agarose or the incorporation of air bubbles. The solution is to melt the sample again directly inside the NMR tube, then to mix it thoroughly and chill again at ambient temperature. Interestingly, agarose is non-interactive with most biomolecules and organic compounds, so DNA, RNA and protein can be recovered from agarose gels.[34,35] Its freezing point is reported to be 260 K and 264 K, routinely reached with H_2O- and D_2O-based 1% agarose gels,[36] respectively, limiting the temperature range for which spin diffusion may occur compared to sucrose solution. Hence, the use of agarose gel will be more dedicated to the study of rigid small or medium-sized molecules in mixtures presenting higher correlation times than very flexible small molecules. Remarkably, due to the rigidity of the agarose gel and therefore the overly active transverse relaxation, all resulting proton and carbon resonances are too broad to be detected in NMR spectroscopy.[36,37] As a result, only the residual proton signal of water has to be removed, compared to the numerous proton signals of non-deuterated sucrose.

The requirement for optimal temperature selection is a compromise between overall spectral resolution and intensity of NOESY cross peaks between nuclei that are not close enough to show a NOE signal in a low viscosity medium. A temperature reduction enhances spin diffusion but also reduces peak height through line broadening caused by a more efficient transverse relaxation process. Sample cooling is therefore required if the NOESY spectrum shows positive NOE responses (diagonal and off-diagonal peaks of opposite signs).

Depending on the complexity of the mixtures, the analysis of ^1H NMR spectra may become intractable due to the overlapping of ^1H resonances. A usual remedy to this issue lies in the spreading of the spectroscopic information along a second axis that encodes chemical shifts of nuclei other than ^1H such as ^{13}C or ^{15}N.[27,28] This approach to mixture analysis is exemplified in sucrose solution by 1D and 2D ^1H–^{15}N and ^1H–^{13}C HSQC-NOESY spectra providing $^1H/^{13}C$ and $^1H/^{15}N$ chemical shift lists for the mixture components.

Results and discussion

Leu–Val, Leu–Tyr, Gly–Tyr and Ala–Tyr mixture in sucrose solution

These four dipeptides in a pure water solution do not reveal any differentiation based on their translational diffusion behaviour, due to their similar molecular weight and shape (Fig. S-1 in the ESI†). This observation encouraged us to investigate the dipeptide mixture dissolved in a viscous sucrose solution by considering the resolving power of homo- and heteronuclear NOESY-based spin diffusion experiments for offering pertinent alternatives to DOSY experiments. The main experimental drawback of our approach was the mandatory elimination of the strong ^{1}H signals of sucrose and water (see Fig. 1a) in order to avoid excessively obscuring solute signals, since deuterated sucrose would be too expensive to produce and deuterated water would give rise to the chemical exchange between the deuterium nuclei of the latter and the amide proton of the dipeptide mixture. The elimination of these signals can be achieved by means of selective excitation and detection pulses when included in an excitation sculpting sequence.[38] The selective pulses invert the equilibrium magnetization of the nuclei of interest and leave untouched that of the solvent nuclei. For this purpose, resonance inversion in the two frequency bands on either side of the solvent signal was successfully realized by means of two consecutive band selective pulses. The spectrum in Fig. 1b proves the quality of solvent suppression that was achieved by band selective detection (see pulse sequence in Fig. 1e).

Since the temperature is an important parameter in spin diffusion experiments (because it is directly related to solvent viscosity and, as a result, to overall rotational correlation times τ_c),[22,26] we have determined the optimal temperature at which NOESY cross peaks were positive (negative NOE enhancements, slow motion regime), well-resolved, and as intense as possible between nuclei that were not supposed to be close enough to present a NOE in a low viscosity medium.

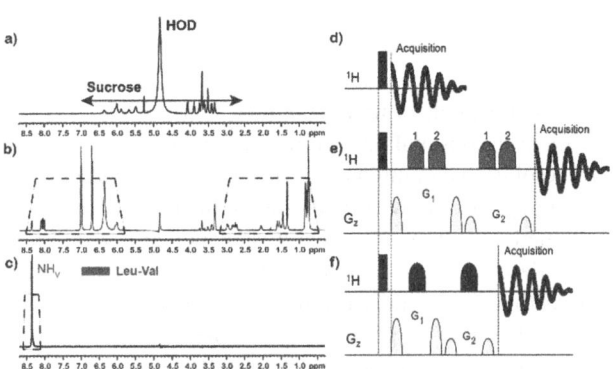

Fig. 1 1D ^{1}H spectra (8 scans), at 600 MHz (^{1}H), and corresponding NMR pulse sequence of the dipeptide test mixture in sucrose solution ((a–c) 283 K). G1 : G2 = 70 : 30. The FIDs (32k points, spectral width = 6.010 Hz) were processed with LB = 0.3 Hz and zero-filled to 32k points. (a and d) Non-selective excitation and detection. (b and e) Selective detection of two resonance bands. The 3 ms I-BURP-2 pulses cover 1.560 Hz (dotted trapezium). The "1" and "2" labels indicate their application to the high and low chemical shift regions, respectively. (c and f) Selective excitation of the valine amide proton doublet of Leu–Val (dotted trapezium) using a 10 ms, 1% truncated, 180° Gaussian pulse.

The optimized temperature of 283 K has been defined by means of band-selective detection NOESY experiments (see Fig. 2a, S-2: amide proton region NOESY spectra at 298, 288, 283, 278 and 273 K, and S-3: full NOESY spectrum at 283 K in ESI†). We have clearly observed that the use of viscous sucrose solution makes full intramolecular magnetization transfer through spin diffusion possible, observed over distances of >14 Å within each very small and flexible dipeptide. In

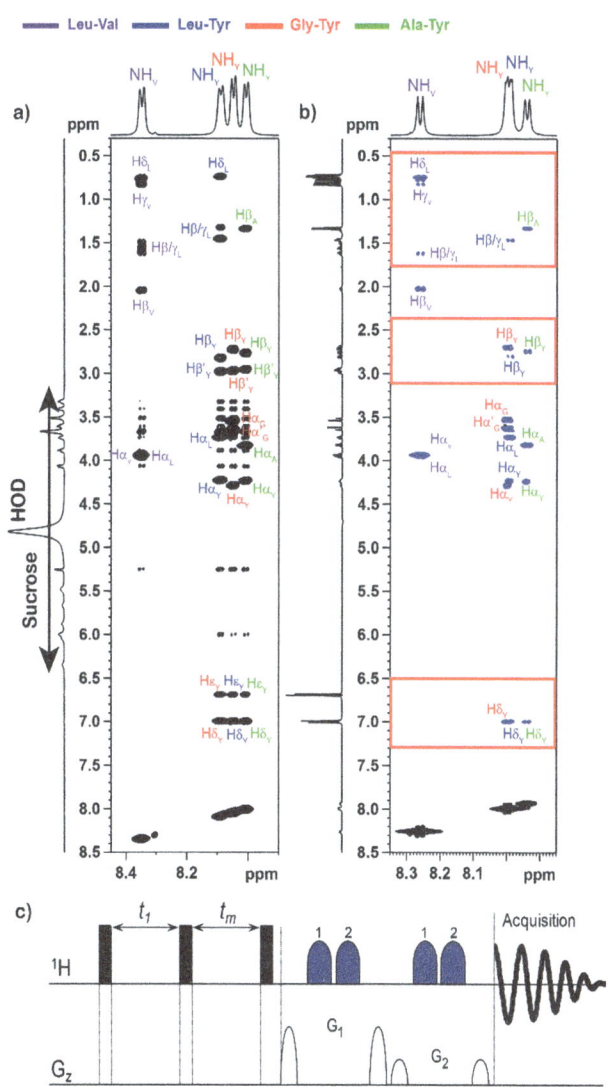

Fig. 2 (a) The amide proton region of the band-selective detection 2D NOESY spectrum of the dipeptide test mixture (10 mM), mixing time (t_m) = 0.5 s, at 600 MHz (^{1}H) using the pulse sequence in part (c), dissolved in sucrose solution, at 283 K. (b) The amide proton region of the 2D NOESY spectrum of the same dipeptide test mixture (10 mM), t_m = 0.5 s, at 600 MHz (^{1}H), using a noesyesgpph pulse sequence, dissolved in H_2O/D_2O (9 : 1, v/v), at 298 K. The red frames correspond to spectral regions of interest in which water as the solvent has a major effect on the number and sign of observable NOESY cross peaks.

comparison, the NOESY spectrum recorded in water at 298 K displays less numerous NOE cross peaks of the opposite sign (positive NOE enhancements, fast motion regime; see Fig. 2b, and the full NOESY spectrum in Fig. S-4 in ESI†). Hence, under viscous conditions, the grouping of proton resonances together is accessible, allowing the individualization of the mixture components since the chemical shift pattern of each dipeptide is predictable. The individualization of the four dipeptides in water would have necessitated the concomitant use of NOESY and TOCSY (and COSY) experiments, according to the common resonance assignment strategy.[39]

For analytical reasons, it appeared relevant to collect supplementary structural information of mixture components by detecting Hα proton resonances in F_2 during signal acquisition, that was not the case by means of band-selective transient NOESY experiments. Incidentally, we implemented a 1D selective NOESY pulse sequence composed of a double pulse field gradient block for the multiplet selective excitation[38,40] of the resonance of interest followed by a mixing time including two wideband adiabatic inversion pulses flanked with gradient pulses, so as to avoid the reintroduction of strong sucrose and water solvent signals that arise from longitudinal relaxation during the mixing time. Fig. 3 undoubtedly proves that all dipeptides are differentiated by spin diffusion in sucrose solution by means of an appropriate set of selectively excited proton resonances. Indeed, the selective excitation of the NH amide proton, at δ 8.35 in sucrose solution, shows a magnetization exchange exclusively with the protons of the Leu–Val dipeptide because the tyrosine Hδ/Hε proton resonances do not appear in the 1D NOESY spectra (see Fig. 3a). The selective excitation of the side chain Hδ and Hγ protons (between 0.7 and 0.9 ppm) shows a magnetization exchange with all protons of the two Leu–Val and Leu–Tyr dipeptides (see Fig. 3b). By comparison with the 1D NOESY spectra in Fig. 3a and b, a complete proton assignment of Leu–Tyr is possible. The selective excitation of the aromatic Hδ/Hε protons of Leu–Tyr, Gly–Tyr and Ala–Tyr reveals all of the proton resonances of Leu–Tyr, Gly–Tyr and Ala–Tyr (see Fig. 3c). One of the three tyrosine Hα protons at δ 4.14 has been selectively excited in order to differentiate all proton resonances from Gly–Tyr and Ala–Tyr (see Fig. 3d). Fig. 3d clearly shows the transfer of the tyrosine Hα magnetization over all protons of Gly–Tyr because of the absence of the side chain proton (H$\beta_A/\beta_L/\gamma_L/\delta_L$) resonances of Leu–Tyr and Ala–Tyr.

Another route to the individualization of the dipeptide mixture, which is less time-consuming than the acquisition of four suitable selective 1D NOESY spectra, has been to focus on the close NH amide resonances as an initial source of magnetization and to resort to a F_1 band-selective F_1 decoupled 2D NOESY experiment,[41,42] since these selected nuclei have their resonances in the same frequency band and they are not scalarly coupled together (see Fig. 4). Carrying out this latter experiment in sucrose solution allows the assignment of all proton resonances of Leu–Val, Leu–Tyr, Gly–Tyr and Ala–Tyr by profiting from spin diffusion acting during the mixing time.

The selective excitation of isolated proton resonances makes the individualization of each peptide within the mixture possible by taking advantage of spin diffusion. Nonetheless, it may happen in other complex mixtures that a component of interest does not present resolved proton resonances due to strong spectral overlap. In such cases, the larger chemical shift dispersion of ^{13}C and ^{15}N nuclei may prove to be helpful. By coupling the HSQC and NOESY experiments,

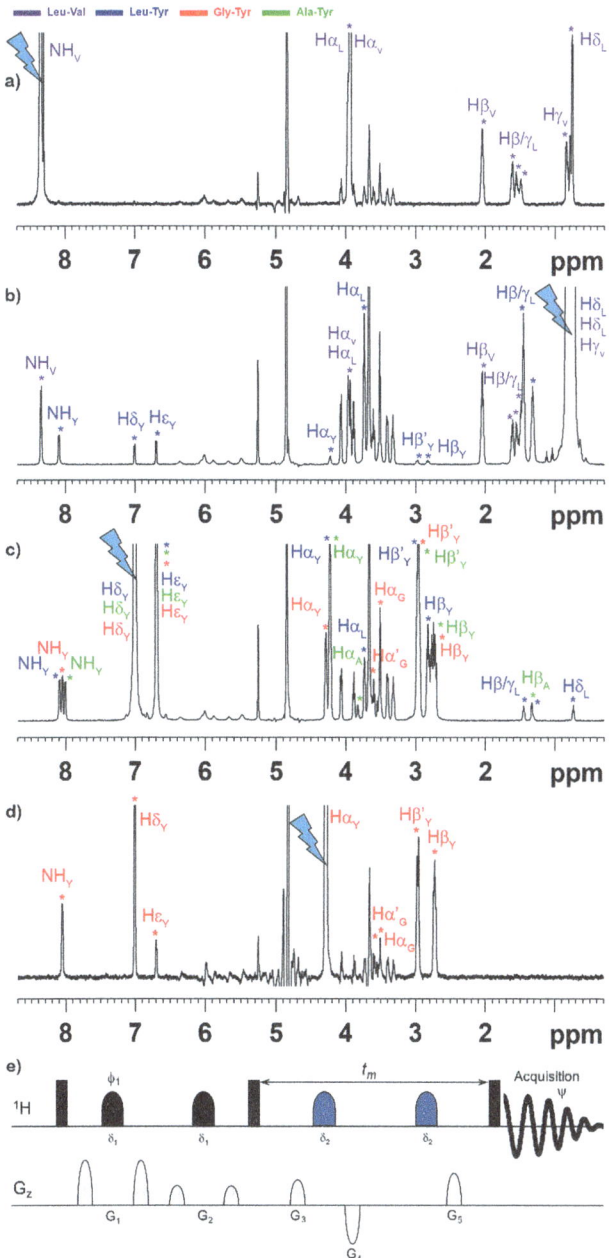

Fig. 3 Multiplet selective excitation 1D ^{1}H NOESY spectra of the dipeptide test mixture (10 mM) dissolved in sucrose solution ((a–d) 283 K), $t_m = 0.5$ s, at 600 MHz (^{1}H). The initial selective inversion pulses excite: (a) the NH$_V$(LV) proton resonance (experiment time (expt) = 134.65 min); (b) the Hδ$_L$(LY)/Hδ$_L$(LV)/Hγ$_V$(LV) proton resonances (expt = 33.67 min); (c) the Hδ$_Y$(LY)/Hδ$_Y$(GY)/Hδ$_Y$(AY) proton resonances (expt = 25.32 min); (d) the Hα$_Y$(GY) proton resonance (expt = 275.07 min). (e) Pulse sequence: $\varphi_1 = x, y, -x, -y$, $\psi = x, -x$.

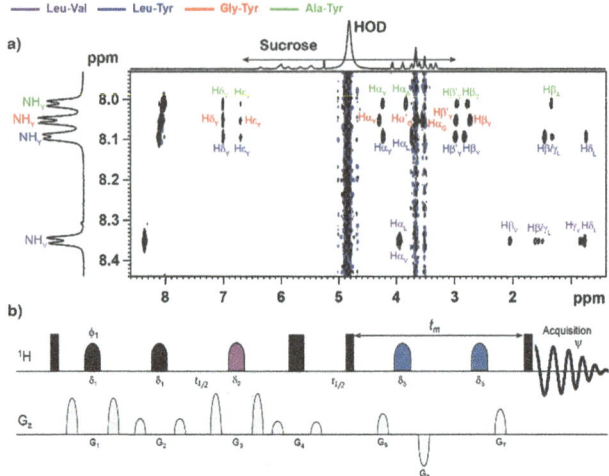

Fig. 4   (a) F_1 band-selective F_1 decoupled 2D NOESY spectrum of the dipeptide test mixture (10 mM) in sucrose solution, at 283 K, at 600 MHz (^{1}H), (64 scans per t_1 value, expt = 222.54 min, t_m = 0.5 s). (b) Pulse sequence: $\varphi_1 = x, y, -x, -y, \psi = x, -x$. The initial selective 180° pulses had a Gaussian shape and were applied to the four NH amide proton resonances.

a complete proton spectrum should be achieved for a molecule starting only from a single carbon or nitrogen resonance.

The 2D ^{1}H–^{13}C and ^{1}H–^{15}N HSQC-NOESY spectra of the dipeptide test mixture have been acquired at 283 K in sucrose solution (Fig. S-5 in ESI† and Fig. 5). Under these viscous conditions, all of the protons of each dipeptide of the mixture are able to propagate their own magnetization with all other protons by spin diffusion. A suitable selection of horizontal rows through carbon resonances at 18.14 and 38.08 ppm (Ala–Tyr), 22.64 and 37.66 ppm (Leu–Tyr), 31.43 and 53.64 ppm (Leu–Val) and 38.11 ppm (Gly–Tyr) allows the extraction of the four complete proton spectra corresponding to Ala–Tyr, Leu–Tyr, Leu–Val and Gly–Tyr. These four spectra have been compared to the conventional 1D ^{1}H spectra (Fig. S-5(b, b′), (c, c′), (d, d′) and (e, e′) in ESI†) and they logically reveal similar peak patterns. In the same way, an appropriate selection of four horizontal slices through nitrogen resonances at 123.93 (Ala–Tyr), 124.49 (Gly–Tyr), 126.01 (Leu–Tyr), and 126.91 ppm (Leu–Val) allows the production of four complete ^{1}H spectra corresponding to Ala–Tyr, Gly–Tyr, Leu–Tyr, and Leu–Val. These four spectra present resonance patterns similar to those of the conventional 1D ^{1}H spectra as well (Fig. 5(b, b′), (c, c′), (d, d′) and (e, e′)). Interestingly, another way is possible to individualize the compounds of the mixture by considering appropriate vertical slices from the 2D ^{1}H–^{13}C HSQC-NOESY (Fig. S-5†). In this case, the resulting spectra should reveal all of the protonated carbons of the four dipeptides in the mixture. As an aside, Fig. S-5(d″, c″, e″ and b″)† presents the four protonated carbon spectra obtained by extracting the column of amide protons at 8.35, 8.09, 8.05 and 8.01 ppm, corresponding to Leu–Val, Leu–Tyr, Gly–Tyr, and Ala–Tyr, respectively. The ability to extract all of the protonated carbon chemical shifts for

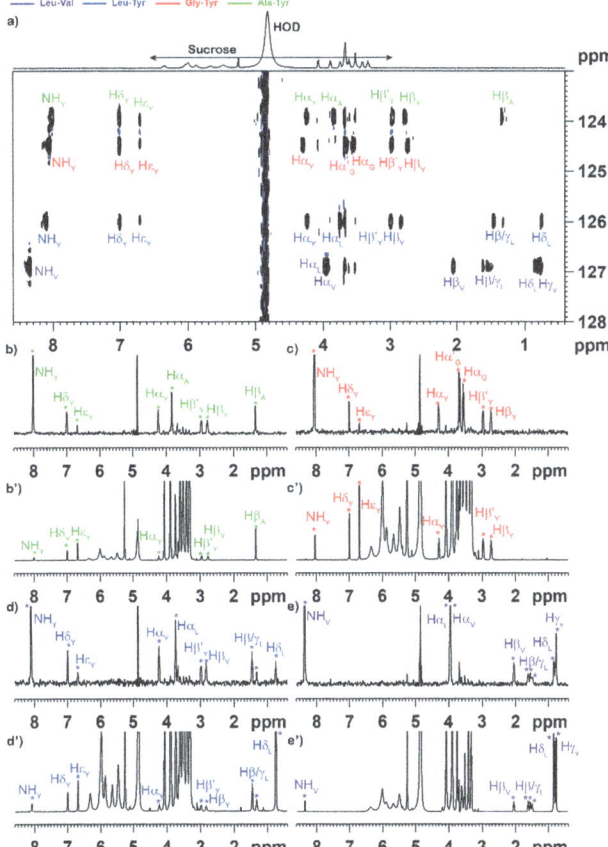

Fig. 5 (a) 2D ^{1}H–^{15}N HSQC-NOESY spectrum of the dipeptide test mixture (20 mM) dissolved in sucrose solution, at 283 K, t_m = 0.5 s, with solvent multiple presaturation, at 600 MHz (^{1}H). Comparison of four ^{1}H horizontal slices extracted from the 2D ^{1}H–^{15}N HSQC-NOESY at 123.93 ((b, b') Ala–Tyr, green row), 124.49 ((c, c') Gly–Tyr, red row), 126.01 ((d, d') Leu–Tyr, blue row), and 126.91 ppm ((e, e') Leu–Val, purple row) with the conventional 1D proton spectra of each pure dipeptide dissolved (20 mM) in sucrose solution, at 283 K, at 600 MHz (^{1}H).

an individual component in a mixture may turn out to be a very convenient tool in the structure assignment of molecules within mixtures.

Leu–Val, Leu–Tyr, Gly–Tyr and Ala–Tyr mixture in agarose gel

As previously published,[36,37] due to the active transverse relaxation of agarose gel, all of the proton resonances from agarose are remarkably not visible (see Fig. 6). This aspect is probably the main advantage of using this viscous medium in the individualization of components within polar mixtures, because that makes the solvent suppression easier.

We also determined the temperature that provides the best compromise between spin diffusion and spectral resolution by means of usual NOESY experiments with water suppression using excitation sculpting (see Fig. S-6 in ESI†).

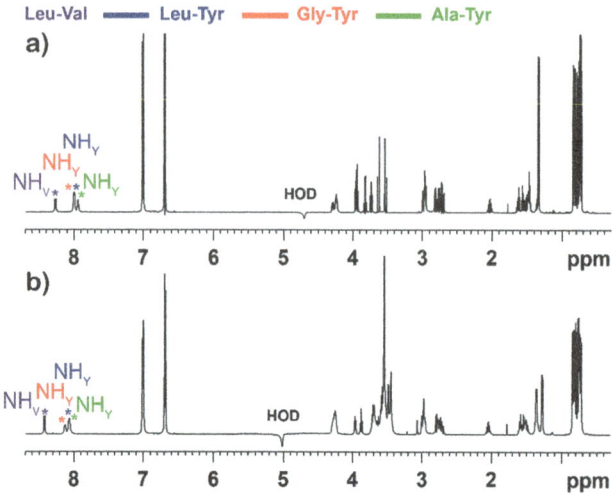

Fig. 6 1D ^{1}H spectra of the dipeptide test mixture (10 mM) dissolved (a) in H_2O/D_2O (9 : 1, v/v), at 298 K, at 600 MHz (^{1}H) and (b) in agarose gel (1%), at 273 K, at 600 MHz (^{1}H) with water suppression using excitation sculpting.

The optimized temperature of 273 K enables the detection of as many negative NOE enhancements (positive NOESY cross peaks) as possible due to the molecular tumbling slowdown. The use of viscous agarose gel yields full intramolecular magnetization transfer through spin diffusion, over distances of >14 Å for Leu–Tyr, Gly–Tyr and Ala–Tyr due to the stiffness of the aromatic moiety. In comparison, the NOESY spectrum acquired in water at 298 K presents fewer NOE cross peaks of the opposite sign (positive NOE enhancements, fast molecular tumbling – see Fig. 2b). However, the remaining flexibility of the Leu–Val peptide in agarose gel at 273 K prevents the observation of full magnetization transfer over the entire molecule by means of spin diffusion. Only positive NOE enhancements are detected. Nonetheless, under sufficiently viscous conditions, the grouping of proton resonances of Tyr-based dipeptides together is achievable, thus allowing the individualization of these three Tyr-based mixture components since the chemical shift pattern of each dipeptide is predictable. By subtraction, the proton resonance pattern of Leu–Val is then deduced.

We sought to highlight the discrepancy of the molecular behaviour of the four dipeptides in agarose gel by only collecting the resonances of interest during the signal acquisition to access supplementary structural information. This approach may turn out to be pertinent in cases of proton overlapping, especially for complex mixtures. Selectively exciting one suitable set of proton resonances illustrates this approach. Incidentally, we reused the 1D selective NOESY experiment previously described in this work. The one HOD signal of agarose gel arising from longitudinal relaxation during the mixing time is more easily suppressed than the multiple ones of sucrose. Fig. 7 illustrates that all Tyr-based dipeptides are differentiated by spin diffusion in agarose gel by means of an appropriate set of selectively excited proton resonances. Indeed, the selective excitation of $CH_{3\beta}$ at δ 1.27 in agarose gel shows a magnetization exchange exclusively with the protons of the Ala–Tyr dipeptide (Fig. 7a). The selective excitation of the aromatic Hδ/Hε protons of Leu–Tyr, Gly–Tyr

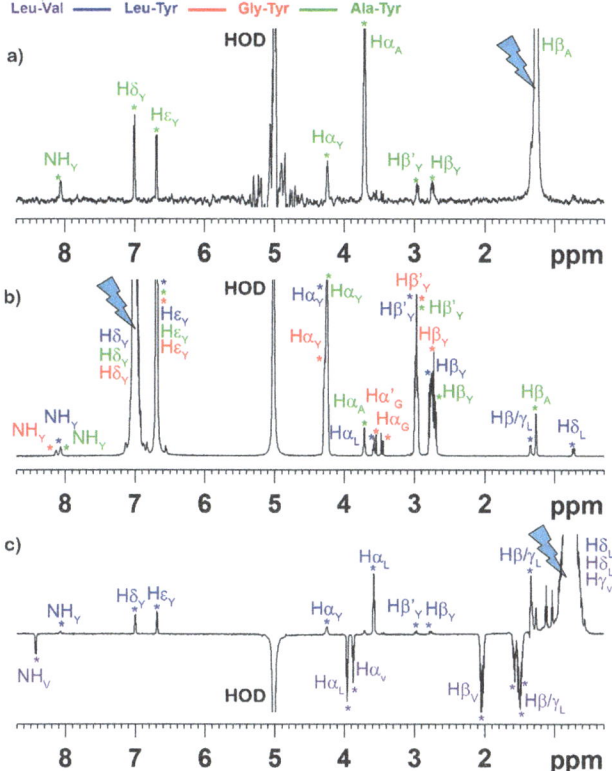

Fig. 7 Multiplet selective excitation 1D ^{1}H NOESY spectra of the dipeptide test mixture (10 mM) dissolved in agarose gel (1%) ((a–c) 273 K), t_m = 0.5 s, at 600 MHz (^{1}H). The initial selective inversion pulses excite: (a) the Hβ$_Y$(AY) proton resonance (expt = 186.25 min); (b) the Hδ$_Y$(LY)/Hδ$_Y$(GY)/Hδ$_Y$(AY) proton resonances (expt = 67.25 min); (c) the Hδ$_L$(LY)/Hδ$_L$(LV)/Hγ$_V$(LV) proton resonances (expt = 67.08 min); Pulse sequence: $\varphi_1 = x, y, -x, -y$, $\psi = x, -x$ (see Fig. 3e).

and Ala–Tyr at δ 7.00 reveals all of the proton resonances of Leu–Tyr, Gly–Tyr, and Ala–Tyr (Fig. 7b). The selective excitation of the side chain Hδ and Hγ protons (between 0.7 and 0.9 ppm) displays a magnetization exchange with all protons of the two Leu–Val and Leu–Tyr dipeptides (Fig. 7c). Due to a different molecular tumbling rate, the NOE enhancements are negative (active spin diffusion, positive peaks) for Leu–Tyr and positive for Leu–Val (negative peaks). As a result, the distinction of the two proton patterns is immediate by considering the change in the sign of the NOE peaks. By comparison with the 1D NOESY spectra in Fig. 7a–c, a complete proton assignment of Gly–Tyr is accessible.

Experimental

Chemical reagents

D$_2$O was purchased from Eurisotop (Gif-sur-Yvette, France). Leu–Val, Leu–Tyr, Gly–Tyr and Ala–Tyr were purchased from TCI Europe (Zwijndrecht, Belgium). Agarose/TBE blend powder (1%) was purchased from Sigma-Aldrich (Saint-Quentin-Fallavier, France). All peptides had 95% or higher purity and were dissolved at a concentration from 10 to 20 mM in sucrose/H_2O (5 : 5, w/w + 10% D_2O, v/v), and agarose gel (1%)/D_2O (9 : 1, v/v).

NMR spectroscopy

All of the NMR experiments on the dipeptide mixture were performed on a Bruker Avance AVIII-600 NMR spectrometer equipped with a 5 mm TCI cryoprobe using the Bruker TOPSPIN Software (Rheinstetten, Germany). Static field gradient pulses were generated by a 10 A amplifier, so that the sample is submitted to a nominal 0.613 Tm^{-1} gradient. Gradient pulses were followed by a 200 µs recovery delay. Temperature control was performed using a Bruker variable temperature (BVT) unit in combination with a Bruker cooling unit (BCU-05) to provide chilled air.

Dipeptide mixture spectra were calibrated so that the tyrosine Hα proton and Cα carbon resonances appeared at 7.00 and 132.00 ppm, respectively. Additional NMR data acquisition and processing parameters for Fig. 1 up to Fig. 7 are described in the ESI S-1 to S-6.†

Conclusions

We have established for the first time that the use of sucrose solution and agarose gel makes the resolution of polar components within complex mixtures possible, by taking advantage of NMR spin diffusion. We have pointed out that using sucrose blends as viscous binary solvents presents valuable advantages compared to the use of agarose gel: spin diffusion is active in a wider temperature range, and NMR sample preparation (no heating) is compatible with thermally fragile compounds and user-friendly. On the other hand, agarose gel presents the considerable benefit of not producing intense residual proton solvent signals due to active transverse relaxation, which is the major drawback of the use of non-deuterated sucrose solution. Mid-sized molecules will only require a small amount of sucrose in water, whereas smaller molecules will require more sucrose up to 50% (w/w) for driving spin diffusion from room temperature to sub-zero temperature. Agarose gel may be a suitable alternative to the use of sucrose solution for the study of small rigid or mid-sized compounds in mixtures presenting quite long correlation times for which spin diffusion should be more active than that of very small and flexible molecules.

The component individualization within a Leu–Val, Leu–Tyr, Gly–Tyr and Ala–Tyr mixture was achieved at 283 K and 273 K in sucrose solution and agarose gel, respectively, by means of selective 1D, 2D ^1H–^1H NOESY. ^1H–^{13}C and ^1H–^{15}N HSQC-NOESY experiments were implemented for sucrose solution since ^{13}C and ^{15}N nuclei were considered as chemical shift markers that increase the spectrum readability, at the expense of a lower sensitivity caused by their low natural abundance.

Future investigations will deal with the study of other polar mixtures of medium-sized molecules for assessing the spin diffusion power of agarose gel and of the solute–solvent interaction.

Conflicts of interest

The authors declare no conflict of interest.

Acknowledgements

We wish to thank CNRS, Conseil Regional Champagne Ardenne, Conseil General de la Marne, and the EU-programme FEDER to the PlAneT CPER project is gratefully acknowledged.

Notes and references

1 K. F. Morris and C. S. Johnson, *J. Am. Chem. Soc.*, 1992, **114**, 3139–3141.
2 K. F. Morris and C. S. Johnson, *J. Am. Chem. Soc.*, 1993, **115**, 4291–4299.
3 K. F. Morris, P. Stilbs and C. S. Johnson, *Anal. Chem.*, 1994, **66**, 211–215.
4 S. Viel, F. Ziarelli and S. Caldarelli, *Proc. Natl. Acad. Sci. U. S. A.*, 2003, **100**, 9696–9698.
5 G. Pages, C. Delaurent and S. Caldarelli, *Angew. Chem., Int. Ed.*, 2006, **45**, 5950–5953.
6 G. Pages, C. Delaurent and S. Caldarelli, *Anal. Chem.*, 2006, **78**, 561–566.
7 S. Caldarelli, *Magn. Reson. Chem.*, 2007, **45**, S48–S55.
8 C. Carrara, S. Viel, F. Ziarelli, G. Excoffier, C. Delaurent and S. Caldarelli, *J. Magn. Reson.*, 2008, **194**, 303–306.
9 M. E. Zielinski and K. F. Morris, *Magn. Reson. Chem.*, 2008, **47**, 53–56.
10 R. Evans, S. Haiber, M. Nilsson and G. A. Morris, *Anal. Chem.*, 2009, **81**, 4548–4550.
11 M. E. Zielinski and K. F. Morris, *Magn. Reson. Chem.*, 2009, **47**, 53–56.
12 J. S. Kavakka, V. Parviainen, K. Wähälä, I. Kilpeläinen and S. Heikkinen, *Magn. Reson. Chem.*, 2010, **48**, 777–781.
13 A. K. Rogerson, J. A. Aguilar, M. Nilsson and G. A. Morris, *Chem. Commun.*, 2011, **47**, 7063–7064.
14 C. Pemberton, R. Hoffman, A. Aserin and N. Garti, *J. Magn. Reson.*, 2011, **208**, 262–269.
15 J. C. Hoch, M. W. Maciejewski and B. Filipovic, *J. Magn. Reson.*, 2008, **193**, 317–320.
16 M. Mobli, M. W. Maciejewski, A. D. Schuyler, A. S. Stern and J. C. Hoch, *Phys. Chem. Chem. Phys.*, 2012, **14**, 10835–10843.
17 N. H. Meyer and K. Zangger, *Angew. Chem., Int. Ed.*, 2013, **52**, 7143–7146.
18 K. Kazimicrczuk and V. Orekhov, *Magn. Reson. Chem.*, 2015, **53**, 921–926.
19 C. Papaemmanouil, C. G. Tsiafoulis, D. Alivertis, O. Tzamaloukas, D. Miltiadou, A. G. Tzakos and I. P. Gerothanassis, *J. Agric. Food Chem.*, 2015, **63**, 5381–5387.
20 K. Zangger, *Prog. Nucl. Magn. Reson. Spectrosc.*, 2015, **86–87**, 1–20.
21 G. D. Poggetto, L. Castañar, G. A. Morris and M. Nilsson, *RSC Adv.*, 2016, **6**, 100063–100066.
22 A. Gierer and K. Wirtz, *Z. Naturforsch., A: Astrophys., Phys. Phys. Chem.*, 1953, **8**, 532–538.
23 M. P. Williamson and D. H. Williams, *J. Chem. Soc., Chem. Commun.*, 1981, **4**, 165–166.
24 A. J. Simpson, G. Woods and O. Mehrzad, *Anal. Chem.*, 2008, **80**, 186–194.
25 H. Farooq, R. Soong, D. Courtier-Murias, C. Anklin and A. Simpson, *Anal. Chem.*, 2012, **84**, 6759–6766.

26 P. Lameiras, L. Boudesocque, Z. Mouloungui, J. H. Renault, J. M. Wieruszeski, G. Lippens and J. M. Nuzillard, *J. Magn. Reson.*, 2011, **212**, 161–168.

27 P. Lameiras and J.-M. Nuzillard, *Anal. Chem.*, 2016, **88**, 4508–4515.

28 P. Lameiras, S. Patis, J. Jakhlal, S. Castex, P. Clivio and J. M. Nuzillard, *Chem.-Eur. J.*, 2017, **23**, 4923–4928.

29 V. R. N. Telis, J. Telis-Romero, H. B. Mazzotti and A. L. Gabas, *Int. J. Food Prop.*, 2007, **10**, 185–195.

30 M. P. Longinotti and H. R. Corti, *J. Phys. Chem. Ref. Data*, 2008, **37**, 1503–1515.

31 C. H. Cho, J. Urquidi, S. Singh and G. W. Robinson, *J. Phys. Chem. B*, 1999, **103**, 1991–1994.

32 M. O. Sonnati, S. Amigoni, E. P. Taffin De Givenchy, T. Darmanin, O. Choulet and F. Guittard, *Green Chem.*, 2013, **15**, 283–306.

33 E. Bismuto and G. Irace, *FEBS Lett.*, 2001, **509**, 476–480.

34 M. Ausubel, R. Brent, R. E. Kingston, D. D. Moore, J. G. Seidman, J. A. Smith and K. Struhl , *Mol. Reprod. Dev.*, 1989, **1**, 146.

35 R. Kim, H. Yokota and S. H. Kim, *Anal. Biochem.*, 2000, **282**, 147–149.

36 A. M. Spring and M. W. Germann, *Anal. Biochem.*, 2012, **427**, 79–81.

37 A. Pastore, S. Salvadori and P. A. Temussi, *J. Pept. Sci.*, 2007, **13**, 342–347.

38 K. Stott, J. Stonehouse, J. Keeler, T.-L. Hwang and A. J. Shaka, *J. Am. Chem. Soc.*, 1995, **117**, 4199–4200.

39 K. Wüthrich, *NMR of proteins and nucleic acids*, Wiley, New York, 1986.

40 K. Stott, J. Keeler, Q. N. Van and A. J. Shaka, *J. Magn. Reson.*, 1997, **125**, 302–324.

41 R. Brüschweiler, C. Griesinger, O. W. Sørensen and R. R. Ernst, *J. Magn. Reson.*, 1988, **78**, 178–185.

42 B. Plainchont, A. Martinez, S. Tisse, J. P. Bouillon, J. M. Wieruszeski, G. Lippens, D. Jeannerat and J. M. Nuzillard, *J. Magn. Reson.*, 2010, **206**, 68–73.

Faraday Discussions

DISCUSSIONS

High resolution techniques: general discussion

Elaine Adair, Carlos Afonso, Nicholle G. A. Bell, Antony N. Davies, Marc-André Delsuc, Ruth Godfrey, Royston Goodacre,  Jeffrey A. Hawkes,  Norbert Hertkorn, Donald Jones, Pedro Lameiras,  Adrien Le Guennec, Anneke Lubben, Mathias Nilsson, Ljiljana Paša-Tolić, Josh Richards, Ryan P. Rodgers, Christopher P. Rüger, Philippe Schmitt-Kopplin, Peter J. Schoenmakers, Philip Sidebottom, Dan Staerk, Stephen Summerfield, Dušan Uhrín,  Pieter van Delft , Justin J. J. van der Hooft, Fleur H. M. van Zelst and Alexander Zherebker

DOI: 10.1039/c9fd90045d

Peter J. Schoenmakers opened discussion of the paper by Ljiljana Paša-Tolić: You state as your objective that you want to achieve spatial resolution on a patch of soil by bringing the sample size down to 1 mm^3. However, given the time required for extraction and (LC) separation, wouldn't it require a horrendous effort to get a good impression of the soil? In light of this, is the very small sample size really a great advantage?

Ljiljana Paša-Tolić responded: This is a question of context, meaning that acquiring spatially resolved data on larger samples using this approach is not practical, but going after specific areas (*e.g.* hot spots) in a targeted fashion is definitely feasible and would provide added value. For instance, the SFE-LC-FTMS approach provides a means to obtain insights into the localization and distribution of organic matter within soil aggregates or on mineral surfaces, as an alternative to the recently demonstrated LDI-FTMS. Faster data acquisition, afforded by a higher magnetic field, frequency multiple detection schemes and data processing, will certainly further improve throughput and our ability to interrogate soil heterogeneity regardless of the method applied.

Christopher P. Rüger said: The presentation nicely pointed out the difference in chemical space addressed by the deployed ionization techniques, such as ESI and LDI. In the context of LDI, it might make sense to work on deploying different matrices (MALDI) also varying the observed chemical space, or inducing certain aspects of sensitivity and selectivity for certain compound classes. Recent developments, such as deploying Proton Sponge (DMAN) for acidic constituents, might be beneficial. Obviously, the benefit of the liquid chromatographic separation has to be taken into account as this information is lost in common LDI approaches/sample preparation.

Ljiljana Paša-Tolić responded: I fully agree.

Ryan P. Rodgers said: If you isolated the CO_2 extract and performed simple C, H, N, O, and S bulk elemental analysis, how representative are the MS results of the bulk elemental ratios? Is the LD biased to more aliphatic or more aromatic species?

Ljiljana Paša-Tolić replied: Unfortunately, we don't have TOC measurements for these samples. However, I do believe that the bias of the SFE extraction is more similar to solvent based extractions than water based extractions. Regardless, the extent of extraction bias is confounded with other biases such as chemical changes during extraction, ionization efficiency, *etc.* Bias is a real issue. We have done several experiments to understand the extraction and measurement bias. Comparing the extractions to C bulk analysis, we find that we don't extract much of the total carbon out of a sample (*i.e.* only 5–10%). So then the question becomes: are our measurements representative of the bulk? If you look under a microscope, you can see all kinds of carbon-containing debris that we are not extracting *e.g.* insect wings, shell, *etc.* The extracts we selected are more representative of the compounds available to microorganisms and diffusion, the more usable subset of chemical compounds in the soil, if you will. We have started to quantify this by spiking "standards" into the soil prior to extraction. Obviously, we find a range of extraction and ionization efficiencies when we do this and rolling this up to an explanation of bias is complicated. We have found the more useful way to discuss bias is to discuss the differences we observe with different extraction protocols, as described herein and elsewhere.

Royston Goodacre asked: How do you actually identify chemical species with FTICR-MS? You claim that you have 25 000 analytes in a typical run, but you can only annotate 6 500 to specific chemical formulae. What do you need to do to identify the 18 500 that are missing? How do we approach 100% annotation in such experiments?

Ljiljana Paša-Tolić responded: Formulae are assigned using a database containing a subset of possible formulae (*e.g.* CHONSP); hence, unassigned analytes are likely not represented in the database employed. Formula assignment is one of the key challenges in this field. The number of possible formulae grows exponentially with increasing mass (especially with the inclusion of heteroatoms). Despite our seemingly endless pursuit of accuracy, resolution and sensitivity, and even with the best performance attainable (21 T FTICR), multiple formulae assignments are unavoidable at present (and in the foreseeable future), requiring additional decisions to be made typically based on chemical feasibility. Improved formula assignment algorithms and better assessment of confidence are desperately needed, as are orthogonal methods for high-throughput molecular characterization of these extremely complex mixtures.

Royston Goodacre enquired: Are there significantly more than 25 000 features that you call analytes? Do you look at adducts and reduce the number of features prior to annotation?

Ljiljana Paša-Tolić replied: The reported number of LCMS features included isotopic peaks detected in at least 3 consecutive spectra in LCMS analyses. Relaxing peak picking parameters and/or LCMS feature definition would undoubtedly produce a higher number of features. However, differentiating between true analytes and data processing artefacts arising from additional low intensity LCMS features would be extremely challenging. Adduct formation was not considered prior to annotation. All peaks were assumed to originate from singly charged deprotonated molecules.

Dušan Uhrín said: You used databases to assist with the formulae assignment. Is this not a limiting factor? Could there be compounds that are not in the databases? Could homologous series help to identify more formulae? An assignment rate of 80% is typically reported these days for (−) ESI. What is the reason for your lower conversion?

Ljiljana Paša-Tolić responded: Yes, that is correct: unidentified analytes are likely not represented in the database used for formula assignment. The typical database represents a limited subset of possible formulae (*e.g.* CHONSP), resulting in limited assignment proficiency. Herein, we used the Compound Identification Algorithm (CIA), with mass accuracy of 1 ppm and a formula propagation with CH_2, H_2 and O building blocks. The lower assignment rate in this case suggests that analytes extracted by SFE are not included in the chemical databases used for formula assignment or include elements not considered in the formula assignment algorithm.

Donald Jones remarked: You have shown in previous studies that different extraction methods predictably provide different results. Does this new methodology result in higher levels of assignments?

Ljiljana Paša-Tolić responded: Correct, we have previously used direct infusion ESI-FTICR to characterize these same peat samples using liquid extraction (LSE) with multiple solvents.[1] This study served as a reference for evaluating the SPE approach introduced herein. In a single SFE-LC-FTICR MS analysis using 8 mg of peat soil, we have detected ~25 000 LCMS features and assigned ~6500 formulas. This compares favorably to the 5800 total formulas that were assigned for the same sample on the same mass spectrometer using LSE separately with four solvents (methanol, acetonitrile, water and hexane). Obviously, the LCMS-based method detected more analytes and assigned more formulae relative to the direct infusion method. However, the point here is that we obtained higher coverage using nearly an order of magnitude smaller soil sample (*e.g.* 8 mg for SFE *vs.* 100 mg for LSE).

1 M. M. Tfaily, R. K. Chu, N. Tolić, K. M. Roscioli, C. R. Anderton, L. Paša-Tolić, E. W. Robinson and N. J. Hess, *Anal. Chem.*, 2015, **87**, 5206–5215.

Stephen Summerfield commented: It is wonderful to see where SFE has gone since I used the method to extract PAHs for the Gas Research Centre from coal gas contaminated soils back in 1994 when I validated the process with Soxhlet Extraction, which took 6 h each time. Did you use any modifiers such as methanol, acetonitrile, acetone, *etc.*? This gives an increase in the range of substances that can be extracted.

Ljiljana Paša-Tolić responded: No, we have not used any modifiers in this study. It is worth mentioning that the SFE-LCMS system we employed was designed for operation at higher temperatures (up to 150 °C) and with modifiers. Hence, using these conditions could potentially extend analytical coverage by changing chemical selectivity. For instance, the addition of water or other polar solvents could improve coverage for tannins, cellulose and amino sugars. This would be an interesting future study.

Marc-André Delsuc opened discussion on Alexander Zherebker's paper: This was a very interesting presentation. With your deuteration protocol you are adding complexity to an already very complex sample, because of unavoidable partial deuteration. How do you deal with the increased complexity and properly assign the spectrum? Have you tried to push the reaction to obtain a 100% deuteration yield to simplify the spectrum this way?

Alexander Zherebker answered: Thank you for the question. We don't directly assign peaks, which correspond to deuterium ions. Instead we juxtapose mass-spectra of parent and labeled samples, and extract exchange series for intense peaks with the mass tolerance 0.5 mDa, which is routinely provided by our FTICR MS instrument. Answering your second question, no, we didn't try to obtain 100% deuteration yield because currently we need a peak distribution to assign exchange series.

Jeffrey A. Hawkes commented: You count the maximum number of deuterations for an accurate mass, and state that number in the data tables (*e.g.* Table 2 in the paper) and in the discussion. From Fig. 3 in your paper, it actually seems that there is a Gaussian-like distribution of the number of deuterations, with maximum peak intensity at 3–4 deuterations, but a maximum number of deuterations at 6 for this example. This seems to fit more with the rest of the literature regarding these complex mixtures: that there is a range of isomers with different chemistries (assuming that all deuterium exchanges are fully complete). Why do you discuss the results only referring to the maximum number of deuterations found?

Alexander Zherebker replied: That is a very good point. Actually, in our experiments with models or humic substances, we don't obtain fully deuterated compounds. That is why we observe a distribution of peaks. Since in that case we can't distinguish between partially deuterated compounds and isomers with a lower amount of active sites, we focused on the maximum number of exchanges.

Dušan Uhrín asked: The conditions of your deuterium exchange are harsh - high temperature and extreme pH values. Did you do a control experiment using non-deuterated substances? What was the outcome? Did you see any chemical modifications?

Alexander Zherebker responded: In order to verify the results of the experiments, we performed blank and labeling experiments on the model compounds and the synthetic humic-like complex mixture.[1,2] We didn't observe any significant modifications of molecular composition. We believe that the usage of sealing tubes for exchange prevents oxygen excess and, consequently, oxidation of humic material. Also, esters are not abundant in alkali-extracted humic fractions. Therefore we don't expect a hydrolysis under basic or acidic conditions.

1 A. Y. Zherebker, D. Airapetyan, A. I. Konstantinov, Y. I. Kostyukevich, A. S. Kononikhin, I. A. Popov, K. V. Zaitsev, E. N. Nikolaev and I. V. Perminova, *Analyst*, 2015, **140**, 4708–4719.
2 A. Zherebker, Y. Kostyukevich, A. Kononikhin, V. A. Roznyatovsky, I. Popov, Y. K. Grishin, I. V. Perminova and E. Nikolaev, *Analyst*, 2016, **141**, 2426–2434.

Christopher P. Rüger commented: In this contribution, selective isotope exchange high-resolution mass spectrometry is deployed for the structural elucidation of coal humic substances. In particular, common structural motifs, such as PAHs, are classified. In this respect, it would make sense to compare the gathered results to standard pyrolysis methods deployed for the description of complex solid sample materials. Pyrolysis gas chromatography mass spectrometry (or FID) might be an excellent method for the validation of the presented techniques, as it is able to degrade the solid material and fully quantify the generated pyrolysis products, in particular, the aromatic constituents.

Alexander Zherebker answered: I agree that pyrolysis gas chromatography mass spectrometry might be an excellent complimentary method to verify the pattern of aromatic ring substitution, and we definitely want to try it. However, particular attention should be paid to secondary transformations which may occur during pyrolysis of functionalized aromatic components comprising coal humic substances.

Donald Jones remarked: In your paper you conclude that the van Krevelen plots do not show significant differences between the samples. The Venn diagrams indicate that there are significant differences between the samples. Given the significant sampling differences between the samples it would be presumed that the differences would be significantly different. Could you expand on this?

Alexander Zherebker answered: The Venn diagram showed that the samples possessed unique molecular assignments. However, in all cases, more than 50% of formulae were common for all samples. This can be related to the common evolutionary processes in coal even of different origins. By usage of selective labeling, we have shown that differences may lay on the structural level and it should be taken into account when comparing natural samples.

Norbert Hertkorn commented further: Commonly we relate visible/obvious similarity to similarity of intrinsic causes. In mass spectra this leads to misconceptions, because *e.g.* any molecular composition/mass within any nominal mass cluster represents entirely different compositions and hence, structures as well. Similarly, even barely visible displacements within van Krevelen diagrams actually means drastic differences in molecular compositions (and structure). It would be nice if we could find more instructive means to describe these molecular differences, although it might be difficult to find pleasing and easily comprehensible diagrams.

Dušan Uhrín asked: Is it possible to automate the analysis of the spectra of deuterated compounds? Is the mass difference of deuterium sufficient to avoid the CHO peaks? How do you relate the parent peaks to the peaks of deuterated compounds?

Alexander Zherebker replied: The major challenge of the labeling experiments presented in this work is the molecular diversity of coal humic substances. The problem is aggravated when high amounts of deuterium atoms are incorporated. Still, in the case of a synthetic humic-like mixture, we could easily distinguish CHO and labeled ions. Unlike natural humic samples, they aren't comprised of CHO compounds exclusively. They contain N and S-containing molecules. Thus, automatic assignment of deuterium peaks requires higher resolution power than provided by a 7 T ICR instrument in broadband acquisition mode. At the same time, we can juxtapose mass-spectra of native and labeled samples and automatically extract peaks, which could lay in exchange series with the error up to 0.5 mDa. Furthermore, we manually select peaks that possess a monomodal distribution, which by our experience is a good approximation for H/D exchange of humic substances. Currently, we are working on the algorithm, which would include evaluation of both intensity and error distributions for automatic enumeration of exchange series.

Dušan Uhrín communicated: When interpreting the spectra of deuterated compounds, do you work with the original spectrum, or do you use a spectrum of treated compounds obtained using protonated substances?

Alexander Zherebker communicated in response: In all cases we worked with the original mass spectrum. Experiments on model compounds and synthetic humic-like mixtures showed a lack of collateral reactions, which could significantly change the molecular composition of coal humic samples. Nevertheless, we believe that even in the case of alteration of humic samples by NaOD or DCl, the calculation of H/D exchange series by comparison with the mass spectrum of the parent sample would enable us to avoid false interpretation.

Antony N. Davies remarked: For molecular weight distribution, we should not lose sight of complex systems being in dynamic equilibrium rather than static systems. How representative is it to identify so many substances (25 000)?

Alexander Zherebker responded: When working with the lab samples, we avoid all processes related to evolution of organic carbon in nature. In that case, we deal with static systems. Nevertheless, the identification of as many compartments as possible is important to study pathways of organic matter transformation, which is particularly important in cases of high rates of such transformation. The most typical example is the study of permafrost. The conserved organic matter is rapidly consumed and transformed by microbial communities upon thaw. Mass-spectrometry, especially FTICR MS, provides insights on organic compartments that are susceptible to biodegradation. Otherwise, I believe that in cases of samples isolated from coal, which is hundreds of millions of years old, the rates of transformation are extremely low and lab samples are relevant to the source.

Adrien Le Guennec opened discussion of the paper by Nicholle G. A. Bell: In the HSQC, since the resolution in the ^{13}C is usually low, some peaks that seem to have the same chemical shift in the ^{13}C dimension could in theory be differentiated by improving the resolution in this dimension. Did you see or suspect several cases like this in your studies?

Nicholle G. A. Bell replied: Indeed, resolution of the carbon-13 dimension can be significantly improved, as demonstrated in previous work.[1] However, for identical C-13 chemical shifts, the methods I presented are a viable solution.

1 L. Castañar and T. Parella, *Magn. Reson. Chem.*, 2015, **53**, 399–426.

Mathias Nilsson asked: You say that as the dimensions go up, the resolution goes down – can you explain this? 3D or 4D experiments normally increase resolution.

Nicholle G. A. Bell answered: The need to sample multiple dimensions in a 3D experiment inevitably reduces the achievable resolution along the individual indirect dimensions. Unless the resolution improvement of a 3D is achieved by moving the cross peaks from a 2D plane to 3D space, the resolution, *e.g.* of the carbon dimension of the 3D experiment, will always be lower than in the corresponding 2D experiment (achieved in the same time).

Norbert Hertkorn said: Natural organic matter (NOM) features abundant carboxylic groups and is subject to relevant interactions between molecules. This causes acceleration of transverse NMR relaxation and concomitant loss of signal amplitude. Plant extracts may behave differently; can you comment on some relevant differences between NMR spectra of dissolved NOM and those of common plant extracts?

Nicholle G. A. Bell answered: Typical plant extracts may contain hundreds of molecules, while NOM may contain many thousands. Therefore the NMR detection threshold for individual compounds is rarely met for unfractionated NOM. To make things worse, the propensity for molecules to interact and therefore lower the T_2 values is naturally larger in more complex mixtures. It may well be that carboxylic groups play a role in this.

Elaine Adair asked: You mention that there is a loss of sensitivity in CLIP-COSY compared to HSQC-TOCSY; what do you think is causing this loss in sensitivity?

Nicholle G. A. Bell responded: The reason for the loss in sensitivity in the HSQC-CLIP-COSY compared to the HSQC-TOCSY is the evolution of passive ^{1}H–^{1}H coupling constants during the CLIP-COSY sequence.

Josh Richards asked: In your (3,2)D ^{1}H ^{13}C HSQC experiment, the number of cross peaks is doubled due to the splitting. In a complex mixture, can you/would it be advantageous to analyse individual carbon signals to resolve the overlap?

Nicholle G. A. Bell answered: The editing process returns two separate spectra with the same number of cross-peaks as the original HSQC. This gives an opportunity to work with either of these spectra and choose the one which removes the overlap of the carbon resonances.

Royston Goodacre opened a general discussion of the papers by Ljiljana Paša-Tolić, Alexander Zherebker and Nicholle G. A. Bell: Can you each briefly discuss sampling error – each of you are using complex sample pretreatment (SFE, HDX) and high resolution instruments, yet you measure a crumb of soil on a millimeter scale, a nugget of coal in a field with seams of coal kilometres deep, and variation in plants (geography and climate) that are extracted with hot water and end up in a cup of green tea. So which is best – more sub-sampling of diverse samples with lower resolution? Or one-off samples with ultra-high resolution modalities?

Ljiljana Paša-Tolić responded: Again, this is a question of context. A soil sample can be homogenized to generate a global chemical profile and this might be enough to answer the question at hand. In other instances, however, spatially resolved measurements will be required *e.g.* to get insights into localization and distribution of organic matter within the heterogeneous soil matrix and better define biotic/abiotic interactions.

Nicholle G. A. Bell answered: The tea was bought from a company and we measured multiple samples from different batches and got identical NMR spectra. Therefore I cannot comment on variations due to geography or climate. When we take soil samples, we take a large number of soil samples from a given site following the standard procedures for field sampling. This takes into account many factors such as topography, vegetation, *etc.* We also maintain the sample integrity by maintaining sample conditions of temperature and atmosphere. We are using low and high resolution techniques to collect data from hundreds of samples to determine the differences. This is done on whole samples and extracts.

When we are developing experimental methods to determine the individual molecules in a complex mixture, it is always best to use a few samples to develop the methodology using the most powerful techniques one has. Once created, one can then refine the methodology and apply it to a higher number of samples, depending, of course, on the research question.

Alexander Zherebker responded: I cannot comment on the sampling error because we didn't do the sampling. But I know that all samples were obtained by following standard analytical procedures including randomization, quartering, *etc.* So, samples are supposed to be well-representative of the source. Nevertheless, I agree that currently complicated treatment and ultra-high resolution methods cannot be applied for a large amount of samples. But it doesn't mean that new state-of-the-art methods shouldn't be developed.

Ruth Godfrey addressed Ljiljana Paša-Tolić, Alexander Zherebker and Nicholle G. A. Bell: The accuracy of sample representation would require measurements *in situ* – offline (away from the site of origin). How accurate are your measurements?

Ljiljana Paša-Tolić responded: I don't really know because there are no means at present to perform these measurements *in situ* (as far as I am aware).

Nicholle G. A. Bell replied: When taking soil samples we preserve them in such a way to maintain the conditions (atmosphere and temperature) and run the measurements as soon as possible. Others have tested the changes in pH and DOC that occur if the samples are not maintained in the ways mentioned above, and they saw differences, which were reflected in the NMR or MS spectra. Therefore, we believe that our measurements are as accurate as possible.

Alexander Zherebker responded: Of course, the development of *in situ* measurement strategies is an important task, and surely the molecular organisation of humic components in nature differs from the lab sample. Nevertheless, we should take into account that components are bonded by noncovalent interactions. Moreover, humic components are small and we don't talk about their conformation. Therefore, samples purified for lab experiments accurately represent a part of the source, which was successfully extracted.

Donald Jones addressed Alexander Zherebker and Ljiljana Paša-Tolić: You have shown a lot of common assignments between samples and I wonder if these assignments represent low-lying fruit, meaning the differences that are unassigned and unique to samples are actually just more difficult to assign.

Ljiljana Paša-Tolić responded: I fully agree.

Alexander Zherebker replied: In our experiments, we were limited to electrospray ionizable compounds, which likely represent only a part of the whole sample. Nevertheless, intensity-weighted average elemental composition collaborated well with the bulk elemental analysis. Moreover, in the present study we aimed to show that even common assignments actually correspond to different structures, which add a new dimension for the comparison of humic components.

Anneke Lubben said: There has been a lot of discussion about the number of ions detected and compounds identified. But why are we doing this? Do we really need to see everything? Can we ever really see everything? I propose that as a community we need to focus on what information is actually needed to answer the research question, and that it is more important to accurately define the research question so that we establish what information is actually needed to answer this question, rather than trying to get more resolution, more dimensions of separation, or to identify more compounds. I would welcome the panel's thoughts on this.

Ljiljana Paša-Tolić replied: I agree. Our goal here was to develop an approach to provide comprehensive chemical representation of a small soil sample and enable more effective studies of hot spots of interest. While the nature of these hot spots will depend on the specific question asked, the need for comprehensive molecular characterization is universal.

Dušan Uhrín asked: Can I be confident that people's formulae assignments based on FTICR MS spectra are correct, especially when heteroatoms such as N, S and P are involved?

Ljiljana Paša-Tolić answered: Formula assignments based on accurate mass only are by definition tentative. Formula assignments were used to broadly classify detected analytes in lipid-like, lignin-like, *etc.* classes based on their O/C and H/C ratios to *e.g.* enable comparison between different extraction approaches. Having the ability to resolve (detect) isotopologues (*i.e.* fine isotopic structure), as offered by ultra-high resolution MS, would improve confidence in formula assignments based on mass only. Additional measurements, such as tandem MS, collisional cross-section, NMR, *etc.* are needed for confident structural identification of analytes and are at present performed only in a targeted fashion.

Stephen Summerfield addressed Nicholle G. A. Bell and Mathias Nilsson: I am a chromatographer and involved in defining substances on behalf of industry for EU regulations such as REACH. The ECHA (European Chemicals Agency) considers NMR one of the required methods for identity (UV-Vis, FT-IR and MS) but not for quantification. People are increasingly using NMR to quantify substances, which is not comparable to chromatography. Can you demystify this? Many of these substances are not thermally stable, they hydrolyse quickly or they suffer from re-arrangements/cyclisation in a mass spectrometer. Additionally, there are substances that are ionised at different pH and this would change NMR shifts, I would assume. Companies have only supplied me with the raw ^{1}H-NMR, ^{13}C-NMR, Si-NMR or P-NMR *etc.*, without peak tables – no 2D-NMR or other methods, and they certainly do not explain how they have interpreted the data to give the quantification of complex mixtures of 3–20 constituents. Do you consider this sufficient to defend their quantification of their substance?

Nicholle G. A. Bell responded: In my mind (and I am not a chromatographer) you need to have a calibration curve for quantification if you are doing chromatography. However, because 1D ^{1}H-NMR is inherently quantitative, quantitative analysis by NMR can be achieved by using a suitable inert compound of known concentration mixed with the sample (external is also possible). This is straightforward if the signal of interest is well-resolved and not overlapping with other resonances. Deconvolution of overlapping signals is possible but this may lower the accuracy. Quantification by 2D NMR methods is inherently more difficult and there is a lot of literature on this topic, but it seems that none of the methods are yet well-established. I agree with you that without knowing the protocols, interpretation, and potentially some error estimates, I would not take such results at face value.

Mathias Nilsson answered: NMR is inherently quantitative, in a correctly set up pulse-acquire experiment: the signal strength (integral) is directly proportional to the number of nuclei contributing to a given resonance. It is therefore possible – and common – to use NMR for quantitation purposes, if appropriate protocols are followed for data acquisition and analysis. However, spectra acquired under routine conditions for identification purposes are not in general suitable for quantitation. To have any evidential value, quantitative spectra need to be accompanied by the details of the data acquisition and analysis methods used.

Marc-André Delsuc said: One word has been missing so far in the current discussion about quantitation: "calibration". In NMR, you don't need calibration because, in a given experiment, all of the signals are proportional to the amount of matter present. There is a "sensitivity coefficient", and this is in contrast with most other methods, which require calibration on a reference sample. This is the reason why quantitativity is a sort of non-issue in NMR.

Antony N. Davies replied: In the real world, inter-laboratory studies have shown that, even for NMR, you need to be very careful in assuming all signals are proportional. Many laboratories have relatively fixed instrument operation parameters that may not be optimized for all samples, even when they are relatively small simple molecules in mixtures. It is also possible when analysing large numbers of samples that the spectroscopist fails to spot when automated peak integration throws up questionable results. I worry about the assumption that it is a non-issue.

Carlos Afonso responded: In mass spectrometry, quantitation is indeed not as direct, as signal response depends on the analyte and on experimental conditions (ionization source, temperature, solvent, *etc.*). For all of these reasons, an external or internal calibration procedure is required. With complex matrices, the so-called "matrix effect" has to be taken into account and is typically solved with internal calibration (*e.g.* isotopically labelled calibrant, standard addition method). On the other hand, MS has other important advantages, the most important being its extreme sensitivity and selectivity.

Peter J. Schoenmakers responded: Is quantitation always a non-issue in NMR? I am willing to accept that signal intensity is directly proportional to the number of protons (in ^{1}H-NMR of liquid solutions), but in cases of complex samples, it is not always easy to distinguish between signals of different constituents of the samples.

Philip Sidebottom opened discussion of the paper by Dan Staerk: In your paper, you identified 17 known compounds by comparison of their spectroscopic data with that published in the literature. Please could you tell us about the dereplication process that led you to these structures?

Dan Staerk responded: Our dereplication process is not automated, and relies on establishing the molecular formula of the material eluted with each peak and searching (in Scifinder, Reaxys or alike) for NMR, MS and UV data to support the identification. This is without doubt a process we could – and should – work on automating in the future.

Marc-André Delsuc commented: I am impressed with the bioassay presented in your last slide, or in Fig. 3 in your paper, for instance. It seems to be very fast and efficient, and you have a very fine time resolution. Is it performed on-line or off-line?

Dan Staerk replied: All of these are performed off-line, which allowed us to optimise the HPLC conditions without taking care of the percentage of organic modifier – which is a problem in direct on-line HPLC-bioassay. It furthermore allowed us to collaborate with whoever has a bioassay implemented in 96-well microplate format.

Nicholle G. A. Bell said: Fig. 5 in your paper shows the HPLC chromatogram monitored at 254 nm and 210 nm, as well as the inhibition profiles. A number of peaks are shown in the inhibition profile that have been assigned to specific compounds. However, there is an unresolved hump just above 25 min retention time, which has a strong HPLC absorbance at 210 nm and is not assigned or commented on in the paper. Can you comment on this?

Dan Staerk replied: Yes, this hump contained unresolved polymeric compounds – most likely tannins – and was therefore not further investigated.

Donald Jones remarked: Fig. 2 in your paper shows that a pattern of responsiveness for the inhibition assays is clearly present. Can you explain these patterns? How are the inhibition cut-offs determined?

Dan Staerk answered: I agree that these threshold levels are somewhat arbitrary, and that could of course have been chosen differently. Thus it is mainly based on our past experience with screening a large number of crude extracts, and the fact that we prefer a higher cut-off for α-glucosidase inhibitors than for PTP1B inhibitors, since we discover the former more frequently and since there are no currently approved drugs for the latter. Therefore, new leads for PTP1B inhibitors are relevant even with a lower inhibition at the chosen concentration.

Philippe Schmitt-Kopplin said: Plant extracts are extremely complex biological molecular mixtures. Can you detail your chromatographic approach and describe what combinations of separations you use to reduce this complexity to single compounds?

Dan Staerk responded: Indeed crude plant extracts are very complex, but even though working with crude extracts creates problems, we rather prefer this than fractionating into many fractions that should be separated with individual separation methods. Thus, as a standard first approach, we develop separation methods using analytical-scale reversed-phase HPLC with 3-micrometer particles (C_{18}). This typically results in prolonged separations of between 45 and 90 min, where the majority of the peaks can be satisfactorily separated. If there are still larger regions with unresolved peaks, or if peaks in the biochromatogram are correlated with unresolved HPLC peaks, we perform semi-preparative or preparative-scale HPLC to isolate these fractions for subsequent separation on the same or a complementary stationary phase with analytical-scale HPLC, *e.g.* often a pentafluorophenyl phase if we expect compounds with aromatic moieties.

Justin J. J. van der Hooft asked: Do you have an idea about how much material you need to isolate from the crude extract to perform successful NMR experiments for complete structural elucidation? And what amount of the isolated molecules do you need for the different bioactivity tests? Do these correspond to each other?

Dan Staerk replied: Depending on the size of the molecule and its behaviour (*i.e.*, relaxation time), the 1.7 mm cryogenic TCI probe allows structure elucidation in the microgram range. This is for many bioassays not enough for medium-active compounds, where you need to be in the high-microgram range to establish IC_{50}-values accurately in triplicate. If you furthermore need to identify the mode of action, *e.g.*, competitive or non-competitive inhibition, you need even more and you need to do semi-preparative or preparative-scale isolation.

Norbert Hertkorn queried: Why do you not use the 1.7 mm Bruker Match tube (employed initially in your 1.7 mm probehead for ^{1}H detected NMR spectra) directly for acquisition of ^{13}C-NMR spectra in 5 mm probeheads? You will have losses when transferring the sample to 2.5 mm tubes as you will have losses when inserting a 1.7 mm tube into an empty 2.5 mm tube. The costs for the exchangeable Match inserts are relatively low.

Dan Staerk replied: Originally we started putting the 1.7 mm tubes in the 2.5 mm tubes because we had the latter available – and because with the 5 mm cryogenic DCH probe we obtained excellent S/N. However, I acknowledge your arguments about the loss of sensitivity and I have now ordered Bruker 1.7 mm inserts to explore the difference between the two methods.

Norbert Hertkorn asked: How often do you discover new compounds that are not previously known? What kind of novelty is more common than another; is there a credible ranking of novel structures?

Dan Staerk responded: On average we discover between 25–50 not previously described ("new") compounds every year. The "novelty" is most often related to *e.g.* different glycosylation, prenylation, and methoxylation/hydroxylation patterns; but of course, there can also be new combinations of known skeletons as well as truly new skeletons – although the latter is rare. To me, it is difficult to rank "novelty": either a compound is "new" (not previously described) or known (previously described) – although I clearly prefer publishing new skeletons rather than new hydroxylation patterns. Moreover, "pharmacological novelty" is even more important for me since my interest is bioactive natural products. Thus, "new" bioactive compounds are preferable, but even previously described compounds with not previously described pharmacological activity is new knowledge and therefore of interest to me.

Justin J. J. van der Hooft enquired: How many of the 200–300 novel molecules you fully elucidate yearly do truly have unique chemistry as compared to molecules that are analogues of ones you elucidated before? As especially for truly novel scaffolds, your platform is very powerful!

Dan Staerk replied: I think you misunderstood. I do not identify 200–300 "new" (not previously described) molecules every year – it amounts to, on average, 25–50 "new" compounds per year. Clearly, as also mentioned in my answer to the previous question, "new" compounds with truly novel skeletons are identified less frequently than "new" compounds that are merely different glycosylation, prenylation, and methoxylation/hydroxylation patterns of known skeletons. Indeed I agree that the hyphenated HPLC-PDA-HRMS-SPE-NMR with 1.7 mm volume-matched cryogenic TCI probe heads is excellent for establishing the structure of compounds with novel skeletons – trapped directly from crude extracts. But it also takes some luck to find the truly novel skeletons.

Mathias Nilsson opened discussion of Fleur H. M. van Zelst's paper: I noted that with your setup, you can only get a simple ^{1}H at the moment, due to restrictions in sensitivity. Is that enough for your current needs, or to what extent would it be useful to be able to acquire mode advanced NMR experiments?

Fleur H. M. van Zelst answered: The microfluidic stripline probe that we use in this paper also allows for 2D NMR. It is a double resonance probe with a ^{1}H and ^{13}C channel. However, due to the concentrations we obtain here, using natural abundance samples, ^{13}C-NMR spectra would take too long to acquire. However, 2D proton NMR experiments are possible. We did acquire COSY and TOCSY spectra for the reference samples (not shown in the manuscript), but for the specific SFC-NMR discussed in the paper it would take a long experimental time. To distinguish between the stereoisomers in this work, more advanced experiments were not needed. Improvements in the experimental setup will make the analysis time shorter (better matched probe detection volume, trapping the sample multiple times, higher external magnetic field, hyperpolarization (DNP)), and therefore will make 2D NMR more feasible at these concentrations.

Adrien Le Guennec asked: The useful volume for NMR acquisition is trapped between two different phases, one liquid and one gaseous, each having different magnetic susceptibilities. How difficult is it to shim the sample in these conditions?

Fleur H. M. van Zelst replied: Since we are using a narrow capillary running over our stripline detection chip, the different phases are not present at the same time on the RF chip. In most of the SFC-NMR experiments, we stop the sample "plug" in the middle of the strip, so the surrounding environment is homogeneous. The shimming is therefore easy to accomplish. Work has been done in our group to further optimise the shimming for stripline chips, since shimming these small volumes can be challenging with regular shim coils.[1]

1 S. G. J. van Meerten, P. J. M. van Bentum and A. P. M. Kentgens, *Anal. Chem.*, 2018, **90**, 10134–10138.

Dušan Uhrín asked: The active volume of your NMR probe is 150 nL. What signal to noise ratio have you achieved? How does it compare with a bigger flow probe? What is the advantage of a smaller volume?

Fleur H. M. van Zelst responded: The absolute limit of detection of the 150 nL stripline detector is 0.1 nmol ^{1}H spins per square-root receiver bandwidth. Practically, this means that the concentration LOD is 67 mM for a spectrum with 10 kHz band-width, in a single scan, for a S/N of 1 in the FID. In 500 scans, the LOD is then 3 mM. This is slightly better than a standard cryoprobe (for the same amount of spins).

The advantage of having such a small volume probe is the enhanced sensitivity due to a better filling factor. Of course, the probe needs to match the sample volume. In our case of sample after in-line concentration, the sample volume is between 5 µL and 40 µL, depending on the amount of co-solvent (polarity of the compound). To only detect 150 nL of this sample is indeed sub-optimal, therefore the detection volume of the stripline chip was scaled up to detect ~1.5 µL. By upscaling, the absolute sensitivity decreases, but since more spins are detected, the S/N improves. We have determined a factor 3.9 signal increase for the 1.5 µL flow probe, compared to the 150 nL probe, simply by detecting more of the sample that was already present. The probe is now ideal for the 5 µL sample volume, but could be upscaled more for a 40 µL sample volume.

Compared to other commercially available flow probes, the sensitivity of the stripline detector is comparable to the solenoid coils often used in these probes. However, the design of the stripline is more practical for microfluidics, and due to the use of a capillary to flow the sample inside the probe, NMR experiments at high pressures can be performed on the stripline chip. The other advantage of using a stripline design is that no special measures (such as susceptibility matching) need to be taken to get a high resolution.

Donald Jones said: It is clear that SFC has advantages for both polar and non-polar compounds. Why isn't it more commonly utilised in more laboratories?

Fleur H. M. van Zelst replied: SFC is often still regarded as the technique it used to be when it was first developed. SFC was then only used with pure CO_2, without co-solvent, and could only separate non-polar compounds. Since then, and especially in the last decade, SFC has improved a lot; not only have the use of (polar) co-solvents and more column chemistries been introduced, but also the equipment (such as the back-pressure regulation) has improved. This means that SFC is now widely applicable, including for more polar compounds.

Also, the use of SFC is often regarded as more complicated then (HP)LC, but it is based on the same principles. However, a lot more publications and examples of separations are known for (HP)LC, so the method development time can be shorter, by looking up the conditions for similar compounds. More and more examples of SFC separations are being published nowadays, so hopefully in the future, more laboratories will consider SFC, since the analysis time is often much shorter.

Norbert Hertkorn asked: Do you have experience with CO_2 directly as a supercritical fluid in NMR? What would you expect?

Fleur H. M. van Zelst replied: In previous experiments we have looked at the expansion behavior of CO_2 in-line, to be able to separate the sample from the CO_2 mobile phase after SFC.[1] Although CO_2 is technically not supercritical below 31 °C, but subcritical, we have done NMR experiments on this (at room temperature). The difference is visible in the relaxation time of molecules in super-/subcritical fluids. In supercritical fluids, faster relaxation occurs, due to the lower viscosity and higher diffusivity compared to fluids. In our group, we are trying to exploit this for Overhauser DNP, to enhance the NMR signal.[2]

1 F. H. M. van Zelst, S. G. J. van Meerten, P. J. M. van Bentum and A. P. M. Kentgens, *Anal. Chem.*, 2018, **90**, 10457–10464.
2 S. G. J. van Meerten, M. C. D. Tayler, A. P. M. Kentgens and P. J. M. van Bentum, *J. Magn. Reson.*, 2016, **267**, 30–36.

Mathias Nilsson opened discussion of the paper by Pedro Lameiras: With respect to efficiency of spin diffusion and resolution, could you comment on the tunability and optimisation? How do you find an optimum?

Pedro Lameiras replied: Thank you for your question. We determined operating conditions at which the best compromise between spin diffusion efficiency and NMR spectral resolution was reached by means of band-selective detection and usual NOESY experiments, respectively, in sucrose solution and agarose gel. Since the temperature is an important parameter in spin diffusion experiments, because it is directly related to solvent viscosity and consequently to overall rotational correlation times τ_c, we determined the optimal temperature at which NOESY cross peaks were positive (negative NOE enhancements and slow motion regime), well-resolved, and as intense as possible between nuclei that were not supposed to be close enough to present a NOE in low viscous medium. The optimized temperatures of 283 K and 273 K were chosen, respectively, in sucrose solution and agarose gel.

Mathias Nilsson asked: Could you comment on the quantitative issues of your experiments?

Pedro Lameiras replied: Thank you for your second question. Until now, we have been taking benefit from spin diffusion by only considering qualitative aspects. Depending on the proton density of molecules of interest, the magnetization exchange may turn out to be drastically altered. The higher the proton density, the more spin diffusion may be efficient under viscous operating conditions. Nonetheless, the integration of proton resonances from the individualized NMR spectrum of each component in mixture will be biased. In the future, we will try to assess the quantitative aspect of spin diffusion in terms of internuclear distances under viscous conditions by considering standard molecules for which internuclear distances will be well-known under non-viscous conditions.

Philip Sidebottom said: You mentioned in your paper that previous work has been published using other solvent systems to give the same type of effect. Under what circumstances would you use the sucrose/agarose gel systems you talked about today rather than those previous systems? I would have expected that a water/DMSO system would have worked for your peptides. DMSO dissolves a very wide range of molecules – it is the solvent of choice for delivery of compounds to biological screens in both the pharmaceutical and crop protection industries.

Pedro Lameiras responded: We will recommend the use of sucrose solution and agarose gel for studying mixtures made of polar compounds. Mid-sized molecules will only require a small amount of sucrose in water whereas smaller molecules will require more sucrose up to 50% (w/w) for driving spin diffusion from room temperature to sub-zero temperature. Agarose gel may be a suitable alternative to the use of sucrose solution for the study of small rigid or mid-sized compounds in mixtures presenting quite long correlation times for which spin diffusion should be more active than that of very small and flexible molecules.

We have published work in the past where viscous solvent blends such as DMSO-d_6/glycerol and DMSO-d_6/water were also dedicated to the investigation of apolar (natural products) and polar compounds (dipeptides) in mixture.[1,2] Fig. 1 in this discussion presents the advantages and drawbacks of viscous solvents (or solvent blends) in mixture analysis by NMR investigated in our research team.

1 P. Lameiras and J. Nuzillard, *Anal. Chem.*, 2016, **88**, 4508–4515.
2 P. Lameiras, S. Patis, J. Jakhlal, S. Castex, P. Clivio and J. Nuzillard, *Chem.–Eur. J.*, 2017, **23**, 4923–4928.

Marc-André Delsuc remarked: Very nice talk, thank you. I am surprised that you don't observe intermolecular NOE. Considering the concentrations at which you are operating (on the order of 10 mM to 100 mM if I am right), there could be some transient molecular interactions which would lead to such a signal. Such intermolecular spin diffusion is sometimes used for instance to explore protein ligand interaction. Can you comment on this?

Pedro Lameiras responded: At a concentration from 10 to 20 mM, we did not observe intermolecular NOE between molecules of interest. For sure, we can expect to detect these later at much higher concentration. Nonetheless, we always observed intermolecular NOE between the molecules of interest and the viscous

	Glycerol	Glycerol carbonate	DMSO-d6/Glycerol	DMSO-d6/Glycerol-d8	DMSO-d6/H$_2$O	Sucrose/H$_2$O	Agarose Gel
Sample preparation	✗	✗	✓	✓	✓	✓	✗
Polar compound dissolution power	✓	✓	✓	✓	✓	✓	✓
Apolar compound dissolution power	✗	✗	✓	✓	✓	✗	✗
Solvent cost	✓	✗	✓	✗	✓	✓	✓
Efficient Spin Diffusion	318 K d > 13 Å ✓	288 K d > 13 Å ✓	288 K d > 13 Å ✓	268 K d > 13 Å ✓	238 K d > 18 Å ✓	283 K d > 13 Å ✓	273 K d > 13 Å ✓
NMR spectral resolution	✓	✓	✓	✓	✓	✓	✓
Labile proton exchange	✓	✓	✓	✗	✓	✓	✓
Routine NMR (lock, shimming)	✗	✗	✓	✓	✓	✓	✓
Solvent suppression	✗	✗	✗	✓	✓	✗	✓

Fig. 1 Advantages (green tick) and drawbacks (red cross) of viscous solvents (or solvent blends) in mixture analysis by NMR previously published by Lameiras *et al.*

solvents such as sucrose (see Fig. 2 of the article, reproduced in this discussion as Fig. 2).

Royston Goodacre queried: How critical is it to control the percentage of gel used? NMR is stable but now with this additional element of diffusion: what is the precision and accuracy like for repeat measurements, especially for database matching?

Pedro Lameiras responded: Thank you for your question. The agarose gel percentage of 1% was chosen since it has been shown that mesh from 1% up to 1.5% was too wide for observing direct effects of confinement on the stability of proteins such as lysozyme.[1] There is limited translational diffusion inside our NMR samples but our method involves the spin diffusion phenomenon corresponding to magnetization transfer through space between nuclei (which are not supposed to be close enough to present a NOE in low viscous medium) such as hydrogen and fluorine. Until now, our approach has been qualitative, but nonetheless we have demonstrated its high reproducibility and robustness by means of several published studies in recent years (DOI: 10.1039/c8fd00226f, ref. 2–4), provided that NMR temperature is precisely regulated. However, we may observe slight chemical shift differences for labile proton resonances.

1 A. Pastore, S. Salvadori and P. A. Temussi, *J. Pept. Sci.*, 2007, **13**, 342–347.
2 P. Lameiras, S. Patis, J. Jakhlal, S. Castex, P. Clivio and J. Nuzillard, *Chem.–Eur. J.*, 2017, **23**, 4923–4928.
3 P. Lameiras and J. Nuzillard, *Anal. Chem.*, 2016, **88**, 4508–4515.
4 P. Lameiras, L. Boudesocque, Z. Mouloungui, J. Renault, J. Wieruszeski, G. Lippens and J. Nuzillard, *J. Magn. Reson.*, 2011, **212**, 161–168.

Dušan Uhrín said: In your protocols you heat and melt the gels at high temperatures, which is not compatible with all samples. To prepare stretched gels, we dry them and then soak them in the solution containing the studied compound in the NMR tube. The compound naturally diffuses into gel. Can you use a similar procedure for your samples?

Pedro Lameiras responded: Thank you for your question. The possible risk of following your procedure is to produce inhomogeneous samples due to partially unmelted agarose or incorporation of air bubbles. As a result, the NMR signal line sharpness resolution may be damaged. The remedy is to melt the sample again directly inside the NMR tube, for instance by means of a hair dryer for a few minutes, then to mix it thoroughly and chill again at ambient temperature. Large molecules such as proteins may suffer from the high temperature exposition (at around 363 K for regular agarose and at around 338 K for low-melting agarose) preventing NMR structural and conformational studies. Nonetheless, the sample preparation is compatible with quite small molecules that are reasonably heat resistant such as nucleic acids, peptides or saccharides, since in practice the heat time only takes a few minutes.

Justin J. J. van der Hooft opened a general discussion of the papers by Dan Staerk, Fleur H. M. van Zelst and Pedro Lameiras: I saw some medical applications during your paper presentations – when considering diagnostic or clinical

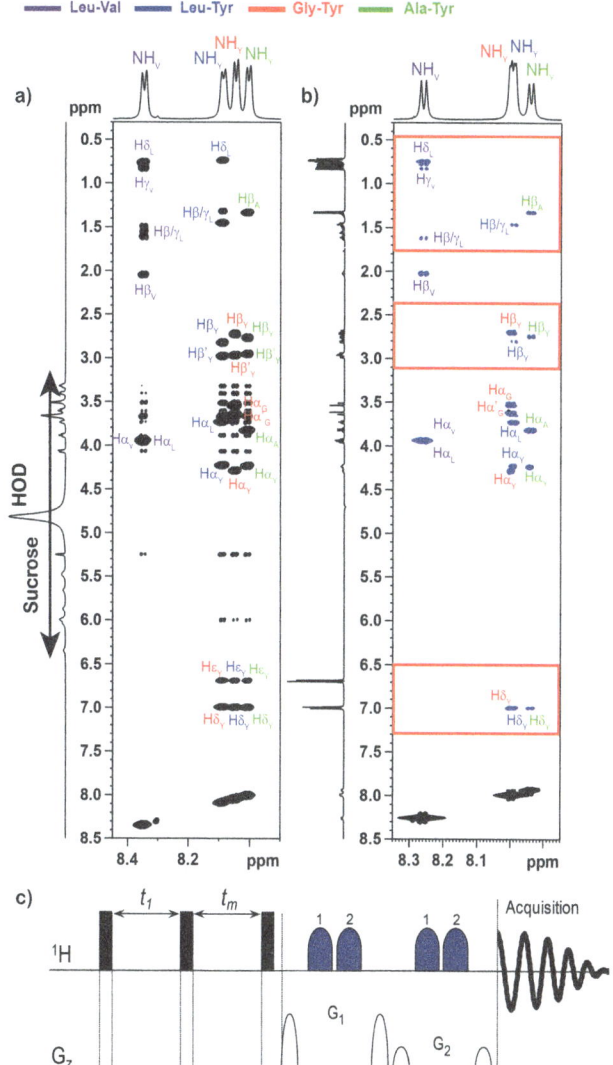

Fig. 2 (a) The amide proton region of the band-selective detection 2D NOESY spectrum of the dipeptide test mixture (10 mM), mixing time (t_m) = 0.5 s, at 600 MHz (^{1}H) using the pulse sequence in part (c), dissolved in sucrose solution, at 283 K. (b) The amide proton region of the 2D NOESY spectrum of the same dipeptide test mixture (10 mM), t_m = 0.5 s, at 600 MHz (^{1}H), using a noesyesgpph pulse sequence, dissolved in H_2O/D_2O (9 : 1, v/v), at 298 K. The red frames correspond to spectral regions of interest in which water as the solvent has a major effect on the number and sign of observable NOESY cross peaks. Reproduced from DOI: 10.1039/c8fd00226f with permission from the Royal Society of Chemistry.

settings, how expensive or cost effective is it to use the platforms that you presented?

Fleur H. M. van Zelst responded: This depends on the type of experiment and information you would like to obtain from the samples. For a fast screening method, this SFC-NMR setup is currently not suitable, and would therefore not be effective. Hyphenation with MS is then more suitable. However, if a structural assignment is needed, then NMR is the best technique to use, and with this set-up complex samples can also be analysed, which is normally difficult in NMR experiments due to overlap in the spectrum. NMR is an expensive technique of course, but you can obtain information from your sample that you cannot get from other, less expensive techniques (full chemical structure and quantitative information).

Pedro Lameiras replied: Thank you for your question. Except for the cost of the NMR spectrometer, our method is very cheap to implement. Table sugar and NMR sample tubes cost a few euros or pounds. Agarose gel costs a little bit more (dozens of euros or pounds) depending on the amount requested.

Pieter van Delft addressed Fleur H. M. van Zelst and Dan Staerk: NMR came up in quantification of well defined chemical entities such as *N*-acetylated carbohydrates. With proteins/antibodies, nature has provided us with biochemical platforms and tools that allow for highly specific and highly sensitive detection and quantitation of small molecules in complex matrices. Should we as analytical chemists sometimes look more into biochemistry/nature before applying techniques such as NMR, in particular when it comes to high throughput (potential) clinical applications?

Dan Staerk replied: I agree that for certain cases where sensitive and specific biochemical probes have been developed, these can be more sensitive – whereas NMR may be more broadly applicable – and they are definitely often needed techniques when dealing with unknown compounds. So as with all other scientific topics, one should look for the best possible technology to answer ones research question/research topic.

Fleur H. M. van Zelst responded: It is always a good idea to develop complementary methods for solving a problem. For screening purposes, antibodies and proteins are a good high throughput method. However, these screening methods rely on certain molecular interactions and binding to specific groups in a molecule. For unknown samples, many antibodies or proteins need to be tested before all molecules in a complex sample are detected. Even then, the full structure of the molecule is not evident, only that a certain binding group is present. For high throughput, known-target screening, using antibodies or proteins would be a good option, also for quantification, but the binding needs to be specific to only bind one molecule (or one stereoisomer). NMR is widely recognised as a tool of choice when it comes to the quality of the information that can be obtained about ligand/target interactions, but NMR suffers from low sensitivity. So if the sensitivity issue can be addressed *e.g.* by integrating nuclear hyperpolarization methods, NMR would be a very interesting technique for unknown targets, for which the full molecular structure is interesting.

Antony N. Davies asked: In the talk, an arbitrary threshold in signal intensity was set to decide only the more intense peaks would be investigated. This might not be the best strategy for the main goal of discovery. In previous literature, systems have been reported for natural product screening – looking at past databases to see if it is a known compound. Would not an alternative way of doing the filtering, looking even at relatively weak responders as interesting compounds to follow if they are new, be more profitable in the long term than simply setting threshold values?

Dan Staerk answered: Indeed, one could add another "filter" of chemical novelty. However, in our experience there is still much new to discover in terms of new bioactivity of known compounds, and we have therefore not yet implemented another selection strategy like the proposed one. But indeed, if over time we fail to discover new interesting bioactivity of known as well as "new" compounds, such a selection criteria as chemical novelty would be an alternative approach.

Conflicts of interest

There are no conflicts to declare.

PAPER

Joint and unique multiblock analysis of biological data – multiomics malaria study†

Izabella Surowiec,[ab] Tomas Skotare, [iD][a] Rickard Sjögren, [iD][ab]
Sandra Gouveia-Figueira,[a] Judy Orikiiriza,[cde] Sven Bergström, [iD][f]
Johan Normark[f] and Johan Trygg [iD] *[ab]

Received 18th December 2018, Accepted 8th February 2019
DOI: 10.1039/c8fd00243f

Modern profiling technologies enable us to obtain large amounts of data which can be used later for a comprehensive understanding of the studied system. Proper evaluation of such data is challenging, and cannot be carried out by bare analysis of separate data sets. Integrated approaches are necessary, because only data integration allows us to find correlation trends common for all studied data sets and reveal hidden structures not known *a priori*. This improves the understanding and interpretation of complex systems. Joint and Unique MultiBlock Analysis (JUMBA) is an analysis method based on the OnPLS-algorithm that decomposes a set of matrices into joint parts containing variations shared with other connected matrices and variations that are unique for each single matrix. Mapping unique variations is important from a data integration perspective, since it certainly cannot be expected that all variation co-varies. In this work we used JUMBA for the integrated analysis of lipidomic, metabolomic and oxylipins data sets obtained from profiling of plasma samples from children infected with *P. falciparum* malaria. *P. falciparum* is one of the primary contributors to childhood mortality and obstetric complications in the developing world, which makes the development of new diagnostic and prognostic tools, as well as a better understanding of the disease, of utmost importance. In the presented work, JUMBA made it possible to detect already known trends related to the disease progression, but also to discover new structures in the data connected to food intake and personal differences in metabolism. By separating the variation in each data set into joint and unique, JUMBA

[a]Computational Life Science Cluster (CLiC), Department of Chemistry, Umeå University, Linnaeus väg 10, 901 87 Umeå, Sweden. E-mail: johan.trygg@umu.se; Tel: +46 730647137
[b]Sartorius Stedim Data Analytics, Tvistevägen 48, 907 36 Umeå, Sweden
[c]Infectious Diseases Institute, College of Health Sciences, Makerere University, P.O. Box 22418, Kampala, Uganda
[d]Department of Immunology, Institute of Molecular Medicine, Trinity College Dublin, St. James's Hospital, Dublin 8, Ireland
[e]Rwanda Military Hospital, P.O. Box: 3377, Kigali, Rwanda
[f]Department of Molecular Biology, Umeå University, 901 87 Umeå, Sweden
† Electronic supplementary information (ESI) available. See DOI: 10.1039/c8fd00243f

reduced the complexity of the analysis and facilitated the detection of samples and variables corresponding to specific structures across multiple data sets, and by doing this enabled fast interpretation of the studied system. All of this makes JUMBA a perfect choice for multiblock analysis of systems biology data.

Introduction

Malaria remains a major global health and economic burden in spite of recent intense preventive measures, with the infection caused by the parasite being one of the primary contributors to childhood mortality and obstetric complications in the developing world.[1] The biochemical mechanisms behind malaria pathogenesis and the impact of the parasite on the host response are still largely unknown. To decipher the underlying mechanisms of malaria, a holistic systems biology approach is crucial. Systems biology aims at a higher level of understanding of organisms by studying them as integrated systems of genetic, protein, metabolic, pathway and cellular events. Analysis of complex biological processes within a systems biology approach is now possible to achieve thanks to the extensive development of a range of 'omics' technologies (genomics, transcriptomics, proteomics, metabolomics, and beyond).[2] All 'omics' technologies provide massive amounts of data, making analysis highly challenging and requiring powerful computational methods.[3,4] Overcoming these challenges and using systems biology to understand the host–parasite relationship in malaria may lead to new ways of treating the malaria infection.

Systems biology requires integrated analysis of multiple data sets. Although analysis of a single data set is a solved problem, integrated analysis of several different types of data sets, also called blocks, is challenging. At the same time, integration reveals previously unknown hidden structures across multiple data sets and detects samples and variables corresponding to them. Combining information from many data sets can also improve interpretation of the trends observed in the studied system. There exists a wide range of methods for integrated analysis, including methods based on network analysis,[5,6] Bayesian factor analysis[7] and multivariate linear projections.[8–10] In this article we used Joint and Unique MultiBlock Analysis (JUMBA),[11,12] which is a multivariate linear projection method. Such methods handle noisy, multicollinear data with many more variables than observations (samples), which is typical for biochemical and biological applications.

JUMBA is based on the OnPLS-algorithm[8,13] and is used to perform unsupervised integration of multiple data sets, so called multiblock analysis. Multivariate linear projection methods such as JUMBA have long been used for both unsupervised analysis, for instance Principal Components Analysis (PCA), as well as supervised analysis, for instance Orthogonal Projects to Latent Structures (O-PLS).[14] Although methods such as PCA can be used to analyse multiple blocks by combining the matrices into one,[15,16] they do not distinguish between variation that is joint between blocks and unique variation, which makes interpretation difficult. To perform unsupervised integration of two blocks, O-PLS was modified into O2-PLS,[17,18] which was later generalized into OnPLS.[8,13] OnPLS is an algorithm that separates data matrices into variation joint between all or only some blocks as well as variation unique to each block. JUMBA is an OnPLS workflow that structures multiblock analysis *via* pre-processing, modelling, validation, visualization and interpretation of the data.[11,12]

There are several examples of the application of multiblock analysis based on multivariate linear projection methods in systems biology. O2-PLS has been used for combined modelling of transcript, protein and metabolite data in plant species,[19,20] for multiblock analysis of fatty acid and lipid profiles in a mice model of familial dysbetalipoproteinemia[21] and for integration of NMR and DIGE data from prostate cancer xenograft mice.[22] OnPLS has been used for the integration of transcriptomic, proteomic and metabolomic data for the global investigation of stress response[23] and secondary cell wall synthesis[24] in *Populus* plants, as well as for transcriptomics, metabolomics, sphingolipids, oxylipins, and fatty acids interrogation of biological interactions in asthma.[25] In all cases, integrated analysis increased the interpretability of the models and enabled important biological conclusions, which would not be possible to achieve with the application of other methods.

We have recently applied metabolomics profiling on plasma from children infected with malaria and showed that a metabolite signature could be used for decision support in disease staging and prognostication, with fatty acids being potential biomarker molecules.[26] The study was later expanded to the analysis of another set of samples and application of an additional two platforms: LCMS lipid[27] and oxylipin profiling.[28] In these studies, we showed that the malaria infection altered the lipid and oxylipin patterns and that these changes could be connected to energy turnover and immune regulation. In the present study we wanted to see if we can confirm the findings from the analysis of separate data sets by application of a faster, integrative approach and to investigate if such an approach can provide more information about unknown trends/structures in the data that can be used for enhanced interpretation of the studied data sets. With this in mind, we have used JUMBA for multiblock analysis on lipidomic, metabolomic and oxylipins data sets obtained from profiling of plasma samples from children infected with *P. falciparum* malaria.

Experimental methods

Samples

Twenty plasma samples from each group of diagnostic categories: healthy controls, mild and severe malaria patients, were chosen from the 690 available, based on the clinical information using a full factorial design, as described before.[26] The research was carried out according to The Code of Ethics of the World Medical Association (Declaration of Helsinki). Ethical clearance was obtained from the Rwanda National Ethics Committee RNEC (no. 279/RNEC/2010) and the Regional Ethical Committee in Umeå (no. 09–064). Written informed consent was provided by the parent or legal guardian of each participant.

Data sets

Samples were extracted and analyzed with the GCMS metabolomics profiling method,[29] with LCMS metabolomics and lipidomics profiling methods[27] and with oxylipin targeted LCMS profiling.[28] For easy interpretation, and due to the large overlap between detected compounds and the fact that they were extracted in the same sample preparation step, metabolomics GCMS and LCMS data (from both ionization modes) were combined into one metabolomic data set. For compounds that were common between platforms or between LCMS ionization modes, the ones with lower relative standard deviations in the pooled samples were kept in the table. The applied analytical procedures are given in ESI, File 1.†

Data normalization, transformation and scaling

For the metabolomic data set, the compound peak areas were normalized using areas of internal standards, according to the following procedure: the PCA (with Unit Variance (UV) scaling without subtraction of the mean) on the peak areas of internal standards was calculated and the first component score value for each sample was used to normalize the resolved data by dividing the peak areas of each sample with the corresponding score value.[30] The oxylipins data set was log-transformed and all three data sets were mean centered and scaled to unit variance before JUMBA analysis.

Joint and unique multiblock analysis – JUMBA

JUMBA is a workflow for analyzing the underlying structures shared between multiple blocks of data measured on the same set of samples. JUMBA is based on the OnPLS-algorithm that decomposes the variation of data sets into joint and unique variation as well as residuals. Joint variation may be in common in all analyzed blocks, globally joint, or joint between some but not all, locally joint. A schematic overview of a three-block JUMBA model is given in Fig. 1. Even though JUMBA summarizes several blocks at the same time, the components can be

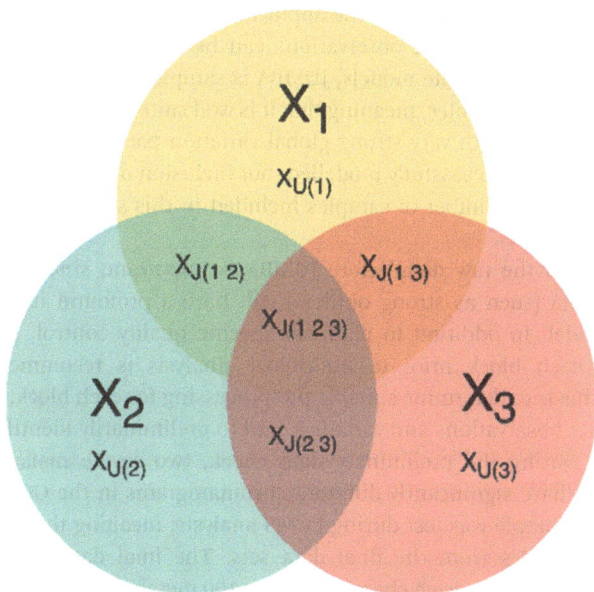

Fig. 1 Overview of a three-block JUMBA model showing possible partition of the variation within three data sets: X_J – joint variation, X_U – unique variation, 1, 2, 3 – number of data block.

investigated using scores and loadings in the same way as single-block methods, such as PLS or PCA. The scores show patterns among samples, and can be used to identify trends and outliers, while loadings show correlations between studied variables and their influence on the distribution of samples.

OnPLS, and therefore JUMBA, uses so called block-scores, meaning that each observation is assigned scores for each block instead of a single score per component across all blocks, which is the case for other multiblock methods, like, for example, JIVE and DISCO.[31,32] The OnPLS-algorithm finds such block-scores that maximize the covariance across all blocks. Separate scores for each block make it possible to investigate how well samples correspond across blocks and may be used to identify observations that correspond poorly between blocks. A single score for each observation gives the impression of artificially strong correspondence across blocks, which limits interpretation.

In this study we used JUMBA[11] to do multiblock analysis; we used the correlation matrix plot, multiblock scatter plot, explained variance plot and external responses correlation plot to evaluate and visualize the model. To visualize loadings, we used the correlation loadings, p(corr), since correlation loadings are scaled to correlations in the range −1 to 1, which simplifies interpretation.

The implementation of the algorithm was done using Mathworks MATLAB. The Pearson correlations were calculated using an in-house script written using the Anaconda Python distribution v. 3.5 (https://continuum.io) and plotted using the Matplotlib library (http://matplotlib.org/). Pathway enrichment analysis was performed on the p(corr) values above 0.2 using MetaboAnalyst 3.0.[33]

Results and discussion

Data analysis and pre-treatment

The fundamental requirement for the application of JUMBA is that there are two or more data sets where the observations can be matched in a 1 : 1 fashion. Similar to other multivariate models, JUMBA is sample efficient and can handle noisy and collinear variables, meaning that it is well suited for, but not limited to, omics data. In data with very strong global variation patterns, as few as twelve observations can be successfully modelled, but inclusion of more observations is recommended. The number of samples included in this study was sufficient to produce good results.

Evaluation of the raw data before JUMBA is important, since disturbances within the data (such as strong outliers) will have a profound impact on the resulting model. In addition to platform specific quality control, single-block analysis of each block prior to multiblock analysis is recommended. The purpose of this is to determine suitable pre-processing for each block, detect and handle outlier observations and variables, and to preliminarily identify trends in each block. During the preliminary data check, two severe malaria samples turned out to have significantly different chromatograms in the GCMS analysis and one mild sample was lost during LCMS analysis, meaning that we removed these three samples from the final data sets. The final data sets consisted, therefore, of 57 samples, each characterized by 100 metabolites, 144 lipids and 37 oxylipins.

Skewness analysis revealed that many variables were not normally distributed, which was especially profound for the oxylipins data set, where no variable passed the normality test. Due to the skewed variable distributions, we chose to use log-transformation for the oxylipins data set before JUMBA. Data transformation influences the interpretation of the results and should therefore be used with caution.[34] In our case, log transformation helped to obtain a less skewed distribution of samples for the oxylipins data set and hence eliminated the need to remove a number of samples due to their what seemed to be deviating behaviour.

JUMBA

General model structure. Using JUMBA, we found three globally joint components explaining 48.0% of variation in the lipids data set (26.6%, 12.9% and 8.5% of variation distributed between the first, second and third globally joint components, respectively), 38.9% variation in the oxylipins data set (16.0%, 17.7% and 5.2%) and 32.7% in the metabolic data set (11.4%, 11.8% and 9.6%). This means that there was a large overlap between the studied data sets. We found two locally joint components between the metabolomic and oxylipins data sets, describing 4.8%, 5.4% and 7.1%, 7.2% of variation, respectively. We also found four unique components in the lipids data set, (11.7%, 9.2%, 8.4% and 4.2% of variation explained). In the metabolomic data set we found five unique components (7.5%, 5.2%, 4.5%, 4.2% and 3.9% of variation explained). For the oxylipins data set we also found five unique components (11.3%, 6.6%, 3.9%, 3.5% and 3.4% of variation explained). The remaining variation corresponded to residuals and was equal to 18.6%, 18.2% and 31.8% of variation for the lipidomic, oxylipins and metabolomic data sets, respectively.

Scores and loading values for all model components are given in ESI, File 2.†

Model evaluation. To evaluate the validity of JUMBA, we inspected the correlation matrix plot (Fig. 2), which is a good tool for the detection of problems in the model.[11] Correlation between different model components may indicate that variation is assigned to joint and unique components in an inappropriate manner. For instance, if a unique component and a joint component correlate significantly, it may indicate that the unique component should actually be part of the joint component. In our case, there were no significant correlations between the different model components, but at the same time there were some correlations between different components (the plot was not too 'clean'), showing that the model was not over fitted, and hence was valid.

Analysis of the globally joint variation. An overview of the sample distribution, as described by the first two globally joint components, is presented on a multi-block scatter plot in Fig. 3. The first globally joint component describes the separation between the controls and the samples from the malaria patients. The second and third globally joint components (not shown) explain within-class variation.

The loadings corresponding to the first globally joint component (Fig. 4) revealed that the main observed compounds that co-varied and were up-regulated in infected individuals were phosphatidylcholines, sphingomyelins and the majority of triacylglycerides (lipids data set), 67% of all detected metabolites in the metabolites data set, as well as the majority of compounds from CYP, four compounds from 5-LOX and one compound (13-oxo-ODE) from the 12/15-LOX synthesis pathway (oxylipins). The main compounds that co-varied and were down-regulated in infected individuals were lysophosphatidylcholines (lipids)

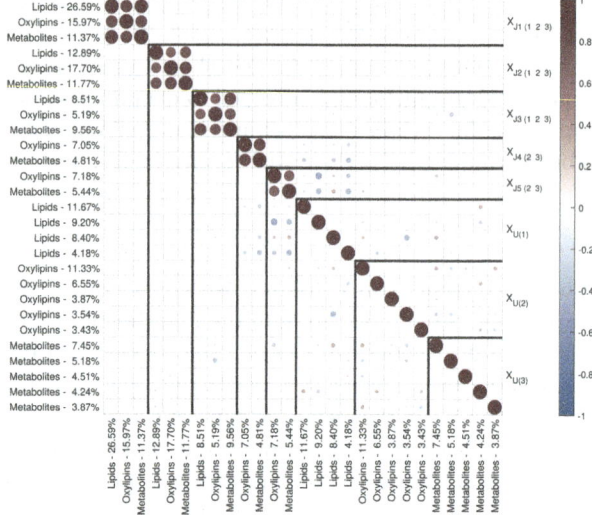

Fig. 2   Correlation matrix plot for the malaria study; the color and area of the circles correspond to the sign and strength of the correlation, with increasing circle size and color intensity indicating increasing correlation (positive (red) or negative (blue)). The thick vertical and horizontal lines are a visual aid to see the distinction between model components and unique component blocks;[11] J – joint component, U – unique component.

and oxylipins from the COX and 12/15-LOX synthesis pathways (oxylipins). These results are consistent with the ones obtained from the OPLS-DA analysis of differences between patients and controls performed for separate data sets.[27,28] This confirms that JUMBA can be used for the fast, successful integration and extraction of relevant information from the multiblock data.

While finding correlations between different data sets corresponding to the known sample groups by studying data sets separately and then combining the results is possible to some extent, analysis of any trends in common between blocks that are not known *a priori* can only be carried out using an integrated approach. In this study we have focused on the analysis of the second and third globally joint, as well as locally joint vectors, to elucidate such trends.

None of the trends observed in the second and third global components could be contributing to the previously known information about the samples. The loadings for the second globally joined vector, Fig. 5, show that the main trend observed with positive scores of the second globally joint component was down-regulation of the majority of phospholipids and triacylglycerides with lower carbon contents and higher degrees of saturation (lipids), down-regulation of amino acids with higher levels of fatty acids (metabolites) and higher levels of practically all oxylipins. The observed metabolic and oxylipin profiles correspond well to the ones observed in human plasma after the intake of a defined meal.[35,36] Also, postprandial dyslipidemia is a well described phenomenon.[37] Since samples included in this study were fed *ad libitum*, it is possible that some of them were taken shortly after a meal, and that the response to food intake represents part of the within-group variation described by the second globally joint component.

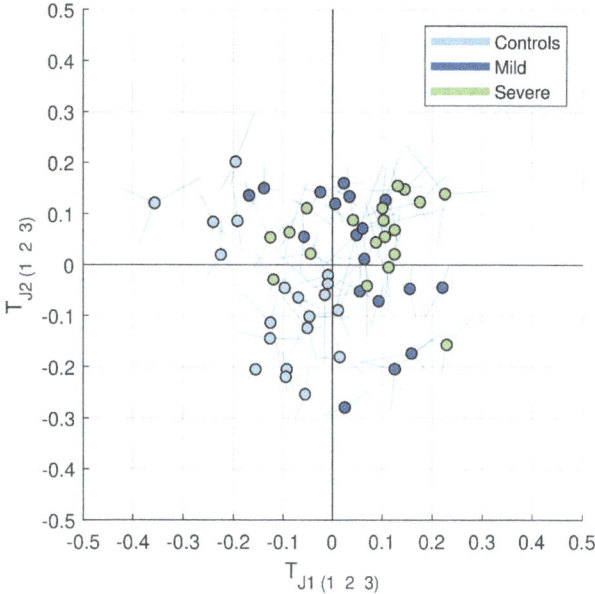

Fig. 3 Multiblock scatter plot made for the JUMBA model showing how the observations varied across several blocks for the globally joint components T_{J_1} and T_{J_2}. The samples are colored according to the class they belong to (controls, mild and severe malaria). The multiblock scatter plot is created by normalizing the score vectors to equal length, averaging the score values across blocks sample-wise and plotting them on a two dimensional plot as a mean score of each observation (sample), with lines drawn to each original block score value.[11] A point with longer lines in comparison to others will correspond to the observation with a large variation in the score values.

Connection of the second joint component to the food intake can also be supported by the size of the variation in the data connected to this component – a similar size for the oxylipins and metabolic data sets and approximately two times lower for the lipidomic data set as compared to the variation connected to the first globally joined vector. It is reasonable to assume that the intake of a meal will have a large impact on the metabolic and lipid composition of plasma. As such, analysis of the second globally joint vector provided additional information about the trends common for the studied data sets, which would be difficult, if not impossible, to gain by the analysis of separate blocks.

Analysis of the locally joint variation. Whereas globally joint components represent trends in the data common for all data sets, the locally joint ones correspond to the variation that is shared between a few but not all data sets. This distinction between different ways of sharing of variation across data sets is difficult to detect with other statistical methods. In our case, JUMBA found a share of the total variation that was joined between the metabolic and oxylipins data sets, split into two components. Since the samples for these data sets were prepared and analysed using different protocols, joint variation is probably connected to the intrinsic properties of biological, rather than experimental, origin. We could not find any clear trends linking to known sample descriptors by inspecting the score plot of joint score vectors. Pathway analysis of the

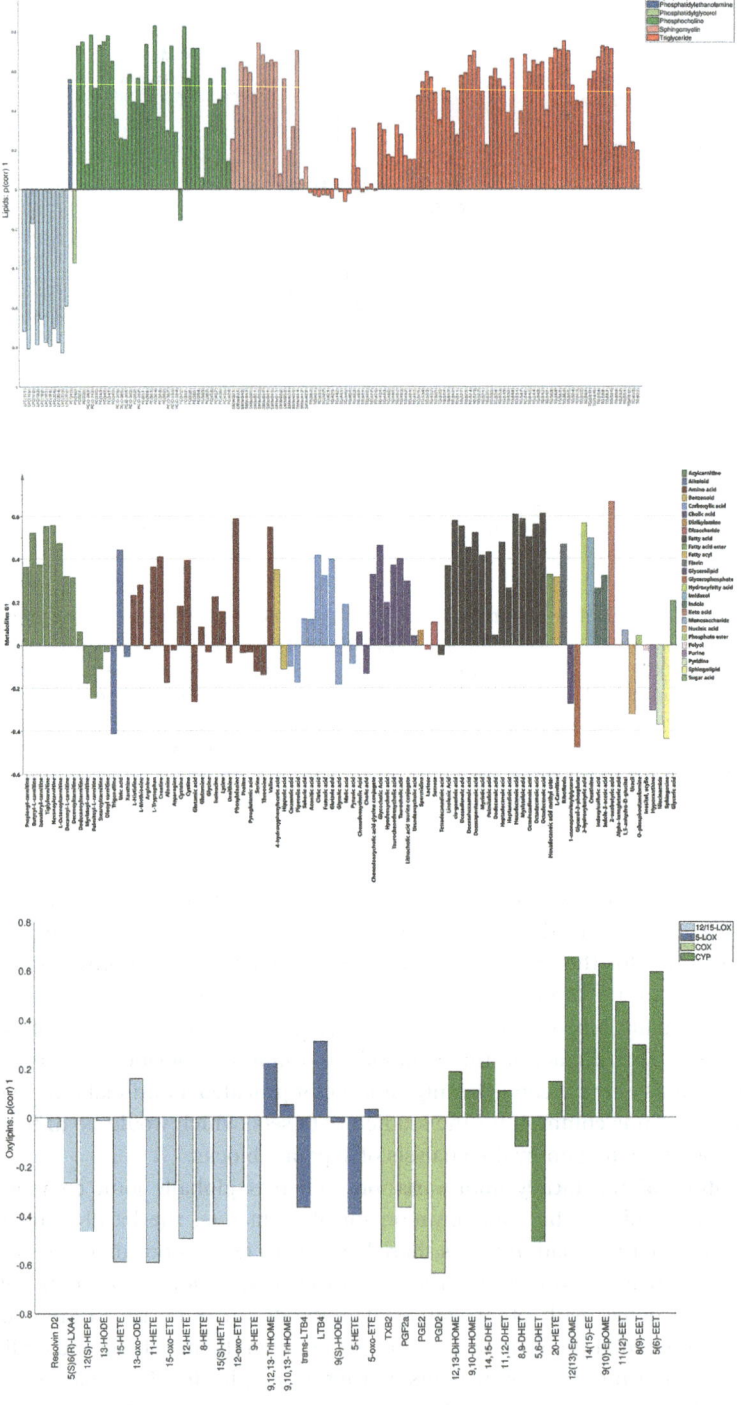

Fig. 4   First globally joint loading vectors (p(corr)) for the lipids (A), metabolites (B) and oxylipins (C) data sets colored according to the chemical classes (metabolites and lipids) and biochemical synthesis pathways (oxylipins).

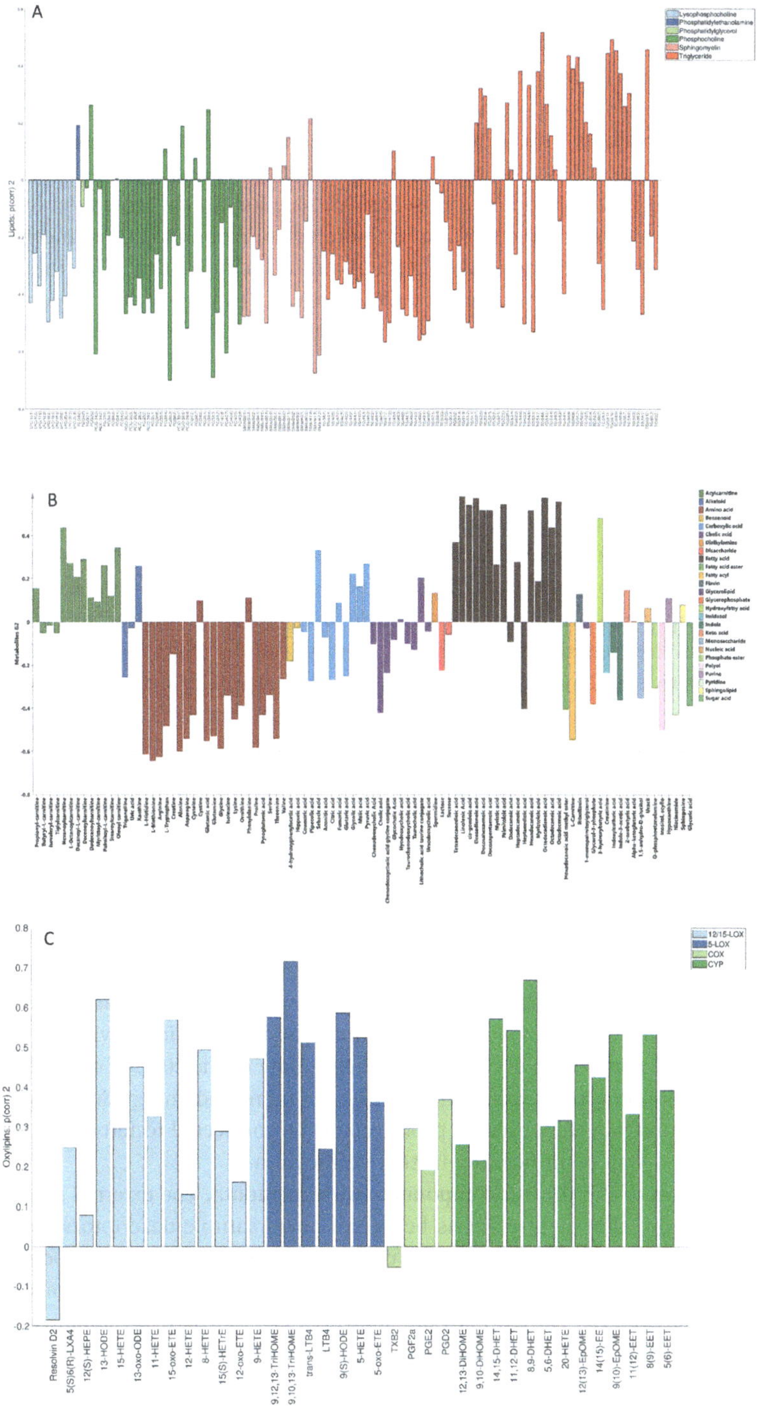

Fig. 5 Second globally joint loading vectors (p(corr)) for the lipids (A), metabolites (B) and oxylipins (C) data sets coloured according to the chemical classes (metabolites and lipids) and biochemical pathways (oxylipins).

corresponding loadings (ESI, File 1, Fig. S1 and S2†) revealed that linoleic and arachidonic acid metabolism as well as amino acid metabolism and the citric acid cycle were the most affected pathways for the first locally joint component. For the second locally joint component, amino acid and arachidonic acid metabolism, primary bile acid biosynthesis and lysine degradation were most affected. At this point, it is not possible to connect observed metabolic/oxylipin trends to specific physiological processes, but the observed connection to the arachidonic acid metabolism suggests changes in the immunological response not connected to the general metabolic response along the main sick–control axis. This indicates that the locally joint information could be related to a personal response to malaria. Since the results of the pathway analysis are highly related and limited to the capabilities of the platform used for the detection of the compounds (for example, types of compounds that can be detected and ones which will be under-represented), as well as the cut-off used for the selection of the compounds used as input for the pathway analysis, interpretation of the loadings presented above would need further verification.

Analysis of the unique variation. Unique variation describes the structured variation characteristic for one data set only and may be connected to experimental and/or biological factors. Loading plots corresponding to the selected four unique components are shown in Fig. 6. The first unique loading for the oxylipins data set and the second for the lipids data set showed the majority of compounds having positive and negative p(corr) values, respectively. Such a trend would rather not be expected from changes in biochemical pathways since this would mean activation of all pathways without any counterbalance effect. As such, trends described by these unique loadings could be most probably connected to experimental bias (for example, extraction errors not fully compensated by normalization of data to internal standards). The first unique loading for the lipids data set was among others characterized by negative p(corr) values of tri-acylglycerides with shorter fatty acid chain lengths and higher levels of saturation, which may be connected to personal differences in lipid metabolism or to specific types of diet. For the metabolomic data set, the first unique loading was characterized by all amino acids having positive p(corr) values and acylcarnitines with fatty acid chain lengths over six carbon atoms and fatty acids having negative p(corr) values. Long-chain acylcarnitines accumulate in the state of dysregulated fatty acid oxidation, especially during periods of increased energy demand from fat. This means that the observed profile could be connected to fatty acid oxidation defects; a statement needing further verification.

Correlation of the JUMBA components with metadata. The external response correlation plot shows correlations between known sample descriptors and the components of the model, and can be used for interpretation of the model. For this study, correlations of the model components against available personal and clinical parameters of the samples are shown in Fig. 7. The first globally joined component was strongly correlated with the class of the sample 'Controls', as well as with several parameters describing the severity of the sickness, like temperature, breathing rate and pulse rate, as well as symptoms describing the severity of the disease, for example loss of consciousness, convulsions, *etc.* This confirmed the previously discussed observation that trends described by the first joint component were related to the differences between the infected individuals and the controls. The first unique component for the metabolic data set correlated

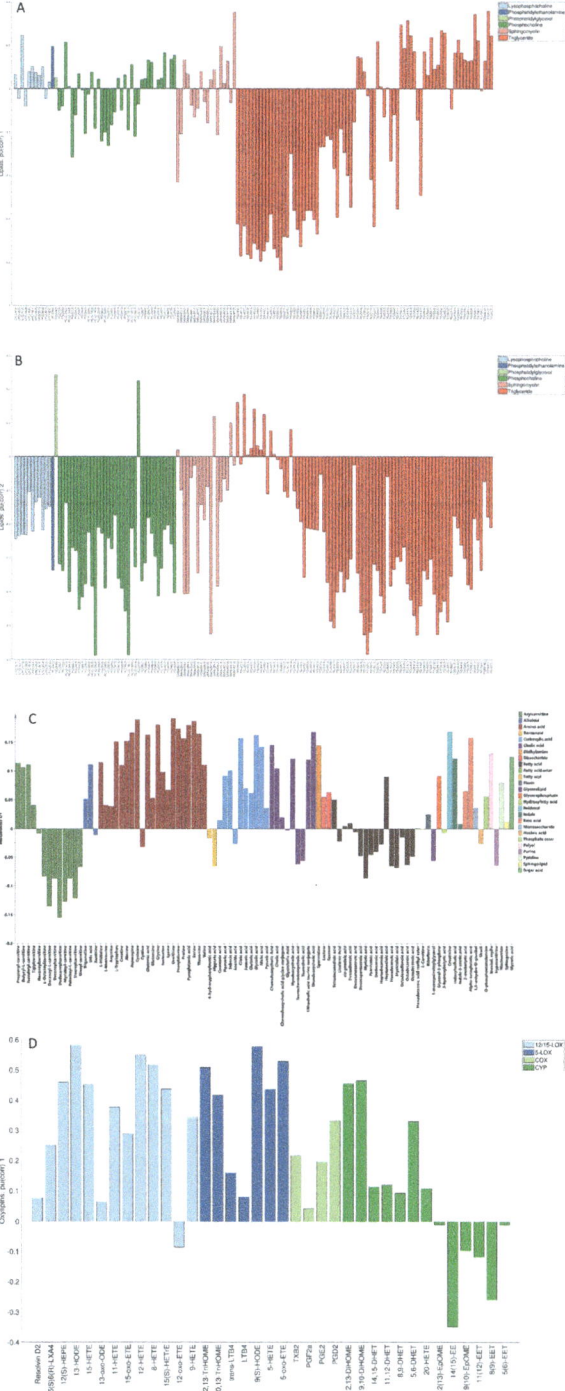

Fig. 6   First unique loading vectors (p(corr)) for the lipids (A, B), metabolites (C) and oxy-lipins (D) data sets coloured according to the chemical classes (metabolites and lipids) and biochemical pathways (oxylipins).

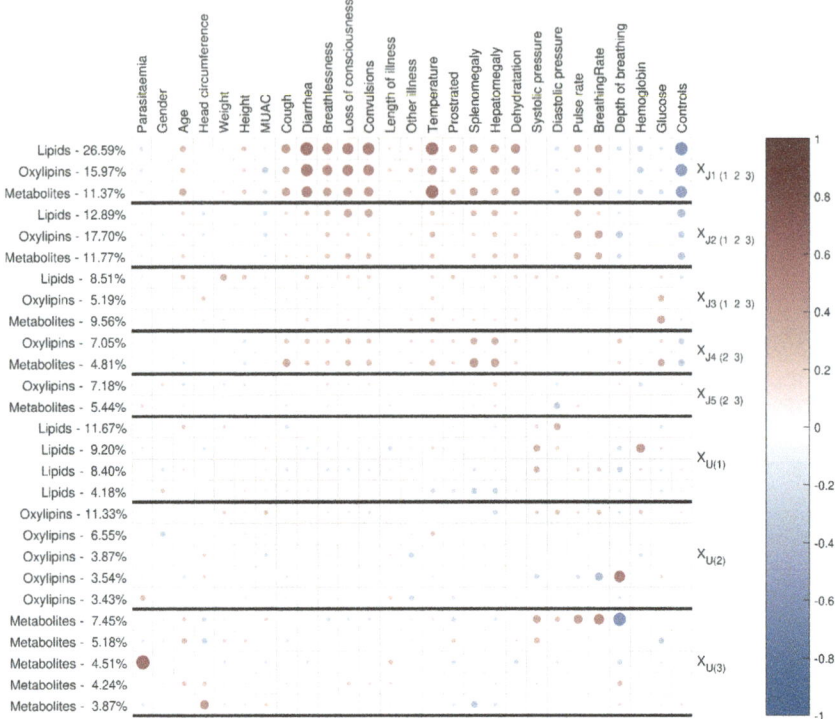

Fig. 7 The external response correlation plot for the malaria study showing correlations between known external response variables (personal and clinical parameters of the samples) and the different components of the model. The colour and size of the circles correspond to the sign and strength of the correlation, with increasing circle size and colour intensity indicating increasing correlation; blue shades are used for negative correlations and red shades for positive correlations.

positively with systolic pressure, pulse rate, breathing rate, hepatomegaly and loss of consciousness. These are parameters related to the severity of the disease, which could be connected to changes in the fatty acid beta-oxidation, as suggested by the analysis of the loading profile of this component.

Conclusions

In this study we have presented an integrative approach for the combined analysis of multiple data sets in clinical settings. Using JUMBA, we were able to reveal and analyse variation shared between lipidomic, metabolomic and oxylipin profiles of plasma that corresponded to intrinsically linked flows of information. Analysis of the joint loadings confirmed previously known trends in data related to the disease biochemistry and helped to immediately elucidate correlations between the studied data sets relating to these trends. We also revealed previously unknown trends that were most probably related to food intake and personal differences in the immunological response and general metabolism. Detection of those trends would not be possible to achieve by analysing each data set separately. Analysis of the relevant vectors from variations unique to each data block provided information specific for each data set that could be interesting both biologically, as well as from the analytical methodology point of view.

We proved that JUMBA can successfully integrate data obtained from different analytical platforms, thanks to its robustness to noise and its compartmentalization of variation into joint and unique parts. Integrated analysis of multiple data sets provides a faster and easier means to visualize and hence interpret the found relationships rather than separate investigation of each data set. Our integrative approach revealed hidden structures in the data not known *a priori*. It also allowed the detection of samples and variables corresponding to the found structures across multiple data sets, such as food intake, personal response to the disease or platform specific variations. To summarize, JUMBA is an easy-to-use workflow to put multiple data sets together, which offers enhanced visualization and allows comprehensive interpretation of multiblock data. JUMBA is a suitable method for handling complex multi-omics data sets and has high potential to improve the biological understanding of the studied systems. However, as in the case of all multivariate modelling methods, also for the application of JUMBA, common sense is needed. In cases where there are very many and very noisy variables, there is always the risk of spurious correlations, meaning that components can be found just by pure chance. One way to alleviate this risk is to investigate each block separately by evaluating its structure using well-established metrics, like, for example, PCA. Since the OnPLS-algorithm finds linear components, it is not suitable for detecting non-linear relationships. In those cases, we recommend finding feature representations of the data that are able to represent non-linear relationships (for instance, polynomial features).

Conflicts of interest

There are no conflicts to declare.

Acknowledgements

We would like to acknowledge the Swedish Metabolomics Centre for support with the chromatographic analysis.

References

1 P. F. Beales, B. Brabin, E. Dorman, L. Loutain, K. Marsh, M. E. Molyneux, *et al.*, Severe falciparum malaria, *Trans. R. Soc. Trop. Med. Hyg.*, 2000, **94**, S1–S90.

2 A. Fukushima, M. Kusano, H. Redestig, M. Arita and K. Saito, Integrated omics approaches in plant systems biology, *Curr. Opin. Chem. Biol.*, 2009, **13**(5–6), 532–538.

3 S. E. Richards, M.-E. Dumas, J. M. Fonbille, T. M. D. Ebbels, E. Holmes and J. K. Nicholson, Intra- and inter-omic fusion of metabolic profiling data in a systems biology framework, *Chemom. Intell. Lab. Syst.*, 2010, **104**(1), 121–131.

4 A. R. Joyce and B. O. Palsson, The model organism as a system: integrating 'omics' data sets, *Nat. Rev. Mol. Cell Biol.*, 2006, **7**(3), 198–210.

5 R. Shen, A. B. Olshen and M. Ladanyi, Integrative clustering of multiple genomic data types using a joint latent variable model with application to

breast and lung cancer subtype analysis, *Bioinformatics*, 2009, **25**(22), 2906–2912.

6   B. Wang, A. M. Mezlini, F. Demir, M. Fiume, Z. Tu, M. Brudno, *et al.*, Similarity network fusion for aggregating data types on a genomic scale, *Nat. Methods*, 2014, **11**(3), 333.

7   R. Argelaguet, B. Velten, D. Arnol, S. Dietrich, T. Zenz, J. C. Marioni, *et al.*, Multi-Omics Factor Analysis—a framework for unsupervised integration of multi-omics data sets, *Mol. Syst. Biol.*, 2018, **14**(6), e8124.

8   T. Lofstedt and J. Trygg, OnPLS-a novel multiblock method for the modelling of predictive and orthogonal variation, *J. Chemom.*, 2011, **25**(8), 441–455.

9   E. F. Lock, K. A. Hoadley, J. S. Marron and A. B. Nobel, Joint and individual variation explained (JIVE) for integrated analysis of multiple data types, *Ann. Appl. Stat.*, 2013, 7(1), 523–542.

10 M. Schouteden, K. Van Deun, S. Pattyn and I. Van Mechelen, SCA with rotation to distinguish common and distinctive information in linked data, *Behav. Res. Methods*, 2013, **45**(3), 822–833.

11 T. Skotare, R. Sjögren, I. Surowiec, D. Nilsson and J. Trygg, Visualization of descriptive multiblock analysis, *J. Chemom.*, 2018, e3071.

12 T. Skotare, D. Nilsson, S. J. Xiong, P. Geladi and J. Trygg, Joint and Unique Multiblock Analysis for integration and calibration transfer of NIR instruments, *Anal. Chem.*, 2019, **91**(5), 3516–3524.

13 T. Lostedt, D. Hoffman and J. Trygg, Global, local and unique decompositions in OnPLS for multiblock data analysis, *Anal. Chim. Acta*, 2013, **791**, 13–24.

14 J. Trygg and S. Wold, Orthogonal projections to latent structures (O-PLS), *J. Chemom.*, 2002, **16**(3), 119–128.

15 S. Wold, N. Kettaneh and K. Tjessem, Hierarchical multiblock PLS and PC models for easier model interpretation and as an alternative to variable selection, *J. Chemom.*, 1996, **10**(5–6), 463–482.

16 J. A. Westerhuis, T. Kourti and J. F. Macgregor, Analysis of multiblock and hierarchical PCA and PLS models, *J. Chemom.*, 1998, **32**, 301–321.

17 J. Trygg, O2-PLS for qualitative and quantitative analysis in multivariate calibration, *J. Chemom.*, 2002, **16**(6), 283–293.

18 J. Trygg and S. Wold, O2-PLS, a two-block (X–Y) latent variable regression (LVR) method with an integral OSC filter, *J. Chemom.*, 2003, **17**(1), 53–64.

19 M. Bylesjo, R. Nilsson, V. Sirvastava, A. Grönlund, A. I. Johansson, S. Jansson, *et al.*, Integrated Analysis of Transcript, Protein and Metabolite Data To Study Lignin Biosynthesis in Hybrid Aspen, *J. Proteome Res.*, 2009, **8**(1), 199–210.

20 M. Bylesjo, D. Eriksson, M. Kusano, T. Moritz and J. Trygg, Data integration in plant biology: the O2PLS method for combined modeling of transcript and metabolite data, *Plant J.*, 2007, **52**(6), 1181–1191.

21 G. M. Kirwan, E. Johansson, R. Kleemann, E. R. Verheji, Å. M. Wheelock, S. Goto, *et al.*, Building Multivariate Systems Biology Models, *Anal. Chem.*, 2012, **84**(16), 7064–7071.

22 M. Rantalainen, O. Cloarec, O. Beckonert, I. D. Wilson, D. Jackson, R. Tonge, *et al.*, Statistically integrated metabonomic-proteomic studies on a human prostate cancer xenograft model in mice, *J. Proteome Res.*, 2006, **5**(10), 2642–2655.

23 V. Srivastava, O. Obudulu, J. Bygdell, T. Löfstedt, P. Ryden, R. Nilsson, *et al.*, OnPLS integration of transcriptomic, proteomic and metabolomic data

shows multi-level oxidative stress responses in the cambium of transgenic hipI-superoxide dismutase *Populus* plants, *BMC Genomics*, 2013, **14**, 893.

24 O. Obudulu, N. Mähler, T. Skotare, J. Bygdell, I. N. Abreu, M. Ahnlund, *et al.*, A multi-omics approach reveals function of secretory carrier-associated membrane proteins in wood formation of *Populus* trees, *BMC Genomics*, 2018, **19**(1), 11.

25 S. N. Reinke, B. Galindo-Prieto, T. Skotare, D. I. Broadhurst, A. Singhania, D. Horowitz, *et al.*, OnPLS-based multi-block data integration: a multivariate approach to interrogating biological interactions in asthma, *Anal. Chem.*, 2018, **90**, 13400–13408.

26 I. Surowiec, J. Orikiiriza, E. Karlsson, M. Nelson, M. Bonde, P. Kyamanwa, *et al.*, Metabolic signature profiling as a diagnostic and prognostic tool in pediatric *Plasmodium falciparum* malaria, *Open Forum Infect. Dis.*, 2015, **2**(2), ofv062.

27 J. Orikiiriza, I. Surowiec, E. Lindquist, M. Bonde, J. Magambo, C. Muhinda, *et al.*, Lipid response patterns in acute phase paediatric *Plasmodium falciparum* malaria, *Metabolomics*, 2017, **13**(4), 41.

28 I. Surowiec, S. Gouveia-Figueira, J. Orikiiriza, E. Lindquist, M. Bonde, J. Magambo, *et al.*, The oxylipin and endocannabidome responses in acute phase *Plasmodium falciparum* malaria in children, *Malar. J.*, 2017, **16**(1), 358.

29 A. Jiye, J. Trygg, J. Gullberg, A. I. Johansson, P. Jonsson, H. Antti, *et al.*, Extraction and GC/MS analysis of the human blood plasma metabolome, *Anal. Chem.*, 2005, **77**(24), 8086–8094.

30 H. Redestig, A. Fukushima, H. Stenlund, T. Moritz, M. Arita, K. Saito, *et al.*, Compensation for systematic cross-contribution improves normalization of mass spectrometry based metabolomics data, *Anal. Chem.*, 2009, **81**(19), 7974–7980.

31 E. F. Lock, K. A. Hoadley, J. S. Marron and A. B. Nobel, Joint and individual variation explained (JIVE) for integrated analysis of multiple data types, *Ann. Appl. Stat.*, 2013, **7**(1), 523–542.

32 M. Schouteden, K. Van Deun, S. Pattyn and I. Van Mechelen, SCA with rotation to distinguish common and distinctive information in linked data, *Behav. Res. Methods*, 2013, **45**(3), 822–833.

33 J. G. Xia, I. V. Sinlenikov, B. Han and D. S. Wishart, MetaboAnalyst 3.0-making metabolomics more meaningful, *Nucleic Acids Res.*, 2015, **43**(W1), W251–W257.

34 C. Feng, W. Hongyue, N. Lu and X. M. Tu, Log transformation: application and interpretation in biomedical research, *Stat. Med.*, 2013, **32**(2), 230–239.

35 S. Gouveia-Figueira, J. Späth, A. M. Zivkovic and M. L. Nording, Profiling the oxylipin and endocannabinoid metabolome by UPLC-ESI-MS/MS in human plasma to monitor postprandial inflammation, *PLoS One*, 2015, **10**(7), e0132042.

36 M. Karimpour, I. Surowiec, J. Wu, S. Gouveia-Figueira, R. Pinto, J. Trygg, *et al.*, Postprandial metabolomics: A pilot mass spectrometry and NMR study of the human plasma metabolome in response to a challenge meal, *Anal. Chim. Acta*, 2016, **908**, 121–131.

37 V. Higgins and K. Adeli, Postprandial Dyslipidemia: Pathophysiology and Cardiovascular Disease Risk Assessment, *eJIFCC*, 2017, **28**(3), 168–184.

Faraday Discussions

Deciphering complex metabolite mixtures by unsupervised and supervised substructure discovery and semi-automated annotation from MS/MS spectra†

Simon Rogers, [iD][a] Cher Wei Ong,[a] Joe Wandy, [iD][b]
Madeleine Ernst, [iD][cd] Lars Ridder [iD][e] and Justin J. J. van der Hooft [iD]*[f]

Received 9th December 2018, Accepted 25th January 2019
DOI: 10.1039/c8fd00235e

Complex metabolite mixtures are challenging to unravel. Mass spectrometry (MS) is a widely used and sensitive technique for obtaining structural information of complex mixtures. However, just knowing the molecular masses of the mixture's constituents is almost always insufficient for confident assignment of the associated chemical structures. Structural information can be augmented through MS fragmentation experiments whereby detected metabolites are fragmented, giving rise to MS/MS spectra. However, how can we maximize the structural information we gain from fragmentation spectra? We recently proposed a substructure-based strategy to enhance metabolite annotation for complex mixtures by considering metabolites as the sum of (bio)chemically relevant moieties that we can detect through mass spectrometry fragmentation approaches. Our MS2LDA tool allows us to discover – unsupervised – groups of mass fragments and/or neutral losses, termed Mass2Motifs, that often correspond to substructures. After manual annotation, these Mass2Motifs can be used in subsequent MS2LDA analyses of new datasets, thereby providing structural annotations for many molecules that are not present in spectral databases. Here, we describe how additional strategies, taking advantage of (i) combinatorial in silico matching of experimental mass features to substructures of candidate molecules, and (ii) automated machine learning classification of molecules, can facilitate semi-

[a]School of Computing Science, University of Glasgow, Glasgow, UK

[b]Glasgow Polyomics, University of Glasgow, Glasgow, UK

[c]Collaborative Mass Spectrometry Innovation Center, Skaggs School of Pharmacy and Pharmaceutical Sciences, University of California San Diego, La Jolla, CA, USA

[d]Skaggs School of Pharmacy and Pharmaceutical Sciences, University of California, San Diego, San Diego, California, USA

[e]Netherlands eScience Center, Amsterdam, The Netherlands

[f]Bioinformatics Group, Wageningen University, Wageningen, The Netherlands. E-mail: justin.vanderhooft@wur.nl

† Electronic supplementary information (ESI) available. See DOI: 10.1039/c8fd00235e

automated annotation of substructures. We show how our approach accelerates the Mass2Motif annotation process and therefore broadens the chemical space spanned by characterized motifs. Our machine learning model used to classify fragmentation spectra learns the relationships between fragment spectra and chemical features. Classification prediction on these features can be aggregated for all molecules that contribute to a particular Mass2Motif and guide Mass2Motif annotations. To make annotated Mass2Motifs available to the community, we also present MotifDB: an open database of Mass2Motifs that can be browsed and accessed programmatically through an Application Programming Interface (API). MotifDB is integrated within ms2lda.org, allowing users to efficiently search for characterized motifs in their own experiments. We expect that with an increasing number of Mass2Motif annotations available through a growing database, we can more quickly gain insight into the constituents of complex mixtures. This will allow prioritization towards novel or unexpected chemistries and faster recognition of known biochemical building blocks.

Introduction

Complex natural mixtures are full of specialized metabolites with diverse structures and functions.[1] In untargeted metabolomics approaches, these molecules give rise to information-rich mass spectral data sets and a key challenge is the interpretation of this data, particularly in terms of identifying chemical structures.[2,3] This process is commonly referred to as metabolite annotation and identification,[4] a highly challenging process that typically enables the assignment of chemical structures to only a very small percentage of the molecules detected.[2,5–7] Consequently, the rapid and automated identification of chemical structures is one of the main obstacles hindering the discovery of novel bioactive molecules addressing global health care threats, such as antimicrobial resistance, cancer or inflammatory diseases.

Recently, we demonstrated how the unsupervised decomposition of fragment (MS2) spectra could aid in the annotation of molecules *via* identifying common fragment and loss patterns that were indicative of particular substructures (termed Mass2Motifs).[8] We showed that through Mass2Motif discovery, we can assign substructures to more than 70% of the fragmented molecules in beer extracts and our approach (MS2LDA) is publicly available through a web application (ms2lda.org).[8] Another widely used tool to organize fragmentation spectra is mass spectral Molecular Networking.[9,10] In combination or as a stand-alone tool, these similarity-based fragment spectra grouping algorithms are the current state-of-the-art in untargeted metabolomics for rapidly obtaining a comprehensive overview of molecular diversity in samples.[11–15] To retrieve chemical structural information for acquired experimental spectra, MS2 fragmentation patterns are matched directly to library reference data or *in silico* by matching substructures of candidate structures,[5,16–18] however only a very low percentage of the molecular features (typically 2–5%, but up to 30% in rare cases) can be confidently assigned to known chemical structures. In comparison to the structural annotation of entire molecules, structural annotation of the Mass2-Motifs is more straightforward and less complex, as Mass2Motifs represent smaller substructures. However, the structural annotation of Mass2Motifs is currently performed *via* a combination of manual peak searching in MS/MS databases such as MetLin[19] and MzCloud[20] as well as expert knowledge, and thus still represents a tedious and time-consuming step, especially for large-scale high-throughput experiments with several hundred discovered Mass2Motifs per experiment. As we and others have shown,[8,17,21,22] the use of reference MS/MS spectra of standards speeds up the annotation process; however, with the increasing size of publicly available MS/MS reference libraries,[9,17] complete manual Mass2Motif annotation and curation is rapidly becoming impractical. Furthermore, with the expected increase in publicly available experimental MS/MS data, the amount of structurally novel Mass2Motifs is expected to steadily rise. This will make structural predictions for Mass2Motifs of non-standards and effective reuse of previously annotated Mass2Motifs essential. Thus, the next step is to semi-automate Mass2Motif annotation and store annotated Mass2Motifs such that they can be used in the future.

In recent years, algorithms that propose chemical substructures and candidate structures for mass features have become available.[23–26] For example, MAGMa maps possible candidate molecules to MS/MS spectra in experimental data by assigning possible substructures from a candidate molecule to the mass fragments, and subsequently ranks different candidate molecules using those annotations based on a relatively simple scoring algorithm.[27] A complementary strategy towards structural annotation is to predict molecular properties such as fingerprints or classification based on spectral features.[28,29] For example, Classy-Fire[30] allows the classification of known molecular structures based on a consistent ontology of chemical descriptors.

In this work, we demonstrate how the integration of both MAGMa and ClassyFire terms within the ms2lda.org application facilitates the structural characterisation of a larger number of discovered Mass2Motifs. The extensions to the original ms2lda.org platform presented here are shown schematically in Fig. 1. MAGMa is used for the automated annotation of mass and neutral loss features within Mass2Motifs discovered from reference spectra, using the known chemical structures as candidates. These Mass2Motifs can then be compared with Mass2Motifs discovered in other experiments, increasing annotation coverage.

ClassyFire terms are used in two ways. Firstly, Mass2Motifs derived from reference spectra are mined for terms enriched in the molecules in which the Mass2Motifs are present. This provides rich structural information about the Mass2Motifs, against which newly discovered Mass2Motifs can be queried. Secondly, using the terms from known reference spectra, we present a machine learning approach (ClassyFirePredict) that predicts terms for spectra from experimental data. Mass2Motifs derived from these experimental data can then be mined for enriched terms based upon the predictions. Using a publicly available annotated MS2LDA experiment, we show how this can guide the user for annotation of fragment-based Mass2Motifs such as flavonoid and saccharide related motifs. Both ClassyFire systems are available at ms2lda.org.

Finally, to effectively reuse previously annotated motifs, we introduce MotifDB (available from ms2lda.org).[31] MotifDB stores annotated Mass2Motifs with their MS/MS features. A number of annotated Mass2Motif sets from various sources including plant extracts, urine, and standards, are already available for matching against Mass2Motifs discovered in new experiments.

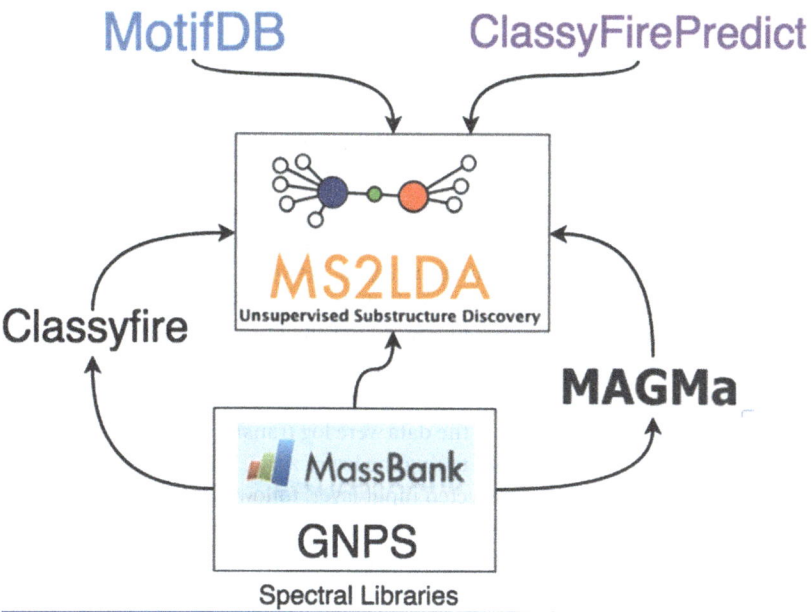

Fig. 1 The extensions to the original MS2LDA model described in this paper. MotifDB provides a platform for storing and re-using annotated Mass2Motifs. MAGMa and ClassyFire are both used with standard datasets to predict substructures corresponding to Mass2Motif features, identify terms enriched within Mass2Motifs and provide insight into the structural makeup of the MassMotifs derived from them. ClassyFirePredict extends this idea to non-standard data by predicting ClassyFire terms directly from the mass spectra.

We expect that the augmentations to the ms2lda.org web app will allow researchers to more rapidly decipher complex mixtures and create annotated and curated sets of Mass2Motifs. Those in turn will be effective in future experiments to more quickly assess the presence of specific molecular types in complex mixtures and assess the chemical diversity of those mixtures based on substructure recognition. We expect these substructure-based annotation strategies to become essential for deciphering complex mixtures and enabling meaningful biochemical interpretation.

Methods

Integrating ClassyFire substituent terms

ClassyFire terms were derived through the ClassyFire API for two of the public standard datasets (massbank_binned_005 and gnps_binned_005 – see Data availability section) stored within ms2lda.org using the ClassyFire web API[30] based on the molecules' InChIKeys. The substituent terms were stored in the database and linked to the relevant molecules such that they are visible when the molecule is explored. Additional functionality was added to ms2lda.org to summarize the terms within a particular Mass2Motif. In particular, based on actual values of the fragment spectra to Mass2Motif probability and overlap score thresholds outputted by MS2LDA,[32] the molecules associated with each Mass2-Motif are extracted, along with their ClassyFire substituent terms. For each term, the proportion of molecules associated with the Mass2Motif that include the term is computed, along with the proportion of molecules in the experiment. Comparing these terms provides evidence as to how unique and concentrated that term is in the Mass2Motif.

When working with new experimental data, exploring ClassyFire terms from standard molecules is useful if a discovered motif closely matches one of those in the standards experiments. To further extend this functionality, we have developed a machine learning approach that can predict putative ClassyFire terms from any mass spectrum. A multilayer neural network was produced that, for a binned mass spectrum, predicts the probability of the presence/absence for each ClassyFire term. The network was built in Python using Keras.[33] Spectral data are currently binned into bins of width 1 Da, with m/z values over 1000 discarded. After normalizing so that the base bin (*i.e.* the most intense bin in a particular spectrum) had intensity of 1000.0, the data were log transformed (after adding 1.0 to avoid problems associated with taking the log of zero). The network consists of a 1000-dimensional densely-connected input layer, followed by two hidden dense layers (of dimension 500 and 200) and then an output layer with dimension equal to the number of ClassyFire substituent terms. Non-linear ReLU (rectified linear unit) activation functions were used for the hidden layers, and a sigmoid function was used for the output layer. The model was optimized using the binary cross entropy loss function. This model represents our initial network design and it is likely that it could be optimized further.

An initial training and validation phase was undertaken using a filtered dataset of 10 038 unique tandem mass spectra with associated chemical structures retrieved from Global Natural Products Social Molecular Networking (GNPS). This dataset was created as follows. First, all public libraries from GNPS were assembled. Subsequently, we used a script in Python (see Code availability section) to sub-select only tandem mass spectra with full chemical structural information in computer readable format (at least SMILES available) to create a dataset in the .MGF data format followed by the selection of 10 105 unique molecules based on the first 14 digits of the InChIKeys with precursor $m/z < 1000$. The ClassyFire API generated classifications for 10 038 of these molecules, resulting in the final dataset.

Ten random splits into training (90%) and validation (10%) were used to assess the performance with respect to each term. Within each split, the area under the receiver operating characteristic curve (AUC) was computed, and these were averaged across the ten splits. Based on this analysis, we selected 444 terms that could be reliably predicted for the final classifier. These 444 terms were chosen *via* two conditions: firstly, all terms with an average AUC across the ten splits of greater than 0.7, and also, terms with an AUC of between 0.6 and 0.7 that appeared in at least 0.5% of the molecules in the dataset. These additional terms were included to increase coverage under the assumption that some false positives can be tolerated for individual molecules, as they are likely to be filtered out when we explore terms at the Mass2Motif level. Finally, the model was re-trained using these 444 terms and all of the available training data.

The predictive model was incorporated into ms2lda.org, allowing users to assign putative ClassyFire terms to any molecules. These terms are then collated at the Mass2Motif level to aid in annotation in exactly the same manner as those linked to the reference molecules.

MAGMa-MS2LDA integration

MAGMa was used to annotate Mass2Motif features as follows. All reference spectra for four data sets of known molecules that were subjected to MS2LDA (massbank_binned_005, gnps_binned_005, 2613 public spectra from various sources in positive ionization mode, and 551 public spectra in negative ionization mode from various sources – see Data availability section) were analyzed and annotated using MAGMa (see Code availability section). Each spectrum was annotated based upon its known structure resulting in the likely molecular substructures being assigned to individual peaks. Only the peaks used in the MS2LDA analysis were included in the MAGMa analysis, of which not all necessarily match with a simple substructure found within the reference molecule. Subsequently, the substructures were matched with the actual features used in the MS2LDA analysis (either fragments or losses within user-defined mass bins). For fragment features, the substructures assigned by MAGMa were stored both as a canonical SMILES, generated by the RDKit software library,[34] and as a mapping (with atom indices) on the original molecule. A SMILES string was generated for the loss features by first removing the MAGMa substructure atoms from the complete molecule and generating a canonical SMILES from the remaining atoms. These SMILES may contain disconnected parts of the molecules (separated by a dot according the SMILES specifications). MAGMa substructure feature annotations were stored in MS2LDA and visualized in the web application with the ChemDoodle package.[35]

As a result, Mass2Motif pages in MS2LDA could now be augmented with the MAGMa substructure annotations as follows. For a given feature explained by a Mass2Motif, all substructures associated to the feature in the corresponding spectra are retrieved and grouped. It is possible that the same fragment or loss in two spectra could be assigned different molecular substructures by MAGMa, a consequence of different molecular structures having the same (or very similar) mass. For example, a methyl carboxylic acid or O-acetyl group could be assigned to a loss of 60.0225 depending upon the parent structure. For a particular Mass2Motif, all unique substructures are presented along with the number of times they occur in the corresponding spectra. Additionally, since the same binned fragment and neutral losses are used as global features across all experiments in MS2LDA.org, annotations for all (and new) features that have corresponding features in MAGMa-annotated experiments can be derived from the existing MAGMa annotations assigned to these shared global features. We show this new information in the Mass2Motif and Document pages of the ms2lda.org web app.

MotifDB

Once Mass2Motifs have been annotated, it is useful to be able to search for them in future MS2LDA experiments. To this end, we have created a new application within MS2LDA.org called MotifDB: a database for annotated Mass2Motifs (http://ms2lda.org/motifdb). This database can be accessed via an API as well as being searchable against other experiments in the ms2lda.org web app. In particular, when an experiment has been run through MS2LDA.org, the user can start a motif matching procedure against Mass2Motifs stored in MotifDB. Where a Mass2Motif discovered in a new experiment exceeds a cosine similarity threshold with a Mass2Motif from MotifDB, the experimental motif can be linked to the MotifDB motif. The MotifDB annotation will now be highlighted in visualizations. It is important to realize that differential fragmentation mechanisms and different choices of collision energies between platforms can result in different fragmentation spectra.[36] As a result, similar substructures discovered in data obtained from different mass spectrometry platforms (*i.e.*, quadrupole time-of-flight, orbitrap, and ion trap) could result in different Mass2Motifs that would still represent the same substructure information. However, as we[8] and others[37] have shown, there are many situations where substructures are represented by diagnostic mass features formed across different platforms or where molecules do have comparable fragmentation spectra. As MotifDB grows by community efforts, more and more Mass2Motifs learnt in experiments under different experimental conditions will be annotated and available to be matched against in the MotifDB database, allowing for more rapid characterization of diverse chemical mixtures.

Code availability

The Python script to generate MAGMa annotations of standards datasets is provided on Github: https://github.com/iomega/motif_annotation.

The Python script to collect all GNPS library molecules including full metadata in .MGF format is provided on GitHub: https://github.com/madeleineernst/EditMGF/blob/master/CompileGNPSMGF_withInChIKey.py for which the following GNPS jobs are needed: https://gnps.ucsd.edu/ProteoSAFe/status.jsp?task=6e22f85aeb0744208e872d1640f508d9, https://gnps.ucsd.edu/ProteoSAFe/status.jsp?task=03fba62d93cb4cbfa3f72106d18f7d2c.

The scripts to prepare the GNPS library molecules for neural networking and perform the neural networking are provided on Github: https://github.com/sdrogers/nnpredict.

The code to perform MS2LDA is available at: https://github.com/sdrogers/lda.

The code for the ms2lda.org visualisation platform is available at: https://github.com/sdrogers/ms2ldaviz.

Data availability

The following public MS2LDA experiments were used in this manuscript.

Reference molecule data sets: massbank_binned_005 – http://ms2lda.org/basicviz/show_docs/190/.

Gnps_binned_005 – http://ms2lda.org/basicviz/show_docs/191/.

2613 public spectra from various sources in positive ionization mode – http://ms2lda.org/basicviz/summary/304/.

551 public spectra in negative ionization mode from various sources – http://ms2lda.org/basicviz/summary/305/.

Complex mixtures: Urine38_POS_mzML_standardLDA_005binned – http://ms2lda.org/basicviz/summary/709.

UrineDrugs_MolNetw_WorkshopSeattle2018 – http://ms2lda.org/basicviz/summary/601/.

Rhamnaceae_plant_extracts_KyoBin_200Motifs_MS1_peaktable – http://ms2lda.org/basicviz/summary/566/.

Results

MAGMa-based annotation of Mass2Motifs

MAGMa-MS2LDA annotations for previously analyzed Mass2Motifs. The integration of MAGMa with MS2LDA resulted in reference MS/MS MS2LDA experiments enriched with available MAGMa annotations for mass fragments and neutral losses for each fragmented molecule (Fig. 2A). MAGMa annotations were evaluated to identify how well they matched with previously (manually) annotated and validated motifs.[8] For example, motif 59 in the GNPS reference set was manually annotated and validated to be related to the phenylalanine minus CHOOH fragment substructure (http://ms2lda.org/basicviz/view_parents/58316/). Indeed, for 79 out of 117 molecules exactly this substructure was annotated by MAGMa for mass fragment 120.0825, with confirmation for the related aromatic fragment 103.0525 for 29 out of 35 appearances. This indicates that indeed this motif is related to [phenylalanine minus CHOOH]; moreover, the MAGMa annotations also provide quick insight in structurally less related molecules in the motif that are included due to isomeric fragments giving rise to the same mass fragment. This highlights the need for manual validation of fragmentation patterns in molecules, which is now supported in the ms2lda.org web application.

Another example is the indole related GNPS motif 25 (http://ms2lda.org/basicviz/view_parents/58017/); here, for 47 out of 110 molecules, MAGMa annotated the 130.0675 mass fragment with a methylindole substructure, and for 11

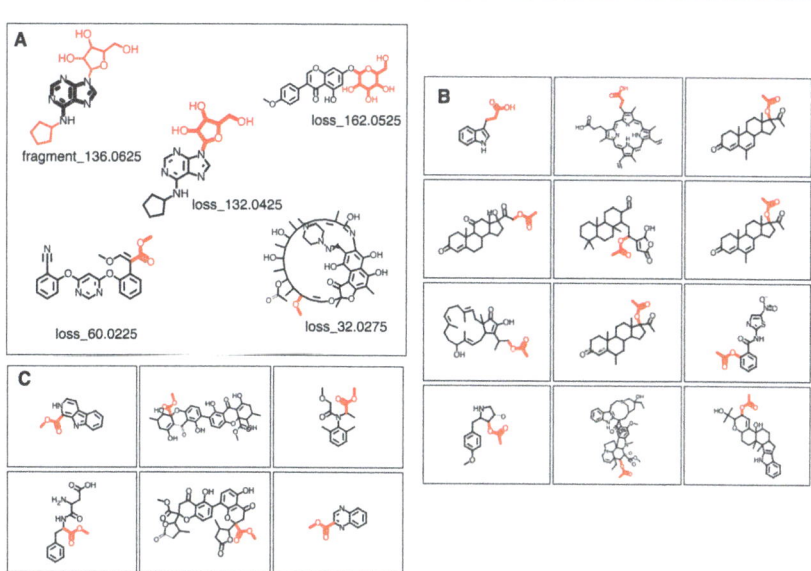

Fig. 2 (A–C) Screenshots of the ms2lda.org web app with (A) MAGMa annotations of Mass2Motif features in 5 motifs discussed in the results section. Annotated fragments are highlighted in black and bold, whereas annotated losses are depicted in red and bold. (B) 12 examples of the 38 molecules for which the loss_60.0225 in GNPS Mass2Motif 49 was annotated with loss (CC(=O)O) in SMILES. (C) 6 examples of the 25 molecules for which the structurally related COC═O loss in SMILES was annotated for the same loss feature in GNPS Mass2Motif 49.

out of 28 molecules, the 118.0675 mass fragment was annotated with the indole substructure. Interestingly, the MAGMa annotations facilitated insight in other isomeric substructures within this motif; for example, MAGMa annotated the 130.0675 fragment for 17 molecules with a 2-aminopropyl-phenyl substructure and for 6 molecules the related 2-aminoethyl-phenyl substructure, indicating that motif 25 is also associated to this aromatic substructure. Other annotations for the 130.0675 fragment included two isobaric substructures with a different elemental formula, the mass of which fell within the 0.005 Da mass bin.

MAGMa also annotated neutral loss-based Mass2Motifs. For example, GNPS Mass2Motif 49 was previously annotated with "Loss possibly indicative of carboxylic acid group with 1-carbon attached" http://ms2lda.org/basicviz/view_parents/58174/. This annotation was confirmed by MAGMa with the loss being annotated as CC(=O)O (in SMILES) in 38 molecules out of 132 (12 of which can be seen in Fig. 2B). 25 of the remaining molecules were annotated with the structurally related COC=O loss (Fig. 2C) and the remainder of the molecules with other isomeric losses. A similar example can be found in the MAGMa annotations for GNPS motif 18 http://ms2lda.org/basicviz/view_parents/58383/ annotated as acetyl loss, as can be seen here: http://ms2lda.org/basicviz/show_doc/273058/. Furthermore, for Massbank Mass2Motif 41, "Loss indicative of [hexose minus H20]" the majority of the MAGMa-annotated losses (50 out of 64) were glucose related http://ms2lda.org/basicviz/view_parents/57676/ (Fig. 3A)

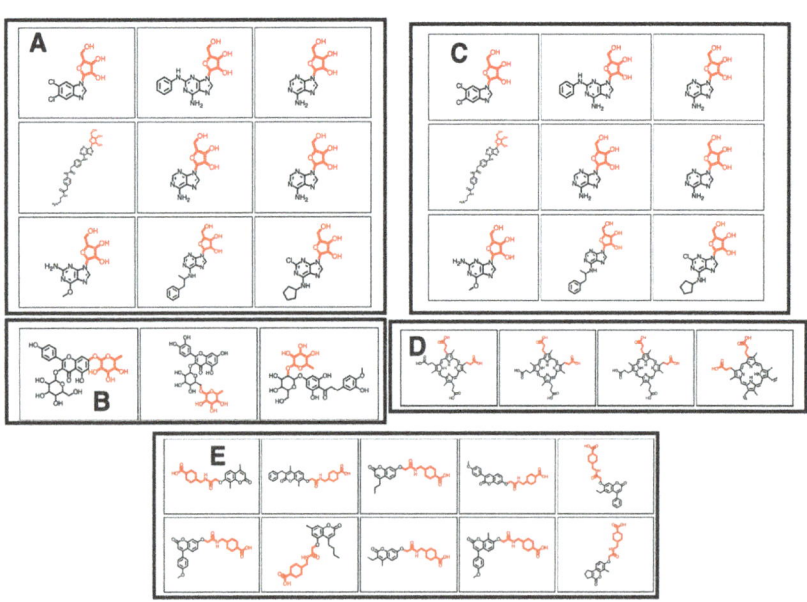

Fig. 3   (A–E) Screenshots of the ms2lda.org web app with (A) 9 different molecules out of the 50 molecules that MAGMa annotated with a hexose moiety for the loss feature in MassBank Mass2Motif 41. (B) 3 examples of the 13 molecules where MAGMa annotated the loss feature in MassBank Mass2Motif 41 with a deoxyhexose moiety. (C) 9 out of the 27 molecules for which MAGMa annotated a pentose moiety for the loss feature in GNPS Mass2Motif 44. (D) Alternative loss annotation of the loss feature in GNPS Mass2Motif 44. (E) Oxyacetyl-amino-methyl-cyclohexane-1-carboxylic acid loss annotated in 10 molecules of GNPS Mass2Motif 439.

with 13 being deoxyhexose moieties (Fig. 3B) that – unusually – included the connecting oxygen atom upon fragmentation of the main scaffold, which normally remains connected to the main scaffold. In the case of GNPS Mass2Motif 44, "[Pentose (C5-sugar)-H_2O] related loss – indicative for conjugated pentose sugar", MAGMa confirmed the pentose loss for 27 out of 56 molecules (Fig. 3C) http://ms2lda.org/basicviz/view_parents/58179/. For this motif, alternative loss annotations were also annotated by MAGMa, as shown in Fig. 3D.

Finally, GNPS motif 54 was annotated as ferulic acid related http://ms2lda.org/basicviz/view_parents/58325/. The MAGMa annotations show how important it is for this motif that the four mass fragments are all present, since 73 molecules contained mass fragment 177.0525. whereas for mass fragment 117.0325. 14 out of 19 molecules contained ferulic acid related substructures. Thus, whereas all GNPS Mass2Motif 54 related fragments have isomeric substructures unrelated to ferulic acid, their combined presence is highly indicative of the presence of ferulic acid.

MAGMa-MS2LDA integration for annotation of yet unexplored Mass2Motifs. In addition to previously annotated motifs, MAGMa annotations of not yet explored Mass2Motifs were analyzed. Fig. 2A shows MAGMa annotations for Mass2Motif fragment and loss features for five of the here described motifs in one of their related molecules. For example, GNPS Mass2Motif 152 could now be easily annotated as methanol loss resulting from the presence of a methoxy group http://ms2lda.org/basicviz/view_parents/58033/. The methoxy related loss could be annotated in 51 out of 58 molecules by MAGMa. Another methoxy group related GNPS Mass2Motif (374) was uncovered, where the loss of 16.0325 was assigned to CH_4 in 33 out the 38 molecules in the motif. In addition, GNPS Mass2Motif 188 could be annotated as related to a 2-dimethylamine-ethanol loss (*m/z* 89.0825), which was present in 9 out of the 14 molecules http://ms2lda.org/basicviz/view_parents/58098/. Other examples where MAGMa facilitated motif annotations include MassBank Mass2Motif 315 (benzyl and phenoxy group containing molecules), where for 77 out of the 84 associated molecules, the benzyl moiety was annotated by MAGMa. Moreover, in 20 molecules the phenoxy group was annotated for the motif fragment *m/z* 95.0475; however, interestingly, in 34 cases this fragment was present in the MS/MS spectrum, while there was no phenoxy group present in the corresponding reference molecule, nor was there any other substructure that could be assigned to this fragment. A possible explanation is that rearrangements are taking place in the mass spectrometer during the fragmentation process leading to the formation of phenoxy fragments as all these molecules do contain benzyl moieties. Here, the MAGMa-MS2LDA integration provides quick insight in assessing the consistency of structural annotations based on the presence/absence of mass fragments. Furthermore, MassBank Mass2Motif 443 could be annotated as "aniline related" due to the fact that 30 of the 32 associated molecules contained an aniline or substituted aniline substructure annotated by MAGMa http://ms2lda.org/basicviz/view_parents/57561/. Finally, GNPS Mass2Motif 439 (http://ms2lda.org/basicviz/view_parents/57921/) was shown by the MAGMa annotation to originate from a specific series of oxyacetyl-amino-methyl-cyclohexane-1-carboxylic acids with a characteristic series of losses (Fig. 3E). Based on the above examples, we show how MAGMa annotations are very helpful during the Mass2Motif annotation process. Our manual analysis of neutral losses is hampered by our inability to detect these generally smaller losses rather than larger scaffolds, which are easier to recognize – and MAGMa annotations are particular helpful here.

Chemical classification-based annotation of Mass2Motifs from standards. With increasing numbers of library MS/MS spectra available, the number of Mass2Motifs that can be extracted from those spectra will steadily increase. An alternative to the MAGMa substructure annotations for annotating this growing number of Mass2Motifs is the use of chemical classification. ClassyFire substituent terms for all of the molecules in the reference MS/MS data set were collected.[30] These substituent terms are based upon more than 5000 SMARTS patterns and are typically used by ClassyFire to organise molecules into a hierarchical chemical ontology. Here, we combined the substituent terms associated with molecules to look for terms that are enriched within Mass2Motifs with respect to their presence across the entire data set. For example, GNPS Mass2Motif 43 was previously annotated as being related to the adenine core structure http://ms2lda.org/basicviz/view_parents/58177/. The enriched substituent terms clearly correlate with this previous annotation: terms like *aminopyrimidine* and *6-aminopurine* are enriched (present in 64.3% and 52.4% of the molecules associated with this Mass2Motif, respectively) as compared to their percentage of occurrence in the entire GNPS data set (2.3% and 0.6%, respectively) (ESI Table S1†). In addition, GNPS Mass2Motif 72 was enriched with *amine* and *tertiary amine* terms (58.3% and 45.2% within the motif, 25% and 14.6% across the experiment), which is consistent with its annotation as diethylamino or dimethylaminoethyl substructure (ESI Table S2†). GNPS Mass2Motif 1 was enriched with oxosteroid related substituent terms *oxosteroid* and *3-oxosteroid* (present at 45.6% and 44.4% within the motif, and 3.9% and 3.3% across the experiment) matching its previous annotation as "sterone related" http://ms2lda.org/basicviz/view_parents/58328/.

The natural product substructure of quinazolinol (4-quinazolinone) was previously assigned to GNPS Mass2Motif 60 http://ms2lda.org/basicviz/view_parents/57956/. Demonstrating the power of the combination of MAGMa and ClassyFire, MAGMa annotated the quinoxaline substructure in 22 out of the 25 molecules (Fig. 4) and the enriched ClassyFire terms confirm this annotation (the *quinoxaline* term is present in 39.2% of molecules within the motif *versus* 0.5% of molecules within the experiment). This example shows that collected substituent terms can be used as guidance for Mass2Motif annotations in reference MS/MS data sets thereby providing consistent and widely-used chemical ontology terms.

With help of MAGMa and ClassyFire a number of novel annotations were made. For example, GNPS Mass2Motif 6 was annotated with the diphenyl-containing substructure following MAGMa annotations for its mass features and its enriched ClassyFire terms http://ms2lda.org/basicviz/view_parents/58331/ (Table 1). The MAGMa annotations of a methoxy group in GNPS Mass2Motif 152 matched with corresponding ClassyFire terms being enriched in this motif, such as *methyl ester* and *carboxylic acid ester* (Table 2). This is remarkable for such a small substructure. Interestingly, for GNPS Mass2Motif 439 (Fig. 3E), amongst the substituent terms ClassyFire did return, there were no helpful terms for Mass2Motif annotation, whilst MAGMa could annotate relevant substructures to guide Mass2Motif annotation, indicating the complementarity of these approaches. Overall, the enriched chemical classification terms confirmed and

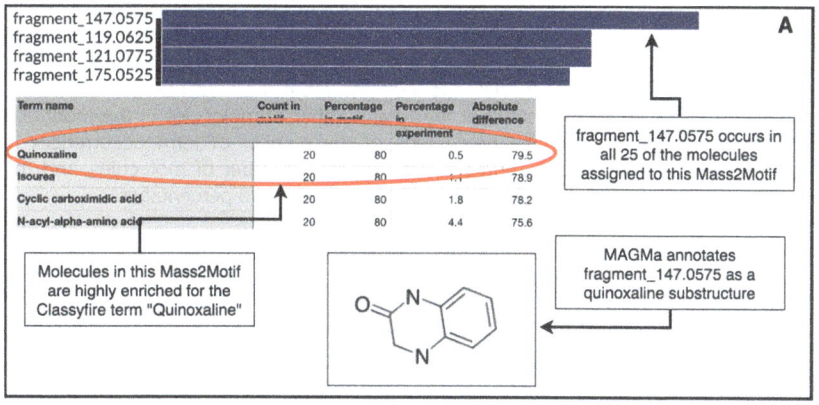

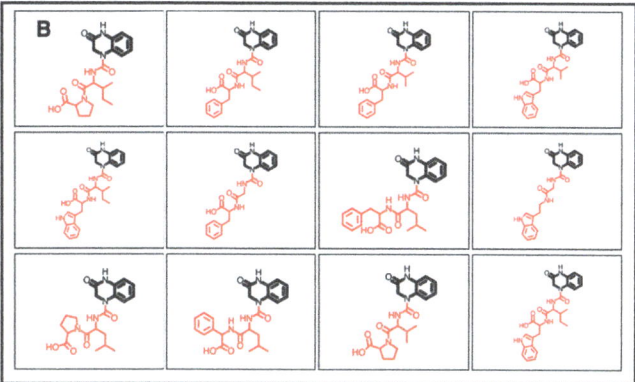

Fig. 4 (A) Top: feature frequency plot for GNPS Mass2Motif 60; middle: most enriched ClassyFire substituent terms in the same motif; bottom: MAGMa assigned the quinazolinol substructure in 22 of the 25 molecules associated to this motif. (B) Screenshot of the ms2lda.org web app with the MAGMa annotated quinazolinol substructure highlighted in 12 of the 22 molecules.

strengthened the manual and MAGMa annotations, and as such they may support and promote the use of consistent chemical terminology during the annotation process.

Chemical classification-based annotations of Mass2Motifs from non-standards. Using more than 10 000 unique GNPS Library reference MS/MS spectra, a neural network was trained to infer 444 ClassyFire substituent terms from fragmentation data (ClassyFirePredict). To evaluate the predictive model, it was applied to a public MS2LDA experiment of 71 Rhamnaceae plant extracts (see Data availability section) in which more than 20 motifs had previously been manually annotated.[14] Terms predicted for each spectrum were collected at the Mass2Motif level and compared with the manual annotations. Rhamnaceae Mass2Motif 33 had been manually annotated with a xylose or arabinose saccharide moiety. The ClassyFire predictions indicated enrichment of *alcohol* and *secondary alcohol* terms as well as *glycosyl* and *O-glycosyl compounds* which are all saccharide related terms http://ms2lda.org/basicviz/view_parents/109416/. Thus, the ClassyFirePredict and manual annotations correspond well for this

Table 1   Top 10 most enriched ClassyFire substituent terms for GNPS Mass2Motif 6, which could in this study be annotated as diphenyl substructure related. The term name represents the ClassyFire substituent term, the count in motif is the number of times the term appeared in a molecule associated to the Mass2Motif, the percentage in motif is the percentage of the count in motif over the total number of molecules in the motif, the percentage in experiment is the percentage of the number of term occurrences in molecules within the entire experiment over the total number of molecules, and the absolute difference is the absolute difference between the two percentages

Term name	Count in motif	Percentage in motif	Percentage in experiment	Absolute difference
Diphenylmethane	23	52.3	2.1	50.2
Tertiary aliphatic amine	21	47.7	13.7	34
Tertiary amine	21	47.7	14.6	33.2
Amine	24	54.5	25	29.5
Heteroaromatic compound	5	11.4	36.8	25.4
Aromatic heteropolycyclic compound	7	15.9	40.3	24.4
Benzenoid	10	22.7	45	22.3
Aromatic homomonocyclic compound	14	31.8	9.6	22.2
Benzylether	8	18.2	0.6	17.5
Dialkyl ether	11	25	7.7	17.3

Mass2Motif, indicating that ClassyFirePredict can assist in Mass2Motif annotations. The unannotated Rhamnaceae Mass2Motif 196 was enriched with overlapping saccharide-related terms, which suggests that this is also a saccharide

Table 2   Top 10 most enriched ClassyFire substituent terms for GNPS Mass2Motif 152 that was annotated with the help of MAGMa as methoxy group related. The term name represents the ClassyFire substituent term, the count in motif is the number of times the term appeared in a molecule associated to the Mass2Motif, the percentage in motif is the percentage of the count in motif over the total number of molecules in the motif, the percentage in experiment is the percentage of the number of term occurrences in molecules within the entire experiment over the total number of molecules, and the absolute difference is the absolute difference between the two percentages

Term name	Count in motif	Percentage in motif	Percentage in experiment	Absolute difference
Methyl ester	14	24.1	2.3	21.8
Carboxylic acid ester	20	34.5	13.9	20.6
Dialkyl ether	15	25.9	7.7	18.2
Enoate ester	11	19	2.9	16.1
Alpha,beta-unsaturated carboxylic ester	11	19	2.9	16.1
Ether	26	44.8	30.9	13.9
Dihydropyridinecarboxylic acid derivative	6	10.3	0.6	9.8
Carboxylic acid	2	3.4	13.3	9.8
Enamine	5	8.6	0.6	8.1
Monocarboxylic acid or derivatives	16	27.6	19.7	7.9

related motif http://ms2lda.org/basicviz/view_parents/109504/. Rhamnaceae Mass2Motifs 3 and 86 were annotated with the 3-hydroxyflavanoid cores myricetin and quercetin, respectively http://ms2lda.org/basicviz/view_parents/109575/ and http://ms2lda.org/basicviz/view_parents/109460/. Indeed, the predicted enriched ClassyFire terms clearly point to flavonoid related terms like *chromone* and *phenol*, which is also reflective of their presence in the training data. Finally, Rhamnaceae Mass2Motif 148 was annotated as a cyclopeptide alkaloid related motif http://ms2lda.org/basicviz/view_parents/109419/. Motif members were previously structurally annotated and found to be cyclic peptides sharing a benzenoid moiety (https://gnps.ucsd.edu/ProteoSAFe/gnpslibraryspectrum.jsp?SpectrumID=CCMSLIB00004679280#%7B%7D).[14] The predicted enriched ClassyFire terms reflect these cyclopeptidic structures well. In particular, *benzenoid* is highly enriched (85.7% present in the motif *versus* 18.5% in the experiment), as is *organonitrogen compound* (60.7% in motif *versus* 29.8% in experiment). Thus, we conclude that ClassyFirePredict can provide annotations that are useful annotations in guiding the analysis of Mass2Motifs from experimental data.

MotifDB. The new motif matching pipeline was used to match newly discovered Mass2Motifs in 5021 mass spectra from a publicly available human urine sample with a set of Mass2Motifs previously manually annotated from urine samples of the same cohort run under the same experimental conditions (http://ms2lda.org/basicviz/manage_motif_matches/709/).[32] Of the 300 Mass2Motifs discovered, 102 could be matched against 82 unique Mass2Motifs from MotifDB with cosine scores of 0.5 or greater, of which 41 had cosine scores greater than 0.9. The distribution of scores is shown in Fig. 5. The ten highest scoring matches are shown in ESI Table S3† along with the annotation and the number of molecules

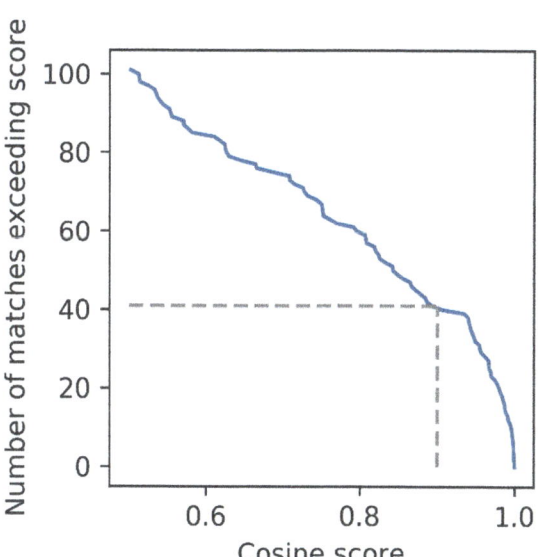

Fig. 5   Distribution of Mass2Motif matching scores for a urine dataset matched against the urine MotifSet in MotifDB. The dashed line shows the number of Mass2Motifs (41) that could be matched against the MotifSet with a cosine score of 0.9 or more.

that are assigned to the discovered motif (at a probability threshold of 0.1 and an overlap threshold of 0.3). These matches include Mass2Motifs related to urine-related substructures such as creatine and carnitine that have large degrees in the 5021 mass spectra, indicating that these substructures are abundant in urine as they are present in many fragmented molecules. In total, across the 102 matched motifs, 3715 unique molecules include at least one of the 102 matched Mass2Motifs (out of a total of 5021 in the experiment; 74%) and 2879 (57%) unique molecules include at least one Mass2Motif matched with a score of >0.9. These percentages indicate the potential of annotating complex mixtures through substructure assignments.

To further evaluate the power of motif matching against MotifDB we compared the urine motif set from MotifDB with Mass2Motifs discovered in fragmentation spectra of 6 urine samples from a different cohort analysed under the same experimental conditions (http://ms2lda.org/basicviz/manage_motif_matches/601/).[22] In this case, of the 200 Mass2Motifs, 55 could be matched at a threshold of at least 0.5 (covering 573 of the 1163 molecules; 49%) and 20 at a threshold of 0.9 (404 molecules; 35%). Although, as expected, the number of matches is lower than in the first example, the ability to immediately match approximately a quarter of the discovered motifs (allowing some level of annotation for half of the molecules) highlights the generalizability of Mass2Motifs across sample sets. This approach aids the discovery and prioritization of novel Mass2Motifs that may well represent xenobiotic-related chemistry (*i.e.*, drugs, food, *etc.*) not previously encountered.

Conclusions and future outlook

In this paper, we have described multiple extensions to the MS2LDA platform (all implemented on the ms2lda.org web app) that enhance the ability of analysts to characterize the makeup of complex mixtures of metabolites. The extensions all make it easier to characterize the Mass2Motifs onto which MS2LDA allows experimental data to be decomposed. These Mass2Motifs often represent chemical substructures and annotating them allows some degree of annotation to all MS2 spectra that include them, as often a relatively small number of annotated Mass2Motifs provide information about a significant proportion of the molecules in an experiment.[8]

The extensions move the platform forward in two general directions. The first, MotifDB, provides a platform that allows for the storage of annotated Mass2-Motifs that can then be accessed *via* an API (details at http://ms2lda.org/motifdb) or used within ms2lda.org by allowing users to match Mass2Motifs discovered within their experiments to those stored in MotifDB. In our experiments with human urine data, we found that roughly 25% of the Mass2Motifs in a urine dataset from a different cohort than the dataset from which the annotated motifs were generated could be matched against Mass2Motifs from MotifDB. These 25% of Mass2Motifs were associated to about 50% of the molecules.

The second direction is the collation of known and predicted molecular properties for individual molecules across Mass2Motifs. Here, we have presented three advances. Firstly, the use of MAGMa on databases of standards that had been analysed with MS2LDA to annotate their fragment spectra with substructures. We show how MAGMa-Mass2Motif annotations provide quick insight in ambiguity of annotations in case of isomeric substructures. These substructure annotations can then be propagated to the features in the Mass2Motifs, providing relevant insight into the substructures they could represent.

The second advance propagates the ClassyFire substituent terms for the same datasets of chemical standards to the Mass2Motif level. Finally, for "unknown" molecules measured in experimental data, we have introduced a machine learning approach based on a neural network that can predict a subset of ClassyFire substituent terms from the spectral data. This model has some limitations: (i) the predictive power is dependent on the chemical diversity present in available training spectra, (ii) the current training set consists of series of structurally correlated molecules, and (iii) very small substructures will be difficult to predict due to their usually widespread presence in molecules with structurally diverse larger scaffolds, making it harder to recognize the specific chemical terms connected to these smaller substructures. Nevertheless, we show that for fragment-based Mass2Motifs from complex mixtures, the predicted terms can guide Mass2Motif annotations. Again, these can be propagated to the Mass2Motif level, providing insight into their structural makeup. We foresee that by annotating more and more Mass2Motifs, the metabolite annotation of yet unknown molecules in complex mixtures – the main bottleneck in untargeted metabolomics data analysis – will become easier. The proposed machine learning approach has the potential for further exploration and optimization. The model can be further augmented by inclusion of neutral loss features as well as mass shifts, which are expected to improve chemical predictions for loss-based motifs such as loss of hexose or deoxyhexose and amino acid related motifs, respectively.

As more Mass2Motifs are extracted and annotated from the growing datasets of standards, MotifDB will grow and the coverage across experiments will increase. We also foresee users including annotated motif sets within their LDA experiment, thereby simultaneously finding known substructure patterns and discovering new ones with the benefit of combining supervised and unsupervised motif discovery in one analysis. Furthermore, users would then also be able to decompose single spectra over these motif sets through an API.

The MAGMa and ClassyFire based annotations can significantly enhance the process of annotation of the rapidly growing (number of) datasets and Mass2Motifs. The expected growth in available fully annotated reference spectra will also increase the training sets available for our ClassyFire predictor, increasing performance and increasing the set of terms that we can confidently predict. Furthermore, the implementation of chemical ontology from ClassyFire assists in more consistent annotations of motifs by using chemical terminology from an ontology.

We expect that substructure-based annotation strategies will prove to be essential to decipher complex mixtures and enable meaningful biochemical interpretation. Our work represents key steps of this workflow by recognizing mass spectral patterns, semi-automated structural annotation and storage of them. An increasing amount of structurally annotated Mass2Motifs will allow metabolomics researchers to gain some structural information on the majority of fragmented molecules. The further closing of the structural annotation gap in metabolomics will make untargeted metabolomics a very powerful tool for studying complex mixtures.

Author contributions

SR, LR, and JJJvdH conceptualized the study. LR and JW designed and implemented MAGMa-MS2LDA integration. ME extracted well-annotated publicly available spectral data from GNPS. SR and JJJvdH designed ClassyFire predictions. CWO and SR built the neural network model for ClassyFire predictions and SR integrated it within MS2LDA. CWO, LR, SR, and JJJvdH analyzed data. All authors contributed to the writing of the manuscript and agreed on the content.

Funding

JJJvdH is supported by an ASDI eScience grant (ASDI.2017.030) from the Netherlands eScience Center (NLeSC). SR is supported by an BBSRC grant BB/R022054/1 and a Carnegie Trust for Scotland grant.

Conflicts of interest

The authors declare there are no conflicts of interest.

Acknowledgements

The authors would like to thank all GNPS contributors who took the efforts to extensively annotate their public spectra including SMILES, which made them reusable in this study. The authors would also like to thank the ClassyFire initiative for sharing the chemical ontology with the scientific community.

References

1 F. Olivon, P.-M. Allard, A. Koval, D. Righi, G. Genta-Jouve, J. Neyts, C. Apel, C. Pannecouque, L.-F. Nothias, X. Cachet, L. Marcourt, F. Roussi, V. L. Katanaev, D. Touboul, J.-L. Wolfender and M. Litaudon, *ACS Chem. Biol.*, 2017, **12**, 2644–2651.

2 J.-L. Wolfender, J.-M. Nuzillard, J. J. J. van der Hooft, J.-H. Renault and S. Bertrand, *Anal. Chem.*, 2019, **91**, 704–742.

3 R. Chaleckis, I. Meister, P. Zhang and C. E. Wheelock, *Curr. Opin. Biotechnol.*, 2019, **55**, 44–50.

4 J. J. J. van der Hooft, R. C. H. de Vos, L. Ridder, J. Vervoort and R. J. Bino, *Metabolomics*, 2013, **9**, 1009–1018.

5 R. R. da Silva, P. C. Dorrestein and R. A. Quinn, *Proc. Natl. Acad. Sci. U. S. A.*, 2015, **112**, 12549–12550.

6 B. Y. L. Peisl, E. L. Schymanski and P. Wilmes, *Anal. Chim. Acta*, 2018, **1037**, 13–27.

7 O. A. H. Jones, *Metabolomics*, 2018, **14**, 101.

8 J. J. J. van der Hooft, J. Wandy, M. P. Barrett, K. E. V. Burgess and S. Rogers, *Proc. Natl. Acad. Sci. U. S. A.*, 2016, **113**, 13738–13743.

9 M. Wang, J. J. Carver, V. V. Phelan, L. M. Sanchez, N. Garg, Y. Peng, D. D. Nguyen, J. Watrous, C. A. Kapono, T. Luzzatto-Knaan, C. Porto, A. Bouslimani, A. V. Melnik, M. J. Meehan, W.-T. Liu, M. Crusemann, P. D. Boudreau, E. Esquenazi, M. Sandoval-Calderon, R. D. Kersten,

L. A. Pace, R. A. Quinn, K. R. Duncan, C.-C. Hsu, D. J. Floros, R. G. Gavilan, K. Kleigrewe, T. Northen, R. J. Dutton, D. Parrot, E. E. Carlson, B. Aigle, C. F. Michelsen, L. Jelsbak, C. Sohlenkamp, P. Pevzner, A. Edlund, J. McLean, J. Piel, B. T. Murphy, L. Gerwick, C.-C. Liaw, Y.-L. Yang, H.-U. Humpf, M. Maansson, R. A. Keyzers, A. C. Sims, A. R. Johnson, A. M. Sidebottom, B. E. Sedio, A. Klitgaard, C. B. Larson, C. A. Boya P, D. Torres-Mendoza, D. J. Gonzalez, D. B. Silva, L. M. Marques, D. P. Demarque, E. Pociute, E. C. O'Neill, E. Briand, E. J. N. Helfrich, E. A. Granatosky, E. Glukhov, F. Ryffel, H. Houson, H. Mohimani, J. J. Kharbush, Y. Zeng, J. A. Vorholt, K. L. Kurita, P. Charusanti, K. L. McPhail, K. F. Nielsen, L. Vuong, M. Elfeki, M. F. Traxler, N. Engene, N. Koyama, O. B. Vining, R. Baric, R. R. Silva, S. J. Mascuch, S. Tomasi, S. Jenkins, V. Macherla, T. Hoffman, V. Agarwal, P. G. Williams, J. Dai, R. Neupane, J. Gurr, A. M. C. Rodriguez, A. Lamsa, C. Zhang, K. Dorrestein, B. M. Duggan, J. Almaliti, P.-M. Allard, P. Phapale, L.-F. Nothias, T. Alexandrov, M. Litaudon, J.-L. Wolfender, J. E. Kyle, T. O. Metz, T. Peryea, D.-T. Nguyen, D. VanLeer, P. Shinn, A. Jadhav, R. Muller, K. M. Waters, W. Shi, X. Liu, L. Zhang, R. Knight, P. R. Jensen, B. O. Palsson, K. Pogliano, R. G. Linington, M. Gutierrez, N. P. Lopes, W. H. Gerwick, B. S. Moore, P. C. Dorrestein and N. Bandeira, *Nat. Biotechnol.*, 2016, **34**, 828–837.

10 J. Watrous, P. Roach, T. Alexandrov, B. S. Heath, J. Y. Yang, R. D. Kersten, M. van der Voort, K. Pogliano, H. Gross, J. M. Raaijmakers, B. S. Moore, J. Laskin, N. Bandeira and P. C. Dorrestein, *Proc. Natl. Acad. Sci. U. S. A.*, 2012, **109**, E1743–E1752.

11 J. Y. Yang, L. M. Sanchez, C. M. Rath, X. Liu, P. D. Boudreau, N. Bruns, E. Glukhov, A. Wodtke, R. de Felicio, A. Fenner, W. R. Wong, R. G. Linington, L. Zhang, H. M. Debonsi, W. H. Gerwick and P. C. Dorrestein, *J. Nat. Prod.*, 2013, **76**, 1686–1699.

12 T. Depke, R. Franke and M. Brönstrup, *J. Chromatogr. B: Anal. Technol. Biomed. Life Sci.*, 2017, **1071**, 19–28.

13 T. Naake and E. Gaquerel, *Bioinformatics*, 2017, **33**, 2419–2420.

14 K. B. Kang, M. Ernst, J. J. J. van der Hooft, R. R. da Silva, J. Park, M. H. Medema, S. H. Sung and P. C. Dorrestein, *bioRxiv*, 2018.

15 M. Ernst, L.-F. Nothias-Scaglia, J. van der Hooft, R. R. Silva, C. H. Saslis-Lagoudakis, O. M. Grace, K. Martinez-Swatson, G. Hassemer, L. Funez, H. T. Simonsen, M. H. Medema, D. Staerk, N. Nilsson, P. Lovato, P. Dorrestein and N. Ronsted, *bioRxiv*, 2018.

16 L. Ridder, J. J. J. van der Hooft, S. Verhoeven, R. C. H. de Vos, J. Vervoort and R. J. Bino, *Anal. Chem.*, 2014, **86**, 4767–4774.

17 M. Vinaixa, E. L. Schymanski, S. Neumann, M. Navarro, R. M. Salek and O. Yanes, *TrAC, Trends Anal. Chem.*, 2016, **78**, 23–35.

18 P.-M. Allard, T. Péresse, J. Bisson, K. Gindro, L. Marcourt, V. C. Pham, F. Roussi, M. Litaudon and J.-L. Wolfender, *Anal. Chem.*, 2016, **88**, 3317–3323.

19 H. P. Benton, J. Ivanisevic, N. G. Mahieu, M. E. Kurczy, C. H. Johnson, L. Franco, D. Rinehart, E. Valentine, H. Gowda, B. K. Ubhi, R. Tautenhahn, A. Gieschen, M. W. Fields, G. J. Patti and G. Siuzdak, *Anal. Chem.*, 2015, **87**, 884–891.

20 *MzCloud Database*, http://www.mzcloud.org.

21 I. Blaženović, T. Kind, J. Ji and O. Fiehn, *Metabolites*, 2018, **8**, 31.

22 J. J. J. van der Hooft, S. Padmanabhan, K. E. V. Burgess and M. P. Barrett, *Metabolomics*, 2016, **12**, 1–15.

23 B. B. Misra and J. J. J. van der Hooft, *Electrophoresis*, 2016, **37**, 86–110.

24 F. Hufsky, K. Scheubert and S. Böcker, *TrAC, Trends Anal. Chem.*, 2014, **53**, 41–48.

25 Y. Wang, G. Kora, B. P. Bowen and C. Pan, *Anal. Chem.*, 2014, **86**, 9496–9503.

26 L. Ridder, J. J. J. van der Hooft, S. Verhoeven, R. C. H. de Vos, R. J. Bino and J. Vervoort, *Anal. Chem.*, 2013, **85**, 6033–6040.

27 L. Ridder, J. J. J. Van Der Hooft, S. Verhoeven, R. C. H. De Vos, R. Van Schaik and J. Vervoort, *Rapid Commun. Mass Spectrom.*, 2012, **26**, 2461–2471.

28 K. Dührkop, H. Shen, M. Meusel, J. Rousu and S. Böcker, *Proc. Natl. Acad. Sci. U. S. A.*, 2015, **112**, 12580–12585.

29 C. Brouard, H. Shen, K. Dührkop, F. d'Alché-Buc, S. Böcker and J. Rousu, *Bioinformatics*, 2016, **32**, i28–i36.

30 Y. Djoumbou Feunang, R. Eisner, C. Knox, L. Chepelev, J. Hastings, G. Owen, E. Fahy, C. Steinbeck, S. Subramanian, E. Bolton, R. Greiner and D. S. Wishart, *J. Cheminf.*, 2016, **8**, 61.

31 J. Wandy, Y. Zhu, J. J. J. van der Hooft, R. Daly, M. P. Barrett and S. Rogers, *Bioinformatics*, 2018, **34**, 317–318.

32 J. J. J. van der Hooft, J. Wandy, F. Young, S. Padmanabhan, K. Gerasimidis, K. E. V. Burgess, M. P. Barrett and S. Rogers, *Anal. Chem.*, 2017, **89**, 7569–7577.

33 F. Chollet, https://keras.io.

34 *RDKit: Open-Source Cheminformatics Software*, https://www.rdkit.org, 2018.

35 M. C. Burger, *J. Cheminf.*, 2015, **7**, 35.

36 J. Sztáray, A. Memboeuf, L. Drahos and K. Vékey, *Mass Spectrom. Rev.*, 2011, **30**, 298–320.

37 H. Oberacher, V. Reinstadler, M. Kreidl, A. M. Stravs, J. Hollender and L. E. Schymanski, *Metabolites*, 2019, **9**(1), 3.

Faraday Discussions

PAPER

Multivariate analysis applied to complex biological medicines†

Timothy R. Rudd, ID *ab Lucio Mauri,*c Maria Marinozzi,c Eduardo Stancanelli,c Edwin A. Yates, ID b Annamaria Naggic and Marco Guerrini ID c

Received 14th January 2019, Accepted 13th March 2019

DOI: 10.1039/c9fd00009g

A biological medicine (or biologicals) is a term for a medicinal compound that is derived from a living organism. By their very nature, they are complex and often heterogeneous in structure, composition and biological activity. Some of the oldest pharmaceutical products are biologicals, for example insulin and heparin. The former is now produced recombinantly, with technology being at a point where this can be considered a defined chemical entity. This is not the case for the latter, however. Heparin is a heterogeneous polysaccharide that is extracted from the intestinal mucosa of animals, primarily porcine, although there is also a significant market for non-porcine heparin due to social and economical reasons. In 2008 heparin was adulterated with another sulfated polysaccharide. Unfortunately this event was disastrous and resulted in a global public health emergency. This was the impetuous to apply modern analytical techniques, principally NMR spectroscopy, and multivariate analyses to monitor heparin. Initially, traditional unsupervised multivariate analysis (principal component analysis (PCA)) was applied to the problem. This was able to distinguish animal heparins from each other, and could also separate adulterated heparin from what was considered bona fide heparin. Taught multivariate analysis functions by training the analysis to look for specific patterns within the dataset of interest. If this approach was to be applied to heparin, or any other biological medicine, it would have to be taught to find every possible alien signal. The opposite approach would be more efficient; defining the complex heterogeneous material by a library of bona fide spectra and then filtering test samples with these spectra to reveal alien features that are not consistent with the reference library. This is the basis of an approach termed spectral filtering, which has been applied to 1D and 2D-NMR spectra, and has been very successful in extracting the spectral features of adulterants in heparin, as well as being able to differentiate supposedly biosimilar products. In essence, the filtered spectrum is determined by

aNational Institute for Biological Standards and Control (NIBSC), Blanche Lane, South Mimms, Potters Bar, Hertfordshire, EN6 3QG, UK. E-mail: tim.rudd@nibsc.org; Tel: +44 (0)1707641120

bDepartment of Biochemistry, Biosciences Building, University of Liverpool, Crown Street, Liverpool, L69 7ZB, UK

cIstituto di Ricerche Chimiche e Biochimiche 'G. Ronzoni', Via G. Colombo 81, 20133 Milano, Italy

† Electronic supplementary information (ESI) available. See DOI: 10.1039/c9fd00009g

subtracting the covariance matrix of the library spectra from the covariance matrix of the library spectra plus the test spectrum. These approaches are universal and could be applied to biological medicines such as vaccine polysaccharides and monoclonal antibodies.

Introduction

Biological medicines (or biologicals) are drugs that are derived from natural sources. They are, by definition, heterogeneous, which can be seen in both their composition and activity. Examples of biological medicines are vaccines, monoclonal antibodies and the family of heparin-based anticoagulants, the latter being amongst the most intrinsically diverse pharmaceutical products on the market.

Many physico-chemical techniques are used to characterise biological medicines. These include HPLC techniques, mass spectrometry and nuclear magnetic resonance (NMR) spectroscopy. Each of these techniques have their own strengths, with NMR spectroscopy being, in the authors opinion, one of the most adaptable. The technique can be used to fingerprint, determine the structure (chemical and physical) and quantify the amount of material present. An event in 2008, the contamination of heparin with oversulfated chondroitin sulfate,[1] further exemplified the usefulness of NMR spectroscopy, with the technique being used to determine the contaminant.[2] Since then, the interest in using NMR spectroscopy to characterise biological medicines has increased even more. The technique is readily applied to the heparin active pharmaceutical product and there is currently great interest in applying NMR spectroscopy to peptide/protein based products, for example, to the qNMR analysis of small peptides,[3] protamine sulfate[4] (reversal of heparin administration), copaxone[5] (glatiramer acetate, an immunomodulator used to treat multiple sclerosis) and monoclonal antibodies (immunotherapies for cancer and autoimmune diseases). These complex molecules are primarily fingerprinted using 1D and 2D-NMR spectroscopy.

The limitation of the manual spectral analysis of these biological medicines is the ability of the analyst to differentiate samples of interest when comparing complex 1D or 2D spectra, and the problem is further compounded when dealing with large datasets, where many samples are compared.

The solution to this is to use multivariate analysis, where complex datasets can be decomposed into a number of key trends that can be used to reconstruct the dataset, as well as where predictions about the sample(s) being analysed are made. These analyses fall into two camps, the first being untaught analysis, where the dataset is blindly analysed and the method differentiates the observations by correlations calculated between the variables. Examples of this type of analysis are principal component analysis or factor analysis. This type of analysis is very informative if the aim is to find the features within the dataset that discriminate the observations. The second type of analysis is taught or supervised analyses, and these are used where various parameters are known about an already existing dataset. This pre-existing dataset can then be used as a reference to compare a test sample against, allowing the parameter of interest to be determined. Analyses that fall into this category include partial least squares-discriminant analysis and orthogonal partial least squares analysis.

As previously mentioned, heparin is a biological medicine,[6,7] principally being derived from the intestines of pigs, but it is also extracted from cows. Heparin has been long established as an anticoagulant drug, which prevents or slows blood clotting, and it is very important for patients undergoing surgery, dialysis and during recovery from surgical procedures. It functions by interacting with a number of proteins of the blood clotting cascade, notably, but not limited to, antithrombin and thrombin.[8] It is composed of a linear, highly sulfated poly-saccharide chain of varying lengths, from 2 to 40 kDa. The carbohydrate is formed of repeating disaccharide units of 1,4 linked α-L-iduronic or β-D-glucuronic acid, and α-D-glucosamine. The predominant substitution pattern comprises 2-O-sulfation of the iduronate residues and N- and 6-O-sulfation of the glucosamine residues. The α-D-glucosamine residue can also be O-sulfated at position 3, and this is important for the molecule's antithrombotic properties.[7] Currently there is no alternative for these applications. It has also been proposed that sheep or camelids could be useful sources of heparin, as well as possibly non-mammalian animals.[9] Its diversity arises from manifold sources; the biosynthesis of heparin is complex involving many enzymes, the extraction method is initially mechanical in nature resulting in material of varying quality, and furthermore, once the mucosa has been extracted many steps of chemical purification, resin capture, precipitation and fractionation take place to produce a pure product, which is then bleached. This process produces a colourless and odourless material that is free from endotoxins, bacteria, mould, viruses and prions.[9,10] The bleaching step can also chemically modify the underlying polysaccharide structure. This diversity means that heparin is a challenging material to analyse, and it was this property that provided the opportunity for heparin to be adulterated with over-sulfated chondroitin sulfate.

NMR spectroscopy was used to identify the adulterant used to contaminate heparin,[2] and it was quickly realised by the research groups working on the problem that manually analysing the data would be inefficient. Principal component analysis (PCA) has been readily used to analyse heparin and model adulterated heparin samples.[11,12] Furthermore, taught analyses have been used to predict the amounts of known heparin contaminants (chondroitin sulfate and dermatan sulfate) present in test samples.[13–16] The techniques can also be applied to the more complex crude heparin, that is composed of heparin as well as other glycosaminoglycans.[17] Novel techniques were also applied, such as spectral filtering, to search for unknown contaminants in heparin.[18–21] The aims of all of these analyses have all been directed to the quality control of heparin, with the goal of detecting heparin samples that contain contaminants, such as chondroitin sulfate/dermatan sulfate, or adulterants, such as oversulfated chondroitin sulfate.

This is not the case for the analysis described within this manuscript. Here, a combination of 2D-NMR spectroscopy and PCA will be used to differentiate heparin from different animal sources. Even though the biosynthesis of heparin in the different animal sources uses the same biosynthetic pathway, the materials have different chemical structures. Normally, the structural differences would be elucidated by enzymic digestion followed by either HPLC or HPLC-MS. The benefit of using a combination of NMR spectroscopy and multivariate analysis is that the sample pre-treatment is minimal; 2 steps of D_2O exchange and lyophilisation and then final resuspension of the material in D_2O or a deuterated buffer containing a chemical shift reference. The experiment used here is a standard HSQC experiment found in the Bruker library.

Historically, the researchers involved in the analysis of heparin were early adopters of NMR spectroscopy, with ^{1}H and ^{13}C spectra successfully being used to characterise the material. One dimensional-NMR measurements of complex materials suffer from many overlapping signals and this problem can be ameliorated by using 2D-NMR experiments. Heteronuclear Single Quantum Coherence (^{13}C–^{1}H HSQC) spectra are two-dimensional containing correlations between ^{13}C atoms and the proton bound to them.

This dispersion in a second dimension means that the problem of overlapping signals is greatly diminished for heparin samples, although the problem is not eradicated entirely due to the heterogeneity of heparin.

The analysis contained within shows that the combination of ^{13}C–^{1}H HSQC NMR spectra and multivariate analysis (PCA) is able to differentiate heparin from different animal sources (porcine intestinal mucosa, bovine intestinal mucosa, ovine intestinal mucosa and bovine lung). Furthermore, if the relationships found within the data are examined, the spectral and therefore the chemical differences of the material can be revealed, thereby providing 2D-spectral fingerprints for the different heparins.

Methods

Materials

Heparin from porcine intestinal mucosa (PMH, 67 samples), bovine intestinal mucosa (BMH, 20 samples), ovine intestinal mucosa (OMH, 13 samples) and bovine lung (BLH, 6 samples) were sourced from different manufacturers. The PMH heparin represents samples from a number of different manufacturers, that have been sourced over many years. The material was lyophilised twice into D_2O. After the final freeze-drying step, the material was resuspended in 600 μL of 20 mM phosphate buffer, which also contained 3-(trimethylsilyl)propionic-2,2,3,3-d_4 acid (TSP) as a chemical shift reference.

NMR spectroscopy

The HSQC (^{13}C–^{1}H) spectra were measured on a Bruker AVANCE III 600 MHz spectrometer (Karlsruhe, Germany), equipped with a TCI 5 mm cryoprobe, using the Bruker hsqcetgpsisp2.2 (Phase-sensitive ge-2D HSQC using PEP and adiabatic pulses for inversion and refocusing with gradients in back-inept) pulse sequence. The experiments were recorded at 298 K using the following acquisition parameters: number of scans 12, number of dummy scans 16, relaxation delay 2.5 s, spectral width 8 ppm (F2) and 80 ppm (F1), transmitter offset 4.7 ppm (F2) and 80 ppm (F1), and $^1J_{CH}$ = 150 Hz.

Multivariate analysis

The spectra were processed so that F2 was comprised of 8 k points and 2 k in the F1 dimension. Importantly, before converting the proprietary HSQC NMR spectra into numerical matrices, the offsets for every spectrum were set as the same values (F1 and F2). These values were for the first spectrum in the dataset, which had been calibrated correctly (TSP set to ^{1}H and ^{13}C equal to 0 ppm). The spectra were processed using Topspin software version 4.0.4 (Bruker BioSpin, Rhein-stetten, Germany). Principal component analysis (PCA) of the HSQC NMR spectra was carried out using R (R: A Language and Environment for Statistical Computing[22]), and the 2D spectra were imported into R using the rNMR package.[23] This involves reading the acquisition (*parseAcqus*) and processing (*parseProcs*) parameters, and the spectra are then converted into the sparky format (*bruker2D*), and then they are finally imported (*ucsf2D*) into R as a matrix. Before the spectra are analysed, they are aligned, normalised for area, and mean centred.[24] PCA is then performed using the prcomp function. All the spectra were assigned the same offset values. Once the spectra were imported into R they were peak picked and then aligned to the signal due to I1(2OH)-A(6S), which can be found between 5.05–4.97 ppm ^{1}H and 105.5–104.5 ppm ^{13}C. This signal was chosen as it is insensitive to its environment, so it is not readily perturbed. The script to perform this task was written in-house. Due to the large size of the dataset involved in the analyses, cross-validation was very time consuming. To this end, a method was used that is an approximation of the leave-one-out cross-validation methods – the general cross-validation.[25] This method is found in the R package FactoMineR,[26] implemented by the function *estim_ncp*.

Results and discussion

The aim of many multivariate analysis techniques is to reduce complex datasets to a number of key trends found within the dataset, that explain the variation within those data. This is the aim of techniques such as principal component analysis, single value decomposition, and factor analysis, to highlight three.

Here, PCA[27] is used to explore the ^{13}C–^{1}H HSQC NMR spectra of heparins from different animal sources. Using ^{13}C–^{1}H HSQC NMR spectra to analyse heparin has one major advantage over ^{1}H NMR spectra. That is signal dispersion, which enables features to be assigned. Furthermore, the ^{13}C–^{1}H HSQC NMR experiment allows information to be gathered regarding the environment surrounding the ^{13}C nuclei present in heparin in less time than a standard 1D-^{13}C NMR experiment.

To avoid artefacts arising in the PCA, a number of steps have to be taken. Firstly, care has to be taken preparing the samples; samples were lyophilised into D_2O to reduce the signal from water, furthermore the samples were reconstituted in a deuterated phosphate buffer, reducing any problems arising from the variations in pH. Secondly, the authors have noted that when preparing the data for analysis, the spectral offset (the furthest limits of the spectra in the F1 and F2 dimension) should be kept constant for the whole dataset. This may change if O1 (the centre of the direct dimension) is allowed to be determined for every experiment and even if the HSQC spectra are calibrated, they may still require internal alignment to avoid artefacts from ghost spectral shifts. The pre-treatment of the dataset that contained all of the HSQC spectra was simple. It was found that normalising the spectra for area and then mean-centring provided the best performance. Previously, the authors have found that when performing multivariate analysis of the 1D-NMR spectra of heparin, the additional normalisation of the data for area and mean centring, as well as Pareto scaling gave the best performance.[11]

PCA of PMH HSQC NMR spectra

Principal component analysis was performed on a dataset containing the ^{13}C–^{1}H HSQC spectra of 67 PMH samples. The analysis decomposed the dataset into 5 components, which explained 60.00% of the variance of the data; component 1 (36.52%), component 2 (8.23%), component 3 (6.22%), component 4 (5.04%) and component 5 (4.00%). Component 1 differentiated samples by the level of sulfation, epimerisation of the uronic acid and the linkage region (Fig. 1A). The linkage region is a tetrasaccharide at the non-reducing end of the carbohydrate that links the polysaccharide to a protein core; the sequence of this tetrasaccharide is GlcA-Gal-Gal-Xyl-serine.[28] The features in red are due to the highly sulfated parts of the polysaccharide, and predominantly show signals from the major trisulfated disaccharide IdoA(2S)–GlcNS(6S). The blue features are due to regions in the heparin chain that contain low sulfation and the linkage region that links the polysaccharide to its protein core. Component 2 (Fig. 1B) is a little subtler. Samples are differentiated in this component by the varying levels of sulfation contained, not the stark differences found in component 1 (Fig. 1A). Interestingly, in both components 1 and 2 (Fig. 1A and B, respectively), signals are seen for the rare disaccharide IdoA(2OH)–GlcA(NH$_2$), as well as signals due to chemical modifications arising in the chain from the manufacturing process, including epoxidation, in component 2. Component 3 (Fig. 1C) shows features specifically due to the linkage region, indicating that the rare GlcA(NH$_2$) residue is correlated with the linkage region, and that this region of the chains adjacent to the linkage tetrasaccharide contains different levels of sulfation. Again, components 4 and 5 (Fig. 1D and E, respectively) differentiate the heparin samples by subtle features in the chain that contain varying levels of sulfation, uronic acid epimerisation and linkage region content. As with component 2, component 4 contains signals arising from chemical modifications in the chain, arising from the manufacturing process, and this time they are from the galacturonic acid residue.

PCA of PMH HSQC NMR spectra compared or HSQC NMR spectra of BMH, BLH and OMH

Principal component analysis was then performed on the PMH HSQC NMR spectra dataset, comparing it to the spectra of BMH, BLH and OMH. Unsurprisingly, the analysis was able to differentiate the other types of heparin from PMH. The individual datasets for BMH, BLH and OMH were not analysed separately as they did not contain sufficient spectra to draw meaningful conclusions. This is a common error made by analysts when PCA is performed.

The comparison of 20 BMH HSQC NMR spectra with the 67 PMH HSQC NMR spectra by PCA found two significant components, one major and one minor (component 1 62.30% and component 2 12.08%, Fig. 2A). The BMH samples are clearly differentiated from the PMH samples in component 1 (Fig. 2B and C). Bovine intestinal mucosal heparin has varying levels of O-sulfation at position 6 and this can clearly be seen in component 1 (Fig. 2C, blue features), as well as signals arising from GlcA-Glc(NAc), GlcA-Glc(NS) and GlcA(2S).

While the PMH samples analysed have higher levels of the standard disaccharide IdoA(2S)–GlcNS(6S), as well as containing more of the linkage region (GlcA-Gal-Gal-Xyl-serine), signals also arose from the trisulfate glucosamine

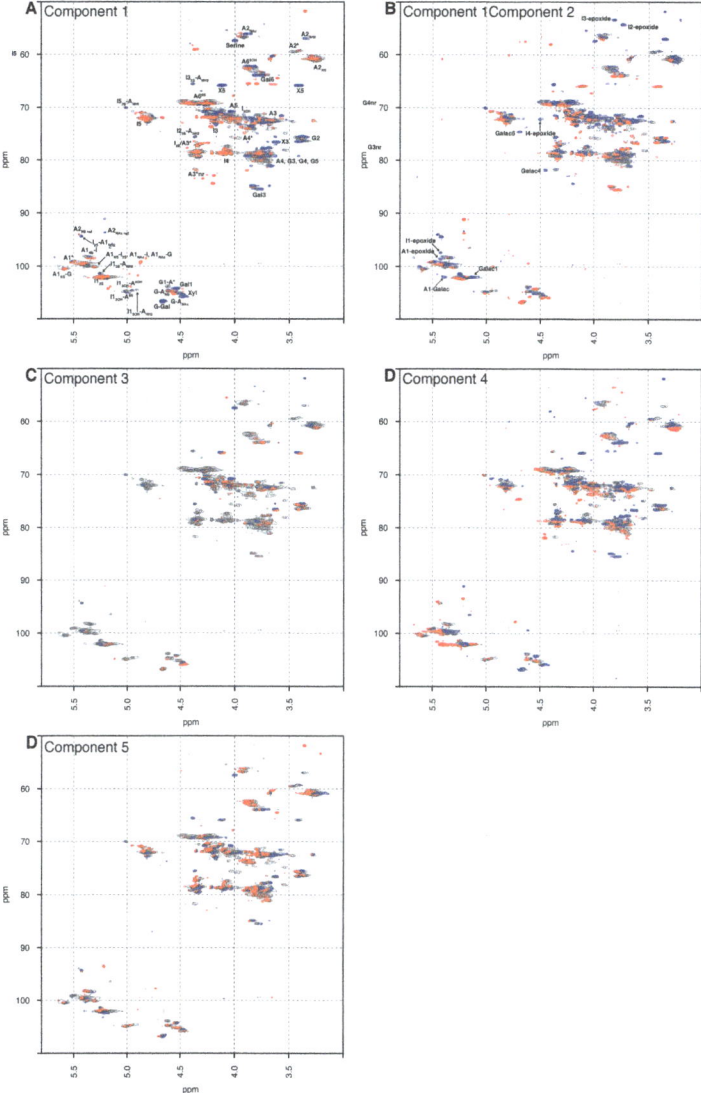

Fig. 1 Principal component analysis of 67 PMH $^{13}C-^{1}H$ HSQC spectra. Prior analysis was performed: the spectra were aligned, normalised for area and mean centred. The analysis decomposed the dataset into 5 major components, and the figure shows the score plots of these 5 components, panels (A) to (E), respectively. The 5 components chosen here explain 60.00% of the variance contained within the dataset. The percentages of variance explained by each component are as follows: 36.52%, 8.23%, 6.22%, 5.40% and 4.00% of the variance, respectively. The scree and score plots can be found in the ESI.† I stands for iduronate, A for glucosamine, and nr indicates that the residue is at the nonreducing end of the molecule. The sub- and superscripts denote the position of sulfation (S) or acetylation (Ac), respectively. AN and IN refer to position N (either C atom or H atom depending on the context) of the glucosamine or iduronate residue, respectively. For example, I_{2S}-A_{NS}^{6S} corresponds to the disaccharide 2-O sulfated iduronic acid linked to 6-O-sulfated N-sulfated glucosamine. A2* signifies position 2 of glucosamine, which is N-sulfated and O-sulfated at positions 6 and 3. IN-epoxide indicates that the iduronate has undergone epoxidation and galac indicates a galacturonic acid residue. Cross-validation of the dataset found that 11 components would explain the variance present in the PMH dataset (see Methods section).

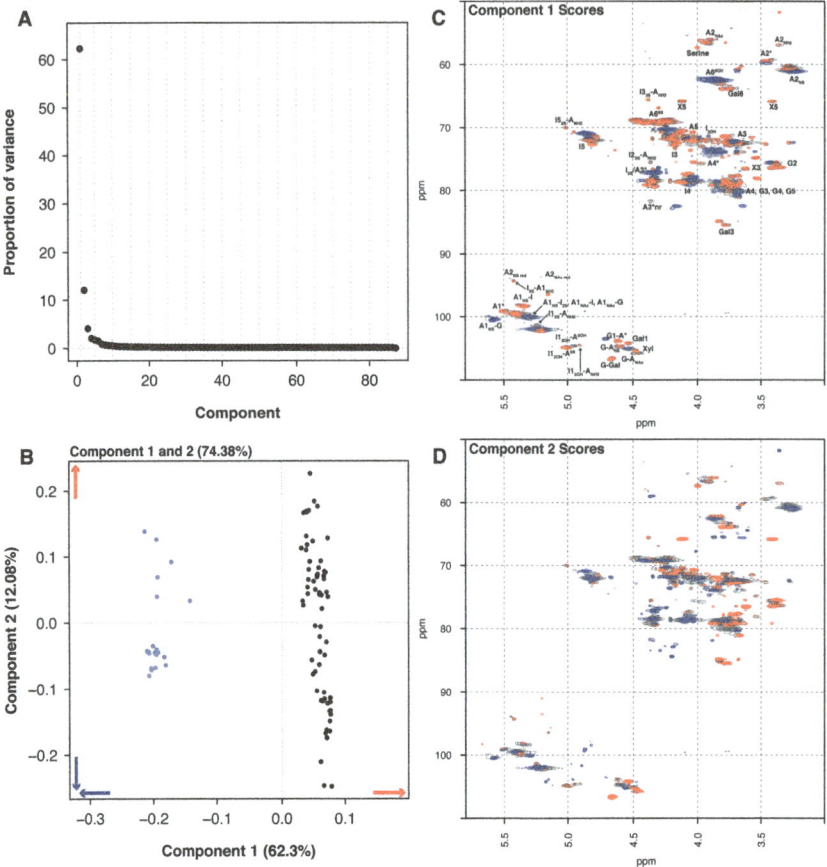

Fig. 2 PMH $^{13}C-^{1}H$ HSQC spectra compared to BMH $^{13}C-^{1}H$ HSQC spectra. Principal component analysis of a dataset composed of 67 PMH and 20 BMH $^{13}C-^{1}H$ HSQC spectra. Prior to analysis, the spectra were aligned, normalised for area and mean centred. The analysis decomposed the dataset into 2 major components explaining 74.38% of the total variance. (A) Scree plot and (B) loading plot (BMH samples are light blue, while the PMH samples are black). The figure shows the score plots of components 1 (62.30%) (C) and 2 (12.08%) (D). Porcine intestinal mucosal heparin is differentiated from BMH by component 1 (B and C). The blue features observed in component 1 (C) are more prevalent in the BMH spectra and the red features are more prevalent in the PMH spectra. Cross-validation of the dataset found that 14 components would explain the variance present in the PMH–BMH dataset (see Methods section).

(Glc(3S,6S,NS)) which is important for the antithrombotic activity of the molecules and disulfate iduronic acid linked to 6-O-sulfated glucosamine (IdoA(2OH)–Glc(6S)). Component 2 differentiated samples based on their overall sulfation level (Fig. 2D), separating both PMH and BMH.

Another source of pharmaceutical heparin that is being considered is sheep. Many regions of the world consume large amounts of lamb or mutton, and therefore a significant amount of ovine mucosa is available. As with the BMH material, OMH is distinct from PMH and PCA of the HSQC NMR spectra can differentiate PMH from OMH. Two significant components are found by PCA, similarly with 1 major and 1 minor component (component 1 52.4% and component 2 9.14%, these two components explain 61.54% of the variance found in the dataset, Fig. 3A). The OMH and PMH samples are differentiated by component 1 (Fig. 3C). The blue features observed in the score plot for component 1 are those that are more prevalent in OMH. The OMH samples have a different amount of the standard IdoA(2S)–GlcA(NS,6S) disaccharide to that seen in PMH. Interestingly, signals due to the trisulfated glucosamine (Glc(3S,6S,NS)) indicate that the antithrombin binding site found in OMH is distinct to that found in PMH. These are signals for positions 1 and 2 of

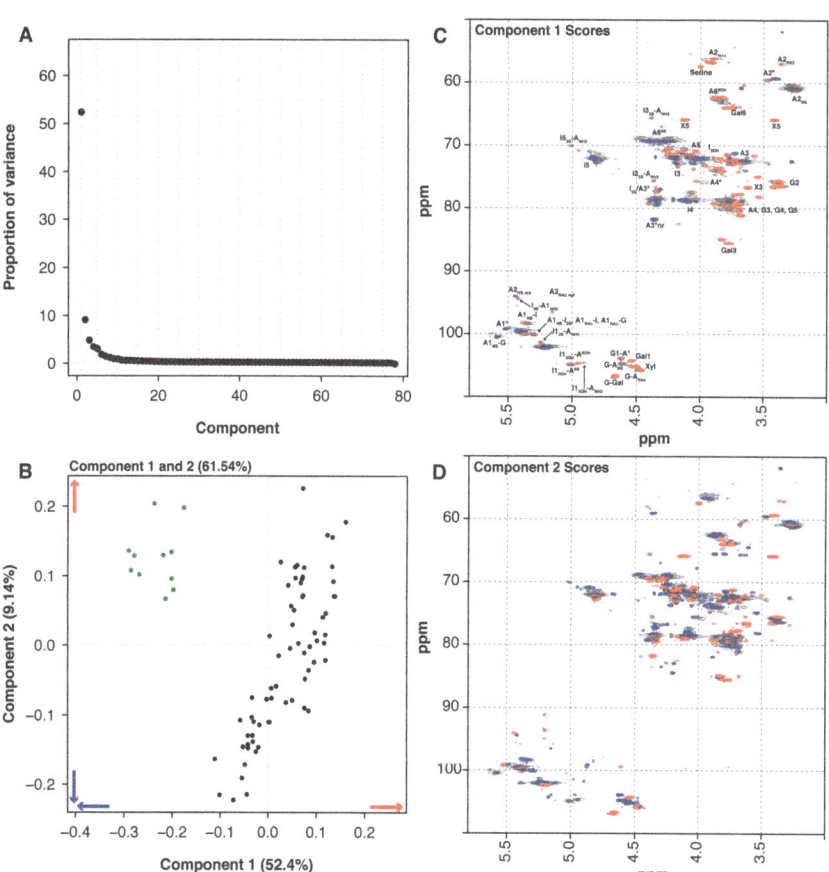

Fig. 3   PMH $^{13}C-^1H$ HSQC spectra compared to OMH $^{13}C-^1H$ HSQC spectra. Principal component analysis of a dataset composed of 67 PMH and 13 OMH $^{13}C-^1H$ HSQC spectra. Before the analysis was performed, the spectra were aligned, normalised for area and mean centred. The analysis decomposed the dataset into 2 major components explaining 61.54% of the total variance. (A) Scree plot and (B) loading plot (OMH samples are green, while the PMH samples are black). The figure shows the score plots of components 1 (52.40%) (C) and 2 (9.14%) (D). Porcine intestinal mucosal heparin is differentiated from OMH by component 1 (B and C). The blue features observed in component 1 (C) are more prevalent in the OMH spectra and the red features are more prevalent in the PMH spectra. Cross-validation of the dataset found that 14 components would explain the variance present in the PMH–OMH dataset (see Methods section).

Glc(3S,6S,NS), as well as position 3 of Glc(3S,6S,NS) located at the non-reducing end of the polysaccharide. As can be seen from the loading plot, the samples from OMH and PMH are not completely orthogonal, so the major variation that differentiates OMH from PMH also arises within the PMH samples. The red features in component 1 (Fig. 3C) are those found more prevalently in the PMH samples and contain signals due to the less sulfate residues, GlcA containing disaccharides and the linkage region. These observations suggest that the OMH samples analysed here have a more homogeneous sequence than the PMH samples. Component 2 disperses the PMH samples (Fig. 3D), with the PMH

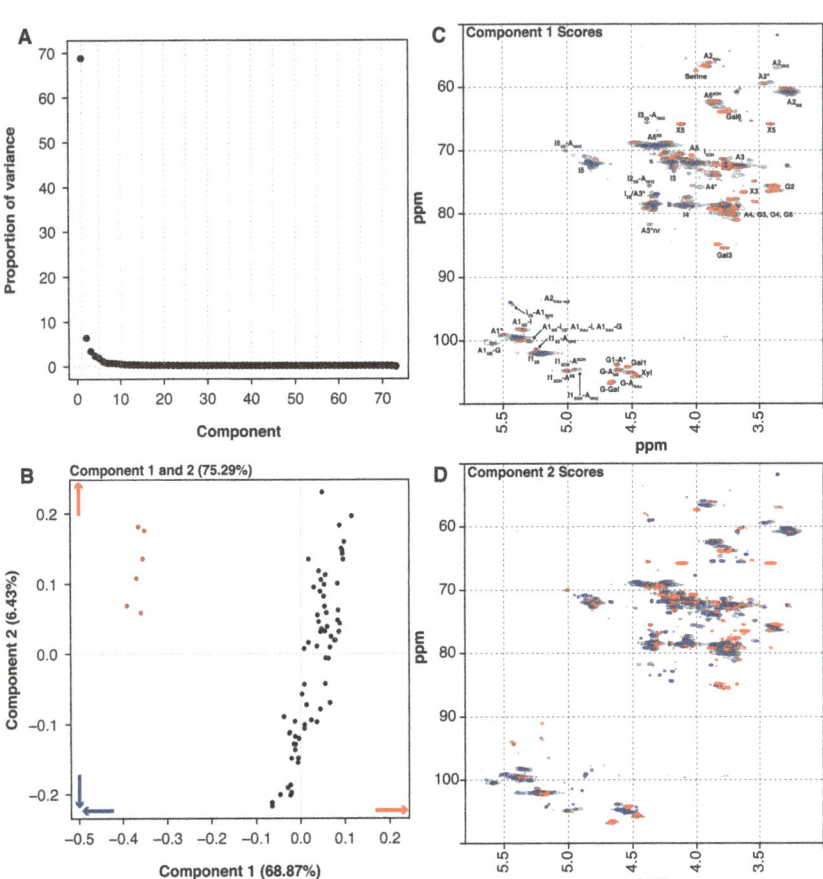

Fig. 4   PMH $^{13}C-^1H$ HSQC spectra compared to BLH $^{13}C-^1H$ HSQC spectra. Principal component analysis of a dataset composed of 67 PMH and 6 BLH $^{13}C-^1H$ HSQC spectra. Before the analysis was performed, the spectra were aligned, normalised for area and mean centred. The analysis decomposed the dataset into 2 components (1 major and 1 nominal minor component), explaining 75.3% of the total variance. (A) Scree plot and (B) loading plot (BLH samples are red, while the PMH samples are black). The figure shows the score plots of components 1 (68.87%) (C) and 2 (6.43%) (D). Porcine intestinal mucosal heparin is differentiated from BLH by component 1 (B and C). The blue features observed in component 1 (C) are more prevalent in the BLH spectra and the red features are more prevalent in the PMH spectra. Cross-validation of the dataset found that 14 components would explain the variance present in the PMH–BLH dataset (see Methods section).

samples containing varying amounts of the component. As can be seen from the loading plot for the analysis (Fig. 3B), the OMH samples only contain the positive features of component 2, which contains signals from the linkage region, as well as signals for the standard IdoA(2S)–GlcA(NS,6S) disaccharide and the trisulfated glucosamine (Glc(3S,6S,NS)). This suggests that the non-reducing end of the OMH samples is, on the whole, more sulfated than the same region found in the PMH samples and, potentially, it also contains a possibly distinct antithrombin binding site.

Historically, heparin was sourced from both cows and pigs, however the emergence of bovine spongiform encephalopathy (BSE) put an end to the use of bovine heparin in most of the world, due to safety concerns. When heparin was widely sourced from cows, the material was extracted from both the intestinal mucosa and lungs. The 6 BLH samples analysed here are distinct from the 67 PMH samples. PCA of the dataset containing the BLH and PMH HSQC NMR spectra isolated 2 significant components, 1 major and 1 minor (component 1 68.87% and component 2 6.43%, these two components explain 75.30% of the variance found in the dataset) (Fig. 4A). The BLH samples have a very homogenous structure, being enriched in the standard IdoA(2S)–GlcA(NS,6S) disaccharide, which is evident in component 1 (Fig. 4C). The PMH samples were dispersed by component 2 (Fig. 4B). The blue signals seen in the score plot for component 2 are the features that separate the PMH samples (Fig. 4D). The PMH samples contain varying levels of the signals originating from the trisulfated glucosamine (Glc(3S,6S,NS)) residue, positions 1, 2 and 4 of Glc(3S,6S,NS), and position 1 of GlcA attached to Glc(3S,6S,NS), indicating that the antithrombin site within PMH is different to that seen in BLH. The BLH samples only contain the red features observed in component 2, the minor signals (Fig. 4C) corresponding with the major repeating disaccharide observed in component 1 (Fig. 4D).

The pairwise approach here allows the differences between PMH and BMH, OMH or BLH to be investigated. This analysis can be expanded to look at global differences between the heparins from 4 difference sources. The ESI† contains the PCA of all the heparin HSQC spectra; components 1 and 2 differentiate the four heparin. Component 1 differentiates PMH and OMH from BLH and BMH, and component 2 differentiates the heparin from the bovine sources.

Conclusions

Multivariate analysis techniques provide a powerful toolbox that can be used to analyse the most complicated mixtures. The material of interest in this paper, heparin, is a highly heterogeneous polysaccharide comprising of chains of varying length, charge and substitution pattern. The application of PCA to the ^{13}C–^{1}H HSQC NMR spectra of heparin allowed heparin from different animal sources and organs to be differentiated. Furthermore, the analysis extracted spectral signatures that are specific to the 4 heparin types (porcine intestinal mucosa, bovine intestinal mucosa, ovine intestinal mucosa and bovine lung). While the ^{13}C–^{1}H HSQC NMR spectrum of heparin provides a great deal of information, the analysis performed here is mainly qualitative, although through the integration of a number of signals in the ^{13}C–^{1}H HSQC NMR spectra of heparin samples, the average disaccharide composition of the polysaccharide can be determined.[29,30] These approaches are much quicker than other traditional methods, such as digestion followed by HPLC or HPLC–MS, and require much less preparation time.

Such approaches are highly valuable to the quality control of the heparin pharmaceutical product; the NMR experiment, spectral processing and subsequent multivariate analysis could all be performed within one working day, with the only barrier being the exchange of the sample into D_2O. This exchange could be circumvented, and the measurement performed in 90% H_2O/10% D_2O. The only drawback would be that the water signal may obscure signals of interest and further complications could be caused by the presence of signals from exchange protons.[31]

The HSQC spectra provide information regarding the average electronic environment surrounding the hydrogen and carbon nuclei present in the carbohydrate. One important piece of data that is lacking is information regarding the sequence/substitution pattern found within the carbohydrate. To provide such information it might be necessary to perform different NMR experiments, possibly analysing datasets of TOCSY or NOESY experiments, or by analysing datasets that contain different experiment types, for example HSQC and TOCSY spectra. The only limitation would be time, since both high quality NOESY and TOCSY spectra take much more time to record than the equivalent HSQC spectrum.

Conflicts of interest

There are no conflicts to declare.

Acknowledgements

The authors would like to acknowledge the contribution of the late Professor Benito Casu, without whom, these developments would not have been possible.

Notes and references

1 T. K. Kishimoto, K. Viswanathan, T. Ganguly, S. Elankumaran, S. Smith, K. Pelzer, J. C. Lansing, N. Sriranganathan, G. Zhao, Z. Galcheva-Gargova, A. Al-Hakim, G. S. Bailey, B. Fraser, S. Roy, T. Rogers-Cotrone, L. Buhse, M. Whary, J. Fox, M. Nasr, G. J. Dal Pan, Z. Shriver, R. S. Langer, G. Venkataraman, K. F. Austen, J. Woodcock and R. Sasisekharan, *N. Engl. J. Med.*, 2008, **358**, 2457–2467.

2 M. Guerrini, D. Beccati, Z. Shriver, A. Naggi, K. Viswanathan, A. Bisio, I. Capila, J. C. Lansing, S. Guglieri, B. Fraser, A. Al-Hakim, N. S. Gunay, Z. Zhang, L. Robinson, L. Buhse, M. Nasr, J. Woodcock, R. Langer, G. Venkataraman, R. J. Linhardt, B. Casu, G. Torri and R. Sasisekharan, *Nat. Biotechnol.*, 2008, **26**, 669–675.

3 C. Li, S. Bhavaraju, M. P. Thibeault, J. Melanson, A. Blomgren, T. Rundlof, E. Kilpatrick, C. J. Swann, T. Rudd, Y. Aubin, K. Grant, M. Butt, W. Shum, T. Kerim, W. Sherwin, Y. Nakagawa, S. Pavon, S. Arrastia, T. Weel, A. Pola, D. Chalasani, S. Walfish and F. Atouf, *J. Pharm. Biomed. Anal.*, 2019, **166**, 105–112.

4 A. C. Gucinski, M. T. Boyne II and D. A. Keire, *Anal. Bioanal. Chem.*, 2015, **407**, 749–759.

5 S. Rogstad, E. Pang, C. Sommers, M. Hu, X. Jiang, D. A. Keire and M. T. Boyne II, *Anal. Bioanal. Chem.*, 2015, **407**, 8647–8659.

6 T. W. Barrowcliffe, *Handb. Exp. Pharmacol.*, 2012, **207**, 3–22, DOI: 10.1007/978-3-642-23056-1_1.

7 D. L. Rabenstein, *Nat. Prod. Rep.*, 2002, **19**, 312–331.

8 B. Mulloy, J. Hogwood, E. Gray, R. Lever and C. P. Page, *Pharmacol. Rev.*, 2016, **68**, 76–141.

9 J. Y. van der Meer, E. Kellenbach and L. J. van den Bos, *Molecules*, 2017, **22**, 1025.

10 R. J. Linhardt and N. S. Gunay, *Semin. Thromb. Hemostasis*, 1999, **25**(suppl. 3), 5–16.

11 T. R. Rudd, D. Gaudesi, M. A. Skidmore, M. Ferro, M. Guerrini, B. Mulloy, G. Torri and E. A. Yates, *Analyst*, 2011, **136**, 1380–1389.

12 T. R. Rudd, M. A. Skidmore, S. E. Guimond, C. Cosentino, G. Torri, D. G. Fernig, R. M. Lauder, M. Guerrini and E. A. Yates, *Glycobiology*, 2009, **19**, 52–67.

13 Q. Zang, D. A. Keire, L. F. Buhse, R. D. Wood, D. P. Mital, S. Haque, S. Srinivasan, C. M. Moore, M. Nasr, A. Al-Hakim, M. L. Trehy and W. J. Welsh, *Anal. Bioanal. Chem.*, 2011, **401**, 939–955.

14 Q. Zang, D. A. Keire, R. D. Wood, L. F. Buhse, C. M. Moore, M. Nasr, A. Al-Hakim, M. L. Trehy and W. J. Welsh, *J. Pharm. Biomed. Anal.*, 2011, **54**, 1020–1029.

15 Q. Zang, D. A. Keire, R. D. Wood, L. F. Buhse, C. M. Moore, M. Nasr, A. Al-Hakim, M. L. Trehy and W. J. Welsh, *Anal. Bioanal. Chem.*, 2011, **399**, 635–649.

16 Q. Zang, D. A. Keire, R. D. Wood, L. F. Buhse, C. M. Moore, M. Nasr, A. Al-Hakim, M. L. Trehy and W. J. Welsh, *Anal. Chem.*, 2011, **83**, 1030–1039.

17 L. Mauri, M. Marinozzi, G. Mazzini, R. E. Kolinski, M. Karfunkle, D. A. Keire and M. Guerrini, *Molecules*, 2017, **22**, 1146.

18 M. Guerrini, T. R. Rudd, L. Mauri, E. Macchi, J. Fareed, E. A. Yates, A. Naggi and G. Torri, *Anal. Chem.*, 2015, **87**, 8275–8283.

19 T. R. Rudd, D. Gaudesi, M. A. Lima, M. A. Skidmore, B. Mulloy, G. Torri, H. B. Nader, M. Guerrini and E. A. Yates, *Analyst*, 2011, **136**, 1390–1398.

20 T. R. Rudd, E. Macchi, L. Muzi, M. Ferro, D. Gaudesi, G. Torri, B. Casu, M. Guerrini and E. A. Yates, *Anal. Chem.*, 2013, **85**, 7487–7493.

21 T. R. Rudd, E. A. Yates and M. Guerrini, New Methods for the Analysis of Heterogeneous Polysaccharides – Lessons Learned from the Heparin Crisis, in *New Developments in NMR*, 2017, pp. 305–334.

22 R Core Team, *R: A language and environment for statistical computing*, R Foundation for Statistical Computing, Vienna, Austria, 2018, https://www.R-project.org/.

23 I. A. Lewis, S. C. Schommer and J. L. Markley, *Magn. Reson. Chem.*, 2009, **47**(suppl. 1), S123–S126.

24 R. A. van den Berg, H. C. Hoefsloot, J. A. Westerhuis, A. K. Smilde and M. J. van der Werf, *BMC Genomics*, 2006, **7**, 142.

25 J. Josse and F. Husson, *Comput. Stat. Data Anal.*, 2012, **56**, 1869–1879.

26 S. Lê, J. Josse and F. Husson, *J. Stat. Softw.*, 2008, **1**(1), DOI: 10.18637/jss.v025.i01.

27 I. T. Jolliffe, *Principal component analysis*, Springer-Verlag, New York, 2002.

28 M. Iacomini, B. Casu, M. Guerrini, A. Naggi, A. Pirola and G. Torri, *Anal. Biochem.*, 1999, **274**, 50–58.

29 M. Guerrini, A. Bisio and G. Torri, *Semin. Thromb. Hemostasis*, 2001, **27**, 473–482.

30 L. Mauri, G. Boccardi, G. Torri, M. Karfunkle, E. Macchi, L. Muzi, D. Keire and M. Guerrini, *J. Pharm. Biomed. Anal.*, 2017, **136**, 92–105.

31 C. N. Beecher and C. K. Larive, *Anal. Chem.*, 2015, **87**, 6842–6848.

PAPER

Resolving complex hierarchies in chemical mixtures: how chemometrics may serve in understanding the immune system

Gerjen Herman Tinnevelt (iD) * and Jeroen Jasper Jansen (iD)

Received 8th January 2019, Accepted 5th February 2019

DOI: 10.1039/c9fd00004f

In immunology, the resolution of complex chemical mixtures familiar from omics, comes with an added layer of hierarchy: bioactive immunological surface markers are embedded on the cell membranes of *e.g.* white blood cells. Therefore, each blood sample actually consists of a comprehensive mixture of cells. The cells need to be resolved based on their surface marker chemistry, to investigate their involvement in an immune response. This mixture may be measured on a single-cell level with Multicolour Flow Cytometry (MFC). Finding such cellular and molecular markers is of the utmost academic and diagnostic importance. Several advanced data analysis methods therefore aim to meet the considerable data challenge of resolving such cell mixtures. These multivariate methods are more resource-efficient than the manual analysis of MFC data, called sequential gating, but also likely provide additional biomedical insight compared to the conventional bivariate approach. To compare such methods more comprehensively than has been done until now, we have developed a list of criteria on how each method recovers the information on both the cell and the underlying molecular levels on an MFC sample of an asthma patient. We compare these methods for the chemometric data analysis commonly used in metabolomics. This shows that all compared methods have their own advantage in recovering the sequential gating results, giving insight into the limitations of sequential gating, providing insight into the chemical relationships between cells within the mixture and resolving information related to chemical heterogeneities between cells. We furthermore show how comparative analyses of different samples may lead to further insight into the subdivision of cells into different types based on their immunological involvement in asthma development, and how sparsity—a currently popular method to enhance the discriminative ability of multivariate models—may reduce the insight into the underlying hierarchical variability in cell chemistry. Although developed for cytometry, the presented chemometrics will be highly valuable to many more chemical systems where hierarchical arrangement of the molecules plays a crucial role.

Radboud University, Institute for Molecules and Materials, (Anaytical Chemistry), Heyendaalsweg 135, Nijmegen, Netherlands. E-mail: chemometrics@science.ru.nl

Introduction

Advances in analytical technology have led to considerably broader and deeper insight into biomedical systems. Omics technologies have greatly increased the breadth of molecular species,[1] simultaneously interrogated for disease involvement. Immunology, however, has focused on the analysis of mixtures of protein molecules expressed on the surface of specific cells,[2] to assess their role within the immune system. The true value of such analytical technologies only comes forward in the translation of the relatively abstract data they provide, e.g. spectra or chemical profiles, into evidence-based biomedical decision support. Such translation affects the interpretation of the result, which is essential for the end-user to understand how a specific model may support research or treatment decisions.

Capture of the considerable diversity in surface protein expression on a mixture of single cells requires separate analysis of the quantitative surface protein expression on each individual cell in e.g. a blood sample. Multicolor Flow Cytometry (MFC) may perform this in high-throughput[3,4] by measuring fluorescently conjugated antibodies specifically attached to the surface proteins on the membranes of white blood cells. A laser then excites every cell, which was previously brought into a laminar flow of thousands to millions of cells. The measured fluorescence is then a quantitative readout of the targeted surface protein expression on each cell.

MFC may be used to detect the pedigrees of white blood cell types that exist within the immune system. Identification takes place by a relatively small number of around 250 surface proteins,[2] where, currently, standard MFC technology allows the simultaneous measurement of eight—although this number is consistently being increased through technological innovations.[5-7] The variability in the quantity and quality of the surface proteins on a cell, however, gives rise to a considerable diversity in both known and unknown cell types, making flow cytometry a potential member of the omics family as a profiling or fingerprinting technology.[8]

Multicolor flow cytometry, therefore, generates data with a hierarchy of information: the measured molecular mixture on each cell determines its identity, 'type', while the number of cells of that type—in combination with all other cell types—determines the activity of the immune system. Understanding the system requires an understanding of how many cells of each type it contains and how much of every surface protein they express, also compared to other samples.

Data from an MFC sample is conventionally analyzed by so-called 'gating': arranging cells into pre-specified types by sequentially setting thresholds on each surface protein expression, either alone or in selected bivariate combinations.[9] This provides fractions of each cell type within each sample that may be tested e.g. between the control and clinical phenotype samples. Manual gating is therefore resource-intensive, potentially subjective, and expertise-dependent. It precludes analyses of more than two proteins simultaneously and thereby limits the discovery of novel, as yet unknown, cellular activity, as is the objective in omics. The discovery of hitherto unknown systematic continua in protein expression in cells that until then were believed to belong to a group of cells with homogeneous protein expression requires more efficient and automated methods that are less-reliant on prior information.

This need has led to a number of high-impact publications on quantitative methodologies for MFC data[10-12] that show the strength of machine learning to cluster the single-cell MFC data into aggregates that align with different cell types. They make extensive use of the recent developments in data science and bioinformatics. However, the connection to the considerable data analysis expertise in metabolomics is sparse. Some authors[13] even explicitly dismiss the contributions of chemometric methods, such as principal component analysis, to the systemic analysis of omics data. The chemometrics expertise in the quantitative processing of mass spectrometry and nuclear magnetic resonance spectroscopy data may, however, be both insightful and powerful in resolving chemical variability. Chemometrics is very strong in validation, visualization and data pre-processing. We have recently introduced the hierarchical perspective essential for MFC in the Discriminant Analysis of MultiAspect Cytometry (DAMACY)[14] method, with which the abundance of cell types can be systematically compared between samples.

In this paper, we explore the ways in which manual sequential gating, machine learning and chemometrics compare, and show complementary strength in the analyses of the hierarchies of MFC data, to resolve molecular and cell mixtures into insightful contributions to the immune system. We compare SPADE, Flow-SOM, t-SNE and PCA biplots, by evaluating how they reproduce sequential gating. Comprehensive assessment and comparison of unsupervised models is challenging. It may be facilitated by an error free 'golden standard'. Such a comparison was done in FlowCAP.[15] However, FlowCAP only quantitatively compares these models based on heuristic measures but not on how well the models are interpreted and how they extract information from the data. We base our analysis on comparisons with sequential manual gating, in terms of resolution, cluster size, recovery of rarer cell types and systematic heterogeneity in surface protein expressions. We use a representative asthma sample from an MFC analysis for the immune response associated to asthma.

Moreover, we evaluate how these methods provide complementary information on the cell composition, in terms of finding a mismatch in the sequential gating where only univariate or bivariate combinations are used instead of the whole multivariate space and in terms of finding new cell subtypes or cellular heterogeneity. Finally, we compare how these methods provide chemical insight in terms of the relationship between proteins or cell clusters, and whether the methods have a heuristic measure of the quality of the model.

In the second part, we compare all samples from the asthma study by calculating the percentage of each cell type found with sequential gating of each individual and subsequently performing multiple two sample t-tests or PCA and compare this with the multivariate method DAMACY and Citrus[16] in terms of finding all the relevant discriminating cell (sub)types and describing the complete cellular heterogeneity.

Material and methods

Resolving cell mixtures in an MFC study involves two separate operations: first, the different cells need to be differentiated based on their chemical surface protein profiles. Then, the role of these mixture constituents in a biomedical study needs to be further studied in a comparison between MFC samples. Several approaches are available for both operations.

Resolving the cells within an MFC sample

(Manual) sequential Boolean gating. Contemporary clinical flow cytometry uses sequential Boolean gating,[9,17] consisting of selecting cells in uni- or bivariate surface protein histograms (Fig. 1). This selection may then be sequentially refined by comparison to expression(s) of other markers, until all discoverable cell types have been separately identified and quantified. This approach requires considerable prior knowledge on the most relevant markers to observe and combine. The gating strategy for the studied representative asthma sample (Fig. 2) provides fractions of fourteen cell types in the sample. The defined gates may be used (or slightly modified to match the individual variance) to unmix other MFC samples from the same study.

SPADE. Spanning-tree Progression Analysis of Density-normalized Events (SPADE)[10] uses agglomerative hierarchical clustering based on Euclidean distances, up to a user-defined number of clusters. As this operation is computationally intensive, it down samples the data while keeping intact the local density: cells from abundant cell types are less-often sampled than cells from less-abundant types. The method removes cells with very low local densities as outliers. The method uses many more clusters than the number of expected cell types, to model systematic continuities in surface marker expression. SPADE represents these many clusters by a minimum spanning tree in combination with the Kamada Kawai layout, in which cluster nodes are connected in a tree with multiple branches[18] that indicate specific changes in surface protein expression that may relate to hematopoietic relations between cell types and continuity within a cell type, see Fig. 3b. Like all other methods we describe hereafter, SPADE observes the multivariate expressions of all markers simultaneously.

FlowSOM. Another popular clustering method in clinical flow cytometry is FlowSOM, based on the Self Organizing Map (SOM).[12,19,20] It arranges all cells within the sample onto a two-dimensional grid of cluster nodes, where proximal nodes are most similar. FlowSOM uses the same minimum spanning tree representation as SPADE, see Fig. 3a. As SOM calculates more efficiently than the clustering of SPADE, FlowSOM does not need to down sample the cells to become computationally feasible.

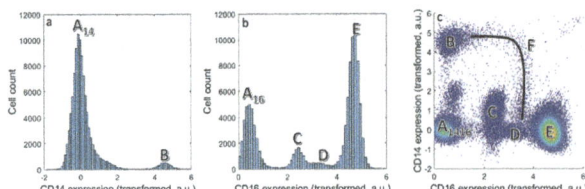

Fig. 1 (a) Histogram of the expression of CD14. (b) Histogram of the CD16 expression. Multiple cell populations can be found. Other cells (A), classical monocytes (B), eosinophils (C), non-classical monocytes/natural killer cells (D) and neutrophils (E). (c) Bivariate plot of CD16 *versus* CD14 expression. Each dot is a single cell. The continuum F describes the intermediate monocytes between B and D.

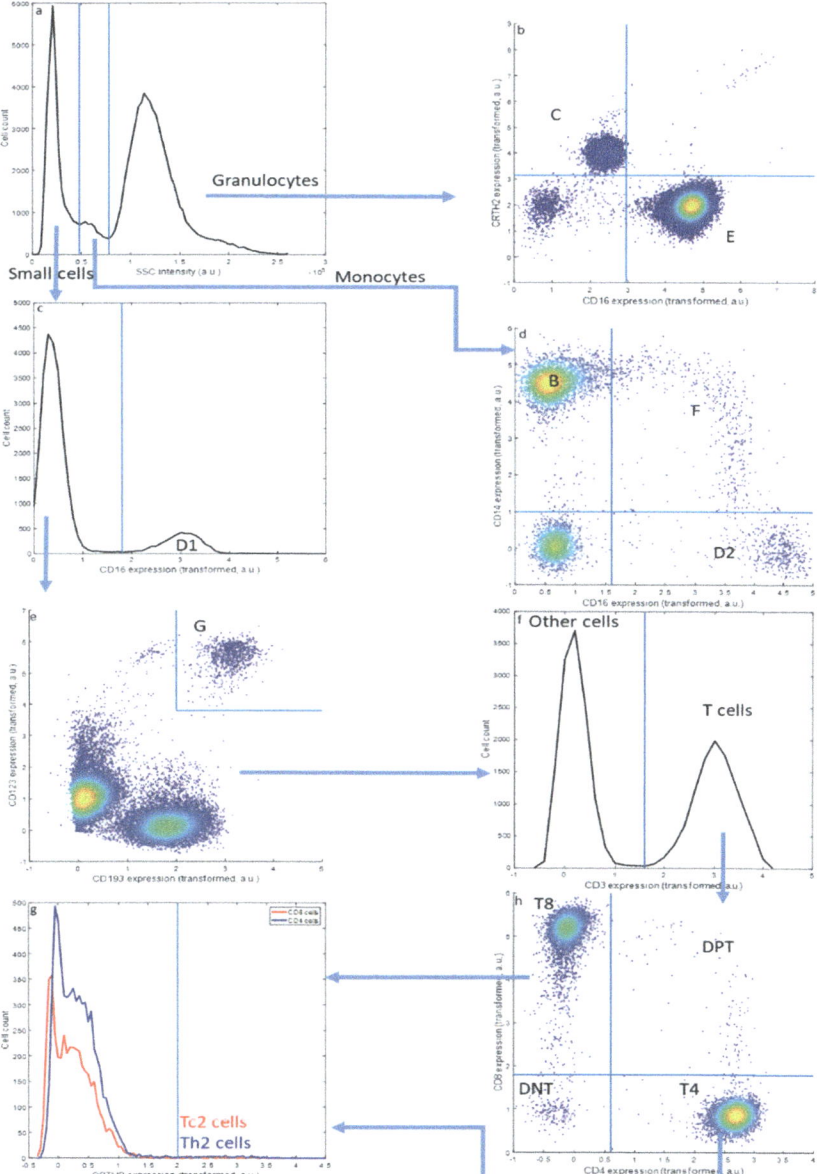

Fig. 2 Bivariate sequential gating. Each step is either setting a threshold in a histogram or in a bivariate plot starting from the top left. The arrows depict the sequence.

t-SNE. The currently most widely used dimension reduction method for MFC data analysis is t-distributed Stochastic Neighborhood Embedding (t-SNE).[11,21] Stochastic neighborhood embedding converts the high-dimensional Euclidean distance between the surface protein expressions of cells into a non-linear map of usually two dimensions (see Fig. 5). In this map, cells with similar high-dimensional expressions are plotted close to each other, while cells with more

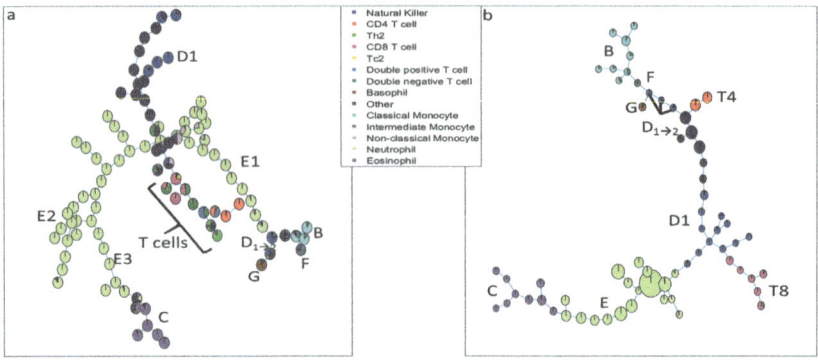

Fig. 3   (a) Minimum spanning tree based on the clusters found with SOM. (b) Minimum spanning tree based on the clusters found with SPADE, a larger number of cells in a cluster node is depicted by a larger node area. The pie charts are colored based on the sequential bivariate gating showing that many nodes contain well-resolved populations from the sequential gating, but that other populations that were well-resolved in sequential gating, end up in the same FlowSOM/SPADE nodes.

diverging expressions are placed much further away. The method aims to mainly represent similarities between cells, but the differences between cell types may be strongly disrupted, such that clusters may appear throughout the map.

Chemometrics: principal component analysis and partial least squares

Chemometrics is the research field that develops quantitative data analyses for chemical analytical technologies.[22] It has proven essential for systematic insight into chromatographic, spectroscopic and otherwise multivariate chemical data and is considered a cornerstone of metabolomics.[23]

Principal Component Analysis (PCA)[24] for exploratory analysis and Partial Least Squares (PLS)[25] for multivariate regression and discrimination are essential tools for metabolomics. A PCA model results in scores of every sample on the

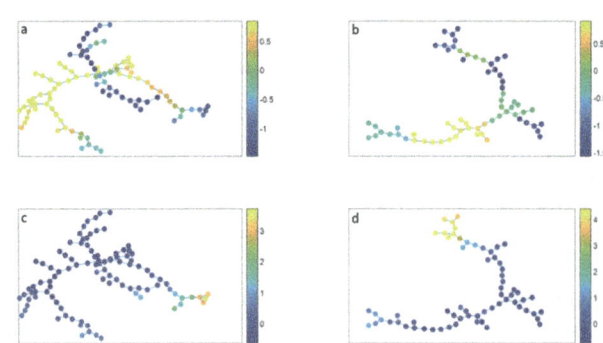

Fig. 4   (a and c) Minimum spanning tree based on the clusters found with SOM. (b and d) Minimum spanning tree based on the clusters found with SPADE. The cluster nodes are colored based on their CD16 expression (a and b) or on their CD14 expression (c and d). The color bar shows the expression (autoscaled, transformed, a.u.), where dark blue is low expression and yellow is high expression. 0 shows the mean expression.

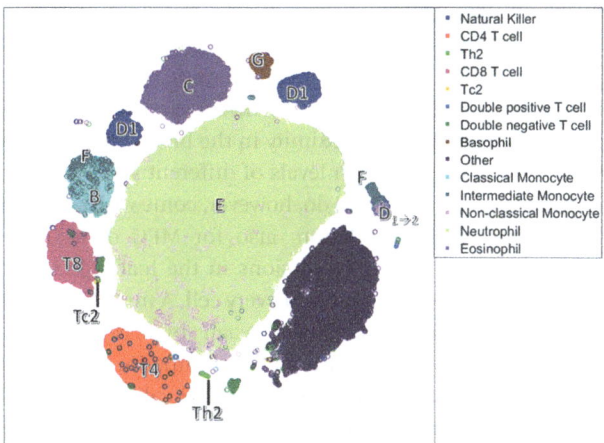

Fig. 5 The t–SNE plot. Each circle responds to a single cell. Cells have been colored based on the sequential bivariate gating.

most prominent multivariate correlations between the chemical features measured on the samples in principal components. The loadings express these correlations and indicate correlations between the features. Combinations between the scores and loadings may indicate how these correlations associate to specific samples. The principal components are orthogonal and fit as much variation in the original data as possible. Partial Least Squares Discriminant Analysis (PLS-DA), together with various helpful methodological extensions, employs a similar approach of dimension reduction to predict a class member-ship (*e.g.* control or clinical phenotype) for every sample.[26]

Like cluster-based methods, PCA and PLS are multivariate. This enables prediction of sample properties from relatively large numbers of correlated predictor features, which provides an alternative to the lasso regularized regres-sion used in Citrus:[16] for PLS the number of samples in the data restricts the maximum number of features that may be simultaneously identified as biomarkers—although the simplicity of the lasso-imposed sparsity may exceed this simplicity further. However, another concomitant advantage that is also highly important to explore the data with PCA, is the increase in resolution between samples that the linear correlations between predictors may bring, compared to *t*-tests of the separate features, the 'multivariate advantage'.[27] This has received relatively little attention in the literature.

PCA biplots. Although already proposed for the analysis of MFC data,[28] and essential in a standard methodology to resolve leukemia,[29] using PCA for MFC data has been largely vocally dismissed. "Principle components analysis has been used classically to calculate linear vectors through all measured parameters, thus identifying those combinations that describe the most variance in the data and relationships between samples. However, this method is not generally useful to immunophenotyping data, because of the general lack of correlations of expres-sion in most surface proteins."[13] This is one of the drivers for the popularity of non-linear methods such as t-SNE[11] for the analysis of MFC data, but disregards the considerable efforts undertaken to linearize the response of MFC technology to the quantitative surface protein expression on the single cell.[30] It also disregards how helpful PCA has been in resolving the complex mixtures in metabolomics and other omics fields. In metabolomics, the non-linearity of the underlying biological system is recognized. The correlation strength between two metabolites may be interpreted as proximity in the biochemical pathways, which may similarly hold for the expression levels of different surface proteins.

Linear methods like PCA (and PLS) do, however, come with significant benefits in interpretation and model validation, also for MFC data. The correlation structure between surface protein expressions in the loadings may be simultaneously represented with the PCA scores of every cell. A model of *e.g.* two PCs may closely resemble the t-SNE map in ordinating each single cell (see Fig. 6) but PCA provides direct feedback to the expression of each protein, and the relationships between proteins, through the loadings that serve as calibrated axes through the biplot map. The loadings may thereby serve as a compass-like guide to show which antibody expressions are most variable and which are highly correlated—indicating co-expressions. They also show on which cells and cell clusters within the MFC sample this co-expression is most prominent. The linearity of PCA also allows the calculation of which percentage of the variation in the original data is represented in the schematic representation, which is at least not-yet available for t-SNE, SPADE and FlowSOM.

Comparing MFC samples

The results from a sequential Boolean gating for a case–control study may be compared by a two sample *t*-test. This then indicates which cell type fractions are significantly different between the two groups, see Table 2. Alternatively, the resulting fractions from the gating may be analyzed with PCA, see Fig. 7.[9] This however relies on sequential Boolean gating with its inherent drawbacks. Therefore, we compared two methods that use the more informative methods to

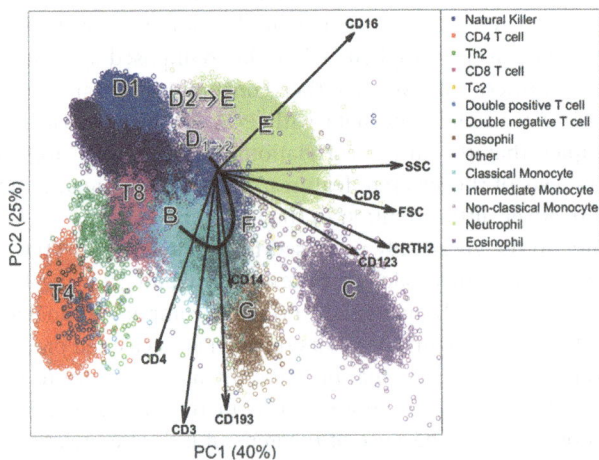

Fig. 6 PC1 *versus* PC2 based on the cells of one individual. Each round shape is a single cell and colored based on the sequential bivariate gating. The arrows show the PCA loadings.

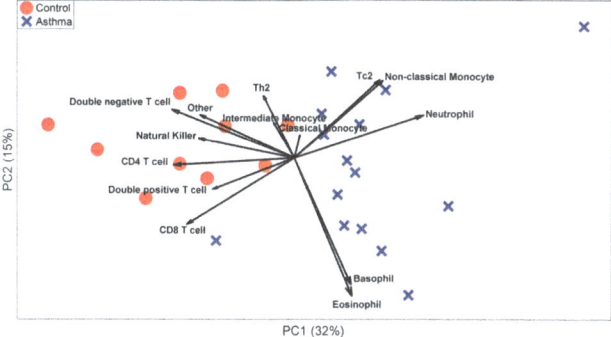

Fig. 7 PCA model on the manual gated data. The red circles represent the control indi-viduals and the blue crosses represent the asthma individuals. The loadings show the percentage of cells in the specific gate.

analyze cell mixtures to find hierarchies through which surface proteins drive cellular disease biomarkers.

Citrus. Citrus[16] uses an agglomerative hierarchical clustering similar to SPADE to arrange the cells into clusters. The clusters must contain at least 5% of the measured events and the hierarchical larger clusters are also included, thus cells may be assigned to multiple clusters. To find differences between sample groups, the clusters are compared between sample groups with a lasso-regularized regression. The lasso provides a 'sparse' result of cluster nodes that are suffi-cient to optimally discriminate between the groups. This sparsity serves the same objective as the dimension reduction of PLS-DA: avoiding collinearity. Note that down sampling by randomly taking 5000 cells per sample is applied to compu-tationally efficiently validate the model with ten-fold double cross validation with twenty iterations.[31] The same minimum spanning tree is used to represent the data as in SPADE and SOM, see Fig. 8.

DAMACY. Chemometrics has shown to be very strong in the application of modeling building blocks to develop novel methods that provide complementary insight for new analytical technologies and new chemical systems (*e.g.* ASCA,[32,33] MCR,[34,35] PARAFAC[36]). We developed Discriminant Analysis of MultiAspect Cytometry (DAMACY) specifically for the quantitative comparison from surface protein to patient population.

The method first builds a PCA model on the cells from all samples, weighting each MFC sample with the number of cells it contains and applying appropriate centering, to avoid dominance of samples with more cells in the model. As far as we know, DAMACY was the first method to apply such a correction. Instead, some methods use down sampling with the risk of losing important rare cell (sub)types.

The single-cell scores per sample are then transformed into 2D smoothed histograms, of which the bins may be compared between samples. This comparison is then performed with OPLS-DA,[26] of which the predictions serve as estimators for class membership and the weights of each bin extracted from the 2D PCA plot may be evaluated for a higher or lower abundance of the corre-sponding cells for either the control or clinical phenotype individuals, see Fig. 9.[14] The loadings from the cell-level PCA biplot may then serve as guides to interpret

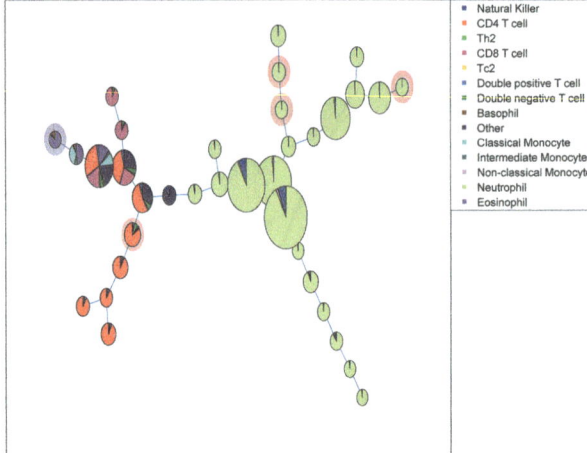

Fig. 8 Minimum spanning tree based on the clusters found with Citrus, a larger number of cells in a cluster node is depicted by a larger node area. The pie charts are colored based on the sequential bivariate gating showing that many nodes contain well-resolved populations from the sequential gating, but that other populations that were well-resolved in sequential gating, end up in the same Citrus nodes. The red shade behind a node means fewer cells in asthmatic patients, and the blue shade means more cells.

the surface marker co-expressions on these differentiating bins. We validated the model with the same double cross validation as used in Citrus.

The asthma MFC data set

The data set contains 15 asthma patients (aged 22–78, $\bar{x} = 57$) and 10 healthy controls (aged 25–57, $\bar{x} = 40$), who were recruited at the respiratory outpatient clinics of the Churchill Oxford University Hospital, UK.[37] The study received ethical approval, and written informed consent was obtained. After inclusion,

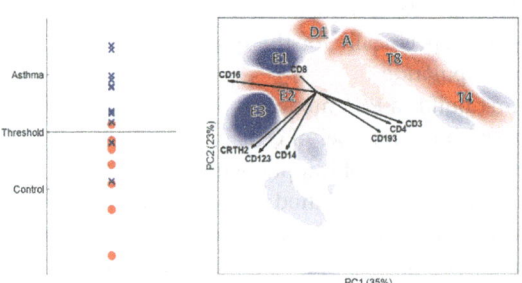

Fig. 9 DAMACY model. The left panel shows the average prediction score of the controls (red circles) and asthma individuals (blue crosses) based on the results of the double cross-validation. If a predicted value is above the threshold, the individual is classified as suffering from asthma. The right panel shows the weights – positive weights are colored blue and belong to cells more represented in the asthma individuals, and negative weights are colored red and belong to cells more present in the controls. The black vectors indicate how each marker contributes to the cell variability in a specific direction.

patients filled out symptom questionnaires, sputum induction was performed, blood was taken, and patients underwent FeNO measurements and lung functional testing. All patients were receiving appropriate asthma treatment at the time of blood withdrawal. The blood cells were stained with a panel of 8 antibodies including CD3, CD4, CD8, CD14, CD16, CRTH2 (CD294), CD123 and CD193. After staining, the red blood cells were lysed using a FACS Lysing solution (Becton Dickinson). The cells were measured on a LSR Fortessa flow cytometer (Becton Dickinson). Only single cells were included by using the correlation between the forward scatter (FSC) maximum height and the FSC area under the curve. Debris was removed by setting a minimum threshold based on FSC. The data was compensated to correct for the fluorophore overlap, using a manually optimized compensation matrix following the principles described by Roederer *et al.*[38] The data was preprocessed by applying an arcsinh transformation with cofactor 150 and subsequently autoscaling either using the mean and standard deviation of the one specific asthma patient or of all individuals together corrected for the number of cells measured in each individual.

Results and discussion

White blood cells from asthma patients (and controls) are analyzed for the quantitative expression of eight surface proteins. Although this data is collected with the current standard in multicolor flow cytometry, the findings may be expected to project directly to *e.g.* mass cytometry[39] and to emerging technologies like single-cell metabolomics.[40] None of the advanced data analysis techniques are intrinsically limited to a specific number of features. The study focuses on the identification of different cell types, differentiated by the immunological activity of the selected surface proteins.

The chemical resolving power of multivariate analysis

We first focus on one set of cell types within the asthma blood samples: the monocytes. These expressions for a specific MFC sample may be represented as histograms (Fig. 1a and b). Expression of CD14 shows division into two cell populations, we call them here A_{14} and B. Expression of CD16 shows a division of the cells into four cell populations by similar, visual distinction of expression levels. These four populations we indicate as A_{16}, C, D and E. Populations A_{14} and A_{16} are the negative populations that do not, or very slightly, express the surface protein. Naturally, the CD14 histogram does not show whether cells in population B also belong to populations A_{16}, C, D or E: it ignores that expression if both surface proteins occur on the same cells.

The bivariate density scatter plot (Fig. 1c) shows the combined expression of both surface proteins on each cell within the sample: the combined expression of both surface proteins resolves the cells A_{16} into a fraction of cells from A_{14} and the classical monocytes B that highly and very reproducibly express CD14. In combination with the expression of CD16, A_{14} may be subdivided into the cells that express very little of CD14 (population A_{1416} in Fig. 1c). A_{14} also has contributions from eosinophils (C), non-classical monocytes/natural killer cells (D) and neutrophils (E) with consistent, yet increasing, CD16 expression. Although somewhat trivial, this shows that the combination between two markers reveals more about the surface protein co-expression in these populations.

Combining CD14 and CD16 also reveals the intermediate monocytes F, a relatively small cell fraction (0.8%) that could not be observed in the expression of either separate surface protein, but resolves itself in the bivariate combination of CD14 and CD16. Contrary to the distinct populations A–E, cells F form a heterogeneous non-linear continuum in CD14 and CD16 expression that ranges between the classical and non-classical monocytes (B and D). The existence of this continuum is known,[41] but its presentation in different models may show their potential to discovery similar, as-yet unknown, continua.[7]

Although cell populations C–E may already be identified as somewhat resolved peaks in CD16, they still overlap to a certain degree—specifically populations C and D. This shows that combining surface protein may also increase the resolution in unmixing cells into different populations. In fact, cell population D contains two distinct populations that may only be resolved by adding information on additional surface proteins in subsequent steps of the sequential gating.

Sequential Boolean gating of the asthma data (Fig. 2) resolves fourteen cell types. The cell types (C–E) may be first distinguished into small cells, monocytes and granulocytes, based on the Sideward SCatter (SSC) of each cell, a parameter representative of cell granularity (Fig. 2a). Subsequently, they can be divided in eosinophils (C), neutrophils (E), NK cells (D1), non-classical monocytes (D2), intermediate monocytes (F) and classical monocytes (B) based on CD16 and CD14 expression. The SSC alone may not perfectly resolve these populations.

This gating strategy results in a subdivision of all cells into the fourteen cell types that could be observed in the asthma data by this approach (Table 3) into fractions for the specific representative sample in Fig. 1 and 2. The sample contains a very high fraction of neutrophils, high fractions of monocytes, eosinophils, CD8 and CD4 T cells, and considerably smaller fractions of basophils, Th2 cells, Tc2 cells, double negative T cells and double positive T cells; all of these cells are, however, important to characterize asthma. The same gating may be performed for all other samples from the same asthma data set, thresholds may even be slightly altered to accommodate a slight shift in thresholds. Of specific relevance here is the sideward scatter, having a much poorer reproducibility between samples than the surface protein expressions. We therefore omitted this feature, as well as the front scatter of each cell, from the quantitative comparisons between MFC analyses presented further.

Although widely used, sequential bivariate gating has serious drawbacks: gating thresholds need to be set manually, using prior biological knowledge or experience, which both introduces arbitrariness and is highly resource-intensive. Gating the asthma data is also limited to interpreting four out of the 28 potential bivariate combinations, which makes it greatly hypothesis driven; even more so when more than eight surface proteins would have been interrogated. Thirdly, the approach does not scale beyond co-expressions of two surface proteins, which limits the resolution of cells defined on the simultaneous expression of more than two surface proteins. Semi-automated methods to define gating thresholds[42] do not necessarily improve resolution between different cell populations, due to this bivariate limitation.

Multivariate resolving of the cell mixture within an MFC sample

The SOM (Fig. 3a) clearly resolves several sequentially gated populations into distinct nodes (eosinophils, natural killer cells, other cells, neutrophils) and the number of nodes associated with each population corresponds to the fraction of those populations in the blood sample. However, as the model aims to describe all cells well, it will focus on describing the most abundant populations. Many nodes contain exclusively neutrophils and are therefore similar, but do show a systematic heterogeneity, namely increasing sideward scatter in E1, E2 and E3 upon further inspection. Several smaller populations (*e.g.* monocytes, T cells), especially those with surface protein expressions that do not differ much from other populations, are not resolved into separate nodes. Moreover, the tree does not show how the intermediate monocytes (F) form the continuum between both other monocyte types (located at B and $D_{1 \rightarrow 2}$, respectively), as observed in Fig. 1c. This may be likely solved by increasing the number of nodes to *e.g.* better resolve the monocyte and T cell branch. However, adding more nodes also makes the tree less-well interpretable and there is no heuristic for the model quality other than compliance to sequential gating.

Although the tree may for large parts be well-interpreted, several inconsistencies appear. A group of natural killer cells ($D_{1 \rightarrow 2}$) is located near the intermediate monocytes (F) and classical monocytes (B) at the right-most end of the tree. Further inspection shows that the current sequential gating strategy may introduce an error caused by the limiting resolving power of the sideward scatter, see Fig. 2a. Several cells identified as natural killer cells (D1) are, in fact, non-classical monocytes (D2). For the same reason, the 'non-classical monocytes' near the neutrophils (E1) may be small neutrophils. The SOM is thus able to discover mismatches in the sequential gating as it is able to use the 'multivariate advantage'.

The minimum spanning tree representation in Fig. 3a focuses on how the method reconstructs the sequential gating and does not give any view on the surface protein expression, but the same tree may be colored for average expressions of specific surface proteins (Fig. 4a and c). The tree may have maxima for specific surface proteins (*e.g.* CD14 for the classical monocytes), but there may be multiple non-connected nodes for which expression is high, like for CD16. Investigating co-expression between proteins is thus limited, because you have to compare the nodes in Fig. 4a with Fig. 4c to find the relationship between CD14 and CD16.

The SOM model needs to describe all surface protein variability simultaneously, which is reflected here in how both branches orient towards each other: basophils (G) and eosinophils (C) are highly similar as they both belong to the granulocyte class, but the minimum spanning tree puts the basophils (G) on the same branch as the monocytes (B and $D_{1 \rightarrow 2}$). In other words, the proximal nodes may be related but the distance between the neighboring nodes may be very large, thus investigating the similarity between nodes is limited.

The SPADE tree of the same representative asthma sample (Fig. 3b) shows that preprocessing with density-dependent down sampling leads to a tree with different characteristics to that from the SOM. The high-abundant neutrophils (E) and eosinophils (C) occupy a much more similar number of nodes as lower-abundant cells (CD8 T cells (T8), NK cells (D1) and monocytes (B and F)) compared with SOM. In SPADE, the number of cells for each cell type is related to the number of nodes and the cluster node area. SPADE resolves cell types well into cell type-specific nodes, even those that are lower-abundant. The continuum of the intermediate monocytes between the classical and non-classical monocytes is reflected in SPADE, although F overlaps with $D_{1 \to 2}$ and the classical monocytes (B). However, the down sampling removes rarer cell populations such as Tc2 and Th2 cells. Not removing these rare cells is also not wanted as that would result in a model which is very sensitive to outliers.

The tree, however, also gives reflection onto the sequential gating performance. Again, one of these NK cell ($D_{1 \to 2}$) groups may be wrongly gated as non-classical monocytes (D2).

The representation of the SPADE cluster nodes with the minimum spanning tree suffers from the same limitations as the tree based on the SOM cluster nodes. Namely, the surface protein co-expressions may also only be observed by separate surface protein-based tree representations, see Fig. 4b and d. Also SPADE deems the similarity of the basophils (G) to the monocytes more important than their high similarity to the eosinophils. Moreover, it splits CD4 (T4) and CD8/CD3DN (T8) into separate branches. Thus, the similarity between cluster nodes is limited.

Unlike the cluster nodes of FlowSOM and SPADE, the t-SNE map (Fig. 5) represents each single cell, although the model then aggregates similar cells into distinct clusters. The large neutrophil cluster (E) is surrounded by clusters of other cell populations. Rare cell types, such as Th2 cells and basophils (G), have their own compact cluster. Tc2 cells are next to CD8 cells (T8). The cluster area in t-SNE may be determined by the number of connected cells and their heterogeneity caused by biological or measurement variation in surface protein expression, and therefore may not be attributed to both the number of and heterogeneity between cells.

The continuum of the intermediate monocytes (F) between classical monocytes (B) and wrongfully gated NK cells ($D_{1 \to 2}$) is represented in two dispersed clusters, corresponding to both continuum endpoints (Fig. 1c). Furthermore, both endpoints are placed at opposite ends of the non-linear map, which does not reflect their similarity in the surface protein expression. Also, in the t-SNE map, surface protein co-expression is not explicitly modelled and may only be revealed by coloring each cell with the expression level of a specific surface protein.

The NK cells split into three clusters, again one cluster ($D_{1 \to 2}$) may be wrongly gated as non-classical monocytes, but a cluster on the left with a medium CD8 expression is also observed. This distinction between NK cells with very low CD8 expression and with a medium CD8 expression is not very common, but has been found earlier in humans after a bout of exercise.[43] Thus, t-SNE may reveal additional information compared to the hypothesis driven approach of sequential bivariate gating. In hindsight, the SOM also describes these NK cells, however these cells are mixed with other cells and thus harder to interpret than in t-SNE.

The biplot shows the first two principal components of the PCA model for this MFC sample (Fig. 6). These two PCs describe 65% of the total variation in surface marker expressions. PCA quantifies how well the model reproduces the original data, unlike the other compared methods. These two principal components show how all sequentially gated populations have distinct locations on the map, and some are well-resolved like CD4 (T4), basophils (G) and eosinophils (C). However, many populations—including the high-abundant neutrophils—overlap with cells from other populations. The size of the clusters observed in PCA is determined by the heterogeneity in surface protein expression, but not by the abundance of the cell population: the area covered by the neutrophils is only slightly larger than that of the eosinophils, although their abundance is much higher. This makes the discovery of rare cell populations like Th2 and Tc2 challenging without further visual aids. The continuum between the classical (B) and non-classical (D2) monocytes by the intermediate monocytes (F) is visible in the model, albeit requiring visual aids to highlight the relevant cells due to overlap. The shape of the continuum is somewhat distorted compared to Fig. 1c, but this may be explained by the non-orthogonal orientation of both PCA loadings: as PCA describes all cells, and co-expressions of CD14 and CD16 with all other markers, the continuum is partially recovered and may be chemically interpreted.

Although most cell populations overlap in the map, their proximity and location with respect to each other indicates linearly increasing surface protein expressions, a multivariate extension of sequential gating (Fig. 2). The model loadings (indicated as arrows in Fig. 6) serve as direction indicators to quantify surface marker co-expressions for each cell. For example, basophils (G) and eosinophils (C), positioned next to each other in the PCA biplot, have similar surface protein expressions on CD123, CRTH2 and CD193 as their loadings direct towards these cells. Eosinophils are higher in sideward scatter (SSC), forward scatter (FSC) and CD16 expression, as these loadings direct from G to C. Basophils (G) have above-average expressions of CD3, CD4 and CD14, although the contribution of CD14 is less, indicated by its considerably shorter loading arrow. Here, the heuristic of the percentage of variance explained (65%) also comes into play. For example, the scores of basophils and CD4 T cells for PC2 are similar, which would suggest that both cell types have above-average expression of *e.g.* CD4, CD3 and CD193. However, CD4 T cells are only high on CD3/CD4 and low/medium on CD193 and basophils vv: principal components 3 and higher will describe such contrasts as they still explain 35% of all variation in surface marker expression.

Table 1 shows an overview of the performance of the methods and Table 2 shows an overview of how each method recovers the sequential gating. The number of nodes in the SOM directly represents the number of cells per cell type. In SPADE, it is a combination between the node area and the number of nodes. In t-SNE, the cluster size is determined by the number of connected nodes and the cellular variability, and in PCA only by the cellular variability. In terms of distinguishing the cell types, t-SNE outperforms other methods as it is optimizing the local structure. SPADE is second best for the larger cell types, but, due to the down sampling, completely detrimental for rare cell types. In SOM and PCA, most cell (sub)types overlap, however this overlap does show the continuous intermediate monocytes F with PCA. The resolving power in t-SNE proved malign for these intermediate monocytes as these cells were counterintuitively split.

All methods were able to find the mismatched NK cells that were in fact non-classical monocytes. SOM was able to find a heterogeneity in eosinophils, t-SNE found two natural killer subsets and PCA finds surface protein co-expression. Thus, multivariate methods are able to find complementary information.

The principal component biplot ordinates the cells by linearly retaining the quantitative surface protein expressions, *i.e.* the cell chemistry, into a map constructed on the largest variation in these expressions. Mutual cell distances in the PCA biplot can be associated to quantitative differences in the surface protein

Table 1 Overview of the performance of the methods, SOM, SPADE, t-SNE and PCA. The methods are scored from poor to reasonable, good and very good. Limited means that it is only possible upon further inspection of the data but not possible to extract from the figures made. In this table we aimed to briefly qualify every method, however for a full description we refer to the respective result sections

Criteria	SOM	SPADE	t-SNE	PCA
Recovery of the sequential gating				
Size of cell type fractions	Very good	Reasonable	Poor	Poor
Resolution of cell types	Reasonable	Good	Very good	Reasonable
Rare cell types recovery	Reasonable	Poor	Very good	Poor
Continuous intermediate monocytes F	Poor	Reasonable	Poor	Good
Complementary to sequential gating				
Sequential gating mismatch discovery	Very good	Very good	Very good	Very good
Complementary information on cell types	Heterogeneity in neutrophils	Not found	NK subtype that express CD8	Surface protein co-expression
Chemical insight				
(Co)-expression between proteins	Limited	Limited	Limited	Very good
Similarity between proximal clusters	Limited	Limited	No relationship	High relationship
Quality of the model	Not present	Not present	Not present	Variance explained per PC

expression throughout the map, *via* the surface protein loadings. This chemical insight is not directly extracted from the figures made by SOM, SPADE and t-SNE, but indirectly extracted when, for each surface marker, a new figure is plotted with the intensity of that surface marker, *e.g.* see Fig. 4.

PCA furthermore stands out in that it provides a heuristic approach to quantify the recovery of the data by the model, *i.e.* the variance explained per PC, which is very useful as a warning for over-interpreting the data.

These observations have been performed on only a single data set, but it is the first comparison of such detail among all these methods that we have come across. Additional criteria may be most relevant, where we have omitted that of calculation speed on purpose. Unmixing the cells into different types is, however, a task with a specific objective: this makes each method appropriate for a specific task.

Comparing MFC samples

The methods that were presented to resolve MFC samples into cell populations based on their multivariate surface protein expressions provide insightful views on the distribution of the cells across different types. However, in most applications of MFC technology, the comparison between samples, in *e.g.* a case–control study, is of primary interest to determine which populations vary between MFC samples and differ between sample groups.

Although variability in cell abundances among both control and asthma samples may be considerable (Table 3), the study shows that neutrophils and

Table 2 Overview of the recovery of the sequential gating with the different methods. Good means no overlap with other cells. Partly means that the majority of cells were recovered but a part overlaps with other cells. Completely means no separation from other cells was possible. Mix means mixed together with many other cell types. Missing means that the cells cannot be found with the method. Good+ means that additional information was found. In t-SNE, natural killer cells could be further distinguished into CD8− and CD8$_{dim}$ cells. In SOM, the neutrophils were differentiated based on sideward scatter

Cell (sub)type		SOM	SPADE	t-SNE	PCA
D1	Natural killer	Good	Good	Good+	Partly A
T4	CD4 T cells	Partly DPT	Good	Good	Partly DPT
Th2	Th2 cells	Partly mix	Missing	Good	Partly DPT/T4
T8	CD8 T cells	Partly DNT	Completely DNT	Partly DNT/Tc2	Partly DNT/B/A
Tc2	Tc2 cells	Completely mix	Missing	Partly T8	Missing
DPT	Double positive T cells	Completely T4/DNT	Missing	Partly T4	Completely T8/Th2
DNT	Double negative T cells	Partly T8	Completely T8	Partly T8	Partly T8
G	Basophils	Partly A	Partly A	Good	Partly B/F
A	Other cells	Partly mix	Good	Good	Partly, D1/T8/D2
B	Classical monocytes	Completely F/D2	Completely F	Completely F	Partly, T8/F/D2/G
F	Intermediate monocytes	Completely B/D2	Completely B/D2	Partly B/D2	Partly, B/D2/G
D2	Non-classical monocytes	Completely mix	Completely F	Completely E/F	Completely F/B/E
E	Neutrophils	Good+	Good	Partly D2	Partly D2
C	Eosinophils	Partly A	Good	Good	Good

eosinophils are significantly higher-abundant in asthma patients. Double negative T cells, CD4 and CD8 T cells, natural killer cells and 'other cells' are relatively lower-abundant in asthma patients. Note that the cell fractions are closed here to 100%: the chemometric expertise in the analysis of such data is available,[44] but beyond the scope of this study.

These *t*-tests give a relevant indication of which cell types are relevant. However, it ignores heterogeneity between individuals or multivariate relationships between the abundances of different cell types. A PCA biplot on the cell-type fractions within all MFC samples (see Fig. 7) already provides much deeper insight into the study. Systematic variation between asthma patients and control samples is the largest source of multivariate variability. The loadings show that most cell types are more abundant for the control than for the asthma samples, but that neutrophils are more abundant for all asthmatic patients, as its loading is directed towards the average of all asthma samples, which agrees with its *t*-test.

The PCA scores also show how the asthma patients are far more heterogeneous than the control individuals. One group of patients scores low on PC2 and therefore has a far above-average abundance of basophils and eosinophils (low on PC2), while another group scores high on PC2 due to a high abundance of Tc2 cells and classical monocytes (which include some erroneously gated small neutrophils). As the sample-level PCA model in Fig. 8 describes 46% of the variance in the data, PCs 3 and higher may contain even more such subgrouping. This

Table 3 Hypothesis testing based on the percentages of the different cell (sub)types found with sequential manual gating using the strategy described in Fig. 2. The table is sorted from low to high p-values. p-values marked * are significantly different between both groups after false discovery rate correction

Cell (sub)type		Specific sample	Asthma	Control	p-Value
DNT	Double negative T cells	0.45%	$0.90 \pm 0.48\%$	$2.41 \pm 0.94\%$	2.00×10^{-5}*
E	Neutrophils	57.29%	$63.96 \pm 7.65\%$	$51.58 \pm 7.80\%$	6.60×10^{-4}*
T4	CD4 T cells	6.03%	$8.29 \pm 3.62\%$	$14.68 \pm 4.62\%$	7.70×10^{-4}*
T8	CD8 T cells	4.14%	$4.56 \pm 3.16\%$	$8.04 \pm 2.14\%$	5.90×10^{-3}*
C	Eosinophils	7.10%	$6.07 \pm 4.40\%$	$2.32 \pm 1.43\%$	0.016*
D1	Natural killer	3.97%	$3.91 \pm 1.65\%$	$6.11 \pm 2.68\%$	0.018*
A	Other cells	14.96%	$5.29 \pm 3.38\%$	$8.52 \pm 2.85\%$	0.021*
DPT	Double positive T cells	0.14%	$0.54 \pm 0.41\%$	$0.96 \pm 0.59\%$	0.044
F	Intermediate monocytes	0.81%	$0.80 \pm 1.32\%$	$1.38 \pm 0.93\%$	0.049
G	Basophils	0.88%	$0.76 \pm 0.38\%$	$0.51 \pm 0.17\%$	0.058
D2	Non-classical monocytes	0.53%	$1.32 \pm 1.98\%$	$0.46 \pm 0.25\%$	0.19
Tc2	Tc2 cells	0.03%	$0.17 \pm 0.31\%$	$0.05 \pm 0.06\%$	0.26
Th2	Th2 cells	0.06%	$0.19 \pm 0.09\%$	$0.21 \pm 0.08\%$	0.64
B	Classical monocytes	3.96%	$3.37 \pm 0.80\%$	$3.44 \pm 1.44\%$	0.9

heterogeneity among asthma patients may be the reason that basophils, non-classical monocytes and Tc2 cells do not show significant elevation for asthma patients (Table 3). However, treating each MFC sample as a fully resolved mixture of cells—analogous to an omics approach—reveals such individualized aspects of the disease.

Although the PCA model in Fig. 7 is very insightful, it still requires the expertise and resources of sequentially gating each MFC sample. A comparison using one of the automatically generated models from the previous section may be much more helpful.

Citrus provided a diagnostic accuracy of 79.4%, using double cross validation. Five out of the 31 cluster nodes were included and highlighted in Fig. 8. One node with basophils and eosinophils was increased in the asthma patients, and a CD4 T cell subset and three neutrophil subsets are decreased in the asthma patients.

Using the same double cross validation as Citrus, we achieved an accuracy of 85.6% for the discriminant analysis using DAMACY (Fig. 9). DAMACY reveals which single cells within the PCA map are more or less abundant in the asthmatic patients. Two cell clusters (indicated as E1 and E3) are higher-abundant and five other cell clusters (E2, D1, A, T8 and T4) are less-occupied. The DAMACY map would suggest that cell cluster E3 could be eosinophils as the loadings CD16, CRTH2 and CD123 are pointing towards this cluster. However, cell clusters E1, E2 and E3 are neutrophils with increasing size by comparing these clusters to the manual gates and their original scatter intensity. Therefore, our hierarchical analysis of the immune system with DAMACY shows additional cell subtyping that could not be resolved by gating separate samples, either sequentially or with the automated methods compared before.

The classification accuracies of 79–86% for both methods are considerable. Although the population-level PCA model (Fig. 7) would allow a seemingly better linear separation between asthma and control samples, both classification accuracies are reported for unseen data and the PCA model validity is limited to the samples in the current data set.

DAMACY shows a direct relationship between surface proteins, cells, cell types and patients: for example, above-average co-expression of CD16 together with CRTH2, CD123 and CD14 is associated with cells that cluster into a 'big neutrophil' cell type (E3). This type is higher-abundant together with the small neutrophils (E1), which together are part of the multicellular biomarker for asthma: this increase is more severe for asthma patients with higher disease predictions (Fig. 9, left panel).

DAMACY showed that asthma patients have more 'small' (E1) and 'big' neutrophils (E3) and less normal-sized neutrophils (E2). Citrus only finds three neutrophil nodes that contain less cells in the asthma patients than in the controls. Table 2 shows that the overall neutrophil population is significantly increased for asthma, but the descent to the single-cell level finds intricacies and heterogeneities within cell populations that aid discrimination between MFC samples and is therefore potentially invaluable to understand asthma.

Apart from the eosinophils (C), DAMACY retrieves all cell types that sequential-based *t*-tests also found. Citrus came to a similar classification accuracy, needing only the populations that decrease in abundance (together with a single basophil node that DAMACY did not find as higher-abundant). This sparsity is directly aligned with Occam's razor and therefore statistically favorable. For the analysis of (any type of) omics data, imposing sparsity should be treated with caution. Especially in binary classifications, highly generic or less-informative features may be sufficiently discriminative to reach a certain classification accuracy. This statistically sound limitation is, however, counterproductive for the objective of omics studies, which is the retrieval of all biomolecules associated with a specific biochemical process.

General discussion

Although the data used throughout this study has been carefully collected, measured and preprocessed, all observations are specific for this data set. The study thereby focuses on immunophenotyping, *i.e.* resolving mixtures of different cell types. Other applications of MFC focus more on the activation of specific cell types, or comparing samples of an individual for different time-points, which might show different results for a similar critical comparison. Secondly, we limited our analyses to the implementation of the methods as they are available in the literature. In principle, quality characteristics could be devised to evaluate how SPADE, FlowSOM and t-SNE recover the original data but this requires further research. Another interesting extension would be the hybridization of different methods, *e.g.* using t-SNE for DAMACY or the SPADE down sampling in the other methods.

In principle, the methods described here are able to analyze every piece of microparticle data. MFC is a high throughput and well established quantitative analytical instrument and is routinely used in hospitals.[45] For the representative asthma sample, 0.03% Tc2 cells in a total of 102 thousand cells were measured. This leads to 35 cells measured and already most methods have trouble detecting this group of cells. Most other single cell omics instruments yield only a fraction of the number of cells measured and are thus unable to detect these rarer cell subtypes.[40] Moreover, these single cell omics data suffer from more technical variability than MFC, which is enhanced by the number of variables measured and thus it will be harder to distinguish the biological relevant variability from the technical variability.

Conclusions

Multicolor flow cytometry is invaluable to chemically characterize cells in a biomedical sample. Manual sequential gating is extremely labor and resource-intensive, such that automated methods to resolve such a sample into different cell types based on their surface protein expression have considerable intrinsic value. This is supported by the diverse methodologies that have been presented in the literature. Developing quality criteria to describe resolving such cell mixtures are context-specific, but we have qualitatively evaluated the methods based on the analysis of an MFC sample obtained from an asthma patient. Each of the four compared methods provided an insightful overview of the mixture, but each method had pre-defined aspects in which it excelled. Although principal component analysis did not resolve all cell types in the mixture well, using it as a basis for hierarchically comparing MFC samples for disease biomarker cells with specific surface marker expressions revealed even more populations than the analysis of a single sample. We also showed how detrimental the implementation of sparsity might be in comprehensively resolving mixtures in high-dimensional biochemistry. Such hierarchies in mixtures become much more prevalent in analytical chemistry, for example in characterizing the complexome of different proteins within a biofluid. Also, in industrial recycling where objects in heterogeneous waste streams need to be individually chemically characterized and separated, the compared technologies may be directly applied.

Conflicts of interest

There are no conflicts to declare.

References

1 O. Beckonert, H. C. Keun, T. M. Ebbels, J. Bundy, E. Holmes, J. C. Lindon and J. K. Nicholson, *Nat. Protoc.*, 2007, **2**, 2692.
2 H. Zola, B. Swart, I. Nicholson, B. Aasted, A. Bensussan, L. Boumsell, C. Buckley, G. Clark, K. Drbal and P. Engel, *Blood*, 2005, **106**, 3123–3126.
3 A. L. Givan, in *Flow Cytometry Protocols*, Springer, 2011, pp. 1–29.
4 J. Picot, C. L. Guerin, C. Le Van Kim and C. M. Boulanger, *Cytotechnology*, 2012, **64**, 109–130.
5 J. Brummelman, K. Pilipow and E. Lugli, *Int. Rev. Cell Mol. Biol.*, 2018, 63–124.

6 D. R. Bandura, V. I. Baranov, O. I. Ornatsky, A. Antonov, R. Kinach, X. Lou, S. Pavlov, S. Vorobiev, J. E. Dick and S. D. Tanner, *Anal. Chem.*, 2009, **81**, 6813–6822.

7 S. C. Bendall, E. F. Simonds, P. Qiu, E.-a. D. Amir, P. O. Krutzik, R. Finck, R. V. Bruggner, R. Melamed, A. Trejo, O. I. Ornatsky, R. S. Balderas, S. K. Plevritis, K. Sachs, D. Pe'er, S. D. Tanner and G. P. Nolan, *Science*, 2011, **332**, 687–696.

8 R. Goodacre, S. Vaidyanathan, W. B. Dunn, G. G. Harrigan and D. B. Kell, *Trends Biotechnol.*, 2004, **22**, 245–252.

9 E. Lugli, M. Roederer and A. Cossarizza, *Cytometry, Part A*, 2010, **77**, 705–713.

10 P. Qiu, E. F. Simonds, S. C. Bendall, K. D. Gibbs Jr, R. V. Bruggner, M. D. Linderman, K. Sachs, G. P. Nolan and S. K. Plevritis, *Nat. Biotechnol.*, 2011, **29**, 886–891.

11 E.-a. D. Amir, K. L. Davis, M. D. Tadmor, E. F. Simonds, J. H. Levine, S. C. Bendall, D. K. Shenfeld, S. Krishnaswamy, G. P. Nolan and D. Pe'er, *Nat. Biotechnol.*, 2013, **31**, 545–552.

12 S. Van Gassen, B. Callebaut, M. J. Van Helden, B. N. Lambrecht, P. Demeester, T. Dhaene and Y. Saeys, *Cytometry, Part A*, 2015, **87**, 636–645.

13 S. C. Bendall, G. P. Nolan, M. Roederer and P. K. Chattopadhyay, *Trends Immunol.*, 2012, **33**, 323–332.

14 G. H. Tinnevelt, M. Kokla, B. Hilvering, S. Staveren, R. Folcarelli, L. Xue, A. C. Bloem, L. Koenderman, L. M. Buydens and J. J. Jansen, *Sci. Rep.*, 2017, **7**, 5471.

15 N. Aghaeepour, G. Finak, H. Hoos, T. R. Mosmann, R. Brinkman, R. Gottardo and R. H. Scheuermann, *Nat. Methods*, 2013, **10**, 228–238.

16 R. V. Bruggner, B. Bodenmiller, D. L. Dill, R. J. Tibshirani and G. P. Nolan, *Proc. Natl. Acad. Sci. U. S. A.*, 2014, **111**, E2770–E2777.

17 C. E. Pedreira, E. S. Costa, Q. Lecrevisse, J. J. van Dongen, A. Orfao and E. Consortium, *Trends Biotechnol.*, 2013, **31**, 415–425.

18 T. Kamada and S. Kawai, *Inf. Process. Lett.*, 1989, **31**, 7–15.

19 J. Friedman, T. Hastie and R. Tibshirani, *The elements of statistical learning*, Springer series in statistics, New York, 2001.

20 R. Wehrens and L. M. Buydens, *J. Stat. Software*, 2007, **21**, 1–19.

21 L. v. d. Maaten and G. Hinton, *J. Mach. Learn. Res.*, 2008, **9**, 2579–2605.

22 D. L. Massart, B. G. Vandeginste, L. Buydens, P. Lewi and J. Smeyers-Verbeke, *Handbook of chemometrics and qualimetrics: Part A*, Elsevier Science Inc., 1997.

23 R. Madsen, T. Lundstedt and J. Trygg, *Anal. Chim. Acta*, 2010, **659**, 23–33.

24 R. Bro and A. K. Smilde, *Anal. Methods*, 2014, **6**, 2812–2831.

25 P. Geladi and B. R. Kowalski, *Anal. Chim. Acta*, 1986, **185**, 1–17.

26 M. Bylesjö, M. Rantalainen, O. Cloarec, J. K. Nicholson, E. Holmes and J. Trygg, *J. Chemom.*, 2006, **20**, 341–351.

27 J. M. Fonville, S. E. Richards, R. H. Barton, C. L. Boulange, T. M. D. Ebbels, J. K. Nicholson, E. Holmes and M.-E. Dumas, *J. Chemom.*, 2008, **24**, 636–649.

28 Y. Kosugi, R. Sato, S. Genka, N. Shitara and K. Takakura, *Cytometry*, 1988, **9**, 405–408.

29 J. Van Dongen, L. Lhermitte, S. Böttcher, J. Almeida, V. Van der Velden, J. Flores-Montero, A. Rawstron, V. Asnafi, Q. Lecrevisse and P. Lucio, *Leukemia*, 2012, **26**, 1908.

30 Y. Saeys, S. Van Gassen and B. N. Lambrecht, *Nat. Rev. Immunol.*, 2016, **16**, 449.

31 E. Szymańska, E. Saccenti, A. Smilde and J. Westerhuis, *Metabolomics*, 2012, **8**, 3–16.

32 J. J. Jansen, H. C. Hoefsloot, J. van der Greef, M. E. Timmerman, J. A. Westerhuis and A. K. Smilde, *J. Chemom.*, 2005, **19**, 469–481.

33 A. K. Smilde, J. J. Jansen, H. C. Hoefsloot, R.-J. A. Lamers, J. Van Der Greef and M. E. Timmerman, *Bioinformatics*, 2005, **21**, 3043–3048.

34 R. Tauler, *Chemom. Intell. Lab. Syst.*, 1995, **30**, 133–146.

35 J. Jaumot, R. Gargallo, A. de Juan and R. Tauler, *Chemom. Intell. Lab. Syst.*, 2005, **76**, 101–110.

36 R. Bro, *Chemom. Intell. Lab. Syst.*, 1997, **38**, 149–171.

37 B. Hilvering, S. Vijverberg, J. Jansen, L. Houben, R. Schweizer, S. Go, L. Xue, I. Pavord, J. W. Lammers and L. Koenderman, *Allergy*, 2017, **72**, 1202–1211.

38 M. Roederer, *Curr. Protoc. Cytom.*, 2002, **22**, 1.14.1–1.14.20.

39 M. H. Spitzer and G. P. Nolan, *Cell*, 2016, **165**, 780–791.

40 R. Zenobi, *Science*, 2013, **342**, 1243259.

41 L. Ziegler-Heitbrock and T. P. Hofer, *Front. Immunol.*, 2013, **4**, 23.

42 M. Malek, M. J. Taghiyar, L. Chong, G. Finak, R. Gottardo and R. R. Brinkman, *Bioinformatics*, 2015, **31**, 606–607.

43 J. P. Campbell, K. Guy, C. Cosgrove, G. D. Florida-James and R. J. Simpson, *Brain, Behav., Immun.*, 2008, **22**, 375–380.

44 Y. Gagnebin, D. Tonoli, P. Lescuyer, B. Ponte, S. de Seigneux, P.-Y. Martin, J. Schappler, J. Boccard and S. Rudaz, *Anal. Chim. Acta*, 2017, **955**, 27–35.

45 J. P. Robinson and M. Roederer, *Science*, 2015, **350**, 739–740.

Faraday Discussions

PAPER

An integrated approach for mixture analysis using MS and NMR techniques†

Stefan Kuhn, ⓘD *a Simon Colreavy-Donnelly, ⓘD a Juliana Santana de Souza b and Ricardo Moreira Borges ⓘD b

Received 4th December 2018, Accepted 7th February 2019

DOI: 10.1039/c8fd00227d

We suggest an improved software pipeline for mixture analysis. The improvements include combining tandem MS and 2D NMR data for a reliable identification of the constituents in an algorithm based on network analysis aiming for a robust and reliable identification routine. An important part of this pipeline is the use of open-data repositories, although it is not totally reliant on them. The NMR identification step emphasizes robustness and is less sensitive towards changes in data acquisition and processing than existing methods. The process starts with LC-ESI-MSMS based molecular network dereplication using data from the GNPS collaborative collection. We identify closely related structures by propagating structure elucidation through edges in the network. Those identified compounds are added on top of a candidate list for the following NMR filtering method that predicts HSQC and HMBC NMR data. The similarity of the predicted spectra of the set of closely related structures to the measured spectra of the mixture sample is taken as one indication of the most likely candidates for its compounds. The other indication is the match of the spectra to clusters built by a network analysis from the spectra of the mixture. The sensitivity gap between NMR and MS is anticipated and it will be reflected naturally by the eventual identification of fewer compounds, but with a higher confidence level, after the NMR analysis step. The contributions of the paper are an algorithm combining MS and NMR spectroscopy and a robust $^{n}J_{CH}$ network analysis to explore the complementary aspects of both techniques. This delivers good results, even if a perfect computational separation of the compounds in the mixture is not possible. All of the scripts are freely available to aid studies such as with plants, marine organisms, and microorganism natural product chemistry and metabolomics, as those are the driving forces for this project.

a De Montfort University, School of Computer Science and Informatics, The Gateway, Leicester LE1 9BH, UK. E-mail: stefan.kuhn@dmu.ac.uk

b Instituto de Pesquisas de Produtos Naturais Walter Mors, Universidade Federal do Rio de Janeiro, Brazil

† Electronic supplementary information (ESI) available: The data in realspectrum.csv and compounds.smi are for *P. boldus*. The spectral data (HSQC and HMBC) of *P. boldus* are provided as well. Additionally, https://github.com/stefhk3/nmrfilter contains all code described in Section 2.2. See DOI: 10.1039/c8fd00227d

1   Introduction

Natural products (NP) are an important source of new pharmacologically active compounds. Regrettably, the rapid extinction of many unexplored plants and other organisms represents losses of a broad range of potential new bioactive and valuable chemicals. An effective and challenge-free method of screening and identifying NP is yet to be well established. Thus, there is a need for new high-throughput approaches to be used as standard procedures to accurately catalog NP. When a biologically relevant spectral feature is identified and it is listed in a given database, the identification process is straightforward, generating a confidence index for the analytical data to a database match. This is routinely done for biological samples, especially in human metabolomics studies, where the broad range of compounds is well-known and well-recorded in various databases.[1] This is not the same for NP, where the chemical diversity is much broader, with varied physicochemical properties. Their available databases are not well organized, comprehensive or publicly available. The complexity of secondary metabolite biosynthesis leads to the opportunity of uncovering additional compounds at different stages of the biosynthetic/metabolic pathway with similar core structures. Within this context, mass spectrometry (MS) and nuclear magnetic resonance (NMR) play a leading role in yielding informative data for the identification of both known and unknown organic chemical compounds. Both have benefits and drawbacks that characterize their complementary usage in terms of the sensitivity, reproducibility and structural information they are able to provide. Whereas MS shows high sensitivity and accuracy but low reproducibility, NMR shows low sensitivity, but high reproducibility and efficiency to unambiguously elucidate complex structures. If a single compound is analysed, the spectrum can be interpreted and will reveal its structure. A molecular structure can be inferred from the peaks in the so-called spectrum acquired from both MS and NMR. In the case of mixtures, the spectrum corresponds to the spectra of all compounds in one analysis. So a direct interpretation as the result of a molecule is not trivial since all signals from each component of the mixture will be shown.

1.1   Background

Mixture analysis is a hot topic today within NP and metabolomics including modern and complex algorithms. Specifically for NP, the analysis of complex mixtures is often referred to as dereplication due to its goal to quickly identify known compounds and prevent replicated results. Dereplication in NP has been extensively reviewed elsewhere.[2] Open-access tools such as MZMine[3] and Open-Chrom[4] enable complex processing of MS data and database matching using open-access or even user-defined databases for dereplication; closed source options from different companies are available as well but under copyright protection. Global Natural Products Social Molecular Networking (GNPS) is an important tool that calculates similarity networks among the fragmentograms and enables a crowdsourcing approach for dereplication.[5] It uses open-access databases for spectra matching and allows the user to submit data from putatively identified compounds to a local database. New workflows are under development for the use of *in silico* fragmentation for compound identification.[6] GNPS now includes a workflow for the use of *in silico* fragmentation of candidate structures, namely Network Annotation Propagation.[7] Regarding NMR compound identification, COLMAR[8] is a broadly used tool for metabolomics mainly focused on primary metabolites. It uses HMDB[1] and BMRB[9] as databases and it offers an interactive web interface. The underlying technique (called DemixC) for NMR analysis uses full high-resolution TOCSY after covariance NMR to deconvolute pure spectra from redundant connectivity information from the cross peaks. Statistical techniques are then used to find correlated changes in the cross peaks and allow separation of the spectra of the individual compounds from the measured spectra.[10] The use of ^{13}C NMR for compound identification is a trend in the last decade,[11-14] probably due to the increasing sensitivity of dedicated (micro- and nano-)probes. Undeniably, the ^{13}C resonances are less affected by external parameters such as solvent, pH or temperature than ^{1}H resonances, but the low sensitivity of direct ^{13}C detection is still prohibitive. The INETA package was designed to use INADEQUATE NMR data using mainly BMRB[9] and assigned ^{13}C resonances as databases for compound identification, but yet it is only feasible for ^{13}C labeled samples. Another interesting approach was developed for a computer-aided ^{13}C profile of NP that uses 1D ^{13}C NMR data and an *in-house* search algorithm based on simulated NMR data from predefined candidates.[15] Later, the same group extended this approach by using HMBC NMR data for compound identification using a community detection algorithm.[16] Differential analysis of 2D NMR spectra (DANS) compares spectra from different biological states. If some signals vary significantly between two spectra, they are assumed to come from compounds unique to a particular sample.[17]

2   Methods

2.1   Overview of the method

The overall method we present here starts with a general compound identification scheme using LC-ESI-MSMS and molecular networks as described elsewhere.[18] We also search a list of expected compounds in the literature assuming chemo-taxonomical relations within the plant species and genus as well as compounds from related biosynthetic pathways. Once we have identified a set of lead compounds using MS data and chemotaxonomic review, we search for similar known compounds through Pubmed and list them together as possible candidates (Fig. 1). Thus, this approach filters compounds by structural variations rather than by the molecular mass only. For the candidates identified, we predict HMBC and HSQC NMR spectra and compare them to the measured spectrum of the mixture for every candidate (see Section 2.3 for details). The goal is to design a process to confirm the MS compound identification using 2D NMR and increase confidence. For this, we have designed output parameters, which calculate how well the simulated spectra fit to the measured data, and the candidates are ranked accordingly.

2.2   NMR network analysis

An NMR network analysis was suggested in ref. 16. This is based on the idea that in an HMBC spectrum, cross peaks originating from one compound should either share the ^{13}C or the ^{1}H chemical shift. So an initial network is built from the long-range proton-carbon couplings. If a complete separation is

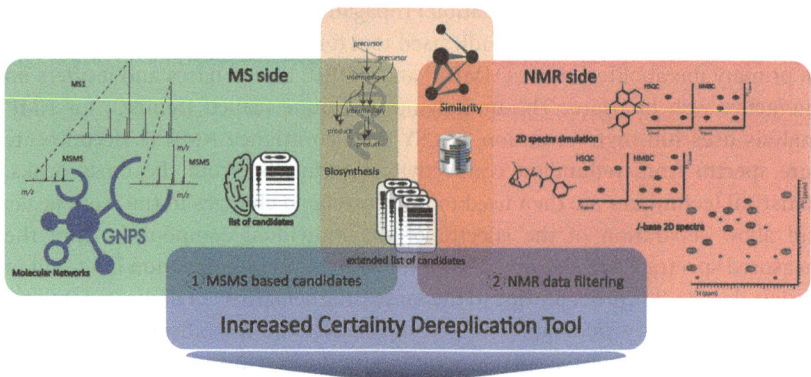

Fig. 1 Overall approach from the MSMS analysis and compilation of the possible candidates to the NMR simulations and matching.

not possible by this, a community clustering algorithm should separate the subnetworks for the individual compounds in the overall network, assuming that there should be more connections between cross peaks within one compound than to the other compounds. In ref. 16 the RBER (Erdös-Rényi null-model) method is used, with the resolution parameter set to 0.2. This method divides the network into clusters based on the density of the connections inside the clusters compared to the density of connections to other clusters. It optimizes the clusters to have many connections within, but few to other clusters. Cross peaks from the same compound should share many chemical shifts. In contrast, different compounds should share chemical shifts rarely, by a combination of similar substructures and not enough resolution in the measurement. Therefore, the clusters should mostly correspond to cross peaks from one structure, not from several structures.

It is reported in ref. 16 that this process yields as many clusters as there are compounds in the mixture, with each cluster corresponding to the cross peaks from one compound. The cross peaks from each cluster can then be matched against a database of compounds to identify the components of the mixture.

Table 1 The clustering achieved for the HMBC cross peaks of a mixture of caffeine and ferulic acid. Various settings for the chemical shift tolerance and the resolution parameter of the RBER algorithm were tested. 1/28, 5/2... means there was 1 cluster with 28 elements, 5 with 2 elements, *etc*.

Tolerance		
RBER resolution	^{13}C: 0, ^1H: 0	^{13}C: 0.2, ^1H: 0.02
0.2	3/7, 2/6, 1/5, 1/2, 15/1	1/28, 1/15, 5/2, 2/1
0.5	3/7, 2/6, 1/5, 1/2, 15/1	1/28, 1/10, 1/5, 5/2, 2/1
1	3/7, 2/6, 1/5, 1/2, 15/1	1/15, 1/13, 1/7, 1/5, 1/3, 5/2, 2/1
10	3/5, 2/4, 3/3, 3/2, 17/1	1/6, 2/5, 3/4, 2/3, 8/2, 5/1

In order to verify the approach, we tested it with the mixture of caffeine and ferulic acid described in Section 4. The HMBC spectrum yields 55 cross peaks. We firstly built a network of HMBC cross peaks using the 0.2/0.02 ppm tolerance and then applied the RBER using the suggested 0.2 resolution parameter. We then varied these parameters. Table 1 shows the numbers of clusters and their size derived with different resolution parameters for the RBER algorithm. As expected, we obtained two large clusters by setting the resolute parameter to 0.2. We listed the observed cross peaks (numbered from 1 to 55) in the clusters, underlined them if there is a matching cross peak for caffeine and overlined them if there is a matching cross peak for ferulic acid. 42 of the 47 predicted chemical shifts are matched to a measured chemical shift. We get the following list:

$[\overline{1},\ \overline{2},\ \overline{3},\ \overline{4},\ \overline{6},\ \overline{7},\ \overline{8},\ \overline{9},\ \overline{10},\ \overline{11},\ \overline{12},\ \overline{13},\ \overline{14},\ \overline{15},\ \overline{16},\ \overline{17},\ \overline{18},\ \overline{20},$
$21,\ \overline{22},\ \overline{23},\ \overline{38},\ \overline{39},\ \underline{45},\ \overline{46},\ \overline{47},\ 50,\ 52]$

$[\overline{5},\ 30,\ \underline{31},\ \underline{32},\ \underline{35},\ \underline{37},\ \underline{41},\ \overline{42},\ 44,\ \overline{48},\ \underline{49},\ \underline{51},\ 53,\ \overline{54},\ \overline{55}]$

$[19,\ 40]$

$[24,\ 25]$

$[26,\ 27]$

$[\underline{28},\ 29]$

$[33,\ 34]$

$[\underline{36}]$

$[\underline{43}]$

We can see that the two large clusters correspond roughly to the two compounds, but the separation is not perfect. Furthermore, when using different parameters for resolution and tolerance, the number of clusters varies. We list the cross peaks for resolution 0.2 and tolerance 0 ppm for both axes in the same fashion as before, with the cross peaks matched marked by underline/overline, and visualize this in Fig. 2 (15 clusters with one cross peak have been left out for clarity):

$[\overline{5},\ \underline{31},\ \underline{32},\ \underline{35},\ \underline{36},\ \underline{37},\ \underline{41}]$
$[\overline{1},\ \overline{10},\ \overline{11},\ \overline{15},\ \overline{16},\ \overline{23},\ \overline{39}]$
$[\overline{4},\ \overline{7},\ \overline{8},\ \overline{9},\ \overline{13},\ \overline{14},\ \overline{17}]$
$[\overline{3},\ \overline{6},\ \overline{9},\ \overline{12},\ \overline{20},\ 21]$
$[\overline{48},\ \underline{49},\ \underline{51},\ 53,\ \overline{54},\ \overline{55}]$
$[\overline{22},\ \overline{38},\ 44,\ \overline{46},\ \overline{47}]$
$[24,\ 25]$

There is no clean separation, but we can still see a clustering pattern: in the clusters, most of the cross peaks belong to one compound.

Overall it is clear that the separation depends on the parameters. Furthermore, with more complex mixtures, the best parameter setting may change. Finally, we found that the data processing and peak picking can influence the separation. On the other hand, even if a complete separation is not possible, the cross peaks for

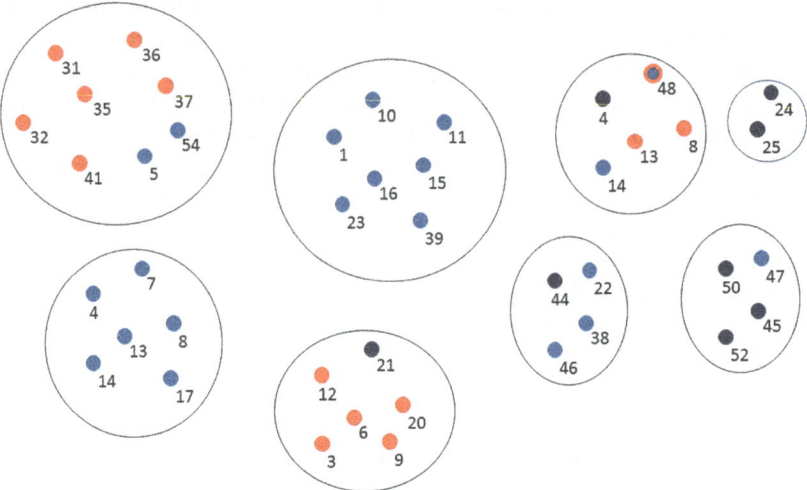

Fig. 2 The clusters derived from the HMBC spectrum of caffeine and ferulic acid with tolerances set to 0 for both chemical shifts and the RBER resolution set to 0.2, with the compounds mapped onto them. Cross peaks for ferulic acid are blue, those for caffeine are red, and cross peaks mapped to both are in both colours. Cross peaks used for none are in black.

the compounds still fall mostly within specific clusters. They are not spread out over all clusters. We have also confirmed that some test compounds other than caffeine and ferulic acid do not show such a clustering pattern when tested against the caffeine and ferulic acid mixture. The cross peaks of those other compounds either do not match any cross peaks or spread out over all clusters.

2.3 NMR filtering by network analysis

Considering our results when using the NMR network analysis, we believe that a reliable separation of compounds is not always possible. If this would be possible, each cluster could be matched against the predicted spectra. Even though a reliable separation is not possible, the clusters still somehow relate to the compounds. Since we have a list of possible candidates from the MS experiments, we have therefore devised a modified algorithm. This does not assume that one cluster represents one compound, but that clusters contain cross peaks belonging to one compound. Even if it would be possible to achieve full separation by fine-tuning the measurement and the data processing and peak picking, our results indicate that the full separation is quite sensitive and is not guaranteed to work. Therefore our method is designed to be more robust and less dependant on the quality of the data. We call our method NMR filtering by network analysis. If the clustering separates the compounds exactly, the simulated spectra should cover exactly one cluster. So the procedure in ref. 16 is actually a special case of our algorithm.

Once a list of candidates has been generated using the MS analysis, the results are ranked according to the likelihood of their occurrence in the 2D NMR spectra. Core ideas presented in ref. 16 are used for this, but they are extended and generalized in the present approach that uses established techniques, but modifies them by introducing new elements. The core steps of our approach are:

• We use HSQC and HMBC spectra, measured and peak-picked as explained in Section 4. The cross peaks of both spectra were put into a single list, with the ^{13}C chemical shift being the first dimension and the ^{1}H chemical shift being the second dimension.

• Every cross peak is a node in the NMR network we built in the next step. An edge between two nodes is added to the network if two cross peaks share a chemical shift on the ^{13}C or ^{1}H axis. A tolerance of 0.2 ppm for ^{13}C chemical shifts and 0.02 ppm for ^{1}H chemical shifts is applied here. These values have been found experimentally and can be changed if desired. As cross peaks from the same compound should share either the ^{13}C or the ^{1}H chemical shift value(s) with other cross peaks from the same compound, this gives an initial network.

• The resulting network is analysed using the RBER algorithm. The resolution parameter is set to 0.2. Again, this can be changed. As explained, inside the clusters produced, the cross peaks should originate predominantly from one compound, even if a complete separation is not possible.

• We then predict the HSQC and HMBC spectra for the candidate structures derived from MSMS. The combined spectra for each of the compounds are then mapped onto the measured spectrum. From the mapping, we calculate two measures: (a) the distance of the simulated spectrum to the measured spectrum and (b) the distribution of the cross peaks matched in the measured spectrum within the clusters calculated in the previous steps. For details of the calculation, see the description of the implementation below.

• We normalize both measures to range from 0 to 1 and use the average of the two measures as the likelihood of a compound to be part of the mixture.

In this algorithm, we map the cross peaks of the simulated spectrum of each candidate onto the whole spectrum and calculate the distribution over all clusters. Ideally, the cross peaks should cover some clusters completely and not have any cross peaks in the remaining clusters. So we have the distribution in the clusters and the distance of the simulated spectrum to the best match in the measured spectrum for each candidate as an indication of how likely the candidate is to occur in the mixture. In order to improve the clustering, we include HSQC and HMBC spectra in our clustering (as opposed to ref. 16, which uses HMBC only). All of these have ^{13}C–^{1}H cross peaks, which are treated the same, forming one network, to which the clustering is applied. The spectrum simulation is also done for HMBC and HSQC spectra and these cross peaks are mapped onto the combined spectra.

Our approach is illustrated in Section 2.2 and Fig. 2. These demonstrate that the cross peaks for the compounds fall mostly within specific clusters. This is true even if a complete separation is not achieved.

We have implemented the described procedure as a set of Python script, including a Java program to perform the prediction, and a shell script to run the overall procedure. Data are transferred between scripts *via* text files. This is primarily intended as proof of concept; a full application is part of the future work. The detailed algorithm for the NMR ranking is as follows:

• For each candidate structure originating from the MSMS step, we simulate the combined HSQC and HMBC spectra using the prediction mechanism of nmrshiftdb2.[19] The Java code in simulate.jar extracts pairs of atoms from the molecule, which are assumed to generate a cross peak in one of the spectra, and writes the pair of chemical shifts of these atoms into a peaklist. For HSQC, cross peaks are built for all atom pairs one bond away, and for HMBC, for all atom pairs two or three bonds away. Couplings and intensities are currently not included; the cross peaks are based on topology only. Experience shows that this gives a sufficient approximation. The chemical shift prediction is based on HOSE codes, uses solvents when possible, and respects wedge bonds if data are available.[20]

• We form a single list of cross peaks out of the HSQC and HMBC spectra measured for the mixture. This list is provided to the clustering.py script, which builds a network as described, using the tolerances from the nmrproc.properties file.

• The network generated is processed by clusterlouvain.py. This applies the RBER algorithm, using the louvain library.[21] The resolution value is taken from the nmrproc.properties file. The result is a list of clusters, containing all cross peaks from the measured spectra in some cluster.

• For every simulated spectrum, we find the nearest matching cross peaks in the measured spectrum. This is done by calculating the distance between each cross peak in the simulated spectrum and each cross peak in the measured spectrum. The formula for this is as follows:

$$\text{Distance}(\text{peak}_1, \text{peak}_2) = (\text{abs}(\text{peak}_{1_x} - \text{peak}_{2_x}) + \text{abs}(\text{peak}_{1_y} - \text{peak}_{2_y}) \times 10)^2 \quad (1)$$

This squares the distance between the two cross peaks on the ^{1}H and ^{13}C axis and adds them. The ^{1}H chemical shift is multiplied by 10 to normalize the range, assuming ^{13}C ranges from 0 to 200 ppm and ^{1}H from 0 to 20 ppm. The factor 10 is commonly used, e.g. in ref. 16 the tolerance for carbon chemical shifts is 1.5 ppm and for hydrogen chemical shifts it is 0.15 ppm. The squaring includes variance and bias, similar to the mean squared error in statistics. This gives us a matrix of size $n \times m$, where n is the number of cross peaks in the measured spectrum and m the number of cross peaks in the simulated spectrum. We then use the function linear_sum_assignment from the scipy.optimize package to find the minimal combination of these costs, which assigns exactly one cross peak to every cross peak in the measured spectrum. The sum of the costs of this minimal combination is the distance of the simulated spectrum to the measured spectrum, which is our first reliability measure. As opposed to other methods,[22] we do not have a fixed limit for cross peaks to match, rather we search for a best match and calculate the distance. Together with the squared distance, this should give us a robust mapping.

• We have previously created the clusters containing the measured cross peaks. In the previous step, we have mapped each simulated cross peak onto one measured cross peak. Therefore, we can now calculate the fraction of cross peaks in each clusters, onto which a simulated cross peak is mapped. Our distance measure will map each simulated cross peak onto some measured peak, even if they are very much apart. For the distance measure this is not a problem, since it will mean a very high distance value in cases of bad matches, which in turn means the compound will not be considered a good match. For the clustering step, we only use mappings where the distance is less than 9, which corresponds to a value of 1.5 ppm for the ^{13}C chemical shift and 0.15 ppm (remember the factor of 10) for the ^{1}H chemical shift, which are the cut offs used in ref. 16. This step gives us *n* decimal numbers between 0 and 1 (since it represents the fraction of mapped peaks, which is between 0 for no mapping and 1 for all peaks mapped), *n* being the number of clusters. We then calculate the standard deviation, using the std method of the numpy package of these numbers for each simulated spectrum. The standard deviation is the second reliability measure.

• In the last step, both reliability measures are normalized to range from 0 (best) to 1 (worst). For each simulated spectrum, and therefore for each candidate compound, they are added and the compounds are ranked by this combined reliability measure, the compound with the lowest value being the most likely candidate.

It should be noted that the current code is not optimized for performance. Running it on a laptop with an Intel Core i5 6300U CPU for the *P. boldus* mixture discussed in the next section takes around 4 minutes. This will be improved in the planned application, but for this type of task a very quick solution cannot be expected, given the amount of information involved.

3 Results

We collected MS data in high resolution under ddMS2 Top3 experiments to yield close to 5000 scans in a more than 2500 Mb file. We converted the raw data to .mzXML for network calculations using the GNPS web system and, then, we used Cytoscape for visualization and further analysis (Fig. 3, section A). From the MS spectra, we could visualize high intensity key features that would indicate well-

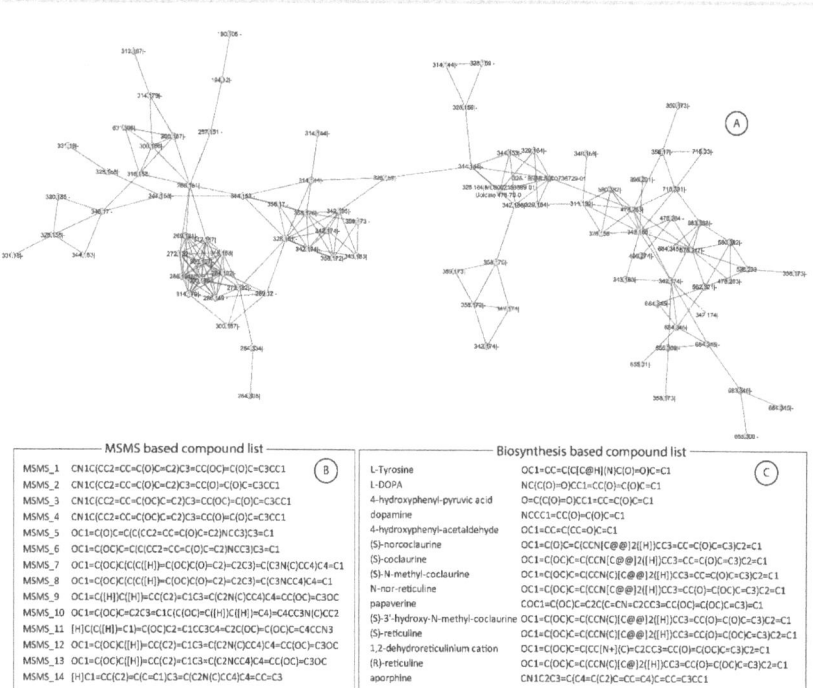

Fig. 3 Main molecular network and list of candidates.

known components of the alkaloidic fraction of *P. boldus*. Boldine (at *m/z* 328.15), coclaurine (at *m/z* 286.14), and norreticuline (at *m/z* 316.15) are well-known components of the aporphine-like pool of alkaloids from this species;[23] other close related (delta-*m/z* 14, 12 and 16) features (at *m/z* 300.15, 342.17 and 358.16) are also displayed at the MS scan (ESI†). Nonetheless, the use of molecular networking for structure elucidation enables the enrichment of the list of candidates, and adding others that are closely related or of expected occurrence. The GNPS processed data can be accessed here.‡ GNPS promptly identified boldine within a network of 94 nodes; this finding enabled us to relay the information and elucidate the possible structure for the closely related nodes. 14 structures were suggested in this stage (Fig. 3, section B). Thus, we identified a core structure as being of an aporphine-like alkaloid and used that to extend the list of candidates with similar known compounds (from Pubmed) and others that play a role in their most accepted biosynthesis pathway (Fig. 3, section C).[24] We took the compounds from the MS side of the method and listed them as SMILES structures as preparation for the NMR filtering. Note that the sensitivity gap between NMR and MS is anticipated and will be reflected naturally by the eventual annotation of fewer compounds after the NMR analysis side of the method. This is due to the peak intensity of MS data, which is structure-dependent and varies highly according to the ionization technique. In contrast, peak intensity in NMR is mainly dependent on the spin concentration, ^{1}H in the case discussed here.

The HMBC and HSQC spectra for *P. boldus* gave 1034 cross peaks. Running the clustering algorithm on these yields 193 clusters. The largest has 257 cross peaks, 75 clusters have one peak, and the other clusters are somewhere in between in size. The average size of the clusters is 4.36, the median is 2. We simulate the spectra for the compounds derived from the MS step. Calculating the similarity and clustering for them gives a ranking for the compounds. The first ten compounds and the last compound are as follows:

(1) OC1=C(OC)C=C(C(CC2=CC=C(O)C=C2)NCC3)C3=C1, distance: 0.21, standard deviation: 1.00; 1-[(4-hydroxyphenyl)methyl]-7-methoxy-1,2,3,4-tetrahydroisoquinolin-6-ol

(2) OC1=C(O)C=C(CCN[C@@]2([H])CC3=CC=C(O)C=C3)C2=C1, distance: 0.23, standard deviation: 0.99; (*S*)-norcoclaurine

(3) OC1=C(O)C=C(C(CC2=CC=C(O)C=C2)NCC3)C3=C1, distance: 0.24, standard deviation: 0.98; norcoclaurine

(4) CC1CN(C)C2CC3=CC=CC=C3C4=C2C1=CC=C4, distance: 0.26, standard deviation: 0.99; 4,6-dimethyl-5,6,6a,7-tetrahydro-4*H*-dibenzo[*de,g*]quinoline

(5)   OC1=C(OC)C=C(CCN[C@@]2([H])CC3=CC(O)=C(OC)C=C3)C2=C1, distance: 0.20, standard deviation: 0.84; (*S*)-norreticuline

(6) OC1=C(OC)C(C(C([H])=C(OC)C(O)=C2)=C2C3)=C(C3N(C)CC4)C4=C1, distance: 0.13, standard deviation: 0.73; (+)-(*S*)-boldine

(7) OC1=C(OC)C=C(CCN[C@@]2([H])CC3=CC=C(O)C=C3)C2=C1, distance: 0.23, standard deviation: 0.82; coclaurine

(8)   CCC1CN(C)C2CC3=CC=CC=C3C4=C2C1=CC=C4, distance: 0.26, standard deviation: 0.81; 4-ethyl-6-methyl-5,6,6a,7-tetrahydro-4*H*-dibenzo[*de,g*]

---

‡ See the link https://gnps.ucsd.edu/ProteoSAFe/status.jsp?task=4275dd938bdf4eea8f30a59afdcfc671.

(9) OC1=C(OC)C=C(CCN(C)[C@@]2([H])CC3=CC(O)=C(O) C=C3)C2=C1, distance: 0.16, standard deviation: 0.67; (S)-3-hydroxy-N-methylcoclaurine

(10) NCCC1=CC(O)=C(O)C=C1, distance: 0.00, standard deviation: 0.47; dopamine

...

(66) O=C(C(O)=O)CC1=CC=C(O)C=C1, distance: 0.99, standard deviation: 0.00; 4-hydroxyphenylpyruvate.

The first hits have high standard deviations (due to normalizing 1 as maximum), meaning they fall into a low number of clusters. They also have relatively high distance to the measured spectrum (0 being optimal). Lower in the list, we get better similarities, but still relatively high clustering. The last hit has high distance and a low standard deviation, making it a highly unlikely candidate. This ranking validates the MS-only based dereplication yielding higher confidence to the result. The successful identification of such aporphine alkaloids (1-[(4-hydroxyphenyl)methyl]-7-methoxy-1,2,3,4-tetrahydroisoquinolin-6-ol, norcoclaurine, 4,6-dimethyl-5,6,6a,7-tetrahydro-4H-dibenzo[de,g]quinoline, norreticuline, boldine, coclaurine, 4-ethyl-6-methyl-5,6,6a,7-tetrahydro-4H-dibenzo[de,g] and 3-hydroxy-N-methylcoclaurine) using both MS and NMR matches the experimental results for *P. boldus* from the previous studies.[23]

4 Experimental section

4.1 Chemicals

HPLC grade methanol and ethyl acetate, LC-MS grade formic acid and HCl and NaOH were acquired from Tedia-Brazil (Rio de Janeiro, RJ, Brazil); D_2O (99.0%), methanol-d^4 and chloroform-d^1 were acquired from Cambridge Isotope Laboratory, Inc. (Andover, MA, USA); caffeine and ferulic acid were acquired from Sigma-Aldrich (St. Louis, MO, United States). Deionized water was purified by a Millipore Milli-Q Gradient A 10 System (Burlington, MA, USA).

4.2 Plant material and sample preparation

Peumus boldus dry leaves from different brands were purchased from different commercial locations in Rio de Janeiro (RJ, Brazil). Samples of 1 g of each were combined and an aliquot (1 g) was saved for the extraction. This representative aliquot was extracted with aqueous HCl 0.02 M (15 mL) at pH 2.5 under ultrasound for 5 minutes. Then, three successive liquid–liquid extractions were performed with 5 mL of ethyl acetate. The pH of the aqueous phase was increased to 9 with 1 mL of aqueous NaOH (1 M) and three successive extractions were again performed with 5 mL of ethyl acetate. The combined organic phases were concentrated to dryness under vacuum to yield 22.6 mg of a crude alkaloid extract. This preparation was made in three replicates. The final samples were divided into 2 aliquots each. 20% of it was resuspended in methanol for LC-MSMS analysis and the remaining 80%, in chloroform-d^1 for the NMR analysis; for the NMR analysis the replicates were combined. The caffeine and ferulic acid mixture was prepared in methanol-d^4 as concentrated samples, centrifuged and transferred to 3 mm NMR tubes.

4.3 Liquid chromatography-tandem mass spectrometry analysis

Ultra-high performance liquid chromatography analysis was performed on a 1260 Infinity Liquid Chromatography system (Agilent) consisting of a quaternary solvent delivery pump and a column oven compartment. Samples (10 μL) were injected using and separated on an Agilent Extend-C18 column (2.1 × 50 mm, 1.8 μm particle size) at 300 μL min^{-1} maintained at 40 °C. The mobile phase consisted of (A) 0.1% formic acid and (B) 0.1% formic acid in methanol in gradient elution mode (0 min 15% B; 0–10 min 100% B; 10–18 min, 15% B; 18.5–25 min 15%). The UHPLC system was coupled to a Q-TOF high resolution and accurate mass spectrometer (Agilent) equipped with an electrospray ion source (Dual ESI; Agilent) operating in positive ionization mode. Source ionization parameters were: spray voltage 3.5 kV; capillary temperature 350 °C; gas flow 10 L min^{-1}; nebulizer 25 psi; skimmer1 65; isolation width MS/MS medium ~4 amu); fixed collision energy for MS/MS 30. Samples were analysed in the scan range of m/z 100 to 1700 (for MS and MS/MS) at a scan rate of 3 spectra per second followed by data-dependent MSMS (ddMS2 Top3 experiments) at a scan rate of 2 spectra per second. The acquired data were converted to mzML or mzXML files using the software MSConvert (ProteoWizard; proteowizard.sourceforge.net/tools.shtml). GNPS network parameters were: MS Fragment Ion Tolerance: 0.02; MS/MS Fragment Ion Tolerance: 0.02; minimum MS/MS peak intensity: 0.0; Run MSCluster: on; minimum consensus cluster size: 1; minimum matched peaks in network edge: 4; minimum MS/MS cosine score in network edge: 0.65; number of neighbors to retain in network: 10; maximum connected component size: 100. The resulted networks were plotted using the Cytoscape software (http://www.cytoscape.org).

4.4 Nuclear magnetic resonance analysis

NMR data was collected using a 800 MHz and a 600 MHz Bruker Avance III equipped with a 1.7 mm TCI cryoprobe and a 5 mm DCH D/H-C carbon cryoprobe, respectively. The pulse sequence hsqcedetgpsp.3 under non-uniform sampling mode (35% of NUS amount and 896 NUS points; 4096 and 5120 points for F2 and F1, respectively; 28.45 points per ppm) was used to acquire the edited HSQC data (24 scans, optimized for $^1J_{CH}$ = 145 Hz; 18 h 15 min), and hmbcetgpl3nd under non-uniform sampling mode (30% of NUS amount and 768 NUS points; 4096 and 5120 points for F2 and F1, respectively; 23.27 points per ppm) for the HMBC data (24 scans, optimized for $^3J_{CH}$ = 8 Hz; 14 h 22 min). For the test sample (caffeine plus ferulic acid), the NMR data was collected using hsqcedetgpsp.3 under non-uniform sampling mode (1024 and 256 points for F2 and F1, respectively) to acquire the edited HSQC data (4 scans, optimized for $^1J_{CH}$ = 145 Hz; 8 min), and hmbcetgpl3nd under non-uniform sampling mode (4096 and 256 points for F2 and F1, respectively) was used for the HMBC data (4 scans, optimized for $^3J_{CH}$ = 8 Hz; 7 min).

5   Conclusion

We have demonstrated a method to infer a list of candidate compounds for complex natural product mixtures. Our method combines MS and NMR techniques to give confidence in the results. The MS step yields a relatively broad result, ensuring coverage of all possible compounds. The NMR step does not rely on predefined libraries, but ranks the suggestions by using their predicted NMR spectra. The prediction can be done for a range of naturally occurring products with a reasonable average error.[20] We found that a full distinction of the compounds in the spectrum is not needed to rank the candidates. Since a full distinction is difficult and in many cases not possible, we consider the combination of prefiltering and ranking a promising approach. It gives reasonable results, even if the peak data are not optimal, due to problems in measurement or data processing. There are indications that the results provide a good match with the actual compounds, but more work to verify this is needed. In particular, larger datasets will be examined by the authors.

We have used an artificial mixture to demonstrate the NMR filtering step and have demonstrated the overall approach using an alkaloid enriched extract of *P. boldus*. A major advantage is that no special sample preparation or experiments are needed. Both the MS and NMR experiments are standard and can be used almost as default. Even though better resolution and higher sensitivity will improve the results, the use of older or less sophisticated equipment is still possible. Furthermore, once the experiments are performed, the processing is relatively quick and will be even more with more automation, which we intend to make possible.

5.1 Future work

The process as presented in this paper is only partially automated. The computational parts are currently done in a command line interface without possibility for user interaction. We aim to increase automation and make the interface more user-friendly in a next step. We consider integrating the program either into a workflow tool like KNIME, or a platform like Bioclipse. The inclusion of $^{n}J_{HH}$ data in the NMR network analysis will be part of this.

Concomitantly, we are applying this approach to other samples and fractions for a broader range of applications; mainly, the NMR filter will benefit significantly by reducing the complexity and the dynamic range with a low-resolution fractionation step. The goal is to establish a source independent tool for dereplication of NP to be used as a driving force towards novelty discovery. The scripts will be made freely available and it will enable data submission to databases as an integral part. New samples of terrestrial plants, marine organisms, microorganisms, fungi and corals are some of the examples to be exploited in the near future.

Conflicts of interest

There are no conflicts of interest to declare.

Acknowledgements

We thank the National Council for the Improvement of Higher Education (CAPES, Brazil) for the financial support. RMB would like to thank Dr Arthur Edison for all of the support in this research topic and for NMR time at the Complex Carbohydrate Research Center (UGA, USA).

References

1 D. S. Wishart, Y. D. Feunang, A. Marcu, A. C. Guo, K. Liang, R. Vazquez-Fresno, T. Sajed, D. Johnson, C. Li, N. Karu, Z. Sayeeda, E. Lo, N. Assempour, M. Berjanskii, S. Singhal, D. Arndt, Y. Liang, H. Badran, J. Grant, A. Serra-Cayuela, Y. Liu, R. Mandal, V. Neveu, A. Pon, C. Knox, M. Wilson, C. Manach and A. Scalbert, *Nucleic Acids Res.*, 2018, **46**, D608–D617.
2 J. Hubert, J.-M. Nuzillard and J.-H. Renault, *Phytochem. Rev.*, 2017, **16**, 55–95.
3 F. Olivon, G. Grelier, F. Roussi, M. Litaudon and D. Touboul, *Anal. Chem.*, 2017, **89**, 7836–7840.
4 P. Wenig and J. Odermatt, *BMC Bioinf.*, 2010, **11**, 405.
5 M. Wang, J. J. Carver, V. V. Phelan, L. M. Sanchez, N. Garg, Y. Peng, D. D. Nguyen, J. Watrous, C. A. Kapono, T. Luzzatto-Knaan, C. Porto, A. Bouslimani, A. V. Melnik, M. J. Meehan, W. T. Liu, M. Crusemann, P. D. Boudreau, E. Esquenazi, M. Sandoval-Calderon, R. D. Kersten, L. A. Pace, R. A. Quinn, K. R. Duncan, C. C. Hsu, D. J. Floros, R. G. Gavilan, K. Kleigrewe, T. Northen, R. J. Dutton, D. Parrot, E. E. Carlson, B. Aigle, C. F. Michelsen, L. Jelsbak, C. Sohlenkamp, P. Pevzner, A. Edlund, J. McLean, J. Piel, B. T. Murphy, L. Gerwick, C. C. Liaw, Y. L. Yang, H. U. Humpf, M. Maansson, R. A. Keyzers, A. C. Sims, A. R. Johnson, A. M. Sidebottom, B. E. Sedio, A. Klitgaard, C. B. Larson, C. A. Boya P, D. Torres-Mendoza, D. J. Gonzalez, D. B. Silva, L. M. Marques, D. P. Demarque, E. Pociute, E. C. O'Neill, E. Briand, E. J. N. Helfrich, E. A. Granatosky, E. Glukhov, F. Ryffel, H. Houson, H. Mohimani, J. J. Kharbush, Y. Zeng, J. A. Vorholt, K. L. Kurita, P. Charusanti, K. L. McPhail, K. F. Nielsen, L. Vuong, M. Elfeki, M. F. Traxler, N. Engene, N. Koyama, O. B. Vining, R. Baric, R. R. Silva, S. J. Mascuch, S. Tomasi, S. Jenkins, V. Macherla, T. Hoffman, V. Agarwal, P. G. Williams, J. Dai, R. Neupane, J. Gurr, A. M. C. Rodriguez, A. Lamsa, C. Zhang, K. Dorrestein, B. M. Duggan, J. Almaliti, P. M. Allard, P. Phapale, L. F. Nothias, T. Alexandrov, M. Litaudon, J. L. Wolfender, J. E. Kyle, T. O. Metz, T. Peryea, D. T. Nguyen, D. VanLeer, P. Shinn, A. Jadhav, R. Muller, K. M. Waters, W. Shi, X. Liu, L. Zhang, R. Knight, P. R. Jensen, B. O. Palsson, K. Pogliano, R. G. Linington, M. Gutierrez, N. P. Lopes, W. H. Gerwick, B. S. Moore, P. C. Dorrestein and N. Bandeira, *Nat. Biotechnol.*, 2016, **34**, 828–837.
6 P.-M. Allard, T. Péresse, J. Bisson, K. Gindro, L. Marcourt, V. C. Pham, F. Roussi, M. Litaudon and J.-L. Wolfender, *Anal. Chem.*, 2016, **88**, 3317–3323.
7 R. R. da Silva, M. Wang, L.-F. Nothias, J. J. J. van der Hooft, A. M. Caraballo-Rodríguez, E. Fox, M. J. Balunas, J. L. Klassen, N. P. Lopes and P. C. Dorrestein, *PLoS Comput. Biol.*, 2018, **14**, 1–26.
8 K. Bingol, L. Bruschweiler-Li, C. Yu, A. Somogyi, F. Zhang and R. Brüschweiler, *Anal. Chem.*, 2015, **87**, 3864–3870.
9 J. F. Doreleijers, S. Mading, D. Maziuk, K. Sojourner, L. Yin, J. Zhu, J. L. Markley and E. L. Ulrich, *J. Biomol. NMR*, 2003, **26**, 139–146.
10 F. Zhang and R. Brüschweiler, *ChemPhysChem*, 2004, **5**, 794–796.
11 A. Bruguière, S. Derbré, C. Coste, M. L. Bot, B. Siegler, S. T. Leong, S. N. Sulaiman, K. Awang and P. Richomme, *Fitoterapia*, 2018, **131**, 59–64.

12 A. Botana, P. W. Howe, V. Caër, G. A. Morris and M. Nilsson, *J. Magn. Reson.*, 2011, **211**, 25–29.

13 A. Mäkelä, I. Kilpeläinen and S. Heikkinen, *J. Magn. Reson.*, 2010, **204**, 124–130.

14 C. S. Clendinen, C. Pasquel, R. Ajredini and A. S. Edison, *Anal. Chem.*, 2015, **87**, 5698–5706.

15 J. Hubert, J.-M. Nuzillard, S. Purson, M. Hamzaoui, N. Borie, R. Reynaud and J.-H. Renault, *Anal. Chem.*, 2014, **86**, 2955–2962.

16 A. Bakiri, J. Hubert, R. Reynaud, C. Lambert, A. Martinez, J. H. Renault and J. M. Nuzillard, *J. Chem. Inf. Model.*, 2018, **58**, 262–270.

17 F. C. Schroeder, D. M. Gibson, A. C. Churchill, P. Sojikul, E. J. Wursthorn, S. B. Krasnoff and J. Clardy, *Angew. Chem., Int. Ed. Engl.*, 2007, **46**, 901–904.

18 L.-F. Nothias, M. Nothias-Esposito, R. da Silva, M. Wang, I. Protsyuk, Z. Zhang, A. Sarvepalli, P. Leyssen, D. Touboul, J. Costa, J. Paolini, T. Alexandrov, M. Litaudon and P. C. Dorrestein, *J. Nat. Prod.*, 2018, **81**, 758–767.

19 S. Kuhn and N. E. Schlorer, *Magn. Reson. Chem.*, 2015, **53**, 582–589.

20 S. Kuhn, B. Egert, S. Neumann and C. Steinbeck, *BMC Bioinf.*, 2008, **9**, 400.

21 *louvain. PyPI*, https://pypi.org/project/louvain/.

22 K. Wolfram, A. Porzel and A. Hinneburg, *Knowledge Discovery in Databases: PKDD 2006*, Berlin, Heidelberg, 2006, pp. 650–658.

23 G. Fuentes-Barros, S. Castro-Saavedra, L. Liberona, W. Acevedo-Fuentes, C. Tirapegui, C. Mattar and B. K. Cassels, *Fitoterapia*, 2018, **127**, 179–185.

24 P. M. Dewick, *Medicinal Natural Products: A Biosynthetic Approach*, John Wiley & Sons, Chichester, 3rd edn, 2009, ch. 6, pp. 291–403.

DISCUSSIONS

Data mining and visualisation: general discussion

Carlos Afonso, Mark P. Barrow, Antony N. Davies,
Marc-André Delsuc, Timothy Ebbels, Francisco Fernandez-Lima,
Caroline Gauchotte-Lindsay, Pierre Giusti, Royston Goodacre,
Norbert Hertkorn, Jeroen J. Jansen, Donald Jones, William Kew,
Stefan Kuhn, Adrien Le Guennec, Anneke Lubben, John Parkinson,
Ljiljana Paša-Tolić, Simon Rogers, Timothy R. Rudd,
Peter J. Schoenmakers, Philip Sidebottom, Stephen Summerfield,
Gerjen H. Tinnevelt, Gianluca Trifirò, Johan Trygg and Justin J. J. van
der Hooft

DOI: 10.1039/C9FD90044F

Gianluca Trifirò opened discussion of the paper by Johan Trygg: With regard to the multivariate analysis of data with different sizes, structures and sources, could the kind of transformation on the raw data affect the final distribution, possibly misleading the interpretation of the correlation matrix? Also, how much do the outliers weigh on the entire multiblock analysis?

Johan Trygg answered: Of course, non-linear data transformations will have an influence on the correlation matrix and hence the resulting multivariate models. The reason why transformation or any preprocessing is made is to correct for "unwanted" shape/variability or to linearize the data. Typically today, data processing workflows include multiple steps and it is hard to understand the impact or interaction between those transformation steps, and how it affects the outcome. Normally, this is not a problem as the workflows are quite established, but otherwise we recommend using a multiblock analysis technique like Joint and Unique MultiBlock Analysis (JUMBA)[1] where the data set in each step in the workflow is assembled and modeled together. Think of principal component analysis (PCA), but on multiple blocks of data, where you get an overview, and can visualize the flow of data and how it changes in each step in a single visualization plot. Multiblock models can also be used for comparing the effect between transformed and non-transformed data. In addition, it is quite common to analyze the same sample on multiple instruments, hence providing multi-block data where JUMBA is great.

1 T. Skotare, R. Sjögren, I. Surowiec, D. Nilsson and J. Trygg, *J. Chemom.*, 2018, e3071.

Gerjen H. Tinnevelt asked: Do you describe unique variations within one block?

Johan Trygg responded: Yes, if any block contains unique variation, the JUMBA model is able to describe that variation as part of that block's unique components. These component(s) can be visualized through plots and used for future data, similar to any other model components. In addition, you have a size measure (R2) that informs its size relative to the other components and the overall variation in that block.

Pierre Giusti enquired: Does the order of the blocks have an influence? How do you order the blocks? What about combining blocks of different sizes? What about a Monte Carlo approach (or other optimization ones) where you put weight on each of the blocks and look for the best correlative model?

Johan Trygg replied: JUMBA is a fully symmetrical method and as such the order of the blocks does not matter when building the model. At this point, we have not implemented specific block weighting in the algorithm.

Simon Rogers queried: Firstly, how does the method scale with the number of features? In your article (DOI: 10.1039/c8fd00243f) you have ~200 features (across the three blocks) whereas a typical LC/MS experiment can easily result in several thousand peaks. Secondly, have you considered incorporating some known relationships into the model? This might be useful where you know something about the features beforehand.

Johan Trygg answered: JUMBA handles many blocks of data matrices and each can contain a different set of features/variables. The number of variables is not a problem, even for thousands of variables, as we utilize different linear algebra techniques in the model building process, *e.g.* kernel calculations. Of course, the code is currently working sequentially and would benefit from being parallelized, which is a matter of implementation. With regard to your second question, this is something we should look into, to go from data driven to a more hybrid modeling that takes into account known structures in the model building process.

Royston Goodacre asked: With your multiblock approach can you handle missing data when samples may be missing in the different input blocks? Also, can you use JUMBA to fuse continuous data (metabolomics) with categorical (qualitative) data that have no ordinal basis?

Johan Trygg responded: Missing data is not implemented in JUMBA. Categorical data can be handled as long as the input data matrix has many variables. If the number of variables is low, then the JUMBA model will be skewed towards the categorical block since each block of data has one "vote" in the model building. We would suggest that if only a few qualitative variables are available, then use them as metadata to "color by" in the model plots and that way you can visualize and highlight the relationship to the other data.

Stephen Summerfield remarked: Analytical chemists traditionally distinguish between samples. It is surprisingly difficult to make substances from different sources the same. Would the multiblock approach be a way of justifying the sameness of a substance made by different companies or from different sources?

The term "Joint Global" seems to suggest this and is a better way to describe similarity and identify the differences with the other terms. My work over the last 7 years for Peter Fisk Associates Ltd has been to determine the sameness between materials produced by different companies with a SIEF (Substance Information Exchange Forum) for over 500 substances, so allowing them to submit joint registrations and share ecotox and toxicology data. The tests are extremely expensive and reduce the amount of animal testing required. Regulators like ECHA (European Chemical Agency) like scores that are validated statistically. Potentially, this could be of use for petrochemicals, natural products and extracts. In addition, this could be useful in the justification of read-across from one substance to another to avoid animal testing and the huge expense.

Johan Trygg replied: Yes, I definitely think that the multiblock approach would be very useful in these areas.

Carlos Afonso opened discussion of the paper by Justin J. J. van der Hooft: You present a nice approach – how does it compare with Global Natural Products Social Molecular Networking (GNPS)?

Justin J. J. van der Hooft answered: GNPS is referring to the overall platform to compare mass fragmentation spectra (molecular networking) and make annotations based on libraries or *in-silico* approaches. I assume you mean the comparison to molecular networking, which compares two spectra and takes into account, after some noise filtering steps, all the mass fragments and compares these across to calculate a score (based on modified cosine scoring). That score equals 1 if the spectra are identical and 0 if the spectra are completely different. By putting a threshold on that score, you can create a network of molecules if you compare many spectra against each other. Using the premise that spectral similarity equals structural similarity, you then create families of structurally similar molecules. MS2LDA, however, searches for co-occurring mass fragments and/or neutral losses – thus also taking into account substructures that are not charged and are directly visible by mass spectrometry – and does not need to look at the entire spectra to determine a probability score. Also, the spectra can have multiple substructures, as discovered by MS2LDA, and one MS2LDA substructure pattern (*i.e.*, Mass2Motif) can be present in multiple spectra. In molecular networking, a spectrum can belong to one molecular family, thereby sometimes masking the structural relationships that approach, as MS2LDA can uncover. The tools are complementary and that is why you can use them together, as we have demonstrated in a few papers now[1,2] and for which a workflow is described in a recent preprint on MolNetEnhancer,[3] where outputs of metabolome mining tools (molecular networking and MS2LDA) and metabolome annotation tools as Network Annotation Propagation and DER-EPLICATOR are combined in one network to facilitate metabolite annotation.

1 M. Ernst, L.-F. Nothias, J. J. J. van der Hooft, R. R. Silva, C. H. Saslis-Lagoudakis, O. M. Grace, K. Martinez-Swatson, G. Hassemer, L. A. Funez, H. T. Simonsen, M. H. Medema, D. Staerk, N. Nilsson, P. Lovato, P. C. Dorrestein and N. Rønsted, *Plant Sci.*, 2019, **10**, 846.
2 K. B. Kang, M. Ernst, J. J. J. van der Hooft, R. R. da Silva, J. Park, M. H. Medema, S. H. Sung and P. C. Dorrestein, *Plant J.*, 2019, **98**, 1134–1144.
3 M. Ernst, K. B. Kang, A. M. Caraballo-Rodríguez, L.-F. Nothias, J. Wandy, M. Wang, S. Rogers, M. H. Medema, P. C. Dorrestein and J. J. J. van der Hooft, *bioRxiv*, 2019, 654459.

Francisco Fernandez-Lima said: How accurate does the MS data need to be, to be processed with your algorithm? Can it be FT-ICR MS1 and/or MS2? Does the algorithm take into account the isotopic pattern?

Justin J. J. van der Hooft responded: The MS2LDA algorithm does need MS/MS (fragmentation) mass spectrometry data in its current set up as it is reporting co-occurring mass fragments and/or neutral losses from the MS/MS spectrum. As such, it does not look at any isotopic pattern as it is not aimed at determining the molecular ion or elemental formula. In terms of mass accuracy, strictly speaking there is no upper or lower limit and we provide bin widths of 0.1 (ion trap) down to 0.005 (orbitrap) to accommodate mass spectrometers along the entire range of mass accuracies. Of course, in the future, MS2LDA could be integrated with other tools that tackle elemental formula assignment and exploit the isotopic pattern (such as SIRIUS and CSI-FingerID), however, this also puts a higher bar on the input data quality.

Ljiljana Paša-Tolić remarked: To my knowledge, there is no definite solution for assigning the confidence to molecular assignments. How do you estimate confidence (or the "false discovery rate")?

Justin J. J. van der Hooft replied: This is indeed a relevant question and whilst there is no "definite" solution to assign confidence to molecular annotations, in fact there are several ways to add metabolite identification levels to metabolomics analysis. For example, the Metabolomics Society Standards Initiative proposed identification levels back in 2007. In the meantime, several groups proposed alternative identification levels as well as a possible way to add false discovery rates to metabolite annotations – all illustrating how relevant and important this topic is. Finally, within the Metabolomics Society Metabolite Identification Task Group, work is ongoing to update the 2007 identification levels to make them technology and method independent as well as more future-proof. Altogether, this shows that this is a vibrant area of metabolomics research and more developments in this area can be expected. A "definite" solution adopted by the entire scientific community will not be available in a very short time-frame but there are already available options to use at the moment.

Timothy Ebbels said: One of your new developments is the MotifDB, which will provide an annotated list of motifs/substructures. This will be a very useful public resource. However, it raises the question of whether the motifs are reproducible between different protocols, instruments and labs. Will two slightly different motifs be recognised as the same substructure?

Justin J. J. van der Hooft answered: I also expect that MotifDB will be a very useful public resource containing annotated Mass2Motifs. Indeed, when using different instruments or fragmentation approaches, the differences in the resulting fragmentation spectra could create different Mass2Motifs for the same substructure. However, we have already shown that when using reference spectra obtained on different machines, MS2LDA was still able to pull spectra from molecules containing the same substructures together. In addition, we can also accommodate different Mass2Motifs for the same substructure in MotifDB and users can choose the relevant Mass2Motif sets that correspond best to the instrument and protocols they used.

Stephen Summerfield queried: Have motif substructures been applied to polymers and surfactants? Note that most surfactants under REACH regulations are considered as polymers so have been excluded from REACH regulations as they have more than 50% of the substance with three or more repeat units. I would consider that the tool could be useful in the future for regulatory applications. There is a data gap on defining a polymer and different types of polymers under the future EU regulations, "REACH for Polymers", that will start in the next couple of years. Peter Fisk Associates Ltd is involved in writing the consultation paper for the EU to formulate these regulations, with a meeting held in May 2019 with regulators, non-governmental organisations (NGOs) and industry. It is considered that "REACH for Polymers" will have as great a constraint upon industry and innovation as REACH has had for substances. From discussion, there seems to be no analytical methodology to reliably define polymers that would be considered polymers of concern where more than 2% of the polymers have a molecular weight less than 1000 daltons, and to distinguish between polymer mixtures, graft co-polymers, *etc.* Size exclusion chromatography is not fit for purpose for determining the weight percent of small molecules compared to the bulk polymer. There are plenty of methods for function and identity but not quantification, as has been proposed in the future regulations. Does anyone in the room have any suggestions? Would you consider this a useful tool to relate motif substructures to toxic effects in toxicology and ecotoxicology in complex substances (referred to under REACH as UVCBs – Unknown or Variable composition, Complex reaction products or Biological materials)?

Justin J. J. van der Hooft responded: This question consists of several sub-questions that could be summarised as: (i) has MS2LDA been applied to polymer-like structures before, and (ii) do you think it would be possible to use it to link toxicity to molecular structures? The answer to the first question is, I do not know as the code is available to use and I do not know all users and what they have done with it. It would be a good idea to try it and I can say that for surfactant-like molecules it could work well as those are lipid-like structures and usually they have recognisable substructure patterns derived from their head groups. As for the second question, I think there is a good opportunity here as instead of focusing on individual molecules you can group molecules together based on the presence of substructures (Mass2Motifs) and correlate the presence/absence of those to toxicity. Another interesting potential application field is doping checks, as quite often new doping molecules are analogues of known molecules that are then slightly modified to escape detection by current assays. However, when we can recognise suspicious Mass2Motifs that are associated with known doping molecules, we can focus on potential novel doping agents previously not seen.

Timothy Ebbels remarked: You mentioned that you often use publicly available data for research. How well described are the public data sets and how often do you need to contact the original authors to help understand the data?

Justin J. J. van der Hooft replied: So far, for the public data sets I used, I did not need the help of the authors to understand the data. However, this was mainly because these data sets were created with the goal in mind to share the data set for future reuse. In many cases, it is indeed hard to reuse data, mainly because sufficient metadata is lacking. In such cases, help from the authors would be needed to unlock the potential of the data.

John Parkinson opened discussion of the paper by Timothy R. Rudd: Thank you for the nice article and the application using 2D [^{1}H, ^{13}C] HSQC and PCA to carry out an assessment of the complex heparin samples. I have attempted to tackle a biofluid mixtures problem myself with a similar approach but struggled to find a good way of actually reducing the data and generating PCA plots. My only software resource at the time was Bruker's AMIX. I have since recognized that there are other software resources available but, at the time, the chosen software struggled to deal with the large size and number of data sets I had available. I am therefore interested in your experience of handling both large data sizes and large numbers of such data sets within the R environment, as you report in your article (DOI: 10.1039/c9fd00009g). I am wondering about some of the detail regarding how you handled your data, how many data sets you were able to handle at once, whether you approached the problem using 2D binning, how you reduced the data, and whether you considered leaving out parts of the data (for instance, regions of the spectra that do not show any cross-peaks) to speed up the data handling process and reduce the CPU time, and how you then computed the PCA analysis?

Timothy R. Rudd answered: My initial experience of performing multivariate analysis on 1D NMR spectra was using SPSS and AMIX, and both were taxing. You mention R and I think this is a brilliant tool. I use the R package rNMR[1] to import NMR spectra (1D and 2D) in R, and once converted into a matrix you are able to manipulate the data as you want. For performing PCA, I normally use the standard functions (*princomp* and *prcomp*), although there are more sophisticated packages, for example factomineR.[2] In the case of HSQC data, I do not bin it. This is because the signals of the carbohydrate I am interested in are very broad due to its heterogenous nature. Binning might be applicable to complex mixtures of small molecules, where the 2D peaks are sharp and well defined. In essence, apart from unwrapping the matrix into a vector, the data is treated in a similar manner to a data set containing 1D spectra, and I found that only mean centering performed best with the 2D data. I have not considered excluding regions from the analysis; I think that this is a good idea, not necessarily for speeding up the analysis, but for improving the performance of multivariate analysis applied to the data set. I find the rate limiting step is importing the data into R; once this is done though, it can be saved as a binary R object, which makes loading the data much quicker.

1 I. A. Lewis, S. C. Schommer and J. L. Markley, *Magn. Reson. Chem.*, 2009, **47**(suppl. 1), S123–S126.
2 S. Lê, J. Josse and F. Husson, *J. Stat. Softw.*, 2008, **1**(1), DOI: 10.18637/jss.v025.i01.

Royston Goodacre said: PCA is not really a classification tool, so I wondered if you had considered using one-class classifiers where you learn to recognise the material, in your case heparin 2D NMR spectra?

Timothy R. Rudd answered: When looking to quantify known features in heparin, we have used techniques such as OPLS-DA, and we were successful in predicting the amount of epoxide modified uronate groups in heparin. Such approaches have been applied to heparin by the FDA to determine the amount of other glycosaminoglycans found in heparin.[1] To date, the approaches have mostly been applied to data sets that contain 1D NMR spectra. I have currently not applied such approaches to 2D NMR spectra, but I would be very interested to apply OPLS or techniques such as SVM to our 2D NMR data sets.

1 Q. Zang, D. A. Keire, L. F. Buhse, R. D. Wood, D. P. Mital, S. Haque, S. Srinivasan, C. M. Moore, M. Nasr, A. Al-Hakim, M. L. Trehy and W. J. Welsh, *Anal. Bioanal. Chem.*, 2011, **401**, 939–955.

William Kew commented: In NMR spectroscopy, sensitivity and limits of detection (LODs) can be a problem in the context of medicines where trace impurities can have negative biological effects. How do the LODs in the HSQC experiments on heparin relate to the required levels of "purity" for the medicine to be safely administered? Will an NMR and statistical analysis technique ever replace bioassays to determine if a sample is safe to administer?

Timothy R. Rudd responded: Heparin is such an old medicine that initially the only method for testing the product's effect and quality would have been a biological assay. The problem with biological assays is that other compounds can be tailored to have the same effect as heparin. This is why, especially for an animal-derived medicine, you would want to couple a biological assay with a physical–chemical test to verify the identity of the compound. We, and other laboratories, have derived 1D NMR tests that can detect small amounts of contaminants in heparin,[1,2] and proton experiments will always outperform HSQC in terms of LOD/LOQ. Where the HSQC experiment provides an advantage is signal dispersion, and in the examples provided here, it allows a large percentage of the signals to be assigned. To answer your final question, compounds such as insulin are now considered a chemical and not a biological compound, it has a defined amino acid sequence and is produced in large amounts by biotechnological means. Heparin is different as the biosynthesis is not template-driven and the enzymes do not run to completion, so a biological assay will always be required to test the activity of the material. This maintains the quality of the medicine and also, by making sure the heparin has a high activity (>180 IU mg^{-1}), you do not have to dose a patient with too much of the drug.

1 T. R. Rudd, D. Gaudesi, M. A. Lima, M. A. Skidmore, B. Mulloy, G. Torri, H. B. Nader, M. Guerrini and E. A. Yates, *Analyst*, 2011, **136**, 1390–1398.
2 Y. B. Monakhova and B. W. K. Diehl, *J. Pharm. Biomed. Anal.*, 2015, **115**, 543–551.

Timothy Ebbels remarked: A key objective seems to be to distinguish the different sources of heparin, and known contaminants. Have you explored the potential of multivariate methods (*e.g.* PCA, PLS, *etc.*) to identify as yet unknown contaminants using the loadings and weights?

Timothy R. Rudd replied: To find unknown contaminants we have tried a filtering approach, which involves using a library of spectra to remove signals from the test sample that are considered *bona fide*. Using this approach, you reveal signals that are not consistent with the library you are using to filter the test sample. This approach has been successfully applied to mono- and di-dimensional NMR data sets. The downside to this approach is that you can not quantify the amount of contaminant, you just know that it is present.[1–3]

1 T. R. Rudd, E. Macchi, C. Gardini, L. Muzi, M. Guerrini, E. A. Yates and G. Torri, *Anal. Chem.*, 2012, **84**, 6841–6847.
2 T. R. Rudd, D. Gaudesi, M. A. Lima, M. A. Skidmore, B. Mulloy, G. Torri, H. B. Nader, M. Guerrini and E. A. Yates, *Analyst*, 2011, **136**, 1390–1398.
3 M. Guerrini, T. R. Rudd, L. Mauri, E. Macchi, J. Fareed, E. A. Yates, A. Naggi and G. Torri, *Anal. Chem.*, 2015, **87**, 8275–8283.

Anneke Lubben opened a general discussion of Johan's, Justin's and Timothy's papers: Research funding is finite, and ultimately we want to maximise impact from our research. We have been talking a lot about the chase for more data and more information. However, it is probably the case that globally there are multiple data sets, by various techniques, which could be combined and mined to gain much deeper insights into specific research questions, sample sets, environmental conditions, *etc.* Do we need to hit the pause button, to take time to really delve deep into existing data with these multivariate analysis techniques for the "big data"? How do we harmonize this data so that it is even useful in this bigger picture? Is this where our resources should be focused? Is this what our funding bodies should be encouraging and funding? Or do we need to hit pause to establish some guidelines/protocols for data collecting so that we can do this synthesis exercise in the future?

Johan Trygg answered: You are touching upon a really important point, and that is data integration, and the reuse of data. This requires primarily that you capture and record the context surrounding the collected data. This is seldom done to a proper degree, or at all. If we, as a community, can agree not only to share data, but also the context and metadata that is needed to reuse that data and integrate with other data, that would be a big win. I know in industry this remains a challenge, and the data infrastructure platforms are being continuously developed to improve this.

Timothy R. Rudd responded: Requesting that the funding bodies include data standards could improve the situation. In the same way that funding bodies request open access publishing, they could request data to be submitted to a database in a predefined manner. This resource could be collated and maintained, which would require funding, but it could be a very valuable resource.

Justin J. J. van der Hooft replied: We need to spend our research funding wisely; however, since instruments and acquisition software constantly evolve, obtaining up-to-date data will remain critical for using our resources optimally. It would also mean that you cannot set up new experiments based on a novel hypothesis that you want to test. A healthy balance between data collection and reuse will be essential to use all the existing infrastructure and data optimally.

Anneke Lubben communicated: Is there any particular application area/ research area/community who would benefit most from a concerted effort of legacy data mining?

Timothy R. Rudd communicated in reply: In my opinion, it would be medically relevant omic data, as such data sets might contain samples that are rare and difficult to obtain again. Therefore, as mentioned previously, it is the associated metadata that is just as valuable as the data itself, especially if the future analysis involves combining data sets produced in different laboratories.

Justin J. J. van der Hooft communicated in response: From complex mixtures such as urine and feral extracts, we can generally annotate less than 5% of the constituents. Having many data files available, we can learn where the same or similar molecules are present, which may help to elucidate their structures or at least aid in identifying the source of those structurally unknown molecules. However, it is very important to have sufficient and good quality metadata available to efficiently reuse the data. Therefore, it is good to see recent efforts in improving metadata collection and standardisation of terminology.

Mark P. Barrow said: In the first talk by Johan Trygg and during the discussion, we have heard about differences between workflows and between instruments. Data sets are often not the same when acquired using different instruments, or even when using the same model of instrument but involving different locations and users. Individual laboratories will also have their own, preferred methods. What are your views on the way forwards for facilitating better comparisons of data? Should we look at greater use of inter-laboratory studies, standardization of workflows, or other approaches?

Johan Trygg answered: Differences between workflows and instruments are inevitable and the only way to validate the results from one laboratory is *via* external validation studies. Regarding standardization of workflows (SOPs), it is a necessity and we think that already there has been a number of initiatives and efforts in the community to standardize, or at least streamline, the way data sets are created and the information that needs to be provided with each data set, for example by setting up metabolomics standards. Here, the FAIR principle should be applied. The follow up initiatives are also on their way (for example, EU COST Integrape action). The use of inter-laboratory studies would be, of course, highly valuable, but this would require access to not only data sets, but also to all relevant metadata, which at this point seems not possible, but we hope it will be in the future when the above mentioned initiatives will create standards for metadata as well.

Timothy R. Rudd responded: One of the primary roles of the National Institute for Biological Standards and Control (NIBSC) is to produce international standards for biological medicines – these are normally activity standards and not structural standards. In the field that I am interested in, complex biological mixtures, the implementation of standards may assist in the comparison of data from different instruments/laboratories. They could be used as an System Suitability Tests (SSTs) before running a set of measurements or actually be included in the set of experiments. I believe that inter-laboratory studies are important, as they allow experimental processes to be disseminated. I do not necessarily think standardised workflows are the way forward as these are always evolving, but all stages of the measurement (material collection and processing, measurement parameters, data processing and analysis, *etc.*) should be detailed in such a manner that it can be repeated by anybody with the relevant expertise. It was mentioned in the group discussions that for the comparison of multiple large data sets that are generated in different laboratories, the metadata might be as valuable as the data itself.

Justin J. J. van der Hooft replied: This is a very relevant question. Let me emphasise that by sharing all information regarding the metabolomics experiment upon publication, including raw data, processed data, and analysed data, as well as the steps taken to get there, including the code, software, and settings used, we can start to process multiple data sets with ever improving batch correction algorithms. In addition, at the level of processed data or identified molecules, data sets could also be analysed together. Having said that, improved concerted efforts to share protocols and standard samples would certainly facilitate better comparisons of data sets from various labs.

Norbert Hertkorn commented: With respect to data generation and dissemination, in the last year, I personally have developed a serious concern about enforced sharing of original data and privacy almost to a level that I see our very future as scientists as we know it at risk. This has to do with the scientific journals, which are mainly commercial enterprises, which increasingly demand deposition of original data that are easily machine-readable. Specifically, every one of us here has specific knowledge from years to decades of scientific experience. This very specific experience is, in all cases, traceable to contributions of single or very few individuals. A dedicated system of artificial intelligence designed to track any single person of interest can bundle the output of this person and combine it with its close collaborators/competitors into trajectories and prediction tools, which can then be transposed into intellectual property owned by these big data companies even without our knowledge. Eventually we would be forced to pay licence fees to use our own ideas and methods to shadowy conglomerates or even forced not to use them. I see analogies of this actually operating at our institution.

Our funding system makes us think small and puts scientists against each other as competitors for limited resources. Thus, we can barely cope and even compete with a conglomerate of big data companies which operate in the uncharted territory of international law for which they themselves largely write and then forcibly impose the rules. These companies are used to thinking in hundreds of billions of dollars, which can be speedily allocated worldwide and which are not subject to credible regulations at all. I see ourselves as rather naïve scientists, ill-prepared for this brave new world, and steps should be taken to protect scientific intellectual property driven by curiosity and desires for applications for the benefit of the entire society, instead of for a (un)happy few. I think that this requires coordinated action at the level of national and, even better, European research institutions and all major funding agencies supported by robust internationally binding legislation.

Justin J. J. van der Hooft responded: I share your concerns with regards to sensitive data, and robust international binding legislation is indeed needed. However, for many data sets, for example bacterial extracts, the shared concerns may(!) be less as one could argue that nobody really "owns" bacteria, especially if they live in the soil all around us. However, the data produced from them may be of value and of course proper accreditation should then go to the data producers. Universities, institutes, and funding bodies should be willing to financially support scientists in case companies violate licences imposed on the data and methods.

Jeroen J. Jansen asked: What about translation from raw data into useful decisions/observations? Where does the reliance on data stop and the model as truth begin? When does our work have independent value?

Johan Trygg answered: Great question. What multivariate data analysis is really good at, is translating raw data into an abstract, yet interpretable, construction we call a model. It is done through a linear combination of the raw data as inputs, and provides a direct $1:1$ link between the raw data and the model. There are so many examples of where uni-variate and bi-variate fail to see the underlying patterns in the raw data, and where it is obvious and easily visible in the multivariate model, *e.g.* score plot visualization showing clusters in the data. Because of the $1:1$ link between the model and raw data, the model interpretation of the meaning of those clusters can be easily interrogated by drilling-down to the raw data. Hence, in effect, the multivariate model is where you either base your decision making on solely, or in combination with raw data, depending on the confidence that the model is correct.

Justin J. J. van der Hooft replied: Most work on modelling complex metabolite mixtures is still in its infancy. Currently, we rely on obtaining (and sharing) good quality data sets to allow models to be trained. For well defined systems or situations, like, for example, the concentration of glucose in plasma, we can translate the acquired information into useful decisions. When markers turn out to be more complex, like the combined presence of a number of molecules, we need sufficient data sets to build reliable models that can be used in diagnostics. Another main challenge that we face is that many spectral features in complex metabolite mixtures are as yet unknown and even if such a signal is statistically significant in making a difference between treatment and control, first its structural identity needs to be solved before it can be used in diagnostics. Having said all this, we are nearing an era where we can obtain much more structural information from metabolomics data from complex mixtures fuelled by the computational developments over the last ten years. That will certainly contribute to translating raw data into actionable decisions.

Antony N. Davies commented: I note the adoption of IUPAC in January of the FAIR data principles. We are struggling with defining generic NMR metadata – without the metadata, archived raw data quickly becomes meaningless. We need better tools to do this. We need to standardise our usage significantly more than we are currently doing.

William Kew said: As more chemists, and other scientists, use increasingly sophisticated statistical analyses, made available through easy-to-use software, how do we ensure non-statistically trained scientists handle and interpret the data correctly? How do we make people more statistically-minded?

Johan Trygg answered: Nothing can substitute good education at the undergraduate level, whether it is through university courses or on-line courses like Coursera, which I find very useful. Also, the open source community brings great support and education tools that help. Also, there are more and more publications (for example, in the metabolomics community) that discuss the applicability scope and limitations of the statistical methods used in the study, which also raises statistical awareness in the community.

Timothy Ebbels responded: The answer must be that non-statistically trained scientists must be statistically trained! We need to offer training at all levels from high school to postdoc and beyond. Some of the most important to reach are actually the generally older, more experienced scientists who may not have been exposed to such training, which has become relatively common at undergraduate and postgraduate levels. Statistical analysis and machine learning/AI should be considered with exactly the same level of scientific judgment and scepticism applied to other areas of science: do you trust it? And, if you do not understand it, how can you trust it? Of course, we do not expect everyone to be an expert statistician. But the same applies to, for example, the use of complex instrumentation: we do not expect everyone to be a mass spectrometry expert but we do expect them to know how to use the instrument to derive reliable measurements. Similarly, we should expect everyone to know how to use algorithms/sophisticated statistics to analyse data reliably, without necessarily being able to explain their inner workings.

Justin J. J. van der Hooft replied: To this question, one word suffices as an answer: education. I think if we add statistical approaches more into chemical, biochemical, and other curricula, scientists will be better equipped to deal with the ever increasing data sets in the future.

Timothy Ebbels remarked: Many comments have been made about the sharing of data. But what about code/data analysis? Do you think journals should require authors to share the analysis scripts so that every step of the data analysis can be reproduced by reviewers or readers?

Johan Trygg answered: Yes, I think that the code/scripts used for data analysis should be made publicly available and nowadays this is becoming a common practice with many journals – all in compliance with the FAIR principle.

Timothy R. Rudd responded: In an ideal world, data and code should be available, and definitely if the work is funded by public money. This becomes more complicated if the work was privately funded. Maybe this would require the submission/reviewing process to change – it could be more of a conversation between the authors and reviewers regarding this aspect. I also think that there is a question of novelty – maybe if the analysis is routine, then journals or reviewers might not want to see the code/data analysis. In an open reviewing process, it could always be requested if it is considered desirable.

Justin J. J. van der Hooft replied: Of course, it would be best if authors share their code and data so that future studies can build on it, especially in the case of method development papers. Having said that, I think that the most important thing to include in papers is a code and data availability statement, where authors explain where the code and data are available and if they are not, why not.

Simon Rogers commented: Authors providing code is obviously good, but looking through (or re-running) code is only one level of reproducibility and it will not detect the bugs that will almost certainly be present in all code. A detailed description of what the code is doing (to the level that it could be re-implemented) is vital.

Timothy Ebbels answered: I agree. At the very least the code needs to be properly commented and versioned. The more complex the software, the more important the documentation needed. Test scripts/data are also useful as a way to ensure code is doing what is expected.

Adrien Le Guennec said: In some groups focused on NMR metabolomics, one way to verify the results from the statistical analysis is to go back to the raw data and check whether the peaks show up as significant for the differentiation of the groups in the overlay of the NMR spectra – is it difficult to do this analysis in mass spectrometry? How can we do a root cause approach for mass spectroscopy or chromatography coupled to MS, or even for MS/MS?

Johan Trygg responded: Nowadays, in any commercial software connected to the instrument where LCMS analysis is done, it is possible to go back to the raw data and drill down and conduct a root cause investigation (*e.g.* based on variable retention time and m/z number and/or MS/MS spectrum) in the same fashion as in NMR analysis.

Justin J. J. van der Hooft replied: Generally, I would argue that for all annotations obtained through *in silico* approaches, it is of utmost importance to validate these results in the raw data. Therefore, in the MS2LDA web app, it is possible to browse through the fragmentation spectra that were uploaded where the mass peaks and/or neutral losses that are part of a Mass2Motif pattern are highlighted. In this way, a user can check whether the substructure pattern discovered by MS2LDA indeed could represent a relevant substructure.

Johan Trygg remarked: In terms of representability, we had 600 people in Africa within a malaria project.[1] The question was, should we perform measurements for all samples or select a subset? Obviously, running all 600 samples provides more data and adds to the statistical power, but that does not mean that you are better off compared to running a subset or a fraction of the total number of samples. The reason is that running all the samples means that you become subject to the inherent bias in the data, linked to sampling. Instead, you should aim for a set of samples that provides balanced and representable data. Our approach has been to use a multivariate design approach, which is a combination of multivariate analysis and design of experiments, to select a subset that is both balanced and representative. We have done this for a number of other studies and this is the way to go. Such a study can then be extended/built up if necessary, both to increase the power of the whole study and also to validate the first sample set.

1 I. Surowiec, S. Gouveia-Figueira, J. Orikiiriza, E. Lindquist, M. Bonde, J. Magambo, C. Muhinda, S. Bergström, J. Normark and J. Trygg, *Malar. J.*, 2017, **16**, 358.

Donald Jones commented: Using subsets can reduce time and instrument time but leaves the study poorly powered and thus susceptible to criticism at review. How are you suggesting to overcome this?

Johan Trygg replied: I can understand the challenge that still exists to convince reviewers to allow a publication using a subset and not all samples. In these cases, we still do subset selection, but we use a set of complementary subsets where each is balanced and representative, and together they form the complete sample set. This ensures that drift and other bias that exist in instruments, time or the technician are minimized. One recent method that works nicely to create such subsets is the Generalized Subset Designs (GSD) published in Analytical Chemistry.[1] It is based on fractional factorial designs, but is more generalizable.

1 I. Surowiec, L. Vikström, G. Hector, E. Johansson, C. Vikström and J. Trygg, *Anal. Chem.*, 2017, **89**, 6491–6497.

Stephen Summerfield communicated: In the EU, like other countries in the world, in the next few years there will be the definition and registration for "Polymers for REACH" focusing on "Polymers of Concern". It is considered that it will have an even greater impact on industry than REACH did for substances. A "polymer molecule" is a molecule that contains a sequence of at least three monomer units, which are covalently bound to at least one other monomer unit or other reactant.[1] Therefore, this includes small molecules of much less than 1000 daltons to macromolecules, especially surfactants, that were outside "REACH for Substances". Furthermore, it is proposed that "Polymers of Concern" will include:

(1) Any substance that has more than 2% w/w of constituents less than 1000 molecular weight that has not already been registered under REACH.

(2) Reactive end groups.

(3) Cationic and anionic polymers.

(4) Water soluble polymers.

(5) Water absorbing polymers, *etc.*

This covers a huge scope. The identification of polymers is relatively straight forward with FT-IR spectroscopy, NIR spectroscopy, NMR spectroscopy and, in some cases, MS. However, the quantification of polymers is not. The traditional method of size exclusion chromatography (SEC) with UV-Vis technology is good for comparing two polymers of the same type to see changes in the distribution.

Question 1: Can anyone suggest techniques and methods to solve this area that is considered by the regulators and my colleagues in Regulatory Affairs as simple?

Question 2: Can graft co-polymers that are covalently linked be distinguished from a mixture of two or more polymers?

Question 3: Bulk analysis has been much neglected. The determination of purity is still very difficult, let alone when performed by different laboratories, companies and analytical methods. Has anyone got any suggestions to close this gap?

1 Guidance for monomers and polymers, European Chemicals Agency, 2012, https://echa.europa.eu/documents/10162/23036412/polymers_en.pdf/9a74545f-05be-4e10-8555-4d7cf051bbed.

Marc-André Delsuc communicated in reply: NMR spectroscopy may answer some of these questions. For instance, NMR spectroscopy is already used as a routine method to estimate the length of PEG chains in anionic detergents. Diffusion ordered NMR spectroscopy (DOSY) can also be used to recognise mixtures from block co-polymers. Finally, it is also possible using DOSY to estimate the polydispersity index (PDI), providing the signals specific to the end group and to the main chain can be observed. Assuming a linear chain, the PDI can be estimated without any reference compound, and barely any calibration.

Peter J. Schoenmakers communicated in response: Question 2 can be very elegantly addressed by comprehensive two-dimensional liquid chromatography (LC × LC). We can choose (gradient elution) conditions in one dimension at which the main chain is retained and the polymer that is grafted is not. We can choose conditions in the other dimensions where this is opposite (the main chain is unretained and the graft is retained). The grafted molecules will stand out, even if they are present at (very) low concentrations. This is the only way I know to rigorously determine the extent of branching in molecules with long grafts that are chemically different from the main chain.

William Kew opened discussion of the paper by Jeroen J. Jansen: Principal components analysis (PCA) is deterministic, whereas t-distributed stochastic neighbour embedding (t-SNE) is a stochastic, parameterised analysis that produces a different output depending on how the analysis is performed. Can you comment on how you can compare and interpret these methods?

Jeroen J. Jansen answered: We used a novel, deterministic implementation of t-SNE. Furthermore, the different non-deterministic t-SNE solutions were largely identical up to rotations, so the results were identical.

Gerjen H. Tinnevelt commented: Do you not always get a different plot when using t-SNE? If you run t-SNE until convergence, the clusters you find in the plot will be the same but differently oriented every time you run t-SNE. There are hyper parameters you can tune and then you will find a different clustering. Prior biological knowledge is then needed to figure out if the clustering makes sense.

Timothy Ebbels said: The flow cytometry data has an interesting hierarchical structure where cells are embedded within samples, which may be further embedded within patients or disease groups. Are there other examples in the complex mixture domain that have a similar hierarchical structure? And could similar models be used in those cases?

Jeroen J. Jansen responded: A very interesting subject would be in recycling, where plastic of different chemical identities is shredded into very small flakes, which then need to be separated. This is a very similar hierarchical structure. Another interesting similar example is in mining geology, where you could study the heterogeneity of a specific ore layer by sampling each layer at different locations.

Philip Sidebottom opened discussion of the paper by Stefan Kuhn: Your method uses predicted NMR data. Have you done any work around how the quality of the prediction influences its success?

Stefan Kuhn replied: The quality of the prediction is crucial and the method relies on this. We can produce a quality measure for the prediction, which can give the user an idea of how valuable a suggestion is. We also intend to make it possible to use experimental or theoretical (simulated by other means) data provided by users as well as the standard data in the near future. This gives us the possibility to improve the prediction for particular classes of compounds.

Justin J. J. van der Hooft commented: You described a number of quality measures to assess the clusters in your paper. It can be expected that there is a difference between the predicted and experimental NMR data. How would this affect the quality of the clusters? And did you also look at how the clusters are formed in relation to the molecules? In other words, are "logical groups" of atoms grouped together in the NMR signal clusters? And could these be used to match to molecular substructures?

Stefan Kuhn answered: The quality of the prediction is clearly crucial. If it is not good for a particular compound, we cannot say much about this compound. We can compensate for this by giving a quality measure for the predictions and we also work on improved predictions, including detecting systematic effects. But generally, the prediction is a crucial component of the method.

Timothy Ebbels said: The clustering technique produces a flat set of clusters, which denote areas of the network with a high connection density. Have you explored other clustering approaches that can detect clusters if there are different densities in different parts of the network, and would this help to detect more structures?

Stefan Kuhn responded: We have not yet tried this. It is definitely worth trying though. Related techniques including hierarchical clustering would be one possible step for improving our method.

Justin J. J. van der Hooft queried: How many different compounds do you recognise out of a complex mixture using your approach compared to what you would expect to measure with 2D NMR spectroscopy? This is useful to know in relation to how well it can help to decipher components from a complex mixture.

Stefan Kuhn replied: Currently our method can only tell how likely it is that a certain compound is in the mixture. We cannot tell how many compounds there are. We did some work in that direction, but this needs more work. So, in the current state, the question for evaluating the method would not be "how many compounds can we identify?", but "how good is the correlation between our in/out likelihood and the actual situation?". Clearly, knowing how many compounds there are would be a good step forwards.

Timothy Ebbels remarked: Your approach combines MS and NMR approaches for identification. Are there any issues due to the very different sensitivities of the two techniques? Are you limited to what is abundant enough to detect by NMR spectroscopy?

Stefan Kuhn answered: We are limited to what is abundant enough to detect by NMR spectroscopy since it is an intrinsic limitation. However, investing NMR time to collect correlations even from minor peaks will allow the user to identify those compounds and annotate the (high sensitive) MSMS-molecular network with increased confidence. Thus, the user is safer to propagate the elucidation of closely related compounds. Another suggestion would be (and we are exploring this for our next paper) to roughly fractionate the raw sample into 5–10 groups to get data for each sample, and so possibly gain better NMR detection of the minor compounds.

Mark P. Barrow opened a general discussion of Jeroen's and Stefan's papers: Researchers are becoming more interested in designing their own data processing and analysis methods, including writing their own code. Instead of relying solely upon commercially available software, people are turning to MATLAB, Python, R, and more. Not long ago, I attended a mass spectrometry conference, which had its own workshop dedicated to programming in R and the room was full to capacity. What recommendations would you have for people getting into this area, particularly with respect to what background expertise is important, what platforms to use, and what collaborations they should consider?

Stefan Kuhn responded: From my point of view, there is a clear increase in researchers interested in programming. Python and R seem very popular, even though they have their problems (as noted previously), and as a more traditional computer scientist I find many aspects not ideal, but it seems to work for many people. My advice would be to practice, because this is still the key with programming, and not to get discouraged quickly. If you want to work in a particular area, look at what the frameworks used in publications are and work with them. That would also offer routes for collaborations. For example, if chemical structures are of interest, then potential frameworks would be *e.g.* CDK or RDkit. Both are widely used and they would offer a good framework and starting point for collaborations.

Jeroen J. Jansen replied: There are a lot of relevant ways to analyse your chemical data, depending on what you want:

(1) Find someone who is experienced in data analysis. Your result depends on your collaboration.

(2) Use a package like Simca, Unscrambler or pls toolbox. These are easy to use, but the results depend on the functionality of the programme used. Most methods available give generic models.

(3) Use a programming language like R, MATLAB or Python. A lot of programmes are available, but they require considerable training. However, this method will allow you to generate a very specific model for your own problem.

Stephen Summerfield asked: What are the user friendly Chemiometric programs that you refer to? I have found Python mind-blowing. MATLAB is slightly better as it is more structured. My experience of well structured statistics programs has been SPSS in a medical statistics area.

Stefan Kuhn answered: As noted previously, there is a clear increase in researchers interested in programming. I do not know much about SPSS, but it might well be that it is better structured, in the same way I would consider Java to be better structured (and better generally) than Python, but lots of people seem to love Python. I do not know why, but I think we have to accept what others do.

Jeroen J. Jansen responded: Simca and Unscrambler are user friendly Chemiometric programs.

Stephen Summerfield commented: My Post-Doc, sponsored by the RSC Analytical Division after my master's level diploma in education, was to create six problem-based case studies to teach analytical chemistry within the context of statistics, environmental science, forensics, laboratory design, pharmaceuticals and industry. This showed that you could take the students anywhere you wished if you gave the students a reason and a safe place to make mistakes, and hence they learnt to problem solve. These have developed over the last two decades as I became a visiting lecturer at various universities, including Loughborough University, MSc in Analytical Chemistry.

Caroline Gauchotte-Lindsay addressed Mark P. Barrow and Jeroen J. Jansen: As we were discussing the integration of advanced analysis to state-of-the-art analytical approaches, I suggest that the discussion with statisticians needs to happen at the very onset of the project, and that the plan for statistical analysis should go hand-in-hand with the experimental design. This also means that the analytical chemists of tomorrow should be trained for advanced statistics (machine learning, AI, *etc.*) and also be able to use the tools, *i.e.* the code. I believe that we should integrated this into our master's programmes. This would enable chemists to have easier conversations with mathematicians.

Mark P. Barrow replied: I very much agree with this. We started with an interest in multivariate analyses and have entered into collaborations with statisticians in more recent years. We have learned a lot, including about the design of experiments, and we are now developing our own algorithms and software, instead of relying upon commercial offerings. The statisticians also emphasize the importance of being involved from the start of a project, rather than only at the end. They often tell people of their dislike of being brought in to "perform the post mortem" on the data!

Jeroen J. Jansen answered: Although you are more than correct that data needs a much more prominent place in the analytical chemistry curriculum, the path chosen by data science is not always the most appropriate. Chemometrics is much older than the current boom in data science. What we have already learned to appreciate is the molecular origin of our data. Neural networks are very appropriate for large amounts of data from non-systematic origin, but we already know a lot about our chemical systems *and* about our measurement technologies. Secondly, we often do not have lots of data. So maintaining an analytical chemistry viewpoint during data design, acquisition and analysis is essential to get the most information out of our limited amounts of high quality data.

Conflicts of interest

There are no conflicts to declare.

Faraday Discussions

PAPER

Focusing on "the important" through targeted NMR experiments: an example of selective $^{13}C-^{12}C$ bond detection in complex mixtures

Amy Jenne,[a] Ronald Soong,[a] Wolfgang Bermel,[b] Nisha Sharma,[c] Antonio Masi,[c] Maryam Tabatabaei Anaraki[a] and Andre Simpson ⓘ *[a]

Received 29th November 2018, Accepted 5th December 2018

DOI: 10.1039/c8fd00213d

Current research is attempting to address more complex questions than ever before. As such, the need to follow complex processes in intact media and mixtures is becoming commonplace. Here, a targeted NMR experiment is introduced which selectively detects the formation of $^{13}C-^{12}C$ bonds in mixtures. This study introduces the experiment on simple standards, and then demonstrates the potential on increasingly complex processes including: fermentation, *Arabidopsis thaliana* germination/early growth, and metabolism in *Daphnia magna* both *ex vivo* and *in vivo*. As signals from the intact ^{12}C and ^{13}C pools are themselves filtered out, correlations are only observed when a component from each pool combines (*i.e.* new $^{13}C-^{12}C$ bonds) in the formation of new structures. This targeted approach significantly reduces the complexity of the mixtures and provides information on the fate and reactivity of carbon in environmental and biological processes. The experiment has application to follow bond formation wherever two pools of carbon are brought together, be it the incorporation of ^{13}C enriched food into a living organism's biomass, or the degradation of ^{13}C enriched plant material in soil.

Introduction

Questions in science research are becoming increasingly complex, with biological and environmental relevance always at the forefront of discussion. Almost every natural sample, including soils, atmospheric particles, plants, aquatic dissolved organic matter, and animals, can be thought of as complex mixtures. Often, isolation or fractionation can reduce complexity and make mixtures more amenable to a wider range of instrumental analyses. Conversely, it can be argued

[a]*Environmental NMR Centre, University of Toronto, 1265 Military Trail, Toronto, ON, Canada, M1C 1A4. E-mail: andre.simpson@utoronto.ca*

[b]*Bruker BioSpin GmbH, Silberstreifen 4, Rheinstetten, Germany*

[c]*Department of Agronomy, Food, Natural Resources, Animals and the Environment, University of Padova, Padova, Italy*

that it is often the synergism between components and phases in complex samples such as soil and living organisms that gives rise to the overall structure and function, and therefore must be kept intact.[1]

Nuclear Magnetic Resonance (NMR) spectroscopy is unique in that it is highly versatile and can be applied to liquids, gels, and solids. Thus it has applications to study both intact mixtures/systems or extracted/isolated sub-components.[2] NMR spectroscopy combines very high resolving capabilities and a diverse range of experiments to measure structure, dynamics, and interactions, while being non-invasive, and non-destructive.[3-6] The high resolving potential of NMR spectroscopy is often underestimated, but summarized beautifully by Hertkorn et al. The authors estimate 1D ^{1}H and ^{13}C NMR spectroscopy have peak capacities of $\sim$3000 and 30 000, while 2D ^{1}H–^{13}C increases to $\sim$2 000 000 and 3D NMR to $\sim$100 000 000.[7] On extrapolation to 7D NMR spectroscopy, which has recently been demonstrated on disordered proteins via sparse sampling approaches, the peak capacity can potentially reach on the order of 10^{18}.[8-10] Of course, in most environmental and biological samples, the limiting factor is a lack of sensitivity when dispersing signals into a larger number of higher dimensions. However, as NMR sensitivity continues to increase and sparse sampling approaches improve,[11] NMR experiments of high dimensionality become more feasible, and will therefore likely become central to the next generation of complex mixture research.[12] Arguably, one of the most impressive applications to date on environmental systems is by Bell et al.[13] Humic substances in soil have been described by many as the most complex known mixture.[14] Bell et al. combined selective labelling and 4D NMR spectroscopy to gain exquisite information on the lignin derived components from soil, with the approach identifying many of the exact structures for the first time.[13]

When performing mixture-based research there are two avenues that can be pursued, namely targeted and non-targeted analysis, for which NMR spectroscopy has strong capabilities in both. However, many analytical approaches are available for targeted analysis, while NMR spectroscopy's high reproducibility, non-destructive, and non-selective nature (i.e. any compounds containing an NMR active nucleus can be detected) make it uniquely suited for non-targeted analysis.[15] Deciding on a targeted or non-targeted approach is most commonly driven by the research question itself. For example, if the question is "what is the molecular composition of aquatic organic matter?", it would make sense that a researcher applies the highest resolution molecular tools available and tries to evaluate what components make up the mixture as a whole in a non-targeted fashion.[16-18] Conversely, if the goal is to follow the fate of specific contaminants in soil, then having an NMR nucleus such as ^{19}F allows targeted monitoring of the contaminant and its interactions, providing information specific to the target of interest.[19] The study of Bell et al. was successful in identifying novel structural units because the ^{13}C enrichment targeted the lignin components in the soil organic matter.[13] In turn, this reduced the overlap with carbohydrates and allowed a wealth of new information to be uncovered. In both of the targeted examples above, a specific NMR nucleus (^{13}C or ^{19}F) was introduced to provide the selectivity. In the current study, a slightly different approach is explored. Instead of monitoring a label selectively, the filtering is built into the NMR experiment itself. Here, a new experiment is developed that specifically targets the formation of ^{13}C–^{12}C bonds, where ^{13}C ^{13}C bonds and ^{12}C–^{12}C bonds are not detected by the experiment. This experiment could have widespread implications for any studies that bring two separate pools of carbon together and complements other experiments that are currently used for isotopic tracing by NMR spectroscopy.[20-22] For example, in 2006 a year-long study was performed following the fate of ^{13}C plant biomass as it degraded using HR-MAS NMR spectroscopy.[23] While humic substances are now thought to be largely complex mixtures,[24,25] at the time humic substances were thought to be chemically distinct, and formed from crosslinking components within soil organic matter.[26] As such, one of the goals of the 2006 study was to see if new bonds were formed. However, as the study followed just the fate of the ^{13}C enriched plant biomass, and as the material was so complex, while it was possible to see general changes (*i.e.* carbohydrates degraded fast, aliphatics accumulated, *etc.*), it was not possible to definitely see if crosslinks formed between the ^{13}C enriched plant biomass and other components in the soil.[23] If this were to be repeated with the experiment introduced here, it should be possible to target only the new bonds between ^{13}C and ^{12}C, which would help unravel how molecules are degraded, and if any novel recalcitrant structures do form within soil during humification. Additionally, the approach could have huge implications for monitoring and understanding the transfer of carbon in food webs. For example, *Daphnia magna* is a keystone aquatic species (a key food source for many fish) but it cannot synthesize many lipids and sterols *de novo*, and is reliant on algae for nutrition.[27] As such, *D. magna* is one of the most studied species in ecology, yet details as to the individual molecular species, their sources, and their impact on growth are still not fully understood.[28] We anticipate that by utilizing targeted experiments that focus on [13]C–[12]C bond formation, then detailed studies, that for example feed ^{13}C enriched *Daphnia* with ^{12}C algae (or *vice versa*), could provide insight on exactly how these organisms utilize food, and likely provide a better understanding of the biochemical pathways involved. Another example of a specific NMR experiment tailored for selective detection would be the amino acid-only experiment developed to selectively observe amino acid-profiles in living organisms.[29]

This study introduces the selective [13]C–[12]C experiment, first on simple standards, and then demonstrates its potential on processes of increasing complexity, including fermentation, plant growth, and *Daphnia* metabolism both *ex vivo* and *in vivo*. We anticipate that due to the versatility of NMR spectroscopy, many more targeted experiments can be developed in the future that examine specific components or processes within complex environmental samples.

Experimental

Ethanol fermentation

Ten milligrams of 1-^{13}C-glucose (Sigma Aldrich) were dissolved into a 1.5 mL 90 : 10 water and D_2O solution. Baker's yeast (produced by ACH Food Companies Inc.) was added to the mixture, which was then vortexed for 30 seconds, and allowed to settle. The remaining suspension was then placed into a 5 mm NMR tube (Norell Inc. NJ, USA) and monitored inside the NMR instrument over 24 hours.

Daphnia magna culturing

D. magna was cultured from a colony originally purchased from Ward's Scientific and maintained at 20 °C, with a water hardness of 124 mg $CaCO_3$ per L, and pH

7.5–8.5, consistent with local freshwater conditions. The cultures were kept under a 16 : 8 light/dark cycle. The species were fed 99% ^{13}C enriched *Chlamydomonas reinhardtii* (purchased from Silantes, GmbH)[30] as their sole food source for 14 days starting at birth, to produce enriched ^{13}C organisms. The daphnids were fed three times a week, and at the same time the water was changed to ensure sufficient oxygen content. The *D. magna* was also provided vitamin B_{12} (Sigma Aldrich, 2 μg L^{-1}) once a week to help with growth. Prior to the NMR studies, the organisms were placed in clean, aged water (dechlorinated *via* bubbling for one week) for 30 minutes to clear off residual algae. During the NMR experiments, the food source was switched to unenriched *Chlamydomonas reinhardtii* cultured in the lab using Bold's Basal Medium and following the Ministry of Ontario's standard operating procedure for algae culturing.

Sterilization and growth of *Arabidopsis thaliana*

The wild-type Columbia (Col-0) *Arabidopsis thaliana* seeds (originally from TAIR, OH, USA) were sterilized by a chlorine gas method. This method does not affect the seed viability, but removes microbial contaminants present on the seed surface.[31] The seeds were placed in Eppendorf tubes and with the cap open, were added to the desiccator. 100 mL of bleach and 6 mL of concentrated HCL (both from Sigma Aldrich) were placed inside the desiccator separately in a beaker. All the processes were performed inside the fume hood. After six hours of sterilization, the seeds were collected and stored in the freezer. The sterilized wild-type Columbia (Col-0) *Arabidopsis thaliana* was grown in sterilized Murashige and Skoog (MS) growth medium containing 1% (w/v) glucose. For the NMR experiments, uniformly labelled $^{13}C_6$-glucose at 99% ^{13}C enrichment (Silantes, GmbH) was used. The seeds were then cold-stratified for three days. The plates with sterilized seeds were transferred to the dark for seven days at 21 °C. Seedlings were collected on day zero, day one, day three, and day seven for NMR analysis.

Sample preparation and extraction of *D. magna* and *A. thaliana*

20 fully labelled *D. magna* were removed from the culture, flash frozen with liquid nitrogen, and lyophilized. The remaining organisms were switched to natural abundance ^{13}C algae as their food source, and sampling was repeated every 24 hours for four days. Metabolites were extracted following the protocol by Nagato *et al.*[32] using 3.2 mg of dried sample in 65 μL of buffer and placed in 1.75 mm capillary tubes (Hirschmann, Eberstadt, Germany) for 2D NMR analysis.

The *A. thaliana* samples were extracted using the protocol above prior to measurement, using 3.2 mg in 65 μL of buffer, and also in 1.75 mm tubes.

NMR spectroscopy

NMR experiments on extractions. NMR experiments on extractions were performed using a Bruker Avance III 500 MHz NMR spectrometer equipped with a ^{1}H–^{13}C–^{15}N TXI 1.7 mm microprobe fitted with an actively shielded gradient. The selective ^{13}C–^{12}C experiment is discussed in the main text. Data were collected with 98 increments, each with 16 scans (1-^{13}C-glucose, 1,2-^{13}C-glucose, and fermentation), and 64 increments with 832 scans (plants), and 896 scans (daphnids), 2048-time domain points, a recycle delay of one second, a 30 ms or 80 ms (see main text) TOCSY mixing time, a ^{1}J ^{1}H–^{13}C coupling of 145 Hz, and GARP-

4 for decoupling. Presaturation (~100 Hz bandwidth) was applied during the recycle delay to help reduce large water signals when required. The 90° pulses were determined in each sample. Data were processed with a sine-squared function phase shifted by 90° in both dimensions and a zero-filling factor of 2.

In vivo **flow system.** A low-volume flow system was utilized for these experiments. This was accomplished using the system and method created by Tabatabaei Anaraki *et al.*[33] 40 *Daphnia*, all 14 days old (fed only ^{13}C algae from birth), were placed in a high-precision, thin-walled 5 mm NMR tube (Wilmad-LabGlass, NJ, USA), with a Teflon plug (machined in-house) placed in the bottom of the NMR tube to prevent the daphnids from swimming outside the coil region. The top plug was created out of Teflon to keep the injection capillary glass and D_2O capillary in place (created in-house), and to keep all the daphnids within the coil region, maximizing the signal. The flow rate was set at 0.25 mL min^{-1}, and the reservoir contained oxygenated water and unenriched algae. This enables the daphnids to survive, and remain in a low-stress environment in the instrument by providing consistent food and oxygen. The entire system was placed inside the NMR spectrometer and kept at 5 °C to slow down the movement of the daphnids, enabling better water suppression and a more stable signal. Separate *in vivo* NMR experiments were conducted for 24 hours on fully enriched ^{13}C *Daphnia*, and *Daphnia* after being fed ^{12}C algae outside of the instrument for nine days.

In vivo **NMR experiments.** Experiments were performed on a Bruker Avance III HD 500 MHz (^{1}H) NMR spectrometer using a ^{1}H–^{13}C–^{15}N TCI Prodigy cryoprobe fitted with an actively shielded z-gradient. The external D_2O capillary lock (~5 μL) was integrated into the flow system and all experiments were run locked. The ^{13}C–^{12}C experiment was performed as with the extracts, with the exception that the reversed editing block was removed to increase the SNR and reduce relaxation. Presaturation (~100 Hz bandwidth) was applied during the recycle delay to help reduce the large water signal. 90° pulses determined in each sample and a TOCSY mixing time of 30 ms was used. A total of 128 increments were collected, each with 480 scans, 2048-time domain points, and a recycle delay of one second. The INEPT transfer was based on ^{1}J ^{1}H–^{13}C coupling of 145 Hz. Data were processed with a sine-squared function phase shifted by 90° in both dimensions and a zero-filling factor of 2.

Spectral assignments. Compound identification and assignment were done using AMIX (analysis of MIXtures software package, version 3.9.15, Bruker Bio-Spin), in combination with the Bruker Bio-reference NMR databases version 2-0-0 through 2-0-5. Spectra were calibrated against the Bruker Bio-reference NMR databases using tyrosine and D-glucose resonances for reference. Assignment was performed using a procedure previously described.[34,35]

Results and discussion

Basic pulse sequences and spectra

Fig. 1 shows the basic sequence used for the selective detection of ^{13}C–^{12}C bonds. In practice, sequences A and B are acquired in an interleaved fashion such that slice one of the 2D is the result from sequence A, and slice two is the result of sequence B. The data sets are split after acquisition to yield separate data sets. As A is essentially a "reference" for B it is important to collect the data sets at similar time points which is the purpose of interleaving the acquisition.

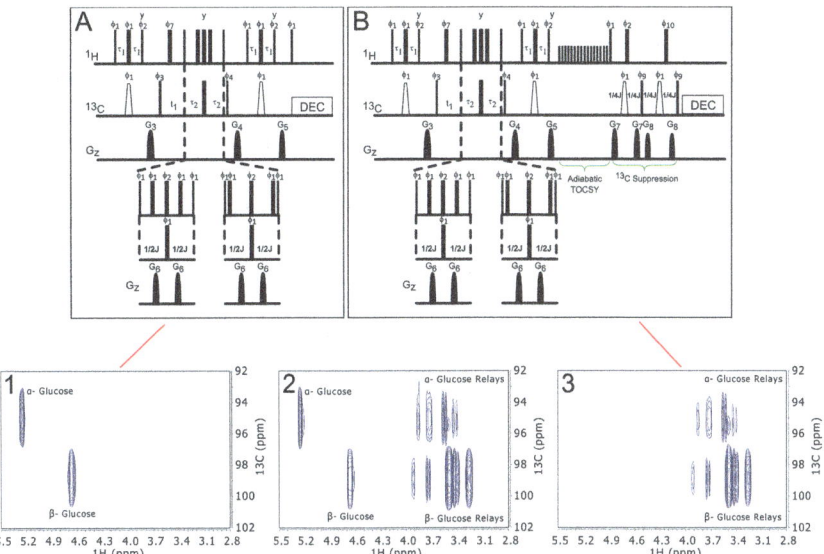

Fig. 1 The basic qualitative sequence. (A) A reverse HSQC spectrum. Narrow rectangles indicate a 90° pulse, whereas wide rectangles indicate a 180° pulse. Unless otherwise stated, pulses are applied along the x-axis. Open trapezoids represent smoothed chirp pulses for inversion with a pulse length of 500 µs, a sweep width of 60 kHz, defined with 1000 points and with a 20% smoothing of the amplitude on either end. The pulsed field gradients are indicated as filled sine envelopes and are 1 ms in length. The amplitudes of the gradient pulses have the following ratio: $G_3 = 60\%$, $G_4 = 40\%$, $G_5 = 31\%$, $G_6 = 19\%$, $G_7 = 23\%$, and $G_8 = 13\%$ (with 100% being 53.5 G cm^{-1}). The pulsed field gradients are applied along the z-axis followed by a gradient recovery delay of 200 µs. The following phase cycling was used for the pulse sequences: $\phi_1 = 0$, $\phi_2 = 1$, $\phi_3 = 0\,2$, $\phi_4 = 0\,0\,0\,0\,2\,2\,2$ 2, $\phi_7 = 0\,0\,0\,0\,2\,2\,2\,2$, $\phi_8 = 0\,0\,2\,2$, $\phi_9 = 0\,1\,2\,3$, $\phi_{10} = 1\,1\,2\,2$, and $\phi_{rec} = 2\,0\,2\,0\,0\,2\,0\,2$. (B) An adiabatic TOCSY which consists of 16 adiabatic processes (ca-WURST, 300 µs, 27.3 KHz pulses) with the following phase cycle, $0\,0\,2\,2\,0\,2\,2\,0\,2\,2\,0\,0\,2\,0\,0\,2$, precedes this sequence, and the additional pulses in panel B are appended. In this part of the sequence the following phase cycling was used: $\phi_1 = 0$, $\phi_2 = 1$, $\phi_9 = 0\,1\,2\,3$, $\phi_{10} = 1\,1\,2\,2$, and $\phi_{rec} = 0\,2\,2\,0\,2\,0\,0\,2$. The bottom panel shows the results of these sequences on 1-^{13}C-glucose. Sequence A leads to panel 1 in which only the one bond ^{1}H–^{13}C bonds are detected. Panel 2 shows the result of data collected after the TOCSY block in B. At this point the data resemble an HSQC-TOCSY experiment (note this data is not actually collected but is included for clarity). Panel 3 shows the result of sequence B. In this case, the ^{1}H–^{13}C bonds are subtracted, leaving only the signal from ^{1}H–^{12}C that is in the same spin system as a ^{13}C. In practice, the data collection is interleaved with sequence A being collected for the first slice and then sequence B being collected for the second slice and so on. The data are then split at the end of the experiment to give panels 1 and panels 3.

The result of sequence A is a standard reverse HSQC spectrum where the CH$_2$ moieties are reflected around the carbon offset (O$_2$P).[36] The advantage is that in complex samples the editing improves the spectral dispersion and helps reduce overlap especially in the aliphatic region where the chemical shift overlap between CH/CH$_2$/CH$_3$ is common. The added dispersion will become clearer in more complex samples shown later in the manuscript (see Fig. 4 as an example).

Fig. 1 (inset 1) shows the results from sequence A on glucose labelled with ^{13}C at the 1-position (1-^{13}C-glucose). As position one is a C–H group, the result from sequence A in this case is the same as a conventional HSQC spectrum and two cross peaks arise, one from α-glucose and one from β-glucose. In Fig. 1 (inset 2) the adiabatic TOCSY block is activated, and protons attached to ^{13}C are now allowed to mix with all other protons in the sample (both those attached to ^{13}C and those attached to ^{12}C). The result is similar to an HSQC-TOCSY experiment, with relays along the horizontal plane. Readers should note that the data set in Fig. 1 (inset 2) is not actually recorded using sequence B, but is included as an additional step to make it easier to visualize how the experiments work (i.e. progressing from Fig. 1 (inset 1) through (inset 3)). An adiabatic TOCSY[37] is used as it shows superior mixing in complex natural samples over MLEV and DIPSI.[3,38]

The result of the full sequence B is shown in Fig. 1 (inset 3). After the TOCSY block and additional ^{13}C filter is applied between Fig. 1 (inset 2) and (inset 3), the signals from any protons attached to ^{13}C are cancelled. The final result is the ^1H atoms attached to ^{12}C are selectively detected. It is important to note that for peaks to occur in the first place they must be in the same $^1H-^1H$ spin system as a ^{13}C nucleus. If a molecule contained only $^1H-^{12}C$ then the protons would not be selected by the first block. If a molecule contains all ^{13}C then the last filter would cancel the signals such that they are not detected. As such, the experiment selectively detects $^{13}C-^{12}C$ in proximity (albeit via their attached protons as ^{12}C is NMR inactive). The maximum number of bonds between the ^{13}C and ^{12}C that can be detected is determined by the TOCSY mixing time. In this first example, 80 ms are used, as such, correlations from the ^{13}C labelled 1-position relay around most of the glucose ring and a number of short (COSY), and long (TOCSY) correlations are seen. However, if the TOCSY mixing time is reduced to 30 ms, mainly COSY type interactions are emphasized. In complex samples this can reduce crowding in data and make it a little easier to interpret. Conversely, additional long range TOCSY correlations can be useful in aiding spectral assignment.

Fig. 2 shows the results of sequence B performed on 1-^{13}C-glucose using a 30 ms mixing time for comparison to Fig. 1, such that only the COSY correlations to ^{12}C show. The result being that only the ^{12}C atoms directly adjacent to ^{13}C atoms are selected. When the results of sequence A (Fig. 2A) and sequence B (Fig. 2B) are overlaid (Fig. 2C), the one bond $^1H-^{13}C$ cross peaks are blue while correlations to the ^{12}C adjacent are shown in red. We find this differential colouring approach

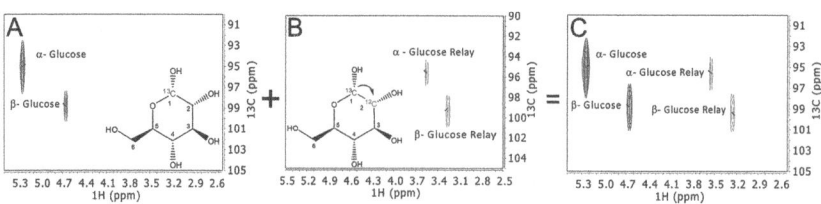

Fig. 2   Examining the process of monitoring $^{13}C-^{12}C$ bonds using 1-^{13}C-glucose. (A) 2D reverse HSQC spectrum that only detects ^{13}C atoms (i.e. the result from sequence 1A), the glucose molecule contains the labelled carbon atom in blue. (B) 2D $^{13}C-^{12}C$ sequence with a 30 ms TOCSY mixing time that selectively detects COSY-like relays and subtracts the signals from ^{13}C (i.e. the result from sequence 1B). (C) The overlaid spectra of 2A and 2B. ^{13}C atoms are show in blue while relays to ^{12}C atoms are shown in red. With a 30 ms TOCSY mixing time, if a blue cross peak is seen on the same row as the red cross peak it means they are directly bonded in the molecule.

very convenient for visualizing data sets, if a red and blue cross peak appear on the same horizontal row, this indicates a ^{13}C and ^{12}C unit are directly bonded. Interpretation from assigned NMR databases is easy, as HSQC cross peaks are commonly labelled with the corresponding structural position (H1/C1, H2/C2, *etc.*). In the example shown in Fig. 2, first the blue cross peaks (one bond correlations of ^{1}H–^{13}C) are identified and then, using the assigned molecular structure, the cross peak (and thus chemical shifts) of the adjacent ^{1}H–^{12}C unit is found. If the molecule selected for assignment is indeed correct, then the proton chemical shift of the relay should match the proton chemical shift at the same position in the molecule of interest. In the case of 1-^{13}C-glucose, the relays match exactly with the proton chemical shifts of carbon 2 in both α-glucose and β-glucose as expected. If assigned molecules are not available in a database, then assignments could be performed by referring to the HSQC and COSY (or TOCSY when a longer mixing time is used) data in combination. The one bond correlation would be identified from the HSQC data and the chemical shift of the proton relay from the COSY data. Whichever way the assignment is performed, the experiment gives a convenient approach to monitor the formation of ^{13}C–^{12}C bonds.

A simple process: fermentation

To demonstrate the concept further, 1-^{13}C-glucose was allowed to ferment in the presence of baker's yeast (*Saccharomyces cerevisiae*) to form ethanol over 24 hours inside the NMR spectrometer. Fig. 3A shows the overlaid one-bond/relay data before the experiment began, which is essentially identical to that shown in Fig. 2C. After 24 hours of fermentation, ethanol has formed. In Fig. 3B, it can be seen that the CH$_3$ group in the ethanol molecule has been derived from the ^{13}C

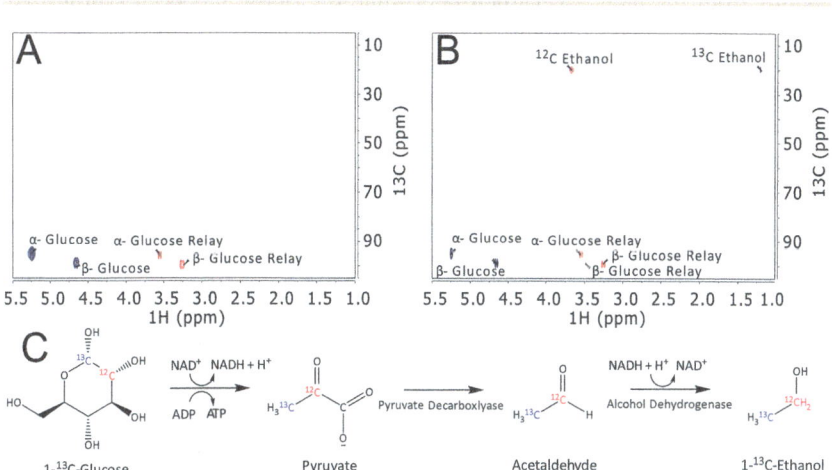

Fig. 3   Following the simple process of fermentation. 1-^{13}C-glucose was mixed with baker's yeast (*Saccharomyces cerevisiae*) for 24 hours inside the NMR spectrometer to monitor the formation of 1-^{13}C-ethanol. Carbons 1 and 2 in glucose are used to form ethanol through fermentation. (A) The same overlaid spectrum from Fig. 2C prior to the fermentation reaction. (B) 24 hours after the addition of yeast. ^{13}C is seen in blue, and the ^{12}C relays are in red. In this case, ethanol is formed with the methylene group from ^{12}C and the methyl group from ^{13}C. (C) Simplified fermentation process with 1-^{13}C-glucose.

labelled 1-position (blue contour), while the CH_2 group is from ^{12}C (red contour). This is consistent with the well understood process of fermentation of glucose. Under anaerobic conditions, glucose breaks down to form ethanol, with the help of ATP and NADP/H. Through this process the six-membered ring breaks, resulting in two 3-carbon pyruvate derivatives, which are then broken down into 2-ethanol. Through this process, the result is a ^{13}C atom in the 1 position, and a ^{12}C atom in the 2 position.[39,40] A simplified mechanism of the fermentation process can be seen in Fig. 3C.

A complex process: *Daphnia magna ex vivo*

Fig. 4 demonstrates the approach in a much more complex system. In this case, *Daphnia magna* have been fed 99% ^{13}C enriched algae two weeks from birth such that the overwhelming majority of carbon in their biomass is ^{13}C. After this they were fed ^{12}C algae (natural abundance ^{13}C) for 96 h. During this time, it is expected that ^{13}C–^{12}C bonds will form as their new ^{12}C food is incorporated into their ^{13}C biomass. A TOCSY time of 80 ms is used in this example such that both long- and short-range correlations are observed. Samples were taken every 24 hours over four days and extracted using a phosphate buffer. Fig. 4A shows the result after 96 hours. The CH_2 groups in the reversed HSQC spectrum are flipped around the carbon offset (100 ppm in this case) and appear in the lower right quadrant of the data set. The CH_3 and CH groups appear as they would in a conventional HSQC experiment. The data are highly complex, and the full analysis is beyond the scope of this paper. However, five regions of interest have been highlighted.

Region one is consistent with a relay from the adenine group to the ribose ring in the energy molecules (*i.e.* ATP, ADP, and AMP). This indicates the adenine is coming from the older ^{13}C carbon pool while the ribose has been synthesized from the ^{12}C biomass introduced over the last 96 hours.[41,42] In this case, the adenine and ribose link *via* a quaternary N center and protons on each side of this (one from adenine and one from ribose) TOCSY correlate due to the 80 ms mixing. This has been confirmed by cross-checking the TOCSY spectra of the energy molecules (ATP, ADP, AMP) in Bruker bio-reference databases, which all show this correlation.[35] The formation of energy molecules from carbon coming from different pools is consistent with the glycolysis and glycogenesis pathways used by *D. magna* for energy formation. These are major energy-creating pathways in crustaceans, producing vital ATP and NAD/H.[43] Pyruvate is formed from the breakdown of the glucose contained within the algae, and given the cyclic nature of the tricarboxylic acid (TCA) cycle, these pyruvate molecules can come from either labelled or non-labelled algae. These pyruvate molecules then form into new energy-containing molecules that are used by *D. magna*. This process has been studied extensively using cell flux analysis with *E. coli* and *S. cerevisiae*. Upon adding labelled glucose into the system, energy metabolism can be monitored and is seen to occur *via* the glycolysis/glycogenesis pathway.[42,44–46] By our estimate, this may be the first time this process has been examined directly in *D. magna*.

Region two is consistent with relays from ^{13}C enriched lipids to ^{12}C lipids containing double bonds. *Daphnia* are known to assimilate most of their lipids from their diet and cannot synthesize many essential fatty acids from scratch.[27,47–49] As the food source is changed to unenriched algae, the *Daphnia*

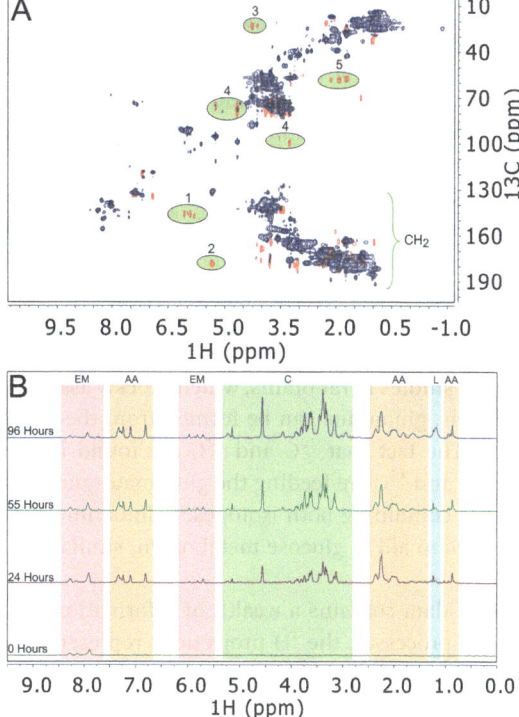

Fig. 4 Sequence applied to *Daphnia magna* extracts. The organisms were enriched for two weeks from birth with 99% ^{13}C enriched algae. They were then switched to natural abundance ^{12}C algae and monitored for four days. Samples were taken every 24 hours and extracted using a phosphate buffer. (A) Overlaid 2D NMR spectrum of the 96 hour sample. Blue is the ^{13}C remaining in the organisms and red is the ^{12}C relays to the new bonds formed. There are five regions of interest highlighted in green. (1) Adenine from ^{13}C bonding with ^{12}C ribose from the algae used in energy formation, (2) ^{13}C lipids containing ^{12}C unsaturation indicative of lipid synthesis, (3) protons in the ^{13}C lipid chains relaying to ^{12}C adjacent to oxygen atoms, (4) carbohydrate components including a mix of carbon enrichment in glucose as it moves through the TCA cycle and glycogenesis, and (5) a mixture of glutamate and glutamine amino acids. (B) 1D ^{1}H projections for all the time points showing the uptake and use of ^{12}C in the *Daphnia*. EM = energy molecules, AA = amino acids, C = carbohydrates, and L = lipids.

begin to form lipids from the new food source. Acetyl CoA, derived from pyruvate and the degradation of glucose, is used as a primer for the carbon chain, and elongated by the repeating condensation of malonyl-CoA yielding lipid pools.[50,51] As well, some dietary fatty acids are able to be desaturated in the carboxyl direction, as opposed to the methyl direction and then elongated to form poly-unsaturated fatty acids.[50] The correlation between ^{13}C and ^{12}C in the lipid chains suggests existing lipids are being modified with units derived from the new ^{12}C food source.

Region three is consistent with protons in the ^{13}C lipid chains relaying to the ^{12}C atoms adjacent to an oxygen atom, such as in a phospholipid head or even an ester group. This follows the above explanation where the formation of phospholipids to lipids occurs through dietary routing, which is the process of linking together multiple fatty acid or lipid chains through digestion, where the chains can be from different carbon pools.[50]

Region four is carbohydrates including glucose breakdown and formation. As the experiment uses an 80 ms TOCSY spinlock, coupling around the entire hexose ring is expected. As such, the correlations indicate that ^{12}C and ^{13}C are being brought together to form the carbohydrates. This is consistent with the process explained above for the breakdown of glucose through glycolysis and then the re-formation through glycogenesis. As there will be pools of ^{13}C pyruvate and ^{12}C pyruvate, the glucose can get re-formed with ^{12}C theoretically in any position, resulting in ^{13}C–^{12}C relays observed within glucose.[52]

Region five is consistent with α-protons in amino acids relaying to the side chains. The peaks are the closest match for a mixture of glutamine and glutamate. This is consistent with studies in rat brains, which quickly assimilate glucose into glutamate.[46,53] After this, glutamine can be formed from the glutamate by glutamine synthetase.[54,55] The fact that ^{13}C and ^{12}C are found in the same system suggests that both ^{12}C and ^{13}C are feeding the glutamate–glutamine cycle, which produces amino acids containing both isotopes.[56] Glutamine and glutamate are used by *Daphnia magna* to aid in glucose metabolism, similar to a wide range of species.[56]

While the 2D NMR data contains a wealth of information down to the exact bonds involved in key processes, the ^{1}H projections represent a complementary source of information. As explained above, the only ^{1}H signal remaining at the end of pulse sequence B is relays to ^{1}H–^{12}C that are in close proximity to ^{13}C. As such, the ^{1}H projection essentially gives a simple visual as to the fate of ^{12}C as a process progresses. Consider, for example, the ^{13}C *Daphnia* feeding on ^{12}C algae. At the start of the experiment, the *Daphnia* are fully ^{13}C enriched so no signal is expected in the ^{13}C–^{12}C experiment or its ^{1}H projection. However, overtime as the *Daphnia* utilize the ^{12}C, new molecules are formed bringing together both ^{13}C and ^{12}C, and a signal will appear in the ^{13}C–^{12}C experiment. In turn, the ^{1}H projection shows the fate of the ^{12}C and what types of molecules it has become incorporated into. Fig. 4B shows the ^{1}H projections over time as the *Daphnia* utilize ^{12}C. At time zero the spectrum is essentially blank, as expected. However, some small signals are present consistent with an adenine group. Prior to lyophilizing, the *Daphnia* were transferred and stored in aged water (which itself contains an algal background) for ∼30 min. It is likely the *Daphnia* used this ^{12}C to start making adenine, the precursor required for energy molecules such as ADP/ATP/AMP.

After 24 hours of exposure to the unenriched food, ^{12}C has been incorporated into a range of structural categories. Signals from energy molecules have become much stronger, and key ribose signals around 6 ppm start to appear indicating a portion of both the adenine and ribose units are being made from ^{12}C substrates and are being brought in proximity to ^{13}C. In addition, a range of amino acid and carbohydrate signals also appear. It is important to remember that to appear in the spectrum a ^{13}C atom and a ^{12}C atom must be bonded together somewhere in the molecule, as such the experiment selectively detects new molecules that had to have been synthesized *de novo* from the food source. After 55 hours the spectral profile is relatively similar with the exception of carbohydrates that have ∼doubled in intensity, consistent with the newly introduced ^{12}C biomass being combined with ^{13}C likely for energy *via* glycolysis.[42,46] After 96 hours a distinct broad signal is noted around 1.2 ppm, consistent with $(CH_2)_n$ in lipids. This indicates the *Daphnia* are modifying ^{13}C lipids using ^{12}C from their food source. This is particularly interesting; as mentioned above, *Daphnia* cannot synthesize many lipids and they rely on their food to assimilate many essential fatty acids. In addition, the lipids incorporated from their diets are key for reproduction. As *Daphnia* reproduce asexually, the lipid content is required for egg formation, and spikes immediately prior to birth, which is on average every 2–3 days.[57] Daphnids without proper lipid content stop producing viable offspring, and will often start to convert to sexual reproduction.[58] Thus, the use of lipids in *Daphnia* is key to their survival. Further, as *Daphnia* are keystone species, the stable reproductive cycles are imperative for many species' survival. *Daphnia* are found in almost every fresh water body worldwide, and are the food source for many other aquatic species.[48,59,60] Therefore, the use of lipids is not only key to *Daphnia* survival, but also has many potential negative consequences to higher trophic levels (*i.e.* fish and higher predators) based on available food sources. As such, the ^{13}C–^{12}C experiment introduced here specifically targets the modification of existing ^{13}C lipids by ^{12}C, and could be quite important in uncovering a better understanding of lipid modification and utilization in key species such as *Daphnia*.

Working in reverse: ^{13}C$_6$-glucose in *Arabidopsis thaliana*

Fig. 5 depicts the germination and early growth of ^{12}C *Arabidopsis thaliana* seeds in the dark using ^{13}C$_6$-glucose as the sole carbon source as previously described.[61] This study has been designed as the reverse of the *Daphnia* study, such that the process begins with ^{12}C, and ^{13}C is added into the system. As the plants are stored in the dark, photosynthesis does not occur, and the seeds grow *via* ^{13}C sorption through their husk and later uptake *via* roots after initial germination. In this situation, the question becomes "what does the plant synthesize from the ^{12}C stored in its seeds?" As the original biomass in the seeds is the only source of ^{12}C, the only way structures can appear is if the carbon from the seed biomass is combined with ^{13}C derived from the glucose, forming a new structure. After one

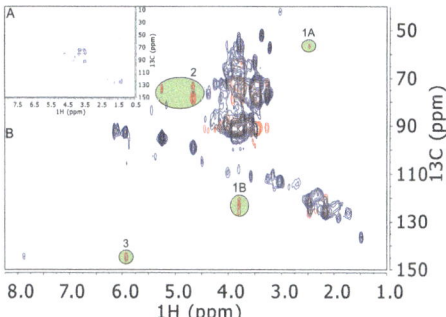

Fig. 5 *Arabidopsis thaliana* grown in the dark with ^{13}C$_6$-glucose as the primary carbon source. (A) One day of growth with ^{13}C in blue and an absence of relays due to the lack of new bond formation. (B) Seven-day growth of the plants with ^{13}C in blue and ^{12}C relays in red. Three areas of interest are highlighted in green. (1A) α^{13}C–β^{12}C glutamate relays, (1B) α^{12}C–β^{13}C glutamate relays, an essential amino acid in plants, (2) relays from glucose from glycolysis and the TCA cycle, and (3) the formation of energy molecules derived from the glucose and TCA cycle.

day of germination/growth, no new signals are seen (see small inset, Fig. 5A). However, after one week of growth, the plant extract shows three main types of signals arising from several different pathways. There are three sections that are worth highlighting in Fig. 5B.

In region one, the new peaks arise from the synthesis of glutamate. This amino acid is crucial for the health of plants as it is used for assimilation and dissimilation of ammonia that is then transferred to all other amino acids in the plant.[62] As well, glutamate has been shown to be the precursor to chlorophyll synthesis in developing leaves.[62] Glutamate, like many of the metabolites under examination, is formed through the TCA cycle in plants and animals, therefore it also has implications in regards to energy formation.[56,63] Given the importance of this amino acid to plant health, it is expected that its formation would occur relatively quickly in the growth of *Arabidopsis*. In this experiment $\alpha^{13}C$–$\beta^{12}C$ (peak 1A) and $\alpha^{12}C$–$\beta^{13}C$ relays (peak 1B) are seen for glutamate, indicating ^{12}C and ^{13}C pools are likely brought together from isotopically different sources *via* the TCA cycle.

Region two represents relays from the ^{13}C portions of glucose to ^{12}C at the 1-position. As a TOCSY mixing time of 80 ms was used for the plant study, long range interactions around the ring will be observed. Plants, like all living things, use glucose as an energy source for ATP formation, and without light, the glycolysis process occurs much like in *Daphnia*. The glucose gets broken down into two pyruvate molecules, and then moves through the TCA cycle to form ATP.[64,65] However, pyruvate can also move 'backwards' through gluconeogenesis to reform glucose, but the two pyruvates can be from different carbon sources. This results in a glucose molecule with C1–3 with one isotope and C4–6 as another.

Region three represents energy molecules such as ATP/ADP/AMP with a ^{13}C adenine connecting to a ^{12}C ribose ring. Much like in the *Daphnia*, the plants utilize the glucose through the TCA cycle to create energy molecules.[44,45] The results here indicate that a portion of the energy molecules is created when adenine derived from the $^{13}C_6$-glucose is combined with ribose derived from the seed (^{12}C).

Daphnia magna in vivo

To further test the applicability of the ^{13}C–^{12}C experiment, the sequence was applied to a flow system containing living *Daphnia* raised on ^{13}C from birth. The main difference here is that the editing step has been removed from the sequence in Fig. 1B such that the CH_2 groups are not flipped around the ^{13}C offset. The reason for this is that in intact samples, such as living *Daphnia*, relaxation is very fast. Therefore, it is prudent to reduce the pulse sequence length as much as possible. When the CH_2 edited step was removed, it was found the signal intensity doubled over the edited version *in vivo*. Due to the additional signal, the non-edited version was applied to the *in vivo* system.

Fig. 6A shows a conventional HSQC experiment for reference, which represents all the ^1H–^{13}C bonds in the sample. Fig. 6B shows the result of the ^{13}C–^{12}C experiment of the *Daphnia* grown solely with ^{13}C algae since birth. As the organisms contain only ^{13}C, no signals appear in the ^{13}C–^{12}C experiment. Note, the vertical streak centered around 5 ppm is the breakthrough of residual water due to the fact that the organisms are swimming in 100% pure H_2O and 90% of

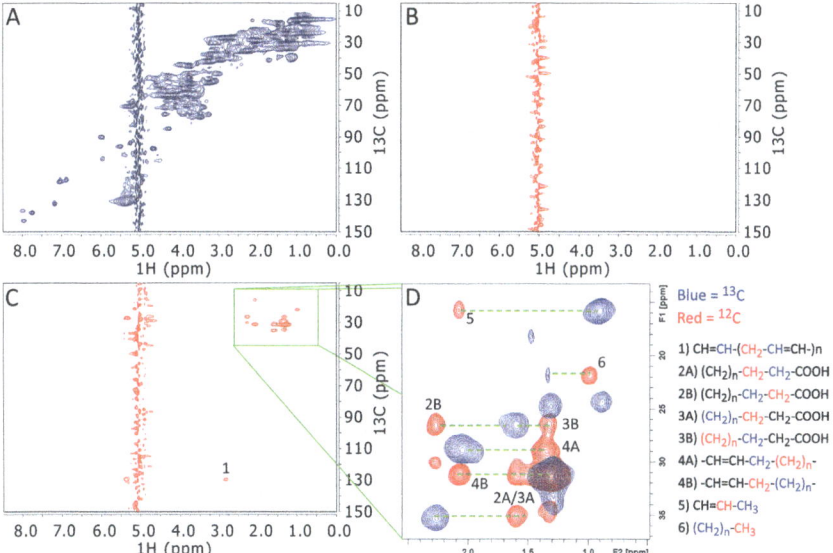

Fig. 6 Testing the $^{13}C-^{12}C$ sequence on an *in vivo* sample of *Daphnia magna*. In this case, the *Daphnia* were cultured from birth for two weeks using 99% enriched ^{13}C algae, and then switched to natural abundance ^{12}C algae for 9 days. (A) 2D HSQC spectrum of day zero *Daphnia* (*i.e.* not ^{12}C fed), (B) $^{13}C-^{12}C$ spectrum at day zero, no relays were detected due to the lack of ^{12}C in the system (note the streak at 5 ppm is residual water), (C) $^{13}C-^{12}C$ relays of *Daphnia* after nine days of exposure to ^{12}C algae as the food source. Relays have started to appear due to the assimilation of the new food into their biomass. (D) An expansion of the lipid region highlighting the new bonds that are formed.

their bodies are also water, making perfect water suppression incredibly difficult. Fig. 6C shows the same culture of organisms after nine days of being fed ^{12}C carbon. As can be seen, ^{12}C has been incorporated into the original ^{13}C biomass at various sites. Interestingly, despite the complexity of the system, all the correlations are relatively easy to assign based on the literature,[66–68] and arise from the modification of lipids (see Fig. 6D). Various structures can be identified, most of which appear twice with the relative locations of the ^{13}C and ^{12}C switched. This suggests that the *Daphnia* can utilize the ^{12}C and ^{13}C pools in similar ways. For example, it indicates the organisms can modify ^{13}C lipids from their own biomass with ^{12}C from their new diet. It also suggests they can modify ^{13}C lipids from their new diet using ^{12}C derived from their own biomass.

As discussed previously, *Daphnia* are limited in the lipids they can make, thus lipid incorporation from their diet is very important.[48,57,58] The expectation is that modification of their food source and lipids from their own biomass will result in new peaks formed using this sequence. For example, as the organism digests the algae, the lipids are broken down into single chain fatty acids that can then be desaturated and elongated using carbon already in the system.[50] Such lipids can be utilized to make more complex chains of lipids with additional carbon coming from either the diet (^{12}C) or biomass (^{13}C) pools.[69,70] This would occur relatively quickly and in higher concentrations compared to other metabolites due to the energy requirement from the lipids, and the inability to synthesize them internally. This experiment is a powerful approach for providing information on a complex *in vivo* system for processes such as lipid modification that are currently not well understood.[28]

The future potential for quantification

The ability to quantify the ratio of $^{12}C/^{13}C$ at a given position in a molecule would be extremely useful in constrained metabolic models, and tracing pathways.[71,72] Using the sequence outlined in Fig. 1B, it is not possible to quantify the $^{12}C/^{13}C$ ratio as the intensity of the cross peaks depends on both the efficiency of the TOCSY transfer and the abundance of ^{12}C in the peaks. Unfortunately, a rigorous implementation of a quantitative approach is beyond the scope of this paper and would likely take years of research to implement thoroughly in complex systems. That said, it is worthwhile to consider the future potential for quantification on a standard that paves the way for development and further discussion.

Fig. 7 shows two pulse sequences that when used in combination, in theory, should permit quantification of site specific $^{13}C/^{12}C$ ratios. The result of Fig. 7B is in fact identical to Fig. 1B (*i.e.* relays to ^{12}C remain, while ^{13}C is subtracted). Fig. 7A is derived from Fig. 7B with the goal to keep the pulses, durations, and timing identical between the two sequences. After the TOCSY proton magnetisation is excited and the CH coupling evolves for 1/2J, either block A or B is executed. In the case of B, the first ^{13}C 90° pulse converts the magnetisation into unobservable zero and double quantum coherence. The remaining proton magnetisation can again evolve into CH antiphase magnetisation, which then is also converted into unobservable magnetisation by a second ^{13}C 90° pulse, thus enhancing the efficiency of the filter. For block A, the position of the second 90° pulse (of B) is changed (moved to the beginning). These two ^{13}C 90° pulses now act as a 180° pulse and reverse the J evolution so that at the end of block A all the magnetisation is back in phase and no filtering takes place. The net result is that the sequence in Fig. 7A now allows both protons attached to both ^{1}H–^{13}C and ^{1}H–^{12}C signals to pass, while the sequence in Fig. 7B blocks the ^{1}H–^{13}C signals. As the timing in both sequences is identical, Fig. 7A acts as a reference for the sequence shown in Fig. 7B. As such, we will refer to the sequence in Fig. 7A as the "quantitative reference" while we will refer to the sequence in Fig. 7B as the "^{12}C-only sequence." The TOCSY transfer in both will be identical, as such, the difference in signal intensity between the reference and the ^{12}C-only sequence will be from the subtraction of the ^{13}C signal. The concept is best explained on a standard. Fig. 7, panel 1 shows the result from 99% 1,2-^{13}C-glucose. Two horizontal bands appear, the upper band representing correlations between the 2-^{13}C position and protons around the ring, with the lower band representing correlations between the 1-^{13}C position and the ring protons. As there is no X filter, both protons attached to ^{12}C and ^{13}C appear in the spectrum. Fig. 8A (top spectrum) shows the proton projection from the experiment. Conversely, Fig. 7, panel 2 shows the result with the X filter turned on, such that the ^{13}C signals subtract. The corresponding projection is shown in Fig. 8A (bottom spectrum). Fig. 8B shows the spectra superimposed and it is clear that the intensities from the protons on ^{12}C are near identical in both experiments, while those from ^{13}C are completely suppressed. As such, the ratio between the two spectra indicates that the ^{13}C positions are essentially fully labelled (99%) while the ^{12}C positions are essentially ^{13}C free. In reality, of course, the ^{12}C position will be at natural

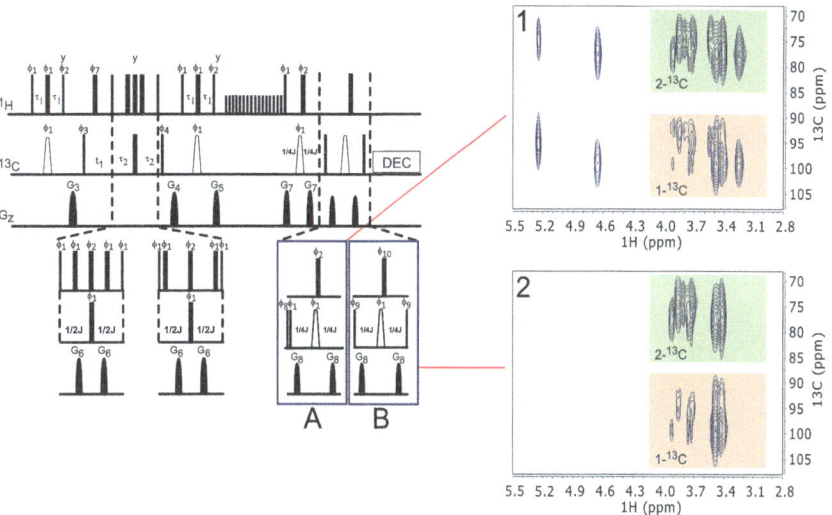

Fig. 7 The basic quantitative sequence. Block B produces the same result as Fig. 1B in that the ^{13}C signals are subtracted to leave only relays to ^{1}H–^{12}C. (A) is derived from (B) with the goal to keep the pulses, durations and timing identical between the two sequences. By reorganizing the X filter element (block B) to become a spin-echo on carbon, and eliminating the carbon phase cycling (*i.e.* $\phi_8 = 0$ in block A), block A essentially does nothing and allows signals from both ^{1}H–^{12}C and ^{1}H–^{13}C to pass. The data from A and B are collected in an interleaved fashion. The narrow rectangles indicate a 90° pulse whereas the wide rectangles indicate a 180° pulse. Unless otherwise stated, pulses are applied along the *x*-axis. The open trapezoids represent smoothed chirp pulses for inversion with a pulse length of 500 μs, a sweep width of 60 kHz, defined by 1000 points and with a 20% smoothing of the amplitude on either end. The pulsed field gradients are indicated as filled sine envelopes and are 1 ms in length. The amplitudes of the gradient pulses have the following ratio: $G_3 = 60\%$, $G_4 = 40\%$, $G_5 = 31\%$, $G_6 = 19\%$, $G_7 = 23\%$, and $G_8 = 13\%$ (with 100% being 53.5 G cm^{-1}). The pulsed field gradients are applied along the *z*-axis followed by a gradient recovery delay of 200 μs. The following phase cycling was used for the pulse sequences: $\phi_1 = 0$, $\phi_2 = 1$, $\phi_3 = 0\ 2$, $\phi_4 = 0\ 0\ 0\ 0\ 2\ 2\ 2\ 2$, $\phi_7 = 0\ 0\ 0\ 0\ 2\ 2\ 2\ 2$, $\phi_8 = 0$, $\phi_9 = 0\ 1$ 2 3, $\phi_{10} = 1$, and $\phi_{rec} = 0\ 2\ 0\ 2\ 2\ 0\ 2\ 0$. The adiabatic TOCSY is comprised of 16 adiabatic processes (ca-WURST, 300 μs, 27.3 KHz pulses) with the following phase cycle, 0 0 2 2 0 2 2 0 2 2 0 0 2 0 0 2. Sequence A results in spectra one, where both the ^{13}C and ^{12}C are acquired. In this case, 1,2-^{13}C-glucose was used. Sequence B subtracts the ^{13}C information leaving only the ^{12}C, as can be seen in spectra two. This allows for quantification by examining signal loss between (A) (the control) and (B) (the ^{12}C-only spectrum), the difference being the % ^{13}C at a specific site.

abundance and should contain ~1.1% ^{13}C. Looking at the peaks, there is a very slight reduction that is consistent with a 1% reduction in signal at the "^{12}C-position", but the accuracy and reproducibility of this would need to be the subject of a much more extensive study. Future work would need to focus on standards with different levels of ^{13}C enrichment to assess how well ratios can be determined and the errors associated with such measurements, before measurements in complex systems would be meaningful. The goal here is simply to introduce one possible route towards isotope ratio quantification in products that are formed when ^{12}C and ^{13}C are bonded together in a complex process, that is a natural extension of the qualitative experiments introduced in the main body of this work.

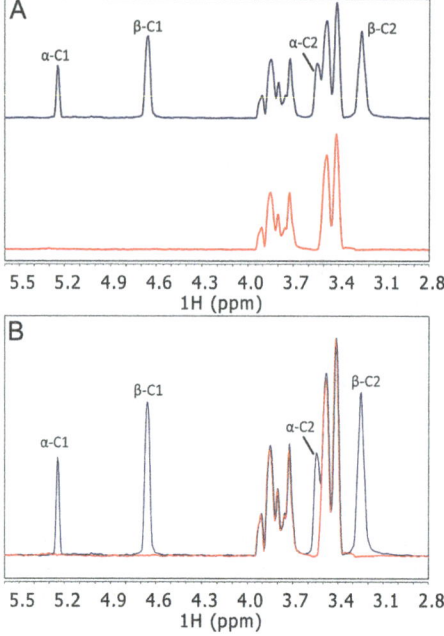

Fig. 8 A simple example of the potential for quantification on 1,2-^{13}C-glucose. (A) The top spectrum is the ^{1}H projection from the quantitative reference (both protons on ^{12}C and ^{13}C appear), and the bottom is the ^{12}C-only sequence (only ^{1}H on ^{12}C are observed). (B) The superimposed spectra show the ^{13}C signals completely cancel, resulting in the red ^{12}C signals. This theoretically allows for quantification by measuring the signal loss in the ^{12}C-only sequence relative to the reference which represents the % ^{13}C at that specific position within the molecule.

Conclusions

A new method for examining new bond formation using targeted 2D NMR spectroscopy has been explored. With this new sequence, ^{13}C–^{12}C bond formation can be selectively observed. To show the concept, processes from simple fermentation to complex metabolic examination in *Daphnia magna* and *Arabidopsis thaliana* were examined both *ex vivo* and *in vivo*. The results provide insight into how these species utilize food and energy sources and synthesize new molecules such as glutamate, ATP, and lipids. Quantification is briefly considered, and the sequence is modified to provide site specific ^{13}C/^{12}C ratios within a simple standard. However, additional further characterization with partially labelled standards is required to determine the error of such measurements before application to complex systems.

In summary, the approach provides a unique insight into the fate and reactivity of carbon in environmental and biological samples. After detailed spectral interpretation, it should be a useful tool for understanding how organisms utilize, store, and transform carbon. Similarly, in environmental research, the transformation and fate of organic matter are of widespread interest. If an enriched substrate (for example, ^{13}C enriched plant biomass, ^{13}C enriched biochar, or a ^{13}C enriched contaminant) is introduced in soil, sediment or water, the experiment should identify when these materials become functionalized, or degraded and recombined with ^{12}C from their environment. In turn, this should provide better insight into carbon sequestration, carbon cycling, humification, and contaminant fate.

Conflicts of interest

There are no conflicts to declare.

Acknowledgements

Andre Simpson would like to thank the Strategic (STPGP 494273-16) and Discovery Programs (RGPIN-2014-05423), the Canada Foundation for Innovation (CFI), the Ontario Ministry of Research and Innovation (MRI), the Krembil Foundation for providing funding, and the Government of Ontario for an Early Researcher Award.

Notes and references

1 A. J. Simpson, Y. Liaghati, B. Fortier-McGill, R. Soong and M. Akhter, Perspective: *in vivo* NMR – a potentially powerful tool for environmental research, *Magn. Reson. Chem.*, 2015, **53**, 686–690.

2 D. Courtier-Murias, H. Farooq, H. Masoom, A. Botana, R. Soong, J. G. Longstaffe, M. J. Simpson, W. E. Maas, M. Fey, B. Andrew, J. Struppe, H. Hutchins, S. Krishnamurthy, R. Kumar, M. Monette, H. J. Stronks, A. Hume and A. J. Simpson, Comprehensive multiphase NMR spectroscopy: Basic experimental approaches to differentiate phases in heterogeneous samples, *J. Magn. Reson.*, 2012, **217**, 61–76.

3 A. J. Simpson, D. J. McNally and M. J. Simpson, NMR spectroscopy in environmental research: From molecular interactions to global processes, *Prog. Nucl. Magn. Reson. Spectrosc.*, 2011, **58**, 97–175.

4 P. R. L. Markwick, T. Malliavin and M. Nilges, Structural Biology by NMR: Structure, Dynamics, and Interactions, *PLoS Comput. Biol.*, 2008, **4**, e1000168.

5 R. Ghose, in *eLS*, John Wiley & Sons, Ltd, Chichester, UK, 2017, pp. 1–20.

6 M. Nilsson, Diffusion NMR, *Magn. Reson. Chem.*, 2017, **55**, 385.

7 N. Hertkorn, C. Ruecker, M. Meringer, R. Gugisch, M. Frommberger, E. M. Perdue, M. Witt and P. Schmitt-Kopplin, High-precision frequency measurements: indispensable tools at the core of the molecular-level analysis of complex systems, *Anal. Bioanal. Chem.*, 2007, **389**, 1311–1327.

8 S. Żerko and W. Koźmiński, Six- and seven-dimensional experiments by combination of sparse random sampling and projection spectroscopy dedicated for backbone resonance assignment of intrinsically disordered proteins, *J. Biomol. NMR*, 2015, **63**, 283–290.

9 X. Yao, S. Becker and M. Zweckstetter, A six-dimensional alpha proton detection-based APSY experiment for backbone assignment of intrinsically disordered proteins, *J. Biomol. NMR*, 2014, **60**, 231–240.

10 S. Hiller, C. Wasmer, G. Wider and K. Wüthrich, Sequence-Specific Resonance Assignment of Soluble Nonglobular Proteins by 7D APSY-NMR Spectroscopy, *J. Am. Chem. Soc.*, 2007, **129**, 10823–10828.

11 M. Mobli, M. W. Maciejewski, A. D. Schuyler, A. S. Stern and J. C. Hoch, Sparse sampling methods in multidimensional NMR, *Phys. Chem. Chem. Phys.*, 2012, **14**, 10835–10843.

12 C. Pontoizeau, T. Herrmann, P. Toulhoat, B. Elena-Herrmann and L. Emsley, Targeted projection NMR spectroscopy for unambiguous metabolic profiling of complex mixtures, *Magn. Reson. Chem.*, 2010, **48**, 727–733.

13 N. G. A. Bell, A. A. L. Michalchuk, J. W. T. Blackburn, M. C. Graham and D. Uhrín, Isotope-Filtered 4D NMR Spectroscopy for Structure Determination of Humic Substances, *Angew. Chem., Int. Ed.*, 2015, **54**, 8382–8385.

14 I. M. Young and J. W. Crawford, Interactions and self-organization in the soil-microbe complex, *Science*, 2004, **304**, 1634–1637.

15 J. L. Markley, R. Brüschweiler, A. S. Edison, H. R. Eghbalnia, R. Powers, D. Raftery and D. S. Wishart, The future of NMR-based metabolomics, *Curr. Opin. Biotechnol.*, 2017, **43**, 34–40.

16 N. Hertkorn, M. Frommberger, M. Witt, B. P. Koch, P. Schmitt-Kopplin and E. M. Perdue, Natural Organic Matter and the Event Horizon of Mass Spectrometry, *Anal. Chem.*, 2008, **80**, 8908–8919.

17 B. P. Koch, M. Witt, R. Engbrodt, T. Dittmar and G. Kattner, Molecular formulae of marine and terrigenous dissolved organic matter detectedby electrospray ionization Fourier transform ion cyclotron resonance mass spectrometry, *Geochim. Cosmochim. Acta*, 2005, **69**, 3299–3308.

18 N. Hertkorn, M. Harir, B. P. Koch, B. Michalke and P. Schmitt-Kopplin, High-field NMR spectroscopy and FTICR mass spectrometry: powerful discovery tools for the molecular level characterization of marine dissolved organic matter, *Biogeosciences*, 2013, **10**, 1583–1624.

19 H. Masoom, D. Courtier-Murias, R. Soong, W. E. Maas, M. Fey, R. Kumar, M. Monette, H. J. Stronks, M. J. Simpson and A. Simpson, From Spill to Sequestration: The Molecular Journey of Contamination *via* Comprehensive Multiphase NMR, *Environ. Sci. Technol.*, 2015, **49**, 13983–13991.

20 I. A. Lewis, R. H. Karsten, M. E. Norton, M. Tonelli, W. M. Westler and J. L. Markley, NMR Method for Measuring Carbon-13 Isotopic Enrichment of Metabolites in Complex Solutions, *Anal. Chem.*, 2010, **82**, 4558–4563.

21 P. N. Reardon, C. L. Marean-Reardon, M. A. Bukovec, B. E. Coggins and N. G. Isern, 3D TOCSY-HSQC NMR for Metabolic Flux Analysis Using Non-Uniform Sampling, *Anal. Chem.*, 2016, **88**, 2825–2831.

22 T. W. M. Fan and A. N. Lane, NMR-based stable isotope resolved metabolomics in systems biochemistry, *J. Biomol. NMR*, 2011, **49**, 267–280.

23 B. P. Kelleher, M. J. Simpson and A. J. Simpson, Assessing the fate and transformation of plant residues in the terrestrial environment using HR-MAS NMR spectroscopy, *Geochim. Cosmochim. Acta*, 2006, **70**, 4080–4094.

24 B. P. Kelleher and A. J. Simpson, Humic Substances in Soils: Are They Really Chemically Distinct?, *Environ. Sci. Technol.*, 2006, **40**, 4605–4611.

25 M. W. I. Schmidt, M. S. Torn, S. Abiven, T. Dittmar, G. Guggenberger, I. A. Janssens, M. Kleber, I. Kögel-Knabner, J. Lehmann, D. A. C. Manning, P. Nannipieri, D. P. Rasse, S. Weiner and S. E. Trumbore, Persistence of soil organic matter as an ecosystem property, *Nature*, 2011, **478**, 49–56.

26 M. H. B. Hayes, *Humic substances II: in search of structure*, J. Wiley, 1st edn, 1989.

27 N. Sengupta, D. C. Reardon, P. D. Gerard and W. S. Baldwin, Exchange of polar lipids from adults to neonates in *Daphnia magna*: Perturbations in sphingomyelin allocation by dietary lipids and environmental toxicants, *PLoS One*, 2017, **12**, 1–25.

28 D. Martin-Creuzburg, E. von Elert and K. H. Hoffmann, Nutritional constraints at the cyanobacteria- *Daphnia magna* interface: The role of sterols, *Limnol. Oceanogr.*, 2008, **53**, 456–468.

29 D. Lane, R. Soong, W. Bermel, W. Maas, S. Schmidt, H. Heumann and A. Simpson, in *57th Experimental NMR Conference*, Pittsburg, 2016.

30 R. Soong, E. Nagato, A. Sutrisno, B. Fortier-Mcgill, M. Akhter, S. Schmidt, H. Heumann and A. J. Simpson, *In vivo* NMR spectroscopy: toward real time monitoring of environmental stress, *Magn. Reson. Chem.*, 2015, **53**, 774–779.

31 B. E. Lindsey, L. Rivero, C. S. Calhoun, E. Grotewold and J. Brkljacic, Standardized Method for High-throughput Sterilization of Arabidopsis Seeds, *J. Visualized Exp.*, 2017, **128**, e56587.

32 E. G. Nagato, B. P. Lankadurai, R. Soong, A. J. Simpson and M. J. Simpson, Development of an NMR microprobe procedure for high-throughput environmental metabolomics of *Daphnia magna*, *Magn. Reson. Chem.*, 2015, **53**, 745–753.

33 M. Tabatabaei Anaraki, R. Dutta Majumdar, N. Wagner, R. Soong, V. Kovacevic, E. J. Reiner, S. P. Bhavsar, X. Ortiz Almirall, D. Lane, M. J. Simpson, H. Heumann, S. Schmidt and A. J. Simpson, Development and Application of a Low-Volume Flow System for Solution-State *in vivo* NMR, *Anal. Chem.*, 2018, **90**, 7912–7921.

34 G. C. Woods, M. J. Simpson, P. J. Koerner, A. Napoli and A. J. Simpson, HILIC-NMR: Toward the Identification of Individual Molecular Components in Dissolved Organic Matter, *Environ. Sci. Technol.*, 2011, **45**, 3880–3886.

35 J. J. Ellinger, R. A. Chylla, E. L. Ulrich and J. L. Markley, Databases and Software for NMR-Based Metabolomics, *Curr. Metabolomics*, 2013, **1**, 28–40.

36 P. Sakhaii and W. Bermel, A different approach to multiplicity-edited heteronuclear single quantum correlation spectroscopy, *J. Magn. Reson.*, 2015, **259**, 82–86.

37 W. Peti, C. Griesinger and W. Bermel, Adiabatic TOCSY for C,C and H,H J-transfer, *J. Biomol. NMR*, 2000, **18**, 199–205.

38 H. Farooq, D. Courtier-Murias, R. Soong, W. Bermel, W. M. Kingery and A. J. Simpson, HR-MAS NMR Spectroscopy: A Practical Guide for Natural Samples, *Curr. Org. Chem.*, 2013, **17**, 3013–3031.

39 D. E. Koshland and F. H. Westheimer, Mechanism of Alcoholic Fermentation. The Fermentation of Glucose-1-C14, *J. Am. Chem. Soc.*, 1950, **72**, 3383–3388.

40 J. L. Galazzo and J. E. Bailey, Fermentation pathway kinetics and metabolic flux control in suspended and immobilized Saccharomyces cerevisiae, *Enzyme Microb. Technol.*, 1990, **12**, 162–172.

41 M. Ikeda, R. Katsumata and O. Zelder, Hyperproduction of tryptophan by Corynebacterium glutamicum with the modified pentose phosphate pathway, *Appl. Environ. Microbiol.*, 1999, **65**, 2497–2502.

42 J. M. Buescher, M. R. Antoniewicz, L. G. Boros, S. C. Burgess, H. Brunengraber, C. B. Clish, R. J. DeBerardinis, O. Feron, C. Frezza, B. Ghesquiere, E. Gottlieb, K. Hiller, R. G. Joncs, J. J. Kamphorst, R. G. Kibbey, A. C. Kimmelman, J. W. Locasale, S. Y. Lunt, O. D. K. Maddocks, C. Malloy, C. M. Metallo,

E. J. Meuillet, J. Munger, K. Nöh, J. D. Rabinowitz, M. Ralser, U. Sauer, G. Stephanopoulos, J. St-Pierre, D. A. Tennant, C. Wittmann, M. G. Vander Heiden, A. Vazquez, K. Vousden, J. D. Young, N. Zamboni and S.-M. Fendt, A roadmap for interpreting (13)C metabolite labeling patterns from cells, *Curr. Opin. Biotechnol.*, 2015, **34**, 189–201.

43 W. M. De Coen, C. R. Janssen and H. Segner, The Use of Biomarkers in *Daphnia magna* Toxicity Testing V. *In Vivo* Alterations in the Carbohydrate Metabolism of *Daphnia magna* Exposed to Sublethal Concentrations of Mercury and Lindane, *Ecotoxicol. Environ. Saf.*, 2001, **48**, 223–234.

44 W. Wiechert, 13C Metabolic Flux Analysis, *Metab. Eng.*, 2001, **3**, 195–206.

45 A. Marx, A. A. de Graaf, W. Wiechert, L. Eggeling and H. Sahm, Determination of the fluxes in the central metabolism of Corynebacterium glutamicum by nuclear magnetic resonance spectroscopy combined with metabolite balancing, *Biotechnol. Bioeng.*, 1996, **49**, 111–129.

46 C. Zwingmann, N. Chatauret, D. Leibfritz and R. F. Butterworth, Selective increase of brain lactate synthesis in experimental acute liver failure: Results of a [^{1}H–^{13}C] nuclear magnetic resonance study, *Hepatology*, 2003, **37**, 420–428.

47 A. Wacker and D. Martin-Creuzburg, Allocation of essential lipids in *Daphnia magna* during exposure to poor food quality, *Funct. Ecol.*, 2007, **21**, 738–747.

48 M. Bastawrous, A. Jenne, M. Tabatabaei Anaraki and A. Simpson, *In Vivo* NMR Spectroscopy: A Powerful and Complimentary Tool for Understanding Environmental Toxicity, *Metabolites*, 2018, **8**, 35.

49 C. E. Goulden and A. R. Place, Fatty acid synthesis and accumulation rates in daphniids, *J. Exp. Zool.*, 1990, **256**, 168–178.

50 L. Ruess and P. M. Chamberlain, The fat that matters: Soil food web analysis using fatty acids and their carbon stable isotope signature, *Soil Biol. Biochem.*, 2010, **42**, 1898–1910.

51 J. A. G. Duarte, F. Carvalho, M. Pearson, J. D. Horton, J. D. Browning, J. G. Jones and S. C. Burgess, A high-fat diet suppresses *de novo* lipogenesis and desaturation but not elongation and triglyceride synthesis in mice, *J. Lipid Res.*, 2014, **55**, 2541–2553.

52 C. Ettenhuber, T. Radykewicz, W. Kofer, H.-U. Koop, A. Bacher and W. Eisenreich, Metabolic flux analysis in complex isotopolog space. Recycling of glucose in tobacco plants, *Phytochemistry*, 2005, **66**, 323–335.

53 R. Gruetter, E. J. Novotny, S. D. Boulware, G. F. Mason, D. L. Rothman, G. I. Shulman, J. W. Prichard and R. G. Shulman, Localized ^{13}C NMR Spectroscopy in the Human Brain of Amino Acid Labeling from D-[1-^{13}C] Glucose, *J. Neurochem.*, 2002, **63**, 1377–1385.

54 M. D. McCoole, B. T. D'Andrea, K. N. Baer and A. E. Christie, Genomic analyses of gas (nitric oxide and carbon monoxide) and small molecule transmitter (acetylcholine, glutamate and GABA) signaling systems in *Daphnia pulex*, *Comp. Biochem. Physiol., Part D: Genomics Proteomics*, 2012, **7**, 124–160.

55 N. R. Sibson, A. Dhankhar, G. F. Mason, K. L. Behar, D. L. Rothman and R. G. Shulman, In vivo 13 C NMR measurements of cerebral glutamine synthesis as evidence for glutamate-glutamine cycling (hyperammonemianeurotransmitter cycledetoxification), *Neurobiology*, 1997, **94**, 2699–2704.

56 L. Hertz, The Glutamate–Glutamine (GABA) Cycle: Importance of Late Postnatal Development and Potential Reciprocal Interactions between Biosynthesis and Degradation, *Front. Endocrinol.*, 2013, **4**, 59.

57 A. Bunescu, J. Garric, B. Vollat, E. Canet-Soulas, D. Graveron-Demilly and F. Fauvelle, In vivo proton HR-MAS NMR metabolic profile of the freshwater cladoceran *Daphnia magna*, *Mol. BioSyst.*, 2010, **6**, 121–125.

58 A. Putman, D. Martin-Creuzburg, B. Panis and L. De Meester, A comparative analysis of the fatty acid composition of sexual and asexual eggs of *Daphnia magna* and its plasticity as a function of food quality, *J. Plankton Res.*, 2015, **37**, 752–763.

59 M. Kariuki, E. Nagato, B. Lankadurai, A. Simpson and M. Simpson, Analysis of Sub-Lethal Toxicity of Perfluorooctane Sulfonate (PFOS) to *Daphnia magna* Using 1H Nuclear Magnetic Resonance-Based Metabolomics, *Metabolites*, 2017, **7**, 1–13.

60 V. Kovacevic, A. J. Simpson and M. J. Simpson, 1H NMR-based metabolomics of *Daphnia magna* responses after sub-lethal exposure to triclosan, carbamazepine and ibuprofen, *Comp. Biochem. Physiol., Part D: Genomics Proteomics*, 2016, **19**, 199–210.

61 H. L. Wheeler, R. Soong, D. Courtier-Murias, A. Botana, B. Fortier-Mcgill, W. E. Maas, M. Fey, H. Hutchins, S. Krishnamurthy, R. Kumar, M. Monette, H. J. Stronks, M. M. Campbell and A. Simpson, Comprehensive multiphase NMR: a promising technology to study plants in their native state, *Magn. Reson. Chem.*, 2015, **53**, 735–744.

62 B. G. Forde and P. J. Lea, Glutamate in plants: metabolism, regulation, and signalling, *J. Exp. Bot.*, 2007, **58**, 2339–2358.

63 V. R. Young and A. M. Ajami, Glutamate: An Amino Acid of Particular Distinction, *J. Nutr.*, 2000, **130**, 892S–900S.

64 R. A. Harris and E. T. Harper, in *eLS*, John Wiley & Sons, Ltd, Chichester, UK, 2015, pp. 1–8.

65 U. Sonnewald, N. Westergaard, B. Hassel, T. B. Müller, G. Unsgård, F. Fonnum, L. Hertz, A. Schousboe and S. B. Petersen, NMR spectroscopic studies of 13C acetate and 13C glucose metabolism in neocortical astrocytes: evidence for mitochondrial heterogeneity, *Dev. Neurosci.*, 1993, **15**, 351–358.

66 L. Lam, R. Soong, A. Sutrisno, R. De Visser, M. J. Simpson, H. L. Wheeler, M. Campbell, W. E. Maas, M. Fey, A. Gorissen, H. Hutchins, B. Andrew, J. Struppe, S. Krishnamurthy, R. Kumar, M. Monette, H. J. Stronks, A. Hume and A. Simpson, Comprehensive Multiphase NMR Spectroscopy of Intact 13C-Labeled Seeds, *J. Agric. Food Chem.*, 2014, **62**, 107–115.

67 A. P. Deshmukh, A. J. Simpson and P. G. Hatcher, Evidence for cross-linking in tomato cutin using HR-MAS NMR spectroscopy, *Phytochemistry*, 2003, **64**, 1163–1170.

68 R. K. Adosraku, G. T. Choi, V. Constantinou-Kokotos, M. M. Anderson and W. A. Gibbons, NMR lipid profiles of cells, tissues, and body fluids: proton NMR analysis of human erythrocyte lipids, *J. Lipid Res.*, 1994, **35**, 1925–1931.

69 M. J. DeNiro and S. Epstein, Mechanism of carbon isotope fractionation associated with lipid synthesis, *Science*, 1977, **197**, 261–263.

70 N. Blair, A. Leu, E. Muñoz, J. Olsen, E. Kwong and D. Des Marais, Carbon isotopic fractionation in heterotrophic microbial metabolism, *Appl. Environ. Microbiol.*, 1985, **50**, 996–1001.

71 T. W.-M. Fan and A. N. Lane, Applications of NMR spectroscopy to systems biochemistry, *Prog. Nucl. Magn. Reson. Spectrosc.*, 2016, **92–93**, 18–53.

72 T. W.-M. Fan and A. N. Lane, Structure-based profiling of metabolites and isotopomers by NMR, *Prog. Nucl. Magn. Reson. Spectrosc.*, 2008, **52**, 69–117.

Faraday Discussions

PAPER

Application of novel solid phase extraction-NMR protocols for metabolic profiling of human urine

Daniel McGill, [ID] *[a] Elena Chekmeneva,[a] John C. Lindon, [ID] [a] Zoltan Takats[a] and Jeremy K. Nicholson[b]

Received 2nd December 2018, Accepted 6th December 2018

DOI: 10.1039/c8fd00220g

Metabolite identification and annotation procedures are necessary for the discovery of biomarkers indicative of phenotypes or disease states, but these processes can be bottlenecked by the sheer complexity of biofluids containing thousands of different compounds. Here we describe low-cost novel SPE-NMR protocols utilising different cartridges and conditions, on both natural and artificial urine mixtures, which produce unique retention profiles useful for metabolic profiling. We find that different SPE methods applied to biofluids such as urine can be used to selectively retain metabolites based on compound taxonomy or other key functional groups, reducing peak overlap through concentration and fractionation of unknowns and hence promising greater control over the metabolite annotation/identification process.

Introduction

Structure elucidation is a necessary and complex task for synthetic chemistry, drug discovery, and natural products research, but it is also a major challenge in areas related to the life sciences. Metabolite structure elucidation is considered a bottleneck in metabolic profiling (the study of low molecular weight metabolite patterns in organisms) – metabolites found within biofluids can be identified and described as biomarkers characteristic of specific phenotypes and disease states. For example, Elliott et al.[1] were able to characterise the biomarkers of adiposity in US adults through analysis of urine samples. Metabolites representative of different metabolite classes and biochemical pathways (such as N-acetylneur-aminate and trimethylamine) were shown to have significant association with BMI. Hence, structure elucidation for metabolic profiling has been demonstrated to indirectly deepen knowledge of metabolic pathways, which may aid the

[a]Division of Computational and Systems Medicine, Department of Surgery and Cancer, Faculty of Medicine, Imperial College London, South Kensington Campus, London SW7 2AZ, UK. E-mail: d.mcgill16@imperial. ac.uk

[b]The Australian National Phenome Center, Research and Innovation, Murdoch University, 50, South Street Murdoch, Perth WA6150, Australia

development of future diagnostic and therapeutic techniques. Within metabolic profiling, ^1H Nuclear Magnetic Resonance (NMR) spectroscopy has become an extremely valuable tool for the characterisation of complex mixtures;[2] statistical spectroscopy tools have also been applied to extract information from complex spectral sets.[3] Posma *et al.*[4] have demonstrated the use of statistical 2D NMR, utilising the RED-STORM probabilistic statistical spectroscopy tool on data acquired from diet-controlled human urine samples, in order to further expand the understanding of dietary biomarkers.

As such, there is currently a high interest in the identification of metabolites. Urine has found widespread use in metabolic profiling due to its non-invasive ease of collection; however, the human urinary metabolome remains only partially mapped. This is in part because of the sheer number and dynamic range of compounds within a given urine sample – and, as a corollary of this, because of peak overlap that frustrates annotation efforts. In a high-throughput NMR study of human urine, it was found by Bingol *et al.*[5] that in a sample ^{13}C–^1H HSQC spectrum, of the 1012 peaks detected, only 437 peaks (belonging to 98 individual metabolites) could be assigned. In 2013, Bouatra *et al.*[6] utilised a variety of analytical platforms (NMR, GC-MS, DFI/LC-MS/MS, ICP-MS, and HPLC) in order to identify 445 and quantify 378 unique urine metabolites, and a literature review led to the identification of an additional 2206 compounds. The Urine Metabolome Database (http://www.urinemetabolome.ca) at the time of writing counts 4237 total metabolites, of which only 1607 are 'detected and quantified', 421 are 'detected but not quantified', and 2209 are 'expected but not detected'.

There is a need to expand the understanding of the human urinary metabolome in order to improve our capacity for characterization of population phenotypes and for discovery of biomarkers related to disease and diet. Currently, the two most powerful analytical techniques used for metabolite annotation (the putative identification of metabolites based on spectral similarity to literature or external data) and identification (confirmation of molecular identity based on 'a minimum of two independent and orthogonal datasets relative to an authentic reference standard')[7] are mass spectrometry (MS) and NMR spectroscopy, as they provide orthogonal qualitative data, as well as absolute and relative quantification, in a very precise and high-throughput manner. In the framework of LC-MS analysis, different column chemistries have been explored to ensure wide metabolome coverage of biofluids, and to take account of the physicochemical diversity of their components. Integration of NMR spectroscopy allows acquisition of complementary information for definitive structure elucidation and confirmation. Despite its reproducibility, ease of use, and quantitative data generation, the limited sensitivity of NMR remains a bottleneck to metabolite identification – low-concentration compounds become indistinguishable from noise, and peak overlap by more concentrated compounds regularly obscures signals from less concentrated compounds.

The use of SPE-NMR of urine has been demonstrated previously by Wilson and Nicholson for retention of specific drug metabolites such as paracetamol, ibuprofen, and naproxen.[8] The fractionation of human urine using SPE to reduce NMR peak overlap has also been demonstrated by Yang *et al.*[9] and Jacobs *et al.*[10] using C18 and HLB cartridges, respectively. All previous SPE-NMR experiments have utilised standard or widely available methods using classic reversed phase sorbents, but have not varied or altered conditions such as pH to generate different retention profiles; additionally, the use of ion exchange SPE cartridges on biofluids for the purpose of metabolite annotation has not been attempted before. The expansion of available SPE methods should promise greater control over metabolite retention, and hence greater insight into the human urinary metabolome – this can be done through studying how different SPE methods can retain not only individual metabolites, but entire compound classes. Hence this approach is highly relevant to the detection and characterisation of unknowns in many complex biological mixtures.

Experimental

Sample collection

Urine was collected from 12 healthy volunteers in 500 mL Corning™ tubes pre-rinsed with ultrapure water – each volunteer provided informed consent in writing. Each urine sample was individually analysed using NMR as a check for polyethylene glycol contamination. Samples were pooled into a pre-rinsed poly-propylene carboy with a stir bar. The pooled sample was homogenised by stirring for five minutes, after which 15 mL aliquots were dispensed into 20 mL Sterilin sample tubes. Samples were labelled sequentially and stored at $-80\,°C$; samples to be used were subsequently transferred to a 4 °C fridge for thawing.

pH-altered urine samples

Concentrated HCl or NaOH solution was added dropwise to 250 mL of pooled urine until the desired pH was achieved. The acidified (pH 2 and pH 5) and basified (pH 11 and pH 9) samples were then stored at 4 °C.

Artificial urine preparation

500 mL ultrapure water was added to a 1 L flask. Constituent compounds and salts were then added (see Results section, Table 2) with constant stirring. An additional 500 mL of distilled water was added to make 1 L of artificial urine.

SPE methods

All solid phase extraction cartridges obtained had a 6 mL capacity and 500 mg bed weight. C18 cartridges were acquired from Thermofisher (Hypersep™) and Agilent (Bond Elut™). C_{18}-ec, NH_2 (Weak Anion eXchange – WAX), CN, and SO_3^- (Weak Cation eXchange – WCX) cartridges were acquired from Biotage (Iso-lute™). C_8, phenyl, and diol cartridges were acquired from Thermofisher (Hypersep™). Hydrophilic–Lipophilic Balance (HLB) cartridges were acquired from Waters (Oasis™). PhenylBoronic Acid (PBA), Strong Anion eXchange (SAX), and Strong Cation eXchange (SCX) cartridges were acquired from Macherey-Nagel (Chromabond™). Sample pretreatment, conditioning, equilibration, wash, and elution steps were tailored for the cartridge and method. Different methods uti-lised a variety of pH levels and solvent systems (Table 1). The use of different conditions altered the retention profiles exhibited by the different cartridges, resulting in elutions composed of different metabolites.

Reversed phase cartridge (C_{18}, HLB, and phenyl), neutral pH sample. A Thermofisher Hypersep™ C_{18} cartridge (6 mL capacity and 500 mg bed weight) was conditioned with methanol (6 mL), then equilibrated with water (6 mL).

Table 1 Descriptions of individual SPE methods

Cartridge	Method name	Modification
C_{18}	Neutral	No modification
	Formic acid	2% formic acid added to conditioning, equilibration, wash, and elution steps
	Weakly acidified	Sample acidified to pH 5
	Strongly acidified	Sample acidified to pH 2
HLB	Neutral	No modification
	Formic acid	2% formic acid added to conditioning, equilibration, wash, and elution steps
	Weakly acidified	Sample acidified to pH 5
	Strongly acidified	Sample acidified to pH 2
Phenyl	Neutral	No modification
	Formic acid	2% formic acid added to conditioning, equilibration, wash, and elution steps
	Weakly acidified	Sample acidified to pH 5
	Strongly acidified	Sample acidified to pH 2
SCX	Neutral	No modification
	Formic acid	2% formic acid added to conditioning and equilibration steps
	Weakly acidified	Sample acidified to pH 5
	Strongly acidified	Sample acidified to pH 2
SAX	Neutral	No modification
	Formic acid	2% formic acid added to conditioning and equilibration steps
	Weakly basified	Sample basified to pH 9
	Strongly basified	Sample basified to pH 11
PBA	Glycine buffer	Glycine-based buffer
	Phosphate buffer	Sodium phosphate-based buffer

Pooled urine (3 mL) was loaded onto the cartridge, which was then washed with water (6 mL) to eliminate interferences. The retained metabolites were then eluted with methanol (6 mL).

Reversed phase cartridge (C_{18}, HLB, and phenyl), 2% formic acid (all steps), acetonitrile elution. A Thermofisher Hypersep™ C_{18} cartridge (6 mL capacity and 500 mg bed weight) was conditioned with 2% formic acid in acetonitrile (6 mL), then equilibrated with 2% formic acid in water (6 mL). Pooled urine (3 mL) was loaded onto the cartridge, which was then washed with 2% formic acid in water (6 mL) to eliminate interferences. The retained metabolites were then eluted with 2% formic acid in acetonitrile (6 mL).

Strong cation exchange, neutral pH sample. A Macherey-Nagel Chromabond™ SCX cartridge (6 mL capacity and 500 mg bed weight) was conditioned with methanol (6 mL), then equilibrated with water (6 mL). Pooled urine (3 mL) was loaded onto the cartridge, which was then washed with 2% formic acid solution (6 mL). Methanol (6 mL) was used to elute the first set of metabolites, followed by 5% NH_4OH in methanol (6 mL) for the second elution.

Strong cation exchange, 2% formic acid. A Macherey-Nagel Chromabond™ SCX cartridge (6 mL capacity and 500 mg bed weight) was conditioned with 2% formic acid in acetonitrile (6 mL), then equilibrated with 2% formic acid in water (6 mL). Pooled urine (3 mL) was loaded onto the cartridge, which was then washed with 2% formic acid solution (6 mL). Acetonitrile (6 mL) was used to elute the first set of metabolites, followed by 5% NH_4OH in methanol (6 mL) for the second elution.

Strong anion exchange, neutral pH sample. A Macherey-Nagel Chromabond™ SCX cartridge (6 mL capacity and 500 mg bed weight) was conditioned with acetonitrile (6 mL), then equilibrated with water (6 mL). Pooled urine (3 mL) was loaded onto the cartridge, which was then washed with 5% NH_4OH solution (6 mL). Acetonitrile (6 mL) was used to elute the first set of metabolites, followed by 2% formic acid in acetonitrile (6 mL) for the second elution.

Phenylboronic acid, sodium phosphate buffer. A Macherey-Nagel Chromabond™ PBA cartridge (6 mL capacity and 500 mg bed weight) was conditioned with a solution of 1% HCl in 70 : 30 water : acetonitrile (6 mL), then equilibrated with sodium phosphate buffer basified to pH 10 with sodium hydroxide (6 mL). Pooled urine (3 mL) was loaded onto the cartridge, which was then washed with sodium phosphate buffer basified to pH 8.5 with sodium hydroxide (6 mL). Water (6 mL) was used to elute the first set of metabolites, followed by a solution of 1% HCl in 70 : 30 water : acetonitrile (6 mL) for the second elution.

NMR sample preparation

Washes and elutions were dried under nitrogen and reconstituted in ultrapure water (3 mL). Buffer containing trimethylsilylpropionate (TMS) as an internal reference standard was added to 540 μL of reconstituted sample, as described by Dona *et al.*[11] 580 μL of the manually vortexed sample was then transferred into 5 mm SampleJet NMR racks.

Samples which required additional 2D NMR experiments were dried under nitrogen and reconstituted in D_2O (3 mL). TMS phosphate buffer (60 μL) was added to 540 μL of reconstituted sample, and 580 μL of the resulting manually vortexed sample was transferred into 5 mm NMR tubes.

NMR data acquisition

All 1D experiments were run using a Bruker Avance III 600 MHz spectrometer equipped with SampleJet. Samples were analysed using one-dimensional water-suppressed ^1H NOESY experiments at 300 K.

Additional ^1H–^1H J-resolved experiments, and 2D-NMR experiments, including ^1H–^1H Total Correlation Spectroscopy (TOCSY), ^1H–^1H Correlation Spectroscopy (COSY), and ^1H–^{13}C Heteronuclear Single Quantum Coherence spectroscopy (HSQC), were utilised for metabolite annotation. The data from the 2D NMR experiments was acquired using a Bruker Avance III 600 MHz spectrometer equipped with a cryoprobe.

Data analysis

NMR datasets were imported into MatLab using the Imperial Metabolic Profiling and Chemometrics Toolbox (IMPaCTS).[12] Water (4.26–5.50 ppm) and formate (8.25–8.63 ppm) regions were removed from the spectra to eliminate interferences; the spectra were then normalised against the TSP region (−0.5 to 0.5 ppm) using a probabilistic quotient normalisation function.[13] Principal Component Analysis (PCA) plots were subsequently constructed with 5 principal components.

Results

Different cartridge chemistries were utilised in order to produce unique retention profiles for different compound classes (as demonstrated in Fig. 1). All samples utilised a sample load incorporating 3 mL of urine as a compromise between substrate retention capacity and spectral resolution. Each elution demonstrated differing retention profiles for each method – replicates of the same method, however, had little difference between spectra. Hence the reproducibility of the SPE methods outlined here can be guaranteed.

Key molecules

Quantifying the extent of retention of different classes of compound can be done using molecules representative of a chemical class. In the Human Urine Metabolome database,[14] chemicals are assigned taxonomically – first to classes, then to subclasses – using the ChemOnt automated taxonomy. As the taxonomy is automated, its class structure can be utilised in order to demonstrate retention profiles for given methods.

Two separate lists of metabolites were utilised to generate a list of representative compounds. One set of metabolites was examined and ranked according to their frequency of occurrence in the human urinary metabolome,[6] such that the metabolites being examined would have a significant chance of being characterised in pooled urine samples. The second set was generated from a method for producing artificial urine.[15] The two lists were combined, and the resulting set of

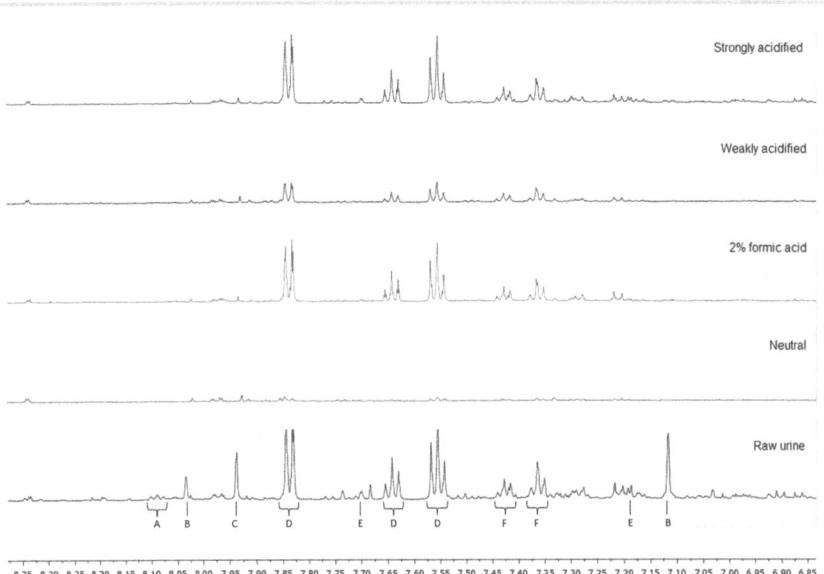

Fig. 1 A comparison of the aromatic regions of 600 MHz NMR spectra utilising different C_{18} SPE methods on natural human urine. From top: sample acidified to pH 2, sample acidified to pH 5, 2% formic acid added to all steps, neutral sample, and pooled urine before SPE treatment. Select metabolites labelled: (A) trigonelline, (B) 3-methylhistidine, (C) histidine, (D) hippurate, (E) 2-furoylglycine, and (F) phenylalanine.

metabolites (Table 2) was then characterised by their assigned subclass from the ChemOnt automated taxonomy.

The intensities of specific peaks corresponding to these metabolites in the elutions were compared to the intensities of the same peaks in the raw pooled urine samples. From this, the percentage retention per compound class could be

Table 2 Artificial urine mixture constituent metabolites

Name	Occurrence in urine (%)[6]	Assigned subclass (HMDB)
trans-Aconitic acid	55	Tricarboxylic acids and derivatives
Tartaric acid	82	Carbohydrates and carbohydrate conjugates
Succinic acid	91	Dicarboxylic acids and derivatives
L-Asparagine	95	Asparagine and derivatives
L-Aspartic acid	95	Aspartic acid and derivatives
2-Furoylglycine	100	*N*-Acyl-alpha amino acids
3-Aminoisobutanoic acid	100	Beta amino acids and derivatives
3-Methylhistidine	100	Histidine and derivatives
Acetic acid	100	Carboxylic acids
Allantoin	100	Imidazoles
Betaine	100	Alpha amino acids and derivatives
Citric acid	100	Tricarboxylic acids and derivatives
Creatine	100	Alpha amino acids and derivatives
Creatinine	100	Alpha amino acids and derivatives
D-Glucose	100	Carbohydrates and carbohydrate conjugates
Dimethyl sulfone	100	Sulfones
Dimethylamine	100	Organonitrogen compounds
Erythritol	100	Carbohydrates and carbohydrate conjugates
Ethanolamine	100	Amines
Formic acid	100	Carboxylic acids
Glycerol	100	Carbohydrates and carbohydrate conjugates
Glycine	100	Alpha amino acids and derivatives
Glycolic acid	100	Alpha hydroxy acids and derivatives
Guanidoacetic acid	100	Alpha amino acids and derivatives
Hippuric acid	100	Benzoic acids and derivatives
Indoxyl sulfate	100	Arylsulfates
L-Alanine	100	Alanine and derivatives
L-Cysteine	100	Cysteine and derivatives
L-Cystine	100	L-Cysteine-*S*-conjugates
L-Glutamic acid	100	Glutamic acid and derivatives
L-Glutamine	100	Alpha amino acids and derivatives
L-Histidine	100	Histidine and derivatives
L-Lactic acid	100	Alpha hydroxy acids and derivatives
L-Lysine	100	Alpha amino acids and derivatives
L-Phenylalanine	100	Phenylalanine and derivatives
L-Serine	100	Serine and derivatives
L-Threonine	100	Alpha amino acids and derivatives
Methanol	100	Alcohols and polyols
Myoinositol	100	Alcohols and polyols
p-Hydroxyphenylacetic acid	100	Phenols
Propylene glycol	100	Alcohols and polyols
Taurine	100	Organosulfonic acids and derivatives
Trigonelline	100	Alkaloids and derivatives
Trimethylamine N-oxide	100	Aminoxides

estimated – hence, insight into which methods can selectively retain different compound classes can be achieved. For example, the peak intensity of retained creatinine (belonging to the subclass 'alpha amino acids and derivatives') from a method utilising a C_{18} cartridge at neutral conditions was measured at 257. When compared to the intensity of the same peak in the 'raw' urine sample (5027), this gives an estimated retention capacity of creatinine using this method of 5%. After accounting for the other members of that subclass, C_{18} under neutral conditions has an overall retention of α-amino acids and derivatives of 0.75%.

The compound class retentions per method were then summed to produce a measure of the total retention capacity out of 100 – where methods ranking 0 would retain nothing, while methods ranking 100 would retain everything (Table 3). It can be expected that the sum of the retention capacity estimates of the elutions and the washes for a given method is equal to 100.

Principal component analysis (PCA)

An alternative approach to quantifying the retention profiles of each SPE cartridge can be achieved using PCA. PCA is an analytical method that can be used to reduce the dimensionality of data and produce visual representations of correlations between datasets – for NMR spectra, it allows for clustering and trends between experiments to be demonstrated, as well as for discovering potential outliers.

A PCA structure built with the datasets from all elution methods utilising natural urine demonstrated that the elutions from acidified ion exchange methods were clearly separated from the other chemistries (Fig. 2). The below

Table 3   Total retention capacity estimates for each SPE method

SPE method	Total retention capacity (%)
C_{18}, neutral conditions	4.57
C_{18}, 2% formic acid	12.91
C_{18}, weakly acidified sample	10.65
C_{18}, strongly acidified sample	15.69
HLB, neutral conditions	8.87
HLB, 2% formic acid	17.40
HLB, weakly acidified sample	13.77
HLB, strongly acidified sample	22.13
Phenyl, neutral conditions	1.68
Phenyl, 2% formic acid	2.05
Phenyl, weakly acidified sample	3.41
Phenyl, strongly acidified sample	5.67
SCX, neutral conditions	0
SCX, 2% formic acid	0
SCX, weakly acidified sample	2.57
SCX, strongly acidified sample	14.67
SAX, neutral conditions	12.11
SAX, 2% formic acid	5.28
SAX, weakly basified sample	2.24
SAX, strongly basified sample	0.35
PBA, phosphate buffer	0.93

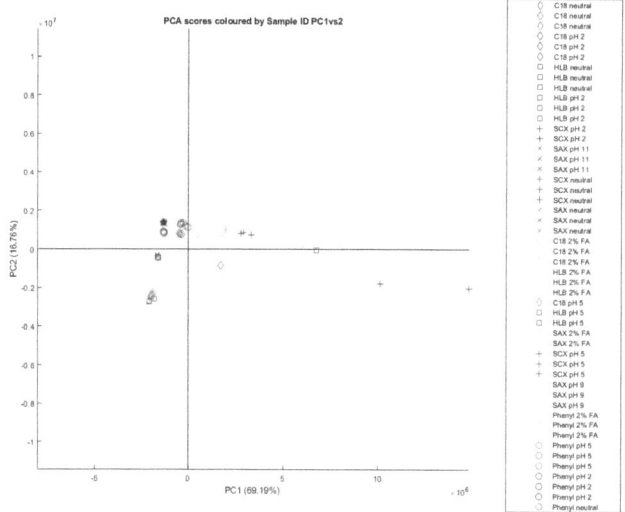

Fig. 2 A PCA scores plot based on all SPE elutions of natural urine, PC1 (69.19%) *vs.* PC2 (16.76%).

table (Table 4) lists the NMR signals visible in the PCA loadings plot, as well as their correlation coefficient and their tentative assignment – all assignments are made with comparison to the reported values in the literature, and hence can be considered annotated to a level 2 standard.[16] It demonstrates that a small number

Table 4 Natural urine all elutions, PC1 assignments

Peak (ppm)	Assignment	Pearson's correlation
7.96 (s)	3-Methylhistidine	0.64
7.90 (s)	Histidine	0.89
7.70 (s)	1-Methylhistidine	0.81
7.10 (s)	Histidine	0.95
7.08 (s)	3-Methylhistidine	0.90
7.02 (s)	1-Methylhistidine	0.92
4.06 (s)	Creatinine	1.00
4.05 (d)	Unknown A	0.79
4.00 (dd)	Histidine	0.90
3.72 (s)	3-Methylhistidine	0.89
3.70 (s)	3-Methylhistidine	0.80
3.29 (s)	**Unknown B**	0.93
3.27 (s)	TMAO	0.96
3.25 (s)	**Unknown B**	0.94
3.24 (s?)	Unknown C	0.94
3.23 (d?)	Unknown D	0.95
3.21 (s)	Unknown E	0.84
3.18 (s)	Unknown F	0.87
3.16 (s)	Unknown G	0.93
3.14 (s)	Unknown H	0.89
3.05 (s)	Creatinine	0.99
3.04 (s)	Creatine	0.83

of peaks (creatinine, histidine, creatine, trimethylamine-N-oxide, and 3-methyl-histidine being the most prominent) contributed to 69.24% of the differences between elutions. These metabolites were all retained by SCX cartridges under acidic conditions, and their spectral peaks are sensitive to pH changes, causing significant chemical shifts even in buffered samples.

Unknowns in bold have both ^{1}H and ^{13}C signals identifiable, but could not be matched to compounds in metabolite databases.

The ion exchange method most separated from other datasets is the one which produces elutions from strongly acidified urine using an SCX cartridge. However, there is also significant differentiation with the SAX elutions with 2% formic acid – under these conditions, the SAX cartridge begins to retain compounds like creatinine and histidine where it otherwise wouldn't under neutral or basic conditions. It is not clear why the SAX retention profile would begin to resemble that of SCX; one possibility could be that the silanol groups from the silica on which the SO_3^- and ammonium modifications are based are more able to retain these compounds under acidic conditions. However, silanols are usually protonated under acidic conditions, and hence would not express this ionic character; the explanation additionally doesn't account for why C_{18} with 2% formic acid – which, similarly, contains silanol groups – does not retain creatinine or histidine.

Besides the separation between ion exchange and reversed phase methods, there is additional separation along the secondary principal component axis between phenyl, C_{18} neutral, SAX neutral, and SCX neutral elutions, and C_{18} acidified and HLB acidified elutions – with HLB neutral elutions found in between the two clusters (Fig. 3). Removing elutions from ion exchange cartridges and reconstructing the PCA affords a similar differentiation with more clarity. The clusters suggest that C_{18} neutral elutions have more in common with phenyl

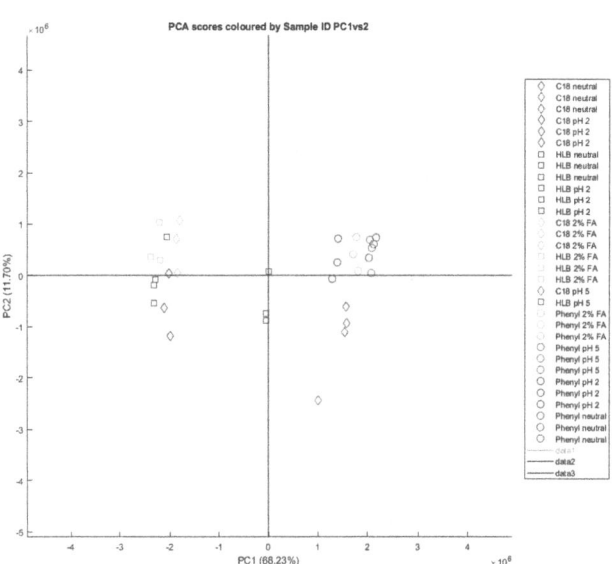

Fig. 3 A PCA scores plot based on reversed phase SPE elutions of natural urine, PC1 (68.23%) vs. PC2 (16.76%).

Table 5 Natural urine reversed phase elutions, PC1 assignments

Peak (ppm)	Assignment	Pearson's correlation	Phenyl correlation
9.13 (s)	Trigonelline	0.78	Positive
8.84 (t)	Trigonelline	0.77	Positive
[8.45 (d)]	Quinolinate	*	Negative
[8.34 (d)]	N-Methyl-2-pyridone-5-carboxamide	*	Negative
8.09 (t)	Trigonelline	0.76	Positive
8.03 (s)	3-Methylxanthine	0.84	Negative
8.00 (s)	Phenylacetyl-glutamine[17]	0.91	Negative
7.97 (dd)	N-Methyl-2-pyridone-5-carboxamide	0.98	Negative
7.94 (s)	**Unknown I**	0.77	Negative
7.91 (d?)	**Unknown J**	0.93	Negative
7.87 (d)	Unknown K	0.90	Negative
7.84 (dd)	Hippurate	0.98	Negative
7.77 (d?)	4-Hydroxyhippurate?	0.92	Negative
7.70 (dd)	2-Furoylglycine	0.92	Negative
7.64 (tt)	Hippurate	0.98	Negative
7.56 (t)	Hippurate	0.99	Negative
7.46 (dd)	Quinolate	0.93	Negative
7.43 (m)	Phenylacetyl-glutamine	0.97	Negative
7.37 (m)	Phenylacetyl-glutamine	0.97	Negative
7.30 (t?)	Unknown L	0.97	Negative
7.24 (d)	Unknown M	0.96	
7.19 (d)	2-Furoylglycine	0.92	Negative
7.17 (d)	4-Hydroxyphenyl-acetate	0.94	Negative
7.12 (ddd)	**Unknown N**	0.96	Negative
7.07 (m)	**Unknown O**	0.98	Negative
6.98 (d)	4-Hydroxyhippurate?	0.94	Negative
6.93 (t)	**Unknown P**	0.89	Negative
6.87 (d)	4-Hydroxyphenyl-acetate	0.95	Negative
6.68 (d)	N-Methyl-2-pyridone-5-carboxamide	0.97	Negative
6.64 (dd)	2-Furoylglycine	0.86	Negative
4.19 (td)	Phenylacetyl-glutamine	0.98	Negative
4.06 (s)	Creatinine	0.55	Positive
4.01 (dd)	Phenylalanine	0.86	Negative
3.97 (d)	Hippurate	0.98	Negative
3.96 (s)	4-Hydroxyhippurate?	0.97	Negative
3.94 (s?)	Unknown Q	0.99	Negative
3.93 (s)	(2-Furoylglycine)	0.93	Negative
3.92 (s?)	Unknown R	0.96	Negative
3.87 (s/t?)	Unknown S	0.96	Negative
3.70 (s)	**Unknown T**	0.96	Negative
3.67 (d)	**Unknown U**	0.97	Negative
3.65 (s)	N-Methyl-2-pyridone-5-carboxamide	0.98	Negative
3.63 (s)	**Unknown T**	0.98	Negative
3.54 (s)	Unknown V	0.92	Negative
3.53 (s)	**Unknown W**	0.99	Negative
3.49 (s)	**Unknown X**	0.88	Negative
3.48 (s)	**Unknown Y**	0.96	Negative
3.45 (s)	Unknown Z	0.98	Negative
3.34 (s)	**Unknown AA**	0.96	Negative
3.32 (s)	**Unknown AB**	0.96	Negative
3.30 (s)	Unknown AC	0.97	Negative
3.27 (s)	Unknown AD	0.67	Positive

Table 5 *(Contd.)*

Peak (ppm)	Assignment	Pearson's correlation	Phenyl correlation
3.17 (s)	**Unknown AE**	0.98	Negative
3.11 (s)	**Unknown AF**	0.79	Positive
3.05 (s)	Creatinine	0.55	Positive
3.00 (s)	Unknown AG	0.91	Negative
2.71 (s?)	Unknown AH	0.93	Negative
2.69 (s?)	Unknown AI	0.87	Negative
2.68 (s?)	Unknown AJ	0.89	Negative
2.67 (s?)	Unknown AK	0.88	Negative
2.63 (d)	**Unknown T**	0.92	Negative
2.35 (s)	**Unknown AL**	0.78	Negative
2.27 (t?)	Phenylacetyl-glutamine	0.98	Negative
2.12 (m?)	Phenylacetyl-glutamine	0.98	Negative
1.93 (m?)	Phenylacetyl-glutamine	0.98	Negative
1.31 (m)	Valerate	0.82	Negative
0.94 (m?)	Unknown AM	0.92	Negative
0.89 (t)	Valerate	0.92	Negative

elutions than with acidified C_{18} or even HLB neutral elutions. The NMR signals responsible for the separation between reversed phase elutions are tabulated in Table 5 – it also notes whether the annotated metabolites or unknowns are positively correlated with phenyl cartridges (and hence retained by phenyl), or are negatively correlated (and hence retained by C_{18}/HLB). Signals marked with an asterisk * are not visible in the loadings plot, but can be observed with manual inspection of the spectra.

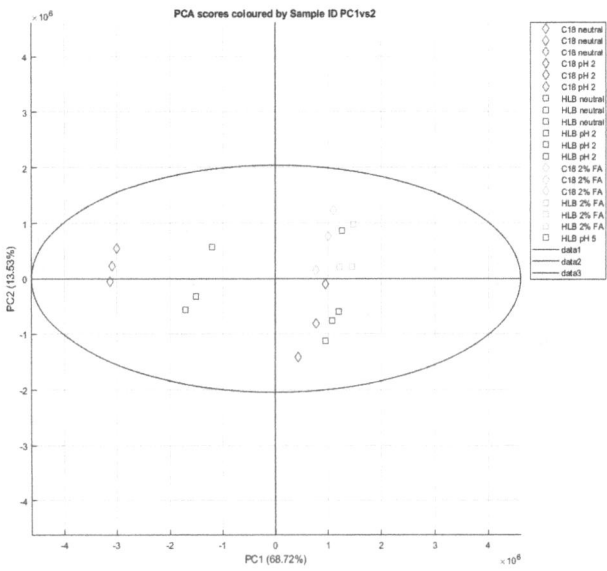

Fig. 4 A PCA scores plot based on C_{18} and HLB SPE elutions of natural urine, PC1 (68.72%) *vs.* PC2 (13.53%).

Table 6 Natural urine C_{18} and HLB elutions, PC1 assignments

Peak (ppm)	Assignment	Pearson's correlation	HLB correlation
8.05 (m?)	Unknown AN	0.87	Positive
7.97 (br s?)	Unknown AO	0.91	Positive
7.97 (dd)	N-Methyl-2-pyridone-5-carboxamide	0.91	Negative
7.96 (dd)	N-Methyl-2-pyridone-5-carboxamide	0.96	Positive
7.94 (s)	**Unknown I**	0.70	Positive
7.93 (s)	Unknown AP	0.83	Negative
7.92 (s)	Unknown AQ	0.82	Negative
7.91 (s)	Unknown AR	0.91	Positive
7.91 (s)	Unknown AS	0.61	Negative
7.90 (s)	Unknown AT	0.84	Positive
7.87 (d)	Unknown K	0.83	Positive
7.84 (dd)	Hippurate	0.91	Positive
7.76 (d?)	4-Hydroxyhippurate?	0.88	Positive
7.70 (dd)	2-Furoylglycine	0.81	Positive
7.64 (tt)	Hippurate	0.99	Positive
7.56 (t)	Hippurate	0.99	Positive
7.43 (m)	Phenylalanine	0.95	Positive
7.37 (m)	Phenylalanine	0.95	Positive
7.31 (s)	Unknown AU	0.83	Positive
7.30 (t)	Unknown L	0.92	Positive
7.19 (d)	2-Furoylglycine	0.80	Positive
7.17 (d)	4-Hydroxyphenyl-acetate	0.85	Positive
7.12 (ddd)	**Unknown N**	0.96	Positive
6.98 (d)	4-Hydroxyhippurate?	0.85	Positive
6.87 (d)	4-Hydroxyphenyl-acetate	0.90	Positive
6.81 (s/t?)	Unknown AV	0.92	Positive
6.68 (d)	N-Methyl-2-pyridone-5-carboxamide	0.92	Positive
6.67 (d)	N-Methyl-2-pyridone-5-carboxamide	0.91	Negative
4.19 (td)	Phenylacetylglutamine	0.96	Positive
4.06 (s)	Creatinine	0.80	Negative
3.97 (d)	Hippurate	0.99	Positive
3.96 (s)	Unknown AW	0.94	Positive
3.94 (d?)	Unknown AX	0.92	Negative
3.93 (d?)	Unknown AY	0.95	Positive
3.93 (s)	Unknown AZ	0.88	Positive
3.87 (s/t?)	**Unknown T**	0.98	Positive
3.70 (s)	**Unknown BA**	0.98	Positive
3.67 (d?)	**Unknown BB**	0.95	Positive
3.65 (s)	N-Methyl-2-pyridone-5-carboxamide	0.92	Negative
3.65 (s)	N-Methyl-2-pyridone-5-carboxamide	0.97	Positive
3.63 (s)	**Unknown T**	0.97	Positive
3.53 (s)	Unknown BC	0.79	Negative
3.53 (s)	3-Methylxanthine?	0.95	Positive
3.49 (s)	**Unknown BD**	0.81	Negative
3.49 (s)	Unknown BE	0.84	Positive
3.48 (s)	Unknown BF	0.90	Positive
3.46 (s)	Unknown BG	0.93	Positive
3.45 (s)	Unknown BH	0.88	Positive
3.36 (s)	Unknown BI	0.87	Negative
3.35 (s?)	Unknown BJ	0.86	Negative
3.34 (s?)	**Unknown BK**	0.95	Positive
3.33 (s?)	Unknown BL	0.83	Negative

Table 6 (*Contd.*)

Peak (ppm)	Assignment	Pearson's correlation	HLB correlation
3.32 (s?)	Unknown BM	0.93	Positive
3.30 (s?)	Unknown BN	0.96	Positive
3.27 (s)	TMAO	0.80	Negative
3.17 (s)	**Unknown BO**	0.88	Negative
3.17 (s)	Unknown BP	0.95	Positive
3.05 (s)	Creatinine	0.80	Negative
2.27 (t)	Phenylacetyl-glutamine	0.97	Positive
2.12 (m)	Phenylacetyl-glutamine	0.97	Positive
0.93 (m)	Phenylacetyl-glutamine	0.97	Positive

Clustering can be observed forming an almost linear scale for C_{18} and HLB cartridges (Fig. 4) – with C_{18} neutral elutions at one end, acidified elutions at the other, and HLB neutral elutions in between. Many of the assigned peaks (Table 6) in the aromatic region are caused by differences in chemical shift between identical compounds (likely due to pH differences) – for example, *N*-methyl-2-pyridone-5-carboxamide (2PY) is significantly correlated both positively and negatively with phenyl elutions (Fig. 5), as its spectral peaks undergo chemical shifting due to pH differences in different experiments.

The PC3 *vs.* PC4 plot in the all-elutions structure also demonstrated clustering (Fig. 6) of the PBA elutions, positively correlated with the PC4 dimension. The PC4 loadings hence closely resemble the averaged spectra from the PBA elutions – the metabolites (Table 7) being mostly represented by mannitol and *N*-methylnicotinamide.

Artificial urine

A similar PCA can be constructed for reversed phase elutions utilising artificial urine – again, there is separation across the first component, demonstrating

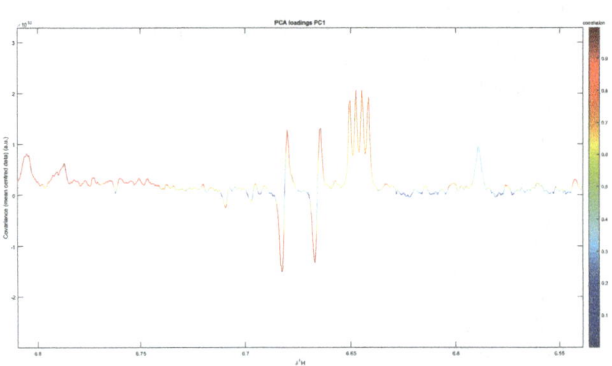

Fig. 5 The PC1 loadings plot for C_{18}/HLB elutions, demonstrating both positive and negative correlation of 2PY.

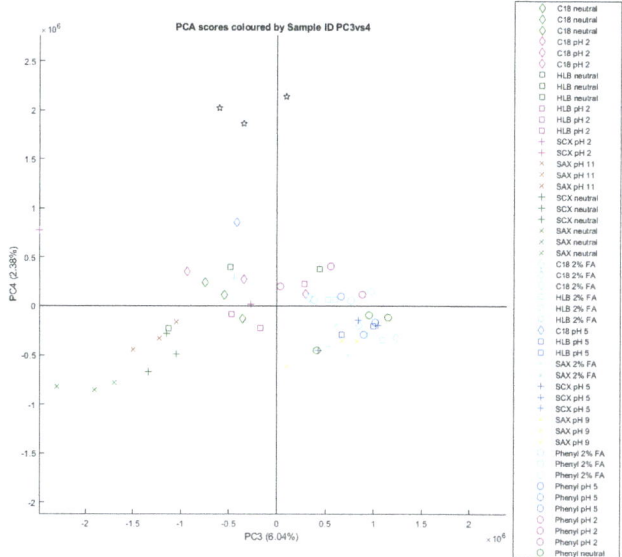

Fig. 6 A PCA scores plot based all SPE elutions of natural urine, PC3 (6.04%) vs. PC4 (2.38%).

a notable difference between phenyl and C_{18}/HLB retention profiles. As with the natural urine elutions, the loadings (Table 8) can be annotated to demonstrate the most important spectral differences separating different methods.

Table 7 Natural urine all elutions, PC4 assignments

Peak (ppm)	Assignment	Pearson's correlation
9.29 (s)	*N*-Methylnicotinamide	0.82
8.97 (d)	*N*-Methylnicotinamide	0.72
8.90 (d)	*N*-Methylnicotinamide	0.71
8.79 (d)	**Unknown BQ**	0.69
[8.53 (t)]	**Unknown BQ**	*
8.06 (t?)	**Unknown BQ**	0.63
7.68 (s)	**Unknown BR**	0.88
5.85 (d)	Unknown BS	0.49
[4.31 (t)]	Unknown BS	*
4.16 (t)	**Unknown BS**	0.53
4.11 (s)	Unknown BT	0.68
4.02 (t)	Unknown BU	0.73
3.89 (d)	Mannitol	0.86
3.87 (d)	Mannitol	0.87
3.82 (s)	Mannitol	0.86
3.80 (s)	Mannitol	0.84
3.77 (m)	Mannitol	0.87
3.67 (d?)	Unknown BV	0.87
3.69 (dd)	Mannitol	0.86
3.20 (s)	Unknown BW	0.79
2.76 (t)	Ethanolamine?	0.76
2.72 (s)	DMA	0.85
2.01 (s)	Acetamide?	0.82

Table 8 Artificial urine reversed phase elutions, PC1 assignments

Peak (ppm)	Assignment	Pearson's correlation	Phenyl correlation
9.13 (s)	Trigonelline	0.95	Positive
8.84 (t)	Trigonelline	0.97	Positive
8.09 (t)	Trigonelline	0.92	Positive
7.97 (d)	Hippurate	0.97	Negative
7.88 (m)	Benzoate	1.00	Negative
7.84 (dt)	Hippurate	1.00	Negative
7.78 (t)	Hippurate	0.97	Negative
7.70 (t)	Hippurate	0.97	Negative
7.64 (tt)	Hippurate	1.00	Negative
7.56 (m)	Hippurate	1.00	Negative
7.49 (m)	Benzoate	1.00	Negative
7.44 (tt)	Phenylalanine	0.99	Negative
7.38 (tt)	Phenylalanine	0.99	Negative
7.34 (m)	Phenylalanine	0.99	Negative
7.17 (dt)	4-Hydroxyphenyl-acetate	0.93	Negative
6.87 (dt)	4-Hydroxyphenyl-acetate	0.93	Negative
6.59 (s)	trans-Aconitate	0.85	Negative
4.00 (q)	Phenylalanine	0.96	Negative
3.97 (d)	Hippurate	0.99	Negative
3.30 (d)	Phenylalanine?	0.97	Negative
3.27 (s)	TMAO	0.91	Positive
3.13 (m)	Phenylalanine?	0.99	Negative
3.05 (s)	Creatine?	0.93	Positive
2.66 (d)	Citrate	0.91	Negative
2.55 (d)	Citrate	0.94	Negative
2.41 (s)	Succinate	0.84	Negative

Here, artificial urine was used to demonstrate that a mixture of representative compounds can be used to estimate the retention capacity of cartridges without using natural urine – as with the natural urine reversed phase elutions, metabolites such as hippurate can be shown to be retained on C_{18}/HLB, but not on phenyl cartridges; similarly, trigonelline can be shown to be retained on phenyl, but not on C_{18}/HLB. This allows for greater control over future SPE experiments aimed at characterising retention profiles of SPE cartridges.

Discussion

The guiding philosophy behind this use of SPE-NMR suggests that not only each cartridge, but each pH and solvent system utilised in a given experiment, would result in different retention profiles. These retention profiles can be classified either through annotation of a selection of common metabolites from different compound classes, or using a more holistic approach, determining the compounds more likely to be retained under different conditions through data treatment. Selective use of methods can then be utilised to reduce peak overlap and aid metabolite identification. One example of this is displayed in Fig. 7; interferences in the 'raw' pooled urine sample are removed by the use of a PBA-based method to clearly reveal mannitol.

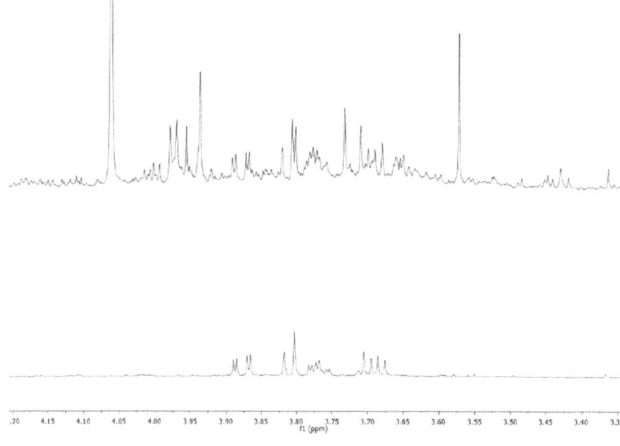

Fig. 7 Comparison of the 3.30–4.15 ppm region of the 600 MHz NMR spectra of the pooled urine sample (top) and PBA SPE-treated urine (bottom), the latter revealing only mannitol peaks.

Annotation of a selection of common metabolites is facile and provides immediate and useful information of individual compounds. Ideally, this could be done using a list of metabolites representative of the compound classes generally found in human urine; unfortunately, the partial identification of the human urinary metabolome hinders the creation of a fully representative sample. Additionally, the natural rate of occurrence of metabolites may make a representation of an 'average' sample difficult or even impossible. Peak intensities can also be impacted by NMR shimming, peak overlap, and pH changes – all of which can affect the intensity recorded. Despite these shortcomings, general trends can be established by considering functional groups and structural commonalities between compound classes.

Clustering in PCA plots can be used to demonstrate substantial differences between datasets. The Strong Cation Exchange (SCX) cartridge – utilising the pseudo-permanently charged phenylsulfonic derivative ($pK_a \approx 2.1$) – provides the greatest separation between clusters when included in PCA structures, due to the ion exchange mechanisms not present in reversed phase chromatography. Ion exchange retention profiles rely heavily on pH control, since all compounds must have at least one positively charged atom in order to have sufficient attraction to the sorbent to be retained; hence, compounds that do not have a positive charge at physiological pH must be in acidic solution for retention to occur. This intrusive sample adjustment will naturally affect the chemistry of the biofluid; using acidified (or basified) conditions is, then, necessarily a trade-off between greater insight into the metabolome through retention, and authenticity of the sample itself. It is additionally feasible that compounds not normally present in the sample may be formed and retained due to the change of conditions, although this was not noted during the course of the experiments.

The importance of pH control is reflected in the retention capacity of SCX cartridges; the neutral pH retention profile is one of the least retaining methods with an estimated retention capacity of 0 – at pH 2, its retention capacity (14.67) is comparable to a more widely recognised reversed phase method, such as HLB with 2% formic acid in all steps (17.40). The compounds best retained on SCX under acidic conditions were predominantly histidine-based – with histidine, 3-methylhistidine, and 1-methylhistidine being well retained at pH 2. Creatinine and TMAO were also present in the elutions. A cationic nitrogen atom, possibly stabilised by electron-donating groups through hyperconjugation, may serve as the most important characteristic uniting the compounds. Compounds without nitrogen-containing functional groups (such as amino acids) were generally not present in the acidified SCX elutions, although the presence of a nitrogen-containing functional group did not necessarily result in retention – for example, of the proteinogenic amino acids that were retained, only histidine was retained in any significant quantity. It may be notable that histidine has a pK_a of 6.04 (pyrrolic nitrogen), far lower than the other positively charged amino acids, arginine (pK_a 12.10) and lysine (pK_a 10.67).

Removing ion exchange elutions from the dataset and reconstructing a PCA plot demonstrates additional separation between C_{18}/HLB and phenyl elutions and allows for further probing into the differences between the reversed phase methods. Previous uses of phenyl cartridges in the literature have remarked on their similar retention capabilities to C_{18}, with slightly better retention for poly-cyclic aromatic compounds,[18] but slightly worse retention for other hydrocarbons.[19] Phenyl cartridges utilise π-stacking on top of hydrophobic forces in order to provide additional retention for aromatic compounds – however, the strength of π–π interactions tends to be bound between around 8–12 kJ mol^{-1} for benzene dimers.[20] For comparison, hydrophobic forces may be up to 4 times stronger;[21] hence, despite having an additional mechanism of action, the actual retention capacity for phenyl cartridges is significantly lower than that of C_{18} or HLB cartridges across all methods due to a weaker hydrophobic retention mechanism.

The two major compounds that were retained selectively by phenyl (but not by C_{18} or HLB) were trigonelline and creatinine, both nitrogen-containing hetero-cycles. Other compounds with phenyl functional groups – such as phenylalanine or hippurate – did not experience greater retention using phenyl cartridges, and in fact retained much less, if at all. Conversely, the cyclic metabolites retained under acidic conditions by C_{18}/HLB – but not by phenyl – are predominantly aromatic, with both heterocycles (2-furoylglycine, quinolinate, and N-methyl-2-pyridone-5-carboxamine), and hydrocarbon rings (hippurate, phenylacetylglut-amine, and p-hydroxyphenylacetate) present. The aromatic heterocycles here do not have the ability to form cationic nitrogen in the rings themselves, unlike the aromatic compounds retained in phenyl elutions, such as trigonelline. Short chain fatty acids such as valeric acid were also retained under non-neutral conditions. It is unclear why charged metabolites like trigonelline would be better retained on phenyl cartridges – although as the charge is positive, it is hypothetically possible that the π-electron clouds located above and below the benzene rings are able to electrostatically attract these metabolites with enough strength that they can be retained.

Out of the remaining reversed phase sorbents, the C_{18} and HLB cartridges are known to have similar retention profiles to each other,[10] with HLB cartridges often being preferred for their tolerance to drying and the possibility for elimination of conditioning and equilibration steps. The differences between the two can be demonstrated through comparison of elutions – while the two have comparable retention, HLB cartridges tend to retain a slightly larger range of compounds in greater quantities. This is especially true under neutral conditions, where C_{18} cartridges retain relatively little. Generally speaking, as with the SCX cartridges, more acidic conditions result in greater retention – possibly due to the deionisation of silanol groups on the surface of the sorbent. However, this again comes with the trade-off of authenticity, as the acidic conditions may cause signal suppression, unwanted reactions between metabolites, or general degradation of the sample itself. Use of 2% formic acid in all steps allows for a balance between the two – while the retention does not extend as deeply as that under pH 2 conditions, the higher pH environment should not be as destructive to sample authenticity, and chemical shifts caused by drastic pH changes should be absent. HLB and C_{18} under acidic conditions are powerful analytical methods that can reveal signals that are otherwise not visible – for example, unknown N, which displays a doublet of doublet of doublets (ddd) signal at 7.12 ppm, is normally obscured by a 3-methylhistidine peak (Fig. 8). The use of reversed phase methods can hence provide additional information about the human urinary metabolome.

C_{18} and HLB cartridges themselves can also be differentiated from each other. On top of the obvious differences in sorbent structure (a hydrocarbon chain, compared to a polymer containing divinylbenzene), HLB cartridges are not silica-based – hence, silanol groups present in C_{18} cartridges are not present in HLB. These silanol groups produce secondary interactions with metabolites, commonly expressed as a weak cation exchanger, which can influence retention. This is reflected in the differences between C_{18} elutions under neutral and acidified conditions – at lower pH, the silanols are generally protonated; at neutral pH, at least some silanols are deprotonated, giving the cartridge the ability to selectively retain some cations. Indeed, observing the PCA results shows that compounds such as creatinine and TMAO are retained under neutral conditions – these compounds also being retained by the SCX cartridge under the appropriate conditions.

The final elutions to consider were those afforded from methods utilising phenylboronic acid (PBA) cartridges. PBA cartridges utilise a unique covalent

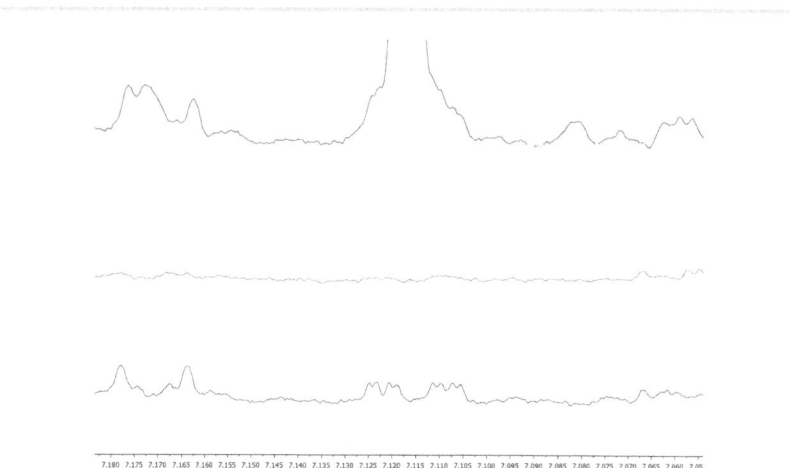

Fig. 8 Unknown N (ddd), normally obscured by 3-methylhistidine (top), is revealed after HLB SPE treatment under acidic conditions (bottom), but not under neutral conditions (middle).

bonding mechanism in order to selectively retain diols, α-hydroxy ketones, or any other functional groups where two unsubstituted heteroatoms are separated by at least one carbon.[22,23] It is not clear whether the diols must adopt a specific isomerism for retention to occur: mannitol is heavily retained in the elutions, but contains both R and S carbon centres, as well as terminal hydroxyls which can rotate to become a given conformer – its retention hence does not give additional insight. Other compounds retained include acetate, acetamide, and N-methyl-nicotinamide, a metabolite of niacin. The presence of adjacent heteroatoms does not guarantee good retention: for example, citric acid is poorly retained, despite having three carboxylate groups. There is also some retention of dimethylamine in both artificial and natural urine samples, despite it not being a diol – however, it could hypothetically be retained through a single substitution of water at the boronate, rather than through a double-substitution, as is normally the case.

Conclusions

We have demonstrated and compared the use of different SPE methods for the retention of different compound classes within the human urinary metabolome. Different retention profiles can give unique insight into the metabolome by revealing metabolite peaks in NMR spectroscopy that had previously been obscured by peak overlap – SPE can hence be used to either remove the suppressing metabolite(s) or to isolate the unknown metabolite itself. These retention profiles can be differentiated based on their retention of not only individual metabolites, but of broader subclasses of compounds united by their structure commonalities, including shared functional groups. Hence, different methods can be utilised in order to give greater control over the annotation process. On top of the metabolites identified by comparison to metabolite databases and available literature, several unknowns were annotated in the elutions – further experiments and comparisons to an authentic chemical reference will be required to identify them to a level 1 standard.[16] There also exists the possibility to study washes from the SPE experiments further, as well as to use multiple SPE methods in series, in order to narrow down a set of retained compounds even further. Finally, it may be possible to transfer these methods to an automated SPE system for more high-throughput analysis. All of these will hopefully broaden our understanding of the human urinary metabolome, and hence our understanding of biomarkers and the disease states and phenotypes that they represent.

Conflicts of interest

The authors register no conflicts of interest.

Acknowledgements

Author Dan McGill is supported by the Bruker Corporation and the Stratified Medicine Graduate Training Programme in Systems Medicine and Spectroscopic Profiling (STRATiGRAD). The authors would like to thank Rose Tolson (Imperial College London) for running the polyethylene glycol screening and Gordon Haggart (Imperial College London) for his assistance with troubleshooting the

IMPaCTS toolbox. Elena Chekmeneva was supported by the NIHR Imperial Biomedical Research Centre.

Notes and references

1 P. Elliott, *et al.*, Urinary metabolic signatures of human adiposity, *Sci. Transl. Med.*, 2015, **7**, 285ra62.

2 J. K. Nicholson and I. D. Wilson, High resolution proton magnetic resonance spectroscopy of biological fluids, *Prog. Nucl. Magn. Reson. Spectrosc.*, 1989, **21**, 449–501.

3 S. L. Robinette, J. C. Lindon and J. K. Nicholson, Statistical Spectroscopic Tools for Biomarker Discovery and Systems Medicine, *Anal. Chem.*, 2013, **85**, 5297–5303.

4 J. M. Posma, *et al.*, Integrated Analytical and Statistical Two-Dimensional Spectroscopy Strategy for Metabolite Identification: Application to Dietary Biomarkers, *Anal. Chem.*, 2017, **89**, 3300–3309.

5 K. Bingol and R. Brüschweiler, NMR/MS Translator for the Enhanced Simultaneous Analysis of Metabolomics Mixtures by NMR Spectroscopy and Mass Spectrometry: Application to Human Urine, *J. Proteome Res.*, 2015, **14**, 2642–2648.

6 S. Bouatra, *et al.*, The Human Urine Metabolome, *PLoS One*, 2013, **8**, e73076.

7 L. W. Sumner, *et al.*, Proposed minimum reporting standards for chemical analysis Chemical Analysis Working Group (CAWG) Metabolomics Standards Initiative (MSI), *Metabolomics*, 2007, **3**, 211–221.

8 I. D. Wilson and J. K. Nicholson, Solid phase extraction chromatography and NMR spectroscopy (SPEC-NMR) for the rapid identification of drug metabolites in urine, *J. Pharm. Biomed. Anal.*, 1988, **6**, 151–165.

9 W. Yang, Y. Wang, Q. Zhou and H. Tang, Analysis of human urine metabolites using SPE and NMR spectroscopy, *Sci. China, Ser. B: Chem.*, 2008, **51**, 218–225.

10 D. M. Jacobs, L. Spiesser, M. Garnier, N. De Roo, F. Van Dorsten, B. Hollebrands, E. Van Velzen, R. Draijer and J. Van Duynhoven, SPE-NMR metabolite sub-profiling of urine, *Anal. Bioanal. Chem.*, 2012, **404**, 2349–2361.

11 A. C. Dona, *et al.*, Precision High-Throughput Proton NMR Spectroscopy of Human Urine, Serum, and Plasma for Large-Scale Metabolic Phenotyping, *Anal. Chem.*, 2014, **86**, 9887–9894.

12 *Imperial Metabolic Profiling and Chemometrics Toolbox (IMPaCTS)*, available at: 10.5281/zenodo.803330.

13 F. Dieterle, A. Ross, G. Schlotterbeck and H. Senn, Probabilistic Quotient Normalization as Robust Method to Account for Dilution of Complex Biological Mixtures. Application in ^{1}H NMR Metabonomics, *Anal. Chem.*, 2016, **78**, 4281–4290.

14 The Human Urine Metabolome Database, available at: http://www.urinemetabolome.ca/.

15 P. G. Takis, H. Schäfer, M. Spraul and C. Luchinat, Deconvoluting interrelationships between concentrations and chemical shifts in urine provides a powerful analysis tool, *Nat. Commun.*, 2017, **8**, 1662.

16 L. W. Sumner, *et al.*, Proposed quantitative and alphanumeric metabolite identification metrics, *Metabolomics*, 2014, **10**, 1047–1049.

17 W. Barton, *et al.*, The microbiome of professional athletes differs from that of more sedentary subjects in composition and particularly at the functional metabolic level, *Gut*, 2017, **67**, gutjnl-2016-313627.

18 R. Marcé and F. Borrull, Solid-phase extraction of polycyclic aromatic compounds, *J. Chromatogr. A*, 2000, **885**, 273–290.

19 M.-C. Hennio, Solid-phase extraction: method development, sorbents, and coupling with liquid chromatography, *J. Chromatogr. A*, 1999, **856**, 3–54.

20 M. O. Sinnokrot, E. F. Valeev and C. D. Sherrill, Estimates of the *Ab Initio* Limit for π–π Interactions: The Benzene Dimer, *J. Am. Chem. Soc.*, 2002, **124**, 10887–10893.

21 R. H. Petrucci, F. G. Herring, J. D. Madura and C. Bissonnette, *General Chemistry: Principles and Modern Applications*, Prentice Hall, 2016.

22 M. Tugnait, F. Y. K. Ghauri, I. D. Wilson and J. K. Nicholson, NMR-monitored solid-phase extraction of phenolphthalein glucuronide on phenylboronic acid and C18 bonded phases, *J. Pharm. Biomed. Anal.*, 1991, **9**, 895–899.

23 P. Martin, B. Leadbetter and I. D. Wilson, Immobilized phenylboronic acids for the selective extraction of β-blocking drugs from aqueous solution and plasma, *J. Pharm. Biomed. Anal.*, 1993, **11**, 307–312.

Faraday Discussions

PAPER

Structural analysis of heavy oil fractions after hydrodenitrogenation by high-resolution tandem mass spectrometry and ion mobility spectrometry†

Johann Le Maître, [ID] abc Marie Hubert-Roux, [ID] ac Benoît Paupy,bc
Sabrina Marceau,bc Christopher P. Rüger, [ID] a Carlos Afonso [ID] *ac
and Pierre Giusti [ID] bc

Received 12th December 2018, Accepted 11th January 2019
DOI: 10.1039/c8fd00239h

Heavy petroleum fractions such as vacuum gas oils (VGOs) are structurally and compositionally highly complex mixtures. Nitrogen species, which have a significant impact on the subsequent refining processes, are generally removed by the hydrodenitrogenation (HDN) catalytic process. The purpose of this study was to identify and characterize compounds that are refractory to the HDN process. This may allow for the examination of the effectiveness of a vacuum distillate hydrotreatment catalytic bed in removing nitrogen-containing compounds before the cracking step. Three different VGO fractions of the same oil before and after HDN processes were analysed in ESI(+) mode by FTICR mass spectrometry and ion mobility spectrometry–mass spectrometry (IMS-MS), in particular compounds containing basic nitrogen, such as quinoline and isoquinoline. Ultra-high-resolution FTICR mass spectrometry provides a sufficiently high mass resolution power to resolve different compounds and attribute a unique molecular formula to each ion. Information on the isomeric content was obtained by use of tandem mass spectrometry (MS/MS) and IMS-MS. The evolution of the fragmentation of the N_1 class of compounds as a function of collision energy allowed for the identification of the molecular nucleus raw formula. From the IMS-MS experiments, it clearly appeared that, based on the IMS peak width, a lower isomeric dispersity was obtained after the HDN process and, based on the drift time and collision cross section determination, species presenting longer alkyl branches are the molecules most refractory to the HDN process.

Normandie Université, COBRA, UMR 6014 et FR 3038, Université de Rouen, INSA de Rouen, CNRS, IRCOF, Mont Saint Aignan Cedex, France. E-mail: carlos.afonso@univ-rouen.fr
bTOTAL Refining & Chemicals, Total Research & Technology Gonfreville, BP 27, 76700 Harfleur, France
cInternational Joint Laboratory – iC2MC: Complex Matrices Molecular Characterization, TRTG, BP 27, 76700 Hurfleur, France
† Electronic supplementary information (ESI) available. See DOI: 10.1039/c8fd00239h

Introduction

Heavy petroleum fractions such as vacuum gas oil (VGO) are some of the most complex mixtures in nature, containing tens of thousands of different organic compounds.[1,2] They consist of saturated and aromatic hydrocarbons, heteroatomic molecules containing mainly nitrogen, oxygen, and sulfur, and organometallic compounds, such as vanadyl and nickel porphyrins.[3-5] The presence of heteroelements drastically affects the activity of the conversion catalyst. Therefore, the removal of nitrogen, sulfur and metals is essential for the petroleum catalytic hydrocracking process. This is generally achieved by use of hydrodemetallation (HDM), hydrodesulfurization (HDS), and hydrodenitrogenation (HDN) steps with catalytic beds in the hydrotreatment process.[6-15] In particular, nitrogen containing species have a significant impact on the refining processes due to poisoning of the HDS catalysts.[14] Hydrodenitrogenation does not occur directly at unsaturated rings, meaning that unsaturated nitrogen-containing heterocycles must first be saturated before the carbon–nitrogen cleavage can occur.[14]

Refractory compounds that resist catalytic hydrodenitrogenation, under given process conditions, are targeted in this study. Knowledge of the molecular composition has become an important challenge in analytical development, particularly in the case of heavy petroleum fractions evaluation. Determining the quantity and structure of molecules containing heteroatoms is essential for these catalytic hydrotreating processes.[16] Due to the complexity of petroleum, ultrahigh-resolution Fourier transform ion cyclotron resonance mass spectrometry (FTICR MS) is an appropriate analytical method[17-19] affording exhaustive molecular formula attribution.[20,21] However, mass spectrometry alone cannot separate isomers, although differences in the isomeric content can yield different molecular properties. Another dimension of separation is required to obtain information on isomeric content. This can be afforded by chromatographic separation,[1] ion mobility spectrometry (IMS)[22] or tandem mass spectrometry (MS/MS).[23] For MS/MS, ions are first selected in the quadrupole and subsequently dissociated, e.g. by collision-induced dissociation (CID) in a collision cell,[23-26] before their detection. It has been demonstrated that weak σ-bonds of alkyl chains connected to an aromatic core structure can be cleaved by CID.[27] Many compositional and structural questions are being investigated for heavy crude oil fractions, such as vacuum gas oil (VGO), bitumen and asphaltene by Kekäläinen et al.,[27] Wittrig et al.,[23] and Chacón-Patiño et al.,[28-31] and petroleum emulsion interfacial materials by Lalli et al.[24] Research on their chemical composition and physical structures includes examining the heteroatom distribution and the structural conformation by the single and multiple nucleus structures with respect to their composition. Two structural models exist concerning the nuclei of oil compounds: archipelago type (a set of nuclei close to each other and bridged by linkers), and island type (a single nucleus).[28,29] Among the different types of oil characterization diagrams issued from FTICR data, the most commonly used is the mapping representing the number of double bond equivalents as a function of the carbon number (DBE/C#) for a given class of heteroatoms.[32-34] These represent the "molecular fingerprint" of a given sample. This type of map makes it possible to visualize the thousands of elemental compositions revealed for a given class, and thus to identify the structural compositions of the various compounds in the oil samples. It is possible to differentiate structural families by studying the DBE/C# maps from the MS/MS spectra.[29,30,35,36]

It has been shown that ion mobility spectrometry coupled to mass spectrometry (IMS-MS) is a valuable tool for the characterization of complex mixtures such as petroleum products. Besides the partial separation of isomers, it gives access to the collision cross section (CCS) that is a descriptor of the ion structure.[37–40] In practice, owing to the very high number of isomers present for one particular molecular formula, one does not expect to obtain separation of the isomeric species and generally a broad unresolved signal is seen. The measured drift time of this signal is related to the average ion CCS whereas the peak width is related to the isomeric content.[37] As in the case of structural determination by CID, many studies are being conducted on the use of IMS-MS, such as for the characterization of crude oil.[22]

The ESI source allows selective ionization of basic or acidic compounds. In petroleum samples, this leads to the detection of compounds containing basic (e.g. quinolines)[15,41] or neutral nitrogen (e.g. carbazole)[42,43] in positive and negative ion mode, respectively, and, thus, the overall observed complexity is reduced.[44] The discriminative nature of electrospray ionization means the nitrogen-containing ions are found to be the most intense signals in both positive and negative ion detection mode. The objective of this work is to characterize the basic nitrogen-containing compounds that are refractory to the catalytic processes. The feed and effluents from the HDN processes were analyzed by ultra-high-resolution FTICR and IMS-MS in positive mode.

Experimental

Sample preparation

Three samples (vacuum gas oil with 1000 ppm N and two effluents with 70 and 10 ppm N) were supplied by Total Research and Technology (Gonfreville, France). The samples were solubilized in toluene and further diluted in methanol/toluene (50/50 v/v) to a final concentration of 1 mg mL^{-1}. The diluted solution was spiked with 1% formic acid for the positive mode analyses.

Instrumentation

A hybrid quadrupole FTICR instrument (solariX XR, Bruker Daltonics, Bremen, Germany) equipped with a 12 T superconducting magnet was operated in the positive electrospray mode. Mass spectra were acquired over a mass range of m/z 147–1300 for 400 scans for the broadband experiments and 20 and 50 scans for the isolation and fragmentation spectra, respectively. The signal was digitalized with 8 M points, resulting in a transient length of 3.4 s. The accumulation time was set to 0.025 s, except for the fragmentation analyses with a 1 s accumulation period, both at a flow rate of 400 μL h^{-1}. The electrospray ionization conditions were set as follows: desolvation gas flow, 4 L min^{-1}; source temperature, 146 °C; source cone, 50 V; capillary voltage, −4500 V; nebulizer pressure, 0.5 bar; octopole energy, 350 V_{pp}; quadrupole lower cut-off, m/z 200; quadrupole collision energy, 1200 V_{pp}; TOF duration, 0.8 ms. A blank spectrum was recorded for 2 min prior to each sample being introduced.

A hybrid quadrupole time-of-flight mass spectrometer, which incorporates a traveling wave (T-Wave)-based mobility separation device (Synapt G2-Si HDMS, Waters Corp., Manchester, UK), was used to obtain ion mobility data. The instrument and the T-Wave device have been described in detail elsewhere.[38] Mass spectra were acquired in positive mode over m/z 50–1200 range for 20 min except for MS/MS (isolation/fragmentation) analyses, which were conducted for 10 min. The high-resolution mode (W reflectron) was used yielding a resolution of 40 000. All analyses were performed in triplicate to validate the repeatability of the experiments. The electrospray ionization conditions were set as follows: desolvation gas flow, 800 L h^{-1}; source temperature, 120 °C; desolvation temperature, 300 °C. The capillary voltage was set at 2.5 kV, the sampling cone at 50 V, and the extraction cone at 5 V. The IMS parameters were: IMS gas flow, 90 mL min^{-1} of N$_2$ (2.96 mbar of N$_2$ IMS cell pressure); IMS traveling wave height and velocity, 40 V and 800 m s^{-1}.

The IMS cell calibration was performed with a solution of D/L-polyalanine at a concentration of 10 ng μL^{-1} in MeCN/H$_2$O, as described by Smith et $al.$[45] The polyalanine CCS values used for the calibration were taken from the Pr. Clemmer CCS database.[46]

The mass spectrometers were externally m/z calibrated using sodium formate solution or sodium trifluoroacetate solution before the sample analyses. The instrument control and data acquisition were provided by DataAnalysis (version 4.4) and MassLynx (version 4.1) software for the FTICR and Synapt G2-Si HDMS, respectively. DriftScope (version 2.2) was used for the display and analysis of the ion mobility data, whereas OriginPro (version 2016) was used to process and visualize the data sets.

From the molecular formulae determined from the accurate mass measurements (typically <0.2 ppm), the number of double bond equivalent (DBE) values can be calculated utilizing eqn (1) (c: carbon number; h: hydrogen number; n: nitrogen number) for a molecular formula of $C_cH_hN_nO_oS_s$.[33,34] Given the resolving power of 0.9×10^6 at m/z 400 for the FTICR instrument, it is possible to separate the N$_1$ class ions from the N$_1$S$_1$ (mass split: 3.4 mDa) class compounds.

$$\text{DBE} = c - \frac{h}{2} + \frac{n}{2} + 1 \tag{1}$$

Results and discussion

Broadband FTMS analysis

The positive ion mode ESI FTICR MS spectra of the feed (VGO) and the two effluents from the HDN process containing 70 and 10 ppm nitrogen are presented Fig. S1.† The first observation is the significant decrease in the abundance and number of ions detected between the VGO and the effluents, as for the deeply denitrogenated sample, the chemical noise signals became predominant. As the positive ESI mode was used, mainly basic nitrogen-containing species were detected. Fig. 1 shows the DBE/C# maps of this N$_1$ family.

The most abundant compounds of the VGO N$_1$ family can be found between C28 and C32 and DBE values between 8 and 11. The predominant components of the N$_1$ class for the 70 ppm effluent are slightly lower and were observed between C24 and C35 and DBE values between 8 and 9. Finally, the prevailing nitrogen

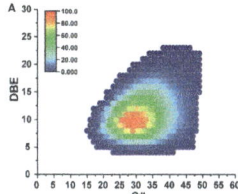 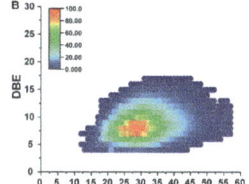 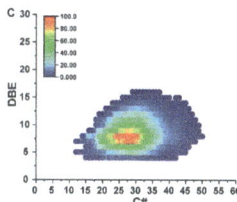

Fig. 1 N_1 class DBE/C# maps of the feed VGO (A), effluent 70 ppm (B) and effluent 10 ppm (C) obtained from the ESI(+) experiments by FTICR MS.

compounds detected in the 10 ppm effluent are between C19 and C30, for DBE values between 7 and 9. Thus, it appears that following the hydrotreating process, the most refractory nitrogen compounds detected in ESI(+) mode correspond to polyaromatic quinoline compounds. The change in the chemical pattern of the N_1 class between the VGO and the 70 ppm and 10 ppm effluents is relatively week and mainly involves the decline of high DBE molecules and a slight diminution of the average mass distribution. More significant differences between the samples may be found in the isomeric composition. For this purpose, MS/MS and IMS techniques can be considered.

Ultra-high-resolution tandem mass spectrometry

In order to obtain further structural information, tandem mass spectrometry (MS/MS) experiments were performed on selected class N_1 ions. The ion at m/z 378.31553 ($C_{27}H_{40}N^+$, DBE 9) was investigated as it is the most abundant refractory compound in the 70 and 10 ppm effluents for the DBE 9 series. The m/z selection is performed with a quadrupole using the smallest achievable window of 1 m/z unit, limited by the hardware itself. Thus, several isobaric ions are isolated within this selection window, belonging to N_1, N_1S_1, and N_1O_1 class constituents. The raw data extracted from the molecular attribution of the peaks of each spectrum have been filtered to focus only on class N_1 species. For the selected nominal masses investigated within this study, two isobaric compounds of class N_1 were detected. Consequently, when selecting the m/z 378 ions, the isobaric ions m/z 378.31553 ($C_{27}H_{40}N^+$, DBE 9) and m/z 378.22163 ($C_{28}H_{28}N^+$, DBE 16) are co-selected simultaneously (Fig. 2); fragmentation will, therefore, occur on these two compounds. In order to visualize these isolation and fragmentation data, DBE/C# maps were drawn (Fig. 2) similar to the approach for the broadband spectra. In this case, the fragment ions of each isobar were detected along two different horizontal lines at a DBE value corresponding to $n + 1$ with respect to the precursor ion, as shown in Fig. 3.[37,47]

Fig. S2† shows the MS/MS spectra of m/z 378 ions (the most abundant ion is m/z 378.31553 ($C_{27}H_{40}N^+$, DBE 9)) at different collision energies: 30 eV, 40 eV, 50 eV and 60 eV. Relatively high collision energies were required to generate fragment ions, which highlights the high stability of these ions. Fig. 3 shows the evolution of this fragmentation as a function of collision energy. A bimodal distribution of fragments is observed starting from the 40 eV spectrum, but more precisely on the 50 and 60 eV spectra, corresponding to the fragmentation pattern of both co-isolated ions m/z 378.31553 ($C_{27}H_{40}N^+$, DBE 9) and m/z 378.22166 ($C_{28}H_{28}N^+$, DBE 16). From the MS/MS experiments, fragment distribution can be used to

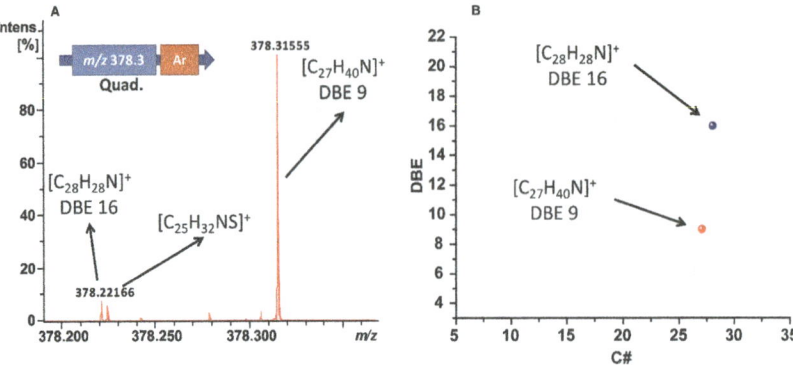

Fig. 2　(A) Expansion of the unit isolated window at *m/z* 378 showing the different isobaric ions that cannot be separated by the quadrupole, and (B) transfer of the isobaric information of class N_1 to a DBE/C# map.

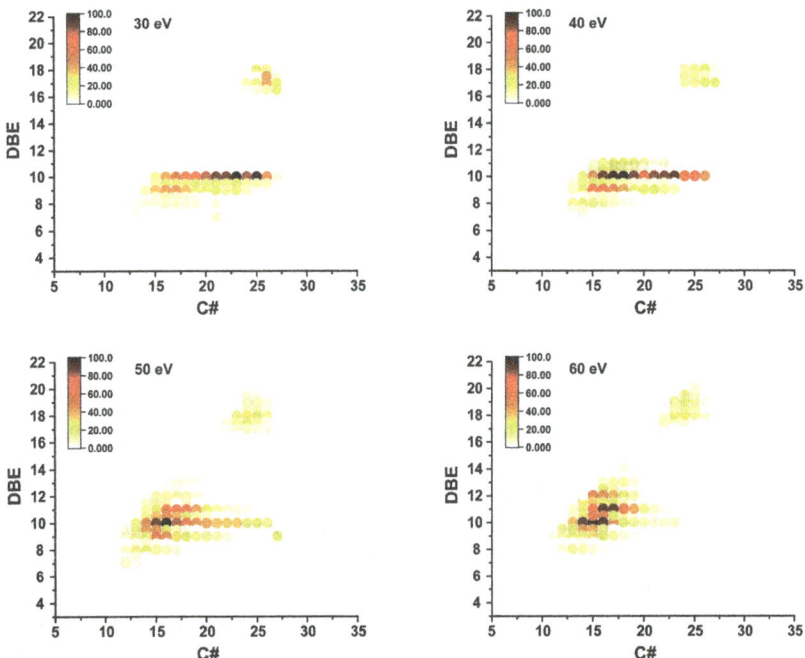

Fig. 3　DBE/C# maps from the MS/MS spectra of *m/z* 378.31553 ($C_{27}H_{40}N^+$, DBE 9) and *m/z* 378.22166 ($C_{28}H_{28}N^+$, DBE 16) as a function of the collision energy value (30–60 eV).

obtain structural information on oil molecules. The main limitation of this work is that the ions of the CID product cannot be directly correlated with the individual precursor ions.[26] It is therefore difficult to monitor the CID mechanisms and draw conclusions about the fragmentation patterns that change with the characteristics of the precursor ions.[23,48]

Fig. 3 visualizes the DBE/C# maps for the respective fragment patterns of the precursor ions *m/z* 378.31553 ($C_{27}H_{40}N^+$, DBE 9) and *m/z* 378.22166 ($C_{28}H_{28}N^+$,

DBE 16) at different collision energies from 30 eV to 60 eV. The main fragment ion series for both precursors correspond to losses of small alkane molecules that involve a hydrogen transfer during the fragmentation mechanism.[37,47] Consequently, this product ion series has DBE values of 10 and 17, involving an increase in the DBE value by one due to the displacement of the proton following the loss of the alkyl chain. It appears, therefore, that at a moderate collision energy value (30 eV), there was mainly a decrease in the carbon number (C#) with no change in the DBE. This is consistent with competitive fragmentation processes involving alkane losses through alkyl chain cleavages.[23] On the other hand, with a higher collision energy value (60 eV), a decrease in the DBE values was observed in addition to a decrease in the carbon number, which is probably related to consecutive processes involving ring opening reactions. The aromatic rings are very stable so this should be possible, in particular, with core structures featuring naphthenic motives. Retro-Diels-Alder processes may be involved among other fragmentation pathways, especially with tetralin core structures. These results imply that the losses in the DBE observed at higher collision energies are consecutive fragmentation processes taking place after the losses of the alkyl chains. This is possible only if we assume island type molecules with fused core structures surrounded by alkyl chains.[23,49–52]

The most intense fragment series from the two N_1 ions isolated in the quadrupole, at DBE values of 10 and 17, are plotted on a histogram graph according to their carbon number (Fig. 4). For a specific DBE value, the series of fragment ions follows the same trend: a monomodal distribution with a specific maximum abundance at a certain carbon number. This maximum corresponds to the most stable fragment before obtaining consecutive fragments with DBE losses. Therefore, it is likely that this point corresponds to the molecular nucleus that presumably lost all its alkyl chains. The diminution of the ion abundance at a lower C# can be rationalized by considering that once all the alkyl chains are lost, consecutive fragmentation takes place involving dissociation of the naphthenic rings on the core.

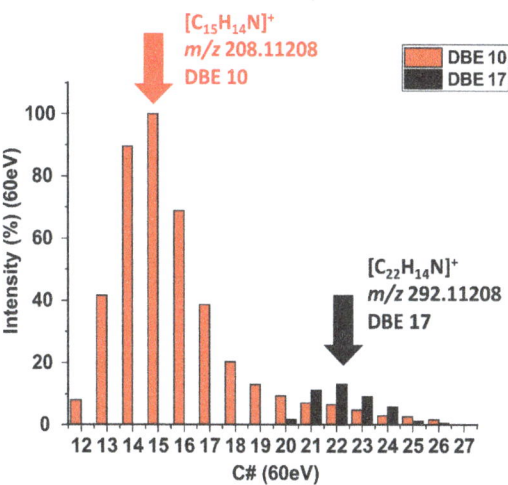

Fig. 4 Histogram of the fragment ions of $C_{27}H_{40}N^+$ and $C_{28}H_{28}N^+$ precursor ions as a function of their relative intensity with a collision energy value of 60 eV.

In this case, the major fragment ions, m/z 208.11208 ($C_{15}H_{14}N^+$, DBE 10) and m/z 292.11208 ($C_{22}H_{14}N^+$, DBE 17), should represent the dominant molecular nuclei of the isobar precursor ions, m/z 378.31553 ($C_{27}H_{40}N^+$, DBE 9) and m/z 378.22163 ($C_{28}H_{28}N^+$, DBE 16), respectively, previously isolated in the quadrupole.

The evolution of the fragmentation of the basic nitrogen protonated molecules revealed a class of compounds allowing the identification of the molecular nucleus of petroleum compounds, as well as an understanding of the fragmentation pathways (by the loss of the alkyl chain in a first step, and the opening and rearrangement of the nuclei thereafter).[26]

Subsequently, the structural determination of the molecular core of the m/z 378.31555 ion ($C_{27}H_{40}N^+$, DBE 9) was also performed on the 70 and 10 ppm effluents. This made it possible to monitor the evolution of the refractory compound after the nitrogen removal treatment, as shown in Fig. 5. Interestingly, the identified fragments are identical to those observed for the VGO, but not with the same proportions. The DBE/C# maps indicate that the most stable fragment found in the effluents is m/z 194.09643 ($C_{14}H_{12}N^+$, DBE 10). This could indicate a new type of molecular core resistant to the hydrodenitrogenation process. Fig. S3† gives hypothetical molecular nucleus structures from the raw formula determined by tandem mass spectrometry. The MS/MS experiments made it possible to determine the raw formula of the molecular nucleus and to tentatively identify the structures of these nuclei: "island" (central nucleus branched by alkyl chains) or "archipelago" (several central nuclei connected by an alkyl chain). In this case, the "island" structures were revealed predominantly. The types of archipelagos, if any, are only visible after the samples have been prepared to

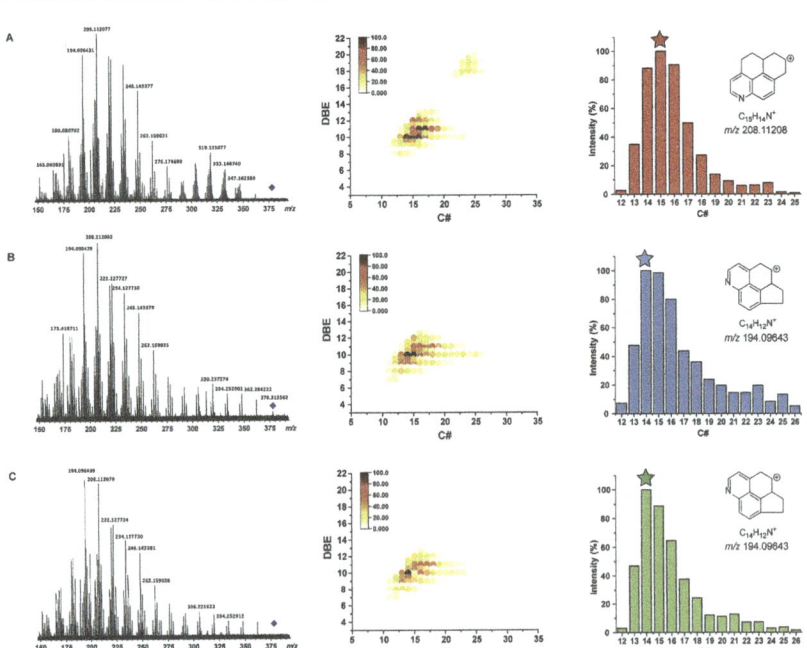

Fig. 5 Fragmentation patterns of the precursor ions m/z 378.31553 ($C_{27}H_{40}N^+$, DBE 9) and m/z 378.22166 ($C_{28}H_{28}N^+$, DBE 16) at a collision energy of 60 eV for (A) VGO, (B) effluent 70 ppm, and (C) effluent 10 ppm.

isolate them, because their ionization efficiency is much lower. The archipelago types form nanoaggregates much more than island types and are therefore not ionized.[28,29]

The addition of an ion mobility separation dimension to separate isomeric compounds/fragments from these complex mixtures would allow further identification of the different conformations and would reveal typical fragments of these molecular nuclei.[22]

Ion mobility spectrometry

The FTICR analysis using the MS/MS method enabled the identification of the molecular core structures based on the high mass resolving power and high mass accuracy. However, the mass spectrometry technique alone limits the understanding of isomers and does not allow one to draw conclusions on the comparison of refractory compounds between the feed VGO and the effluents.

This is where ion mobility spectrometry is beneficial as it allows the separation of isomers.[40,53-55] This technique can be applied on both precursor and product ions yielding various factors such as the full width at half maximum (FWHM), drift time (t_D) and CCS (collision cross section).[45,56] CCS represents the ability of an ion to be involved in collisions, so the more compact the conformation of an ion, the less likely it is to have collisions, and therefore the lower its CCS will be. Conversely, an unfolded conformation ion will have a higher CCS because it will have a greater collision probability. A compact conformation ion is, therefore, characterized by a low CCS value, and will present a lower drift time as it reaches the detector faster compared to a less compact ion.

The resolution of the ion mobility is far too low to separate all isomers and, thus, the FWHM can be used as a descriptor of the isomeric content of a particular molecular formula.[37] The larger the width at half height, the greater the isomeric dispersity. Another descriptor of the isomeric dispersity is the drift time of this ion mobility profile that is representative of the average CCS of the detected isomers.

The ion mobility spectra of the precursor ion at m/z 378.31553 ($C_{27}H_{40}N^+$, DBE 9) for the VGO before (A) and after mild (B) and deep (C) HDN processes are presented in Fig. 6. A significant change in the ion mobility peak profile is observed between the VGO and the deep HDN effluent as the latter presents a significantly higher drift time and lower FWHM. The effluent with 70 ppm (mild HDN) presents only a slight change in the signal profile with a lower amount of low drift time species. Interestingly, there is a significant increase in the drift time and CCS values for the 10 ppm refractory ion, indicating the presence of fewer compact compounds. Illustrated differently, Fig. S4† shows the evolution of the isobar ion mobility spectra at m/z 378 for the VGO before and after mild and deep HDN processes. This same trend is observed for other examples of ions of class N_1 – m/z 380.331177 ($C_{27}H_{42}N^+$, DBE 8) and m/z 382.346827 ($C_{27}H_{44}N^+$, DBE 7), illustrated in Fig. S5.† As the HDN process progresses, the FWHM of the precursor mobility peak decreases, indicating a decrease in the number of isomers. Furthermore, an increase in drift time is observed, reflected in the peak apex value, indicating the presence of compounds with higher CCS values.

Table 1 reports the results of the drift time and FWHM measurements after a Gaussian fit, as well as the CCS values calculated from the drift time values of

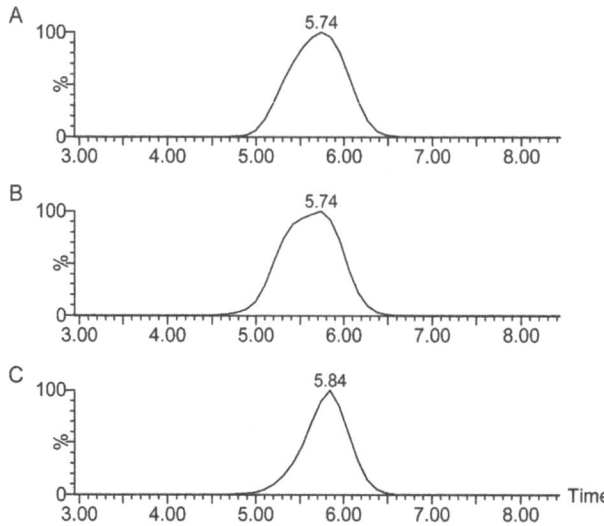

Fig. 6 Ion mobility spectra for m/z 378.31553 ($C_{27}H_{40}N^+$, DBE 9) selected in the quadrupole for (A) VGO, (B) effluent 70 ppm, and (C) effluent 10 ppm.

Table 1 Ion mobility descriptors of the refractory precursor ion isolated in the quadrupole

| | $C_{27}H_{40}N^+$ m/z 378.31553 | | |
	t_D (ms)	FWHM (ms)	CCS ($\mathring{A}^2$)
VGO	5.70 ± 0.01	0.63 ± 0.01	137.9 ± 0.3
Effluent 70 ppm	5.62 ± 0.02	0.66 ± 0.01	136.5 ± 0.3
Effluent 10 ppm	5.81 ± 0.02	0.49 ± 0.01	140.3 ± 0.1

the precursor refractory ions, selected in the quadrupole for the three samples. In each case, a decrease in the FWHM between the VGO and the 10 ppm effluent is observed. This indicates a decrease in isomeric variability after the HDN treatment.[37,57,58] On the other hand, there is an increase in the values of CCS and drift time for each refractory precursor. This may indicate that basic nitrogen molecules with larger conformations are more resistant to the hydrodenitrogenation processes. This can be due to the presence of different core structures or due to longer alkyl chains, as shown previously in the discussion of the FTMS results.[37]

In a second stage, we were interested in the ion mobility spectrometry study of the refractory core fragments previously determined *via* the structural study by CID on the FTICR instrument. The ion mobility spectra of the main fragment ion, m/z 208.11208 ($C_{15}H_{14}N^+$, DBE 10), resulting from the fragmentation of the precursor ion at m/z 378.31553 ($C_{27}H_{40}N^+$, DBE 9) for the VGO before and after mild and deep HDN processes are presented Fig. S6.† The refractory precursor in the VGO and effluents was fragmented in the collision cell (trap cell) located before the ion mobility cell. This operating mode makes it possible to study the ion mobility of the CID fragments.

Table 2 reports the results of the drift time and FWHM measurements after a Gaussian fit, as well as the CCS values calculated from the drift time values of

Table 2 Ion mobility descriptors of the fragment ion at m/z 208.11208 ($C_{15}H_{14}N^+$, DBE 10) from the refractory precursor ions fragmented in the trap cell

	$C_{15}H_{14}N^+$ m/z 208.11208		
	t_D (ms)	FWHM (ms)	CCS ($\mathring{A}^2$)
VGO	2.81 ± 0.02	0.17 ± 0.01	87 ± 1
Effluent 70 ppm	2.83 ± 0.01	0.17 ± 0.01	87.6 ± 0.2
Effluent 10 ppm	2.84 ± 0.02	0.17 ± 0.01	86 ± 1

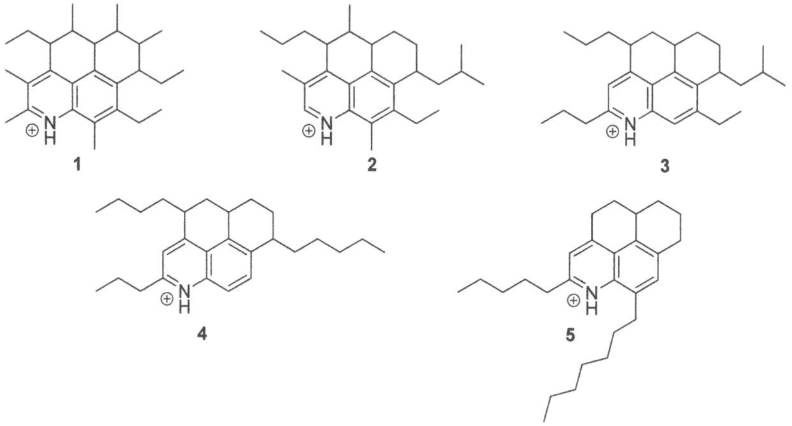

Fig. 7 Examples of compact (1–3) and branched (4–5) structures for m/z 378.31553 ($C_{27}H_{40}N^+$, DBE 9).

the fragment ion m/z 208.11208 ($C_{15}H_{14}N^+$, DBE 10), representing the molecular core of the refractory m/z 378.31553 ($C_{27}H_{40}N^+$, DBE 9), which has been studied earlier in this article. These calculations indicate that there is no significant difference in CCS for the fragment $C_{15}H_{14}N^+$ between the feed and its effluents. The compact fragment has a few conformational possibilities, which would explain the small variation in CCS between samples. Differences in CCS for the molecules are mainly due to a change in the organization of the alkyl chains (position and length). From these IMS data, it can be postulated that the high CCS value refractory molecules present long alkyl chains. Examples of structures are shown in Fig. 7. Such long chains may, depending on their position in the molecule relative to the nitrogen atom, prevent efficient access to the catalyst active site.

Conclusions

The aim of this study was to identify and characterize compounds refractory to hydrodenitrogenation through the use of ultra-high-resolution tandem mass spectrometry and ion mobility spectrometry. The broadband FTICR MS data allowed the determination of the basic compound family refractory to the HDN process, which mainly presents DBE values of 7, 8, and 9, and involves the partial removal of high DBE molecules. In fact, when comparing species with the same carbon number, those with a low DBE might exhibit a smaller molecular core and a larger number of carbon atoms as alkyl chains. From the tandem mass spectrometry experiments, a putative structure of the ion refractory core was proposed based on the evolution of alkane losses from the protonated precursor ion as a function of the collision energy. This simplification facilitated a better understanding of the correlation between precursor and fragment ions and allowed differentiation between single and multi-core structures. Here, the obtained data are consistent with island core structures. A significant change in the IMS profiles for the refractory precursor indicated that the isomers presenting high CCS values are more resistant to the HDN processes. This should correspond to species presenting long branches compared to ramified molecules that should present lower CCS values. The combination of IMS-MS and ultra-high-resolution mass spectrometry opens up interesting and promising prospects for studies into the structural determination of complex mixtures, in particular, compounds that are problematic in different refining processes. It would be interesting in the future to add the ion mobility spectrometry separation dimension to ultra-high-resolution mass spectrometry (TIMS-FTICR) to separate isomeric compounds from these resistant species, as well as to investigate more complex mixtures and other ionisation techniques.

Conflicts of interest

There are no conflicts to declare.

Acknowledgements

This work was supported by the European Regional Development Fund (ERDF) NHN0001343, the European Union's Horizon 2020 Research Infrastructures program (Grant Agreement 731077), the Région Normandie, and the Laboratoire d'Excellence (LabEx) SynOrg (ANR-11-LABX-0029). Financial support from the National FT-ICR network (FR 3624 CNRS) for conducting the research is also gratefully acknowledged.

Notes and references

1 R. P. Rodgers and A. M. McKenna, *Anal. Chem.*, 2011, **83**, 4665–4687.
2 A. G. Marshall and R. P. Rodgers, *Proc. Natl. Acad. Sci. U. S. A.*, 2008, **105**, 18090–18095.
3 C. A. Hughey, R. P. Rodgers and A. G. Marshall, *Anal. Chem.*, 2002, **74**, 4145–4149.
4 C. A. Hughey, C. L. Hendrickson, R. P. Rodgers and A. G. Marshall, *Energy Fuels*, 2001, **15**, 1186–1193.
5 C. A. Hughey, R. P. Rodgers, A. G. Marshall, C. C. Walters, K. Qian and P. Mankiewicz, *Org. Geochem.*, 2004, **35**, 863–880.
6 C. M. Celis-Cornejo, D. J. Pérez-Martínez, J. A. Orrego-Ruiz and V. G. Baldovino-Medrano, *Energy Fuels*, 2018, **32**, 8715–8726.
7 A. Miyata, S. Omi and M. Nagai, *J. Jpn. Pet. Inst.*, 1995, **38**, 251–257.
8 M. Lewandowski, *Appl. Catal., B*, 2015, **168–169**, 322–332.

9 P. Benigni, R. Marin and F. Fernandez-Lima, *Int. J. Ion Mobility Spectrom.*, 2015, **18**, 151–157.

10 H. Farag, M. Kishida and H. Al-Megren, *Appl. Catal., A*, 2014, **469**, 173–182.

11 S. Albersberger, J. Hein, M. W. Schreiber, S. Guerra, J. Han, O. Y. Gutiérrez and J. A. Lercher, *Catal. Today*, 2017, **297**, 344–355.

12 X. Fan, G.-F. Liu, Z.-M. Zong, X.-Y. Zhao, J.-P. Cao, B.-M. Li, W. Zhao and X.-Y. Wei, *Fuel Process. Technol.*, 2013, **106**, 661–665.

13 J. Ancheyta, *Deactivation of Heavy Oil Hydroprocessing Catalysts: Fundamentals and Modeling*, John Wiley & Sons, Inc., 2016DOI: 10.1002/9781118769638.

14 I. Mochida and K.-H. Choi, *J. Jpn. Pet. Inst.*, 2004, **47**, 145–163.

15 Q. Shi, C. Xu, S. Zhao, K. H. Chung, Y. Zhang and W. Gao, *Energy Fuels*, 2010, **24**, 563–569.

16 A. A. Al-Hajji, H. Muller and O. R. Koseoglu, *Oil Gas Sci. Technol.*, 2008, **63**, 115–128.

17 L. A. Stanford, S. Kim, R. P. Rodgers and A. G. Marshall, *Energy Fuels*, 2006, **20**, 1664–1673.

18 J. Maillard, N. Carrasco, I. Schmitz-Afonso, T. Gautier and C. Afonso, *Earth Planet. Sci. Lett.*, 2018, **495**, 185–191.

19 R. P. Rodgers and A. G. Marshall, *Petroleomics: Advanced Characterization of Petroleum-Derived Materials by Fourier Transform Ion Cyclotron Resonance Mass Spectrometry (FT-ICR MS)*, Springer, New York, 2007, 63–93DOI: 10.1007/0-387-68903-6_3.

20 A. G. Marshall and T. Chen, *Int. J. Mass Spectrom.*, 2015, **377**, 410–420.

21 J. M. Purcell, C. L. Hendrickson, R. P. Rodgers and A. G. Marshall, *Anal. Chem.*, 2006, **78**, 5906–5912.

22 F. A. Fernandez-Lima, C. Becker, A. M. McKenna, R. P. Rodgers, A. G. Marshall and D. H. Russell, *Anal. Chem.*, 2009, **81**, 9941–9947.

23 A. M. Wittrig, T. R. Fredriksen, K. Qian, A. C. Clingenpeel and M. R. Harper, *Energy Fuels*, 2017, **31**, 13338–13344.

24 P. M. Lalli, J. M. Jarvis, A. G. Marshall and R. P. Rodgers, *Energy Fuels*, 2017, **31**, 311–318.

25 J. Bergquist, G. Baykut, M. Bergquist, M. Witt, F. J. Mayer and D. Baykut, *Int. J. Genom. Proteonomics*, 2012, **2012**, 342659.

26 A. Vetere, W. Alachraf, S. K. Panda, J. T. Andersson and W. Schrader, *Rapid Commun. Mass Spectrom.*, 2018, **32**, 2141–2151.

27 T. Kekäläinen, J. M. H. Pakarinen, K. Wickström, V. V. Lobodin, A. M. McKenna and J. Jänis, *Energy Fuels*, 2013, **27**, 2002–2009.

28 M. L. Chacón-Patiño, S. M. Rowland and R. P. Rodgers, *Energy Fuels*, 2017, **31**, 13509–13518.

29 M. L. Chacón-Patiño, S. M. Rowland and R. P. Rodgers, *Energy Fuels*, 2018, **32**, 314–328.

30 M. L. Chacón-Patiño, S. M. Rowland and R. P. Rodgers, *ACS Symposium Series*, 2018, Vol. 1282, pp. 113–171.

31 M. L. Chacón-Patiño, C. Blanco-Tirado, J. A. Orrego-Ruiz, A. Gómez-Escudero and M. Y. Combariza, *Energy Fuels*, 2015, **29**, 6330–6341.

32 C. A. Hughey, C. L. Hendrickson, R. P. Rodgers, A. G. Marshall and K. Qian, *Anal. Chem.*, 2001, **73**, 4676–4681.

33 S. Kim, R. W. Kramer and P. G. Hatcher, *Anal. Chem.*, 2003, **75**, 5336–5344.

34 E.-K. Kim, M.-H. No, J.-S. Koh and S.-W. Kim, *Mass Spectrom. Lett.*, 2011, **2**, 41–44.

35 C. P. Rüger, C. Grimmer, M. Sklorz, A. Neumann, T. Streibel and R. Zimmermann, *Energy Fuels*, 2018, **32**, 2699–2711.

36 C. P. Rüger, A. Neumann, M. Sklorz, T. Schwemer and R. Zimmermann, *Energy Fuels*, 2017, **31**, 13144–13158.

37 M. Farenc, B. Paupy, S. Marceau, E. Riches, C. Afonso and P. Giusti, *J. Am. Soc. Mass Spectrom.*, 2017, **28**, 2476–2482.

38 D. N. Mortensen, A. C. Susa and E. R. Williams, *J. Am. Soc. Mass Spectrom.*, 2017, **28**, 1282–1292.

39 J. G. Forsythe, A. S. Petrov, C. A. Walker, S. J. Allen, J. S. Pellissier, M. F. Bush, N. V. Hud and F. M. Fernandez, *Analyst*, 2015, **140**, 6853–6861.

40 S. Hupin, H. Lavanant, S. Renaudineau, A. Proust, G. Izzet, M. Groessl and C. Afonso, *Rapid Commun. Mass Spectrom.*, 2018, **32**, 1703–1710.

41 X. Li, J. Zhu and B. Wu, *Bull. Korean Chem. Soc.*, 2014, **35**, 165–172.

42 K. Qian, W. K. Robbins, C. A. Hughey, H. J. Cooper, R. P. Rodgers and A. G. Marshall, *Energy Fuels*, 2001, **15**, 1505–1511.

43 Q. Shi, S. Zhao, Z. Xu, K. H. Chung, Y. Zhang and C. Xu, *Energy Fuels*, 2010, **24**, 4005–4011.

44 M. Farenc, Y. E. Corilo, P. M. Lalli, E. Riches, R. P. Rodgers, C. Afonso and P. Giusti, *Energy Fuels*, 2016, **30**, 8896–8903.

45 D. P. Smith, T. W. Knapman, I. Campuzano, R. W. Malham, J. T. Berryman, S. E. Radford and A. E. Ashcroft, *Eur. J. Mass Spectrom.*, 2009, **15**, 113–130.

46 http://www.indiana.edu/~clemmer/Research/CrossSectionDatabase/Peptides/polyaminoacid_cs.htm.

47 J. D. Ciupek, D. Zakett, R. G. Cooks and K. V. Wood, *Anal. Chem.*, 1982, **54**, 2215–2219.

48 C. S. Ling Liu, S. Tian, Q. Zhang, X. Cai, Y. Liu, Z. Liu and W. Wang, *Fuel*, 2018, 130.

49 K. Miyabayashi, Y. Naito, K. Tsujimoto and M. Miyake, *J. Jpn. Pet. Inst.*, 2004, **47**, 326–334.

50 Z. H. Linzhou Zhang, S. R. Horton, M. T. Klein, Q. Shi, S. Zhao and C. Xu, 2013.

51 D. J. Porter, P. M. Mayer and M. Fingas, *Energy Fuels*, 2004, **18**, 987–994.

52 L. Zhang, Y. Zhang, S. Zhao, C. Xu, K. H. Chung and Q. Shi, *Sci. China Chem.*, 2013, **56**, 874–882.

53 A. A. Shvartsburg and R. D. Smith, *Anal. Chem.*, 2008, **80**, 9689–9699.

54 K. Giles, J. P. Williams and I. Campuzano, *Rapid Commun. Mass Spectrom.*, 2011, **25**, 1559–1566.

55 A. M. Hamid, Y. M. Ibrahim, S. V. Garimella, I. K. Webb, L. Deng, T. C. Chen, G. A. Anderson, S. A. Prost, R. V. Norheim, A. V. Tolmachev and R. D. Smith, *Anal. Chem.*, 2015, **87**, 11301–11308.

56 C. Kune, J. Far and E. De Pauw, *Anal. Chem.*, 2016, **88**, 11639–11646.

57 A. L. Mendes Siqueira, M. Beaumesnil, M. Hubert-Roux, C. Loutelier-Bourhis, C. Afonso, Y. Bai, M. Courtiade and A. Racaud, *J. Am. Soc. Mass Spectrom.*, 2018, **29**, 1678–1687.

58 A. L. Mendes Siqueira, M. Beaumesnil, M. Hubert-Roux, C. Loutelier-Bourhis, C. Afonso, S. Pondaven, Y. Bai and A. Racaud, *Analyst*, 2018, **143**, 3934–3940.

Faraday Discussions

PAPER

Understanding the structural complexity of dissolved organic matter: isomeric diversity†

Dennys Leyva, [ID] [ab] Lilian V. Tose,[a] Jacob Porter,[a] Jeremy Wolff,[c] Rudolf Jaffé[b] and Francisco Fernandez-Lima [ID] *[ad]

Received 2nd December 2018, Accepted 4th February 2019

DOI: 10.1039/c8fd00221e

In the present work, the advantages of ESI-TIMS-FT-ICR MS to address the isomeric content of dissolved organic matter are studied. While the MS spectra allowed the observation of a high number of peaks (e.g., PAN-L: 5004 and PAN-S: 4660), over $4\times$ features were observed in the IMS-MS domain (e.g., PAN-L: 22 015 and PAN-S: 20 954). Assuming a total general formula of $C_x H_y N_{0-3} O_{0-19} S_{0-1}$, 3066 and 2830 chemical assignments were made in a single infusion experiment for PAN-L and PAN-S, respectively. Most of the identified chemical compounds ($\sim$80%) corresponded to highly conjugated oxygen compounds (O_1–O_{20}). ESI-TIMS-FT-ICR MS provided a lower estimate of the number of structural and conformational isomers (e.g., an average of 6–10 isomers per chemical formula were observed). Moreover, ESI-q-FT-ICR MS/MS at the level of nominal mass (i.e., 1 Da isolation) allowed for further estimation of the number of isomers based on unique fragmentation patterns and core fragments; the later suggested that multiple structural isomers could have very closely related CCS. These studies demonstrate the need for ultrahigh resolution TIMS mobility scan functions (e.g., $R = 200-500$) in addition to tandem MS/MS isolation strategies.

Introduction

Dissolved organic matter (DOM) is a highly complex mixture of organic compounds that is ubiquitous in aquatic ecosystems, resulting mainly from the degradation of aquatic and terrestrial primary producers.[1] It is mainly composed of carbon, hydrogen and oxygen, with the other elements being at relatively lower abundance. The biogeochemical functions of natural DOM are extremely important because of their influence on many environmental processes, including the fate and transport

[a]Department of Chemistry and Biochemistry, Florida International University, Miami, FL 33199, USA. E-mail: fernandf@fiu.edu

[b]Southeast Environmental Research Center, Florida International University, Miami, Florida 33199, USA

[c]Bruker Daltonics, Inc., Billerica, Massachusetts 01821, USA

[d]Biomolecular Sciences Institute, Florida International University, Miami, FL 33199, USA

† Electronic supplementary information (ESI) available. See DOI: 10.1039/c8fd00221e

of contaminants, ecological processes and water treatment.[1] Despite the important role of DOM in global carbon cycling, and while tens of thousands of molecular formulas have been reported in DOM,[2,3] and many structural features identified,[4] the molecular structures of most components in this complex mixture remain largely unknown.[5] This is primarily due to the fact that DOM components are highly variable in volatility, polarity, molecular structure, functionality and elemental composition, leading to serious challenges in their separation and identification.[3] However, the combination of multiple analytical approaches[2,6] and the utilization of advanced analytical techniques have moved this field forward. In particular, Fourier Transform Ion Cyclotron Resonance Mass Spectrometry (FT-ICR MS) and Quadrupole Time-of-Flight Mass Spectrometry (Q-TOF-MS) have aided much in the characterization of DOM due to their high resolution capabilities and flexibility toward coupling with separation techniques. While FT-ICR MS has been widely and successfully used to assess the molecular composition of DOM, limitations with regards to isomer characterization, an important aspect of DOM complexity, still remain. A recent report focused on characterizing DOM complexity and composition in a highly variable set of DOM samples using FT-ICR MS in combination with advanced statistical methods,[7] confirmed the notion that a significant component of DOM seems to be molecularly indistinguishable between samples and is thus ubiquitous in the environment.[8] Not only the co-occurrence of thousands of identical molecular formulae, but also, a remarkable similarity of fragment ion intensities among samples, and thus molecular structure commonalities, were reported.[7] Using a modeling approach, the authors estimated the isomers associated with the large number of identified molecular formulas. However, constraining isomerization aspects in DOM characterization continues to be challenging, and such information might be most accurately achieved by Ion Mobility Spectrometry (IMS) in tandem with mass spectrometry.[9]

During the last few decades, several attempts have been made to utilize IMS in tandem with mass spectrometry for the analysis of complex mixtures.[9] A common trend is towards the possibility to separate chemical classes by their IMS-MS trend lines, measurement of ion-neutral collision cross sections, shorter analysis time, easy coupling to other separation techniques (e.g., gas and liquid chromatography), increased peak capacity and reduction of the chemical noise. With the advent of high resolution mobility analyzers ($R > 80$), there is a natural push for their integration with high resolution mass analyzers for the analysis of complex mixtures.[10-19] Our team pioneered the integration of trapped IMS (TIMS) with FT-ICR MS in 2015,[20] and several reports have shown the unique advantages of TIMS-FT-ICR MS.[9,21-27]

In the present work, we discuss the advantages and current challenges during ESI-TIMS-FT-ICR MS/MS analysis of complex mixtures. The goal is to address the analytical advantages of ESI-TIMS-FT-ICR MS and ESI-q-FT-ICR MS/MS for two freshwater DOM samples in assessing their isomeric diversity and future challenges provided from MS/MS experiments at nominal mass.

Experimental

Sample preparation

Surface water was collected from Pantanal (PAN) National Park – SE Brazil, one of the largest subtropical and biodiverse freshwater wetlands in the world. The PAN samples were collected from the Paraguay River (PAN-L) and a wetland channel in Pantanal National Park (PAN-S). For further details on sampling and sample preparation, see ref. 2. The DOM and the individual standards were dissolved in 50 : 50 v/v methanol/water to a final concentration of 1 ppm. Prior to analysis, all samples were spiked with 5% (v/v) of the Tuning Mix calibration standard. All solvents used were of Optima LC-MS grade or better, obtained from Fisher Scientific (Pittsburgh, PA).

Sample ionization

An electrospray ionization (ESI) source based on the Apollo II ESI design (Bruker Daltonics, Inc., MA) was used in negative ion mode for all experiments. Sample solutions were introduced into the nebulizer at a rate of 360 μL h^{-1} using a syringe pump. Typical operating conditions were 3000–3500 V capillary voltage, 10 L min^{-1} dry gas flow rate, 1.0 bar nebulizer gas pressure, and a dry gas temperature of 180 °C.

Trapped ion mobility spectrometry analysis

The concept behind TIMS is the use of an electric field to hold ions stationary against a moving gas, so that the drag force is compensated by the electric field and ion packages are separated across the TIMS analyzer axis based on their mobility.[28-30] During mobility separation, a quadrupolar field confines the ions in the radial direction to increase trapping efficiency. The mobility, K, of an ion in a TIMS cell is described by:

$$K = \frac{v_g}{E} \cong \frac{A}{(V_{elution} - V_{out})} \tag{1}$$

where v_g, E, $V_{elution}$ and V_{out} are the velocity of the gas, applied electric field, elution voltage and tunnel out voltage, respectively. Mobility spectra were calibrated using a Tuning Mix calibration standard (Tunemix, G2421A, Agilent Technologies, Santa Clara, CA) with the following reduced mobility (K_0) values: m/z 301 $K_0 = 1.909$, m/z 601 $K_0 = 1.187$, m/z 1033 $K_0 = 0.776$, and m/z 1333 $K_0 = 0.710$ cm^2 V^{-1} s^{-1}.[31,32]

The mobility values (K) can be correlated with the ion-neutral collision cross section (Ω, Å^2) using the Mason–Schamp equation:

$$\Omega = \frac{(18\pi)^{1/2}}{16} \frac{z}{(k_B T)^{1/2}} \left(\frac{1}{m_I} + \frac{1}{m_b}\right)^{1/2} \frac{1}{K} \frac{760}{P} \frac{T}{273.15} \frac{1}{N^*} \tag{2}$$

where z is the charge of the ion, k_B is the Boltzmann constant, N^* is the number density, and m_I and m_b refer to the masses of the ion and bath gas, respectively.[33]

ESI-TIMS-FT-ICR MS/MS analysis

All experiments were performed on a custom built ESI-TIMS-q-FT-ICR MS 7T Solarix spectrometer equipped with an infinity ICR cell (Bruker Daltonics Inc., MA). The TIMS analyzer is controlled using in-house software, written in National Instruments Lab VIEW, and synchronized with the 7T Solarix FT-ICR MS acquisition program. TIMS separation was performed using nitrogen as a bath gas at ca. 300 K, $P_1 = 2.2$ and $P_2 = 0.9$ mbar, and a constant rf (2200 kHz and 140–160

Vpp). A nonlinear stepping scan function was used,[27] with a gate width of 3 ms. The TIMS cell was operated using a fill/trap/elute/quench sequence 9/3/9/3 ms, using an average of 1000 IMS scans per MS spectrum and a voltage difference across the ΔE gradient of 5.0 V. The ramp voltage gradient was stepped by 0.25 V per frame with a ΔV_{ramp} range of -160 to -60, for a total of 400 steps. The deflector (V_{def}), funnel entrance (V_{fun}), analyzer base (V_{out}) and gating lens (V_{gate}) voltages were $V_{def} = -180/180$ V, $V_{fun} = -90$ V, $V_{out} = -50$ V and $V_{gate} = -80$ V/ 80 V. TIMS-FT-ICR MS spectra were processed using sine-squared apodization followed by Fast Fourier Transform (FFT), in magnitude mode, resulting in an experimental MS resolving power of $R \sim 400\,000$ at m/z 400. ESI-q-FT-ICR MS/MS experiments were performed using quadrupole isolation at nominal mass and typical CID energies of 15–20 eV.

Data processing

The ESI-TIMS-FT-ICR MS spectra were externally calibrated for mass and mobility using the Agilent ESI-L mass calibration standard. The formulae calculations from the exact mass domain were performed using Composer software (Version 1.0.6, Sierra Analytics, CA) and confirmed with Data Analysis (Bruker Daltonics v 4.2) using formula limits of $C_x H_y N_{0-3} O_{0-19} S_{0-1}$, and odd and even electron configurations were allowed. The TIMS spectrum for each molecular formula was processed using a custom-built Software Assisted Molecular Elucidation (SAME) package – a specifically designed 2D TIMS-MS data processing script written in Python v2.7.[34] SAME package utilizes noise removal, mean gap filling, "asymmetric least squares smoothing" base line correction, peak detection by continuous wavelet transform (CWT)-based peak detection algorithm (SciPy package), and Gaussian fitting with non-linear least squares functions (Levenberg–Marquardt algorithm). The SAME final outcome is [m/z; chemical formula; K; collision cross section (CCS)] for each TIMS-MS dataset. The 2D TIMS-MS contour plots were generated in Data Analysis (Version v. 5.1, Bruker Daltonics, CA) and all the other plots were generated using matplotlib and OriginPro 2016 (Originlab Co., MA). The MetFrag CL software was used for *in silico* determination of potential candidate structures using the PubChem database.[35]

Results and discussion

ESI-TIMS-FT-ICR MS analysis

The analysis of the PAN complex dissolved organic matter using ESI-TIMS-FT-ICR MS resulted in a single, broad trend line in the IMS-MS domain composed of singly charged species (Fig. 1). Inspection of the MS domain leads to the observation of a similar profile of a single, broad Gaussian distribution centered around m/z 400, regardless of the sample.

Closer inspection of the MS spectra allowed a comparison of the number of MS peaks (*e.g.*, PAN-L: 5004 and PAN-S: 4660), with the number of IMS-MS features (*e.g.*, PAN-L: 22 015 and PAN-S: 20 954). Assuming a total general formula of $C_x H_y N_{0-3} O_{0-19} S_{0-1}$, we found 3066 and 2830 for PAN-L and PAN-S compounds, respectively. Most of the identified chemical compounds ($\sim$80%) corresponded to highly conjugated oxygen compounds (O_1–O_{20}), in good agreement with previous

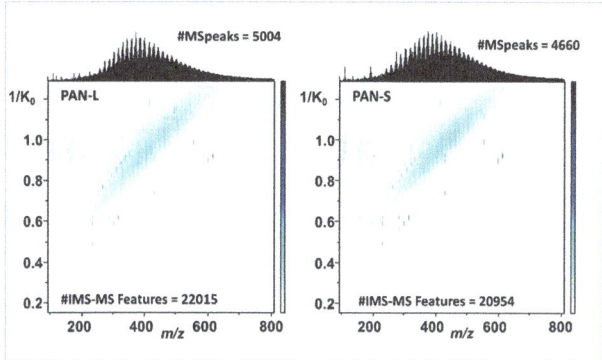

Fig. 1 Typical 2D-IMS-MS contour plots for the cases of the PAN-L and PAN-S complex dissolved organic matter.

reports.[36] This complexity can be visualized at the level of nominal mass (see example in Fig. 2) for 391 *m/z*.

ESI-q-FT-ICR MS/MS analysis

While a large isomeric diversity is observed at the level of nominal mass and per chemical formula, complementary information on the nature of the sample constituents can be obtained by performing ESI-q-FT-ICR MS/MS. At the level of nominal mass, several *m/z* signals are observed (*e.g.*, over 7 at 391 *m/z*). When subjected to Collision Induced Dissociation (CID), several common neutral losses are observed (see Fig. 3 and Tables S1–S3†).

If we assume that the neutral losses can be directly associated with functional groups and the overall structure of the parent ion, a number of potential structural isomers can be estimated for a given chemical formula; under this assumption, conformational isomers will present the same fragmentation

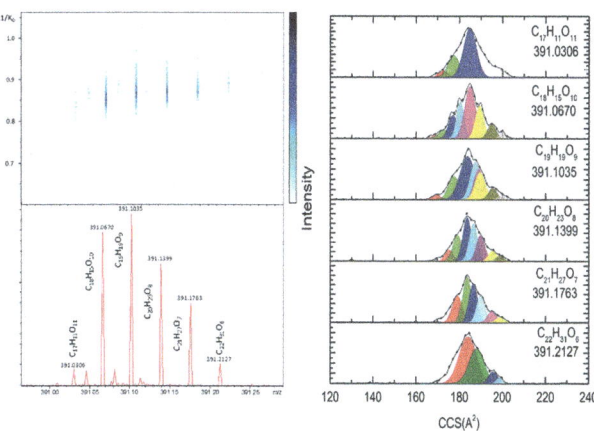

Fig. 2 Typical 2D-IMS-MS, as well the MS and IMS projections at nominal mass (*i.e.*, 391 *m/z*). Different bands are annotated in the IMS projections based on the SAME algorithm.

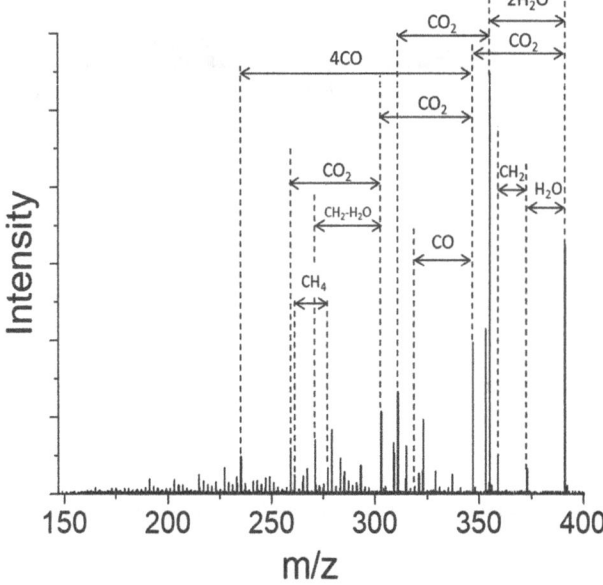

Fig. 3 A typical FT-ICR MS/MS spectrum from a 391 m/z precursor ion isolated at nominal mass and subjected to CID prior to injection in the ICR cell.

pathway and are not considered. For example, CO_2 can be associated with carboxyl groups and H_2O loss with the presence of hydroxyl groups. In addition, we observed the CO, CH_2, and CH_4 neutral losses (see Table S2† for all neutral loss fragments observed), in good agreement with previous FT-ICR MS/MS reports.[37] Taking advantage of the high mass accuracy of the FT-ICR MS measurements, neutral loss assignments can be easily identified. For example, the fragmentation pathways for 391.1031 m/z ($C_{19}H_{19}O_9$, Table S3†) were generated utilizing the fragmentation data obtained at nominal mass (Table S1†) and all possible combinations of neutral loss fragments (Table S2†) with a mass tolerance error of 1 mDa. Duplicate fragmentation pathways with the same syntaxes were eliminated (*e.g.*, $2CH_2$–$3CO$ is the same as $3CO$–$2CH_2$), since sequential fragmentation was not performed. Inspection of the fragmentation pathway shows a total of 16 end core fragments, each of them with multiple neutral loss pathways (see Table 1 and S3†). Since each pathway denotes the number and type of functional groups that were lost during fragmentation, the number of pathways could provide an upper estimate of the number of structural isomers. For instance, $3CO_2$–$2CO$–$2CH_4$ is one of the fragmentation pathways ending in the core formula $C_{12}H_{11}O$ (m/z 171.0814). That is, the parent ion ($C_{19}H_{19}O_9$, 391.1031) presumably experienced losses of three carboxylic groups, two carbonyl groups and two methane groups, suggesting that one isomer structure contains an arrangement of these functional groups. Conversely, for the same ending core formula ($C_{12}H_{11}O$), another fragmentation pathway involved consecutive losses of two hydroxyl groups, one carboxyl group, two methylene groups and four carbonyl groups ($2H_2O$–CO_2–$2CH_2$–$4CO$), indicating the presence of a different structural isomer.

Table 1 Core fragments and number of neutral loss pathways observed for 391.1031 m/z ($C_{19}H_{19}O_9$) during ESI-q-FT-ICR-MS/MS with isolation at nominal mass

Precursor ion m/z	Core fragment m/z	Structural isomers
391.1031 $C_{19}H_{19}O_9$	161.0607 $C_{10}H_9O_2$	13
	163.0763 $C_{10}H_{11}O_2$	7
	165.0192 $C_8H_5O_4$	3
	165.056 $C_9H_9O_3$	2
	167.0349 $C_8H_7O_4$	1
	171.0814 $C_{12}H_{11}O$	23
	173.0607 $C_{11}H_9O_2$	23
	175.0400 $C_{10}H_7O_3$	15
	183.0450 $C_{12}H_7O_2$	40
	183.0814 $C_{13}H_{11}O$	25
	185.0607 $C_{12}H_9O_2$	29
	187.0400 $C_{11}H_7O_3$	25
	201.0192 $C_{11}H_5O_4$	25
	202.9984 $C_{10}H_3O_5$	15
	205.0140 $C_{10}H_5O_5$	7
	241.0140 $C_{13}H_5O_5$	7

A parallel analysis performed using *in silico* fragmentation of 391.1031 m/z ($C_{19}H_{19}O_9$) with the MetFrag CL software across PubChem, that included the MS/MS information at nominal mass, resulted in 96 hits (see Fig. S1†). That is, 96 candidate structures were obtained based on the accurate mass of the precursor and fragment ions with 1 mDa mass tolerance.

While the ESI-q-FT-ICR MS/MS analysis with nominal mass quadrupole isolation is suggested as a rapid way to estimate an upper limit of the structural diversity and complexity of DOM, it is important to consider, that because the isolation was only performed at the level of nominal mass, potential over-estimation of the number of pathways is possible due to rearrangements of the fragments during CID. That is, interferences from fragments from other isobaric parent ions with similar chemical compositions (*i.e.*, $C_cH_hO_o$) may be a limitation in this approach (see Fig. 2). Nevertheless, the data summarized in Table 1 suggest the presence of up to 260 structural isomers. When compared to IMS data and MetFrag output, we can speculate that there are multiple structural isomers that share the same IMS band (only seven bands separated by the SAME algorithm).

Conclusions

In the present work, we illustrated the advantages of ESI-TIMS-FT-ICR MS/MS to address the isomeric content of DOM. The MS analysis permitted the identification of chemical components based on mass accuracy. When complemented with IMS measurements, an estimate of structural and conformational isomers can be obtained (*e.g.*, an average of 6–10 isomers were observed). While the MS spectra allowed the observation of a large number of peaks (*e.g.*, PAN-L: 5004 and PAN-S: 4660), over 4× features were observed in the IMS-MS domain (*e.g.*, PAN-L: 22 015 and PAN-S: 20 954). Assuming a total general formula of $C_xH_yN_{0-3}O_{0-19}S_{0-1}$, 3066 and 2830 for PAN-L and PAN-S chemical assignments were found in a single infusion experiment, respectively. Most of the identified chemical compounds ($\sim$80%) corresponded to highly conjugated oxygen compounds (O_1–O_{20}). Moreover, when ESI-q-FT-ICR MS/MS is performed at the level of nominal mass, further estimation of the number of structural isomers is possible based on unique neutral loss fragmentation patterns and core fragments. The data provided shows that multiple structural isomers could have very closely related CCS, which will demand the use of ultrahigh resolution TIMS mobility scan functions in tandem with MS/MS. Future studies can further push the analytical boundaries of ESI-TIMS-FT-ICR MS by mobility selective ESI-TIMS-FT-ICR MS/MS, and applying the correlated harmonic excitation field (CHEF)[37] on the quadrupole 1 Da isolated parent ions.

Conflicts of interest

There are no conflicts to declare.

Acknowledgements

This work was supported by the National Science Foundation Division of Chemistry, under CAREER award CHE-1654274, with co-funding from the Division of Molecular and Cellular Biosciences to FFL. DL acknowledges the fellowship provided by the National Science Foundation award (HRD-1547798) to Florida International University as part of the Centers for Research Excellence in Science and Technology (CREST) Program. This is contribution number 906 from the Southeast Environmental Research Center in the Institute of Water & Environment at Florida International University and a contribution from the Florida Coastal Everglades LTER. The authors would like to acknowledge the Advance Mass Spectrometry Facility at Florida International University and CAPES (process 88881.135156/2016-01) for their support. RJ acknowledges the George Barley Endowment in support of this research.

Notes and references

1 L. A. Kaplan and R. M. Cory, in *Stream Ecosystems in a Changing Climate*, ed. J. J. a. E. Stanley, Academic Press, 2016, pp. 241–320.
2 N. Hertkorn, M. Harir, B. Koch, B. Michalke and P. Schmitt-Kopplin, *Biogeosciences*, 2013, **10**, 1583–1624.
3 M. Zark, J. Christoffers and T. Dittmar, *Mar. Chem.*, 2017, **191**, 9–15.
4 N. Hertkorn, M. Harir, K. M. Cawley, P. Schmitt-Kopplin and R. Jaffé, 2016.
5 T. Dittmar and A. Stubbins, *Treatise on Geochemistry*, Elsevier, Oxford, 2nd edn, 2014, pp. 125–156.
6 R. Jaffé, Y. Yamashita, N. Maie, W. Cooper, T. Dittmar, W. Dodds, J. Jones, T. Myoshi, J. Ortiz-Zayas and D. Podgorski, *Geochim. Cosmochim. Acta*, 2012, **94**, 95–108.
7 M. Zark and T. Dittmar, *Nat. Commun.*, 2018, **9**, 3178.
8 P. E. Rossel, A. V. Vähätalo, M. Witt and T. Dittmar, *Org. Geochem.*, 2013, **60**, 62–71.

9 L. V. Tose, P. Benigni, D. Leyva, A. Sundberg, C. E. Ramírez, M. E. Ridgeway, M. A. Park, W. Romão, R. Jaffé and F. Fernandez-Lima, *Rapid Commun. Mass Spectrom.*, 2018, **32**, 1287–1295.

10 E. W. Robinson and E. R. Williams, *J. Am. Soc. Mass Spectrom.*, 2005, **16**, 1427–1437.

11 E. W. Robinson, D. E. Garcia, R. D. Leib and E. R. Williams, *Anal. Chem.*, 2006, **78**, 2190–2198.

12 E. W. Robinson, R. D. Leib and E. R. Williams, *J. Am. Soc. Mass Spectrom.*, 2006, **17**, 1470–1480.

13 E. W. Robinson, R. E. Sellon and E. R. Williams, *Int. J. Mass Spectrom.*, 2007, **259**, 87–95.

14 J. Saba, E. Bonneil, C. Pomiès, K. Eng and P. Thibault, *J. Proteome Res.*, 2009, **8**, 3355–3366.

15 Y. Xuan, A. J. Creese, J. A. Horner and H. J. Cooper, *Rapid Commun. Mass Spectrom.*, 2009, **23**, 1963–1969.

16 G. Bridon, E. Bonneil, T. Muratore-Schroeder, O. Caron-Lizotte and P. Thibault, *J. Proteome Res.*, 2012, **11**, 927–940.

17 W. Schrader, Y. Xuan and A. Gaspar, *Eur. J. Mass Spectrom.*, 2014, **20**, 43–49.

18 F. A. Fernandez-Lima, C. Becker, A. M. McKenna, R. P. Rodgers, A. G. Marshall and D. H. Russell, *Anal. Chem.*, 2009, **81**, 9941–9947.

19 M. Fasciotti, P. M. Lalli, C. F. Klitzke, Y. E. Corilo, M. A. Pudenzi, R. C. L. Pereira, W. Bastos, R. J. Daroda and M. N. Eberlin, *Energy Fuels*, 2013, **27**, 7277–7286.

20 P. Benigni, C. J. Thompson, M. E. Ridgeway, M. A. Park and F. Fernandez-Lima, *Anal. Chem.*, 2015, **87**, 4321–4325.

21 P. Benigni and F. Fernandez-Lima, *Anal. Chem.*, 2016, **88**, 7404–7412.

22 Y. Pu, M. E. Ridgeway, R. S. Glaskin, M. A. Park, C. E. Costello and C. Lin, *Anal. Chem.*, 2016, **88**, 3440–3443.

23 M. E. Ridgeway, J. J. Wolff, J. A. Silveira, C. Lin, C. E. Costello and M. A. Park, *Int. J. Ion Mobility Spectrom.*, 2016, **19**, 77–85.

24 P. Benigni, K. Sandoval, C. J. Thompson, M. E. Ridgeway, M. A. Park, P. Gardinali and F. Fernandez-Lima, *Environ. Sci. Technol.*, 2017, **51**, 5978–5988.

25 P. Benigni, R. Marin, K. Sandoval, P. Gardinali and F. Fernandez-Lima, *J. Visualized Exp.*, 2017, **121**, e55352.

26 P. Benigni, C. Bravo, J. M. E. Quirke, J. D. DeBord, A. M. Mebel and F. Fernandez-Lima, *Energy Fuels*, 2016, **30**, 10341–10347.

27 P. Benigni, J. Porter, M. E. Ridgeway, M. A. Park and F. Fernandez-Lima, *Anal. Chem.*, 2018, **90**, 2446–2450.

28 D. R. Hernandez, J. D. DeBord, M. E. Ridgeway, D. A. Kaplan, M. A. Park and F. Fernandez-Lima, *Analyst*, 2014, **139**, 1913–1921.

29 F. A. Fernandez-Lima, D. A. Kaplan and M. A. Park, *Rev. Sci. Instrum.*, 2011, **82**, 126106.

30 F. Fernandez-Lima, D. Kaplan, J. Suetering and M. Park, *Int. J. Ion Mobility Spectrom.*, 2011, **14**, 93–98.

31 E. R. Schenk, M. E. Ridgeway, M. A. Park, F. Leng and F. Fernandez-Lima, *Anal. Chem.*, 2014, **86**, 1210–1214.

32 E. R. Schenk, V. Mendez, J. T. Landrum, M. E. Ridgeway, M. A. Park and F. Fernandez-Lima, *Anal. Chem.*, 2014, **86**, 2019–2024.

33 E. W. McDaniel and E. A. Mason, *Mobility and diffusion of ions in gases*, John Wiley and Sons, Inc., New York, 1973.

34 P. Benigni, K. Sandoval, C. J. Thompson, M. E. Ridgeway, M. A. Park, P. Gardinali and F. Fernandez-Lima, *Environ. Sci. Technol.*, 2017, **51**, 5978–5988.

35 C. Ruttkies, E. L. Schymanski, S. Wolf, J. Hollender and S. Neumann, *J. Cheminf.*, 2016, **8**, 3.

36 A. Stubbins, R. G. M. Spencer, H. M. Chen, P. G. Hatcher, K. Mopper, P. J. Hernes, V. L. Mwamba, A. M. Mangangu, J. N. Wabakanghanzi and J. Six, *Limnol. Oceanogr.*, 2010, **55**, 1467–1477.

37 M. Witt, J. Fuchser and B. P. Koch, *Anal. Chem.*, 2009, **81**, 2688–2694.

Faraday Discussions

PAPER

Automatised pharmacophoric deconvolution of plant extracts – application to *Cinchona* bark crude extract†

Laure Margueritte,‡[a] Laura Duciel,‡[bc] Mélanie Bourjot,[d] Catherine Vonthron-Sénécheau[a] and Marc-André Delsuc 🆔 *[c]

Received 17th December 2018, Accepted 24th January 2019

DOI: 10.1039/c8fd00242h

We present a development of the *"Plasmodesma"* dereplication method [Margueritte *et al.*, *Magn. Reson. Chem.*, 2018, **56**, 469]. This method is based on the automatic acquisition of a standard set of NMR experiments from a medium sized set of samples differing by their bioactivity. From this raw data, an analysis pipeline is run and the data is analysed by leveraging machine learning approaches in order to extract the spectral fingerprints of the active compounds. The optimal conditions for the analysis are determined and tested on two different systems, a synthetic sample where a single active molecule is to be isolated and characterized, and a complex bioactive matrix with synergetic interactions between the components. The method allows the identification of the active compounds and performs a pharmacophoric deconvolution. The program is freely available on the Internet, with an interactive visualisation of the statistical analysis, at https://plasmodesma.igbmc.science.

1 Introduction

During their evolution, plants have acquired various adaptive characteristics in order to survive and to grow in their environment. In particular, small-sized

[a]*Laboratoire d'Innovation Thérapeutique (LIT), UMR CNRS 7200, LabEx Medalis, Faculté de Pharmacie, Université de Strasbourg, Illkirch-Graffenstaden, France*

[b]*CASC4DE Le Lodge, 20, Avenue du Neuhof, 67100 Strasbourg, France*

[c]*Institut de Génétique et de Biologie Moléculaire et Cellulaire (IGBMC), INSERM U596, CNRS UMR 7104, Université de Strasbourg, Illkirch-Graffenstaden, France. E-mail: madelsuc@unistra.fr; Fax: +33 368 85 46 88; Tel: +33 368 85 47 18*

[d]*Institut Pluridisciplinaire Hubert Curien (IPHC), UMR CNRS 7178, Faculté de Pharmacie, Université de Strasbourg, Illkirch-Graffenstaden, France*

† Electronic supplementary information (ESI) available: The ^{1}H and ^{13}C chemical shifts of artemisinin, quinine, quinidine, cinchonine, and cinchonidine. Sensitivity/selectivity graph for all conditions in the artemisinin series. See DOI: 10.1039/c8fd00242h

‡ These authors contributed equally to this work.

molecules called "specialized metabolites" have an important role in the defence against biotic or abiotic pathogens or an attractive role to facilitate reproduction, and the pressure of evolution has generated the huge chemodiversity of these metabolites in nature.[1,2] These natural products (NPs) are often bioactive[3] and represent a non-negligible resource in the drug discovery field. Indeed, over the last thirty years, 33% of new approved medicines in the small molecule category have a natural origin and an additional 27% are synthesized molecules which mimic natural compounds. NPs provide, therefore, high chemodiversity and new chemical entities for the pharmaceutical industry.[4]

The major difficulty in the study of natural products is the complex matrix of extracts in which they are located. The identification of bioactive compounds in these natural matrices is classically performed by bioactivity-guided isolation. The bioactive crude extract is fractionated by chromatographic methods and all fractions are submitted to bioassays until isolation and purification of molecules. Then, these molecules are structurally characterized by NMR, HR-MS and chiroptical spectroscopic methods. This approach is laborious, slow and costly but still remains effective today. Its chance of success is reduced by a possible loss of activity during isolation, possibly due to sample degradation, or molecular synergy.[5]

In recent years, the use of analytical methods for metabolomic studies is increasing, in particular for dereplication, a process to quickly identify known compounds in a mixture.[6] In natural product research, the analysis of tandem mass spectrometry data by molecular networking is attracting a lot of interest, and access to this strategy is eased by the web platform of GNPS (Global Natural Products Social Molecular Networking). This platform is a collaborative community database of MS/MS spectra where users participate to its enrichment. This approach clusters molecules into molecular families based on their MS/MS spectral similarities. The identification and annotation of unknown metabolites is permitted by the propagation from structurally close known compounds in the database.[7,8]

The use of NMR spectroscopy for dereplication or new compound identification in complex mixtures has seen less developments because of the traditional poor sensitivity of the method.[9] Nowadays, this problem has been, for a large part, solved by improvements such as cryogenic probes, reduced sample volumes and increased magnetic fields,[10] and many applications are now developed.[11-13] In metabolomic studies, ^{1}H NMR experiments are usually preferred because of their high sensitivity,[14] however ^{13}C NMR experiments, while less sensitive, are also used in dereplication strategies because of their higher analytical power through the use of chemical databases.[15]

Besides clustering or networking approaches, the tentative use of machine learning technology has recently been developed.[16] Applied to NMR spectroscopy, it allows the early detection of known compounds in complex plant extracts.[17-19] The present work explores the use of machine learning applications for the molecular fingerprinting of active molecules from unrefined fractions of a crude extract.

Automatic large scale analysis of big datasets requires the development of machine learning algorithms which are based on models and multiple parameters.[20] These parameters are of great importance and can highly impact the results and performances of the program. These parameters are called hyperparameters of the algorithm and should be refined and optimized to get the better results in the given application.[21] Indeed, as in any scientific research, the chosen model is as important as the dataset on which the analysis is applied. To perform this optimization, some iterations have to be done manually and the results must be interpreted and evaluated by experts in the domain of the application. In some general cases, algorithms have been developed to help non-experts, either in computing or in the field, choosing an optimized set of parameters.[22] For example, this has been well developed for Neural Network learning, which is a machine learning technique wide-spread in the domain.[23]

We recently presented the automatic processing of a large series of 2D NMR spectra obtained from a set of samples, and the reduction of this large amount of data to tables amenable to statistical analysis.[18] These tables were further analysed using machine learning approaches, and it was shown that molecular information on the active compound can be efficiently extracted from this large dataset.[24]

In the present work, we explore and optimize the efficiency of this process with respect to the hyperparameters of the process, such as bucket size, screening parameters, or response curve (hyperparameters are parameters set prior to the analysis process, and determine the quality of the result). We then apply the optimized process to the study of *Cinchona pubescens* (Red* *Cinchona* tree) bark crude extracts and explore the complexity of the activity pattern of the different molecules sharing the quinine skeleton. Finally, we present a web server which implements this process and which can be freely accessed by end users who desire to try this approach.

2   Principles of the analysis

In the present work we extend the initial approach, nick-named "*Plasmodesma*", introduced recently[18] for automatic dereplication and active compound identification. The principle of the *Plasmodesma* analysis is based on the study of a series of related natural extracts presenting a variation of the measured biological activity. This series can be obtained either by a fractionation of some starting material, or from different extracts obtained from different parts of the plant, or from plants harvested at different times of the year.

Each sample is then analyzed using exactly the same set of standardized 1D and 2D NMR experiments, including COSY, TOCSY, DOSY or HSQC experiments. This large set of data is then automatically processed and a bucketing procedure is applied. Buckets are small spectral zones of constant size which cover the whole spectrum. For each bucket, quantities such as the mean intensity of the signal, the minimum and maximum values over the bucket, or the signal standard deviation are computed. Each of these quantities (to the exception maybe of the minimum) can be used as a proxy of the signal intensity in the zone for the statistical tests. This bucketing operation reduces the size of the dataset to analyse, allows a rich analysis thanks to the local statistical values, and protects against local small spectral variations which appear when comparing different samples.

Buckets retain the geometry of the spectra from which they are extracted. As a consequence, they can be displayed in a mode resembling NMR spectra, that a seasoned user will analyse as actual NMR spectra. For the same reason, standard spectral corrections, such as symmetrisation or t_1-noise removal, can readily be applied to the bucket list. This is used to apply the statistical analysis on curated bucket lists rather than on the one directly created by the processing step. It should be noted that for this analysis, peak lists are not considered; only the bucket lists are considered as data matrices, or pictures, and handled as spectra. Nevertheless, peak lists are computed by the processing pipeline, and can be used independently.

Two procedures are proposed. In a first step, a one-to-one sample comparison is performed, which outlines salient differences between the two selected samples. This procedure can be seen as a difference map, even though the applied mathematical procedure is different and more robust.

In a second approach, the complete series can be analysed at once, by detecting signal variations which correlate with the activity. This is allowed by the proportionality of the NMR signal to the concentration of the molecule, and assumes that the activity is directly related to the active molecule concentration. Details of the mathematical procedures are presented in the Materials & methods section. In both cases, the goal is to extract the spectrum of the compound of interest from the crowded NMR spectrum of the mixture. For this reason, the results are presented in a graphic interactive context, in a spectral format that resembles the spectrum of the compound, and this allows rapid recognition of a possible molecular fingerprint.

3   Results

3.1   Artemisinin series – optimisation of the processing parameters

The first step of the analysis is an automatic processing of the NMR data, starting from the raw FIDs, as obtained from the spectrometer, and performing Fourier analysis and data reduction to the bucket lists and peak lists. Starting from the raw data rather than from processed spectra ensures a constant format of the final data, and protects from subtle variations introduced by the user, which might introduce a bias in the experiment. However, automatic processing cannot be as precise and optimized as a manual one, and a lot of care was taken in the initial work to optimize this step.[18] Here, we have analysed the impact of the bucket shape and content to the final results.

For this, we analysed a data series obtained from a synthetic set of samples. A crude algal extract was supplemented with a small amount of the artemisinin molecule. The supplemented extract was then fractionated by flash chromatography to yield six fractions, and the antimalarial activity was evaluated for each fraction using the pLDH bioassay. Activities were measured from 0% to 98% of growth inhibition of *P. falciparum*.

This series closely resembles the case of a bioguided isolation task, usually used to identify bioactive compounds from a crude plant extract. The set of fractions was then dissolved in D-methanol and NMR experiments were carried out. A 1D ^{1}H spectrum, as well as COSY, TOCSY, HSQC and DOSY experiments, were performed for a total time of about 24 hours of acquisition, and the processing was performed as previously described.

For the processing, the various parameters of the Fourier transform had been optimized previously (zero-filling, denoising, baseline correction, *etc.*), and we rapidly checked that the set-up was optimal. We then decided to check the effect of the size of the buckets, which was varied systematically from 0.02 ppm to 0.1 ppm along the ^{1}H axis, and from 0.5 ppm to 2.0 ppm along the ^{13}C axis. These buckets are then stored as arrays for 1D experiments or matrices for 2D experiments.

For each set of experiments processed in this manner, the complete statistical analysis was performed, and the results were analysed. In this series, the result is known in advance, as the spectrum of the active compound is known. It is thus possible to compare the results of the analysis with the theoretical ones. So, for each bucket size and each experiment type (COSY, TOCSY and HSQC), using the standard deviation descriptor, we counted the number of correctly detected signals (True Positive), the number of peaks not detected (False Negative), and the number of wrongly detected peaks (False Positive). The values are presented in Fig. 1.

From these values it is possible to estimate the quality of the analysis by plotting, for each set of hyperparameters and all detected peaks, the ratio of True Positive peaks to the ratio of False Positive ones (Fig. 2). Such a curve can be seen as a partial Receiver Operating Characteristic (ROC) curve as the hyperparameters are only sampled on a few conditions and not for the full range of possible values. On such a curve, the further the points are from the diagonal, the better the analysis. The condition which has been selected for the rest of the study is marked with a star and corresponds to buckets of 0.04 ppm in ^{1}H and 1.0 ppm in ^{13}C.

Another possibility is to choose different bucket descriptors for the analysis. From the original proposition which was, for each bucket, the total area (A), the standard deviation of the signal (σ), the minimum and the maximum values (min, max), we added the number of isolated peaks within the bucket (N_{pk}), and the combinations: ($A \times \sigma, A/\sigma, \max - \min, \log(A), \log(\sigma), \log(\max - \min), N_{pk} \times \sigma, N_{pk} \times \log(\sigma)$). The two descriptors which present the best results for this series are σ and $\log(\sigma)$. The other descriptors tend to present poorer results, but eventually present good results in certain cases.

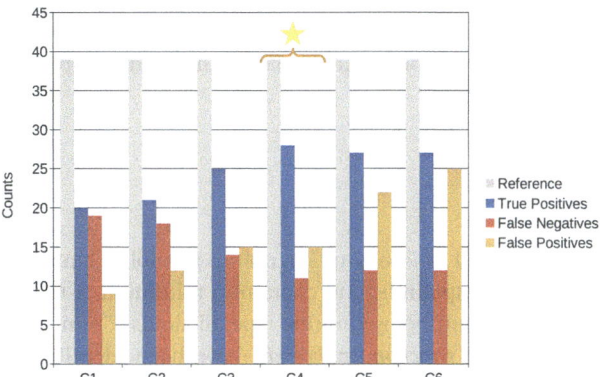

Fig. 1   Variation of the True Positive (artemisinin peaks detected), False Positive (peaks not from artemisinin but wrongly proposed) and False Negative peaks (artemisinin peaks not detected) obtained during the Recursive Feature Elimination (RFE) analysis of the artemisinin series, while varying the bucket size. Reference bars show the expected number of signals for artemisinin. The x-axis goes from smaller to larger buckets (see the Materials & methods section for details). The star corresponds to the parameters chosen for the further studies.

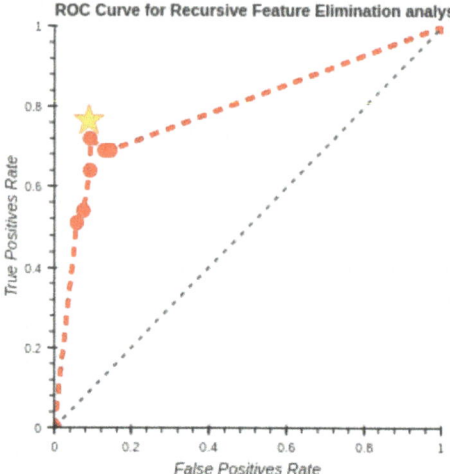

Fig. 2 Partial Receiver Operating Characteristic (ROC) curve obtained with the different bucket sizes used in Fig. 1; the vertical axis is the proportion of artemisinin peaks detected by the analysis, the horizontal axis is the proportion of peaks not from artemisinin but wrongly proposed by the analysis nevertheless. The further the points are from the diagonal, the better the quality of the analysis. The star corresponds to the parameters chosen for the further studies.

Finally, the effect of the cleaning step used to create the curated bucket lists has been investigated. This cleaning is applied to all descriptor matrices, and it consists of two sequential operations. First, for each descriptor matrix, the noise is evaluated and the bucket values below twice the noise level are zeroed in a hard thresholding operation. This operation is performed either all at once on the whole matrix (mild condition), or on each F1 column independently either with a hard thresholding (medium condition) or with a soft thresholding (harsh condition). Finally, homonuclear experiments are symmetrized by replacing each symmetric value with the smallest of both. The effect of these various processes is very dependent on the quality and nature of the spectra studied.

3.2 *Cinchona* series

The dried bark of the Red *Cinchona* tree is rich in phenolic compounds, organic acids and triterpenic compounds, and contains a minimum of 6% of total alkaloids, 30% to 60% of them being quinine type alkaloids.[25] Almost thirty of the alkaloids extracted from *Cinchona pubescens* have been identified and several show an *in vitro* antimalarial activity.[26,27] In addition, the bark contains phenolic compounds.[28] Due to the number of *Cinchona* alkaloids, their structural similarities and their antimalarial activities, the crude extract of *Cinchona* bark is a very complex case-study.

Six fractions have been obtained from a hydroalcoholic extract of dried bark of *Cinchona pubescens* containing the four main quinoleic alkaloids among many other alkaloids as well as tannins. The *in vitro* antimalarial activities of these

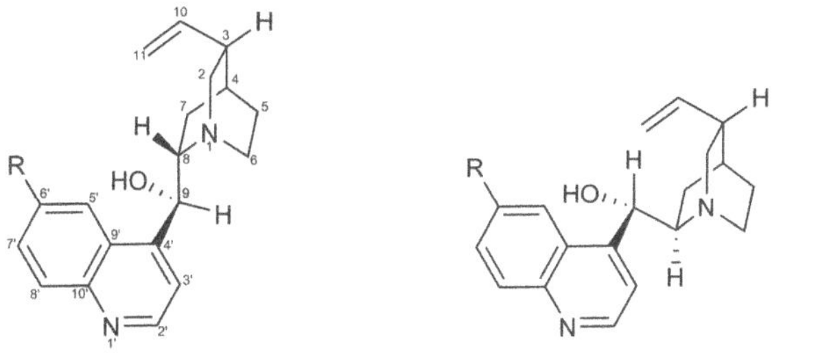

R = OCH₃ : (-)-quinine (8S, 9R)   R = OCH₃ : (+)-quinidine (8R, 9S)

R = H : (-)-cinchonidine (8S, 9R)   R = H : (+)-cinchonine (8R, 9S)

Fig. 3   Structures of the four major active compounds in *Cinchona pubescens*.

fractions and of reference molecules were measured. According to our activity tests, quinidine and cinchonine are the most active molecules (see Fig. 3).

All fractions show an *in vitro* antimalarial activity, and fractions 2 and 3 are the most bioactive.

TOCSY and HSQC NMR spectra were recorded for these samples. The aromatic zone of the spectra does not vary much among the reference compounds. However, the proton chemical shifts of the quinuclidine nucleus, as seen from the TOCSY experiments, significantly vary over the different molecules. These chemical shifts are important to identify the different *Cinchona* alkaloids.

Linear regression is applied between bucket intensities or standard deviations and the bioactivity. The observed spectral fingerprint is characteristic of a *Cinchona* alkaloid. From the TOCSY experiment, the proton chemical shifts of the quinuclidine nucleus give non-ambiguous information allowing the identification of the *Cinchona* alkaloids. When compared to reference spectra, the larger number of correlation peaks are observed to belong to the cinchonine and quinidine molecules, the most bioactive references.

3.3   *Plasmodesma* server

A program has been developed which provides the *Plasmodesma* service as a Web server. This approach has many advantages. First, the user does not need to download and install any particular software as the whole analysis takes place on the Internet, using a regular Internet browser. Then, the software can be easily improved, for instance by adding new functions, without any burden for the users. The server is freely available, and its output is considered public. Due to resource limitations, some usage limit may apply and the results are kept on the server only for a short period of time (typically 15 days).

The program provides most of the features described in the present publication, and is available at https://plasmodesma.igbmc.science. This server provides the service as two independent procedures. First, the user creates a zip archive of the set of NMR experiments to analyse, and uploads the data to the server, along with an E-mail address. No particular parameter is required from the user at this level, and care should only be taken to set-up the zip file correctly. This action launches the processing of the raw data contained in the zip archive, using the

Plasmodesma set-up, running on the top of the SPIKE library.[29] If present, DOSY datasets are processed using the PALMA program available in our laboratory[30] and are joined to the analysis. Once the processing is terminated, the results of the analysis (mostly bucket and peak lists, as well as some reports) are stored on the server disk, and an E-mail is sent to the user, which contains a link to a private page from which the analysis takes place in an interactive session.

When the user connects to the interactive session, the preprocessed datasets are loaded into the program according to the parameters provided by the user at that time (measured activity, type of experiment to look at, type of cleaning, *etc.*).

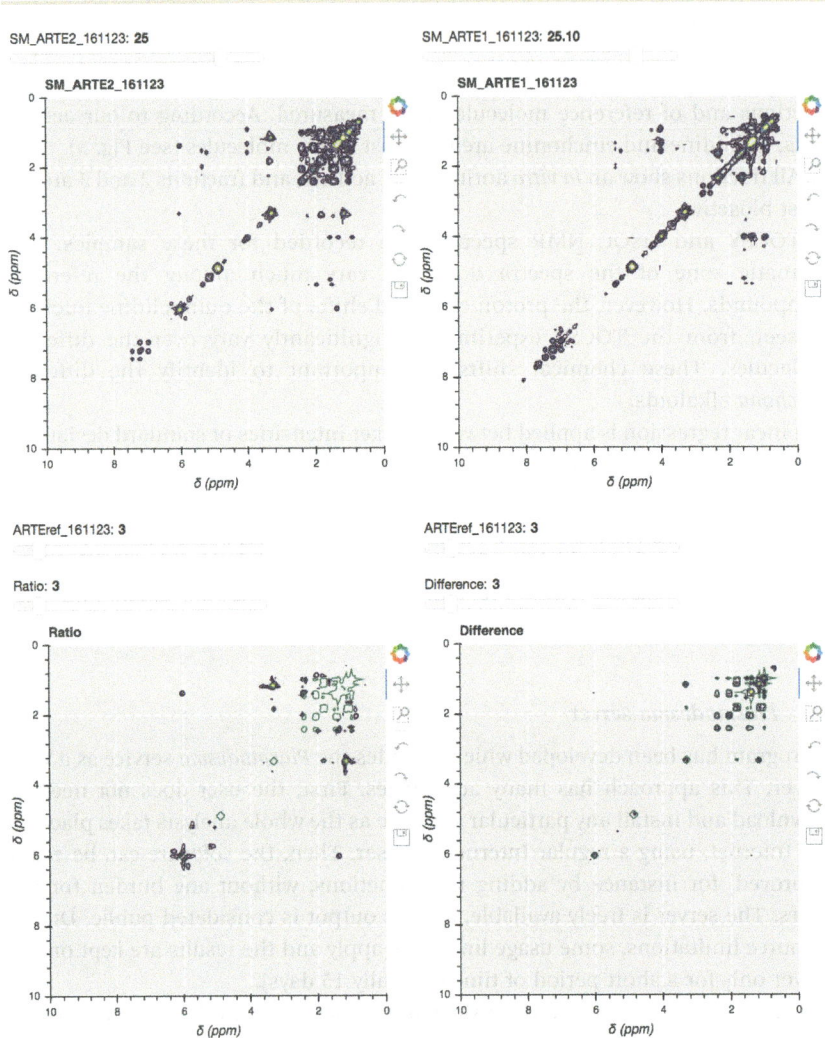

Fig. 4 Interactive display of the *Plasmodesma* program results; a comparison between two selected samples from the artemisinin series, using bucket standard deviations from TOCSY experiments. The top view presents the curated bucket lists of the two samples, the bottom view presents the ratio (left) and the difference (right) differential maps. In green, the reference map of the reference compound artemisinin is overlaid. Each image is interactively controlled with a slider, which allows the level of the contouring to be chosen.

With this data and from these parameters, a program is launched that interactively displays the results of the statistical analysis. Fig. 4 and 5 present an example of such an interactive session. Each panel on the figures is a graphic representation of a data matrix, either showing bucket lists, or the results of the statistical analysis.

In Fig. 4, the one-to-one analysis is shown. The top two images present the bucket lists of the two selected samples, usually one with no activity and the most active one, and the bottom panels present two differential analyses, from the ratio (left) or the difference (right) of the values. Each image is controlled with a slider, which allows the level of the contouring to be chosen. It can be seen on the bottom right image that most of the selected spectral channels fall correctly into

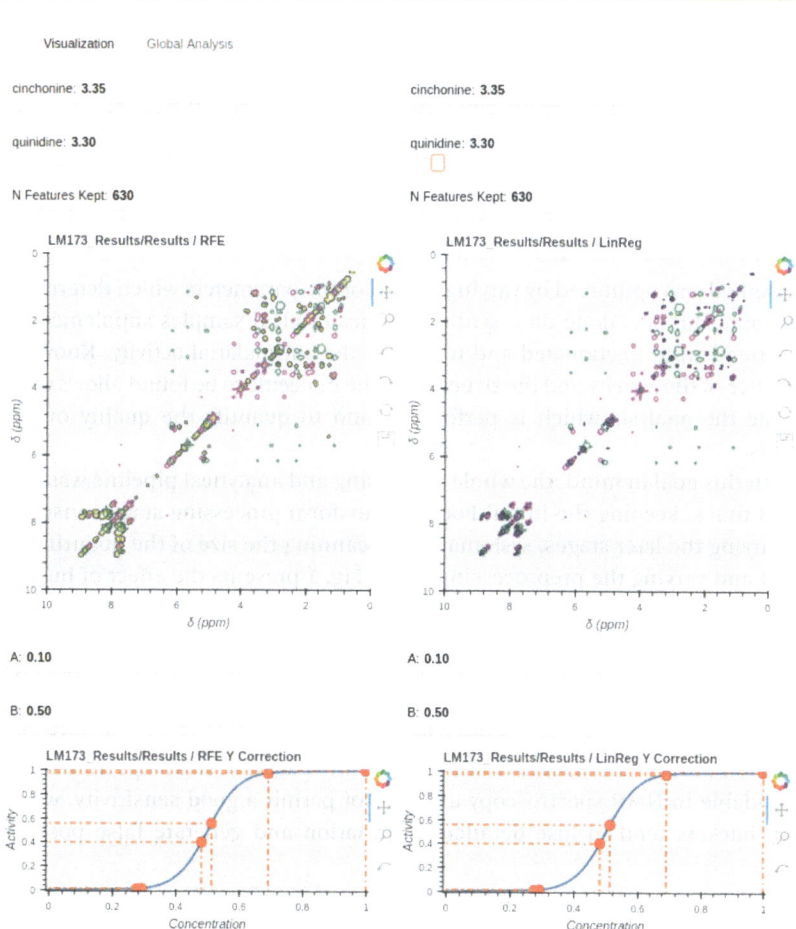

Fig. 5 Interactive display of the correlation analysis; the analysis obtained from the whole set of data from the measured activity response as obtained on the *C. pubescens* set of experiments. The top view presents the results of the Recursive Feature Elimination method (RFE) (left) and of the linear regression (right), a cursor allows the number of selected features to be varied, thus varying the number of displayed spectral channels. The bottom view presents the activity to concentration estimator function, with controls for the slope and offset. In red and green, two possible active compounds are superimposed (here, cinchonine and quinidine).

the green reference contours, indicating the successful extraction of the artemisinin fingerprint from the mixture spectrum.

Fig. 5 shows the results of the analysis from the whole set. Two methods are shown here in an interactive manner: Recursive Feature Elimination (RFE) and Linear Regression. The response curve, transforming activities to estimated relative concentrations, is shown at the bottom, with the actual values displayed, along with interactive cursors to vary the slope and the offset of the response curve.

4 Discussion

We explore here the capabilities of the *Plasmodesma* method to extract the spectral fingerprint of a molecule of interest from a set of NMR experiments, in a pharmacophoric deconvolution, and to find the optimal parameters of the method. The method is meant to be applied to complex samples, and used by people from the pharmacognosy community. For this reason, it has to present its results in an easily readable manner and be robust against noise, artefacts and user errors.

4.1 Optimisation of the processing pipeline

The method was optimized by varying several of the parameters which determine its output. This was done on a synthetic series of algal samples supplemented with artemisinin, fractionated and tested for the antimalarial activity. Knowing the source of the activity and the structure of the molecule to be found allows us to evaluate the analysis which is performed, and to quantify the quality of the analysis.

With this goal in mind, the whole processing and analytical pipeline was run several times, keeping the initial Fourier transform processing stage constant, and varying the later stages, systematically scanning the size of the quantifying bucket and varying the preprocessing steps. Fig. 1 presents the effect of bucket size variation. It can be seen that there is a trade-off between the sensitivity of the method, measured by the number of signals effectively detected with respect to all possible signals, and the selectivity of the analysis evaluated from the number of wrongly selected signals. The sensitivity increases for larger buckets, but the selectivity decreases for the same increase. Small buckets conserve the resolution of the NMR spectra but are less robust against small spectral shifts unavoidable in NMR spectroscopy and do not permit a good sensitivity, while larger buckets tend to lose detailed information and generate false positive signals.

The ROC curve (Fig. 2) allows us to present this information in a clear and synthetic manner, and allows us to choose the optimal point for a given analysis. It can be seen that all bucket sizes give a point far from the diagonal, indicating that the conditions for a correct analysis are present regardless of the detail of the parameter. This shows that the method is not very sensitive to this value, within the range tested here. For this reason, we decided to keep the optimal values determined here for all subsequent analyses (0.05 ppm in ^{1}H and 1.0 ppm in ^{13}C).

The rather large bucket size is compensated by the possibility to compute additional descriptors. More than a dozen are currently defined in the program. In our test experiment, the standard deviation of the signal within the bucket, or its logarithm, has shown to give the best results. This is certainly dependent on the nature of the sample and of the spectrometer, and all possibilities are available.

Because of the non-optimized automatic measurement of the NMR experiment implied in this approach, and the nature of the samples (barely filtered plant crude extracts automatically measured), the quality of the analysed spectra can be expected to be suboptimal. For this reason, a cleaning step is applied to the buckets generated from the first step to produce curated lists. This cleaning step is proposed in three levels (mild, medium and harsh) followed by an optional symmetrisation of the homonuclear experiments.

The cleaning steps are quite efficient in removing noise, in particular t_1-noise (for the medium and harsh conditions), and may not always be needed, or may even cause deleterious effects on good quality spectra. The symmetrisation is also very efficient, in part because it is applied on a synthetic matrix, obtained from bucketing, with symmetric dimensions. However, as with any other symmetrisation, it may introduce spurious peaks, in particular in the presence of strong t_1-ridges. In the synthetic series, the medium cleaning conditions associated with symmetrisation gave the best results and are presented here (the full results are presented in the ESI†).

These curated lists are the material used by the following statistical step. Two approaches are proposed, the difference maps obtained from a one to one comparison, and the analysis of the whole series at once. The difference maps can be quite efficient when the different samples present a strong contrast in their activities but do not vary much in their preparation and composition, for instance consecutive fractions of the fractionation. This is the case in the synthetic series, where the artemisinin molecule is mostly present in one of the fractions (see Fig. 4). Two mathematical methods are provided for this operation, one with a better selectivity but missing the diagonal peaks of the homonuclear experiments, and the other one with a better sensitivity, but probably more False Positives. The two methods are provided to the user for a simple analysis.

Of course, the whole series analysis is more powerful, and is meant to extract information from low concentration species in complex mixtures. It is expected that the intensity of the signals of the compound of interest varies along the different samples in correlation with the activity of the different samples. This is ensured thanks to the NMR spectroscopy linearity to concentration, and to the constant parameters used for data acquisition and processing. A simple linear regression is efficient to extract the spectral fingerprint from the coefficients with the larger positive values. A recursive feature elimination, allowing us to reject the features irrelevant to the analysis in a first step, and to concentrate on the best ones, proved to be efficient in our hands in certain cases, but not systematically. For this reason, both computations are left to the user (see Fig. 5). Thanks to the efficiency used in the implementation, and to the simplicity of this analysis, the computation is nearly instantaneous, and the number of features to select can be chosen interactively by the user.

To give easy access of this method to users in laboratories, it has been made available on the Internet, and can be freely accessed at the address: https://plasmodesma.igbmc.science as an interactive site. The whole program is written in Python, as this specific programming language is conceived to be powerful, versatile, and also easy to learn and apply, offering a lot of programming possibilities, and it is well-adapted to scientific processing.[31] The processing and statistical parts are developed thanks to the well-established machine learning library *scikit-learn*[32] as well as the data analysis *pandas*[33] program and the NMR processing *SPIKE* library.[29] Interactive graphics over the Internet are made available thanks to the *Bokeh* server.[34]

4.2 Cinchona

The bark of the Red *Cinchona* tree (*Cinchona pubescens*) is still used as anti-malarial medication, thanks to the numerous quinoleic alkaloids. Many of the alkaloids present in this sample share a common quinine skeleton, which is nevertheless diversified with the presence of different asymmetric carbon configurations, and different substitutions. These alkaloids are eventually substituted at C-6′, and bound through a hydroxylated secondary carbon to a quinuclidine moiety. The most important are the quinine with its ᴅ-isomer quinidine and the cinchonine with its ʟ-isomer cinchonidine (Fig. 3). Non-related alkaloids are also present, as well as polyphenols and tanins, and these species are present in the hydroalcoholic extract studied here. The six fractions present non-negligible antimalarial activities below 10 μg mL^{-1}, and two fractions are extremely active (see the IC$_{50}$ values in Table 1). This activity is certainly due to the known synergic effect presented by the active quinoleic alkaloids.[35]

The method is thus facing the challenge of a real bioactive extract here, with many secondary metabolites, sharing similar structures with carbon isomers, and synergic biological effects.

This is a real-life sample, which would be extremely difficult to fully interpret even with more involved techniques such as molecular networking or bio-guided isolation. Nevertheless, despite the complexity of the mixture and the structural similarities of the studied compounds, the obtained fingerprint led to the correct identification of the most bioactive compounds.

On the experimental side, it should be noted that the *Plasmodesma* method presents advantages over more involved approaches. Acquisition and processing should introduce as little bias and artefacts as possible. For this reason, we operate in a fully automatised manner, and do not manually optimise the parameter for each measurement. The obvious drawback is that the experimental

Table 1 Activity measured in the *Cinchona* series

Fraction	IC$_{50}$, μg mL^{-1}
F1	0.23
F2	0.095
F3	0.095
F4	0.166
F5	6.17
F6	8.53
Chloroquine	0.1
Quinine	0.24
Quinidine	0.06
Cinchonine	0.06
Cinchonidine	0.24

Table 2   Bucket sizes used in this work

	Bucket size (ppm)					
	C1	C2	C3	C4	C5	C6
^{1}H 2D	0.02	0.02	0.03	0.05	0.04	0.1
^{13}C 2D	0.5	1	1	1	2	2

part is suboptimal in terms of field homogeneity, pulse calibration, processing details, *etc.*, however, in contrast, the measurement time and the burden of user interactions are minimized. By the same token, one should note that in the case of a relatively costly activity test, such as the one used here that implies the use of parasite cultures, the possibility to work with a relatively small number of samples is certainly an advantage.

5 Conclusion

We have presented the optimisation of the *Plasmodesma* tool and explored some of its capabilities. Starting with a small number of fractions presenting contrasting biological characteristics, it allows us to extract a NMR spectral fingerprint in only a few steps.

This tool handles the NMR measurements as statistical entities, and uses curated bucket lists rather than peak lists for the analysis, presenting its results in a graphical manner in which a NMR spectroscopist easily recognises molecular patterns. In addition, it does not rely on the use of molecular databases, and can thus be used for the analysis of unknown systems.

It was tested on the analysis of bioactive natural extracts with the prospect of analysing the original sample and characterizing its active compounds. It was presented in two different scenarios. The first one, performed on a synthetic sample with only one known active compound, artemisinin at 2.7% w/w, allowed its rapid characterisation, as a first step over a complete dereplication. The second one, applied on a *Cinchona pubescens* bark extract, allowed the identification of several active molecules, probably acting in synergy, in a pharamacophiric deconvolution of the active species.

In its general approach, its use can be generalized to different types of studies, where a given property is modulated over a set of samples. For instance, one may think of metabolite analysis for sets of samples obtained under different conditions, or from patients under clinical monitoring; but also in impurity control with the study of sets of samples obtained from different sampling points or at different time periods.

Finally, the program has been made available on the Internet, with an interactive visualisation of the statistical analysis. It provides most of the features described in the present publication, and is available at https://plasmodesma.igbmc.science.

6 Materials & methods

6.1 Samples

Chemicals. Artemisinin 98% was purchased from Sigma-Aldrich and deuterated methanol (10 × 0.75 mL) from Eurisotop (Saint Aubin, France). The bark powder of *Cinchona pubescens* L. was purchased from Herboristerie CAILLEAU (Chemillé-en-Anjou 49120 France).

Algae collection and identification. The alga *Sargassum muticum* was collected in June 2006 in Cap Lévy (Manche), France. Taxonomic determination was performed by Dr A.-M. Rusig and a voucher specimen was deposited in the Herbarium of the University of Caen. Extraction was realized as in Vonthron-Sénécheau *et al.*[36]

Artemisinin study. A *Sargassum muticum* hydroalcoholic extract was supplemented with artemisinin according to natural concentrations observed in *Artemisia annua* and its extract.[37] 9.4 mg of artemisinin was added to 350 mg of extract (2.7% w/w). This supplemented extract was fractionated by flash chromatography using a normal phase column (PuriFlash PF-30SiHP/40G, Interchim). The sample was eluted with a heptane/dichloromethane mixture of increasing polarity (50/50 to 0/100 in 60 min), then with a dichloromethane/methanol mixture (100/0 to 70/30 in 60 min), at a flow rate of 16 mL min^{-1}. The column was washed with a CH_2Cl_2/acetone/MeOH/formic acid mixture (5/3/0.5/0.5). Six fractions were obtained, each 30 min (F1 to F4), F5 was the washing, and F6 was a 50/50 mix of the F2 and F3 fractions. All the samples were lyophilized and dissolved in 750 µl of methanol-d4, and then put in 5 mm NMR tubes. A sample of pure artemisinin was prepared in methanol-d4 and studied by NMR spectroscopy for reference. Activities were measured as described in ref. 18.

Cinchona **samples.** The bark powder (15 g) of *Cinchona pubescens* was extracted with 70% EtOH (10% w/v) during 24 h. The filtrate was dried under vacuum at 35 °C and the residues were stored at 4 °C until testing. 3.6 g of hydroalcoholic extract of *C. pubescens* was fractionated by flash chromatography using a normal phase column (PuriFlash PF-30SiHP/40G, Interchim). The sample was eluted with a dichloromethane/methanol mixture containing 1% of diethylamin (98/2 to 70/30 in 90 min) at a flow rate of 18 mL min^{-1}. Six fractions were monitored by TLC (Sigma-Aldrich silica gel on TLC aluminum plates) and the spots were visualized by UV light and by spraying with iodoplatinate reagent (Sigma-Aldrich). All samples were lyophilized and dissolved in 750 µL of methanol-d4, and then put in 5 mm NMR tubes.

6.2 Bioassays

Plasmodium falciparum 7G8 chloroquine-resistant strain was cultured following Trager and Jensen's method[38] using O+ red blood cells and complete media RPMI1640 medium supplemented by 1 µg mL^{-1} hypoxanthine (Sigma-Aldrich), 110 µg mL^{-1} Na pyruvate (Sigma-Aldrich), 20 µg mL^{-1} gentamycin (Euromedex), 1 mM L-glutamine (Gibco) and 10% alpha-calf serum (HyClone) at 37 °C under reduced oxygen conditions (candle jar). The *in vitro* antimalarial activity was evaluated by the *Plasmodium* lactate dehydrogenase (pLDH) immunodetection assay against the *P. falciparum* 7G8 chloroquine-resistant strain, with a commercially available sandwich enzyme-linked immunosorbent assay kit (Advanced Practical Diagnostics BVBA, Turnhout, Belgium), as reported previously.[39] The assays were performed in a 96-well culture plate with cultures mostly at ring stages at 1% parasitaemia (2% hematocrit). Screening of the artemisinin series was carried out at two different concentrations (100 µg mL^{-1} and 10 µg mL^{-1}). To do so, the parasite culture was incubated with the corresponding concentration of the tested extract for 96 h at 37 °C under reduced oxygen conditions. The positive control, composed of chloroquine diphosphate (Sigma-Aldrich), was evaluated at the same concentrations (99% of growth inhibition at 100 μg mL^{-1} and 92% at 10 μg mL^{-1}). Each experiment was performed two times in triplicate.

IC$_{50}$ measurements of the *Cinchona* series and *Cinchona* alkaloid references were carried out with increasing concentrations of the extract (0.01 to 10 μg mL^{-1}, 16 concentrations) for 96 h at 37 °C under reduced oxygen conditions (candle jar). Each experiment was performed 3 times in duplicate, and chloroquine was used as a positive control. The IC$_{50}$ values were calculated with GraphPad Prism 6 software.

6.3 NMR spectroscopy

Acquisitions were performed on a Bruker Avance-III spectrometer operating at 700 MHz, and equipped with a TCI cryoprobe and a standard Bac60 sample changer. The prepared NMR tubes were spun with a small bench centrifuge to help sedimentation of insoluble parts, and placed in the NMR sample changer. Each sample was automatically inserted into the spectrometer, tuned and shimmed after a stabilization delay of 120 s. All experiments were automatically run on each sample. The spectral parameters (π/2 pulses, receiver gain, *etc.*) were optimized on one sample and used for the whole series without further checks.

6.4 Data analysis

The data processing. The data processing was realised with the *Plasmodesma* Program,[18] written in Python. *Plasmodesma* is based on the *SPIKE* Processing Library, developed for the analysis of NMR and MS spectral datasets.[29] *Plasmodesma* produces peak lists and bucket lists, which are analysed mainly with *scikit-learn*, a specific Python library providing tools for data analysis and application of machine learning techniques.[32] In order to ensure sustainable computational science,[40] the program is deposited on an open-source repository, and is freely available at https://github.com/delsuc/plasmodesma.

For each bucket zone, the total signal volume is computed, as well as the minimum and maximum values, the standard deviation of the signal and the number of isolated peaks found in the bucket zone. In order to enrich the analysis, additional composite descriptors are computed from these values as sums, differences or logarithms of the former. The bucket spectra are then cleaned against t_1-noise and artefacts, and homonuclear 2D experiments are eventually symmetrized. In Fig. 1, the bucket sizes as listed in Table 2 were used.

The one-to-one spectral difference. The one-to-one spectral difference was performed by computing either the ratio or the difference of the two compared spectra. The standard deviation of the bucket value or its logarithm have been shown to provide the best results.

The correlation analysis. The correlation analysis between the spectral features and the activity levels was performed using the set of regression tools from the *scikit-learn* library. The procedure consists of applying a linear regression of each descriptor with respect to the estimated concentration, and selecting the buckets with the largest positive coefficients. We mostly use the Linear Regression (sklearn.linear_model.LinearRegression) and the Recursive Feature Elimination (sklearn.feature_selection.RFE) methods of scikit-learn. In this second approach, linear regression is repeated several times, each time selecting a given ratio of buckets (for instance 50%) and ignoring the non-selected variables for the next iterations, until the required number of selected buckets is reached. In both cases, the final number of selected spectral channels is chosen by the user. Other methods have been tested: ridge regression, Huber and passive–aggressive methods, but they were not selected.

For this analysis, the concentration has to be estimated from the activity levels. This is done by inverting a sigmoid function, for which the user can interactively modify the slope and the offset. In the case of an activity level indicated as an IC_{50} value, the inverse of the IC_{50} value is directly considered as the compound concentration.

The graphic interface. The graphic interface was developed using the *Bokeh* Python library for interactive visualization on web devices,[34] and the Flask framework.[41] The code is open-source and is available at https://github.com/LauraDuciel/plasmodesma.

In the present work, we used the following programs: Python version 3.6.5 as distributed by Continuum Analytics (Austin, TX), *Pandas* version 0.23.4, *scikit-learn* version 0.20.1; *SPIKE* version 0.99.0 Revision Id 369, and *Bokeh* version 1.0.1.

Conflicts of interest

There are no conflicts to declare.

Acknowledgements

We thanks M. Hibert for the scientific discussion dealing with the pharmacophoric deconvolution method and his constructive criticism. The authors are very grateful to Labex Medalis and Région Alsace for a fellowship (LM). We are also grateful to A.-M. Rusig for the collection and the identification of the algal material and to F. Nardella and Pr. E. Candolfi for their help with the bio-assays. We thanks J. Seiler and all the IGBMC IT team for the support in deploying the program on their server.

References

1 E. Grotewold, *Trends Plant Sci.*, 2005, **10**, 57–62.
2 E. Pichersky and E. Lewinsohn, *Annu. Rev. Plant Biol.*, 2011, **62**, 549–566.
3 E. Kellenberger, A. Hofmann and R. Quinn, *Nat. Prod. Rep.*, 2011, **28**, 1483–1492.
4 D. J. Newman and G. M. Cragg, *J. Nat. Prod.*, 2016, **79**, 629–661.
5 K. Hostettmann, J. L. Wolfender and C. Terreaux, *Pharm. Biol.*, 2001, **39**(1), 18–32.
6 J. Hubert, J.-M. Nuzillard and J.-H. Renault, *Phytochem. Rev.*, 2017, **16**, 55–95.
7 M. Wang, J. J. Carver, V. V. Phelan, L. M. Sanchez, N. Garg, Y. Peng, D. D. Nguyen, J. Watrous, C. A. Kapono, T. Luzzatto-Knaan, C. Porto, A. Bouslimani, A. V. Melnik, M. J. Meehan, W.-T. Liu, M. Crüsemann, P. D. Boudreau, E. Esquenazi, M. Sandoval-Calderón, R. D. Kersten, L. A. Pace, R. A. Quinn, K. R. Duncan, C.-C. Hsu, D. J. Floros, R. G. Gavilan, K. Kleigrewe, T. Northen, R. J. Dutton, D. Parrot, E. E. Carlson, B. Aigle, C. F. Michelsen, L. Jelsbak, C. Sohlenkamp, P. Pevzner, A. Edlund, J. McLean, J. Piel, B. T. Murphy, L. Gerwick, C.-C. Liaw, Y.-L. Yang,

H.-U. Humpf, M. Maansson, R. A. Keyzers, A. C. Sims, A. R. Johnson, A. M. Sidebottom, B. E. Sedio, A. Klitgaard, C. B. Larson, C. A. Boya P., D. Torres-Mendoza, D. J. Gonzalez, D. B. Silva, L. M. Marques, D. P. Demarque, E. Pociute, E. C. O'Neill, E. Briand, E. J. N. Helfrich, E. A. Granatosky, E. Glukhov, F. Ryffel, H. Houson, H. Mohimani, J. J. Kharbush, Y. Zeng, J. A. Vorholt, K. L. Kurita, P. Charusanti, K. L. McPhail, K. F. Nielsen, L. Vuong, M. Elfeki, M. F. Traxler, N. Engene, N. Koyama, O. B. Vining, R. Baric, R. R. Silva, S. J. Mascuch, S. Tomasi, S. Jenkins, V. Macherla, T. Hoffman, V. Agarwal, P. G. Williams, J. Dai, R. Neupane, J. Gurr, A. M. C. Rodríguez, A. Lamsa, C. Zhang, K. Dorrestein, B. M. Duggan, J. Almaliti, P.-M. Allard, P. Phapale, L.-F. Nothias, T. Alexandrov, M. Litaudon, J.-L. Wolfender, J. E. Kyle, T. O. Metz, T. Peryea, D.-T. Nguyen, D. VanLeer, P. Shinn, A. Jadhav, R. Müller, K. M. Waters, W. Shi, X. Liu, L. Zhang, R. Knight, P. R. Jensen, B. O. Palsson, K. Pogliano, R. G. Linington, M. Gutiérrez, N. P. Lopes, W. H. Gerwick, B. S. Moore, P. C. Dorrestein and N. Bandeira, *Nat. Biotechnol.*, 2016, **34**, 828–837.

8 P.-M. Allard, T. Péresse, J. Bisson, K. Gindro, L. Marcourt, V. C. Pham, F. Roussi, M. Litaudon and J.-L. Wolfender, *Anal. Chem.*, 2016, **88**, 3317–3323.

9 S. P. Gaudêncio and F. Pereira, *Nat. Prod. Rep.*, 2015, **32**, 779–810.

10 H. Schwalbe, *Angew. Chem., Int. Ed.*, 2017, **56**, 10252–10253.

11 N. Aligiannis, M. Halabalaki, E. Chaita, E. Kouloura, A. Argyropoulou, D. Benaki, E. Kalpoutzakis, A. Angelis, K. Stathopoulou, S. Antoniou, M. Sani, V. Krauth, O. Werz, B. Schütz, H. Schäfer, M. Spraul, E. Mikros and L. A. Skaltsounis, *ChemistrySelect*, 2016, **1**, 2531–2535.

12 G. Dal Poggetto, L. Castañar, M. Foroozandeh, P. Király, R. W. Adams, G. A. Morris and M. Nilsson, *Anal. Chem.*, 2018, **90**, 13695–13701.

13 A. Bakiri, J. Hubert, R. Reynaud, C. Lambert, A. Martinez, J.-H. Renault and J.-M. Nuzillard, *J. Chem. Inf. Model.*, 2018, **58**, 262–270.

14 J. J. J. van der Hooft and N. Rankin, in *Modern Magnetic Resonance*, ed. G. A. Webb, Springer International Publishing, Cham, 2017, pp. 1–33.

15 J. Hubert, J.-M. Nuzillard, S. Purson, M. Hamzaoui, N. Borie, R. Reynaud and J.-H. Renault, *Anal. Chem.*, 2014, **86**, 2955–2962.

16 H. Mohimani and P. A. Pevzner, *Nat. Prod. Rep.*, 2016, **33**, 73–86.

17 A. Bakiri, B. Plainchont, V. de Paulo Emerenciano, R. Reynaud, J. Hubert, J.-H. Renault and J.-M. Nuzillard, *Mol. Inf.*, 2017, **36**, 1700027.

18 L. Margueritte, P. Markov, L. Chiron, J.-P. Starck, C. Vonthron-Sénécheau, M. Bourjot and M.-A. Delsuc, *Magn. Reson. Chem.*, 2018, **56**, 469–479.

19 J.-L. Wolfender, J.-M. Nuzillard, J. J. J. van der Hooft, J.-H. Renault and S. Bertrand, *Anal. Chem.*, 2019, **91**(1), 704–742.

20 I. Goodfellow, Y. Bengio and A. Courville, *Deep Learning*, MIT Press, 2016.

21 G. Luo, *Netw. Model. Anal. Health Inform. Bioinform.*, 2016, 5, 18.

22 J. Bergstra, R. Bardenet, Y. Bengio and B. Kégl, *NIPS'11 Proceedings of the 24th International Conference on Neural Information Processing Systems*, USA, 2011, pp. 2546–2554.

23 T. Domhan, J.-T. Springenberg and F. Hutter, *Proceedings of the 24th International Conference on Artificial Intelligence (IJCAI'15)*, 2015, pp. 3460–3468.

24 L. Margueritte, *Développement d'une méthode de déconvolution pharmacophorique pour accélérer le processus de découverte de molécules antipaludiques d'origine algale*, Université de Strasbourg, 2018.

25 M. Willcox, G. Bodeker and P. Rasoanaivo, *Traditional medicinal plants and malaria*, Taylor & Francis Ltd, 2004, p. 552.

26 J. M. Karle and A. K. Bhattacharjee, *Bioorg. Med. Chem.*, 1999, **7**, 1769–1774.

27 D. V. McCalley, *J. Chromatogr. A*, 2002, **967**, 1–19.

28 G. Nonaka and I. Nishioka, *Chem. Pharm. Bull.*, 1982, **30**, 4268–4276.

29 L. Chiron, M.-A. Coutouly, J.-P. Starck, C. Rolando and M.-A. Delsuc, arXiv, 1608.06777, 2016.

30 A. Cherni, E. Chouzenoux and M.-A. Delsuc, *Analyst*, 2017, **142**, 772–779.

31 H. P. Langtangen, *A primer on scientific programming with Python*, Springer, Dordrecht, The Netherlands, New York, 2009.

32 F. Pedregosa, G. Varoquaux, A. Gramfort, V. Michel, B. Thirion, O. Grisel, M. Blondel, P. Prettenhofer, R. Weiss, V. Dubourg, J. Vanderplas, A. Passos, D. Cournapeau, M. Brucher, M. Perrot and E. Duchesnay, *J. Mach. Learn. Res.*, 2011, **12**, 2825–2830.

33 W. McKinney, *Pandas, Python Data Analysis Library*, 2015, Reference Source, 2014.

34 Bokeh Development Team, *Bokeh: Python library for interactive visualization*, 2018.

35 P. Druilhe, O. Brandicourt, T. Chongsuphajaisiddhi and J. Berthe, *Antimicrob. Agents Chemother.*, 1988, **32**, 250–254.

36 C. Vonthron-Sénécheau, M. Kaiser, I. Devambez, A. Vastel, I. Mussio and A.-M. Rusig, *Mar. Drugs*, 2011, **9**, 922–933.

37 D. L. Klayman, A. J. Lin, N. Acton, J. P. Scovill, J. M. Hoch, W. K. Milhous, A. D. Theoharides and A. S. Dobek, *J. Nat. Prod.*, 1984, **47**, 715–717.

38 W. Trager and J. B. Jensen, *Science*, 1976, **193**, 673–675.

39 P. S. Atchade, C. Doderer-Lang, N. Chabi, S. Perrotey, T. Abdelrahman, C. D. Akpovi, L. Anani, A. Bigot, A. Sanni and E. Candolfi, *Malar. J.*, 2013, **12**, 279.

40 N. P. Rougier, K. Hinsen, F. Alexandre, T. Arildsen, L. A. Barba, F. C. Benureau, C. T. Brown, P. de Buyl, O. Caglayan, A. P. Davison, M.-A. Delsuc, G. Detorakis, A. K. Diem, D. Drix, P. Enel, B. Girard, O. Guest, M. G. Hall, R. N. Henriques, X. Hinaut, K. S. Jaron, M. Khamassi, A. Klein, T. Manninen, P. Marchesi, D. McGlinn, C. Metzner, O. Petchey, H. E. Plesser, T. Poisot, K. Ram, Y. Ram, E. Roesch, C. Rossant, V. Rostami, A. Shifman, J. Stachelek, M. Stimberg, F. Stollmeier, F. Vaggi, G. Viejo, J. Vitay, A. E. Vostinar, R. Yurchak and T. Zito, *PeerJ Computer Science*, 2017, **3**, e142.

41 Flask Development Team, *Flask is a microframework for Python based on Werkzeug, Jinja 2 and good intentions*, 2018.

Faraday Discussions

PAPER

Challenges in the decomposition of 2D NMR spectra of mixtures of small molecules

Afef Cherni,[a] Elena Piersanti,[b] Sandrine Anthoine, [ID][a]
Caroline Chaux, [ID]*[a] Laetitia Shintu, [ID]*[b] Mehdi Yemloul [ID][b]
and Bruno Torrésani [ID][a]

Received 6th February 2019, Accepted 7th March 2019

DOI: 10.1039/c9fd00014c

Analytical methods for mixtures of small molecules require specificity (is a certain molecule present in the mix?) and speciation capabilities. NMR spectroscopy has been a tool of choice for both of these issues since its early days, due to its quantitative (linear) response, sufficiently high resolving power and capabilities of inferring molecular structures from spectral features (even in the absence of a reference database). However, the analytical performances of NMR spectroscopy are being stretched by the increased complexity of the samples, the dynamic range of the components, and the need for a reasonable turnover time. One approach that has been actively pursued for disentangling the composition complexity is the use of 2D NMR spectroscopy. While any of the many experiments from this family will increase the spectral resolution, some are more apt for mixtures, as they are capable of unveiling signals belonging to whole molecules or fragments of it. Among the most popular ones, one can enumerate HSQC-TOCSY, DOSY and Maximum-Quantum (MaxQ) NMR spectroscopy. For multicomponent samples, the development of robust mathematical methods of signal decomposition would provide a clear edge towards identification. We have been pursuing, along these lines, Blind Source Separation (BSS). Here, the un-mixing of the spectra is achieved relying on correlations detected on a series of datasets. The series could be associated with samples of different relative composition or in a classically acquired 2D experiment by the mathematical laws underlying the construction of the indirect dimension, the one not recorded by the spectrometer. Many algorithms have been proposed for BSS in NMR spectroscopy since the seminal work of Nuzillard. In this paper, we use rather standard algorithms in BSS in order to disentangle NMR spectra. We show on simulated data (both 1D and 2D HSQC) that these approaches enable us to accurately disentangle multiple components, and provide good estimates for the concentrations of compounds. Furthermore, we show that after proper realignment of the signals, the same algorithms are able to disentangle real 1D NMR spectra. We obtain similar results on

[a]Aix Marseille Univ, CNRS, Centrale Marseille, I2M, Marseille, France. E-mail: caroline.chaux@univ-amu.fr
[b]Aix Marseille Univ, CNRS, Centrale Marseille, iSM2, Marseille, France. E-mail: laetitia.shintu@univ-amu.fr

2D HSQC spectra, where the BSS algorithms are able to successfully disentangle components, and provide even better estimates for concentrations.

1 Introduction

Nuclear magnetic resonance (NMR) is a powerful spectroscopy that provides comprehensive information on molecular structure and is well suited for the detection and identification of small molecules. The simplest NMR experiment yields a potentially informative one-dimensional spectrum that typically results in overlapping signals for complex mixtures and hinders the identification and the quantification of components. Several developments, occurring at different stages of the NMR analysis (from pulse sequence implementation to data processing), have been proposed to differentiate between multiple peaks in extensively crowded spectra. One powerful approach to increase the information content of NMR spectra is to acquire two-dimensional data. In that regard, HSQC,[6] TOCSY,[1,7] Maximum-Quantum (Max-Q)[3,8] or Diffusion Ordered SpectroscopY (DOSY)[2,9] NMR experiments are very interesting for mixture analysis since they allow the direct identification of a whole molecule or fragments of it. Indeed, DOSY experiments represent a commonly used pseudo-2D NMR experiment that allows the differentiation of molecules in a mixture according to their diffusion coefficients. This technique is efficient when applied to relatively simple mixtures (less than ten molecules), leading to the extraction of the NMR spectrum of each compound. However, a limiting factor of its applicability is the requirement of a mathematical treatment capable of distinguishing molecules with similar spectra or diffusion constants. Similarly, 2D NMR experiments, such as $^1H-^1H$ COSY, $^1H-^1H$ TOCSY and $^1H-^{13}C$ HSQC, are also performed routinely since they are necessary for the assignments of the mixture's molecules.[10] However, their use for the analysis of a large number of samples, such as in metabolomics, is difficult since the acquisition of a 2D NMR spectrum can be extremely time-consuming, and therefore, 2D NMR experiments are only performed on representative samples. In addition, molecule assignment relies on a very exhausting and time-consuming process for which each signal of each 2D NMR spectrum must be thoroughly peak-picked and gathered according to the molecule they characterize. In order to overcome the issue of an extensive acquisition time, ultrafast and fast NMR methods such as single-scan[11] or non-uniform sampling (NUS)[12] techniques have been developed, making the use of 2D NMR experiments for high-throughput study of complex mixtures possible.[13] Regarding the signal assignments and in the case of well-studied samples such as blood plasma or cerebrospinal fluid, automated methods for the metabolite identification from 2D experiments are available online.[14,15] However, their efficiency depends on strict sample preparation protocols that limit their use for a wider range of samples. Consequently, the development of robust mathematical methods that would perform signal decomposition is of prime importance for the analytical study of complex mixtures.[5] The mathematical "demixing" of 1D or 2D NMR spectra would thus provide a clear edge towards identification, with a non-negligible gain of time. In a previous study, our group presented a processing strategy for DOSY experiments based on the synergy of two high-performance Blind Source Separation (BSS) techniques: Non-negative

Matrix Factorization (NMF) using additional Sparse Conditioning (SC), and the Joint Approximate Diagonalization of Eigenmatrices (JADE) declination of independent component analysis (ICA).[4,16,17] Both approaches enabled us to improve the processing of DOSY experiments, in cases of mildly overlapping species. For mixtures with strong overlapping signals of moieties with similar diffusion coefficients, such as a mixture of sucrose and maltotriose, NMF-SC provided a very good method for molecule separation, although not perfect and needing improvement to make it suitable for the processing of more complex mixtures.[18]

In this paper, we address the un-mixing problem in a more general setting, with the aim of processing 1D as well as nD mixtures. Sticking to the family of non-negative matrix factorization approaches to BSS, we consider several algorithms and apply them to simulated and real mixture spectra. These are presented in a unified framework, that includes classical NMF approaches such as alternate least squares (ALS)[19] and sparsity-penalized versions,[20] proximal approaches,[21] and wavelet-based[22] variants. Interestingly enough, the framework also includes algorithms for un-mixing nD spectra. We then provide objective performance evaluations for un-mixing algorithms, using quality indices that allow for assessing the quality of the estimation of the source spectra and concentrations in the mixtures.

The results are given for a dataset that has been prepared on purpose, for which pure spectra and concentrations in solutions are available, which allows for computing the above indices. The results on 1D simulated data (*i.e.* mixtures mathematically generated from pure spectra and concentrations) show that the algorithms under consideration are indeed able to recover the ground truth. The results on real 1D mixtures do not reach the same level of quality even after correcting alignment biases, which raises concerns regarding the mathematical mixture model. In the case of the 2D HSQC spectra, the performances of the algorithms are again fairly good on the simulated data. The results on real 2D mixtures are of weaker quality in terms of the objective performance evaluation indices. However, the increased sparsity of the 2D spectra allows a good identification of the components of mixtures. In addition, the concentrations appear to be better estimated than in the 1D case, which may also be interpreted as a consequence of the sparsity of the 2D spectra. It is worth mentioning that the computational burden is significantly increased in the 2D case, which may be a limitation. The 2D case then represents an important challenge, this is presumably true for higher dimensional spectra.

Besides the objective evaluations, visual inspection of the spectra shows that the mathematical un-mixing algorithms are able to identify pure compounds in the solutions under study, in several situations. This is clearly the case in simulated situations, which show that the algorithms are able to identify compounds when mixtures have been generated under a well defined model. This is also the case, to a smaller extent, in the case of ^{1}H real data, provided the pre-processing steps have been carefully performed (in particular, shift correction). Our results, however, raise a number of important questions that include, among others, the validity of the mathematical mixture model, the possibility of performing some pre-processing steps simultaneously with un-mixing, and also the relevance of the quantitative assessment measures in the context of NMR spectroscopy un mixing.

2 Problem statement

2.1 Blind source separation, the linear instantaneous mixture model

Blind souis the value of therce separation (BSS) aims for the separation of a set of pure signals called sources from a set of mixed signals, called mixtures, with limited information on the sources or the mixing process. The sources are generically represented as an $M \times L$ matrix, $S = \{s_{m\ell}\} \in \mathbb{R}^{M \times L}$, and mixtures by an $N \times L$ matrix, $X = \{x_{n\ell}\} \in \mathbb{R}^{N \times L}$. N is the number of mixtures, M is the number of sources, and L is the number of observations. In the context of NMR spectroscopy un-mixing, N is the number of observed spectra, M is the number of compounds, and L is the number of points of the spectra. As an example, the number $x_{n\ell}$ is the value of the ℓ-th sample of mixture n, $i.e.$ its value at frequency l.

Among the BSS problems, the simplest instance originates from the Linear Instantaneous Mixture (LIM) model, where the observed mixtures are linear combinations of the sources. The mixing process is then expressed mathematically as

$$X = AS + B \approx AS, \tag{1}$$

more explicitly

$$x_{n\ell} = \sum_{m=1}^{M} a_{nm} s_{m\ell} + b_{n\ell}, \quad n = 1, \dots N, \quad \ell = 1 \dots L, \tag{2}$$

where $B = \{b_{n\ell}\} \in \mathbb{R}^{N \times L}$ is some residual noise, and $A = \{a_{nm}\} \in \mathbb{R}^{N \times M}$ is called the mixing matrix. The BSS problem is to identify jointly the mixing matrix A and the source matrix S from the sole observation matrix X.

Different assumptions or models have led to different identification algorithms, among which we may mention statistics based approaches such as ICA and SOBI, or non-negative matrix factorizations (NMF), which will constitute our approach. Thorough descriptions can be found in reference textbooks,[16,17] and we refer to ref. 4 for a review of applications in NMR spectroscopy.

2.2 The LIM model for 2D spectra

In the case of 2D data such as the HSQC discussed below, observations X and pure signals S are not matrix-shaped anymore, but take the form of three-way arrays: $X \in \mathbb{R}^{N \times L_1 \times L_2}$ and $S \in \mathbb{R}^{M \times L_1 \times L_2}$. The LIM model can still formally be written as in (1), provided that the matrix $\times$ tensor product is suitably defined, in the sense

$$x_{n\ell_1\ell_2} = \sum_{m=1}^{M} a_{nm} s_{m\ell_1\ell_2}, \quad n = 1, \dots N, \ \ell_1 = 1 \dots L_1, \ \ell_2 = 1 \dots L_2, \tag{3}$$

where ℓ_1 and ℓ_2 label the two spectral dimensions.

By rc-organizing the ℓ_1 and ℓ_2 spectral indices into a single one (of length $L_1 L_2$), $i.e.$ transforming the three-way arrays into matrices, one can be back to model (1) (in significantly higher dimension). We call this approach data matricization. However, matricization is not always suitable, as this reshaping procedure breaks the 2D structure, which can be exploited by some algorithms.

2.3 Indeterminacies

Quite obviously, the solution of such a general problem is not unique, as for any solution (A,S) and any invertible $M \times M$ matrix Λ, one can also write $X = AS = A'S'$ where $A' = A\Lambda$ and $S' = \Lambda^{-1}S$, which produces infinitely many other solutions. Therefore, additional assumptions are necessary to solve the problem. Among these indeterminacies, the following two play a special role (and correspond to two specific types of matrices Λ):

• Scale indeterminacy (Λ diagonal): sources can only be identified up to a constant factor (in other words, multiplying a row of S by a constant and dividing the corresponding column of A by the same constant do not change X).

• Order indeterminacy (Λ a permutation matrix): estimated sources are not ordered, therefore comparison of estimated sources with reference sources has to be preceded by an ordering step.

Two solutions (A,S) and (A',S') that only differ by these two transformations are generally considered equivalent.

To overcome indeterminacy problems, additional assumptions have to be made, either on the sources, mixing matrix or both. The non-negativity assumptions made in NMF approaches described below turn out to resolve a part of these problems, nevertheless scale and order indeterminacies remain. These will have to be accounted for in the NMR un-mixing algorithms, as we shall see later.

2.4 Non-negative matrix factorization (NMF)

Non-negative matrix factorization techniques address situations where both the source coefficients $s_{m\ell}$ and mixing matrix coefficients a_{nm} are non-negative. In NMR spectroscopy un-mixing problems, such assumptions are relevant, since mixing matrix coefficients represent concentrations, and source coefficients represent spectrum values.

Many approaches to NMF have been proposed since the early works of Paatero and Tapper[19] and Lee and Seung.[23,24] Most of them are based on so-called variational formulations, where numerical algorithms are used to minimize some objective function, which involves a data fidelity term (which forces the product AS to be close to the data matrix X) and possibly additional terms that may encode prior information on the mixing matrix and/or the source matrix. The mathematical formulation takes the form

$$\min_{A,S} F(X|A, S), \quad \text{under constraints} \quad A \geq 0, \ S \geq 0. \tag{4}$$

Here F is the objective function, that depends on the data matrix X and the unknown matrices A and S. Moreover, the non-negativity constraints are imposed entry-wise, *i.e.* all matrix elements have to be non-negative.

The most classical choices for the objective function are penalized versions of the standard quadratic objective function

$$F(X|A, S) = \frac{1}{2}\|X - AS\|_F^2 + f_A(A) + f_S(S), \tag{5}$$

where the first term, called the squared Frobenius norm, is simply the sum of the squares of the coefficients of the matrix $X - AS$, and f_A and f_S are regularizations that can encode prior information on A and/or S. Standard choices involve the so-called ℓ^p norms denoted by $\|\cdot\|_p$ (where $p \geq 0$), for example

$$f_S(S) = \lambda\|S\|_p^{\,p} = \lambda\sum_{m,\ell} S_{m\ell}^{\,p}. \tag{6}$$

with $\lambda \geq 0$ the regularization parameter. The case $p = 2$ is the widely used Tikhonov regularization. We shall use $p = 1$, which tends to enforce sparse solutions.

Alternative choices for data fidelity terms include the Kullback–Leibler divergence $F_{KL}(X|A,S)$, used for example in ref. 23 and 24, which is (as well as the squared Frobenius norm) a special case of the family of so-called β-divergences.[25]

2.5 Evaluation criteria

Un-mixing results on real data have to be evaluated by experts. However, performance evaluation for separation algorithms can be done using numerical simulations, in which case objective assessment is possible. We briefly describe here some evaluation metrics that are routinely used in BSS problems.

In numerical simulations, one starts with a pre-defined source matrix S and mixing matrix A, and the mixture $X = AS + B$ can then be formed (where either $B = 0$, *i.e.* no noise, or B is a matrix containing random Gaussian white noise with prescribed variance). Separation algorithms yield estimates, denoted by $\hat{S}$ and $\hat{A}$, to be compared with ground truth S and A.

To assess the quality of the mixing matrix estimate, a relevant quantity is the matrix product $G = A^{\dagger}\hat{A}$, with $A^{\dagger}$ the pseudo-inverse of A. G equals the identity matrix when the estimation is perfect, *i.e.* $\hat{A} = A$. The departure from that perfect situation may be quantified by the Amari index

$$I = \frac{1}{2M(M-1)}\sum_{m=1}^{M}\left[\frac{\sum_{m'=1}^{M}|g_{mm'}|}{\max_{m'}|g_{mm'}|} + \frac{\sum_{m'=1}^{M}|g_{m'm}|}{\max_{m'}|g_{m'm}|} - 2\right], \tag{7}$$

which ranges between 0 and 1, and is equal to 0 when $\hat{A} = A$ and 1 in the case where $\hat{A}$ and A are maximally different. It is worth noticing that the Amari index I is insensitive to source ordering and normalization, so that no action is required to compensate for the order and normalization indeterminacies described in Subsection 2.3.

To assess the quality of the estimated sources, we rely on indices implemented in the software toolbox BSSEval.[26] In a nutshell, the estimation error $S - \hat{S}$ is split as a sum of several terms, that are used to evaluate various types of error. Of particular interest to us here are the Signal to Distortion Ratio (SDR), which provides a global measure of the distortion introduced by mixing and separation, and the Signal to Interference Ratio (SIR), which provides a quantitative evaluation of crossover terms after separation (in our case, peaks from a given source that could be completely or partially found in the estimate of another source). Both indices are graded on a logarithmic scale, and expressed in dB, as for the traditional SNR (Signal to Noise Ratio). Contrary to the Amari index, these ratios are sensitive to order and normalization, and therefore suitable re-ordering and normalization steps are mandatory prior to computing the SDR and SIR.

2.6 Changing the representation domain, wavelets

In the above approaches, mixtures and sources are represented by point values, $x_{n\ell}$ and $s_{m\ell}$, respectively. The objective functions that are optimized in NMF algorithms are separable, in the sense that for a given mixing matrix A, columns of S are processed independently of each other, so that possible correlations in the spectral domain (represented by index ℓ) are not exploited. It is, however, possible to describe the spectral domain using a different representation, based upon an expansion of a set of L-dimensional vectors, that form a basis of the L-dimensional space, and to introduce a regularization on the corresponding coefficients rather than the source matrix itself. Denoting these vectors by $\{\psi^{(k)}$, $k = 1,...L\}$, and concatenating them in a square matrix denoted by Ψ (columns of Ψ are the vectors $\psi^{(k)}$), it may be shown that the coefficients of the source matrix S on this basis are given by the matrix $\Gamma = \Psi^T S$ (where T stands for matrix transposition).

The variawhere the coefficients of matrix Γ are denoted by tional formulations described above can be adapted to this new setting by introducing adapted objective functions of the generic form

$$F(X|A, S) = \frac{1}{2}\|X - AS\|_F^2 + f_A(A) + f_\Gamma(\Psi^T S), \qquad (8)$$

where f_Γ is a suitable penalty function. As before, we will choose an ℓ_1 penalization with a regularization parameter $\lambda \geq 0$, i.e. $f_\Gamma(\Gamma) = \lambda \sum_{m,\ell} |\gamma_{m,\ell}|$, where the coefficients of matrix Γ are denoted by $|\gamma_{m,\ell}|$. This will have the effect of promoting sparse coefficient matrices, i.e. matrices having a very large number of coefficients equal or close to zero.

Among the possible bases, here we will use bases of orthonormal wavelets.[27] The use of wavelets for representing NMR spectra has been advocated by several authors,[28,29] the main argument being the ability of wavelet expansions to compress signals.[30]

3 Algorithms

3.1 Generic algorithm

In this section, we describe a generic algorithmic approach to the NMR un-mixing problem. All algorithms to be described here aim at solving problem (4), i.e. a joint minimization problem with respect to the source matrix S and mixing matrix A. This problem is addressed through an alternate optimization algorithm, i.e. we optimize alternately with respect to A and S. Various instances of the algorithm are proposed, depending on the choice of the objective function, optimization strategy and source representation domain (spectral domain or wavelet domain). In all cases, the generic structure is given in Algorithm 3.1, and differs only by the update rules for A and S, which we will generically denote by

$$\text{Upd}_A : (A, S) \rightarrow \text{Upd}_A(A, S) \in \mathbb{R}^{N \times M},$$

$$\text{Upd}_S : (A, S) \rightarrow \text{Upd}_S(A, S) \in \mathbb{R}^{M \times L}.$$

Data: X (data matrix); iter_max; ε; A_{init}; S_{init}; Crit $= +\infty$;
Result: Non-negative matrix factors A and S
Initialization: $A^{(0)} = A_{\text{init}}$, $k = 0$, $S^{(0)} = S_{\text{init}}$;
While $k \leq$ iter_max **and** Crit $> \varepsilon$ **do**
Update of A: $A^{(k+1)} = \text{Upd}_A(A^{(k)}, S^{(k)})$;
Update of S: $S^{(k+1)} = \text{Upd}_S(A^{(k+1)}, S^{(k)})$;
Optional: normalization of A and/or S;
Evaluation of stopping criterion Crit$(k + 1)$
Evaluation of the objective function $F(X|A^{(k+1)}, S^{(k+1)})$
$k = k + 1$
end

Algorithm 1: Generic structure of the alternate optimization algorithm for non-negative matrix factorization (starting by updating A is an arbitrary choice).

This generic algorithm requires additional ingredients/options, some of which are listed below.

(1) Initialization: initial source and mixing matrices S_{init} and A_{init} are necessary to start the iterations (some algorithms require only one of these). Usual choices include random initialization, or deterministic ones (based upon SVD, ICA or other classical methods).

(2) Stopping criterion: the algorithm stops when a prescribed maximal number of iterations is reached, or preferably when some precision criterion reaches a small enough value. Possible choices include the absolute value of the objective function's gradient, or normalized norms of differences between two consecutive iterates of A and S.

(3) To account for normalization indeterminacy, it is possible to normalize rows of S and columns of A at each iteration, so as to enforce a certain normalization property, without changing the product AS. This makes sense only when the objective function is itself invariant under renormalization.

(4) Some algorithms require non-negative data. In such cases, it is necessary to project the data matrix X accordingly, *i.e.* to set all negative matrix elements $x_{n\ell}$ to zero.

3.2 Projected alternate least squares (PALS)

The objective function here is the most classical one, *i.e.* the sum of the squares of the matrix coefficients (termed the squared Frobenius norm) of the discrepancy $X - AS$ between data X and the LIM model AS, and reads

$$F(X|A, S) = \frac{1}{2}\|X - AS\|_F^2. \tag{9}$$

The corresponding update rules are given by

$$\text{Upd}_A(A, S) = \Pi_+[(AS - X)S^T], \quad \text{Upd}_S(A, S) = \Pi_+[A^T(AS - X)],$$

where Π_+ denotes the operator that sets all negative matrix coefficients of its argument to zero.

3.3 Soft thresholded projected alternate least squares (STALS)

To enforce sparsity of the sources, a common practice is to add an ℓ_1 penalization to the above quadratic objective function, namely the sum of the absolute values of the source terms, denoted by $\|S\|_1$

$$F(X|A, S) = \frac{1}{2}\|X - AS\|_F^2 + \lambda\|S\|_1, \tag{10}$$

where λ is a positive constant that tunes the strength of the penalty. A commonly used approach for solving this problem is to replace the projection Π_+ (of S) onto non-negatives with the non-negative soft thresholding operator $\mathbb{S}_\lambda^+$, which sets all matrix coefficients smaller than the threshold λ (including negative values) to zero. The update rules become

$$\text{Upd}_A(A, S) = \Pi_+\left[(AS - X)S^T\right], \quad \text{Upd}_S(A, S) = \mathbb{S}_\lambda^+\left[A^T(AS - X)\right].$$

Notice that PALS coincides with STALS in the case $\lambda = 0$.

3.4 Proximal alternating linearized minimization (PALM) and pre-conditioned version (BC-VMFB)

The objective here is to deal with the cost function defined previously in (10). The idea is to propose an algorithm which intertwines the minimization of the quadratic part and the regularization part. This can be done by using either the PALM (proximal alternating linearized minimization) algorithm[32] or its preconditioned version named the BC-VMFB (Block-Coordinate Variable Metric Forward–Backward) algorithm.[33] Both are based on a projected gradient descent algorithm and an optional preconditioning step that allows for increasing the convergence speed. The update rules in this case are defined by

$$\text{Upd}_A(A,S) = \Pi_+[A - \gamma(AS - X)S^T],$$

$$\text{Upd}_S(A, S) = \mathbb{S}_{\lambda/\gamma}^+\left[S - \gamma A^T(AS - X)\right].$$

where γ stands for the gradient descent stepsize.

3.5 Wavelet-based PALM and BC-VMFB

These algorithms address the case where sparsity is imposed on the wavelet coefficients of the spectra rather than the spectra themselves. The considered objective function is a special case of (8), namely

$$F(X|A, S) = \frac{1}{2}\|X - AS\|_F^2 + \lambda\|\Psi^T S\|_1,$$

with $\|\Gamma\|_1$ being the sum of the absolute values of the coefficients $\gamma_{m\ell}$. The PALM and BC-VMFB algorithms can be adapted to this new setting. The wavelet-based PALM and BC-VMFB algorithms thus reduce to PALM and BC-VMFB algorithms, except that the thresholding operation is done on wavelet coefficients rather than spectrum coefficients. The update rules become

$$\mathrm{Upd}_A(A,S) = \Pi_+[A - \gamma(AS - X)S^T],$$

$$\mathrm{Upd}_S(A, S) = \Pi_+\left[\Psi\left(\mathbb{S}_{\lambda/\gamma}\left[\Psi^T\left(S - \gamma A^T(AS - X)\right)\right]\right)\right]$$

where $\mathbb{S}_\lambda$ sets only the values whose absolute value is smaller than λ to zero.

3.6 Processing 2D spectra

In the case of 2D spectra, most of the algorithms described above can still be used on matricized data (see Section 2.2). However, wavelet-based algorithms are not compatible with data matricization, as the latter breaks the 2D structure that is exploited by 2D wavelets. Nevertheless, the algorithms given in Section 3.5 can still be used, with Ψ now being a two-dimensional wavelet transform, similar to the transform used in the JPEG2000 image compression standard, which has been shown to be extremely good at compressing 2D NMR spectra.[30] The same procedure would apply to higher dimensional spectra as well.

4 Experimental results

In this section, we present and discuss numerical results obtained using the algorithms described above, on real and simulated NMR mixtures.

Throughout this section, we term real mixtures as the spectra of the solutions that have been acquired by NMR spectroscopy. By simulated mixtures we mean spectra that have been computed using the mathematical LIM model (1), using the spectra of pure compounds measured by NMR spectroscopy in S, and the concentrations that have been used to produce the solutions in A.

We describe the datasets before discussing the results.

4.1 Data acquisition

4.1.1 Data description. Four commercially available solutions of terpenes were purchased from Sigma-Aldrich (Merck) and Saint Quentin Fallavier, France: (R)-(+)- limonene, nerol, α-terpinolene, and $(-)$-*trans*-caryophyllene. The initially pure compounds were dissolved in 600 µL of $CDCl_3$ at respective concentrations of 181 mM, 36.5 mM, 26.6 mM and 43.7 mM and then transferred to 5 mm NMR tubes which were sealed to prevent loss of solvent at the operating temperature. The samples were then stored at -4 °C until the NMR characterization. Five synthetic mixtures of the four terpenes were prepared, varying the concentrations of each compound as reported in Table 1. All the concentrations were recalculated using the ERETIC method (Electronic REference To access In vivo Concentrations) after the NMR tubes were sealed.[31]

Table 1 Concentrations of each component of the proposed terpenes

	Limonene	Nerol	α-Terpinolene	β-Caryophyllene
Solution 1	23.3 mM	26 mM	8.78 mM	10.87 mM
Solution 2	17.1 mM	11.93 mM	15.5 mM	15 mM
Solution 3	9.05 mM	14.23 mM	18.89 mM	4.67 mM
Solution 4	20.99 mM	6.86 mM	13.54 mM	11.96 mM
Solution 5	4.88 mM	9.01 mM	10.81 mM	13.15 mM

4.1.2 NMR spectroscopy. All experiments were performed on a Bruker Avance III spectrometer operating at 600 MHz for ^{1}H and equipped with a triple resonance high-resolution probe, using a SampleJet with a pre-cooling rack refrigerated at 4 °C. A standard 1D pulse sequence was applied to each sample: zg ^{1}H 1D (90°-Taq), with a spectral width of 6600 Hz, 96 scans, and a relaxation delay of 10 s. The 90° pulse length was automatically calibrated for each sample at around 9.5 μs. Subsequently, the spectra were pre-processed: phased and baseline corrected automatically and referenced to CDCl$_3$ at δ 7.27 ppm using the inbuilt software TOPSPIN 3.5 version (Bruker BioSpin, Germany).

2D ^{1}H–^{13}C HSQC spectra were recorded with phase sensitive sequence "hsqcetgpsi" using Echo/Antiecho-TPPI gradient selection with an INEPT delay adjusted to one-bond ^{1}H–^{13}C coupling constant of 145 Hz. 256 t1 points were acquired for the indirect dimension and 32 scans were acquired for each point with TD2 = 4096. The acquisition time for the direct period was 142 ms and the resolution was 7.045 Hz. For the indirect period, the acquisition time was 4.2 ms and the resolution was 235.80 Hz. No linear prediction was used for these experiments. Both ^{1}H and ^{13}C axes were calibrated using the chloroform peak at 7.27 and 77.2 ppm, respectively.

The spectra of the pure compounds are presented in the stacking plot, Fig. 1. The studied mixtures are well adapted to evaluate the source separation algorithms. Indeed, terpenes are natural molecules present in plants and have highly crowded spectra between 1.5 and 2.5 ppm.

4.2 Algorithm validation on 1D simulated mixtures

We first report on the simulation results. The goal is to validate the algorithms in a situation where a ground truth is available, in the framework of the LIM model described in Section 2.1. The ground truth is provided by: (1) the spectra of the pure compounds that were obtained as described in Section 4.1.2, and are collected in a source matrix S, and (2) the concentrations given in Table 1, organized in a 5 × 4 mixing matrix A.

From these, simulated mixtures of the form

$$X_m = AS + B$$

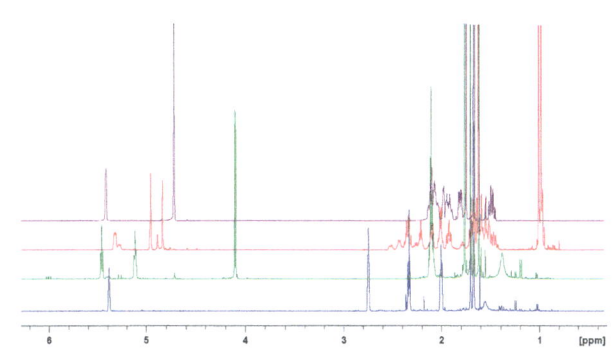

Fig. 1 ^{1}H NMR spectroscopy: a stacked plot of the spectra of the pure compounds. Top to bottom: limonene, β-caryophyllene, nerol, and α-terpinolene.

were generated according to the LIM model (1), involving the linear mixtures given by the matrix product AS, and a zero mean Gaussian white noise B. The standard deviation σ of the latter was set to the standard deviation of the experimental noise, estimated in a signal-free segment of the real mixtures X.

The six algorithms above (PALS, STALS, PALM, BC-VMFB and PALM, BC-VMFB using wavelets) were run on the simulated dataset. For the stopping criterion, we used the relative size of the objective function update from one iteration to the next, *i.e.*

$$\text{Crit}(k) = \left| \frac{F\left(X|A^{(k+1)}, S^{(k+1)}\right) - F\left(X|A^{(k)}, S^{(k)}\right)}{F(X|A^{(k)}, S^{(k)})} \right|,$$

where $A^{(k)}$ and $S^{(k)}$ are the estimates at iteration k. The algorithms also require an initial estimate. Several choices are possible, and here we used estimates obtained using the JADE ICA algorithm.[34] More precisely, running JADE on the mixture matrix yields an estimate for the un-mixing matrix, denoted by D, so that DX provides an estimate of the sources. Also, the pseudo-inverse $D^{\dagger}$ yields an estimate for the mixing matrix. Estimates obtained with JADE contain both positive and negative values which is incompatible with the non-negativity assumption. For the initialization, we thus use the absolute value of these estimates given by $S_{\text{init}} = |DX|$ and $A_{\text{init}} = |D^{\dagger}|$.

JADE only requires the number of sources to be estimated. A Principal Component Analysis (PCA) on the observation matrix shows that only 4 out of the 5 corresponding latent variables are significant, which suggests we set the number of sources to estimate to 4 (which turns out to be the actual number of terpenes present in the solutions).

STALS, PALM and BC-VMFB require choosing a thresholding parameter λ, for which five choices were tested, namely 0.01σ, 0.1σ, σ, 10σ and 100σ. Similar choices are made for the wavelet-based versions of PALM and BC-VMFB.

The un-mixing results on simulated data are globally very good for most (if not all) algorithms. The best results seem to be obtained by the STALS and PALS approaches, with various values of the thresholding parameter λ (we recall that PALS is a special case of STALS with $\lambda = 0$). The best estimate for the concentrations (mixing matrix A) was obtained by STALS with the threshold value set to 10σ (σ being the standard deviation of the noise). This corresponds to the relative errors reported in Table 2. This relative error has been computed as follows: let $\hat{A}$ be the mixing matrix estimate of A. Then the relative error (in %) is given by $(\hat{A} - A)/A \times 100$ where the quotient is computed element-wise.

Table 2 ^{1}H NMR spectra (simulated case): the relative errors in the estimated concentrations (in %) using STALS with $\lambda = 10\sigma$. The corresponding Amari index equals 0.008

	Limonene	Nerol	α-Terpinolene	β-Caryophyllene
Solution 1	0.12%	−0.67%	3.82%	0.09%
Solution 2	−0.29%	0.72%	0.07%	−0.07%
Solution 3	1.29%	−0.74%	−1.93%	0.52%
Solution 4	−1.31%	2.06%	−0.38%	−0.04%
Solution 5	3.68%	0.61%	0.64%	−0.15%

Table 3 ^{1}H NMR spectra (simulated case): the numerical results, using all algorithms, for $\lambda = 10\sigma$ (numbers between parentheses indicate the source number and m stands for the mean value)

	Algorithms					
	PALS	STALS	PALM	BC-VMFB	PALM (wav)	BC-VMFB (wav)
Amari	0.019	0.008	0.022	0.025	0.036	0.031
SIR (1)	26.7	52.4	22	24.4	20.7	22.2
SIR (2)	32.5	31.2	29.8	28.5	30.1	32.5
SIR (3)	19.5	45.7	41.6	25.2	24.3	23.5
SIR (4)	47.6	29.3	19.9	24	21.4	21.7
SIR (m)	31.6	39.6	28.3	25.5	24.1	25
SDR (1)	26.7	51.4	21.7	24.4	20.4	21.3
SDR (2)	32.5	31.2	29.6	28.5	29.7	29.5
SDR (3)	19.5	44.9	40.8	25.2	24	22.8
SDR (4)	47.4	28.6	19.6	23.9	21	20.6
SDR (m)	31.5	39	27.9	25.5	23.8	23.6

In Table 3 we provide the values of the evaluation indices for all the algorithms, λ being fixed to 10σ. For STALS, the SIR and SDR values globally range between 30 dB and 55 dB, which is generally considered very good.

The separation quality is exemplified in Fig. 2 where we can see good correspondences for the limonene spectrum while underlining some extra peaks coming from another source. As could be expected, the spurious peaks are significantly reduced for high values of the thresholding parameter λ.

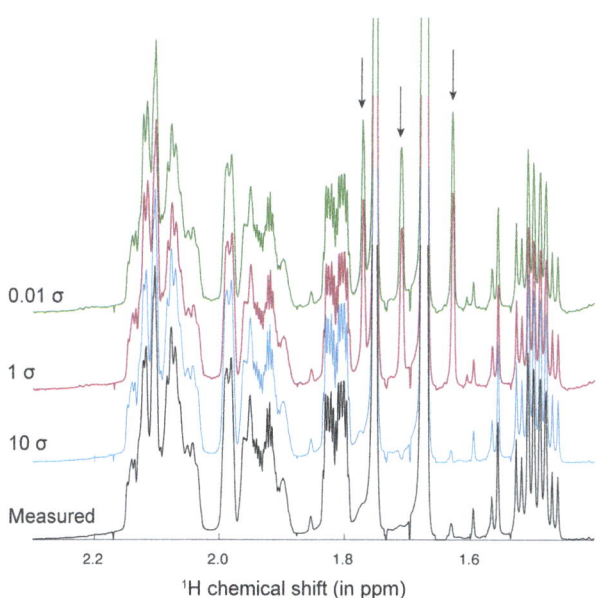

Fig. 2 Comparison between the ^{1}H NMR measured spectrum of limonene and the calculated spectra estimated with STALS for $\lambda = 0.01\sigma$, σ, 10σ (simulated case). The three arrows show the presence of 3 extra peaks that are not present when $\lambda = 10\sigma$. These extra peaks are residual signals from nerol.

Table 4 ^{1}H NMR spectra (simulated case): the numerical results for the STALS algorithm, with various values of the thresholding parameter λ numbers between parentheses indicate the source number and m stands for the mean value

	λ					
	0	$\sigma/100$	$\sigma/10$	σ	10σ	100σ
Amari	0.019	0.019	0.019	0.019	0.008	0.018
SIR (1)	26.7	26.7	26.7	26.8	52.4	42.5
SIR (2)	32.5	32.3	32.2	31.9	31.2	27.6
SIR (3)	19.5	19.5	19.5	19.6	45.7	39.3
SIR (4)	47.6	45.9	44.6	42.4	29.3	18.3
SDR (1)	26.7	26.7	26.7	26.8	51.4	39.7
SDR (2)	32.5	32.3	32.2	31.9	31.2	27.3
SDR (3)	19.5	19.5	19.5	19.6	44.9	35.4
SDR (4)	47.4	45.8	44.5	42.3	28.6	17.1

A closer look at the results shows that the best results are obtained with STALS with $\lambda = 10\sigma$ for two sources ((1) and (3)), and PALS for the other two. This indicates that the optimal thresholding parameter may be source dependent. This is further confirmed when looking at the results obtained with STALS using different regularization parameters λ (see Table 4).

Note that the other algorithms also yielded good un-mixing results on the simulated data.

4.3 Real 1D mixtures

The same algorithms as above were run on the real mixtures, and the results were, at first glance, very disappointing: the algorithms failed to separate the pure compound spectra from the mixture spectra. A closer analysis revealed that the main reason was the presence of spurious shifts between the pure and mixed spectra, the peaks of these two families of spectra were not correctly aligned. The

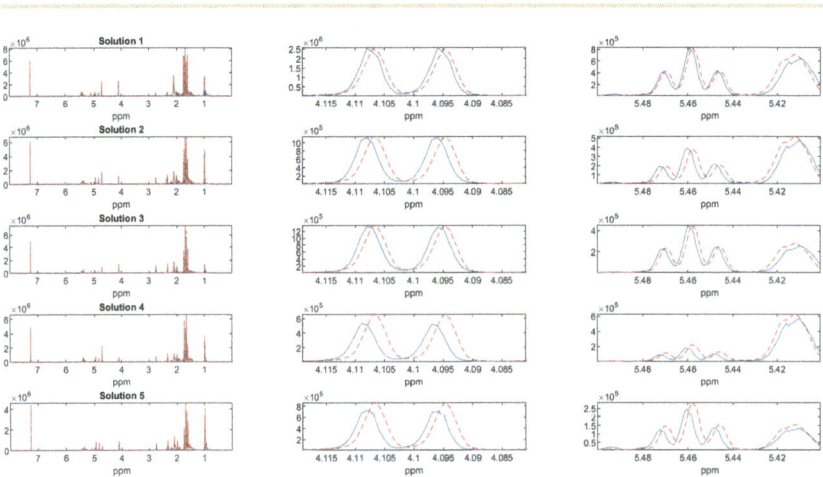

Fig. 3 ^{1}H NMR spectra: the comparison of X (blue) and X_m (red) before alignment preprocessing (whole spectra (left), expanded regions (middle, right)).

algorithms under consideration are extremely sensitive to such issues, and obtaining reliable un-mixing results without proper alignment is hopeless. In Fig. 3 we display the spectra of the real mixtures (*i.e.* the five rows of the X matrix) and the simulated mixtures (the rows of $X_m = AS$). The first column displays the complete spectra, while the second and third columns zoom in on specific regions. As can be seen, there is a clear shift between peaks, that may even be significantly location dependent (third column). This could be the result of a slight variation in pH value due to the decomposition of $CDCl_3$ or, of a difference in ionic strength between samples. Molecular interactions between the individual compounds could also contribute to the observed chemical shift variations.

To overcome this effect, all spectra were aligned using a standard tool. Several approaches of the peak alignment problem exist,[35] based upon various approaches such as correlation analysis, least squares, dynamic time warping, parametric time warping and several others. Here, the online tool NMRProcFlow was used to provide re-aligned spectra, on which the un-mixing algorithms could be suitably tested. This alignment method is based on a least squares algorithm for which a reference spectrum (here, the average spectrum) was calculated. Each region was re-aligned by shifting it to match the reference spectrum. For the sake of further comparison, the mixture spectra were aligned together with the pure spectra, which will not be possible in general situations where the latter will not be available.

The re-aligned spectra are displayed in Fig. 4, which again shows the complete spectra and zoomed in areas of the same regions as in Fig. 3. The procedure allowed us to fix the alignment problems quite successfully, as exemplified by columns 2 and 3. However, the latter also exhibited slight amplitude modulations, which suggests that the departure of real mixtures from the mathematical model includes more than peak shifts. To get a quantitative assessment of the adequacy of the model, we computed values of the SIR and SDR indices, that measure discrepancies between X and X_m. The results are given in Table 5. The SIR values are fairly acceptable, which tends to indicate that the peaks are located

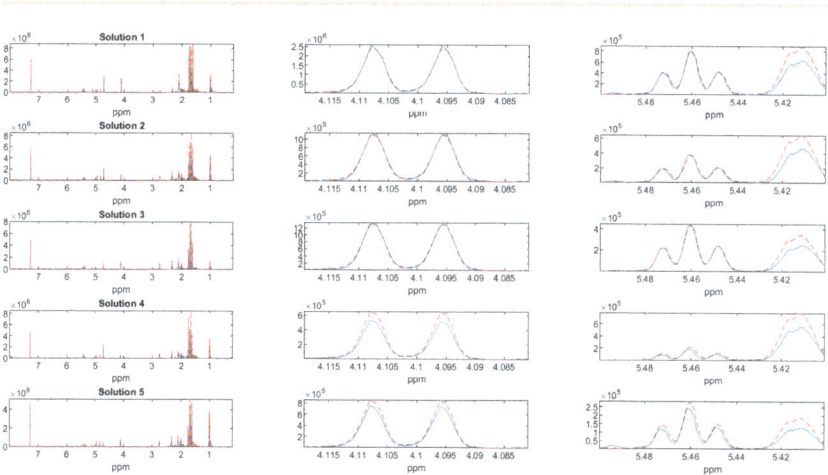

Fig. 4 ^{1}H NMR spectra: the comparison of X (blue) and X_m (red) after alignment preprocessing (whole spectra (left), expanded regions (middle, right)).

Table 5 ¹H NMR spectra: the SIR and SDR indices (in dB), comparing measured and simulated mixture spectra

	Sol. 1	Sol. 2	Sol. 3	Sol. 4	Sol. 5
SIR	16.12	12.96	18.89	19.95	17.15
SDR	10.86	8.84	12.90	9.76	9.72

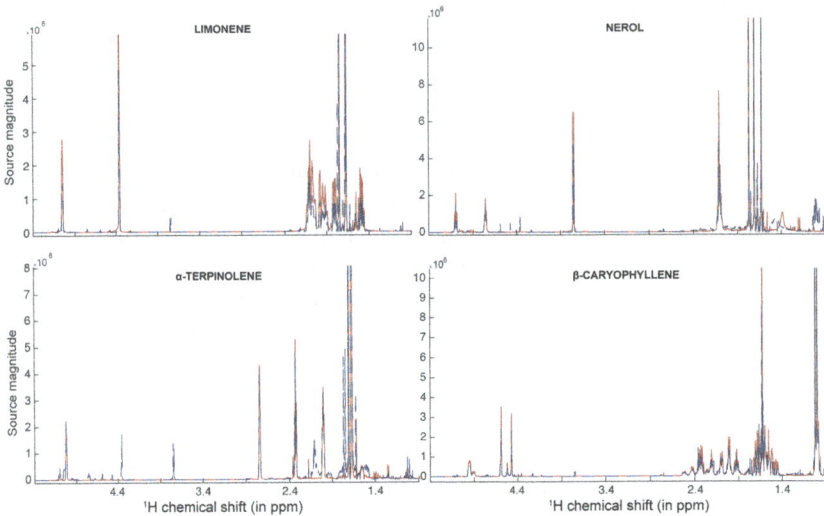

Fig. 5 ¹H NMR spectra (real case): the spectra of the 4 sources estimated using BC-VMFB with $\lambda = \sigma$ (blue) compared to the spectra of the real sources (red).

at the right place, but the SDR values are significantly lower, which we interpret mainly as a consequence of amplitude modulations. This suggests that a more adequate model for describing real mixtures should involve both shift and amplitude modulation, in combination with the linear instantaneous mixing model (1).

The un-mixing algorithms were run on aligned real mixture spectra. The resulting un-mixed spectra for the BC-VMFB algorithm with $\lambda = \sigma$ are displayed in Fig. 5, and the corresponding evaluation indices are given in Table 6.

β-caryophyllene was extracted with great accuracy while limonene presented 4 extra peaks that belonged to nerol. Although the major signals from the nerol spectrum were well recovered, an extra signal coming from β-caryophyllene was also observed at about 1 ppm. The worst result was obtained for the α-terpinolene

Table 6 ¹H NMR spectra (real case): the SIR and SDR indices (in dB) comparing the true and estimated source spectra with the BC-VMFB algorithm with $\lambda = \sigma$

	Limonene	Nerol	α-Terpinolene	β-Caryophyllene
SIR	13.1	15.0	20.2	14.8
SDR	9.7	9.9	4.8	6.8

Table 7 ^{1}H NMR spectra (real case): the relative errors in the estimated concentrations (in %) using BC-VMFB with $\lambda = \sigma^a$

	Limonene	Nerol	α-Terpinolene	β-Caryophyllene
Solution 1	12.34%	25.61%	−9.03%	−4.33%
Solution 2	4.05%	−6.68%	1.11%	26.27%
Solution 3	−47.70%	1.91%	−0.53%	−32.53%
Solution 4	16.36%	−69.78%	1.54%	−11.15%
Solution 5	−55.03%	−14.94%	4.74%	−4.70%

^{a} The corresponding Amari index equals 0.081.

spectrum where signals from this compound were mixed with some signals from nerol and limonene. However, despite the presence of these artifacts, the major signals of each source spectrum were accurately found, allowing the identification of the corresponding molecule without ambiguity.

In Table 7 we report the relative errors on the mixing matrix estimate.

4.4 Results on the HSQC mixtures

The algorithms were also tested on HSQC data, and here we report on the results.

4.4.1 Numerically simulated HSQC mixtures. The first tests were on simulated mixtures, *i.e.* mixtures generated using the mathematical formulas described in Section 2.2, using the measured source matrix and a mixing matrix corresponding to the true concentrations in solutions. In Table 8 we provide a summary of the results obtained using the BSS algorithms with the thresholding parameter set to $\lambda = 100\sigma$, with σ being the standard deviation of the noise measured in a signal-free part of the spectra. The results are globally very good, in particular the SIR values (remember that the SIR provides a measure of the cross-talk between sources, in other words the presence in an estimated source of spurious components originating from the others), which are quite high. The SDR values are lower, which indicates that distortions are present in the estimated

Table 8   2D ^{1}H–^{13}C HSQC spectra (simulated case): the numerical results for $\lambda = 100\sigma$ (numbers between parentheses indicate the source number and m stands for the mean value)

	Algorithms					
	PALS	STALS	PALM	BC-VMFB	PALM (wav)	BC-VMFB (wav)
Amari	0.135	0.012	0.015	0.11	0.081	0.016
SIR (1)	21.6	50.1	39.6	39.8	20.9	42.8
SIR (2)	12.4	39.1	31.7	10.3	21.8	34.6
SIR (3)	22.0	29.0	28.0	9.7	31.5	31.4
SIR (4)	8.3	41.4	45.4	20.8	34.2	28.6
SIR (m)	16.1	39.9	36.2	20.2	27.1	34.4
SDR (1)	20.5	25.9	25.7	24.6	19.8	24.9
SDR (2)	11.6	16.0	17.2	9.3	16.4	16.4
SDR (3)	15.1	13.6	13.5	8.6	14.1	13.2
SDR (4)	7.1	9.9	10.1	10.5	10.4	9.8
SDR (m)	13.6	16.3	16.6	13.3	15.2	16.1

Table 9   2D ^{1}H–^{13}C HSQC spectra (real case): the SIR and SDR indices (in dB), comparing pure and estimated source spectra using wavelet-based BC-VMFB with $\lambda = 10\sigma$

	Limonene	Nerol	α-Terpinolene	β-Caryophyllene
SIR	46.3	25.2	22.7	32.1
SDR	8.5	13.1	9.4	9.4

Table 10   2D ^{1}H–^{13}C HSQC spectra (real case): the relative errors in the estimated concentrations (in %) using wavelet-based BC-VMFB with $\lambda = 10\sigma$. The corresponding Amari index equals 0.042

	Limonene	Nerol	α-Terpinolene	β-Caryophyllene
Solution 1	−0.39%	3.16%	−7.02%	9.53%
Solution 2	−8.56%	−3.43%	−2.66%	10.90%
Solution 3	5.84%	8.42%	2.94%	6.68%
Solution 4	5.74%	−14.40%	3.47%	−12.84%
Solution 5	−3.65%	−6.89%	0.03%	−11.00%

sources. Even though it is not easy to draw clear conclusions, the best results seem to be obtained with the STALS algorithm. Notice that the wavelet-based BC-VMFB algorithm yields quite good results as well. The latter algorithm turns out to be the most effective on real mixtures. We do not display the graphical comparison of the real and estimated HSQC spectra, as we prefer to focus on the real mixtures (see below). However, let us mention that the results are visually excellent, the fingerprint of each terpene is perfectly recovered.

4.4.2   **Real HSQC mixtures.** The algorithms were also tested on real mixtures (the same solutions as the ones reported in the section devoted to ^{1}H NMR spectroscopy). The results in this case too are quite good. We do not report on the performances of all algorithms, and we focus on the best performing one, which turns out to be the wavelet-based BC-VMFB (mentioned above), with the thresholding parameter set to $\lambda = 10\sigma$. The corresponding SIR and SDR indices are provided in Table 9, and the Amari index equals 0.042, meaning that the concentrations have been correctly estimated (see Table 10 for the relative errors on the estimated concentrations). As can be seen, the SIR values are again very good (around 32 dB on average), which indicates that even in crowded regions of the spectra, no significant cross-talk between sources is observed. The SDR values remained significantly weaker, meaning that the estimated spectra were significantly perturbed. As in the 1D situation, this may be interpreted as a departure from the 2D LIM model. Again, as in the 1D case, a visual inspection of the estimated sources *versus* the true one in Fig. 6 shows that the results are of sufficient quality to identify the four terpenes in these solutions.

Remark: As in the 1D case, the adequacy of the LIM model for HSQC spectra can be assessed by looking at quality indices. The SIR and SDR indices computed from the measured and simulated mixture spectra are provided in Table 11. The conclusions that can be drawn are essentially similar to the previous ones. The SIR values are acceptable, at least for solutions (1–3), and are lower for solutions 4

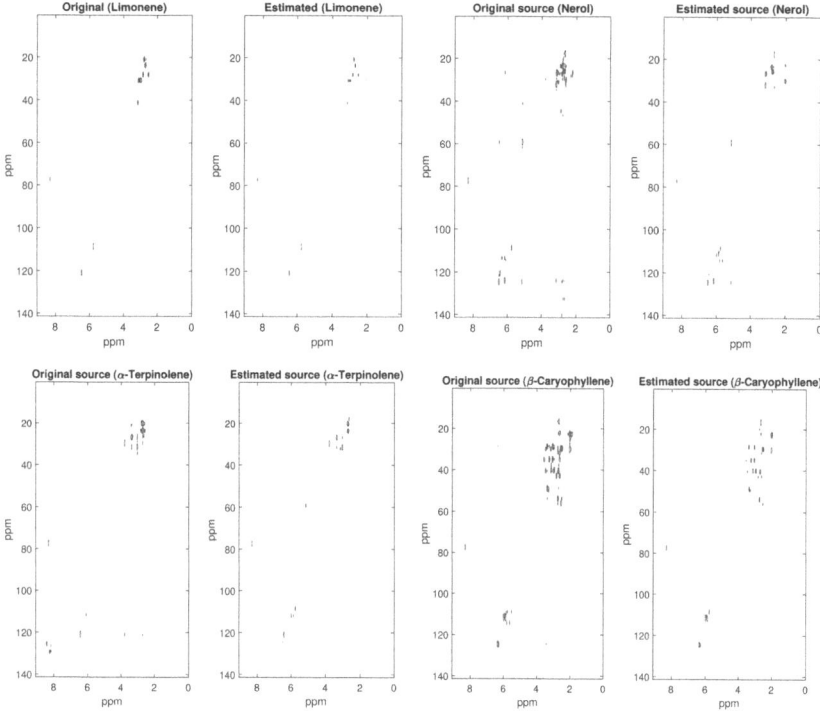

Fig. 6 2D ^{1}H–^{13}C HSQC spectra (real case): a comparison of the pure HSQC spectra (left) and the HSQC spectra estimated from the real mixtures (right), according to the two-dimensional LIM model.

Table 11 2D ^{1}H–^{13}C HSQC spectra: the SIR and SDR indices (in dB), comparing measured and simulated mixture spectra

	Sol. 1	Sol. 2	Sol. 3	Sol. 4	Sol. 5
SIR	15.83	14.04	19.01	13.93	11.29
SDR	8.42	9.33	8.02	8.36	7.11

and 5. The SDR values are very low, which may be interpreted in terms of departures from the LIM model.

5   Discussion and conclusions

In this paper we have presented a general approach for the blind identification of compounds from solutions using NMR spectroscopy and blind source separation algorithms. It is worth recalling that these algorithms are blind in the sense that they attempt to jointly estimate pure compound spectra and concentrations in solutions. From the mathematical point of view, we have provided a general algorithmic approach that includes, as special instances, a number of different algorithms, which we could test and compare. We also considered, for the sake of quantitative performance evaluation, some quantities (Amari index, various forms of the signal to noise ratio) that had been introduced long ago in the blind source separation literature. Numerical tests were performed on data specifically generated for this work, including 1D as well as 2D HSQC spectra.

The results presented here show that blind source separation algorithms have the potential to perform successfully. For the studied dataset, the results on the simulated data range from good to excellent, depending on the algorithm and parameter values. For the real data, good results could be obtained provided some important pre-processing steps could be done carefully.

However, our results also raise a number of questions, some of which we list below, that should be addressed before drawing more complete conclusions.

• The results on the 1D data show that pre-processing is a crucial step. In the case considered here, alignment of the spectra turned out to be fundamental. However, even after careful alignment, the Linear Instantaneous Mixture model turned out to be incompletely satisfactory, with the aligned data showing significant departures from that model. Therefore, this may suggest that the model is not 100% adequate, and that it would be worth considering more complex models, that could include, for instance, amplitude modulations as we observed on the terpene data we studied. One may also imagine including spectral shift in the model, provided the phenomenon could be sufficiently well understood and therefore modelled.

• In the same spirit, this would also suggest modifying the un-mixing algorithms to estimate the amplitude modulations (and shifts) at the same time as the concentrations and pure compound spectra. There are situations in signal and image processing where one faces similar problems, it might be possible to transpose corresponding approaches to the case under consideration here.

• In the considered datasets, alignment was not problematic for the HSQC spectra, for which the un-mixing algorithms could be run without pre-processing. The un-mixing results on the HSQC data turned out to be of very good quality, with the algorithms being able to clearly identify 2D fingerprints of the four terpenes, as well as the concentrations. This is confirmed by the satisfactory values for the interference index (SIR) and Amari index. Besides, the distortion index (SDR) is significantly weaker, which may indicate (as in the 1D case) that additional distortions should be taken into account in the model. However, as such indices have not been used so far (to the best of our knowledge) in NMR spectroscopy, such conclusions must be taken cautiously. More experiments are needed to validate the use of such tools in this context. It is also worth pointing out that the computational burden is significantly higher in the 2D case, and can be expected to grow fast when the dimension of the spectra increases. This will be an important problem to address if one wants to proceed to higher dimensional spectra, where sparsity is expected to be higher, and further facilitate the separation.

• Whatever the model, blind separation problems are always ill-posed problems, and in the framework of variational formulations, they result in non-convex minimization problems. This means that the objective function to be optimized can have (and as a matter of fact, has) several (and often, many) local minima, and algorithms are extremely sensitive to initialization. We have stuck here to a simple choice, that was advocated in ref. 18 for different algorithms. It is not clear that this choice is the most relevant for the family of approaches studied in this paper, and this point clearly deserves an in-depth study. Very much in the same spirit, the fact that different algorithms that aim at optimizing the same objective function actually yield significantly different results and performances raises questions, even though there is no guarantee that they should give the same result, given the non-convexity of the problem. Again, further investigations are necessary at this point.

Conflicts of interest

There are no conflicts to declare.

Acknowledgements

The project leading to this publication has received funding from the Excellence Initiative of Aix-Marseille University – A*Midex, a French "Investissements d'Avenir" program. The research presented in this article is part of a project launched in collaboration with our friend and colleague Stefano Caldarelli, who passed away in late 2018. His knowledge, ideas and enthusiasm have been at the heart of this work, and we miss him very much. This article is dedicated to his memory.

Notes and references

1 D. Marion, P. C. Driscoll, L. E. Kay, P. T. Wingfield, A. Bax, A. M. Gronenborn and G. Clore, *Biochemistry*, 1989, **28**, 6150–6156.
2 C. S. Johnson, *Prog. Nucl. Magn. Reson. Spectrosc.*, 1999, **34**, 203–256.
3 M. G. N. Reddy and S. Caldarelli, *Anal. Chem.*, 2010, **82**, 3266–3269.
4 I. Toumi, S. Caldarelli and B. Torrésani, *Prog. Nucl. Magn. Reson. Spectrosc.*, 2014, **81**, 37–64.
5 D. Nuzillard, S. Bourg and J.-M. Nuzillard, *J. Magn. Reson.*, 1998, **133**, 358–363.
6 R. D. Boyer, R. Johnson and K. Krishnamurthy, *J. Magn. Reson.*, 2003, **165**, 253–259.
7 A. Bax and D. G. Davis, *J. Magn. Reson.*, 1985, **65**, 355–360.
8 G. N. M. Reddy and S. Caldarelli, *Chem. Commun.*, 2011, **47**, 4297–4299.
9 K. F. Morris and C. S. Johnson, *J. Am. Chem. Soc.*, 1992, **114**, 3139–3141.
10 O. Beckonert, H. C. Keun, T. M. D. Ebbels, J. Bundy, E. Holmes, J. C. Lindon and J. K. Nicholson, *Nat. Protoc.*, 2007, **2**, 2692.
11 L. Frydman, A. Lupulescu and T. Scherf, *J. Am. Chem. Soc.*, 2003, **125**, 9204–9217.
12 M. Mobli, M. W. Maciejewski, A. D. Schuyler, A. S. Stern and J. C. Hoch, *Phys. Chem. Chem. Phys.*, 2012, **14**, 10835–10843.
13 A. L. Guennec, P. Giraudeau and S. Caldarelli, *Anal. Chem.*, 2014, **86**, 5946–5954.
14 J. Xia, T. C. Bjorndahl, P. Tang and D. S. Wishart, *BMC Bioinf.*, 2008, **9**, 507.
15 K. Bingol and R. Brueschweiler, *Anal. Chem.*, 2011, **83**, 7412–7417.
16 A. Cichocki and S.-I. Amari, *Adaptive Blind Signal and Image Processing: Learning Algorithms and Applications*, John Wiley & Sons, Inc., 2002.
17 P. Comon and C. Jutten, *Handbook of Blind Source Separation: Independent Component Analysis and Applications*, Academic Press, 1st edn, 2010.
18 I. Toumi, B. Torrésani and S. Caldarelli, *Anal. Chem.*, 2013, **85**, 11344–11351.
19 P. Paatero and U. Tapper, *Environmetrics*, 1994, **5**, 111–126.

20 P. O. Hoyer, *J. Mach. Learn. Res.*, 2004, **5**, 1457–1469.

21 P. L. Combettes and J.-C. Pesquet, *Fixed-point algorithms for inverse problems in science and engineering*, Springer Verlag, 2010, pp. 185–212.

22 C. Chaux, P. L. Combettes, J.-C. Pesquet and V. R. Wajs, *Inverse Probl.*, 2007, **23**, 1495–1518.

23 D. D. Lee and H. S. Seung, *Nature*, 1999, **401**, 788–791.

24 D. D. Lee and H. S. Seung, *Proc. Ann. Conf. Neur. Inform. Proc. Syst.*, 2001, pp. 556–562.

25 C. Févotte and J. Idier, *Neural Comput.*, 2011, **23**, 2421–2456.

26 E. Vincent, R. Gribonval and C. Févotte, *IEEE Trans. Audio, Speech, Lang. Process.*, 2006, **14**, 1462–1469.

27 S. Mallat, *A Wavelet Tour of Signal Processing, Third Edition, The Sparse Way*, Academic Press, 3rd edn, 2008.

28 I. Kopriva, I. Jeric and V. Smrecki, *Anal. Chim. Acta*, 2009, **653**, 143–153.

29 X. Shao, H. Gu, J. Wu and Y. Shi, *Appl. Spectrosc.*, 2000, **54**, 731–738.

30 J. C. Cobas, P. G. Tahoces, M. Martin-Pastor, M. Penedo and F. J. Sardina, *J. Magn. Reson.*, 2004, **168**, 288–295.

31 S. Akoka, L. Barantin and M. Trierweler, *Anal. Chem.*, 1999, **71**, 2554–2557.

32 J. Bolte, S. Sabach and M. Teboulle, *Math. Program.*, 2014, **146**, 459–494.

33 E. Chouzenoux, J.-C. Pesquet and A. Repetti, *J. Optim. Theory Appl.*, 2014, **162**, 107–132.

34 J.-F. Cardoso, *Proc. IEEE*, 1998, **86**, 2009–2025.

35 T. Vu and K. Laukens, *Metabolites*, 2013, **3**, 259–276.

Faraday Discussions

PAPER

Systems biology approach to elucidation of contaminant biodegradation in complex samples – integration of high-resolution analytical and molecular tools†

Caroline Gauchotte-Lindsay, 📍 *a Thomas J. Aspray,‡b Mara Knappc and Umer Z. Ijaz 📍 a

Received 14th February 2019, Accepted 21st March 2019

DOI: 10.1039/c9fd00020h

We present here a data-driven systems biology framework for the rational design of biotechnological solutions for contaminated environments with the aim of understanding the interactions and mechanisms underpinning the role of microbial communities in the biodegradation of contaminated soils. We have considered a multi-omics approach that employs novel *in silico* tools to combine high-throughput sequencing data (16S rRNA amplicons) with chemical data including high-resolution analytical data generated by comprehensive two-dimensional gas chromatography (GC × GC). To assess this approach, we have considered a matching dataset with both microbiological and chemical signatures available for samples from two former manufactured gas plant sites. On this dataset, we applied the numerical procedures informed by ecological principles (predominantly diversity measures) as well as recently published statistical approaches that give discriminatory features and their correlations by maximizing the covariances between multiple datasets on the same sample space. In particular, we have utilized *sparse projection to latent discriminant analysis* and its derivative to multiple datasets, an N-integration algorithm called DIABLO. Our results indicate microbial community structure dependent on the contaminated environment and unravel promising interactions of some of the microbial species with biodegradation potential. To the best of our knowledge, this is the first study that incorporates with the microbiome an unprecedented high-level distribution of hydrocarbons obtained through GC × GC.

aInfrastructure and Environment Research Division, School of Engineering, University of Glasgow, Glasgow G12 8QQ, UK. E-mail: caroline.gauchotte-lindsay@glasgow.ac.uk

bSchool of Energy, Geoscience, Infrastructure and Society, Heriot-Watt University, Edinburgh EH14 4AS, UK

cDepartment of Civil and Environmental Engineering, University of Strathclyde, Glasgow G1 1XQ, UK

† Electronic supplementary information (ESI) available. See DOI: 10.1039/c9fd00020h

‡ Now at: ERS Ltd, Westerhill Road, Bishopbriggs, Glasgow G64 2QH, UK.

1. Introduction

Industrialisation has left behind a legacy of pollution that presents a risk to human health and the environment. Disposal to landfill has long been the preferred approach for disposal of contaminated soils, however rising landfill tax costs drive the development of novel, cheaper and safer remediation technologies.

Newly developed sustainable and safe remediation approaches include *ex situ* (such as biopiles) and *in situ* (monitored natural attenuation (MNA) and enhanced natural attenuation (ENA)) bioremediation. MNA consists of the monitoring and testing of the progress of natural processes that degrade contaminants *in situ*, and biopiles and ENA consist of stimulating *in situ* microbial processes by either introducing new microorganisms (bioaugmentation) or modifying the environmental conditions to stimulate the growth of degrading organisms (biostimulation).[1,2] They are advantageous practices because they produce minimal waste and disturbance to the site but also minimise contact between remediation engineers and the contaminated site.

The microbial organisms and catabolic genes involved in the biodegradation of organic contaminants have been well characterised for a large variety of contaminants including polycyclic aromatic hydrocarbons (PAHs). PAHs are recognised lipophilic legacy organic pollutants present in crude oil and are produced during the combustion of coal and organic matter – they are therefore ubiquitous soil contaminants. They are usually present in their thousands in complex environmental samples but only a few, such as naphthalene, phenanthrene or benzo(*a*)pyrene, are recognized as carcinogenic or mutagenic and are currently on priority pollutant lists. Therefore, the biodegradation of PAHs by microbes has up until now been almost exclusively studied *in vitro* in microcosms and is usually demonstrated for a single compound or a very limited number of compounds. This is, however, an overly simplistic view of what happens *in situ* where multiple contaminants are present and their fate likely to be interdependent.

The design of smart and efficient bioremediation solutions in soil requires extracting from the exhaustive knowledge on the degradation of complex samples the information that will enable enhancement of the degradation of listed contaminants. Eventually, the aim is also to optimise analysis carried out during site investigation to provide remediation practitioners with the necessary information to design remediation approaches. Before this is achievable, however, understanding what the relevant factors governing biodegradation of PAHs in soil are is crucial. These factors will include: soil characteristics, physicochemical characteristics (*e.g.* pH and organic matter content), microbial ecology and the nature of the contaminations (PAH distribution and concentration, but also the presence of organic and inorganic co-contaminants).

The recent rapid technological advances in nucleic acid sequencing have enabled the high-resolution characterisation of microbial communities in PAH contaminated soils. Specifically, metataxogenomic and metagenomic approaches have demonstrated that contamination induces a reduction in species richness and enriches the population with species that are adapted to hydrocarbon degradation.[3] Consequently, initial microbial ecology could also be linked to the potential for PAH degradation,[4] demonstrating that microbial ecology carries meaningful information on the potential for bioremediation in soil. However, without high resolution information on the contamination profile, notably the presence of other contaminants, this information might not be sufficient for the design of an efficient site-specific bioremediation. Comprehensive two-dimensional gas chromatography coupled with mass spectrometry (GC × GC-MS) has allowed the near comprehensive characterisation of semi-volatile organic carbons (SVOCs) in tars and contaminated soils from former manufactured gasworks (FMGs).[5-7] The coupling of two columns allows for a two-dimensional separation across a retention plane rather than along a retention line. The peak capacity of GC × GC is several orders of magnitude higher than traditional techniques and thousands of compounds can be resolved; coupled with the resolution power of mass spectrometry, it generates ultra-resolution chemical signatures of environmental samples. It has been used for source apportionment,[8] monitoring of bioremediation across SVOC classes[6] and estimation of ecotoxicity.[9]

Integration of metataxogenomics and GC × GC analysis has the potential to unveil information on the interactions between complex mixtures of environmental contaminants and microbial ecology never accessed before. Integrative multivariate statistical approaches for multi-omics data are being developed in biomedical applications.[10] For instance, the DIABLO[10] method builds on the Generalised Canonical Correlation Analysis, which integrates multiple datasets by finding principal components (latent variables) that maximize the covariance of scores between different datasets and the categorical outcome of interest. The resulting loading vectors are then constrained to give discriminants that correlate between these datasets.

Here, through the comparison of soils from two different FMGs, we present an example of statistical integration of chemical and molecular data in contaminated soils. Chemical and DNA extracts were analysed and processed through robust methods and pipelines[1-4] previously developed in our laboratories and statistical analysis was first carried out independently. Multivariate analysis enabled forensic characterisation of the site by exploring intra- and inter-site variations in distributions of PAHs and other compounds.[2] 16S rRNA sequencing data were explored for alpha and beta diversity analyses,[5] inferred metabolic pathway analysis[6] and differential abundance analysis of both species and metabolic pathways between sites.[5] Then, the N-integration algorithm DIABLO[7] was used to classify and discriminate features that correlate between the microbiome and chemical metadata including, for the first time, high-resolution distribution of SVOCs obtained by GC × GC-MS.

2. Experimental

2.1. Samples

Samples from two different former manufactured gas plants were provided by collaborators. Very little information was provided about the samples so no spatial resolution has been attempted in this work. The samples were stored in plastic tubes at 4 °C. Eighteen samples came from a site in the United Kingdom (COV site) and nine samples from a site in Switzerland (CH site). The COV soil samples, generally dark, compact and sticky in nature, were each sieved once with a 10 mm mesh. The CH samples, brown in colour and drier in nature, were each sieved through 10, 2.36 and 1.70 mm meshes. For each individual sample, the moisture content in percentage was measured by placing a subsample in an oven at 105 °C for 24 h, and loss on ignition (LOI), also in percentage, was measured by placing a subsample in a furnace at 550 °C for 2 h. No further soil characterisation was carried out at this stage.

2.2. Semivolatile organic compound analysis

Approximately 0.25 g of each soil sample was extracted *via* pressurised liquid extraction (PLE) with on-line silica gel clean-up.[5] The first fraction (hexane) and second fraction (hexane : toluene) (v/v) (8 : 2) were collected together and concentrated to 1.15 mL. Quantification of 14 PAHs along with four surrogates was carried out using GC-MS. It was carried out in TIC or SIM mode depending on the concentrations. We first validated the PLE method by extracting two of the COV samples six times and quantifying the 14 PAHs in duplicates. The relative standard deviation for the quantification of an individual PAH varied between 1% and 16.2%, with only one value over 10% demonstrating that the extraction was repeatable. Consequently, all other samples were only extracted once.

To optimise chromatographic resolution, we also employed a GC × GC method we had previously developed for the exhaustive characterisation of environmental coal tar samples.[5] The method ensured optimal separation of PAH isomers but also separated alkanes and alkylated benzenes. Comprehensive semi-volatile signatures of the extracts[5] were obtained using a LECO (St. Joseph, Michigan) time of flight mass spectrometer, model Pegasus 4D, connected to an Agilent 7890A gas chromatograph equipped with a LECO thermal modulator. The column set-up employed is known as reversed phase, with a first dimension capillary column (TR50-MS; 30 m × 0.25 mm i.d. × 0.25 μm film thickness; Thermo) that was more polar than the second capillary column (Rxi-5Sil; 2 m × 0.25 mm i.d. × 0.25 μm film thickness; Thames Restek). In this set-up, PAHs have a short retention time in the second dimension while alkanes are retained for longer. Alkylbenzenes, alkynes and alkenes, in order of polarity, elute in the retention space between PAHs and alkanes.[5] To be input into multivariate analysis, the GC × GC data must first be aligned between samples. Alignment can be carried out either by aligning chromatograms or peak tables (the output of GC × GC data processing). Here, we employed a combination of peak picking using the LECO ChromaTOF software and peak alignment using the R code R2DGC.[11] We optimised the data processing and R2DGC parameters but decided not to carry out any manual tidying up of the peak tables, which were left. Indeed, the peak tables contained over 500 compounds each, and a manual check of each peak for truly exhaustive analysis was not feasible. Two instrumental duplicates for each sample were analysed. For one COV soil sample, five replicate PL extracts were analysed in duplicate. Additional instrumental replicates were included for some samples because the second-dimension capillary column needed replacing during the study, shifting both retentions; hence samples were re-analysed after the change in columns. All peak tables were aligned to the CH peak table that contained the most peaks (top left in Fig. 1). The alignment of 68 peak tables was carried out twice with missing value limits equal to 0.1 *i.e.* a compound had to appear in at least 10% of the peak tables to be included in the alignment table, and 1 where a compound had to be present in all peak tables.

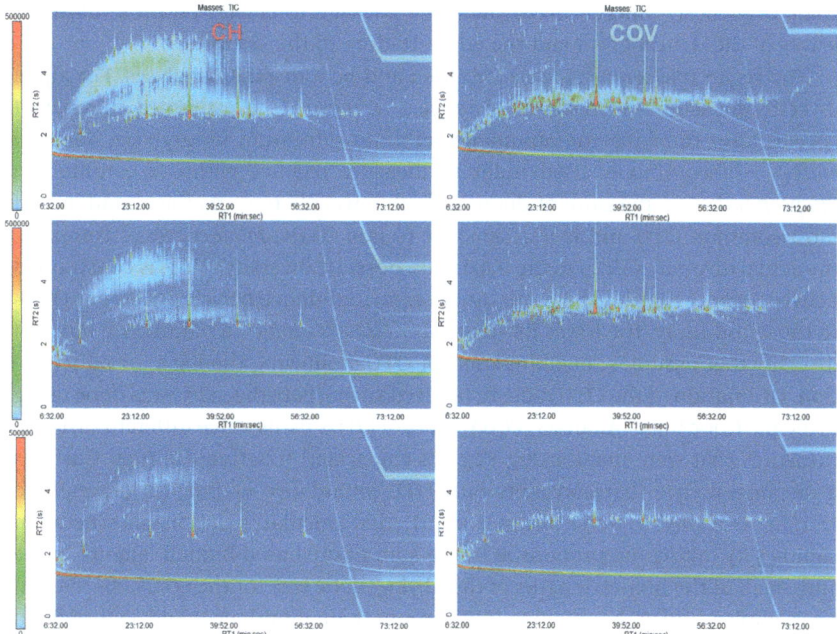

Fig. 1 Representative GC × GC chromatograms for both sites. CH is on the left and COV is on the right.

All methods are described in detail in the ESI.†

2.3. Transition metal analysis

The concentrations of lead, iron, cadmium, chromium, zinc, copper, and nickel were determined using inductively coupled plasma optical emission spectrometry (ICP-OES) and the methods are discussed in the ESI.†

2.4. Genomic DNA analysis

Genomic DNA was extracted twice from soil samples (0.25 g fresh weight); once for quantitative PCR (qPCR) for the alkB[12] and PAH RHD GN and PAH RHD GP genes, and a second time for 16S (V3 and V4 regions) metataxogenomic library preparation and sequencing. The detailed methods and bioinformatics workflow are presented in the ESI.†

2.5. Statistical analysis

All statistical analyses were performed in R.

Hierarchical clustering analysis (HCA) of the GC × GC alignment table was performed by calculating Manhattan distance between the samples each comprising 961 metabolites. Prior to analysis, zero values were replaced by one third of the smallest value in the table and each peak area was normalised to the calculated total peak areas for a given sample. Afterwards, we performed agglomerative clustering using complete linkages utilising R's hclust() function. For visualisation, we used R's dendextend package.[13] We then used the color_branches() function from the same package to cluster both the terminal leaves of the dendrogram and the edges leading to the samples.

The vegan package was used for alpha and beta diversity analyses. For alpha diversity measures we have used: *rarefied richness* – the estimated number of species after rarefying the abundance table to minimum library size; *Shannon entropy* – a commonly used index to measure balance within a community; *Simpson index* – a measure of dominance that weighs towards the abundance of most common OTU and is less sensitive to rare OTUs, *Pilou eveness* – compares the actual diversity values to the maximum possible diversity value, is constrained between 0 and 1.0, and the more variation in abundance between different OTUs in the community, the lower its value; and *Fisher's alpha* – a parametric index of diversity that assumes the abundance of OTUs following the log series distribution. Ordination of the OTU table in reduced space (beta diversity) was done using Principal Coordinate Analysis (PCoA) plots of OTUs using three different distance measures that were made using Vegan's cmdscale() function: (1) Bray–Curtis is a distance metric that considers only OTU abundance counts, (2) unweighted UniFrac is a phylogenetic distance metric that calculates the distance between samples by taking the proportion of the sum of unshared branch lengths in the sum of all of the branch lengths of the phylogenetic tree for the OTUs observed in two samples, and without taking into account their abundances and, (3) weighted UniFrac is a phylogenetic distance metric combining phylogenetic distance with relative abundances. This places emphasis on dominant OTUs or taxa. UniFrac distances were calculated using the phyloseq package.[14]

Analysis of the variance for explanatory variables (or sources of variation) was performed using Vegan's adonis() against distance matrices (Bray–Curtis/unweighted UniFrac/weighted UniFrac). This function, referred to as PERMANOVA, fits linear models to distance matrices and uses a permutation test with pseudo-F ratios. To give an account of environmental filtering (phylogenetic overdispersion *versus* clustering), phylogenetic distances within each sample were further characterised by calculating the nearest taxa index (NTI) and net relatedness index (NRI). This analysis aimed to determine whether the community structure was stochastic (overdispersion and driven by competition among taxa) or deterministic (clustering and driven by strong environmental pressure). The NTI was calculated using mntd() and ses.mntd(), and the mean phylogenetic diversity (MPD) and NRI were calculated using mpd() and ses.mpd() functions from the picante package.[15] NTI and NRI represent the negatives of the output from ses.mntd() and ses.mpd(), respectively. Additionally, they quantify the number of standard deviations that separate the observed values from the mean of the null distribution (999 randomisation using null.model-'richness' in the ses.mntd() and ses.mpd() functions and only considering taxa as either present or absent regardless of their relative abundance). Based upon the recommendations,[16] only the top 1000 most abundant OTUs were used for the calculations.

Discriminant analyses between the two sites were considered using microbiome data alone, and then together with the meta data (Table 1). For the former case, we used Sparse Projection to Latent Structure-Discriminant Analysis (sPLS-DA) with the R's mixOmics package.[10] The procedure constructs artificial latent components of the predicted dataset (OTUs table denoted $X(N \times P)$) and the response variable (denoted Y with categorical information of samples, *e.g.* CH and COV) by factorizing these matrices into scores and loading vectors in a new space

Table 1 Summary of analyses (meta means metadata for DIABLO analysis)

Chemical analysis	Molecular biology	Statistical analysis
Moisture content (A)	qPCR: alkB and PAH RHD GN and PAH RHD GP (I)	HCA of GC × GC signatures (meta: E)
LOI (B)	16S metataxogenomic sequencing (II)	Alpha and beta diversity (rDNA: I)
Quantification of 14 PAHs by GC-MS (C)		Tax4Fun KEGG pathways analysis (rDNA: I)
Quantification of 7 transition metals by ICP-OES (D)		*Discriminant analysis:*
GC × GC-TOFMS signatures (missing values limit = 10%) (E)		PLS-DA (rDNA: I)
GC × GC-TOFMS signatures (missing values limit = 100%) (F)		DIABLO 1 (rDNA: I; meta: II + A + B + C + D). DIABLO 2 (rDNA: I; meta: II + A + B + C + D + F).

such that the covariance between the scores of these two matrices $\mathrm{cov}(X_h a_h, Y_h b_h)$ in this space is maximized under two constraints: $\|a_h\|_2 = 1$; and $\|a_h\|_1 \leq \lambda$, where a_h and b_h are the corresponding loading vectors for X and Y, and h represents the number of components (akin to PCA analysis). To integrate meta data further, we utilised DIABLO from R's mixOmics package. We have combined $M = 2$ datasets denoted $X^{(1)}(N \times P_1)$, $X^{(2)} (N \times P_2)$ where $X^{(1)}$ represents the microbiome data, and $X^{(2)}$ represents meta data (moisture content, LOI, 14 PAHs concentrations and GC × GC signatures whether considered or not) (Table 1). An additional matrix $X^{(3)}$ required for the algorithm to enable the discriminant analysis is a dummy matrix of the classes the samples belong to (whether COV or CH, and equivalent to matrix Y in the sPLS-DA case). The algorithm then constructs artificial latent components of the datasets by factorizing the datasets into scores and loading vectors in new space such that the covariance between the scores of these matrices in this space is maximized, *i.e.*, for $q = 1, 2, ..., Q$, DIABLO solves for each component $h = 1, ..., H$:

$$\begin{array}{c} \arg\ \max \\ a_h^{(1)},\ ...,\ a_h^{(Q)} \end{array} \sum_{q,j=1,q \neq j}^{Q} c_{q,j}\, \mathrm{cov}\left(X_h^{(q)} a_h^{(q)}, X_h^{(j)} a_h^{(j)} \right) \quad \text{s.t.}\ \|a_h^{(q)}\|_2 = 1\ \text{and}\ \|a_h^{(q)}\|_1 \leq \lambda^{(q)}$$

where $\lambda^{(q)}$ is the penalization parameter, $a_h^{(q)}$ is the loading vector on component h associated to the (deflated) matrix $X_h^{(q)}$ of the data set $X^{(q)}$, and $C = \{c_{q,j}\}_{q,j}$ is the design matrix, where $Q = 3$. C is a $Q \times Q$ matrix that specifies whether datasets should be correlated and includes values between zero (datasets are not connected) and one (datasets are fully connected). The first constraint $\|a_h^{(q)}\|_2 = 1$ (similar to sPLS-DA) ensures the loading vector has unit magnitude (requirement of the procedure) and the second constraint $\|a_h^{(q)}\|_1 \leq \lambda^{(q)}$ (also called l_1 penalty) ensures that for the features that do not vary between the categories, the corresponding loading vector coefficients go to zero. This is done using the sparsity control parameter $\lambda^{(q)}$ in the above equation, and adjusting it enforces shrinkage of the loading vector coefficients. According to the recommendations given in the mixOmics package (http://www.mixomics.org), before applying the sPLS-DA and DIABLO procedures, we pre-filter 1% of the lowest OTUs and then perform TSS +

CLR (Total Sum Scaling followed by Centralised Log Ratio) normalisation. For the design matrix in DIABLO, mixOmics suggests that a *full weighted design* where $c_{q,j} = 0.1$ between data matrices and 1 for the outcome leads to a trade-off between maximizing correlation between datasets and maximizing the discrimination of the outcome, and therefore we used this, *i.e.*, $C = \begin{bmatrix} 0.1 & 0 & 1 \\ 0 & 0.1 & 1 \\ 1 & 1 & 1 \end{bmatrix}$.

To predict the number of latent components (associated loading vectors) and the number of discriminants, for sPLS-DA, we used the perf.plsda() and tune.splsda() functions, whereas for DIABLO, block.splsda() and tune.-block.splsda() functions were used, respectively. In both cases, we fine-tuned the model using leave-one-out cross-validation by splitting the data into training and testing sets and then finding the classification error rates employing two metrics, overall error rates and balanced error rates (BER), between the predicted latent variables with the centroid of the class labels (categories considered in this study) using the max.dist (which gave the minimal classification rate for the scenarios considered in this study). BER accounts for differences in the number of samples between different categories. Other than TSS + CLR normalisation for the abundance table, $\log_{10}$ normalisation for qPCR data was used, and GC $\times$ GC signatures were normalised using pareto scaling. When displaying boxplots, pair-wise ANOVA or Kruskal–Wallis were performed taking two categories at a time, and where significant ($p \leq 0.05$), they were joined together by a line and significance was plotted on top (*: $0.01 \leq p < 0.05$; **: $0.05 \leq p < 0.001$; ***: $p \leq 0.001$).

3. Results & discussion

3.1. Comprehensive SVOCs characterisations of contamination

Comprehensive two-dimensional analysis coupled with time-of-flight mass spectrometry analysis was carried out on the soil samples. GC $\times$ GC-TOFMS enables an exhaustive characterisation of the distribution of semi-volatile organic compounds (SVOCs) in the samples as a novel "omics" dimension in the study of bioremediation of complex contaminated soils. Fig. 1 shows the GC $\times$ GC chromatograms for representative samples of CH and COV. Visually, COV samples appeared more heavily contaminated with PAHs than the CH samples, which was confirmed using the one dimensional GC quantification (Fig. 2). The proportion of substituted PAHs was also higher in the COV samples. By comparison, the CH samples had higher proportions of alkanes, alkenes, alkynes and alkylbenzenes. This could possibly be explained by a difference in the processes involved in the coal gasification as we demonstrated previously in a forensic study of coal tar samples from FMGPs.[8] Higher proportions of PAH parents and lower proportions of alkanes were associated with high temperature gasification, while high proportions of petroleum-like hydrocarbons can be associated with carburetted water gas plants.

While the potential of GC $\times$ GC for comprehensive analysis of SVOCs in contaminated samples is evident in terms of the resolution of the analytical method, the availability of fully comprehensive, automated and reproducible data processing and alignment for GC $\times$ GC data is arguably the biggest bottleneck to its use in omics-like studies. The accuracy of the chosen processing method was evaluated using replicates and multivariate analysis. Alignment of 68 peak tables

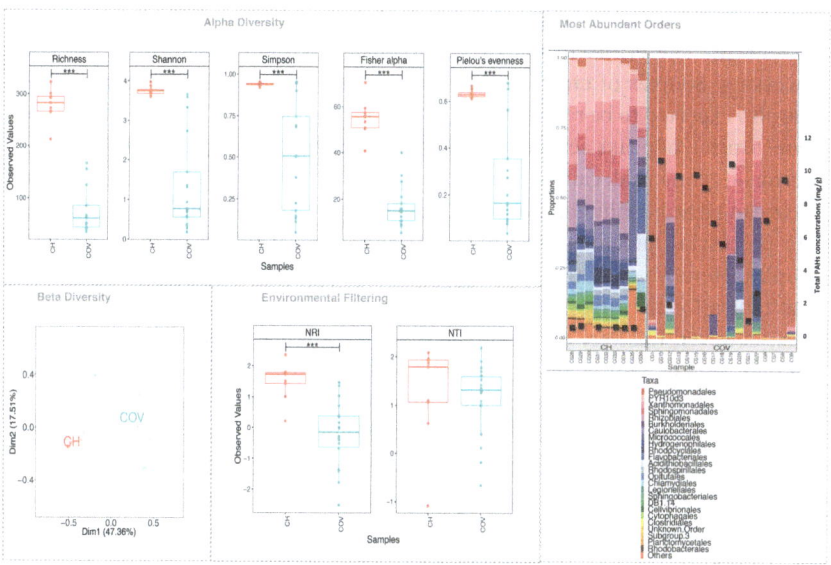

Fig. 2 Microbial diversity and function. (a), (c), and (d) represent alpha diversity, beta diversity and environmental filtering indices, respectively. For beta diversity analysis, PERMANOVA suggested 39%, 42%, and 58% variability (R^2; all significant $p < 0.001$) between COV and CH for Bray–Curtis [shown in (c)], UniFrac, and weighted UniFrac distances, respectively. In (c), the ellipses are drawn at a 95% confidence interval of standard error with the group labels located at the mean of the ordination. Note that in (d), the positive value of NTI indicates that species co-occur with more closely related species than expected by chance, with negative values suggesting otherwise. (b) shows community structure based on the relative abundance of the top-25 most abundant orders from across each site, where 'others' refers to all orders not included in the 'top-25'. The grey bands indicate replicates of the same soil sample. The total concentration (mg g^{-1}) of the 14 PAHs quantified was superimposed to this data.

with a missing values limit of 10% returned 961 compounds. To evaluate the robustness of the data processing, HCA was carried out on the samples using the alignment table (Fig. S1†). The clustering clearly separated the two sites, with only one COV sample (both instrumental replicates) clustering with CH samples. The GC × GC chromatogram for this extract is presented in the bottom right of Fig. 1. One CH sample clustered away from both sites' clusters; its chromatogram is presented in the top left of Fig. 1. In most cases, a sample nearest neighbour was its instrumental replicate. The five extracts from the same soil sample (COV-O5 in Fig. S1†) clustered altogether. Replicate runs before and after the column change, however, did not cluster near each other (*e.g.* see COV-18 and COV-17 on Fig. S1†). The data alignment was therefore deemed successful although it appeared somewhat susceptible to retention shift. Alignment with the missing values limit of 100% returned 58 compounds (Table S1†). While we were able to ascertain during GC-MS quantification that all of the samples contained the 14 quantified PAHs, the alignment only returned seven of those. Similarly, while five surrogates and one internal standard were added to each sample, only three were in the final alignment table and their peak areas showed great variations. Recovery quantification for the surrogates was carried out using GC-MS and showed good reproducibility (relative standard deviation around 10% for all surrogates) (data not shown). The errors in the GC × GC data, therefore, came from the data processing. Sources of errors could be related to a degree of misalignment but were also expected to be related to the peak finding algorithm. Deconvolution and reconstitution of peaks across multiple modulation times might lead to two types of errors: (1) one peak can be incorrectly split into two peaks and (2) two or more peaks can be artificially combined as one. The first error is acknowledged in most data processing pipelines for GC × GC data[11,17] as it occurs in metabolomics samples. Here we used the PrecompressFiles function of R2DGC to correct the peak tables for the first error prior to alignment. The concentrations of surrogates, however, was very high compared to any other compound in the samples and their signals saturated the detectors, which increased the possibility of peak splitting and might explain the non-repeatability of their peak areas.

The second error is less common in metabolomics as compounds have more distinct mass spectra. In hydrocarbon analysis, however, position isomers are very common, have very similar spectra and elute near each other and are therefore more likely to be integrated as a single peak. This can be minimised by optimising the parameters of data processing such as minimum peak widths in both dimensions and signal-to-noise ratio; it, however, remains an issue in samples with large dynamic ranges between compounds. The results therefore indicated that the probability of one of these errors to occur for any one compound in one of the 68 peak tables was high. Further optimisation of the data processing needs to be carried out for accurate and precise exhaustive characterisation of SVOCs.

3.2. Bacterial diversity in CH and COV soils

While 16 different samples were collected for the COV site and 9 for the CH sites, not all extracted DNA was successfully amplified. Additionally, some samples were extracted in replicates (see Fig. 2b) and variations in microbial communities between replicates was lower than the inter-sample variations.

Five measures of alpha diversity clearly showed that the CH samples had significantly higher species richness than COV samples had (Fig. 2). While considering the top 25 most abundant orders, we can notice that the proportions of gammaproteobacteria, alphaproteobacteria and betaproteobacteria were consistent in the CH samples but varied in the COV samples. Comparison with the total concentrations of 14 PAHs suggested a possible connection between the levels of PAH contamination and the abundance of the dominant species in the samples (Fig. 2b), particularly the proportions of gammaproteobacteria. Two samples, however, showed contrary trends. One of them presented low total PAH concentration and a low proportion of gammaproteobacteria. In this sample, recovery values for the PAH surrogates were very low (between and 20 and 35%) indicating an issue during PLE extraction and therefore an underestimation of total PAH concentration.

NMDS (using Bray–Curtis distance) of sample dissimilarities showed separation between the two sites with more inter-sample variability in COV (Fig. 2c). PAH concentrations in COV were one order of magnitude higher on average than in CH samples and also presented a greater variance. This may explain the dispersion in the beta diversity space resulting in several ecological niches. The pathways analyses (not shown here) indicated that a significantly higher proportion of

OTUs in the COV samples were available in the reference SilvaMod 123 database (for which reference pathways are available in the Tax4Fun package) than in CH samples, which suggested that the OTUs that inhabited the more contaminated sites are more ubiquitous in nature and thus well characterised with their pathways available. The gene content for each OTU in each sample was inferred from the closest sequenced genome using the Tax4Fun package and we investigated particular pathways from the KEGG database linked to contaminant degradation[4] and related metabolic pathways. Out of the 12 tested pathways, seven presented significantly different average relative abundance between COV and CH. COV presented statistically more potential for degradation of hydrocarbon contaminants *via* the toluene (ko00623) and naphthalene (ko00626) pathways and CH *via* the drug metabolism (other pathways) (ko00983) and PAH (ko00624) degradation pathways (Fig. 3). Incidentally, COV samples also showed significantly more potential for the propanoate and pyruvate metabolisms and for glycolysis.

Next, we explored the influence of the environment on the assembly of microbial communities. We applied two phylogenetic alpha diversity indices, NTI and NRI (Fig. 2d), to explore phylogenetic clustering at both local and global scales. Positive values of NTI indicate that species co-occur with more closely related species than expected by chance, with negative values suggesting otherwise. This is mainly because NTI measures tip-level divergences (putting more emphasis on terminal clades and akin to "local" clustering) in phylogeny while NRI measures deeper divergences (akin to "global" clustering). For both NTI and NRI, values >+2 indicate strong environmental pressure, values <−2 indicate strong competition among species as the driver of community structure, and the values in between are a gradient between the two. The results indicated that the communities in CH samples are more deterministic, and influenced by the environment, whereas COV sites are driven by the competitive exclusion principle where the species that can outcompete others in a given niche dominate, leading to more dispersion in the phylogenetic tree. This phenomenon also supports the much higher variability in beta diversity space for COV samples (Fig. 2c).

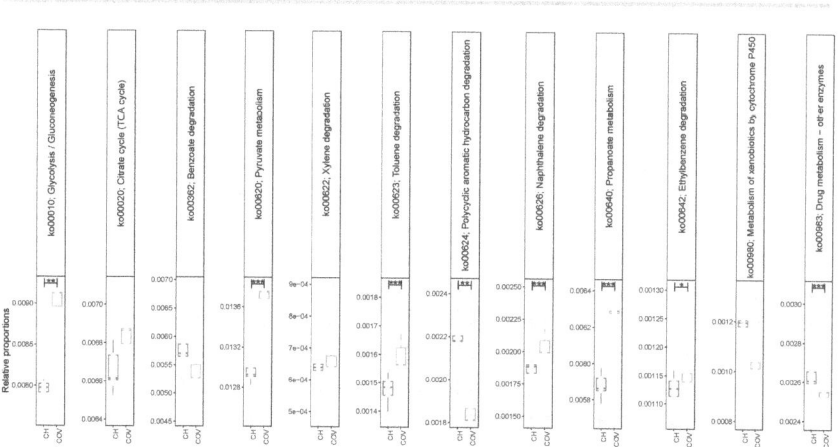

Fig. 3 Relative abundance of KEGG pathways for hydrocarbon degradation and related metabolic pathways inferred using the Tax4Fun package. Lines connect two categories where the differences were significant (Kruskal–Wallis) with *** ($p < 0.001$).

3.3.　Discriminant analysis of microbiome and chemical data

Through discriminant analysis based on the microbiome alone (sPLS-DA), we found a total of 10 discriminating OTUs (five for the first two components each). PLS-DA (Fig. 4a) shows ordination of the samples using all of the OTUs and sPLS-DA (Fig. 4b) shows ordination of the samples using only the discriminating OTUs. By comparing these two figures, it can be seen that in the reduced space, the sample dissimilarities (between COV and CH) are conserved. Three out of the ten OTUs were greater in abundance for CH, while the rest were greater for COV. Clustering of the samples according to these OTUs showed that four out of the five discriminating features from the first component (OTUs 146, 1214, 384 and 333; this group is henceforth referred to as Group A) were significantly more abundant in COV than in CH, and the last OTU (OTU_14) was significantly more abundant in the CH site. OTUs 333, 384 and 1214 belong to the Clostridiales order. OTUs 333 and 384 are both plant biopolymer degraders of the Ruminococcaceae family, which might relate to the organic matter present on the site, while OTU_1214 belongs to the genus *Bacillus*, which is known for both PAH and alkane degradation.[18] OTU_146 belongs to the *Desulfotomaculum* genus, in which some species have been found to degrade cresol in sulfate-degrading condition[19] and OTU_4 belongs to the genus *Sulfuritalea*, oxygen independent aromatic compound degrading bacteria.[20] While none of the Group A OTUs are amongst the most abundant in the samples, OTU_14 is amongst the most abundant in the CH samples (results not shown). These 5 OTUs might be markers of the difference in hydrocarbon degradation mechanisms between the two sites because of differences in concentration and in soil properties and quality.

A first DIABLO analysis (hitherto referred to as DIABLO 1) (Fig. 5) was carried out by integrating additional metadata to the microbiome data: the concentrations of 14 PAHs, LOI, moisture, heavy metal concentrations and qPCR data for Gram negative PAH degrader (PAH RHD α GN), Gram positive PAH degrader (PAH RHD α GP) and alkane degrader (alkB) gene abundance (Table 1). We found two components to reduce the classification error rates resulting in 10 discriminating OTUs and 10 discriminating meta data features (five each for the two components). Group A represented four of the five OTUs that were selected in the first component (Table 2). This confirmed that in both cases, the first component isolated the OTUs that were the most representative of the differences between COV and CH and particularly the ones that were much more abundant in COV than in CH. The final OTU, OTU_147, was identified as the actinobacteria, *Micromonospora* sp. WMMB 894, for which little information is available in the literature. These five OTUs are further referred to as Group A'.

Consequently, this overlap between the first component OTUs for the sPLS-DA and DIABLO 1 allowed us to interpret the discriminating metadata selected for the first component in DIABLO 1 as the ones that were the most representative of the differences between CH and COV. These included moisture, naphthalene and dibenzo(*a*,*h*)anthracene, cobalt and iron. Given that moisture content between the two sites is significantly different – the average moisture in COV samples was 19.5% ± 0.8 (95% confidence interval) and the CH average moisture was 8.1% ± 1.8 (95% confidence interval) – it is likely to be a strong driver for bacterial community composition. Naphthalene and dibenzo(*a*,*h*)anthracene were not the PAHs with the highest concentrations in the samples but were selected

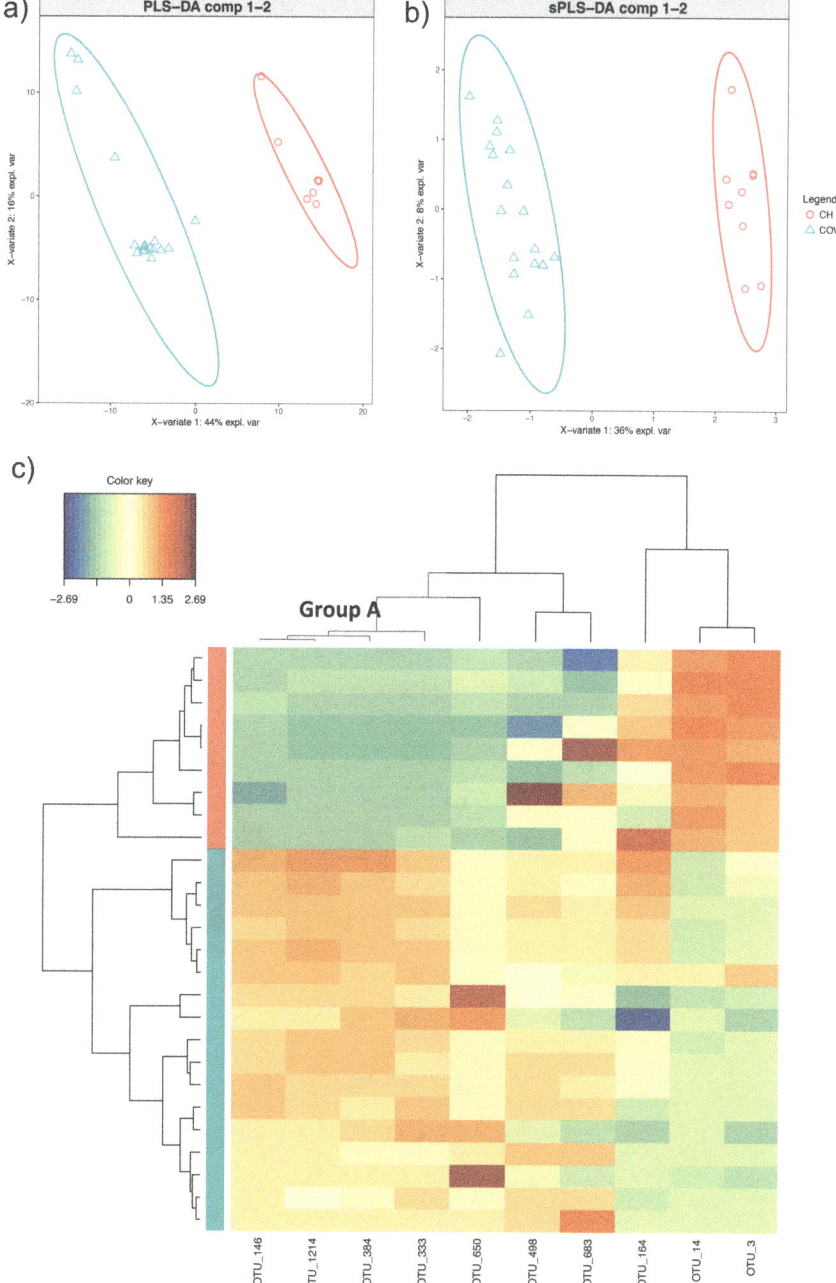

Fig. 4   sPLS-DA of the microbiome data. (a) PLS-DA discriminant analysis and (b) sPLS-DA discriminant analysis – the ordination of the data is conserved when only the discriminating OTUs are employed. (c) shows the heatmaps of the 10 discriminant features selected in the sPLS-DA (see Table 2), with both rows and columns ordered using hierarchical (average linkage) clustering to identify blocks of features of interest. Group A (see Table 2) is indicated in a grey box.

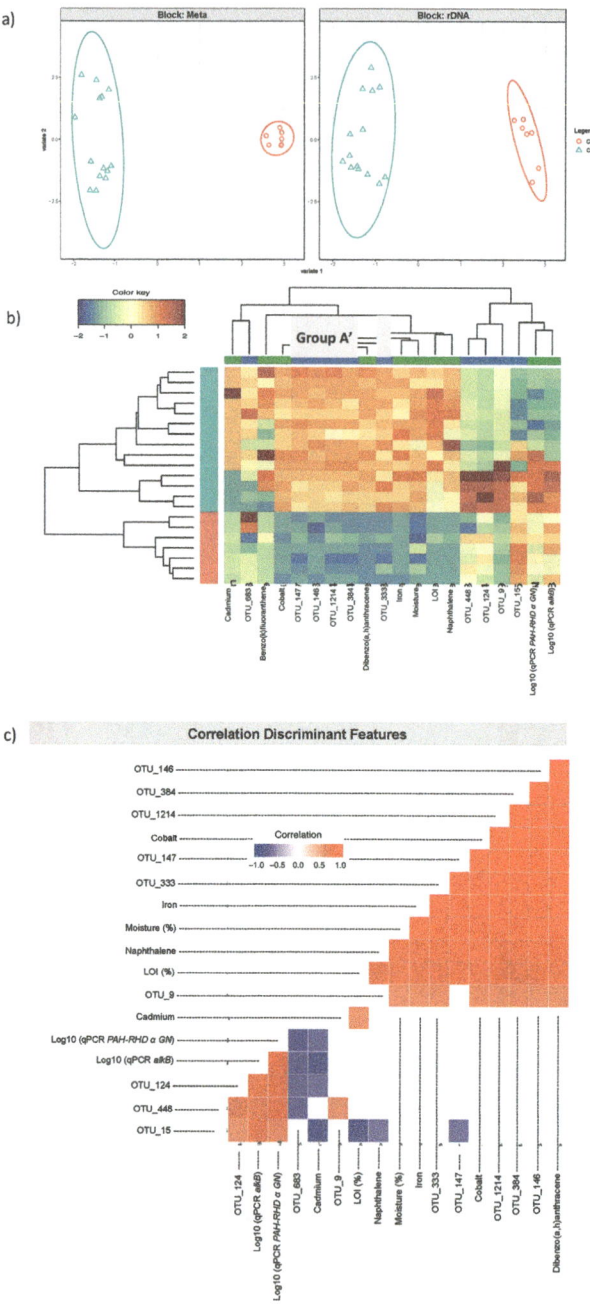

Fig. 5 DIABLO 1. (a) The algorithm found two components reducing the classification error rates in the DIABLO algorithm and shows the ordination of samples with ellipses representing 95% confidence interval and percentage variations explained by these components in axes labels for both microbiome (Block: rDNA) and meta data (Block: Meta). (b) shows the heatmaps of the discriminant features in DIABLO 1 (see Table 2), with both rows and columns ordered using hierarchical (average linkage) clustering to identify blocks of features of interest. Group A' (Table 1) is indicated in a grey box. (c) shows the significant correlations ($-0.6 < R > 0.6$) between the features as calculated by the algorithm.

Table 2 Discriminating features for the first and second components of the three discriminant analyses. OTUs in Group A are highlighted in bold and Group A' in italic

Multivariate analysis		ID	Kingdom	Phylum	Class	Order	Family
		Discriminating OTUs					
sPLS-DA (OTUs)	**First Component**	**OTU_146**	**Bacteria**	**Firmicutes**	**Bacilli**	**Bacillales**	**Bacillaceae**
		OTU_333	**Bacteria**	**Firmicutes**	**Clostridia**	**Clostridiales**	**Ruminococcaceae**
		OTU_384	**Bacteria**	**Firmicutes**	**Clostridia**	**Clostridiales**	**Ruminococcaceae**
		OTU_1214	**Bacteria**	**Firmicutes**	**Clostridia**	**Clostridiales**	**Peptococcaceae**
	Second Component	OTU_14	Bacteria	Proteobacteria	Alphaproteobacteria	Sphingomonadales	Sphingomonadaceae
		OTU_164	Archaea	Euryarchaeota	Methanomicrobia	Methanosarcinales	Methanosarcinaceae
		OTU_683	Bacteria	Proteobacteria	Alphaproteobacteria	Rhizobiales	Methylobacteriaceae
		OTU_650	Bacteria	Proteobacteria	Gammaproteobacteria	Xanthomonadales	Xanthomonadaceae
		OTU_498	Bacteria	Proteobacteria	Deltaproteobacteria	Bdellovibrionales	Bdellovibrionaceae
		OTU_3	Bacteria	Proteobacteria	Betaproteobacteria	Rhodocyclales	Rhodocyclaceae

Multivariate analysis		Genus	OTUs	Discriminating Metadata Features ID
		Discriminating OTUs		
sPLS-DA (OTUs)	**First Component**	**Bacillus**		n.a.
		Ruminiclostridium 1	[Clostridium] termitidis CT1112	n.a.
		Acetivibrio	Acetivibrio cellulolyticus CD2	n.a.
		Desulfotomaculum		n.a.
		Sandaracinobacter	Y	n.a.

Table 2 (Contd.)

Multivariate analysis	Discriminating OTUs		Discriminating Metadata Features
	Genus	OTUs	ID
Second Component			n.a.
	Stenotrophomonas		n.a.
	Bdellovibrio		n.a.
	Sulfuritalea		n.a.
			n.a.

Multivariate analysis	Discriminating OTUs					
	ID	Kingdom	Phylum	Class	Order	Family
DIABLO 1 First Component	OTU_146	Bacteria	Firmicutes	Bacilli	Bacillales	Bacillaceae
	OTU_384	Bacteria	Firmicutes	Clostridia	Clostridiales	Ruminococcaceae
	OTU_333	Bacteria	Firmicutes	Clostridia	Clostridiales	Ruminococcaceae
	OTU_147	Bacteria	Actinobacteria	Actinobacteria	Micromonosporales	Micromonosporaceae
	OTU_1214	Bacteria	Firmicutes	Clostridia	Clostridiales	Peptococcaceae
Second Component	OTU_15	Bacteria	Proteobacteria	Gammaproteobacteria	Xanthomonadales	Xanthomonadaceae
	OTU_683	Bacteria	Proteobacteria	Alphaproteobacteria	Rhizobiales	Methylobacteriaceae
	OTU_448	Bacteria	Proteobacteria	Betaproteobacteria	Burkholderiales	Alcaligenaceae
	OTU_9	Bacteria	Proteobacteria	Betaproteobacteria	Burkholderiales	Alcaligenaceae
	OTU_124	Bacteria	Proteobacteria	Alphaproteobacteria	Rhizobiales	Rhizobiaceae

Table 2 (*Contd.*)

		Discriminating OTUs		Discriminating Metadata Features
Multivariate analysis		Genus	OTUs	ID
DIABLO 1	First Component	*Bacillus*	*Acetivibrio cellulolyticus CD2*	Dibenzo(a,h)anthracene
		Acetivibrio	*[Clostridium] termitidis CT1112*	Cobalt
		Ruminiclostridium 1	*Micromonospora sp. WMMB 894*	Iron
		Plantactinospora		Moisture
		Desulfotomaculum		Naphthalene
	Second Component	*Pseudoxanthomonas*	Pseudoxanthomonas spadix BD-a59	Log10 (qPCR PAH-RHD _ GN)
				Log10 (qPCR alkB)
		Achromobacter		Cadmium
		Pusillimonas		Benzo(k)fluoranthene
		Shinella	Sinorhizobium sp. enrichment culture clone Van49	LOI

		Discriminating OTUs					
Multivariate analysis		ID	Kingdom	Phylum	Class	Order	Family
DIABLO 2	First Component	*OTU_384*	*Bacteria*	*Firmicutes*	*Clostridia*	*Clostridiales*	*Ruminococcaceae*
		OTU_146	*Bacteria*	*Firmicutes*	*Bacilli*	*Bacillales*	*Bacillaceae*
		OTU_1214	*Bacteria*	*Firmicutes*	*Clostridia*	*Clostridiales*	*Peptococcaceae*
		OTU_333	*Bacteria*	*Firmicutes*	*Clostridia*	*Clostridiales*	*Ruminococcaceae*
		OTU_147	*Bacteria*	*Actinobacteria*	*Actinobacteria*	*Micromonosporales*	*Micromonosporaceae*
	Second Component	*OTU_116*	*Bacteria*	*Proteobacteria*	*Alphaproteobacteria*	*Sphingomonadales*	*Erythrobacteraceae*
		OTU_255	*Bacteria*	*Proteobacteria*	*Alphaproteobacteria*	*Sphingomonadales*	*Erythrobacteraceae*
		OTU_635	*Bacteria*	*Proteobacteria*	*Alphaproteobacteria*	*Rickettsiales*	*Holosporaceae*
		OTU_344	*Bacteria*	*Proteobacteria*	*Betaproteobacteria*	*Burkholderiales*	*Comamonadaceae*
		OTU_1397	*Bacteria*	*Proteobacteria*	*Alphaproteobacteria*	*Sphingomonadales*	*Sphingomonadaceae*

Table 2 (Contd.)

Multivariate analysis		Discriminating OTUs		Discriminating Metadata Features
		Genus	OTUs	ID
DIABLO 2	First Component	Acetivibrio	Acetivibrio cellulolyticus CD2	Dibenzo(a,h)anthracene
		Bacillus		Cobalt
		Desulfotomaculum		Iron
		Ruminiclostridium 1	[Clostridium] termitidis CT1112	Moisture
		Plantactinospora	Micromonospora sp. WMMB 894	Naphthalene
				1-Butyloctylbenzene
				1-Butylheptylbenzene
				1-Methylundecylbenzene
	Second Component	Extensimonas		2,4-Di-tert-butylphenol
		Sphingopyxis		1-Pentyloctylbenzene

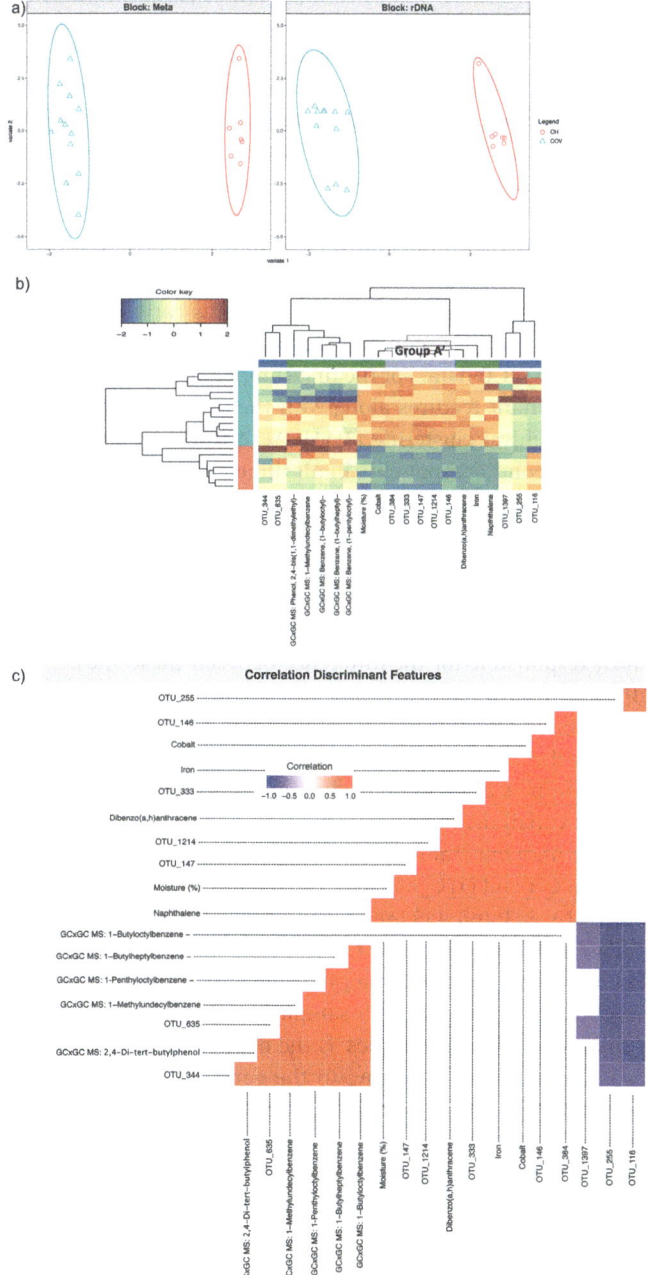

Fig. 6 DIABLO 2. (a) The algorithm found two components reducing the classification error rates in the DIABLO algorithm and shows the ordination of samples with ellipses representing 95% confidence interval and percentage variations explained by these components in axes labels for both microbiome (Block: rDNA) and meta data (Block: Meta). (b) shows the heatmaps of these discriminant features, with both rows and columns ordered using hierarchical (average linkage) clustering to identify blocks of features of interest. Group A′ (Table 2) is indicated in a grey box. (c) shows the significant correlations ($-0.6 < R > 0.6$) between the features as calculated by the algorithm.

statistically to represent low molecular weight and high molecular weight PAHs, respectively, as markers of the difference in PAH concentration between the two sites. Similarly, the heavy metal concentrations were much higher in COV samples than in CH samples and this is reflected in the presence of iron and cobalt as significant contributors to the first component. Noteworthy was the fact that D8-naphthalene was the surrogate with the lowest repeatability and therefore, although naphthalene in an aged contaminated soil is less likely to evaporate than a freshly spiked surrogate during extraction, the concentration of naphthalene was likely to be underestimated in our sample and this should be addressed in the future considering its importance in the analysis.

A second DIABLO analysis (DIABLO 2) (Fig. 6) was carried out introducing to the previous dataset, the 58 compounds that were found in all samples by alignment of the GC × GC data. For the first component, once again Group A' was selected as discriminant features along with the same metadata features. Two-dimensional hierarchical clustering (2D-HCA) of the samples and variables for the first and the second components in all three cases (Fig. 4c, 5b and 6b) presented ideal site separation of the samples and the Group A' OTUs and the metadata that got selected for the first component drove this clustering.

Pairwise correlations (Fig. 5c and 6c) showed that the discriminating features from the first component (in both DIABLO 1 and DIABLO 2) were also highly correlated to each other, demonstrating the link between the microbiome and the stressors most responsible for the difference between these sites.

Since all three discriminatory analyses above rank the components based on percentage variability explained by the components, the features selected in lower components (i.e. the second component) serve as a cue to elucidate finer differences between samples as opposed to the first component.

In the sPLS-DA analysis, while two of the OTUs amongst the five OTUs selected for the second component (Table 2) are clearly more abundant in one site than the others (OTU_3 in CH and OTU_650 in COV), the three others have more nuanced distribution. OTUs 683 and 498 are more abundant in COV overall but are also present in high abundance in some CH samples, while OTU_164 is more abundant in CH on average but is abundant in some COV samples. These three OTUs that are significantly present in both samples might be indicative of change of principal function of the communities as the hydrocarbon distribution changes. OTU_650 was identified as belonging to the genus of *Stenotrophomonas* (gammaproteobacteria), which have been found to be PAH degraders.[21] OTU_3 belongs to the genus *Sandaracinobacter*, a genus of aerobic anoxygenic phototrophic extremophiles.[22] OTU_683 is a member of the Methylobacteriaceae family, some strains of which are known to exhibit high tolerance to heavy metal contamination and have been used beneficially in the bioremediation of contaminated environment.[23,24]

OTU_683 was the only OTU from the second component of the sPLS-DA to also be selected for the second component in DIABLO 1. Three other OTUs (124, 15 and 448) selected for the second component in DIABLO 1 were significantly and positively correlated with each other and with the ($\log_{10}$) qPCR results for Gram-negative PAH and alkane degraders. The same three OTUs correlated negatively with OTU_683 and with cadmium (Fig. 5c). The abundance of the aforementioned OTUs seems to drive the clustering within the COV samples in the 2D-HCA, demonstrating two possible regimes of hydrocarbon degradation, which, however, do not appear to be related to the concentration of PAHs. OTUs 124, 15 and 448 all belong to known aromatic hydrocarbon degrading genera of *Achromobacter*, *Pseudoxanthomonas* and *Sinorhizobium*.[25–29] The abundance of these OTUs was low in samples from COV where cadmium concentration was high, which appeared to favour OTU_683. In the DIABLO 2 analysis, the discriminating features of the second component were entirely different than those of DIABLO 1. The five OTUs formed two distinct groups negatively correlated to each other: OTUs 635 and 344 lying in one group and OTUs 116, 255 and 1397 lying in the other group. The latter all belong to the Sphingomonadales and were also negatively correlated to the discriminating metadata features: four long chain branched alkylbenzenes and 2,4-di-*tert*-butylphenol (Fig. 6c). The abundance of these chemicals also appeared to upregulate the abundance of OTUs 635 and 344. OTU_344 was identified as belonging to the *Extensimonas* genus, aerobic chemoorganotrophs that have been isolated before from wastewater.[30] These highlighted again two different regimes in the COV samples, depending this time on the abundance of hydrocarbons that were not PAHs. Samples in COV with high abundance of these hydrocarbons clustered closer to the CH samples (in which the abundances were high too) than to other COV samples. The integration of the GC × GC data does not affect the first component, demonstrating that the differences in moisture level, heavy metals and PAH concentrations explains the differences between the two sites more than the differences in the exhaustive SVOC signatures. It highlighted, however, through the second component, the influence on the autochthonous microbiome of compounds that would not have normally been measured or taken in consideration during site investigation.

4. Conclusions

Our results demonstrated the usefulness of "multi-omics" approaches in the context of contaminated soils to correlate the abundance of chemical contaminants to the abundance of microbial OTUs. At the same time, utilising the ecological principles, we have highlighted the deterministic nature of microbial communities, dependent on the presence of chemical contaminants. Whilst the discriminant analysis approaches, sPLS-DA and DIABLO, adopted in this study give a reduction of large microbial feature space to the subset of species that form an association with the chemical contaminants and other meta data, care must be taken to interpret causality, primarily because we have only considered sites with spatial variabilities and the patterns found were predominantly inter-site discriminants. It was already successful, however, in demonstrating the efficacy of qPCR analysis, where causality is already known, as a rapid screening tool for hydrocarbon biodegradation potential on a site. This preliminary study has been useful in firstly validating our experimental methods such as exhaustive SVOC and DNA extraction from highly contaminated soil samples and evaluating our *in silico* pipelines for data processing, highlighting notably the potential of the R2DGC free package for alignment of GC × GC peak tables but also its limitations and those of peak picking algorithms, which will need improving for true comprehensive studies. To give the mechanistic underpinnings, a thorough exploration such as in the context of time series microcosms is required as well as subsequent carefully designed experiments that are informed by the findings of this study. Furthermore, the metabolic potential of the microbial community in this study is explored through a proxy method (Tax4Fun) dependent on the availability of the reference pathway database. Although approaches such as shotgun metagenomics give the actual metabolic potential, we have demonstrated that with high representation of taxa from contaminated soils in the reference database, Tax4Fun offers a cost-effective solution with reasonable resolution for biodegradation pathways. Thus, for contaminated sites, coupling microbial community surveys using 16S rRNA with GC × GC MS offers a whole that is greater than the sum of its parts.

Conflicts of interest

There are no conflicts to declare.

Acknowledgements

The authors would like to thank Felipe Sepúlveda Olea and Gillian MacKinnon from the Scottish Universities Environmental Research Centre for transition metal quantification and Diana Guillen Ferrari from Heriot-Watt University for carrying out the qPCR analysis. We are also thankful to Scottish Crucible and the RSE for funding through the project "Describing multiple contaminants biodegradation in soil using comprehensive two-dimensional gas chromatography coupled with intelligent data analysis" and to NERC for Umer Z. Ijaz's Independent Research Fellowship NE/L011956/1. EPSRC is also acknowledged for financial support with grant EP/K038885/1.

References

1 T. J. Aspray, D. J. C. Carvalho and J. C. Philp, Application of soil slurry respirometry to optimise and subsequently monitor *ex situ* bioremediation of hydrocarbon-contaminated soils, *Int. Biodeterior. Biodegrad.*, 2007, **60**, 279–284.

2 T. Aspray, A. Gluszek and D. Carvalho, Effect of nitrogen amendment on respiration and respiratory quotient (RQ) in three hydrocarbon contaminated soils of different type, *Chemosphere*, 2008, **72**, 947–951.

3 S. Yang, *et al.*, Hydrocarbon degraders establish at the costs of microbial richness, abundance and keystone taxa after crude oil contamination in permafrost environments, *Sci. Rep.*, 2016, **6**, 37473.

4 M. Crampon, J. Bodilis and F. Portet-Koltalo, Linking initial soil bacterial diversity and polycyclic aromatic hydrocarbons (PAHs) degradation potential, *J. Hazard. Mater.*, 2018, **359**, 500–509.

5 L. A. McGregor, *et al.*, Ultra resolution chemical fingerprinting of dense non-aqueous phase liquids from manufactured gas plants by reversed phase comprehensive two-dimensional gas chromatography, *J. Chromatogr. A*, 2011, **1218**, 4755–4763.

6 D. Mao, *et al.*, Detailed analysis of petroleum hydrocarbon attenuation in biopiles by high-performance liquid chromatography followed by comprehensive two-dimensional gas chromatography, *J. Chromatogr. A*, 2009, **1216**, 1524–1527.

7 C. Gauchotte-Lindsay, P. Richards, L. A. McGregor, R. Thomas and R. M. Kalin, A one-step method for priority compounds of concern in tar from former industrial sites: Trimethylsilyl derivatisation with comprehensive two-dimensional gas chromatography, *J. Chromatogr. A*, 2012, **1253**, 154–163.

8 L. A. McGregor, C. Gauchotte-Lindsay, N. Nic Daeid, R. Thomas and R. M. Kalin, Multivariate Statistical Methods for the Environmental Forensic Classification of Coal Tars from Former Manufactured Gas Plants, *Environ. Sci. Technol.*, 2012, **46**, 3744–3752.

9 D. Mao, *et al.*, Estimation of ecotoxicity of petroleum hydrocarbon mixtures in soil based on HPLC–GC × GC analysis, *Chemosphere*, 2009, **77**, 1508–1513.

10 F. Rohart, B. Gautier, A. Singh and K.-A. L. Cao, mixOmics: An R package for 'omics feature selection and multiple data integration, *PLoS Comput. Biol.*, 2017, **13**, e1005752.

11 R. C. Ramaker, E. R. Gordon and S. J. Cooper, R2DGC: threshold-free peak alignment and identification for 2D gas chromatography-mass spectrometry in R, *Bioinformatics*, 2018, **34**(10), 1789–1791.

12 K. Kloos, J. C. Munch and M. Schloter, A new method for the detection of alkane-monooxygenase homologous genes (alkB) in soils based on PCR-hybridization, *J. Microbiol. Methods*, 2006, **66**, 486–496.

13 dendextend: an R package for visualizing, adjusting and comparing trees of hierarchical clustering|Bioinformatics|Oxford Academic, available at https://academic.oup.com/bioinformatics/article/31/22/3718/240978, accessed 13th February 2019.

14 phyloseq: An R Package for Reproducible Interactive Analysis and Graphics of Microbiome Census Data, available at https://journals.plos.org/plosone/article?id=10.1371/journal.pone.0061217, accessed 13th February 2019.

15 S. W. Kembel, *et al.*, Picante: R tools for integrating phylogenies and ecology, *Bioinformatics*, 2010, **26**, 1463–1464.

16 J. C. Stegen, X. Lin, A. E. Konopka and J. K. Fredrickson, Stochastic and deterministic assembly processes in subsurface microbial communities, *ISME J.*, 2012, **6**, 1653–1664.

17 X. Wei, *et al.*, MetPP: a computational platform for comprehensive two-dimensional gas chromatography time-of-flight mass spectrometry-based metabolomics, *Bioinformatics*, 2013, **29**, 1786–1792.

18 D. Ghosal, S. Ghosh, T. K. Dutta and Y. Ahn, Current State of Knowledge in Microbial Degradation of Polycyclic Aromatic Hydrocarbons (PAHs): A Review, *Front. Microbiol.*, 2016, **7**, 1369.

19 K. L. Londry, P. M. Fedorak and J. M. Suflita, Anaerobic Degradation of *m*-Cresol by a Sulfate-Reducing Bacterium, *Appl. Environ. Microbiol.*, 1997, **63**, 6.

20 M. Sperfeld, C. Rauschenbach, G. Diekert and S. Studenik, Microbial community of a gasworks aquifer and identification of nitrate-reducing *Azoarcus* and *Georgfuchsia* as key players in BTEX degradation, *Water Res.*, 2018, **132**, 146–157.

21 S. Gao, *et al.*, Multiple degradation pathways of phenanthrene by *Stenotrophomonas maltophilia* C6, *Int. Biodeterior. Biodegrad.*, 2013, **79**, 98–104.

22 V. Yurkov and E. Hughes, Aerobic Anoxygenic Phototrophs: Four Decades of Mystery, in *Modern Topics in the Phototrophic Prokaryotes: Environmental and Applied Aspects*, ed. P. C. Hallenbeck, Springer International Publishing, 2017, pp. 193–214, DOI: 10.1007/978-3-319-46261-5_6.

23 D. P. Kelly, I. R. McDonald and A. P. Wood, The Family Methylobacteriaceae, in *The Prokaryotes: Alphaproteobacteria and Betaproteobacteria*, ed. E. Rosenberg, E. F. DeLong, S. Lory, E. Stackebrandt and F. Thompson, pp. 313–340, Springer Berlin Heidelberg, 2014, DOI: 10.1007/978-3-642-30197-1_256.

24 P. De Marco, C. C. Pacheco, A. R. Figueiredo and P. Moradas-Ferreira, Novel pollutant-resistant methylotrophic bacteria for use in bioremediation, *FEMS Microbiol. Lett.*, 2004, **234**, 75–80.

25 J.-S. Seo, Y.-S. Keum, R. M. Harada and Q. X. Li, Isolation and Characterization of Bacteria Capable of Degrading Polycyclic Aromatic Hydrocarbons (PAHs) and Organophosphorus Pesticides from PAH-Contaminated Soil in Hilo, Hawaii, *J. Agric. Food Chem.*, 2007, **55**, 5383–5389.

26 A.-M. Tanase, R. Ionescu, I. Chiciudean, T. Vassu and I. Stoica, Characterization of hydrocarbon-degrading bacterial strains isolated from oil-polluted soil, *Int. Biodeterior. Biodegrad.*, 2013, **84**, 150–154.

27 Y.-S. Keum, J.-S. Seo, Y. Hu and Q. X. Li, Degradation pathways of phenanthrene by *Sinorhizobium* sp. C4, *Appl. Microbiol. Biotechnol.*, 2006, **71**, 935–941.

28 J. M. Kim, *et al.*, Influence of Soil Components on the Biodegradation of Benzene, Toluene, Ethylbenzene, and *o*-, *m*-, and *p*-Xylenes by the Newly Isolated Bacterium *Pseudoxanthomonas spadix* BD-a59, *Appl. Environ. Microbiol.*, 2008, **74**, 7313–7320.

29 D. R. Nielsen, P. J. McLellan and A. J. Daugulis, Direct estimation of the oxygen requirements of *Achromobacter xylosoxidans* for aerobic degradation of monoaromatic hydrocarbons (BTEX) in a bioscrubber, *Biotechnol. Lett.*, 2006, **28**, 1293–1298.

30 A. Willems, The Family Comamonadaceae, in *The Prokaryotes: Alphaproteobacteria and Betaproteobacteria*, ed. E. Rosenberg, E. F. DeLong, S. Lory, E. Stackebrandt and F. Thompson, Springer Berlin Heidelberg, 2014, pp. 777–851, DOI: 10.1007/978-3-642-30197-1_238.

DISCUSSIONS

Future challenges and new approaches: general discussion

Carlos Afonso, Caroline Chaux, Antony N. Davies, Marc-André Delsuc, Francisco Fernandez Lima, Caroline Gauchotte-Lindsay, Pierre Giusti, Royston Goodacre,  Jeffrey A. Hawkes,  Norbert Hertkorn, Jeroen J. Jansen, William Kew,  Stefan Kuhn,  Anneke Lubben, Daniel McGill, Mathias Nilsson, John Parkinson, Ryan P. Rodgers, Simon Rogers, Philippe Schmitt-Kopplin, Peter J. Schoenmakers, Laetitia Shintu, Ronald Soong, Stephen Summerfield, Andrew Surman, Dušan Uhrín  and Justin J. J. van der Hooft

DOI: 10.1039/c9fd90046b

Stephen Summerfield opened discussion of the paper by Ronald Soong: This is an ecotoxicology question.

(1) How do you create ^{13}C algae that are then fed to *Daphnia*? I understand that the ^{13}C *Daphnia* are fed to shrimp and so these become ^{13}C labelled.

(2) The concept of labelling the organism rather than the chemical is very exciting for me. We often deal with complex substances that are referred to under REACH as UVCBs (chemical substances of unknown or variable composition, complex reaction products and biological materials). These include petrochemicals and those derived from biological sources. We cannot therefore label the constituents due to cost, complexity and timeliness.

(3) Have *Daphnia* been used for toxicology/bioaccumulation studies? Are many other laboratories taking this concept up?

(4) On the regulatory side, more and more chemicals are becoming restricted, possibly unfairly, because current technology cannot prove they are degrading. The current methods are not fit for purpose in proving whether the substance is bioaccumulative.

(5) The concept of being able to see metabolism working in real-time will I hope give us a better understanding of our environment than we have now.

Ronald Soong replied:

(1) Thank you so much for your comment. The ^{13}C algae is created from feeding the organisms with ^{13}C (99%) enriched CO_2. For more information on this process please see the work of Akhter *et al.*[1] In regard to the second point: yes, we also label *Hyalella azteca* (freshwater shrimp); please see the work of Mobarhan *et al.*[2] for more information specific to shrimp.

(2) Thank you very much for the comment. We think that the *in vivo* NMR of isotopically enriched organisms will be very important in the future to understand living processes and toxic responses.

(3) Thanks for your comment. The *in vivo* NMR of enriched organisms is still relatively new, and the specific technique introduced at the Faraday Discussion has not yet been applied beyond the bioaccumulation of lipids shown in the paper. However, we have already applied NMR to follow toxic impacts *in vivo*. For more information on 2D *in vivo* screening and its potential for toxicity studies please see the following review article written by Bastawrous *et al.*[3]

(4) Thank you for the comment. We agree that bioaccumulation and conversion are very difficult to monitor, and by continuing to develop *in vivo* NMR techniques, we hope to provide a complementary tool that can be used to understand binding, bioaccumulation, and bioconversion *in vivo*, which is information that is desperately needed to assist with regulatory decisions.

(5) Thank you for your comment – we agree.

1 M. Akhter, R. D. Majumdar, B. Fortier-McGill, R. Soong, Y. Liaghati-Mobarhan, M. Simpson, G. Arhonditsis, S. Schmidt, H. Heumann and A. J. Simpson, *Anal. Bioanal. Chem.*, 2016, **408**, 4357–4370.
2 Y. Liaghati Mobarhan, B. Fortier-McGill, R. Soong, W. E. Maas, M. Fey, M. Monette, H. J. Stronks, S. Schmidt, H. Heumann, W. Norwood and A. J. Simpson, *Chem. Sci.*, 2016, **7**, 4856–4866.
3 M. Bastawrous, A. Jenne, M. T. Anaraki and A. J. Simpson, *Metabolites*, 2018, **8**, 35.

Philippe Schmitt-Kopplin remarked: Your approach is excellent but is deduced on NMR spectroscopy. I'm convinced that complementing mass spectrometry and stable isotopes would increase the number of metabolites annotated to follow-up the integration of the isotopic probes.

Ronald Soong responded: I would like to thank you for your comments. While our approach mainly focused on NMR spectroscopy, we believe that with other complementary techniques, such as MS, an increased number of metabolites will be found. As such, we agree with the comments.

Marc-André Delsuc addressed Ryan P. Rodgers and Carlos Afonso: This is a naive question to all MS spectroscopists in the audience. We have seen in the first example (fermentation of glucose to ethanol) that it is possible using NMR to follow individual atoms during the metabolism, and see from which initial carbons from the glucose ring the CH_2 and the CH_3 of ethanol originated. This gives hints on the metabolic pathways and on the enzymes involved, and there are several examples in the literature. Can you track in the same manner ^{13}C in the metabolism using mass spectrometry?

Carlos Afonso replied: Mass spectrometry is certainly a tool very complementary to NMR for this. First, the higher sensitivity and selectivity of MS can allow the detection of low concentration metabolites. Second, tandem mass spectrometry can provide evidence of specific groups presenting the isotopic label. However, depending on the fragmentation pathways for specific ions, some ambiguity concerning the position of the ^{13}C may remain.

Ryan P. Rodgers replied: Yes – stable isotope labeling is very useful in mass spectrometry and has been exploited for years.

Mathias Nilsson addressed Ronald Soong: Can you comment on the quantitative aspects of this experiment?

Ronald Soong answered: Thank you for the question. We have discussed the quantitative aspects of the experiment towards the end of the paper – we included a sequence specifically for quantification, which performs well on simple standards. However, as outlined in the paper, we recommended further confirmation with standards of differing complexity before moving onto more complex real-world systems.

Stephen Summerfield opened discussion of Daniel McGill's paper: In ecotoxicology testing, there has over the last few years been much interest in passive sampling, where a receptacle is in the medium for the duration of the test. SPE (Solid Phase Extraction) to me is a similar technology. Previously I have been restricted to GC-MS, which is not suitable for some constituents. So can we use NMR for determining the constituent identity and quantification from passive sampling?

Daniel McGill replied: NMR is an excellent reliable and non-destructive method for quantification and structural identification, and the minimal sample preparation and lack of need for derivatisation makes it, in some circumstances, a powerful technique on par with mass spectrometry. However, it should be noted that the sensitivity of NMR is much lower than that of MS (the exact degree of which is method-dependent), such that some trace compounds may become indistinguishable from noise – even with the use of SPE, there may be competing pressures from time (how long should the sampling take, at a maximum?) and the extent to which compounds may retain on a given cartridge. Hence it may be worth trialling the use of NMR in passive sampling, but whether it will produce useful results at the sample concentrations you may be used to may depend on your research question.

Peter J. Schoenmakers commented: You describe solid phase extraction (SPE) as a form of liquid chromatography. That is not incorrect, but it has little separation power (plates, in chromatographic jargon) and relies heavily on developing materials with good selectivity, which is a difficult and time-consuming process. You want as few compounds as possible in each of your fractions and, thus, you want as much separation as possible. Why do you opt for SPE and not for more powerful methods, such as liquid chromatography (LC) or supercritical fluid chromatography? The latter was explained by Fleur van Zelst at this meeting and in ref. 1.

1 F. H. M. van Zelst, S. G. J. van Meerten, P. J. M. van Bentum and A. P. M Kentgens, *Anal. Chem.*, 2018, **90**, 10457–10464.

Daniel McGill replied: Thank you for your question. It's undeniable that solid phase extraction – the so-called "poor man's chromatography" – is often of inferior build quality, with the sorbents suffering from greater impurities (often metals) and irregular structure. For some cases, it would certainly be a good idea to utilise alternative methods like HPLC to take advantage of its greater separation power. However, HPLC is not always suitable – applying raw, untreated urine to a HPLC column will not give good results, for example.

The aim of the project is to aid metabolite identification and structural elucidation efforts through "cleaning up" of spectra, by concentrating unknowns and eliminating peak overlap in an untargeted manner. The "holy grail" might be to develop a method that selectively retains all compounds that exhibit one specific functional group – of course, in practice this is almost impossible regardless of the technique used. SPE offers several advantages over HPLC to this end – SPE cartridges are relatively cheap, methods can be run in series when automated, and the sorbents have a high retention capacity. Hence our aims are not necessarily aligned with using the technique with the most separating power – if we have a urine sample with an unknown that we have annotated to a level 3 confidence (*i.e.* we have putative information as to the compound class of our unknown), we might utilise SPE methods in order to aid identification efforts without needing to invest in the rep-scale equipment, which would be necessary for using urine samples on HPLC. Ultimately, our methods do not need to have separation power comparable to HPLC, nor do they need to be applicable in every circumstance – they exist in order to add valuable approaches to the analyst's "toolbox", in order to gain more information about biofluids and other complex mixtures.

Dušan Uhrín commented: Table 3 in your paper is about the total retention capacity. Despite different conditions tested, the highest retention is around 20% – is that right? Others have retained 80% of NOM (natural organic matter). Why is yours so low?

Daniel McGill answered: Thank you for your question. I believe that the discrepancy may come from differences in the way that the retention capacity is being described for the purposes of the work, and hence in ways that retention capacity is calculated. For example, in the work of Dittmar *et al.*,[1] the "extraction efficiency" is calculated at about 62% (although this could be as low as 43%); the authors arrive at this number by dividing the concentration of DOC in the methanol SPE extract by the concentration factor and the DOC concentration in the seawater sample. Hence there is a holistic understanding of the total dissolved matter in the seawater, with a higher extraction efficiency generally considered a plus. By comparison, my aim is not to broadly retain as many compounds as possible, but to selectively retain specific compound classes in an untargeted manner in order to aid structural elucidation and annotation efforts – hence a lower retention capacity may be overall better suited if it is more selective for a given functional group or structure. It is also worth noting that the retention capacity here is calculated by comparing the peak intensities of a limited list of model compounds before and after retention; while a good model list will naturally reflect the total retention of all compounds, in practice there are usually discrepancies

1 T. Dittmar, B. Koch, N. Hertkorn and G. Kattner, *Limnol. Oceanogr.: Methods*, 2008, **6**, 230–235.

John Parkinson said: Dan, thanks for your presentation and for the paper. Can you comment on the potential applicability of this approach for other biofluids such as plasma and serum? The motivation for the question is whether and how SPE would handle the mix of large and small molecule components in these fluids including large molecular weight protein and lipid components compared with urine samples, which are chiefly composed of complex mixtures of low molecular weight metabolites?

Daniel McGill responded: Thank you for your kind words, John. I personally have limited experience of using other biofluids besides urine with the SPE methods described, but I understand that SPE has been used in a targeted manner to extract specific compounds from plasma, for example, without any deleterious effect from the proteins you described. Hence, while I can't be perfectly certain, there should be no reason why the same principles shouldn't apply to other biofluids with both large and small molecules.

Mathias Nilsson asked: From a metabolomics perspective, what are the future challenges with NMR?

Daniel McGill answered: Thank you for your question Mathias. Firstly, it goes without saying that greater resolution never goes amiss; stronger magnets (as well as cheaper/more widely available spectrometers) will always be welcome when doing metabolic profiling. However, stronger magnets are obviously not helpful in deconvoluting spectra. Hence, I would suggest that there is a need for more methods – such as hyphenated techniques, statistical approaches, or novel pulse programs – that can either be broadly useful (or improvements on current standards), or have unique applicability to a specific focus or situation, such as in structural elucidation.

Stephen Summerfield queried:
(1) Have you considered performing multiple passes on the SPE cartridge? It is about residence time – if you pull it off too quickly, it does not actually stick.
(2) Have you looked at SPE disks if you are dealing with high liquid volumes?
(3) Have you considered using a co-solvent to improve the extraction efficiency?

Daniel McGill responded: Thank you for your suggestions. For context, the longer term goal of developing the SPE-NMR methods is to migrate them onto an automated system for quick and efficient fractionation of samples, primarily for aiding metabolite identification efforts with respect to biofluids such as urine. This goal has guided our research in several ways. The machine is adapted to use 3 mL and 6 mL SPE cartridges, and hence the use of disks has not been examined in much detail. The use of co-solvents hasn't been studied, and could be useful, but the nature of urine – as well as that of the cartridges being utilised – suggests limited room for improvement when considering solvent systems. Having said that, one of the most retentive methods trialled on reversed-phase cartridges has led to recommending the use of 2% formic acid in all SPE steps.

Multiple-pass SPE is an interesting concept and could hypothetically reveal compounds not absorbed in the SPE wash the first time around. This is especially relevant when considering the effects of compound breakthrough. It may be possible that more information (and cleaner spectra) may become available using multiple passes.

Andrew Surman opened discussion of the paper by Carlos Afonso: Would you mind commenting on your experience of the usefulness of CCS prediction/ modelling (Mobcal), and algorithm selection?

Carlos Afonso answered: CCS prediction accuracy is very dependent on the type of molecule. The calculation of 3D structures of small molecules with low degree of freedom, such as PAHs, is very simple and can be done with very standard molecular modeling methods. The 3D files are then sent to Mobcal or other CCS calculation programs. This can be much more complicated with very flexible molecules such as lipids. We generally tend to evaluate our CCS prediction based on the analysis of standard molecules. In the case of natural organic matter such as petroleum, the absence of real standard molecules representative of heavy fractions is obviously an issue.

Philippe Schmitt-Kopplin said: Involving ion mobility is a great approach – did you think about adding yet another dimension with chromatography for in-depth sample characterization? How do you face the challenges in the multidimensional data produced?

Carlos Afonso responded: The addition of a separation method with IMS and MS is a very powerful tridimensional separation approach. This is however only possible with time-of-flight (TOF) mass spectrometers thanks to its high acquisition speed. It should be pointed out that in the case of natural organic matter, the existence of many isomers can typically yield unresolved chromatographic profiles for each molecular formula that are not always simple to rationalize. The IMS separation is very simple and the results can yield direct structural information through the determination of collision cross section. For the most complex mixtures such as petroleum, Fourier transform mass spectrometers can only be used with trapped ion mobility spectrometry (TIMS), and because of the low acquisition speed, TIMS-FTMS cannot be used with on-line chromatographic separation.

Pierre Giusti remarked: From an industrial point of view, it is important to keep in mind that following this very exhaustive approach, once the targeted chemistry or compounds have been identified or understood, simpler analytical protocols/instruments are needed. Of course, we also need automated data treatment processes.

Dušan Uhrín opened discussion of the paper by Francisco Fernandez Lima: Can the separation power of IMMS be enhanced? *E.g.* can you use more anisotropic gases such as CO_2?

Francisco Fernandez Lima responded: Yes, the separation in IMS changes with the collision partner. We currently mostly operate in N_2 for simplicity and lower cost in the case of TIMS-MS. We have shown the advantages on using bath gas modifiers in TIMS. Other gases (*e.g.* CO_2, Ar, *etc.*) are also possible and will likely provide better separations.

Dušan Uhrín asked: In your fragmentation approach, do you do MS-MS on a nominal mass selection? Did you consider looking at individual peaks? Do you rely on databases in interpreting the MS-MS data?

Francisco Fernandez Lima replied: The example presented focuses on the fragmentation at nominal mass, as a way to practically analyze a DOM sample by scanning the MS1 (typically 150–650 *m/z*) with a 1 *m/z* isolation window in the quadrupole prior to the FT-ICR MS analysis. We also discussed and will explore the possibility of performing MS/MS in the ICR cell using a correlated harmonic excitation field (CHEF).

Two approaches were described to interpret the MS/MS data. One is based on accurate mass database search of parent and fragment ions (*e.g.* using MetFrag). The second approach uses a molecular generator based on the core and neutral loss to predict 2D and 3D candidate structures to be validated in the CCS domain.

Norbert Hertkorn remarked: I'm intrigued that the ion mobility derived mass peaks show fine structure that invites fitting of "components" as done by you. It appears tempting at first to attribute (numbers of) compounds to these peaks. As an example: when a molecule is composed of a single benzene ring, a *tert*-butyl group and a carboxylic group, each of these units will have an individual cross section, which is probably additive (in this case) for these small aromatic molecules that do not have a complex 3D shape. What kind of meaning can you extract from these peak shapes in terms of numbers of compounds/isomers and where are improvements of resolution conceivable?

Francisco Fernandez Lima answered: These are very good questions. In TIMS, we measured the averaged momentum collision cross section. It has been shown for the case of peptides that the CCS can be averaged by the sum of the CCS of the AA units and the backbone. In principle, this can be extrapolated to other chemical classes, but typically complementary information based on MS/MS or other structural tools can lead to a better candidate structure assignment.

There is a search for better and more sensitive IMS devices. With TIMS, we have shown resolving powers of $R \sim 400$–450, and other IMS devices (*e.g.* overtone IMS) have shown $R \sim 1000$. We continue to push for higher R in the TIMS architecture, but most of the upcoming breakthroughs will be based on the use of complementary, orthogonal separations (*e.g.* GC, LC, and non-linear IMS) in tandem with TIMS-MS/MS.

Jeffrey A. Hawkes commented: You suggest that each molecular formula is constituted by a diverse range of isomers, and that these can be distinguished by their fragmentation patterns. The isomer count can be investigated by considering the number of neutral loss pathways that lead to common core structures. You also suggest that each isomer has a narrow CCS profile (modeled in Fig. 2 in your paper), while the whole molecular formula isomer distribution has a broader drift profile. Can these two theories be evaluated by plotting the drift profile of single core structures? Shouldn't a core structure that supposedly comes from one isomer have a drift profile that is narrower than a core structure that supposedly comes from 40 isomers?

This would correspond to data in a recent paper[1] that showed this in chromatographic separations (shown in Fig. 1 in this discussion), in which fraxin was added as a known isomer of $C_{16}H_{18}O_{10}$, and a unique neutral loss for Fraxin ($-C_6H_{10}O_5$) was found to have a narrow profile, while a common loss ($-3 \times CO_2$) had a broad, unresolved profile.

1 J. A. Hawkes, C. Patriarca, P. J. R. Sjöberg, L. J. Tranvik and J. Bergquist, *Limnol. Oceanogr. Lett.*, 2018, **3**, 21–30.

Francisco Fernandez Lima replied: This is a very good point. At this point we are only performing TIMS on the parent ion. By adding the TIMS of the fragments (*e.g.* core structures and intermediates), we will have a much better guess of the structure of the parent ion, and the number of potential isomers in the IMS profile per chemical formula.

Royston Goodacre opened a general discussion of the papers by Ronald Soong, Daniel McGill, Carlos Afonso and Francisco Fernandez Lima: These Faraday Discussion meetings are often repeated but some considerable time in the future (often a decade). As we are in the "Future challenges and new approaches" session, what does each panel member believe will have been solved in 10 years time?

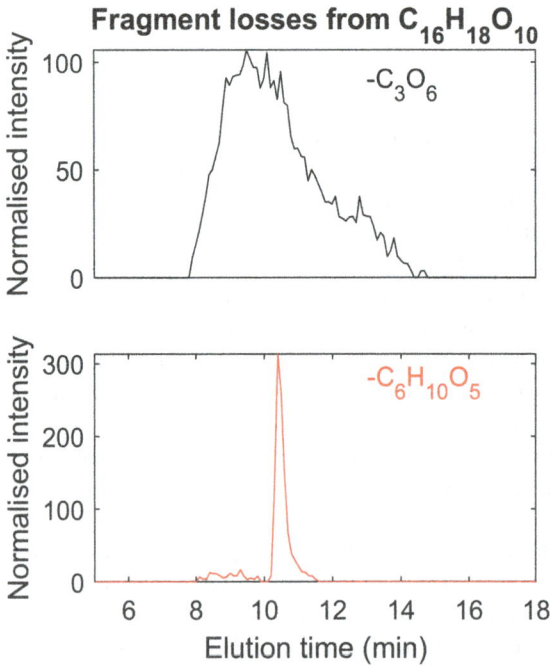

Fig. 1 Tandem MS chromatograms showing neutral losses from pre-cursor ion $C_{16}H_{18}O_{10}$. The loss of 3 CO_2 groups is common to numerous isomers, so gives a broad hump, while the loss of $C_8H_{10}O_5$ is essentially unique to a specific isomer called Fraxin, which was added to the complex mixture.[1]

Daniel McGill answered: Thank you for your question. I would suggest that – given the nature of chemistry – for every question answered, two more immediately spring up; hence, a question in response to this one (!) could be "what tools might we have to solve problems in 10 years?". As I have mentioned elsewhere, stronger magnets in NMR and more comprehensive databases will never go amiss, but perhaps more important is the general expansion of our "toolbox" in order to extract more information from complex mixtures than current methods allow. The further integration with other scientific fields will also hopefully provide unique perspectives, accelerating our ability to solve problems. Finally, we would hope that education efforts at every stage of learning keep up with the trends moving forward, in order to keep students and researchers alike at the cutting edge.

Perhaps most importantly – if not directly relevant to complex mixture analysis – we will have made a significant dent in the climate crisis through the application of renewable and sustainable technologies and practices, so that we will be able to continue to hold such productive meetings.

Ronald Soong answered: This is an insightful question; we believe that one of the key challenges moving forward will require the better education of the next generation of environmental researchers in NMR approaches and using NMR spectroscopy's full potential. Ultimately the field will require a combination of mass spectrometry (for example molecular formula information) and NMR spectroscopy (isomeric information) to truly assign and deconvolute complex mixtures, which is often a necessary precursor to understanding and predicting environmental reactivity.

Carlos Afonso replied: Our instruments are becoming more powerful allowing us to routinely tackle what was a challenge just a few years ago. With high field FTICR, together with technologies such as absorption mode and 2ω detection, the analysis of ultra-complex natural matter by on-line chromatography should become simpler.

The combination of IMS with FTICR mass spectrometry is most likely an important tool for the characterization of natural mixtures as it will give access to valuable information on isomeric content. The coupling of trapped ion mobility (TIMS) with FTICR is nowadays only available as a prototype, but a commercial solution may be available soon, making this technology more widespread.

Francisco Fernandez Lima responded: In our case, we are making strides on the utilization of complementary separations and characterization tools based on mass spectrometry. I believe that IMS-MS will be a standard tool for any analysis within the next 10 years. Relative to the analysis of complex mixtures, I believe we will be able to have well curated databases and protocols for the characterization of DOM (and others) based on complementary GC/LC/-IMS/IMS-MS/MS and their combinations.

Anneke Lubben commented: Combining the information from NMR and MS is not a new idea. Don't forget that undergraduate organic chemists do this on a daily basis; we are training them in exactly that way. Additionally, most pharmaceutical companies have had fully integrated LC-NMR-MS systems for their development and QC processes, *e.g.* Pfizer had this set-up about 20 years ago, when this was already very much in vogue. Even though these are generally simpler chemical systems, we need to revisit and use these "ancient" approaches for the benefit of our understanding of more complex samples. It will require further software and community developments, but all of the individual techniques and expertise seem to be out there already.

Justin J. J. van der Hooft replied: You are right that multiple groups – including myself – have been working on integrating MS and NMR data. I have no knowledge of large companies doing this – this is typically not shared widely in the scientific community – but how come this is not mainstream yet? I think this is because (i) running both an MS and NMR system in one lab is expensive, (ii) most solutions are not as high-throughput as you may wish, and iii) the data analysis is still relatively cumbersome and needs expertise. Your point on developing further software and community efforts is spot on, but it might need a large funding initiative where groups are forced to work together to make it work.

Stephen Summerfield remarked: Sample storage and suitable preparation are essential considerations that are ignored by my fellow lecturers teaching analytical chemistry, such as the use of inappropriate sampling materials that add to the contamination of the sample, or where the analytes adsorb. For example: (1) PET bottles are not suitable for determining organics. (2) The use of glass for the analysis of metals. (3) Sometimes samples decompose before they are analysed where the sample has sat on the shelf for a week.

Another area is to look at solvent compatibility before it becomes an issue where the analytes selectively dissolve, precipitate or react, often causing a blockage of a column. If we do not get the fundamentals correct then it does not matter how advanced our equipment is, the validity of the data is compromised. This has come out of three decades of being in academia and industry. After reviewing over 3 500 ecotoxicology and toxicology tests, I now know more ways of doing the analysis poorly.

Antony N. Davies responded: Hopefully not ALL fellow lecturers teaching analytical chemistry! There are some excellent, well documented examples of where poor sampling has caused outcry when erroneous results have prematurely reached the public domain to keep students interested and focused on the magnitude of the problem. We only need to look at contaminated baby-food scares of the past.

Antony N. Davies commented: In 10 year's time, we should be saying "why we have done this", and "here is the result". We are missing the "why"; we currently focus on tools and technology. How are we driving science forwards? This is what will drive funding.

Caroline Gauchotte-Lindsay remarked: We were discussing what the future looks like and what we should be working on. I think that if we want analytical chemistry to be more integrated and much more recognised than it is nowadays, we need to be at the centre of the big challenges our world is currently facing. Climate change, pollution and health issues will not be addressed without excellent analytical chemistry as part of interdisciplinary teams. I do recognise the value of investigating things because they are interesting, as this allows us to make the breakthrough that leads to the development of more powerful tools. However, we can't just give the tools to the world and move on; we need to also be at the heart of applications.

Justin J. J. van der Hooft answered: This is a fair point and also taking into account the earlier question on how we can ensure that developed tools are being used and adopted by the community. I think this requires efforts from both sides: users will need to try out novel tools and assess their use for their user case and developers will need to bridge their tools to others and put them in a framework that improves their ease of use. You are right in saying that we cannot just give the tools to the world, but to integrate analytical chemistry into practical solutions to the big challenges also requires the time and effort from users to test novel tools that they may think are relevant and provide feedback to the developers and community. If they stay in their silos and keep using what they are using since they know how it works, things will not change. Another important thing to note is that developers can (should) organise hands-on workshops with conferences to show and teach their tools to the community. This will certainly make them easier to use, as often the initial start-up is most of the hurdle into using a new tool (where to find what, which settings to use, *etc.*) – I guess there is not a short and simple answer to this question but I believe there are things that developers and users can do to streamline this process a little bit. As Caroline puts it, many current big challenges require excellent analytical chemical solutions that may well be in development now!

Stefan Kuhn replied: From my perspective as somebody working on computational methods, this is a very relevant point. Lots of good methods are developed, but not many make their way into mainstream labs. This partly has to do with standards and formats, but also with integration into workflows. Any improvement here would be very welcome. I could imagine that the problem is similar in analytical tools. I think this is a very relevant and open question – how do we make sure tools are used and stay relevant beyond the development phase?

Dušan Uhrín addressed Ryan P. Rodgers and Marc-André Delsuc: Where are we with 2D mass spectrometry? Is there potential in this technique?

Ryan P. Rodgers replied: This has been extensively discussed in the literature and in online articles.[1-4]

1 https://www.labnews.co.uk/features/mass-spec-goes-2d-19-01-2017/
2 M. A. van Agthoven, Y. P. Y. Lam, P. B. O'Connor, C. Rolando and M. Delsuc, *Eur. Biophys. J.*, 2019, **48**, 213–229.
3 M. A. van Agthoven, A. M. Lynch, T. E. Morgan, C. A. Wootton, Y. P. Y. Lam, L. Chiron, M. P. Barrow, M. Delsuc and P. B. O'Connor, *Anal. Chem.*, 2018, **90**, 3496–3504.
4 M. A. van Agthoven, M. Delsuc, G. Bodenhausen and C. Rolando, *Anal. Bioanal. Chem.*, 2013, **405**, 51–61.

Marc-André Delsuc answered: From my NMR spectroscopist background, my gut feeling is that 2D mass spectrometry could have a strong impact in the domain, in a similar way to the impact 2D had in NMR in the 1980s. 2D MS is ideal for complex mixtures, as it naturally provides such a huge spectroscopic space and does not require the use of any chromatographic separation before analysis, and does not require any assumption on the sample in a purely Data Independent Acquisition (DIA) approach.[1]

1 M. A. van Agthoven and P. B. O'Connor, *Rapid Commun. Mass Spectrom.*, 2017, **31**, 674–684.

Of course, we are facing many technical and theoretical challenges, but in my view, the major one is to catch up with several decades of development of highly optimized coupling techniques in MS. However, no doubt that in some particular applications, it will rapidly overtake the classical hyphenated approach.

One point that could greatly assist the deployment of 2D would be to allow 2D MS to be run on a simpler and cheaper instrument than the large FT-ICR machines on which it has been shown so far. In this respect, the developments of the Warwick group, aiming to have 2D MS on linear trap, are extremely promising.

William Kew opened discussion of the paper by Marc-André Delsuc: It is great to see that you are making the software reproducible, free and open source, and deploying an easy-to-use online tool for its use. This should be common practice in methods development for data analysis.

In this example, you used only five samples/fractions of a crude extract. This is convenient for NMR time, but it seems to be a small number. How many samples would be needed for more complex samples, or is five sufficient?

Marc-André Delsuc replied: The purpose of this approach is to speed-up the dereplication question, and to allow a rapid identification of potential new molecules in the mixture. More samples would mean more NMR instrument time (or a lesser variety of experiment type for the same time) and more bioassays, and is not always preferable.

As the method is based on the correlation between activity and signal intensity, it is important that the activity is dispersed in the samples, with high and low activities present, and that several samples present non-null activities. This is the reason why a very efficient separation is not chosen for sample preparation.

We believe that an optimal number of samples, at least in the conditions we have conducted, is around 4 to 8 samples.

Philippe Schmitt-Kopplin remarked: Your approach experimentally integrates the NMR information with the analyzed bioactivities – would you win in refining your model by also integrating experimental NMR information of non-active compounds?

Marc-André Delsuc responded: The approach is based on the analysis of a set of samples that are different but similar, and with varying activities. In that respect, you are right, and inactive and partially active samples are as important as the active ones. In all of the examples we have tested so far, we have be careful to integrate the whole range of activities. However, I do not think that adding a loosely related, inactive sample to the series would help the analysis – but this has not been really tested. An assessment of the best protocol in terms of number and quality of samples is yet to be done.

John Parkinson said: Marc, thanks for the paper – I welcome the fact that you are making this software application publicly available and open source. My question concerns how you treat the data. Are you filtering out real data here that could be telling you something and are therefore risking losing information? Unless I'm mistaken or have misunderstood, the process appears to "clean up" the data, which would therefore possibly clear out real information that may be being treated as artefactual.

Marc-André Delsuc answered: We indeed "clean-up" the initial NMR signal, however we try to mitigate the loss of information. The point is that NMR signals are dependent on the environment (solvent, other solutes, *etc.*) in terms of exact position and shape of the lines. Following previous studies, we solve this with a bucketing approach, integrating over a small portion of the spectra. Here the samples are much more diverse than in metabolomics, and the bucket size has to be larger – this is also required by the use of 2D experiments. We compensate for the loss of detailed information by creating a richer content for the bucket and adding descriptors to the integral value, such as standard deviation, minima, maxima and number of peaks.

Jeroen J. Jansen asked: If you find an interesting peak with your slider, how are you going to communicate this finding in a scientific paper? There is a desire for *p*-values: what is the degree of significance for the observation? It is unclear how many plants you actually analyse – I figured only the least and most active. If you include all plants, you could find links and relationships also perhaps with a non-linear trend if you use a nonparametric statistic.

Marc-André Delsuc replied: The primary goal of this tool is to provide a simple and interactive tool for the pharmacognosist to rapidly determine the interest of a given sample. The figures that are created, despite showing composite statistical features, are presented as NMR spectra, and as such can be stored and used in reports and publications as with any other NMR spectra.

About the improvements you mention, the platform has been proposed as a first draft and is meant to be developed, with the addition of new tools and controls. For instance, the exploitation of cross-correlations between different kinds of spectrum has to be developed.

About the *p*-value, this is indeed a nice addition. The sliders in the interactive maps are directly related to a level of confidence, and can be directly related to a global *p*-value. We are currently developing an estimate of this *p*-value, based on a randomisation approach; this should be added in the next version of the tool. In a similar manner, the use of nonparametric statistics is also investigated.

Mathias Nilsson remarked: DOSY is one type of spectrum you include. These spectra are fundamentally different from conventional 2D spectra and I suspect that they are less suitable for automated analysis. Could you comment on how you find such spectra helpful in your automated setup? What DOSY processing are you using? Do you make the source code available for this processing? Can you comment on which pulse sequences you can use for the automated analysis?

Marc-André Delsuc responded: We use our own method for the analysis of DOSY called PALMA, available on-line at http://palma.labo.igbmc.fr. Based on numerical Laplace inversion, and not assuming monodisperse species, this approach, while probably less resolutive, is quite robust and suitable for automatism.

About the DOSY experiment itself, we found that for a small cost in experimental time, it could bring unique information on the 1D ^{1}H spectrum, by filtering out small irrelevant metabolites but also large aggregates. As such, we chose to systematically use it in our studies so far.

Jeroen J. Jansen opened discussion of the paper by Caroline Chaux and Laetitia Shintu: You are treating a well-known research area in chemometrics. Over decades, algorithms have been developed for visible/NIR spectroscopy called Multivariate Curve Resolution-Alternating Least Squares (https://scholar.google.nl/scholar?hl=nl&as_sdt=0%2C5&as_vis=1&q=MCR-ALS&btnG=). However, the focus on NMR of this method has been limited (and on 2D NMR there has been even less[1]). I think there is a lot of possibility to include more NMR-specific constraints (*e.g.* peak split-up) into the MCR framework, which would be a very relevant line of research. Removing shifts is also something which has received a lot of attention in chemometrics, but is usually separately treated as warping or alignment (*e.g.* Dynamic Time Warping and Correlation Optimized Warping). One method that integrates it into the analysis itself is IDLE by Martens (https://www.idletechs.com/#Products).

Maybe except the last part, all of these things are widely known in chemometrics, so if you submit a paper about this and a chemometrician reads it, they will immediately respond to this. I really enjoyed your presentation!

1 R. Huo, R. Wehrens, J. van Duynhoven and L. M. C. Buydens, *Anal. Chim. Acta.*, 2003, **490**, 231–251.

Laetitia Shintu and **Caroline Chaux** replied: Thank you very much for the references you gave us. We will definitely look into it. We developed ALS algorithms in order to establish state-of-the-art results to which we can compare. However, our objective is to develop new algorithms that are more robust, faster and that can handle more sophisticated constraints, as is the case for the BC-VMFB algorithm.

Mathias Nilsson remarked: NMR spectra are never perfect, and may require pre-processing and correction before more advanced data analysis, such as you are presenting here. What sort of pre-processing did you use for these samples?

There is significant sample-to-sample variation in *e.g.* shimming, peak position and phase. You can correct for many such systematic errors using reference deconvolution. I think that this could be very helpful to prepare spectra for your analysis.

Laetitia Shintu and **Caroline Chaux** responded: We manually corrected the phase and the baseline of each spectrum (1D and 2D), and we calibrated them on the $CDCl_3$ signal, for both ^{1}H and ^{13}C frequencies, using TOPSPIN software. The alignment of some signals was also necessary. Since the number of analysed samples was low, it was not necessary to correct the lineshape of the signals in this study. However, it is something that we will take into account for further work and we will then consider using reference deconvolution as advised.

Stephen Summerfield said: Being that my PhD was in fluorescence, and after a decade dealing with water analysis, I'm intrigued. You commented that you use chemometrics on environmental fluorescence? Was this in regards to river water? What species were you picking up with the fluorescence spectroscopy?

Laetitia Shintu and **Caroline Chaux** answered: Our previous paper dealing with fluorescence spectroscopy[1] is available at https://hal.archives-ouvertes.fr/hal-01387439/document. N. Thirion-Moreau and E. Carstea will be more competent than me to answer to your questions. The application was demonstrated effectively in the context of river water monitoring.

1 X. Vu, C. Chaux, N. Thirion-Moreau, S. Maire and E. Carstea, *J. Chemom.*, 2017, **31**, e2859.

William Kew remarked: I have some experience applying independent component analysis (ICA) to NMR datasets, following the ideas of Monakhova *et al.*[1] However, the problem with ICA is that you have to define the number of components to calculate, limited only by the sample set size. The output generated therefore depends on the number of components defined, and possibly the sample size, and thus it is impossible to know if the output model is valid and reflects the components genuinely without an *a priori* understanding of the mixture.

In other words, how do you validate the model without knowing the mixture components to begin with? Or, can you generalise any blind source separation techniques so that a validated approach can be applied to an unknown mixture?

1 Y. B. Monakhova, A. M. Tsikin, T. Kuballa, D. W. Lachenmeier and S. P. Mushtakova, *Magn. Reson. Chem.*, 2014, **52**, 231–240.

Laetitia Shintu and **Caroline Chaux** responded: As mentioned in the paper (page 12, second paragraph), a PCA on the mixing matrix was sufficient to determine the number of sources. However, this study was limited as we were in an overdetermined problem and we only considered model mixtures. In the case of the 5 mixtures of 4 terpenes and 7 mixtures of 6 amino acids, a PCA was sufficient to determine the number of sources.

When considering real samples with an unknown number of compounds, the task will be harder but we think that it would be possible looking at the NMR spectrum to approximate the number of molecules according to the number of signals.

Simon Rogers queried: In the experiments you present the situation is very controlled – a mixture of only 4 compounds. How does the data requirement scale as the number of compounds is increased, and, in a realistic scenario where you do not know what is present, how will you set the number of sources?

Laetitia Shintu and **Caroline Chaux** replied: This question is related to the previous question – please see the answer to this question.

Marc-André Delsuc commented: Thank you for the very nice talk. In Table 10 in your paper, where you present the results for a set of HSQC experiments, you show error bars on the estimated concentration on the order of 10%. This is not very far from the precision with which you can prepare a given solution, in particular considering you are using chloroform, which is a fast evaporating solvent. I believe these numbers are not very representative of the quality of your result.

It would be better to have some kind of descriptor that is able to qualify the quality of the separation in a global manner for the experiment, sensitive to additional and missing signals. You present SIR and SDR but not being global, they are not very handy. Can you comment on this?

Laetitia Shintu and **Caroline Chaux** responded: We took into account the evaporation of the solvent, and all of the concentrations were recalculated using the ERETIC method after the NMR tubes were sealed. This information was included in the paper.

Concerning the quality measure, as we were working in a totally controlled context, we computed the SDR and SIR source-wise and globally. Computing the mean of the obtained individual SDR and SIR allows us to evaluate globally the separation quality.

Concerning the question of additional and missing signals, this information is contained in the SDR value, which is computed as described in the work of Vincent *et al.*[1]

1 E. Vincent, R. Gribonval, and C. Févotte, *IEEE Trans. Audio, Speech, Language Process.*, 2006, **14**, 1462–1469.

John Parkinson commented: I'm pleased to see the successful interface and communication between a mathematician and an analyst in this work, and the honesty expressed in the difficulties of finding a common language and vocabulary through which it is possible to communicate from the perspective of different disciplines. From my point of view, this is encouraging and makes me realize I'm not alone in experiencing the same frustrations, having attempted something similar myself over a long period of time, in finding common language through which to communicate the problems of analysing a very similar type of system to the one presented in the paper. My dialogues have led to the exploration of different concepts about which I know almost nothing, so in a similar way, I'm looking for a partner and expert that understands how to use those concepts to address the analytical problem that I have. We have talked in the meeting about data and about dealing with it as a pattern recognition problem. The terms AI and machine learning have emerged in several discussions. I don't know about how these processes work or even where to start to access such approaches to data interrogation. How do we therefore begin to try to integrate these potentially game-changing methods into our data analysis procedures? How do we draw on these methods and draw down knowledge and information from the wealth of pre-existing data to help us understand the current complex systems we are studying and how do we create a vocabulary that will help the different disciplines communicate effectively with one another? These are open ended questions that don't necessarily have an answer, but that could be helpful in shaping future thought about developments at the interface between disciplines.

Laetitia Shintu and **Caroline Chaux** replied: Thank you for your encouraging remarks.

Stephen Summerfield opened discussion of the paper by Caroline Gauchotte-Lindsay: For me it is wonderful to revisit the work that I did for the Gas Research Centre in Loughborough, that was part of British Gas before it was divided by privatisation. Here I was looking at contamination and remediation of old coal gas sites such as Beckton in London. Like with this work, this included phenols and PAHs.

(1) Sampling is always a problem – where is the pollution from? How many samples have you actually looked at? From experience, there can be huge differences within a meter square, let alone over a whole site. This could be three or more orders of magnitude.

(2) Are you looking at culturing the bacteria or nurturing the native bacteria? From our experience, the latter was far more successful. Possibilities are the use of molasses, seaweed, essential metals *etc.* to stimulate the bacteria, and aeration to promote aerobic bacteria.

(3) Have you distinguished between chemical and physical degradation of petrochemicals, PAHs and phenols? The turning and aeration of the soil did a huge amount of this work.

(4) Could you monitor the success of remediation by the species of bacteria?

Caroline Gauchotte-Lindsay answered:

(1) We had 16 samples for one site and 8 for another. We know little about the different sampling points as the samples were given to us by industrial collaborators. We have looked at the differences between the samples (such as LOI, moisture content, PAHs and heavy metal concentration) intra-site and they were actually more similar to each other than expected. Our ultimate aim is to run microcosms, so we were more interested in getting contaminated soils than looking at intra-site variations.

(2) We are mostly interested in understanding the potential of the autochthonous bacteria: what is the diversity, what are their functions. Eventually, by studying several sites, we want to investigate whether we can use our integrative tools to predict and potentially optimise the local biodegradation.

(3) We have not done this yet. Working in microcosms in the laboratory at first will enable us to control or turn off some of these effects. Using the GC × GC will also be very useful in delineating the different effects, as physico-chemical removal can be predicted by/matched to GC × GC retention.[1]

(4) One of our aims as discussed above is to establish whether the local microbial ecology can help us predict the success of biodegradation.

1 G. D. Wardlaw, J. S. Arey, C. M. Reddy, R. K. Nelson, G. T. Ventura and D. L. Valentine, *Environ. Sci. Technol.*, 2008, **42**, 7166–7173.

Dušan Uhrín asked: In GC × GC, there are problems with aligning the spectra. Is this common for 2D GC?

Caroline Gauchotte-Lindsay replied: The success of peak finding and data alignment is crucial for accurate non-targeted analysis by GC × GC. These are processes that are very well catered for in 1D GC but are still issues in GC × GC. Here we wanted to test a newly published alignment code R2DGC. Our experience with peak finding and data alignment from the proprietary software is that manual verification and correction is usually needed. We decided for this piece of work not to carry out any manual verification as it is unrealistic to check all peaks when you have thousands of peaks in several dozens of samples. We aimed here to illustrate some of the common issues that GC × GC users face and that are holding back the use of GC × GC for omics studies.

Peter J. Schoenmakers queried: During biodegradation, your contaminants (such as alkanes and PAHs) become more polar. To what extent are you paying attention to derivatization of polar analytes prior to your analysis by GC × GC? How does the derivatisation affect the interpretation of your results.

Caroline Gauchotte-Lindsay responded: For this particular paper we did not carry out derivatisation, but as our eventual aim is to predict the end-products of biodegradation, we acknowledge that derivatisation should be carried out to be able to see and identify oxidation products. Our experience is that classic silylation, while theoretically possible, is not successful for the dialcohols formed by dioxygenase enzymes in complex samples. We are currently developing ultrasound assisted derivatisation to access difficult to derivatise molecules.

Dušan Uhrín asked: There is an example in the paper combining different types of data in chemometric analysis. What is the best way to go about it and what are the tools?

Caroline Gauchotte-Lindsay answered: Other PLS algorithms (akin to PCA) could be used. They factorize the original feature matrix into score and loading matrices in such a way that the scoring matrix has the highest variability both in terms of the data spread but also in terms of the separation between the categories the data points belong to. The weights of the loading vector can then serve as indicators for the importance of features. The reason why we have chosen sPLS-DA and DIABLO specifically is mainly because of the additional "sparsity" constraint that enforces some of the loading weights to go to zero and thus serves two purposes, dimensionality reduction as well as discriminant analysis. The latter functionality is missing in the majority of the PLS implementations. Other integration algorithms that have been applied to microbiome data and other omics data were the ones used in the R STATegra package (http://www.stategra.eu/), however these rely on factorizing the data matrices into submatrices that contain the common and joint variations (such as SCA, DISCO-SCA, JIVE, *etc.*) focusing more on visual cues without emphasizing on discriminating attributes.

Norbert Hertkorn opened a general discussion of the papers by Marc-André Delsuc, Caroline Chaux and Laetitia Shintu, and Caroline Gauchotte-Lindsay: We need to think about what specifically is happening on the atomic/molecular level within the analytical method being used and the capacity of these methods to display distinctive features in the face of ill-constrained interferences. For example, a complex array of molecular processes operates in mass spectrometry of mixtures, ranging from volatilization, distinct concentration gradients within droplets from differential surface activity, primary ionization and ion-molecule reactions, rearrangements/fragmentation of high-energy species, and so on. Using fractionation before mass spectrometry is easily available to reduce complexity and to decrease effects of projection within mass spectra. Specialists using different methods should speak to each other in an open-minded way and concede limitations intrinsic to methods.

Mathias Nilsson commented: I think that open source, providing source code for methods – including pulse sequences – is a very useful path to get our work to potential end-users. Ideally, code should be annotated well so that others can easily understand it.

Conflicts of interest

There are no conflicts to declare.

PAPER

The blind men and the elephant: challenges in the analysis of complex natural mixtures

Royston Goodacre (iD)

Received 30th May 2019, Accepted 13th June 2019

DOI: 10.1039/c9fd00074g

The identification of molecules from complex mixtures is difficult and full structure determination of the complete chemical milieu is yet to be achieved. Thus the comprehensive analysis of complex natural mixtures continues to challenge physical and analytical chemistry. Over the last 50 years or so, many research laboratories have strived to invent better analytical techniques with complementary physicochemical properties and improved resolving power, and to investigate upfront sample pre-treatments, which are necessary to enhance sample coverage from complex mixtures. The purpose of this Concluding remarks article is to try to capture the recent developments in high-resolution mass spectrometry and nuclear magnetic resonance spectroscopy applied to complex mixtures that were presented and debated, the parallel progress in chemometrics, data processing and machine learning approaches, as well as capturing and highlighting future challenges that still need to be addressed. The summary begins with a brief contextual overview and explains that the title – the blind men and the elephant – reflects that no single method measures everything and that multiple 'tricorders' are needed in order to understand complex systems. Next, the meeting highlights are provided, and I hope those that were present are happy that this captures the many diverse areas of research that were discussed and that this article may act as a yardstick to indicate where complex natural mixture analysis stands today.

Introduction

Researchers with highly diverse interests came together on the 13th May 2019 in the John McIntyre Conference Centre at the University of Edinburgh for three days for the 304th Royal Society of Chemistry *Faraday Discussion* on "Challenges in analysis of complex natural mixtures". Whilst the sun shone brightly outside the conference venue, with Arthur's Seat looking very tempting for some outside rambling activity, inside the lecture room science glowed as scientists delivered

Department of Biochemistry, Institute of Integrative Biology, University of Liverpool, Biosciences Building, Crown Street, Liverpool L69 7ZB, UK. E-mail: roy.goodacre@liverpool.ac.uk; Web: http://www.twitter.com/ roygoodacre

their latest analytical and data processing developments and findings, with intense discussions and debate warming up the atmosphere.

The meeting was chaired by Dušan Uhrin (University of Edinburgh, UK), with excellent help from the conference organisers, including Mark Barrow (University of Warwick, UK), Timothy Ebbels (Imperial College London, UK), Ruth Godfrey (Swansea University, UK), Donald Jones (University of Leicester, UK) and Mathias Nilsson (University of Manchester, UK). There were 93 delegates from some 13 different countries attending the meeting, with a total of 25 oral presentations and 26 posters. One particularly NICE feature was the lightning poster presentations session, which took place in the late afternoon on the first day. These presentations were a mere 45 seconds each with automatic slide advances. Presenters assembled into an orderly queue, and even within just 3/4 of a minute, the audience was treated to NICE summaries that were both Novel and Interesting, and Clearly delivered with great Enthusiasm! The meeting then had its poster session where in depth discussions were held.

The meeting was split into four different sections, which are detailed below, but before getting into this it is worth setting the scene as to why the time was right for a *Faraday Discussion* to be devoted to the diverse and therefore multidisciplinary sciences behind unraveling complex natural mixtures.

The blind men and the elephant

The reader may be familiar with the proverb of *the blind men and the elephant*, which has its origins in India. This fable is about six blind men who stumble across a strange creature and they try to understand what it is.[1] To do this, they each feel a different part of the animal and come to a conclusion based on their limited experience; this is depicted in Fig. 1A. The first blind man feels the elephant's body and comes to the conclusion that the creature is in fact a wall, while his friend feels the tusk and declares that the elephant is a spear. Another shakes his head after feeling the trunk of the elephant and claims with some anxiety that it's actually a snake. The fourth blind man, whilst feeling the elephant's leg, states that they are incorrect and that it's indeed a tree. The next man has got hold of the elephant's large ear and announces that it's a fan, whilst the last blind man, who has hold of its tail, declares that his friends are all wrong and that this elephant is in fact a rope.

Of course, all of the men are wrong, and only if they had shared their interactions would they have come to the correct conclusion about the elephant. The meaning of this parable is often used to illustrate that what people perceive as truth or fallacy is based on one's all too often subjective and narrow experience(s). We can readily extend this to the analysis of complex natural systems.

Fig. 1B illustrates that multiple approaches are used for the analysis of chemical systems, and in this example how multiple physicochemical techniques may be used to identify a specific molecule (*e.g.* ref. 2). These are based on:

• Sample pretreatment that may involve fractionation or chemical/enzyme reactions.

• Chromatographic separation may then be employed, which uses different physical characteristics to effect separation of molecules in mixtures: *viz.*, polarity, volatility or charge; and combinations of these. This can be combined

The Blind Men and the Elephant — What is it?

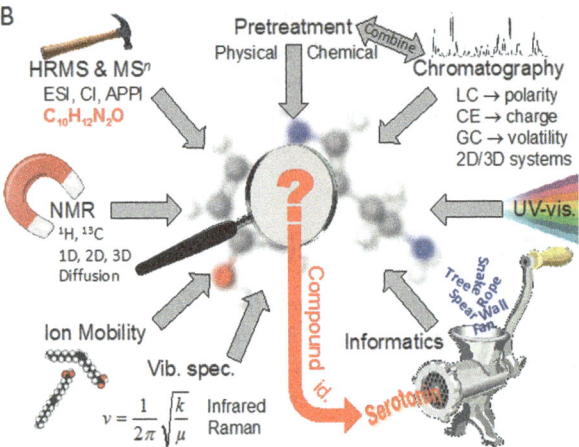

Fig. 1 (A) *The blind men and the elephant* illustrates that personal perception of a situation is subjective if one only measures a small part of the whole. This can be extended to the analysis of complex natural systems as shown in (B), where many different analytical approaches, which have different physicochemical properties, are used. These methods need informatics in order to synthesise and combine this information (parts of the 'elephant') in order to identify a molecule. Abbreviations: HRMS, high resolution mass spectrometry; ESI, electrospray ionisation; CI, chemical ionisation; APPI, atmospheric pressure photoionisation; NMR, nuclear magnetic resonance; LC, liquid chromatography; CE, capillary electrophoresis; GC, gas chromatography; Vib. Spec., vibrational spectroscopy. The image in (A) is under a free Pixabay license from https://pixabay.com.

with pretreatment processes, and would yield some information on a yet to be identified molecule, be it log P, pK_a or volatility.

• For some analyses, additional separation can be performed using ion mobility where molecules are separated on the basis of their size, shape, and charge.

• Detection of molecules can involve many different approaches. These are also highlighted in Fig. 1B and include:

○ Simple UV that may provide information on specific absorbing species in the UV part of the electromagnetic spectrum.

○ Mass spectrometry (MS) may be used to infer chemical formulae (in the example here, accurate mass would suggest: $C_{10}H_{12}N_2O$), and with MS-MS or MS^n one can narrow down potential arrangements of atoms within a molecule.

○ Nuclear magnetic resonance (NMR) spectroscopy is accepted as *the* tool for structural assignments, and provided there is enough sample with sufficiently high purity, it is the tool of choice.

○ Infrared (IR) or Raman spectroscopy[3] can also be used, though rarely, and they can provide information about functional groups due to their vibrational fingerprints.

This *Faraday Discussion* meeting therefore illustrates that systems chemical analysis for understanding complex systems is only really achieved by combining many different methods as they supply complementary information needed to identify an unknown substance. In addition to human interpretation, integration of multiple analytical approaches with appropriate informatics is needed for molecular identification. This is depicted by the 'mincer' in Fig. 1B and the informatics used may include statistics, chemometrics or some machine learning approach, in order to reveal in an objective fashion what the molecules may be within a mixture of diverse chemicals.

Thus we can see that chemical analysis is a multidisciplinary subject practiced by many scientists with diverse interdisciplinary skills. All of these disciplines were represented at this *Faraday Discussion*, and only with cooperation and integration can the whole molecular picture be 'seen': in Fig. 1B, this would be for the identification of serotonin. In reality, complex systems are much more complicated!

Chicken tikka masala

In this *Faraday Discussion* meeting, many different complex systems were analysed; these included plants, plant products (traditional medicines and essential oils), soils, coal, soot and petroleum fractions (petroleomics) as well as human derived samples such as urine.

Particularly complex systems are the food that we eat.[4] If we take a plate of chicken tikka masala and maybe have side accompaniments of raita and roti, then the meal we eat is very complex. This meal would (for example; other recipes do exist!) contain chicken marinated in tikka masala paste containing oil or butter, onion, ginger, garlic, cumin, turmeric, coriander, paprika, chilli powder, tomato, cream and coriander. This would then be skewered on bamboo or wood and cooked (which may release chemicals in the wood into the food) and then this chicken tikka added to a curry gravy containing yogurt, lemon juice, garlic, ginger, salt, cumin, garam masala and paprika. The raita may contain (*e.g.*) yoghurt, cucumber and mint, and the roti (*e.g.*) flour, salt and oil. With the exception (perhaps) of salt, each of these individual ingredients are highly complex mixtures and so the ensemble on the plate is an incredibly diverse mixture of chemicals. The analysis of this concoction would be very detailed and multifaceted.

If we consider just two of the ingredients – lemon and mint – we recognise these as having distinct aromas and flavours, yet the chemicals that give rise to these characteristics are very simple. As illustrated in Fig. 2, the distinct lemony flavour comes from (*S*)-(−)-limonene, whilst its enantiomer (*R*)-(+)-limonene is

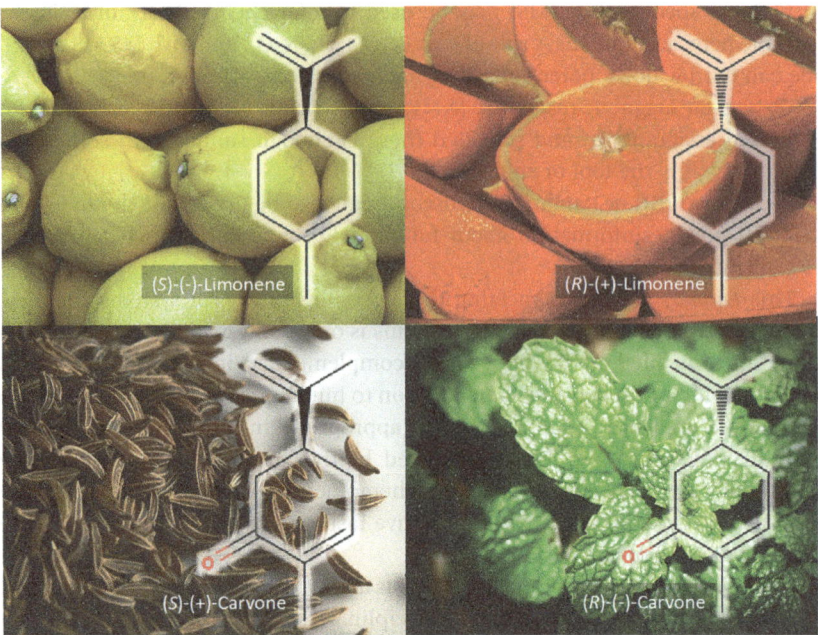

Fig. 2 The chiral limonene monoterpenes that give lemons and oranges their characteristic smell and flavour. Also shown are the two chiral forms of carvone that are responsible for the distinct smell and taste of caraway and spearmint. The images are under a free Pixabay license (https://pixabay.com). The chemical structures were generated in MolView (http://molview.org/).

found in oranges and is responsible for their aroma.[5] The simple addition of a carbonyl group to the benzene ring of limonene gives rise to either the mint or caraway aroma and the flavour from (R)-$(-)$-carvone or (S)-$(+)$-carvone, respectively.[6] These four very simple monoterpenes give highly diverse flavours and this highlights the importance of chirality in molecules, and in particular the interaction of such molecules with our taste receptors.

Whilst clearly important, the chiral nature of analysis was not really explored within this *Faraday Discussion*, and neither were positional isomers and their importance. By way of further example, if the food prepared above had used olive oil, then the major component would be $(9Z)$-octadec-9-enoic acid (oleic acid). This *cis*-isomer is considered healthy, while the *trans*-isomer (E)-octadec-9-enoic acid is not.[7] The point being made here is that the analysis of complex systems requires careful analysis and the analyst needs to decide which chemical resolution is sufficient to report.

Dealing with complexity

The first session discussed the latest advances in high resolution mass spectrometry (HRMS) and chromatography, and their hyphenation. Philippe Schmitt-Kopplin gave a fascinating Introductory lecture that set the scene perfectly. He discussed complexity and diversity and remarked that the former is subjective and often hard to define. In terms of the diversity of chemicals,

Philippe used small molecule chemistry (DOI: 10.1039/c9fd00078j) and high-lighted that there are currently 97.3×10^6 different molecules in PubChem (https://pubchem.ncbi.nlm.nih.gov), of which 113 000 are detailed in the Human Metabolome Database (http://www.hmdb.ca) and if we confine this to lipids there are around 43 000 entries in Lipid Maps (https://lipidmaps.org). He also horrified the audience by explaining that for linear molecules of 700 amu containing just carbon, hydrogen, and oxygen, there were 10^{46} possible isomers!

My own calculations for linear peptides show even more degrees of freedom. A simple peptide containing a mere 20 amino acids has $20^{20} = 10^{26}$ possible amino acid sequences. If we extend this to the average protein in archaea, bacteria or eukaryotes, which contain 283, 311 and 438 residues, respectively,[8] then the complexity by numbers becomes astronomical with 10^{368} archaeal proteins, 10^{404} bacterial ones and 10^{569} eukaryotic proteins! Whilst during the discussion, we learnt that 21 Tesla FT-ICR-MS has enough resolving power to resolve two analytes that differ by the mere mass of an electron (DOI: 10.1039/c9fd00005d) and with 7 dimensional NMR spectroscopy 10^{18} analytes can be resolved (ref. 9, DOI: 10.1039/c8fd00213d), we are likely to run out of time before all of these proteins are measured as the lifetime of the universe is 10^{17} s (ref. 10), and you've probably used several of those reading this far!

For the analysis of any system, what is needed is an ideal detector, along with upfront sample separation or preparative chromatography (DOI: 10.1039/c8fd00234g), and of course like Father Christmas, the Easter Bunny or an honest politician, there is no such thing; there is no magic 'tricorder'! If it were to exist, then the ideal detector (DOI: 10.1039/c8fd00233a) would be fast, have good orthogonality, provide uniform ionization, allow simplified data analysis and have improved (perhaps absolute) quantification.

For MS, electrospray ionisation (ESI) dominates most LC-MS and direct infusion-MS analyses, the latter mainly employing FT-ICR-MS. To a degree, this is a rather crude ionisation technique as ions are generated by squirting a con-ducting liquid through a needle to which a high voltage is then applied. As compound identification requires two orthogonal features,[11] MS-MS or MS^n is needed. This is however also rather crude and uncontrolled as the ions are often bombarded with an inert gas and the resulting fragmentation is akin to hitting a nut with a hammer and working out which nut was destroyed in the experiment. Thus library matching with standards is the key to compound identification along with more orthogonal techniques like NMR spectroscopy, which provides detailed structural analysis.

Other ionisation techniques are thus needed and applied, and within this session and elsewhere in the meeting, electron ionisation (EI) and various chemical ionisation (CI) methods as well as atmospheric pressure photoioniza-tion (APPI) were discussed. Each ionisation method has a bias to specific chemical classes and thus only part of the chemical milieu is ionized. The discussion of MS and fragmentation highlighted that any MS detector will be compromised in terms of having enough scanning speed, high enough mass resolution and enough duty cycle time to perform MS-MS or higher. Thus, for analyses with MS, there are always some concessions to be made.

All in all, this first session was interesting and perfectly set the scene. However, the most memorable thing that was elegantly and rather terrifyingly illustrated by

Ryan Rodgers (DOI: 10.1039/c9fd00005d), was that all analyses were only scraping the top of the iceberg in terms of the comprehensiveness of analysis. For the analysis of petroleum fractions, the routine approach was to perform ESI FT-ICR-MS of aminopropyl silica (APS) extracts of bitumen. However, this revealed only a very small fraction of peaks compared to the same MS approach on six different modified aminopropyl silica (MAPS) fractions (Fig. 3). This relatively simple pre-fractionation revealed that the standard analysis had failed to ionise so many of the components within bitumen, and advocates for the use of prior separation and increased ionisation methods. Up front separation was also used by Jeffrey Hawkes who used exclusion chromatography coupled to MS in order to reveal dark matter that could be detected by UV but was seemingly invisible to MS (DOI: 10.1039/c8fd00222c).

Within this context, the words 'brutal' and 'depressing' were used with reference to the above analysis and to the realisation that so much 'dark matter' was missing. Whilst da Silva and colleagues refer to dark matter in metabolomics as instances where there are MS data but no reference structure,[12] we consider here dark matter to also include small and large molecules that are not even measured by the analytical method and so go undetected.[13] The worrying thing is that there is no real way to estimate the level of dark matter when a complex sample is analysed.

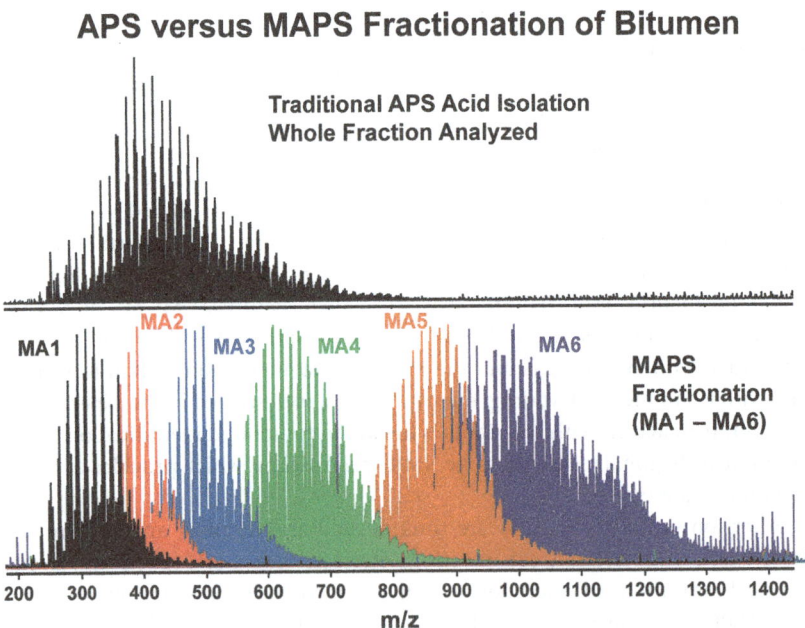

Fig. 3 This figure illustrates that many analytes are not ionised in ESI and so a lot of information is missing. The top spectrum is a broadband negative ion mode ESI FT-ICR mass spectrum of the aminopropyl silica combined acids extract obtained from MacKay bitumen. When modified aminopropyl silica extracts from the same material are analysed in 6 fractions (MA1 (black)–MA6 (purple)), the broadband ESI FT-ICR mass spectra shown in the bottom spectra contain considerably more information, and this 'dark matter' missing from the first analysis is revealed. Reproduced from DOI: 10.1039/c9fd00005d with permission from the Royal Society of Chemistry.

High resolution techniques

In this session, high resolution NMR spectroscopy was highlighted as a complementary technique to high resolution MS. NMR spectroscopy is very powerful, as in contrast to MS it has the capability to solve structures, but often struggles with complexity in systems where there are many analytes present that vary in their concentration over large dynamic ranges. It was thus highly appropriate to hear about the latest developments and challenges with high resolution NMR and MS.

In this session, much consideration was also given to upfront sample preparation and separation. Supercritical fluid extraction (SFE) coupled to both MS (DOI: 10.1039/c9fd00011a) and in-line sample concentration for NMR (DOI: 10.1039/c8fd00237a) were illustrated and discussed in terms of the selection of analytes extracted in this manner. The audience learnt that modifications of the supercritical fluid can readily allow for the analysis of both non-polar[14] and polar analytes with NMR spectroscopy (DOI: 10.1039/c8fd00237a): by simply adding methanol to CO_2, one can shift the mobile phase from non-polar to be more polar. A different approach highlighted for NMR analysis was to use viscous materials such as sucrose or 1% agarose gels to enable spin-diffusion during NMR acquisitions (DOI: 10.1039/c8fd00226f), an alternative to the popular DOSY-NMR approach.[15,16]

We were to learn later in the meeting that 7D NMR is possible, but even with modest 3D NMR, time is a limiting factor. In Nicholle Bell's paper and presentation (DOI: 10.1039/c9fd00008a), (3,2)D NMR was introduced as a method of reducing the dimensionality of hyphenated NMR whilst still keeping the information content of 3D spectra, but offering the speed advantages of 2D NMR measurements.

Combining different methods was also highlighted in this session in terms of high resolution NMR and MS, but also combining these along with bioassays to decide which fractions from Chinese medicinal plants contained pharmacologically active substances, and hence which fractions to concentrate on for structural elucidation (DOI: 10.1039/c8fd00223a). This is an essential component in the analysis of highly complex mixtures, such as those derived from plant sources.

Finally in this session, the use of hydrogen–deuterium exchange (HDX), which in ambient conditions normally only occurs on exchangeable protons such as –OH and –NH, was extended to labeling protons on aromatic rings and –CH side chains (from substances found in coal) for their identification with FT-ICR-MS (DOI: 10.1039/c9fd00002j). The conditions used for this HDX were somewhat harsh as they involved treatments with 4 M NaOD or 16% DCl and heating to 120 °C for 40 h, so would only be useful for non-labile chemical species such as the components found within lignin.

Optimisation of sample pretreatment prior to high resolution analyses featured heavily in the discussion and it was suggested that this process needed to be done for each individual scenario – there was 'no free lunch'. It would seem that most of this optimisation was done by brute force and with tongues maybe firmly in cheek by armies of PhD students. It is possible that this could be performed better by improved design of experiments and this would feature in the next session.

Data mining and visualisation

Collecting data on complex mixtures is only the start of the journey, and the next session of discussion was on chemometrics along with data mining, multivariate calibration or multi-way analysis, and how best these can be applied to different types of complex mixture.

We were reminded by Johan Trygg (DOI: 10.1039/c8fd00243f) that:

"The challenge is not in data collection but in maximising information in data and transforming data into information, knowledge and wisdom."

Johan Trygg, *Faraday Discussion* on *Challenges in analysis of complex natural mixtures*, 2019

This had perhaps been borrowed from an early quote by Henry Nix who was discussing national geographic information systems:

"Data does not equal information; information does not equal knowledge; and, most importantly of all, knowledge does not equal wisdom. We have oceans of data, rivers of information, small puddles of knowledge, and the odd drop of wisdom."

Henry Nix, *Keynote address*, *AURISA*, 1990

Of course, this processing is important, but Johan also reminded us that the design of the experiment was vital in order to maximise the extraction of knowledge about a complex natural system, and hence become a wiser person after the data have been collected and analysed.

As had already been discussed, the analysis of chemical systems using more than one tool is important but the challenge is then what to do with such data. Multiblock analysis was suggested as one potential approach (DOI: 10.1039/c8fd00243f) and with JUMBA (Joint and Unique MultiBlock Analysis) this would allow for the extraction of variation in the systems under analysis at three different levels: (i) globally joint level that would provide information on common features across all data sets; (ii) locally joint information that would provide knowledge within one particular block (analytical technique); and (iii) unique features that may be specific to (*e.g.*) lipidomics rather than metabolomics or oxylipin analyses; in the example given for differentiation between people with mild or severe malaria from control populations.

The need for comparison of multiple data analysis algorithms on the same set of data also featured in this session, and this is always necessary when a new algorithmic approach is proposed. This was exemplified in one paper (DOI: 10.1039/c9fd00004f) where immunological markers from cells were measured using flow cytometry and the conclusion was again that there was 'no free lunch' as the performance of the algorithm depended on the nuances of the multivariate approach used.[17]

Structural analysis of molecules that have not already been measured and thus do not feature in databases of known substances is a challenge. This was addressed in two papers (DOI: 10.1039/c8fd00235e and DOI: 10.1039/c8fd00227d) that used GNPS libraries and data sets (https://gnps.ucsd.edu/, ref. 18). A series of algorithms was developed that allowed in a semi-automated fashion for sub-structural analysis and annotation of MS^n data (DOI: 10.1039/c8fd00235e). In the future, more molecules will be identified using *in silico* predictions,[19] as it will not be possible to acquire or make all possible standards for confirmatory MS^n testing.

I reflected on this significant challenge that Justin van der Hooft was addressing and was reminded of the quote by a famous French philosopher:

"Science is built up with facts, as a house is with stones. But a collection of facts is no more a science than a heap of stones is a house"

Jules Henri Poincaré, *La Science et l'hypothèse*, 1854–1912

If we are to measure a complex mixture of molecules in order to understand the whole system, then currently this is like doing a jigsaw puzzle with only some of the pieces (or for Jules he can only find some of the bricks). This process is illustrated in Fig. 4. We can ask: what does the system do? Who interacts with whom? And in this example: what is in the picture? This is especially complex when we have three types of jigsaw piece with no idea as to whether we have a large or small proportion of the identified components (*identified matter*), and

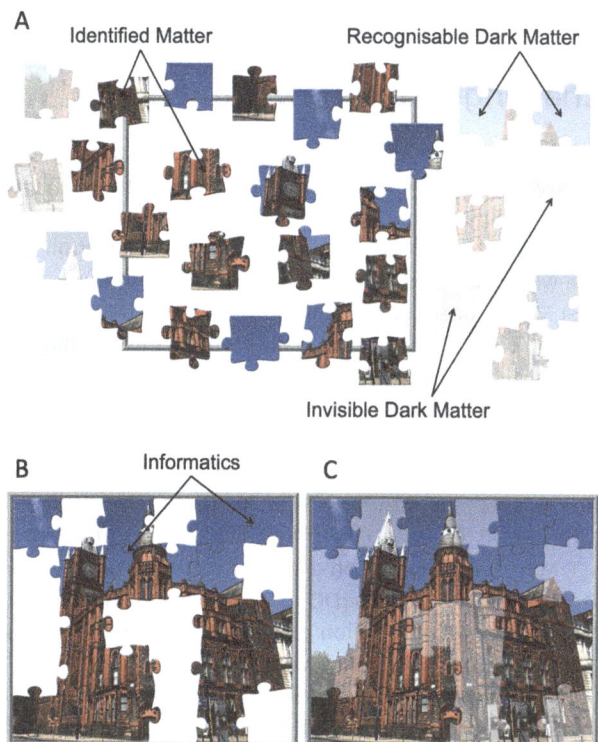

Fig. 4   The complexity jigsaw: what is the picture? After analysis of a complex mixture of chemicals there are (A) three types of jigsaw piece that need to be stitched together: *identified matter*, where a structure is assigned to the analyte being measured; *recognisable dark matter*, where the analyte can be recognised by (*e.g.*) its retention time and accurate mass in LC-MS, but can not be assigned to a chemical structure; whilst *invisible dark matter* consists of analytes that are not detected in the experiment and so are completely unknown to the analyst. Using (B) informatics, the links (edges) between the *identified matter* jigsaw pieces can be joined together and part of the picture is revealed. Finally (C) using bibliometrics, informatics and inference, the full *systems chemical analysis* reveals the picture is indeed the Victoria Gallery & Museum at the University of Liverpool. The jigsaw was generated using the free online software I'm a Puzzle (https://im-a-puzzle.com), using a picture taken by the author.

what we need to do to uncover the dark matter. Informatics is key to fitting the jigsaw pieces together. These computational approaches must infer what goes where, where and what the gaps are, and how to fill them; a central feature in any systems chemistry or biology analysis.[20,21] In Fig. 4 some of the pieces can be put together and after gap filling we can see that the picture is revealed as the Victoria Gallery & Museum at the University of Liverpool; the VGM resides within the Victoria Building, which was constructed in 1892 and was the inspiration for the term 'red brick university'.[22]

In this session, the discussion was directed towards having transparency in the data analysis process, standardisation in data collection and reporting the full informatics analysis pipeline. It was agreed that this could be enabled by suitable training activities where the ambition is to ingrain objectivity into the whole process. In order to enable this transparency, this community agreed that it is desirable to have more people make their data and code freely available. For discussions on this process within the metabolomics community, the interested reader is directed here.[23-25]

Future challenges and new approaches

The final day of the meeting was spent discussing new sample processing and instrumental advantages and what these may offer in the future, along with how data from different sources can be fused to understand more of the whole (DOI: 10.1039/c8fd00242h). Novel approaches to feature extraction from high-resolution data were also discussed (DOI: 10.1039/c9fd00014c).

Over the last 5 years or so, ion mobility spectrometry (IMS) has been coupled with mass spectrometry.[26,27] IMS is potentially useful as this technique adds a new dimension to analyte separation as ionised molecules are separated in the gas phase based on their mobility within a carrier gas; the orthogonal characteristics are therefore based on drift time in the carrier gas and these can be represented as collision cross sections (CCS), which can be computationally predicted.[28] Two papers detailed IMS coupled with MS: the first with a 12 T FT-ICR-MS system for the analysis of heavy oil (DOI: 10.1039/c8fd00239h), and a further study exploited trapped IMS (TIMS) for the elucidation of isomeric species from dissolved organic matter (DOM) from aquatic systems (DOI: 10.1039/c8fd00221e). It was clear from both studies that this extra dimension of separation offered by IMS was highly useful for resolving components within complex mixtures.

As shown in other sessions, complex natural systems analysis must embrace sample pretreatment and separation. This was shown for human urine where solid phase extraction (SPE) was assessed using diverse types of column chemistry (DOI: 10.1039/c8fd00220g). This allowed for enrichment of specific molecular fractions and, perhaps as expected, the matching of the SPE type to the polarity or anion/cation enrichment is predictable and thus enhances the molecular content of the analyses.

Finally, a particularly elegant approach highlighted in one paper (DOI: 10.1039/c8fd00213d) was to use substrates with stable isotopes to label organisms prior to NMR spectroscopy. As was mentioned in the discussion session after this paper, "you are what you eat" and the novel aspect of feeding *Daphnia magna* with ^{13}C labeled algae (*Chlamydomonas reinhardtii*) enhanced the ability to investigate metabolism as ^{13}C–^{12}C bond formation could be selectively observed. With further work, this could lead to accurate quantification of *in vivo* processes as these measurements could be made inside the NMR instrument. A mind boggling memory of the discussion on this paper was when the presenting author Ronald Soong admitted to spinning whole shrimp inside the instrument; I'm sure the poor creature also had its mind boggled!

Conclusions and prospects

This *Faraday Discussion* certainly showcased the current state-of-the-art for the analysis of complex natural systems. These analyses are in themselves complex and require diverse analytical techniques in combination with robust data analysis, interpretation and visualisation. I hope those that were present would agree that the above captures the main outcomes of the meeting and what was discussed.

Within any research area there is always room for improvement. In addition to improvements in high resolution instrument hardware and informatics software, there were three main areas that would make the analysis of complex natural mixtures more complete. Each of these is readily achievable and these are based on sound analytical chemistry.

The first is that in many studies presented at this *Faraday Discussion* meeting, there seemed to be a lack of ownership and suitable level of background knowledge of the sample under interrogation. Some fantastic analyses were performed using HRMS and NMR, but little was mentioned in terms of where the sample had come from and whether the sample was relevant to answer the question under examination. Sampling and experimental design should be considered as an essential aspect of the analysis of complex systems[29] and this was discussed at some length throughout the meeting. Collecting material may be achieved by a grab sample, though this single sample may under-represent the chemical diversity. It was considered that passive sampling over many days would generate a more comprehensive specimen from a heterogeneous environment, as well as the ability to take miniaturised analytical equipment into the field for on-site point and shoot analyses.[30-32]

The second was that there was in general a lack of figures of merit presented. With the exception of a few posters presented by early career researchers, no one discussed the precision in the amount of material measured, nor the accuracy of these levels. Here I am referring to the ordinate (MS ion count or NMR intensity) and not the abscissa, where for HRMS and NMR there certainly was excellent precision in m/z or δ (ppm); hence the high-resolution terminology. Similarly, where analytes were identified in complex mixtures, very few data were presented in terms of limits of detection (LOD) and quantification (LOQ), or limits of linearity (LOL). All of these are useful metrics and vital to appreciate the robustness of an analytical technique.[33]

The final challenge, which brings both of the above together, is the need to have validation in the measurement process. The 4 Rs of any analytical experiment are that it should have excellent *reproducibility* and *robustness*, and this is only achieved by *resampling* and *repeating* the experiment. Statistics within the sampling process are paramount. One would never perform a highly detailed and intricate analysis of a single grain of sand and then declare that this could be extrapolated to the point where every beach, sand dune or desert on the planet were fully understood!

In the extensive discussion sessions, which are a huge benefit of these *Faraday Discussion* meetings, it was debated how, as a community, analyses could be improved over the next few years. Two main conclusions arose from these deliberations: the first was that this was a friendly, welcoming community with diverse interests, and that people should talk, be listened to, learn the common language, and collaborate; the second was that one should address the question first and not the technology – no one really thought that the tail should wag the dog.

I would hope that if in 10 years or so I were to read a follow up *Faraday Discussion* volume on "challenges in analysis of complex natural mixtures", there would be more thought into the background of the sample and whether it is representative of the problem, along with experimental design being used by many. In addition, I would like to see more confidence in the reproducibility and robustness of the process, with proof, and that suitable statistical figures of merit were presented alongside the data. Only then will we know whether we are addressing the challenge of analysing complex natural mixtures.

As was clear from these three days in Edinburgh, chemical systems analysis is vibrant and has brought together scientists from many different disciplines. I believe the future of this field is sunny and bright, and that the next decade will see further improvements in analytics and data processing that will allow for even more comprehensive analysis of complex natural mixtures. The pinnacle of these analyses is to reach the top of the complexity mountain, just like those that climbed to the peak of Arthur's Seat after this most memorable *Faraday Discussion* meeting was over!

Conflicts of interest

There are no conflicts to declare.

Acknowledgements

The inspiration for the title – *the blind men and the elephant* – came from seeing my colleague David Ellis at the University of Manchester use this in a recent talk. I would also like to thank in particular Dušan Uhrin, Tim Ebbels, and the organising committee for their very kind invite to spend three days at this very stimulating *Faraday Discussion*.

References

1 E. B. Goldstein, *Encyclopedia of Perception*, SAGE Publications Inc., 2009.
2 W. B. Dunn, D. Broadhurst, H. J. Atherton, R. Goodacre and J. L. Griffin, Systems level studies of mammalian metabolomes: the roles of mass spectrometry and nuclear magnetic resonance spectroscopy, *Chem. Soc. Rev.*, 2011, **40**, 387–426.
3 D. I. Ellis and R. Goodacre, Metabolic fingerprinting in disease diagnosis: biomedical applications of infrared and Raman spectroscopy, *Analyst*, 2006, **131**, 875–885.

4 D. I. Ellis, V. L. Brewster, W. B. Dunn, J. W. Allwood, A. P. Golovanov and R. Goodacre, Fingerprinting food: current technologies for the detection of food adulteration and contamination, *Chem. Soc. Rev.*, 2012, **41**, 5706–5727.

5 P. Laszlo, *Citrus: A History*, The University of Chicago Press, Chicago, 2007.

6 C. C. C. R. De Carvalho and M. M. R. Da Fonseca, Carvone: Why and how should one bother to produce this terpene, *Food Chem.*, 2006, **95**, 413–422.

7 A. Kiritsakis, *Olive Oil*, American Oil Chemists Society, Champagin. IL, USA, 1991.

8 L. P. Kozlowski, Proteome-pI: proteome isoelectric point database, *Nucleic Acids Res.*, 2017, **45**, D1112–D1116.

9 S. Hiller, C. Wasmer, G. Wider and K. Wüthrich, Specific Resonance Assignment of soluble non-globular proteins by 7D APSY-NMR Spectroscopy, *J. Am. Chem. Soc.*, 2007, **129**, 10823–10828.

10 J. D. Barrow and J. Silk, *The left hand of creation: the origin and evolution of the expanding universe*, Penguin, London, 1995.

11 L. W. Sumner, A. Amberg, D. Barrett, R. Beger, M. H. Beale, C. Daykin, T. W.-M. Fan, O. Fiehn, R. Goodacre, J. L. Griffin, N. Hardy, R. Higashi, J. Kopka, J. C. Lindon, A. N. Lane, P. Marriott, A. W. Nicholls, M. D. Reily and M. Viant, Proposed minimum reporting standards for chemical analysis, *Metabolomics*, 2007, **3**, 211–221.

12 R. R. da Silva, P. C. Dorrestein and R. A. Quinn, Illuminating the dark matter in metabolomics, *Proc. Natl. Acad. Sci. U. S. A.*, 2015, **112**, 12549–12550.

13 O. A. H. Jones, Illuminating the dark metabolome to advance the molecular characterisation of biological systems, *Metabolomics*, 2018, **14**, 101.

14 F. H. M. van Zelst, S. G. J. van Meerten, P. J. M. van Bentum and A. P. M. Kentgens, Hyphenation of supercritical fluid chromatography and NMR with in-line sample concentration, *Anal. Chem.*, 2018, **90**, 10457–10464.

15 H. Barjat, G. A. Morris, S. Smart, A. G. Swanson and S. C. R. Williams, High-resolution diffusion-ordered 2D spectroscopy (HR-DOSY) – a new tool for the analysis of complex-mixtures, *J. Magn. Reson., Ser. B*, 1995, **108**, 170–172.

16 M. Nilsson, The DOSY Toolbox: A new tool for processing PFG NMR diffusion data, *J. Magn. Reson.*, 2009, **200**, 296–302.

17 P. S. Gromski, H. Muhamadali, D. I. Ellis, Y. Xu, E. Correa, M. L. Turner and R. Goodacre, A tutorial review: Metabolomics and partial least squares-discriminant analysis – a marriage of convenience or a shotgun wedding, *Anal. Chim. Acta*, 2015, **879**, 10–23.

18 M. Wang, J. J. Carver, V. V. Phelan, L. M. Sanchez, N. Garg, Y. Peng, D. D. Nguyen, J. Watrous, C. A. Kapono, T. Luzzatto-Knaan, C. Porto, A. Bouslimani, A. V. Melnik, M. J. Meehan, W.-T. Liu, M. Crüsemann, P. D. Boudreau, E. Esquenazi, M. Sandoval-Calderón, R. D. Kersten, L. A. Pace, R. A. Quinn, K. R. Duncan, C.-C. Hsu, D. J. Floros, R. G. Gavilan, K. Kleigrewe, T. Northen, R. J. Dutton, D. Parrot, E. E. Carlson, B. Aigle, C. F. Michelsen, L. Jelsbak, C. Sohlenkamp, P. Pevzner, A. Edlund, J. McLean, J. Piel, B. T. Murphy, L. Gerwick, C.-C. Liaw, Y.-L. Yang, H.-U. Humpf, M. Maansson, R. A. Keyzers, A. C. Sims, A. R. Johnson, A. M. Sidebottom, B. E. Sedio, A. Klitgaard, C. B. Larson, C. A. P. Boya, D. Torres-Mendoza, D. J. Gonzalez, D. B. Silva, L. M. Marques, D. P. Demarque, E. Pociute, E. C. O'Neill, E. Briand, E. J. N. Helfrich, E. A. Granatosky, E. Glukhov, F. Ryffel, H. Houson, H. Mohimani,

J. J. Kharbush, Y. Zeng, J. A. Vorholt, K. L. Kurita, P. Charusanti, K. L. McPhail, K. Fog Nielsen, L. Vuong, M. Elfeki, M. F. Traxler, N. Engene, N. Koyama, O. B. Vining, R. Baric, R. R. Silva, S. J. Mascuch, S. Tomasi, S. Jenkins, V. Macherla, T. Hoffman, V. Agarwal, P. G. Williams, J. Dai, R. Neupane, J. Gurr, A. M. C. Rodríguez, A. Lamsa, C. Zhang, K. Dorrestein, B. M. Duggan, J. Almaliti, P.-M. Allard, P. Phapale, L.-F. Nothias, T. Alexandrov, M. Litaudon, J.-L. Wolfender, J. E. Kyle, T. O. Metz, T. Peryea, D.-T. Nguyen, D. VanLeer, P. Shinn, A. Jadhav, R. Müller, K. M. Waters, W. Shi, X. Liu, L. Zhang, R. Knight, P. R. Jensen, B. Ø. Palsson, K. Pogliano, R. G. Linington, M. Gutiérrez, N. P. Lopes, W. H. Gerwick, B. S. Moore, P. C. Dorrestein and N. Bandeira, Sharing and community curation of mass spectrometry data with Global Natural Products Social Molecular Networking, *Nat. Biotechnol.*, 2016, **34**, 828–837.

19 T. De Vijlder, D. Valkenborg, F. Lemière, E. P. Romijn, K. Laukens and F. Cuyckens, A tutorial in small molecule identification *via* electrospray ionization-mass spectrometry: The practical art of structural elucidation, *Mass Spectrom. Rev.*, 2018, **137**, 607–629.

20 D. B. Kell, Metabolomics, modelling and machine learning in systems biology towards an understanding of the languages of cells. The 2005 Theodor Bücher lecture, *FEBS J.*, 2006, **273**, 873–894.

21 D. B. Kell and J. D. Knowles, The role of modeling in systems biology, in *System modeling in cellular biology: from concepts to nuts and bolts*, ed. Z. Szallasi, J. Stelling and V. Periwal, MIT Press, Cambridge, 2006, pp. 3–18.

22 A. L. Mackenzie and A. R. Allan, *Redbrick University Revisited*, Liverpool University Press, 1996.

23 P. Rocca-Serra, R. M. Salek, M. Arita, E. Correa, S. Dayalan, A. Gonzalez-Beltran, T. M. D. Ebbels, R. Goodacre, J. Hastings, K. Haug, A. Koulman, M. Nikolski, M. Oresic, S.-A. Sansone, D. Schober, J. Smith, S. Steinbeck, M. R. Viant and S. Neumann, Data standards can boost metabolomics research, and if there is a will, there is a way, *Metabolomics*, 2016, **12**, 14.

24 R. J. M. Weber, T. N. Lawson, R. M. Salek, T. M. D. Ebbels, R. C. Glen, R. Goodacre, J. L. Griffin, K. Haug, A. Koulman, P. Moreno, M. Ralser, C. Steinbeck, W. B. Dunn and M. R. Viant, Computational tools and workflows in metabolomics: An international survey highlights the opportunity for harmonisation through Galaxy, *Metabolomics*, 2017, **13**, 12.

25 B. Burla, M. Arita, M. Arita, A. K. Bendt, A. Cazenave-Gassiot, E. A. Dennis, K. Ekroos, X. Han, K. Ikeda, G. Liebisch, M. K. Lin, T. P. Loh, P. J. Meikle, M. Orešič, O. Quehenberger, A. Shevchenko, F. Torta, M. J. Wakelam, C. E. Wheelock and M. R. Wenk, MS-based lipidomics of human blood plasma – a community-initiated position paper to develop accepted guidelines, *J. Lipid Res.*, 2018, **59**, 2001–2017.

26 A. B. Kanu, P. Dwivedi, M. Tam, L. Matz and H. H. Hill, Ion mobility–mass spectrometry, *J. Mass Spectrom.*, 2008, **43**, 1–22.

27 F. Lanucara, S. W. Holman, C. J. Gray and C. E. Eyers, The power of ion mobility-mass spectrometry for structural characterization and the study of conformational dynamics, *Nat. Chem.*, 2014, **6**, 281–294.

28 V. Gabelica, A. A. Shvartsburg, C. Afonso, P. Barran, J. L. P. Benesch, C. Bleiholder, M. T. Bowers, A. Bilbao, M. F. Bush, J. L. Campbell, I. D. G. Campuzano, T. Causon, B. H. Clowers, C. S. Creaser, E. De Pauw,

J. Far, F. Fernandez-Lima, J. C. Fjeldsted, K. Giles, M. Groessl, C. J. Hogan Jr, S. Hann, H. I. Kim, R. T. Kurulugama, J. C. May, J. A. McLean, K. Pagel, K. Richardson, M. E. Ridgeway, F. Rosu, F. Sobott, K. Thalassinos, S. J. Valentine and T. Wyttenbach, Recommendations for reporting ion mobility Mass Spectrometry measurements, *Mass Spectrom. Rev.*, 2019, **38**, 291–320.

29 L. Leardi, Experimental design in chemistry: A tutorial, *Anal. Chim. Acta*, 2009, **652**, 161–172.

30 L. Li, T.-C. Chen, Y. Ren, P. I. Hendricks, R. G. Cooks and Z. Ouyang, Mini 12, Miniature Mass Spectrometer for Clinical and Other Applications— Introduction and Characterization, *Anal. Chem.*, 2014, **86**, 2909–2916.

31 X. Meng, X. Zhang, Y. Zhai and W. Xu, Mini 2000: A Robust Miniature Mass Spectrometer with Continuous Atmospheric Pressure Interface, *Instruments*, 2018, **2**, 2.

32 D. I. Ellis, H. Muhamadali, S. A. Haughey, C. T. Elliott and R. Goodacre, Point-and-shoot: rapid quantitative detection methods for on-site food fraud analysis – moving out of the laboratory and into the food supply chain, *Anal. Methods*, 2015, **7**, 9401–9414.

33 J. C. Miller and J. N. Miller, *Statistics for Analytical Chemistry*, Ellis Horwood, London, 1988.

Poster titles

Ion mobility spectrometry for rapid online monitoring of fermentation and degradation, **S. Murrell**, *University of South Wales, UK*

Molecular-level characterization of raised and blanket bogs under restoration, **G. Trifirò**, *University of Edinburgh, UK*

Automated assignment of carbohydrate NMR spectra, **E. Adair**, *University of Edinburgh, UK*

A novel unbiased method links variability of co-expression between multiple markers on single cells to a clinical phenotype, **G. Tinnevelt**, *Radboud University, Netherlands*

Authentication of geographical origin of crude palm oil by GC-IMS fingerprinting and different analytical classification scenarios, **K. Goggin**, *University of South Wales, UK*

Probing the complexity of water disinfection by ^{15}N and ^{19}F NMR, **A. Smith**, *University of Edinburgh, UK*

Use of non-uniform sampling (NUS) for 2D NMR metabolomics, **A. Le Guennec**, *King's College London, UK*

Robustifying a PLS property prediction workflow on NMR spectra with optimized processing, **L. Duval**, *IFP Energies nouvelles/Université Paris-Est, France*

Absorption mode spectral processing of dissolved organic matter samples measured by FT-ICR MS, **M. P. Da Silva**, *Helmholtz Centre for Environmental Research - UFZ, Germany*

Capabilities and limitations of quadrupole-selected dissociation FTMS for complex mixtures, **J. Maillard**, *Université de Rouen, Laboratoire COBRA, France*

Molecular-level monitoring of Maillard reaction networks, **D. Hemmler**, *Analytical Food Chemistry, Technical University of Munich, Germany*

Supercritical fluid chromatography hyphenated to electron ionization mass spectrometry to enhance identification reliability of semi-volatile compounds, **D. Sciarrone**, *University of Messina, Italy*

Linear retention indices in combination with MS libraries for automated and reliable characterization of psoralens in citrus beverages, **D. Sciarrone**, *University of Messina, Italy*

This journal is © The Royal Society of Chemistry 2019

Development of an automated offline aerosol mass spectrometry technique to measure dissolved particulate matter composition, **J. Cash**, *University of Edinburgh/Centre for Ecology and Hydrology, UK*

Combining two worlds - integration of IMS and UHRMS information on the example of HDN refractory species, **J. Le Maître**, *TOTAL - Laboratoire COBRA - Université de Rouen, France*

Molecular characterization of organosulfur species in snow from North China using FT-ICR MS, **S. Su**, *Tianjin University, China*

Advanced identification of bioactivity hotspots via global screening of the metabolome of entire ecosystems, **C. Mueller**, *HMGU, Germany*

Characterisation of algal feedstock and algae-derived bio-crudes by GC-MS hyphenated methods and NMR spectroscopy, **D. Nowakowski**, *Aston University, UK*

Extreme compositional complexity of organic carbon in atmospheric aerosol particles revealed via 2D-LC fractionation and FT-ICR mass spectrometry, **O. J. Lechtenfeld**, *Helmholtz Centre for Environmental Research - UFZ, Germany*

Data processing in petroleomics: petroinformatics approaches and CERES - a Matlab based platform, **C. P. Rüger**, *University of Rouen-Normandy, France*

Comparison of extraction and derivatization methods for sample preparation of algal oils prior to FAME determination, **B. Brennan**, *Dublin City University, Ireland*

The suite of capabilities for the analysis of complex mixtures in the environmental molecular science laboratory, **M. Lipton**, *Pacific Northwest National Laboratory, USA*

Characterization of unextracted natural organic matter by ultra-high resolution mass spectrometry, **W. Kew**, *Pacific Northwest National Laboratory, USA*

Investigating various soil parameters pertaining to soil health status in Irish soils and potential impacts on biodiversity, **A. Lee**, *Dublin City University, Ireland*

Reassembling fracking fluids - novel technologies for identifying the unknown-unknowns using LC-IMS-Q-TOF and advanced data mining, **D. Blyth**, *Imperial College London, UK*

Sorption and column leaching of neonicotinoid insecticides in contrasting soil from 5 locations in Britain, **A. Aseperi**, *Kingston University, UK*

Please note that only the presenting author (and their affiliation) is listed for each poster.

The Faraday Discussions Poster Prize for the best poster was jointly awarded to Mrs Sheri Murrell of the University of South Wales, UK, for her poster on ion mobility spectrometry for rapid online monitoring of fermentation and degradation, and Mr Gianluca Trifirò of the University of Edinburgh, UK, for his poster on molecular level characterization of raised and blanket bogs under restoration.

List of participants

Miss Elaine Adair, *University of Edinburgh, United Kingdom*
Professor Carlos Afonso, *University of Rouen-Normandy, France*
Mr Adeniyi Aseperi, *Kingston University, United Kingdom*
Dr Mark Barrow, *Department of Chemistry, University of Warwick, United Kingdom*
Dr Nicholle Bell, *School of Chemistry, University of Edinburgh, United Kingdom*
Mr David Blyth, *Imperial College London, United Kingdom*
Dr Adolfo Botana, *JEOL (UK) Ltd, United Kingdom*
Mr Brian Brennan, *Dublin City University, Ireland*
Professor Eleanor Campbell, *University of Edinburgh, United Kingdom*
Mr James Cash, *Centre for Ecology and Hydrology, University of Edinburgh, United Kingdom*
Dr Caroline Chaux, *CNRS I2M Aix-Marseille Université, France*
Miss Maria Paula Da Silva, *Helmholtz Centre for Environmental Research - UFZ, Germany*
Professor Antony N. Davies, *Expert Capability Center Deventer, Nouryon, Netherlands*
Dr Marc-André Delsuc, *Institut de Génétique et de Biologie Moléculaire et Cellulaire (IGBMC)/CNRS, France*
Mr Sylvain Demanze, *Eli Lilly, United Kingdom*
Dr Claire Dickson, *University of Edinburgh, United Kingdom*
Mr Adam Drummond, *University of Glasgow, United Kingdom*
Mr Dumitru Duca, *University of Lille, France*
Dr Laurent Duval, *IFP Energies nouvelles/Université Paris-Est, France*
Dr Timothy Ebbels, *Imperial College London, United Kingdom*
Professor Dr Francisco Fernandez-Lima, *Florida International Unversity, USA*
Professor Cristian Focsa, *University of Lille, France*
Dr Caroline Gauchotte-Lindsay, *School of Enginering, University of Glasgow, United Kingdom*
Dr Pierre Giusti, *TOTAL, France*
Dr Ruth Godfrey, *Swansea University, United Kingdom*
Miss Kirstie Goggin, *University of South Wales, United Kingdom*
Professor Royston Goodacre, *University of Liverpool, United Kingdom*
Dr Jeffrey Hawkes, *Uppsala University, Sweden*
Mr Daniel Hemmler, *Analytical Food Chemistry, Technical University of Munich, Germany*
Dr Norbert Hertkorn, *Helmholtz Zentrum Muenchen, Germany*
Dr Suzanne Howson, *Royal Society of Chemistry, United Kingdom*
Dr Jeroen Jansen, *Radboud University, Netherlands*
Professor Dr Renault Jean-Hugues, *University of Reims Champagne-Ardenne, France*
Dr Myron Johnson, *Johnson Matthey, United Kingdom*
Professor Donald Jones, *University of Leicester, United Kingdom*
Dr William Kew, *Pacific Northwest National Laboratory, USA*
Mr Stefan Kuhn, *De Monfort University, United Kingdom*

Dr Pedro Lameiras, *Reims Institute of Molecular Chemistry*, *France*
Dr Adrien Le Guennec, *King's College London*, *United Kingdom*
Dr Johann Le Maître, *TOTAL, Laboratoire COBRA, Université de Rouen*, *France*
Dr Oliver Lechtenfeld, *Helmholtz Centre for Environmental Research - UFZ*, *Germany*
Dr Alan Lee, *Dublin City University*, *Ireland*
Dr Mary Lipton, *Pacific Northwest National Laboratory*, *USA*
Dr Anneke Lubben, *University of Bath*, *United Kingdom*
Mr Julien Maillard, *Laboratoire COBRA, Université de Rouen*, *France*
Miss Elizabeth Marsden, *University of Southampton*, *United Kingdom*
Mr Dan McGill, *Imperial College London*, *United Kingdom*
Dr Constanze Mueller, *Helmholtz Zentrum Muenchen*, *Germany*
Mrs Sheri Murrell, *University of South Wales*, *United Kingdom*
Professor Mathias Nilsson, *University of Manchester*, *United Kingdom*
Dr Daniel Nowakowski, *Aston University*, *United Kingdom*
Dr John Parkinson, *University of Strathclyde*, *United Kingdom*
Dr Ljiljana Paša-Tolić, *Pacific Northwest National Laboratory*, *USA*
Dr David Peggie, *National Gallery*, *United Kingdom*
Miss Elena Piersanti, *Aix-Marseille Université*, *France*
Miss Aurélie Pirayre, *IFP Energies nouvelles*, *France*
Mr Bill Plumb, *LECO*, *United Kingdom*
Mr Colin Potter, *Bangor University*, *United Kingdom*
Dr Josh Richards, *Aptuit (Evotec UK Ltd)*, *United Kingdom*
Dr Joanne Roberts, *Glasgow Caledonian University*, *United Kingdom*
Dr Ryan Rodgers, *National High Magnetic Field Laboratory*, *USA*
Dr Paul Rodwell, *Merck*, *United Kingdom*
Dr Simon Rogers, *University of Glasgow*, *United Kingdom*
Dr Timothy Rudd, *National Institute for Biological Standards and Control*, *United Kingdom*
Dr Christopher Rüger, *University of Rouen-Normandy*, *France*
Miss Rachel Sanig, *Micromass UK Ltd*, *United Kingdom*
Professor Dr Philippe Schmitt-Kopplin, *Helmholtz Zentrum Muenchen*, *Germany*
Professor Peter Schoenmakers, *University of Amsterdam*, *Netherlands*
Professor Dr Danilo Sciarrone, *University of Messina*, *Italy*
Dr Laetitia Shintu, *Aix-Marseille Université*, *France*
Dr Philip Sidebottom, *Syngenta*, *United Kingdom*
Mr Alan Smith, *University of Edinburgh*, *United Kingdom*
Dr Ronald Soong, *University of Toronto*, *Canada*
Dr Maria Southall, *Royal Society of Chemistry*, *United Kingdom*
Professor Dan Staerk, *University of Copenhagen*, *Denmark*
Mr Marc Stockwell, *University of Edinburgh*, *United Kingdom*
Miss Sihui Su, *Tianjin University*, *China*
Dr Stephen Summerfield, *Peter Fisk Associates*, *United Kingdom*
Dr Andrew Surman, *King's College, London*, *United Kingdom*
Dr Izabella Surowiec, *Corporate Research, Sartorius AG*, *Sweden*
Mrs Sarah Thirkell, *Royal Society of Chemistry*, *United Kingdom*
Mr Gerjen Tinnevelt, *Radboud University*, *Netherlands*
Mr Gianluca Trifirò, *University of Edinburgh*, *United Kingdom*
Professor Johan Trygg, *Computational Life Science Cluster (CLiC),Umeå University*, *Sweden*
Professor Dušan Uhrín, *University of Edinburgh*, *United Kingdom*
Dr Pieter van Delft, *Corbion*, *Netherlands*
Dr Justin J. J. van der Hooft, *Wageningen University*, *Netherlands*

Miss Fleur van Zelst, *Institute for Molecules and Materials (IMM), Radboud University, Netherlands*
Mr James Watt, *University of Edinburgh, United Kingdom*
Dr Susan Weatherby, *Royal Society of Chemistry, United Kingdom*
Miss Ella Wren, *Royal Society of Chemistry, United Kingdom*
Mr Richard York, *University of Edinburgh, United Kingdom*
Dr Alexander Zherebker, *Skolkovo Institute of Science and Technology, Russia*